MICROBIOLOGY OF
Fruits and Vegetables

Edited by

Gerald M. Sapers
James R. Gorny
Ahmed E. Yousef

Taylor & Francis
Taylor & Francis Group

Boca Raton   London   New York

A CRC title, part of the Taylor & Francis imprint, a member of the
Taylor & Francis Group, the academic division of T&F Informa plc.

Published in 2006 by
CRC Press
Taylor & Francis Group
6000 Broken Sound Parkway NW, Suite 300
Boca Raton, FL 33487-2742

International Standard Book Number-10: 0-8493-2261-8 (Hardcover)
International Standard Book Number-13: 978-0-8493-2261-7 (Hardcover)
Library of Congress Card Number 2005046298

Library of Congress Cataloging-in-Publication Data

Microbiology of fruits and vegetables / edited by Gerald M. Sapers, James R. Gorny, Ahmed E. Yousef.
    p. cm.
  Includes bibliographical references and index.
  ISBN 0-8493-2261-8
  1. Fruit--Microbiology. 2. Vegetables--Microbiology. I. Sapers, Gerald M. II. Gorny, James R. III. Yousef, Ahmed Elmeleigy.

QR115.M495 2005
664'.8'001579--dc22   2005046298

Preface

Fruits and vegetables represent an important part of the human diet, providing essential vitamins, minerals, and fiber, and adding variety to the diet. In their Food Guide Pyramid, the U.S. Department of Agriculture encourages consumption of 3–5 servings of vegetable items, and 2–4 servings of fruit items per day. In today's global economy, fresh fruits and vegetables are available year round.

In the U.S. and other technologically advanced countries, high-quality fresh and processed fruits and vegetables are widely available. Fresh-cut fruits and vegetables represent a large and rapidly growing segment of the fresh produce industry. These commodities have an excellent safety record with respect to incidence of foodborne illness. Nevertheless, surveillance statistics compiled by the U.S. Centers for Disease Control and Prevention indicate that significant and increasing numbers of outbreaks have been associated with fresh fruits and vegetables, or their products. The presence of human pathogens in fresh produce is borne out by U.S. Food and Drug Administration product recall data, and by microbiological surveys of domestically produced and imported commodities. Increased recognition of a food safety problem with produce may reflect greater consumption of fruits and vegetables, more frequent eating out, greater reliance on imports of out-of-season fruits and vegetables from "third world" producers, and improved surveillance and reporting methods by public health agencies.

In addition to safety concerns, microbial spoilage of fresh produce represents a source of waste for consumers, and an economic loss to growers, packers, and retailers. Post-harvest decay, bacterial soft rot, and microbial spoilage of fresh-cuts and processed juices are continuing problems.

In recent years, extensive research has been conducted on microbiological problems relating to the safety and spoilage of fruits and vegetables. Active areas of research include incidence of human pathogen contamination, sources of microbial contamination, microbial attachment to produce surfaces, intractable spoilage problems, efficacy of sanitizing treatments for fresh produce, novel interventions for produce disinfection, and methodologies for microbiological evaluation of fruits and vegetables.

In this book, we have attempted a comprehensive examination of these topics, focusing on issues, rather than attempting an encyclopedic compilation of information about all commodities, classes of microorganisms, or categories of spoilage. We have not included certain topics, such as preharvest diseases of produce or production of fermented vegetables, which are adequately covered elsewhere. We have selected chapter authors who are active researchers in their respective fields, and thus bring a working knowledge of current issues, industry practices, and advances in technology.

The book is divided into five sections: (I) Contamination and State of Microflora on Fruits and Vegetables; (II) Microbial Spoilage of Fruits and Vegetables; (III) Food Safety Issues; (IV) Interventions to Reduce Spoilage and Risk of Foodborne Illness; and (V) Microbiological Evaluation of Fruits and Vegetables. Within each section we have grouped chapters that cover specific issues related to the overall topic. For example, Section I contains chapters on sources of microbial contamination, attachment of microorganisms to fresh produce, internalization and infiltration of microorganisms in produce, and stress adaptation by microorganisms and safety of produce.

I wish to thank the individual chapter authors for the authoritative and comprehensive coverage of their respective topics, and my co-editors, Dr. James R. Gorny and Dr. Ahmed E. Yousef, for their assistance in developing the concept and organizational structure of the book, identifying suitable chapter authors, reviewing the completed chapters, and helping me assemble the manuscripts into a form suitable for publication. I also thank Susan Lee, Food Science Editor at Dekker/CRC Press and her editorial staff for their guidance, invaluable help, and patience in working with us on this project. I thank my employer, the USDA Agricultural Research Service's Eastern Regional Research Center, for allowing me the time, and providing the resources, that enabled me to participate in this project. Finally, I must thank my wife for her unlimited patience and understanding during the many long hours when I was attached to the computer and unavailable to meet her needs.

Gerald M. Sapers

Editors

Gerald M. Sapers received his Ph.D. in food technology from MIT in 1961. He joined the USDA's Eastern Regional Research Center (ERRC) in 1968, after 2 years at the U.S. Army Natick Laboratories, and 6 years in private industry. He has conducted research on dehydrated potato stability, apple volatiles, safety of home canned tomatoes, utilization of natural pigments, pigmentation of small fruits, cherry dyeing, control of enzymatic browning in minimally processed fruits and vegetables, mushroom washing, and microbiological safety of fresh produce, which is his current area of research. He has been a Lead Scientist at ERRC since 1991. Dr. Sapers has published 110 scientific papers, 3 book chapters and 5 patents. He is an active member of the Institute of Food Technologists' Fruit and Vegetable Products Division, and the International Fresh-cut Produce Association.

James R. Gorny received his Ph.D. in plant biology from the University of California, Davis, and his M.S. and B.S. degrees in food science from Louisiana State University in Baton Rouge. He is currently vice president of Technology and Regulatory Affairs for the International Fresh-cut Produce Association, and has been the author and editor of numerous scientific publications including: Editor-In-Chief of the IFPA Food Safety Guidelines for the Fresh-cut Produce Industry and a contributor to the chapter on "Produce Food Safety" in the recently revised *U.S. Department of Agriculture Handbook 66*. His research has focused on the effects of modified atmospheres on the quality and safety of whole and fresh-cut fruit produce. He has been actively involved in the fresh-cut produce industry since 1986, and has worked extensively as a consultant on food safety, packaging, quality assurance, operations, and general management issues, both nationally and internationally.

Ahmed E. Yousef received his Ph.D. in food science from the University of Wisconsin (UW)-Madison in 1984. Subsequently, he served as a post-doctoral researcher at the Department of Food Science and the Department of Food Microbiology and Toxicology, UW. Dr. Yousef joined The Ohio State University (OSU) as an assistant professor in 1991. At OSU, Dr. Yousef investigated food biopreservation using bacteriocins, explored new applications of ozone in food processing, and addressed the safety of foods processed by novel technologies such as pulsed electric field, high pressure processing and ohmic heating. He is currently a professor at the Department of Food

Science and Technology and the Department of Microbiology, teaching the main food microbiology course at OSU. Dr. Yousef has published 2 books, 10 book chapters, and 70 scientific papers and review articles, and a patent. He is an active member of the Institute of Food Technologists, the American Society for Microbiology, and the International Association of Food Protection.

Contributors

Bassam A. Annous
Eastern Regional Research Center
Agricultural Research Service
U.S. Department of Agriculture
Wyndmoor, Pennsylvania

Jerry A. Bartz
Department of Plant Pathology
University of Florida
Gainesville, Florida

Robert B. Beelman
Department of Food Science
Pennsylvania State University
University Park, Pennsylvania

Larry R. Beuchat
Center for Food Safety
Department of Food Science and
Technology
University of Georgia
Griffin, Georgia

Maria T. Brandl
Western Regional Research Center
Agricultural Research Service
U.S. Department of Agriculture
Albany, California

F. Breidt, Jr.
Agricultural Research Service
U.S. Department of Agriculture
and Department of Food Science
North Carolina State University
Raleigh, North Carolina

Naveen Chikthimmah
Department of Food Science
Pennsylvania State University
University Park, Pennsylvania

Pascal Delaquis
Food Safety and Quality
Agriculture and Agri-Food Canada
Summerland, British Columbia,
Canada

Mary Ann Dombrink-Kurtzman
National Center for Agricultural
Utilization Research
Agricultural Research Service
U.S. Department of Agriculture
Peoria, Illinois

Elazar Fallik
Department of Postharvest Sciences
of Fresh Produce
ARO-The Volcani Center
Bet-Dagan, Israel

William F. Fett
Eastern Regional Research Center
Agricultural Research Service
U.S. Department of Agriculture
Wyndmoor, Pennsylvania

Daniel Y.C. Fung
Department of Animal Sciences and
Industry
Kansas State University
Manhattan, Kansas

Jim Gorny
International Fresh-cut Produce
Association
Davis, California

Lisa Gorski
Western Regional Research Center
Agricultural Research Service
U.S. Department of Agriculture
Albany, California

Dongsheng Guan
Department of Animal and Food
Sciences
University of Delaware
Newark, Delaware

Yingchan Han
Department of Food Sciences
Purdue University
West Lafayette, Indiana

Dallas G. Hoover
Department of Animal and Food
Sciences
University of Delaware
Newark, Delaware

J.H. Hotchkiss
Department of Food Sciences
Cornell University
Ithaca, New York

William C. Hurst
Department of Food Science and
Technology
University of Georgia
Athens, Georgia

Lauren Jackson
Center for Food Safety and
Applied Nutrition
U.S. Food and Drug Administration
Bedford, Illinois

Susanne E. Keller
National Center for Food Safety
and Technology
U.S. Food and Drug
Administration
Summit Argo, Illinois

Michael F. Kozempel
Eastern Regional Research Center
Agricultural Research Service
U.S. Department of Agriculture
Wyndmoor, Pennsylvania

Ching-Hsing Liao
Eastern Regional Research Center
Agricultural Research Service
U.S. Department of Agriculture
Wyndmoor, Pennsylvania

Richard H. Linton
Center for Food Safety Engineering
Purdue University
West Lafayette, Indiana

Robert E. Mandrell
Western Regional Research Center
Agricultural Research Service
U.S. Department of Agriculture
Albany, California

Pamela G. Marrone
AgraQuest, Inc.
Davis, California

Julien Mercier
AgraQuest, Inc.
Davis, California

Arthur J. Miller
U.S. Food and Drug
Administration
Center for Food Safety and
Applied Nutrition
College Park, Maryland

J.-M. Monier
Laboratoire d'Ecologie Microbienne
Université Claude Bernard Lyon 1
Villeurbanne, France

Philip E. Nelson
Department of Food Sciences
Purdue University
West Lafayette, Indiana

Ynes R. Ortega
Center for Food Safety
Department of Food Science and
Technology
University of Georgia
Griffin, Georgia

Mickey E. Parish
Citrus Research and Education
Center
University of Florida
Lake Alfred, Florida

Luis A. Rodriguez-Romo
Department of Food Science and
Technology
The Ohio State University
Columbus, Ohio

Gerald M. Sapers
Eastern Regional Research Center
Agricultural Research Service
U.S. Department of Agriculture
Wyndmoor, Pennsylvania

Travis L. Selby
Department of Food Sciences
Purdue University
West Lafayette, Indiana

Charles R. Sterling
Department of Veterinary Science
and Microbiology
University of Arizona
Tucson, Arizona

Dike O. Ukuku
Eastern Regional Research Center
Agricultural Research Service
U.S. Department of Agriculture
Wyndmoor, Pennsylvania

B.G. Werner
Department of Food Sciences
Cornell University
Ithaca, New York

Ahmed E. Yousef
Department of Food Science and
Technology
The Ohio State University
Columbus, Ohio

Contents

SECTION IV Interventions to Reduce Spoilage and Risk of Foodborne Illness

SECTION V Microbiological Evaluation of Fruits and Vegetables

Section I

Contamination and State of Microflora on Fruits and Vegetables

1 Microbial Contamination of Fresh Fruits and Vegetables

Jim Gorny

CONTENTS

1.1 INTRODUCTION

Fresh fruits and vegetables are perceived by consumers to be healthful and nutritious foods because of the plethora of scientifically substantiated and documented health benefits derived from consuming fresh fruits and vegetables [1]. However, recent foodborne illness outbreaks in the U.S. and throughout the world have been increasingly linked epidemiologically to consumption of fresh fruits, vegetables, and unpasteurized juices. These incidents have caused growers, shippers, fresh-cut produce processors, distributors, retailers, importers, and government public health officials to re-evaluate the risk of contracting foodborne illness from consumption of fresh fruits and vegetables and to re-evaluate current production and handling practices.

While the probability of contracting a foodborne illness via consumption of fresh fruits or vegetables is very low, a small probability does exist. Because fresh fruits and vegetables are often consumed uncooked so that there is no "kill" step, prevention of produce contamination with human pathogens is the only practical and effective means of ensuring that these food products are wholesome and safe for human consumption. This means that a complete supply chain approach to prevent contamination at any point in the produce continuum is essential to ensuring public health by minimizing the incidence of foodborne illness associated with produce consumption. Ensuring the integrity of produce from field to fork is the responsibility of everyone in the produce continuum, including growers, shippers, processors, distributors, retailers, and consumers. It must also be remembered that the health benefits derived from eating at least five servings of fresh fruits and vegetables daily far outweigh the very small probability of contracting a foodborne illness.

A meaningful assessment of the risk associated with contracting a foodborne illness from consumption of fresh fruits and vegetables involves understanding the microbiology of fresh fruits and vegetables as well as field production, processing, and handling practices. As such, the fresh produce industry is extraordinarily diverse and complex in the number of products produced, how the products are grown and handled, and the geographic areas from which these products are sourced. A typical retail grocer in North America will have available on a daily basis upwards of 300 different produce items for sale. The morphological characteristics of a produce item may also contribute to its propensity for contamination, since produce items may be derived from the leaves, stems, stalks, roots, fruits, and flowers of plants. Because the produce continuum represents such diversity, it is only possible to describe broad generalities about current practices of the produce continuum and the food safety risk associated with them, as an in-depth analysis of this plethora of products would be encyclopedic in volume.

1.2 PRODUCE CONTAMINATION

Contamination of fruits and vegetables by human pathogens can occur anywhere in the farm to table continuum including contamination of seed stocks and during production, harvesting, postharvest handling, storage, processing, transport distribution, retail display, and/or preparation (foodservice or home). Produce contaminated with human pathogens cannot be completely disinfected by washing or rinsing the product in an aqueous solution, and low sporadic levels of human pathogens can be found on produce [2,3]. In 2004 the Alliance for Food and Farming [4] analyzed Centers for Disease Control and Prevention (CDC) data sets [5,6] and summarized information regarding foodborne illness outbreaks that have been associated with produce consumption. The study's objective was to analyze likely sources of produce contamination and categorize the most likely place that the contamination occurred, that being either during production/ growing or during postproduction handling. The "postproduction" category included produce-associated foodborne illnesses that were most likely due to improper handling at the foodservice, retail, or consumer level, while the "grower" category included foodborne illnesses associated with produce that were most likely attributable to the farm, packing, shipping, or other agricultural postharvest handling. Analysis of CDC data indicated that improper handling of fruits and vegetables at foodservice establishments or by consumers caused 83% of produce-associated foodborne illness outbreaks, while "grower"-implicated cases comprised 17% of produce-associated

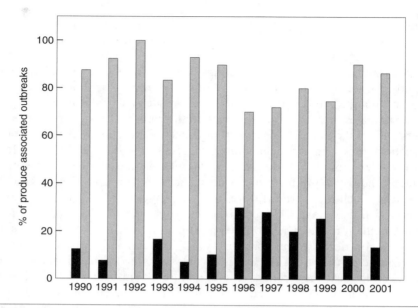

% of produce outbreaks associated with growing, packing, shipping and/or processing
% of produce outbreaks associated with improper handling after leaving the farm

FIGURE 1.1 Produce-associated outbreaks due to suspected farm contamination versus postproduction handling. (Adapted from *Analysis of Produce Related Foodborne Illness Outbreaks*, Alliance for Food and Farming, April 2004, www.foodandfarming.info/documents/85876_produce_analysis_604.pdf.)

foodborne illness outbreaks. Data from this report presented in Figure 1.1 show that the percentage of "grower"-related contamination incidents as a percent of all produce related outbreaks has been declining since 1996, and this trend is most likely due to implementation of good agricultural practices (GAPs) by grower/shipper/packers.

The Alliance for Food and Farming 2004 report and the CDC [7] both indicate that about 12% of foodborne illnesses occurring in the U.S. between 1990 and 2001 has been associated with consumption of fresh fruits and vegetables. This figure of 12% of outbreak cases associated with produce consumption represents a greater proportion of foodborne illness burden being represented by fresh fruits and vegetables than was reported in the past (Table 1.1). CDC data also indicate that produce-related outbreaks have become larger, involving more individuals and increasing in frequency. Foodborne illness outbreak reports related to produce consumption have most likely increased due to:

- Better detection and diagnostic methods for human pathogens which can epidemiologically associate produce consumption with illness (PulseNet, SODA salmonella outbreak detection algorithm, etc.).

TABLE 1.1
Trends in Burden: Foodborne Outbreaks Related to Fresh Produce,
1973–1997

	1970s	1990s
No. outbreaks per year	2	16
Median cases per outbreak	21	43
Outbreaks of known vehicle (%)	0.7	6
Outbreak associated cases (%)	0.6	12

Adapted from Sivapalasingham, S., Friedman, C.R., Cohen, L., and Tauxe, R.V., *J. Food Prot.* 67, 10, 2004.

- Increased surveillance for human pathogens by public health agencies.
- Increased per capita consumption of fresh fruits and vegetables in North America.
- Increased awareness that produce may be a potential vehicle for human pathogens, thus leading to increased epidemiological investigations of produce as a potential vector.
- Increased global sourcing of produce items to ensure year around supply of the broad diversity of produce available in modern grocery stores.
- Longer postharvest storage and longer shipment times that may also contribute to increased potential for illness by allowing for proliferation of an initial low number of human pathogens to an infectious or disease-causing dosage.

1.3 MICROORGANISMS OF CONCERN

Harris *et al.* [8] extensively reviewed outbreaks associated with fresh produce and reported that the most common human pathogens associated with produce foodborne illness outbreaks are: *E. coli* O157:H7, *Salmonella* spp., *Shigella* spp., *Listeria monocytogenes*, *Crytosporidium* spp. [9], *Cyclospora* spp., *Clostridium botulinum*, hepatitis A virus, Norwalk virus, and Norwalk-like viruses. These microorganisms can be categorized as follows:

- Soil-associated pathogenic bacteria (*Clostridium botulinum*, *Listeria monocytogenes*).
- Feces-associated pathogenic bacteria (*Salmonella* spp., *Shigella* spp., *E. coli* O157:H7, and others).
- Pathogenic parasites (*Cryptosporidium*, *Cyclospora*).
- Pathogenic viruses (hepatitis A, enterovirus, Norwalk-like viruses).

Many of these pathogens are spread via a human (or domestic animal) to food to human transmission route. Handling of fruits and vegetables by infected field-workers or consumers, cross contamination, use of contaminated water, use of inadequately composted manure, or contact with contaminated soil are just a few of the ways that transmission of human pathogens to food can occur.

Data from the CDC foodborne outbreak surveillance system show that from 1988 to 1998 the two most commonly reported microorganisms associated with fresh produce foodborne illness outbreaks were *Salmonella* spp. and *E. coli* O157:H7 with 45% and 38% of the fruit and vegetable linked outbreaks, respectively, being attributed to these two microorganisms. However, recent foodborne illness outbreaks associated with produce consumption have been caused by viruses (hepatitis A) and parasites (*Cyclospora* spp.). CDC data demonstrate that the majority of reported foodborne illnesses in the U.S. are of unknown etiology and are most likely caused by viruses such as Norwalk-like viruses [10,11]. Unfortunately, diagnostic tools for detection and enumeration of viruses that may cause foodborne illnesses are severely lacking.

1.4 INCIDENCE AND ASSOCIATION OF HUMAN PATHOGENS WITH PRODUCE

1.4.1 FDA IMPORTED PRODUCE SURVEY

In March 1999 the U.S. Food and Drug Administration (FDA) initiated a 1000-sample survey of imported fresh produce raw agricultural commodities from 21 countries and included: broccoli, loose-leaf lettuce (radicchio, escarole, endive, chicory leaf, mesclun, and others), cantaloupe, celery, strawberries, scallions/green onions, tomatoes, parsley, culantro (a herb), and cilantro [12]. Loose-leaf lettuce products included radicchio, escarole, endive, chicory and others. These high-volume imported fresh produce raw agricultural commodities were selected by the FDA for the imported produce sampling assignment based on the following risk factor criteria: epidemiological outbreak data, structural characteristics of the produce item, growing conditions, processing and consumption rates. Raw agricultural commodities are defined in the Federal Food, Drug, and Cosmetic Act as "any food in its raw or natural state, including all fruits that are washed, colored, or otherwise treated in the unpeeled natural form prior to marketing." These raw agricultural commodities were analyzed for the presence of *Salmonella* spp. and *E. coli* O157:H7. All commodities except for cilantro, culantro, and strawberries also were analyzed for *Shigella* spp. Produce imported from Mexico, Canada, Costa Rica, Guatemala, the Netherlands, Honduras, Belgium, Italy, Israel, Chile, Peru, Colombia, Trinidad and Tobago, New Zealand, Nicaragua, the Dominican Republic, France, Argentina, Ecuador, Haiti, and Korea were sampled. Six countries provided 25 or more samples for analysis: Mexico, Canada, Costa Rica, Guatemala, the Netherlands, and Honduras.

TABLE 1.2
Number of Samples Collected and Analyzed and the Number of Samples Confirmed Positives for Human Pathogens Per Each Type of Imported Produce

Produce item	No. of samples	No. human pathogen positive samples (% positive per produce item)	*Salmonella* spp.	*Shigella* spp.
Broccoli	36	0 (0.0%)	0 (0.0%)	0 (0.0%)
Cantaloupe	151	11 (7.3%)	8 (5.2%)	3 (2.0%)
Celery	84	3 (3.6%)	1 (1.2%)	2 (2.4%)
Cilantro	177	16 (9.0%)	16 (9%)	N/A
Culantro	12	6 (50.0%)	6 (50%)	N/A
Lettuce	116	2 (1.7%)	1 (0.9%)	1 (0.9%)
Parsley	84	2 (2.4%)	1 (1.2%)	1 (1.2%)
Scallions	180	3 (1.7%)	1 (0.6%)	2 (1.1%)
Strawberries	143	1 (0.7%)	1 (0.7%)	N/A
Tomatoes	20	0 (0.0%)	0 (0.0%)	0 (0.0%)
Total	1003	44 (4.5%)	35 (3.5%)	9 (1.3%)

Note: N/A = not analyzed.
Adapted from FDA Survey of Imported Fresh Produce, U.S. Food and Drug Administration Center for Food Safety and Applied Nutrition, Office of Plant and Dairy Foods and Beverages, January 30, 2001, www.cfsan.fda.gov/~dms/prodsur6.html.

Data presented in Table 1.2 show that of 1003 samples that were collected and analyzed, 35 samples (3.5% of the total number of samples) were found to have detectable levels of *Salmonella* spp., 9 samples (0.9% of the total number of samples) were found to have detectable levels of *Shigella* spp., and no samples (0%) were found to have detectable levels of *E. coli* O157:H7. These 44 samples positive for the presence of a human pathogen represent approximately 4.4% of the total number of product samples tested.

The three produce items with the greatest incidence of pathogen contamination were cilantro, cantaloupe, and culantro, accounting for 1.6, 1.1, and 0.6%, respectively, of the overall contamination (4.4%). The remaining produce items each contributed 0.3% or less to the overall contamination.

1.4.2 FDA DOMESTIC PRODUCE SURVEY

In March 2000 the FDA initiated a 1000-sample survey of domestic fresh fruit and vegetable raw agricultural commodities [13]. Cantaloupe, celery, cilantro, loose-leaf lettuce, parsley, scallions (green onions), strawberries, and tomatoes were collected and analyzed for *Salmonella* spp. and *E. coli* O157:H7. Cantaloupe, celery, parsley, scallions, and tomatoes were also analyzed for *Shigella* spp. This survey was the domestic complement to the FDA imported produce survey.

TABLE 1.3

Number of Samples Collected and Analyzed and the Number of Confirmed Positives for Human Pathogens Per Each Type of Domestic Produce Sampled

Produce item	No. of samples	No. human pathogen positive samples (% positive per produce item)	*Salmonella* spp.	*Shigella* spp.
Cantaloupe	164	5 (3.1)	4 (2.4%)	1 (0.6%)
Celery	120	0 (0.0%)	0 (0.0%)	0 (0.0%)
Cilantro	85	1 (1.2%)	1 (1.2%)	N/A
Lettuce	142	1 (0.7%)	1 (0.7%)	N/A
Parsley	90	1 (1.1%)	0 (0.0%)	1 (1.1%)
Scallions	93	3 (3.2%)	0 (0.0%)	3 (3.2%)
Strawberries	136	0 (0.0%)	0 (0.0%)	N/A
Tomatoes	198	0 (0.0%)	0 (0.0%)	0 (0.0%)
Total	1028	11 (1.1%)	6 (0.6%)	5 (0.8%)

Note: N/A = not analyzed.

Adapted from FDA Survey of Domestic Fresh Produce, U.S. Department of Health and Human Services, U.S. Food and Drug Administration Center for Food Safety and Applied Nutrition, Office of Plant and Dairy Foods and Beverages, January 2003, www.cfsan.fda.gov/~dms/prodsu10.html.

Data presented in Table 1.3 show that of 1028 domestic samples that were collected and analyzed, 6 samples (0.58% of the total number of samples) were found to have detectable levels of *Salmonella* spp., 5 samples (0.49% of the total number of samples) were found to have detectable levels of *Shigella* spp., and no samples (0%) were found to have detectable levels of *E. coli* O157:H7. One or more samples of cantaloupe, cilantro, lettuce, parsley, and scallions were found to have detectable levels of human pathogens. Cantaloupes had the highest number of positive samples (5), followed by scallions (3), cilantro, lettuce, and parsley (1 each).

When adjusted to account for the number of samples of each commodity collected, scallions had the highest detectable rate of human pathogens (3.2%) of the total 93 samples collected. Cantaloupe had a 3.1% rate of detectable human pathogens with 5 out of 164 samples collected testing positive. One of 85 cilantro samples tested positive for the presence of a human pathogen giving a detection rate of 1.2%. One of 90 parsley samples (1.1%) was found to have detectable levels of *Shigella* spp. and one of 142 (0.7%) lettuce samples was found to have detectable levels of *Salmonella* spp.

1.4.3 USDA MICROBIOLOGICAL DATA PROGRAM (MDP)

In 2001 the U.S Department of Agriculture (USDA) implemented a program to collect information regarding the incidence, number, and species of important foodborne pathogens and indicator organisms on domestic and imported fresh fruit and vegetable raw agricultural commodities. USDA's Agricultural Marketing Service (AMS) was appointed to undertake the program that is currently known as the microbiological data program (MDP). MDP was primarily designed to provide data on microbial presence in order to establish a microbial baseline to assess the risks of contamination, if any, in the domestic food supply.

In 2002 USDA MDP analyzed a total 10,317 samples of five raw agricultural commodities: cantaloupe, celery, leaf lettuce, romaine lettuce, and tomatoes [14]. Samples were collected in commerce at wholesale and/or distribution centers, and 86% of the samples came from domestic sources; 11% of the samples were imported; and no country of origin information was obtained for 3% of the samples.

Samples were analyzed for generic *E. coli* and *Salmonella* spp. with *E. coli* isolates being further analyzed for the presence of the following virulence factors: enterohemorrhagic shiga-toxins SLT-1 and SLT-2, hemolysin HlyA, invasive trait (intimin-eae) enterotoxigenic toxins (heat stable STa, STb; heat labile LT), enteropathogenic — invasive character (intimin eae-α), enteroaggregative — gene associated with the virulent plasmid, necrotizing cytotoxic — cytotoxic necrotizing factor (CNF-1 and 2), enteroinvasive — IpaH gene known to be associated with EIEC, and K1 capsular antigen. The presence of virulence factors does not necessarily mean that the strains isolated from the produce items are pathogenic to humans, but may have pathogenic potential.

Data presented in Table 1.4 show that of the 10,315 USDA MDP samples that were collected and analyzed for *Salmonella* spp., only 3 samples (0.03% of the total number of samples) were found to have detectable levels of *Salmonella* spp. Of the 10,276 USDA MDP samples that were collected

TABLE 1.4
Summary of the USDA MDP Analysis for *Salmonella* spp. and *E. coli* with Associated Virulence Factors for Cantaloupe, Celery, Leaf Lettuce, Romaine Lettuce, and Tomatoes

Produce item	No. of samples tested for *Salmonella* spp.	No. of samples tested for *E. coli*	No. and % of samples testing positive for *Salmonella* spp.	No. and % of samples testing positive for *E. coli* with a virulence factor
Cantaloupe	1,077	1,077	0 (0.0%)	2 (0.19%)
Celery	2,175	2,174	0 (0.0%)	3 (0.14%)
Leaf lettuce	2,180	2,161	3 (0.14%)	27 (1.25%)
Romaine lettuce	2,177	2,158	0 (0.0%)	29 (1.34%)
Tomatoes	2,706	2,706	0 (0.0%)	3 (0.11%)
Total	10,315	10,276	3 (0.03%)	64 (0.62%)

Adapted from USDA, *Microbiological Data Program Progress Update and 2002 Data Summary,* www.ams.usda.gov/science/mpo/MDPSumm02.pdf, 2004.

and analyzed for *E. coli*, 64 samples (0.62% of the total number of samples) were found to have detectable levels of *E. coli* with associated virulence factors. Twenty-seven (1.25%) of 2161 leaf lettuce and 29 (1.34%) of 2158 romaine lettuce samples were found to have detectable levels of *E. coli* with associated virulence factors. Cantaloupe, celery, and tomato had incidence rates for the presence of *E. coli* with associated virulence factors of 0.19, 0.14, and 0.11%, respectively.

Follow-up FDA farm investigations and other information from both the agency's imported and domestic produce surveys indicated that failure to follow GAPs was often associated with the findings of pathogen contamination. In particular, inadequate manure management and lack of appropriate field and transport sanitation practices were most frequently associated with overall contamination. Specific problems included fields that were open to domestic animals or were fertilized by untreated animal manure, equipment and tools that were not being sanitized, unsanitary harvesting and/or packing equipment or practices (e.g., woven plastic bags to collect cilantro after harvest), and unsanitary methods of transportation (e.g., trucks washed with nonchlorinated water and/or cleaned infrequently) [12].

1.4.4 PRODUCE-ASSOCIATED FOODBORNE ILLNESS TRACEBACK INVESTIGATION RESULTS

Traceback investigations have yielded no definitive information as to the causes of recent produce-associated foodborne illness outbreaks. The inability to identify clearly where contamination occurred and the actual causes of recent foodborne illness outbreaks associated with produce consumption is frustrating to the industry and regulators alike and is a significant hurdle to developing a means of ensuring that similar outbreaks do not recur. Without science-based data that clearly identify the cause of recent foodborne illnesses associated with produce consumption, only speculation and opinion can be used to hypothesize about what may have gone wrong. It is imperative that industry, academia, government, and consumers collaborate and take an active role by working together on developing and implementing measures that enhance produce food safety. Guzewich [15] reported in a summary of produce-related outbreak farm investigations that the practices most likely to have contributed to numerous recent outbreaks related to produce consumption are:

- Questionable practices regarding safe water use.
- Inadequate animal management (domestic and/or wild animals).
- Unsanitary facilities and equipment.
- Inadequate employee health and hygiene practices.

It is important that future investigations do not simply focus on the suspected primary causes of produce contamination in the supply chain, but allow for identification of hitherto unidentified actual causes of produce contamination. Regulatory agency traceback investigations of facilities suspected of being involved in a foodborne illness outbreak must focus on determining the efficiency and effectiveness of the facilities' GAP program and attempt to identify clearly if the contamination occurred due to non-compliance with GAPs or due to deficiencies in GAPs as they are currently formulated.

1.5 POTENTIAL SOURCES OF PRODUCE CONTAMINATION BY HUMAN PATHOGENS

While produce quality can be judged by outward appearance based on such criteria as color, turgidity, and aroma, food safety cannot. Casual inspection of produce cannot determine if it is in fact safe and wholesome to consume. Most fresh fruits and vegetables are grown in nonsterile environments, and conventional fruit and vegetable growers have less control over conditions in the production field as compared to an enclosed production or food preparation facility. The surfaces of produce have natural microflora composed of microorganisms that are generally benign. However, low-level contamination of produce with pathogenic microorganisms may sporadically occur. Production, harvesting, washing, cutting, slicing, packaging, transporting, and preparation all offer opportunities for produce contamination. While it is well established from the data presented above that the vast majority of produce contamination with human pathogens occurs in postproduction situations (Figure 1.1), if contamination does occur during growing and initial postharvest handling of produce, the consequences can be far greater. This is due to the potential for amplification of human pathogens throughout distribution and the increased risk of cross contamination posed by handling a food product contaminated with a human pathogen.

1.5.1 FOOD SAFETY RISK FACTORS ASSOCIATED WITH PRODUCTION OF FRESH PRODUCE

Management of growing conditions is of paramount importance in preventing the contamination of fresh produce by human pathogens. There are risk factors to consider such as growing conditions, agricultural practices used by specific growers, the time of year, growing region/environment, and management practices that may change over the course of a season. Climate, weather, water quality, soil fertility, pest control, as well as irrigation, and other management practices are difficult to integrate towards the development and implementation of microbial risk prevention and reduction programs on the farm [16].

Organic foods including organic fresh fruits and vegetables are one of the fastest-growing segments of the U.S. food industry, and there are many product claims among organic producers and handlers that organic products are safer and more nutritious. Only a limited number of studies have been conducted comparing conventional versus organic fruit and vegetable production practices and the effects on product food safety risk. There is currently no scientific evidence to support claims that organically grown fruits and vegetables are either safer or pose a greater food safety risk than conventionally grown produce [17–19].

1.5.1.1 Land Use

The safety of fruits and vegetables grown on any given piece of land is not only influenced by the current agricultural practices but also by former land use practices. Human pathogens may persist in soils for long periods of time [20–22]. There may be increased risk of soil contamination if production land was previously used as a feedlot or for animal grazing since fecal contamination of the soil may be extensive. However, it is difficult to determine exactly the magnitude of the risk as the persistence of human pathogens in soil varies by the pathogen in question, soil type, climate, irrigation regimes, initial pathogen population numbers, etc. [23].

1.5.1.2 Soil Amendments

Soil amendments are commonly but not always incorporated into agricultural soils used for fruit and vegetable production to add organic and inorganic nutrients to the soil as well as to reduce soil compaction. Human pathogens may persist in animal manures for weeks or even months [24,25]. Proper composting via thermal treatment will reduce the risk of potential foodborne illness. However, the persistence of many human pathogens in untreated agricultural soils is currently unknown and under extensive investigation [26–28].

1.5.1.3 Wild and Domestic Animal Control

Wild and domestic animals such as birds, deer, dogs, rodents, amphibians, insects, and reptiles are known to be potential reservoirs for human pathogens and their feces may facilitate the spread of human pathogens in agricultural production settings, packinghouses, processing, and during distribution [29–31]. Food processing, warehousing, and distribution facilities routinely have animal control programs in place to prevent contamination of fruits and vegetables. However, production agriculture in open fields is challenged by infestation of wildlife and has only a limited number of remedies available to deal with periodic infestations by these pests. There is little or no data available for production agriculture operations to assess the risks associated with the presence of a particular wild animal species in production fields, field harvesting equipment, and/or in an adjacent field. While a zero tolerance for the presence of wild animals in production environs would potentially eliminate the risk of produce contamination, such operating procedures are simply impractical if not impossible to implement.

1.5.1.4 Irrigation Water

Irrigation water is another potential vector by which contaminants may be brought in contact with fruits and vegetables. Well water is perceived to be less likely to be contaminated with human pathogens than surface water supplies, due to the limited access to sources of potential contamination. Production agriculture operations routinely test irrigation water sources for the presence of human pathogens and/or indicator microorganisms. However, such testing is of only limited value, particularly for flowing surface water sources, since water tested at any given point in time will not necessarily be the same water used to irrigate crops in the future. Whenever water comes in direct contact with edible portions of fruits and vegetables, particular care should be taken to ensure that the water does not contain human pathogens. Pesticide application with contaminated water is thought to be the cause of the 1996 cyclosporosis outbreak associated with fresh raspberries grown in Guatemala [32–34], and recent research has demonstrated that commonly used pesticides and fungicides do not significantly affect the survival or growth of human pathogens [35].

Irrigation water if contaminated with human pathogens may contaminate soils, and splashing of soils by irrigation or heavy rain may facilitate produce contamination [36]. A number of recent studies have also indicated that fresh produce may be contaminated by root uptake of human pathogens during irrigation with contaminated water [37,38]. Other research reports have indicated that this phenomenon does not occur [39,40]. It is currently unclear if root uptake of human pathogens is a significant source of contamination of fresh produce. However, direct contact of contaminated water with edible potions of crops is an obvious means of produce contamination by human pathogens.

1.5.1.5 Harvest Operations

During harvesting operations field personnel may contaminate fresh fruits and vegetables by simply touching them with an unclean hand or knife blade. Portable field latrines as well as hand wash stations are routinely made available and used by harvest personnel. Monitoring and enforcement of field worker personal hygiene practices such as hand washing after use of field latrines are critical to reduce the risk of human pathogen contamination on fresh produce. Due to the potential for contamination, produce once harvested should not be placed on bare soils before being placed in clean and sanitary field containers [41]. Field harvesting tools should be clean, sanitary, and when possible not be placed directly in contact with soil. Harvest ladders are commonly used to harvest tree fruit and may serve as a potential source of contamination, if soiled ladder rungs are handled by pickers to move the ladder. Therefore, ladders should be constructed in a sanitary manner so as to allow the easy movement of the ladder without the fruit picker having to grip the ladder rungs. Reusable field harvest containers must also be cleaned and sanitized on a regular basis to reduce the potential for cross contamination.

1.5.2 FOOD SAFETY RISK FACTORS ASSOCIATED WITH POSTHARVEST HANDLING OF PRODUCE

Depending upon the commodity, produce may be field packaged in containers that will go all the way to the destination market or may be temporarily placed in bulk bins, baskets, or bags that will be transported to a packing shed. Employees, equipment, cold storage facilities, packaging materials, and any water that directly or indirectly contacts harvested produce must be kept clean and sanitary to prevent contamination.

1.5.2.1 Employee Hygiene

Human beings are a significant reservoir for human pathogens and therefore gloves, hairnets, and clean smocks are routinely worn by packinghouse employees and field harvest crews to reduce the potential for contamination of fresh produce during handling. The cleanliness and personal hygiene of employees handling produce at all stages of production and handling must be managed to minimize the risk of contamination. Availability of adequate restroom facilities and hand washing stations and their proper use are critical to preventing contamination of produce by employees. Shoe or boot cleaning stations may also be in place to reduce the amount of field dirt and potential contamination from field operations that may enter packing sheds, processing plants, and distribution centers. Employee training regarding sanitary food handling practices, in a language in which employees are fluent, is essential to reducing the potential for employees contaminating food products that they are handling. This is particularly difficult in the produce industry as employees are often seasonal or temporary contract employees; thus a strategy of repetitive training is often needed.

1.5.2.2 Equipment

Recent research has demonstrated that unsanitary packinghouse facility equipment may play a major role in contaminating fresh fruits and vegetables if packinghouse facility food contact surfaces such as conveyor belts and dump tanks that convey produce are not cleaned and sanitized on a regularly scheduled basis with food contact surface approved cleaning compounds [42,43]. Sanitizers to be effective should be used only after thorough cleaning with mechanical action to remove organic materials such as dirt or plant materials. Food processing plants and equipment associated with them are normally designed with wash-down sanitation in mind. However, sanitary design of facilities and equipment used to handle raw agricultural commodities has received only limited attention. Therefore, there are currently no universally accepted standards for equipment or sanitary design for facilities that handle raw agricultural commodities. Rough postharvest handling at packinghouse facilities should be avoided to reduce mechanical damage and punctures to fruit which may allow for the introduction of plant spoilage pathogens via these wounds, as this has been demonstrated to enhance the potential for growth and survival of some human pathogens [44].

1.5.2.3 Wash and Hydrocooling Water

All water that comes in contact with produce for drenching, washing, hydrocooling, or vacuum cooling must be of sufficient microbial quality to prevent contamination. Recirculated water should have sufficient quantities of an approved wash water disinfectant to reduce the potential for cross contamination of all produce in the drenching, washing, or hydrocooling system. Wash water disinfectants are not capable of sterilizing the surface of produce. Research has demonstrated that washing produce in cold chlorinated water will reduce microbial populations by two or three log units (100- to 1000-fold), but complete elimination of microbes is never achieved because microorganisms adhere so tenaciously to the surface of produce and may be present in microscopic hydrophobic areas on the produce surface [2,3] or in inaccessible attachment sites (stomata, lenticels, punctures). Rinsing produce with water that contains a wash water disinfectant will significantly reduce the number of microorganisms present on the produce but it will not remove or inactivate all bacteria. Human pathogens cannot be completely removed from produce by washing in cold chlorinated water [20,45]. (See Chapter 17 for more details.)

It is particularly important that water used for hydrocooling produce be free of pathogenic microorganisms, as when warm produce is placed in cold water, intercellular air spaces within fruits and vegetables contract, creating a partial vacuum (pressure differential). This has been demonstrated to facilitate infiltration of water, which may contain human pathogens, into fresh produce items. While this phenomenon is known to be an important source of plant pathogen infections during postharvest handling of fruit and vegetables [46–49], only recently has direct evidence been brought forward to show that human pathogens may enter produce by this same mechanism. In a follow-up investigation of potential sources of imported mango contamination, Penteado et al. [50] provided evidence that Salmonella spp. may be internalized in fresh mangoes during simulated postharvest hot water insect disinfestation procedures which included a water bath cooling step [51]. However, Richards and Beuchat [52] demonstrated that adhering to or infiltrating of S. Poona cells into cantaloupe tissue via the stem scar is not dictated entirely by the temperature differential between the melon and the immersion solution containing salmonella cells, but it is also influenced by properties unique to tissue surfaces.

1.5.2.4 Cold Storage Facilities

Cold storage facilities and, in particular, refrigeration coils, refrigeration drip pans, forced air cooling fans, drain tiles, walls, and floors are potential harborages for human pathogens and as such should be cleaned and sanitized on a frequent and regular basis. *Listeria monocytogenes* can proliferate quite slowly at refrigerated temperatures and may contaminate cold stored produce if condensation from refrigeration units or the ceiling drips onto produce. Placing warm produce with field heat into a cold room with insufficient refrigeration capacity will cause a temperature rise in the room and, as the room cools, a fog or mist may occur. As the water condenses out of the air and onto surfaces of walls and ceilings that harbor human pathogens, contaminated condensate may end up dripping onto the stored produce. Therefore, it is imperative that sufficient cooling capacity is available when cooling produce.

1.5.2.5 Packaging Materials

Since packaging materials come in direct contact with fresh fruits and vegetables, they may serve as a potential source of contamination. Packages such as boxes and plastic bags require storage in such a manner as to protect them from insects, rodents, dust, dirt, and other potential sources of contamination. All packaging materials cannot be stored inside enclosed facilities due to space constraints. However, if packaging materials are stored outside an enclosed building, sufficient precautions should be taken to reduce the probability of rodent/animal infestation, and measures should be taken to allow for easily identifiable indicators of an infestation. Plastic field bins and totes are preferred to wooden containers, since plastic surfaces are more amenable to cleaning and sanitizing, which should be done after every use to reduce the potential for cross contamination. Wooden containers or field totes are almost impossible to surface sanitize since they have a porous surface. Cardboard field bins if reused should be visually inspected for cleanliness and lined with a polymeric plastic bag before reuse to prevent the potential risk of cross contamination.

1.5.2.6 Modified Atmosphere Packaging of Fresh Produce

The risk of *Clostridium botulinum* on ready-to-eat modified atmosphere packaged (MAP) fresh-cut fruits and vegetables has been investigated extensively by a number of research groups in recent years [53–57]. *C. botulinum* is a spore-forming bacterium commonly found in agricultural environs. Under suitable environmental conditions (temperatures above 5°C, low oxygen conditions, and a pH above 4.6) this microorganism may produce a deadly toxin. Recent research efforts have examined *C. botulinum* risk factors for various fresh-cut MAP produce. In general, overt gross spoilage of fresh-cut produce occurs well before toxin is produced on shredded cabbage, shredded lettuce, broccoli florets, sliced carrots, and rutabaga. The endemic microflora on fresh-cut produce play an important role in signaling the end of shelf life and are also believed to suppress toxin production by *C. botulinum*

[58]. However, some products such as butternut squash and onions have been demonstrated under temperature abuse conditions to have the potential of appearing acceptable although containing botulinal toxin [53]. The important interaction between MAP and microbial food safety must always be considered, and continued research efforts to understand fully these relationships are currently underway. An in-depth assessment of the risk of botulism contributed by MAP of fresh-cut produce may be found in Gorny et al. [59]. Several studies at research institutes have found that MAP technologies commonly used in the fresh-cut industry have varying effects on the survival and growth of E. coli O157:H7, Salmonella spp., Shigella spp., and L. monocytogenes [60–63]. While some pathogenic strains may be inhibited, others are unaffected, weakly inhibited, or even stimulated. Because L. monocytogenes can grow at refrigeration temperatures, there is concern that low inoculum levels, coupled with extended shelf life obtained by the use of MAP, may allow L. monocytogenes to proliferate to infectious dosages late in shelf life. The FDA recently reviewed the risk associated with consumption of fresh fruits and vegetables as well as 20 other ready-to-eat food categories and published, as a draft, a risk assessment on the relationship between foodborne L. monocytogenes and human health (www.fda.gov). Risk from human pathogens due to the use of MAP must be assessed on a per product basis. This is due to the complex interactions between the produce, the indigenous microflora, the pathogen, and its environment. An excellent example of this interaction is the inhibitory effect of carrot extract on growth of L. monocytogenes [64]. Due to these complex interactions, broad generalities cannot be drawn regarding the risk of specific human pathogens on various fresh-cut fruits or vegetables and interactions with MAP.

1.5.2.7 Refrigerated Transport, Distribution, and Cold Storage

Produce is best shipped in temperature-controlled refrigerated vehicles. Maintaining perishables at their appropriate temperature when being transported to destination markets will extend shelf life. When appropriate, holding fresh fruits and vegetables at or below 5°C will significantly reduce the growth rate of microbes including human pathogens. However, cold temperatures and high relative humidity conditions which are often optimal for shelf life extension of fresh fruits and vegetables may actually help favor the viability of some human pathogens such as viral particles.

Trucks used during transportation are also a potential source of contamination from human pathogens. Therefore, trucks should be routinely cleaned and sanitized on a regular basis, and trucks that have been used to transport live animals, animal products, or toxic materials should not be used to transport produce or used only after effective cleaning and sanitation.

1.5.3 FOOD SAFETY RISK FACTORS ASSOCIATED WITH FOODSERVICE, RESTAURANT, AND RETAIL FOOD STORES HANDLING OF PRODUCE

In 2003 the FDA collected data via site visits to over 900 establishments representing nine distinct facility types including restaurants, institutional foodservice operations, and retail food stores. Direct observations of produce handling practices were supplemented with information gained from discussions with management and food workers and were used to document the establishments' compliance status based on provisions in the 1997 Model FDA Food Code [65].

Failure to control product holding temperatures, poor personal hygiene, use of contaminated equipment/failure to protect food handling equipment from contamination, and risk of potential chemical contamination were the risk factors found to be most often out of compliance with the 1997 FDA Model Food Code. The percentages of "out of compliance" observations for each of these risk factors were found to be: improper holding time/temperature (49.3%), poor personal hygiene (22.3%), contaminated equipment (20.5%), and chemical contamination (13.5%). Specifically, for the improper holding time and temperature risk factor, it was found that maintaining cold holding temperatures at or below 5°C (41°F) for produce items that are classified as potentially hazardous foods (PHFs) did not occur in 70.2% of the observed situations. Holding PHFs at or below 5°C (41°F) is critical to preventing the potential growth of human pathogens, which may rapidly proliferate on inadequately refrigerated PHFs. Date marking of refrigerated ready-to-eat PHFs is also an important component of any food safety system, and it is designed to promote proper food rotation and limit the growth of *L. monocytogenes* during cold storage. However, appropriate date marking of ready-to eat PHF produce items made on-site did not occur in 34% of the observations.

The personal hygiene risk factors associated with produce that are most in need of attention at retail and foodservice operations include adequate, available, and accessible hand washing facilities. These personal hygiene risk factors were found by the survey to be not in compliance with the 1997 FDA Model Food Code 33.3, 26.2, and 20.6% of the time, respectively. Hands are very common vehicles for the transfer of human pathogens to food products, and food handlers' hands may become contaminated when they engage in activities such as handling raw meat products, using the lavatory, coughing, or handling soiled tableware.

Food safety procedures for cleaning and sanitizing food contact surfaces and utensils for handling produce were found to be not in compliance with the 1997 FDA Model Food Code in 44.4% of the observations in this study. Proper cleaning and sanitization of food contact surfaces is essential to preventing cross contamination. The 2004 FDA report clearly indicates that foodservice and retail operators must ensure that their produce food safety management systems are designed to achieve active managerial control over the risk factors associated with handling produce identified in the report.

1.5.4 CONSUMER HANDLING OF PRODUCE FROM PURCHASE TO PLATE

Li-Cohen and Bruhn [66] in 2002 published the most extensive consumer handling study of fresh produce from the time of purchase to the plate. Via a national mail survey of 624 respondents these researchers quantified consumer produce handling practices as they relate to food safety risk. Six percent of consumer respondents replied that they never or seldom wash fresh produce before consumption, and greater than 35% of respondents do not wash melons before consumption. Approximately half of all respondents did not wash their hands before handling fresh produce. Ninety-seven percent of all respondents reported that they always washed food preparation surfaces after contact with raw meat products. However, 5% of respondents only dry wipe, and 24% of respondents wash these potentially contaminated food preparation surfaces only with water (without soap or a disinfectant). This survey also found that many respondents did not separate produce from raw meat, poultry, or fish in their refrigerators. These limited observations clearly indicate the need for educational outreach to consumers that must emphasize safe handling practices of produce from purchase to consumption.

1.6 EFFECTIVE MANAGEMENT STRATEGIES: CONTAMINATION PREVENTION AND INTERVENTION

Every foodborne illness outbreak is a tragic event, and an approach that prevents contamination and possible amplification of human pathogens in the produce supply chain is the most effective means of ensuring fresh produce safety. However, the complexity of effectively implementing this strategy is stated concisely by the FDA [16]:

> Although the available scientific literature is adequate to identify sources of contamination and estimate microbial persistence on plants, the specific influence and interactions among the production environments and crop management practices are not sufficiently understood to provide detailed guidance to growers and shippers. Also, the diversity of cropping systems, scale of operation, use and design of equipment, regional and local practices, environmental influences, specifics of on-farm soil related factors, and many other production factors defy any attempt to develop an encompassing assignment of microbial risk to commodities or to crop management practices.

Sampling produce is not an effective means of ensuring product safety. Data from the USDA MDP and FDA domestic and imported produce sampling surveys indicate that human pathogens are found on fresh produce infrequently and in low numbers. Because of this fact increased sampling for the presence of human pathogens by either private enterprises or government regulators will not effectively reduce foodborne illnesses associated with produce consumption because it is simply an ineffective strategy. Increased produce sampling or surveillance would also potentially take valuable limited resources away from potentially more productive research efforts that identify risk factors and mitigation strategies.

Approaches that prevent contamination are warranted and these strategies include effective management and intervention strategies for growing, handling, distributing, and preparing fresh produce that include but are not limited to:

- Good Agricultural Practices (GAPs)
- Good Manufacturing Practices (GMPs)
- Hazard Analysis Critical Control Point (HACCP) programs

1.6.1 GOOD AGRICULTURAL PRACTICES (GAPs)

The FDA published *Guidance for Industry: Guide to Minimize Microbial Food Safety Hazards for Fresh Fruits and Vegetables* in 1998 which has since come to be referred to as Good Agricultural Practices (GAPs). Although this document carries no regulatory or legal weight, due diligence requires producers to take prudent steps to prevent contamination of their crops. GAPs have been widely implemented by the fresh fruit and vegetable industry and as formulated provide the produce industry with an excellent description of broad prescriptive actions that may be taken to enhance produce food safety. Numerous retail and wholesale buyers have made compliance to GAPs, and subsequent independent third-party audits to ensure compliance with GAPs, a requirement for the purchase of fresh fruits and vegetables.

The guide identifies eight principles of food safety within the realms of growing, harvesting, and transporting fresh produce and suggests that the reader "use the general recommendations in this guide to develop the most appropriate good agricultural and management practices for your operation." The application of these principles is aimed at preventing contamination of fresh produce with human pathogens. The eight principles are listed below followed by areas of implementation:

1. Prevention of microbial contamination of fresh produce is favored over reliance on corrective actions once contamination has occurred.
2. To minimize microbial food safety hazards in fresh produce, growers or packers should use GAPs in those areas over which they have a degree of control while not increasing other risks to the food supply or the environment.
3. Anything that comes in contact with fresh produce has the potential of contaminating it. For most foodborne pathogens associated with produce, the major source of contamination is associated with human or animal feces.

4. Whenever water comes in contact with fresh produce, its source and quality dictate the potential for contamination.
5. Practices using manure or municipal biosolid wastes should be closely managed to minimize the potential for microbial contamination of fresh produce.
6. Worker hygiene and sanitation practices during production, harvesting, sorting, packing, and transport play a critical role in minimizing the potential for microbial contamination of fresh produce.
7. Follow all applicable local, state, and federal laws and regulations, or corresponding or similar laws, regulations, or standards for operators outside the U.S. for agricultural practices.
8. Accountability at all levels of the agricultural environment (farms, packing facility, distribution center, and transport operation) is important to a successful food safety program. There must be qualified personnel and effective monitoring to ensure that all elements of the program function correctly and to help track produce back through the distribution channels to the producer.

It is currently unclear if recent outbreaks associated with consumption of produce are due to lack of compliance with GAPs or if there are deficiencies in GAPs as they are currently formulated. Little scientifically based data exist regarding the risk associated with many of the production and postharvest handling practices commonly used in production agriculture and in postharvest handling situations or what the most effective risk management strategies may be.

1.6.2 Current Good Manufacturing Practices (cGMPs)

The cGMPs are set forth in 21CFR110 and provide guidelines that ensure that food for human consumption is safe and has been prepared, packed, and held under sanitary conditions. The cGMPs provide food processors, such as fresh-cut produce processors, with the core principles of sanitary food handling, and they serve as well-recognized and agreed upon standards of conduct and operation. The cGMPs are well written in that they provide general guidance regarding regulatory expectations of performance and conduct without being overly specific or prescriptive, and this aspect of the cGMPs accommodates the many diverse specific situations that are encountered in the food industry today. The regulations as currently written provide flexibility for the diverse formats under which these regulations are applied, by use of terminology such as "adequate facilities," "where appropriate," "necessary precautions," and "adequate controls." This flexibility allows the cGMPs to be applied to the plethora of situations encountered during the production, handling, and distribution of food products. Also, and very importantly, by not being overly prescriptive the cGMPs allow for incorporation of new technologies and innovation without the need to revise the regulations. The cGMPs are the commonly agreed upon and scientifically based standards by which industry and regulators effectively and harmoniously communicate the standards of performance and conduct when food products are being prepared, packed, or held. As such the cGMPs are centrally important in reducing the risk of product adulteration and food safety risk to consumers.

21CFR110.19 specifically exempts raw agricultural commodities from compliance with cGMPs, and raw agricultural commodity safe production and postharvest handling practices are not as clearly defined and commonly agreed upon as cGMPs and HACCP in the food processing industry. Therefore, raw agricultural commodities producers and handlers do not have the advantage of simply adopting long-standing food safety programs that exist in the food processing industry, as they must modify these programs on a site-specific basis.

1.6.3 HAZARD ANALYSIS CRITICAL CONTROL POINT (HACCP)

Hazard Analysis Critical Control Point (HACCP) is a systems approach method to ensure the safety of a food product. The terms HACCP and food safety program are often used interchangeably and synonymously. However, HACCP is not the equivalent of a food safety program, as HACCP is merely a component of an overall food safety program. The terms food safety program and HACCP are not interchangeable and should not be used synonymously. A HACCP plan cannot be established without prerequisite programs such as GAPs, cGMPs, and sanitation standard operating procedures (SSOPs) being in place. HACCP is a food safety system pioneered by the Pillsbury Co. to reduce the risk associated with the food eaten by astronauts for manned space flights. HACCP is a systems approach that:

- Identifies potential sources of contamination in food production systems.
- Establishes methods for detecting the occurrence or prevention of contamination.
- Clearly prescribes what corrective actions will be taken to prevent consumption of contaminated food items.

The National Advisory Committee on the Microbiological Criteria for Foods (NACMCF) has clearly defined what HACCP is in a 1997 document entitled *HACCP Principles and Application Guidelines* (available on line at: www.fst.vt.edu/haccp97/). In this document, NACMCF clearly defines seven criteria that must be met by a HACCP program [67]. The seven basic principles of HACCP are:

1. Assessment of hazards.
2. Determine critical control points (CCPs) to control the identified hazards.

3. Establishment of limits at each CCP.
4. Establishment of CCP monitoring procedures.
5. Establishment of corrective actions to be taken when CCPs exceed set limits.
6. Establishment of record keeping systems to document the HACCP program.
7. Establishment of procedures to verify that the HACCP is functioning properly.

HACCP is described as a management system — designed for use in all segments of the food industry from growing, harvesting, processing, manufacturing, distributing, and merchandising to preparing food for consumption. The NACMCF committee endorsed HACCP as an effective and rational means of ensuring food safety from harvest to consumption [67]. However, if all of the above criteria cannot be met, then a HACCP plan cannot be established and HACCP may not be the appropriate food safety solution for the process under consideration. This does not mean that process hazards should be ignored but simply that the risks and hazards associated with a process need to be dealt with via an alternative mechanism. Another important aspect of any HACCP program is prerequisite ability to monitor quantitatively critical control points. If one cannot monitor and control important process critical control points then HACCP is not appropriate. Food safety programs such as HACCP and cGMPs are well defined and may function well within the control environs of a food processing plant; however, these food safety program components may not be appropriate in production agriculture situations. For example, as food handling operations move from a confined four-walled food processing facility to a three-walled packinghouse operation and/or back to an open agricultural growing operation, it is obvious that not all cGMPs and/or HACCP requisites could possibly be implemented.

The fresh-cut produce industry strongly believes that HACCP is an effective means of enhancing food safety by control of chemical, physical, and biological hazards that are reasonably likely to occur in the absence of controls. HACCP systems may be considered for intact and fresh-cut produce only when sufficient information and data have been gathered to establish appropriate preventive control measures (FDA, 1998).

It is unclear if HACCP can or should be used as a component of a food safety program for production agriculture. HACCP as currently formulated by NACMCF cannot be used as a food safety program for production agriculture. However, risk reduction and mitigation must be evaluated and implemented in production agriculture to enhance produce food safety. See Chapter 15 for a comprehensive discussion of HACCP.

1.7 RESEARCH NEEDS

Everyone in the produce handling continuum must understand the food risks that they are facing, because if these risks are not clearly understood then they cannot be appropriately addressed and managed. Speculative actions that attempt to reduce produce food safety risks, if incorrect, potentially take limited food safety resources away from actual risks which have not been addressed while adding to the perception that the issue has been addressed, and raising expectations. Enhanced research efforts and financial support are needed to identify clearly means of intervention and quantify how much risk is reduced by specific actions, so that limited food safety resources can most effectively be deployed.

There are a number of food safety issues related to fresh and fresh-cut produce production and handling that warrant further investigation to gain a better basic understanding of how human pathogens and produce interact. A better understanding of this interaction will aid in the development of intervention strategies and increase the safety of the food supply. Five areas of research that are of high priority for the fresh and fresh-cut produce industries are discussed in the following.

1.7.1 MICROBIAL ECOLOGY OF HUMAN PATHOGENS IN THE AGRICULTURAL PRODUCTION ENVIRONMENT

Human pathogens in agricultural/farm environs are typically present in low numbers and frequency, making their investigation difficult if not impossible. Preventing human pathogen contamination of produce is currently the most effective means of reducing foodborne illness risk. However, there is a significant lack of information regarding human pathogens on the farm and in postharvest produce environments. Understanding the microbial ecology, persistence, niches, harborages, life cycle, and factors affecting survival and growth of human pathogens in an agricultural/farm environment, including water and soil amendments, is essential to developing and implementing intervention and control measures to reduce the risk of contaminating fresh produce.

1.7.2 AGRICULTURAL WATER

GAPs rely on management practices that prevent contamination of produce on the farm and during postharvest handling operations. Water is a significant potential source of human pathogens in the farm environment. Ensuring that agricultural water is of sufficient microbial quality for its intended purpose is critical in ensuring the safety of produce. Therefore, identification of better methods to determine the food safety risk associated with a particular irrigation water source for a particular use warrants further investigation. Potential lines of investigation include identification of indicator organisms that highly correlate with the presence/absence of viable human pathogens.

1.7.3 SOIL AMENDMENTS

Identification of better methods to determine the food safety risk associated with a particular lot of composted manure to be used as a soil amendment is warranted. Identification of indicator microorganisms that correlate well with the presence/absence of viable human pathogens is needed. Determination of the time/temperature history and other composting variables that affect the survival of human pathogens in compost is also needed.

1.7.4 PROXIMITY RISK OF POTENTIAL CONTAMINANT SOURCES

No produce operation is an island unto itself. Therefore it is important to assess risks posed by adjacent agricultural and nonagricultural operations that are known to be potential sources of human pathogens. Greater understanding and quantification of risk posed by such adjacent operations is needed to formulate strategies to reduce risk. Simply put, how close is too close? What factors should be contemplated when assessing the risk of adjacent operations to agricultural production and postharvest handling operations, and what mitigation steps would be effective to reduce risk?

1.7.5 INTERVENTION STRATEGIES TO REDUCE THE RISK OF HUMAN PATHOGEN CONTAMINATION OF FRESH PRODUCE

Aqueous-based wash water disinfectants do not achieve significant reductions in microbial populations of human pathogens on fresh produce. Investigation of alternative nonaqueous-based disinfectants on produce, such as the use of vapor phase ozone and chlorine dioxide disinfection technologies, warrants further investigation.

1.8 SUMMARY

Produce contamination by a multitude of human pathogens can occur anywhere in the produce continuum from field to fork, and once contamination occurs, no effective interventions exist to eliminate human pathogens from fresh fruits and vegetables. Although there are many potential scenarios for produce contamination to occur, no science-based risk assessment has clearly identified and quantified the risk associated with various produce handling steps from field to fork. A better understanding of risk factors associated with produce handling practices is needed, so that more effective intervention strategies may be developed to enhance produce food safety and reduce the incidence of foodborne illnesses associated with fresh fruit and vegetable consumption. To date, a preventative approach to contamination of fresh fruits and vegetables by the use of GAPs, cGMPs, and HACCP has proven to be the most effective means of ensuring produce food safety. It is imperative that public health officials and industry establish standardized metrics and baseline data regarding produce-associated foodborne illnesses and the risks associated with various handling practices. Data detailing foodborne illnesses associated with produce consumption must be indexed and standardized to ensure that the data that are being reported, accurately reflect actual illness incidence trends, and are not simply reporting anomalies due to increased surveillance, improved detection techniques, or increased per capita consumption of a specific commodity. Without the ability to quantify accurately foodborne illness and compare data over a prolonged period of time, it will be impossible to measure accurately progress and the efficacy of enhanced produce safety activities and tactics that are being implemented to reduce the incidence of produce contamination with human pathogens.

REFERENCES

1. Produce for Better Health Foundation, 2004 Research Summaries. www. 5aday.com/html/research/summaries.
2. Brackett, R.E., Antimicrobial effect of chlorine on *Listeria monocytogenes*, *J. Food Prot.*, 50, 999, 1987.
3. Seo, K.H. and Frank, J.F., Attachment of *Escherichia coli* O157:H7 to lettuce leaf surface and bacterial viability in response to chlorine treatment as demonstrated by using confocal scanning laser microscopy, *J. Food Prot.*, 62, 3, 1999.
4. Alliance for Food and Farming, *Analysis of Produce Related Foodborne Illness Outbreaks*, Alliance for Food and Farming, April 2004, www.foodandfarming. info/documents/85876_produce_analysis_604.pdf.
5. CDC, U.S. Foodborne Disease Outbreaks Report Annual Databases for 1996–2001, www.cdc.gov/foodborneoutbreaks/us_outb.htm, 2001.
6. CDC, U.S. Foodborne Disease Outbreaks Report Searchable Database 1990–1995, www2.cdc.gov/ncidod/foodborne/fbsearch.asp, 1996.
7. Sivapalasingham, S., Friedman, C.R., Cohen, L., and Tauxe, R.V., Fresh produce: a growing cause of outbreaks of foodborne illness in the United States, 1973–1997, *J. Food Prot.*, 67, 10, 2004.
8. Harris, L.J., Farber, J.N., Beuchat, L.R. Parish, M.E., Suslow, T.V., Garrett, E.H., and Busta, F.F., Outbreaks associated with fresh produce: incidence, growth and survival of pathogens in fresh and fresh-cut produce, *Compr. Rev. Food Sci. Food Saf.*, 2, 78, 2003.
9. Monge, R. and Chinchilla, M., Presence of *Cryptosporidium* oocysts in fresh vegetables, *J. Food Prot.*, 59, 202, 1996.
10. Williams, K.E. and Jaykus, L., Norwalk-like viruses and their significance to foodborne disease burden, *J. Assoc. Food Drug Officials*, 4, 28, 2002.
11. Cliver, D.O., Virus transmission via food, *Food Technol.*, 51, 71, 1997.
12. FDA Survey of Imported Fresh Produce, U.S. Food and Drug Administration Center for Food Safety and Applied Nutrition, Office of Plant and Dairy Foods and Beverages, January 30, 2001, www.cfsan.fda.gov/~dms/prodsur6.html.
13. FDA Survey of Domestic Fresh Produce, U.S. Department of Health and Human Services, U.S. Food and Drug Administration Center for Food Safety and Applied Nutrition, Office of Plant and Dairy Foods and Beverages, January 2003, www.cfsan.fda.gov/~dms/prodsu10.html.

14. USDA, *Microbiological Data Program Progress Update and 2002 Data Summary*, www.ams.usda.gov/science/mpo/MDPSumm02.pdf, 2004.
15. Guzewich, J., Hazards identified in FDA farm investigations, International Association of Food Protection 90th Annual Meeting, New Orleans, LA, August 10–13, 2003.
16. FDA, Guidance for Industry: Guide to Minimize Microbial Food Safety Hazards for Fresh Fruits and Vegetables, U.S. Food and Drug Administration, U.S. Department of Agriculture, Centers for Disease Control and Prevention, October 26, 1998, http://vm.cfsan.fda.gov/~dms/prodguid.html.
17. Mukherjee, A., Speh, D., Dyck, E., and Diez-Gonzalez, F., Preharvest evaluation of coliforms, *Escherichia coli*, *Salmonella*, and *Escherichia coli* O157:H7 in organic and conventional produce grown by Minnesota farmers, *J. Food Prot.*, 67, 894, 2004.
18. McMahon, M.A. and Wilson, I.G., The occurrence of enteric pathogens and *Aeromonas* species in organic vegetables, *Int. J Food Microbiol.*, 70, 155, 2001.
19. Sagoo, S.K., Little, C.L., and Mitchell, R.T., The microbiological examination of ready to eat organic vegetables from retail establishments in the United Kingdom, *Lett. Appl. Microbiol.*, 33, 434, 2001.
20. Beuchat, L.R., Survival of enterohemorrhagic *Escherichia coli* O157:H7 in bovine feces applied to lettuce and the effectiveness of chlorinated water as a disinfectant, *J. Food Prot.*, 62, 845, 1999.
21. Islam, M., Morgan, J., Doyle, M.P., and Jiang, X., Fate of *Escherichia coli* O157:H7 in manure compost-amended soil and on carrots and onions grown in an environmentally controlled growth chamber, *J. Food Prot.*, 67, 574, 2004.
22. Kudva, I.T., Blanch, K., and Hovde, C.J., Analysis of *Escherichia coli* O157:H7 survival in ovine and bovine manure and manure slurry, *Appl. Environ. Microbiol.*, 33, 131, 1998.
23. Ingham, S., Losinski, J., Andrews, M., Breur, J., Wood, T., and Wright, T., Bacterial contamination associated with application of non-composted bovine manure to soils in low chemical input vegetable gardening, International Association of Food Protection Annual Meeting, Phoenix, AZ, August 8–11, 2004.
24. Fukushima, H., Hoshina, K., and Goymoda, M., Long-term survival of Shiga toxin-producing *Escherichia coli* O26, O111 and O157 in bovine feces, *Appl. Environ. Microbiol.*, 65, 5177, 1999.
25. Gagliardi, J.V. and Karns, J.S., Leaching of *Escherichia coli* O157:H7 in diverse soils under various agricultural management practices, *Appl. Environ. Microbiol.*, 66, 877, 2000.
26. Jiang, X., Morgan, J., and Doyle, M.P., Fate of *Escherichia coli* O157:H7 during composting of bovine manure in a laboratory-scale bioreactor, *J. Food Prot.*, 66, 25, 2003.
27. Jiang, X., Morgan, J., and Doyle, M.P., Thermal inactivation of *Escherichia coli* O157:H7 in cow manure compost, *J. Food Prot.*, 66, 1771, 2003.
28. Islam, M., Doyle, M.P., Phatak, S.C., Millner, P., and Jiang, X., Persistence of enterohemorrhagic *Escherichia coli* O157:H7 in soil and on leaf lettuce and parsley grown in fields treated with contaminated manure composts or irrigation water, *J. Food Prot.*, 67, 1365, 2004.
29. Beuchat, L.R., Pathogenic microorganisms associated with fresh produce, *J. Food Prot.*, 59, 204, 1996.

30. Fenlon, D.R., Wild birds and silage as reservoirs of *Listeria* in the agricultural environment, *J. Appl. Bacteriol.*, 59, 537, 1985.
31. Janisiewicz, W.J., Conway, W.S, Brown, M.W., Sapers, G.M., Fratamico, P., and Buchanan, R.L., Fate of *Escherichia coli* O157:H7 on fresh-cut apple tissue and its potential for transmission by fruit flies, *Appl. Environ. Microbiol.*, 65, 1, 1999.
32. CDC, Update: outbreaks of cyclosporiasis: United States, *MMWR*, 46, 461, 1997.
33. CDC, Update: outbreaks of cyclosporiasis: United States and Canada, *MMWR*, 46, 521, 1997.
34. CDC, Outbreak of cyclosporiasis: northern Virginia, Washington D.C., Baltimore, Maryland metropolitan area, *MMWR*, 46, 690, 1997.
35. Vlahovich, K.N., Bihn, E.A., Gravani, R.B., Worobo, R.W., and Churney, J.J., The detection and survival of *Salmonella, Escherichia coli* and *Listeria monocytogenes* in selected pesticide sprays used on fresh produce, International Association of Food Protection Annual Meeting, Phoenix, AZ, August 8–11, 2004.
36. Wachtel, M.R., Whitehand, L.C., and Mandrell, R.E., Prevalence of *Escherichia coli* associated with a cabbage crop inadvertently irrigated with partially treated sewage wastewater, *J. Food Prot.*, 65, 471, 2002.
37. Solomon, E.B., Pang, H., and Mathews, K.R., Persistence of *Escherichia coli* O157:H7 on lettuce plants following spray irrigation with contaminated water, *J. Food Prot.*, 66, 2198, 2003.
38. Wachtel, M.R., Whitehand, L.C., and Mandrell, R.E., Association of *Escherichia coli* O157:H7 with preharvest leaf lettuce upon exposure to contaminated irrigation water, *J. Food Prot.*, 65, 18, 2002.
39. Beuchat, L.R., Scouten, A.J., Allen, R.I., and Hussey, R.S., Potential of a plant-parasitic nematode to facilitate internal contamination of tomato plants by *Salmonella, J. Food Prot.*, 66, 1459, 2003.
40. Jablasone, J., Brovko, L.Y., and Griffiths, M.W., A research note: the potential for transfer of *Salmonella* from irrigation water to tomatoes, *J. Sci. Food Agric.*, 84, 287, 2004.
41. Guo, X., Chen, J., Brackett, R.E., and Beuchat, L.R., Survival of *Salmonella* on tomatoes stored at high relative humidity, in soil, and on tomatoes in contact with soil, *J. Food Prot.*, 65, 274, 2002.
42. Castillo, A., Mercado, I., Lucia, L.M., Martinez-Ruiz, Y., Ponce de Leon, J., Murano, E.A., and Acuff, G.R., *Salmonella* contamination during production of cantaloupe: a binational study, *J. Food Prot.*, 67, 713, 2004.
43. Gagliardi, J.V., Millner, P.D., Lester, G., and Ingram, D., On-farm and postharvest processing of bacterial contamination to melon rinds, *J. Food Prot.*, 66, 82, 2003.
44. Wade, W.N. and Beuchat, L.R., Proteolytic fungi isolated from decayed and damaged raw tomatoes and implications associated with changes in pericarp pH favorable for survival and growth of foodborne pathogens, *J. Food Prot.*, 66, 911, 2003.
45. Beuchat, L.R., *Surface Decontamination of Fruits and Vegetables Eaten Raw: A Review. Food Safety Issues*, World Health Organization, 1998.
46. Bartz, J.A. and Showalter, R.K., Infiltration of tomatoes by aqueous bacterial suspensions, *Phytopathology*, 71, 515, 1981.

47. Bartz, J.A., Infiltration of tomatoes immersed at different temperatures to different depths in suspensions of *Erwina carotovora* subsp. *carotovora*, *Plant Dis.*, 66, 302, 1982.

48. Bartz, J.A., Potential for postharvest disease in tomato fruit infiltrated with chlorinated water, *Plant Dis.*, 72, 9, 1988.

49. Bartz, J.A., Relation between resistance of tomato fruit to infiltration by *Erwinia carotovora* subsp. carotovora and bacterial soft rot, *Plant Dis.*, 75, 152, 1991.

50. Penteado, A.L., Eblen, B.S., and Miller, A.J., Evidence of *Salmonella* internalization into fresh mangos during simulated postharvest insect disinfestation procedures, *J. Food Prot.*, 67, 181, 2004.

51. FDA, *Potential for Infiltration, Survival and Growth of Human Pathogens Within Fruits and Vegetables*, Food and Drug Administration, Washington D.C., 1999.

52. Richards, G.M. and Beuchat, L.R., Attachment of *Salmonella poona* to cantaloupe rind and stem scar tissues as affected by temperature of fruit and inoculum source, *J. Food Prot.*, 67, 1359, 2004.

53. Austin, J.W., Dodds, K.L., Blanchfield, B., and Farber, J.M., Growth and toxin production by *Clostridium botulinum* on inoculated fresh-cut packaged vegetables, *J. Food Prot.*, 61, 324, 1998.

54. Hao, Y.Y., Brackett, R.E., Beuchat, L.R., and Doyle, M.P., Microbial quality and the inability of proteolytic *Clostridium botulinum* to produce toxin in film-packaged fresh-cut cabbage and lettuce, *J. Food Prot.*, 61, 1148, 1998.

55. Larson, A.E., Johnson, E.A., Barmore, C.R., and Hughes, M.D., Evaluation of the botulism hazard from vegetables in modified atmosphere packaging, *J. Food Prot.*, 60, 1208, 1997.

56. Solomon, H.D., Kautter, D.A., Lilly, T., and Rhodehamel, E.J., Outgrowth of *Clostridium botulinum* in shredded cabbage at room temperature under modified atmosphere, *J. Food Prot.*, 53, 831, 1990.

57. Farber, J.M., Microbiological aspects of modified atmosphere packaging: a review, *J. Food Prot.*, 54, 58, 1991.

58. Larson, A.E. and Johnson, E.A., Evaluation of botulinal toxin production in packaged fresh-cut cantaloupe and honeydew melons, *J. Food Prot.*, 62, 948, 1999.

59. Gorny, J.R., Brandenburg, J., and Allen, M., *Packaging Design for Fresh-cut Produce*, International Fresh-cut Produce Association, Alexandria, VA, 2003.

60. Beuchat, L.R. and Brackett, R.E., Survival and growth of *Listeria monocytogenes* on lettuce influenced by shredding, chlorine treatment, modified atmosphere packaging and temperature, *J. Food Sci.*, 55, 755, 1990.

61. Jacxsens, L., Devlieghere, F., Falcata, P., and Debevere, J., Behavior of *Listeria monocytogenes* on fresh-cut produce packaged under equilibrium-modified atmosphere, *J. Food Prot.*, 62, 1128, 1999.

62. Kallander, K.D., Hitchins, A.D., Lancette, G.A., Schmieg, J.A., Garcia, G.R., Solomon, H.M., and Sofos, J.N., Fate of *Listeria monocytogenes* in shredded cabbage stored at 5 and 25°C under modified atmospheres, *J. Food Prot.*, 54, 302, 1991.

63. Omary, M., Testin, R., Barefoot, S., and Rushing, J., Packaging effects on growth of *Listeria innocua* in shredded cabbage, *J. Food Sci.*, 58, 623, 1993.

64. Beuchat, L.R. and Brackett, R.E., Inhibitory effects of raw carrots on *Listeria monocytogenes, Appl. Environ. Microbiol.*, 56, 1734, 1990.

65. FDA, Occurrence of Foodborne Illness Risk Factors in Selected Institutional Foodservice, Restaurant, and Retail Food Store Facility Types, report, 2004, www.cfsan.fda.gov/~dms/retrsk2.html#execsum.

66. Li-Cohen, A.E. and Bruhn, C.M., Safety of consumer handling of fresh produce from the time of purchase to the plate: a comprehensive consumer survey, *J. Food Prot.*, 65, 1287, 2002.

67. NACMCF, Hazard analysis and critical control point: principles and application guidelines as adopted on August 14, 1997 by the National Advisory Committee on Microbiological Criteria for Foods (NACMCF), *J. Food Prot.*, 61, 1246, 1998, http://vm.cfsan.fda.gov/~comm/nacmcfp.html.

2 Attachment of Microorganisms to Fresh Produce

Robert E. Mandrell, Lisa Gorski, and Maria T. Brandl

CONTENTS

2.1   INTRODUCTION

Microbes are crucial to plant life, and, therefore, to the successful production of produce as a commodity. Plant microbes can be beneficial as symbionts [1–3], competitors of plant pathogens for biocontrol [4,5], and for promoting plant growth [6–8]. Indeed, most of the fundamental knowledge of the biology, microbial ecology, and genetics of plants has been obtained as a result of studies to understand and prevent plant disease. However, plants also are vulnerable during growth to microbial pathogens from the environment (e.g., soil, water, air, amendments). The links between fresh produce/produce dishes with more than 300 outbreaks in the U.S. since 1990 [9,10], and the obvious vulnerability of preharvest produce to pathogens in the production environment, have stimulated similar basic studies of the biology of enteric pathogens on produce.

The sources of the microorganisms exposed to plant surfaces may be from the plant seed itself [11–13], and through the initial contact with soil, irrigation water and air. The microbial communities of the rhizosphere (roots and the part of the soil affected by contact with roots) and the phyllosphere (leaves and the environment in contact with leaves, e.g., water, air) of produce are in constant change due to factors that affect microbes, such as humidity, temperature, nutrients, UV radiation, insects, and wild animals. Plant tissues are in close contact with potentially thousands of different species of bacteria, viruses, and other microorganisms [14]. Fruit and vegetable crops also have a rich microbial flora, including in many cases coliforms and fecal coliforms that are unavoidable considering the presence of domestic and wild animals near production environments [15–18]. Indeed, 190 produce-associated outbreaks have been documented in the U.S. for the years 1973–1997 [19]. Plant bacteria have evolved multiple mechanisms suitable for initiating interactions essential for successful colonization of plants [3,20,21]. However, a major interest of those working on microbial food safety of produce is whether there are equivalent, or similar, mechanisms of attachment used by human pathogens that contaminate produce commodities.

Considering the analogous or common secretion systems, outer surface proteins, and polysaccharides among the plant and human pathogens, and similarities among disease-associated genes of humans and plants (e.g., *Arabidopsis thaliana*), this area of investigation perhaps offers more promise than many plant and animal pathogen researchers perceived initially [22]. An ultimate goal of studies of attachment of human pathogens to plants is the development of intervention methods to minimize attachment and survival of human pathogens.

Our goal in this chapter is to review current knowledge of perhaps the most important event that initiates the association between most microorganisms and plants: *attachment*. We focus our attention on studies that provide insight into fundamental molecular plant–microbe interactions. The most definitive knowledge on plant–microbe interactions involving attachment has been provided in two areas: (1) mechanisms of disease of bacterial plant pathogens, and (2) molecular mechanisms involved in the symbiotic relationship between nitrogen-fixing bacteria and plants. These studies provide a context for assessing the potential mechanisms of attachment of enteric human pathogens to produce.

2.2 BASIC ANATOMY AND BIOCHEMISTRY OF ROOTS AND LEAVES

2.2.1 RHIZOPLANE

The rhizoplane is the area of the root interfacing with soil. Root tips and root hairs are immersed in mucigel, a substance created by the combination of plant-secreted mucilage (composed of pectins and hemicelluloses) and complex polysaccharides produced by bacteria that also degrade mucilage [23]. Bacteria are exposed to the mucilage when the cuticle covering it on root hairs is punctured or degraded physically or chemically. Depending on the plant, root exudates contain a variety of substances that can act as chemoattractants for microorganisms and/or substrates for growth. Sugars, amino acids and other amino compounds, organic acids, fatty acids and sterols, growth factors, nucleotides, and other compounds are produced from the aging epidermal cells [24]. Most natural rhizosphere bacteria attach to specific regions of the root; root hairs are a common site of attachment of the rhizobial bacteria involved in nitrogen fixation [25,26], and, possibly, attachment of human enteric pathogens [27]. Figure 2.1 shows a general schematic of a region of a plant root illustrating an emerging root hair (Figure 2.1A). As noted above, epidermal cells comprise the surface of the root tissue as it develops with specific epidermal cells becoming root hair cells; cortex cells compose a second layer under epidermal cells (Figure 2.1G). After attachment of certain nitrogen-fixing bacteria such as the rhizobia, root hair cells are induced

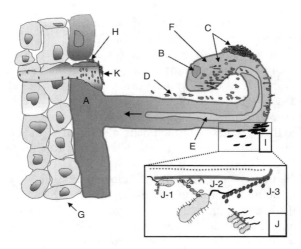

FIGURE 2.1 (Color insert follows page 594) Anatomy of a root hair. Schematic representation of structures that are part of the anatomy of a plant root hair with attached bacteria. (A) Root hair epidermal cell; (B) nucleus; (C) bacteria bound to epidermal cell surface in aggregates and as biofilm; (D) rhizobial bacteria; (E) root hair infection thread initiated by rhizobial bacteria; (F) curling root hair tip; (G) cortex cells; (H) junction between root epidermal cells with attached bacteria; (I) bacteria binding as single cells, then aggregating; (J) magnification of I (not drawn to scale): J-1, single bacterial cell binding by pili/fimbriae; J-2, plant lectins interacting with bacterial carbohydrate (e.g., EPS, LPS, CPS, cellulose fibrils); J-3, bacterial flagellin interacting with plant receptor (e.g., polysaccharide); (K) lesion produced by plant pathogen.

to curl, initiating the complex process of thread formation within the root hair (Figure 2.1E).

2.2.2 PHYLLOPLANE

The phylloplane is the interface between the leaf and the environment. Epidermal cells compose the upper and lower surfaces of a leaf (Figure 2.2B), and are covered by the cuticle, which is composed of a polymer matrix (cutin), polysaccharides, and associated waxes. The cuticle acts as a barrier and prevents water loss from the leaf (Figure 2.2A). The cuticular waxes are lipophilic long-chain fatty acids (20 to 40 carbons); some fatty acids are oxygenated forming aldehydes, ketones, sterols, and esters [28,29]. The waxes in leaves are mostly saturated and, therefore, highly resistant to degradation by most microorganisms. However, an important part of the microbial ecology of plants, and of the phyllosphere in particular, are fungi that secrete cutinases that degrade leaf waxes [30]. Lesions in the cuticle can expose potential sites of attachment for other microorganisms (Figure 2.1K and Figure 2.2L). In addition, some bacteria, including epiphytic bacteria, probably adhere to the cuticle and interact with the plant and obtain nutrients

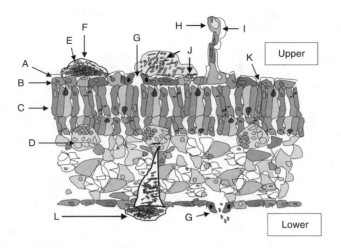

FIGURE 2.2 (Color insert follows page 594) Anatomy of the cross-section of a leaf. Schematic representation of structures that are part of the anatomy of most plant leaves and attached microorganisms. (A) Cuticle layer; (B) upper epidermis; (C) palisade parenchyma; (D) vascular bundle composed of phloem and xylem; (E) biofilm composed of bacteria and other microorganisms; (F) EPS; (G) stomates within upper and lower epidermis; (H) trichome; (I) cuticle; (J) free bacteria and other microorganisms within water droplet; (K) recessed area between epidermal cells; (L) biofilm on underside of leaf forming a lesion into the vascular system.

without damaging the surface [31]. Depressions formed at junctions of epidermal cells appear to have thinner cuticles, with microorganisms often residing at these sites (Figure 2.2B).

The upper and lower surfaces of plants are not considered favorable to microbes because of the cuticle and the rapid and repetitive fluctuations in the physicochemical conditions (e.g., humidity, temperature, leachates) to which microbes must adapt [32,33]. Many of the nutrients present in root exudates noted above have been detected also in leaf exudates [34]. Some of the chemicals in leaf exudates/leachates are likely chemoattractants, inducing movement of the microbe towards a closer interaction that may involve attachment [35].

2.3 MICROBIAL FLORA OF PLANTS

The unique physical and biochemical qualities of each plant surface as a result of the plant genotype and of responses to environmental stimuli (light, temperature, humidity, atmosphere, pH, soil) are major determinants of the plant microbial community. Large differences in the types and numbers of bacteria can occur on different plants, leaf-to-leaf of the same plant [36], and leaf-to-leaf seasonally and even daily [32,37]. Bacteria usually colonize and/or are present in areas of a leaf that retain water and are protected from UV light [38].

Early studies to quantify the populations of bacteria on vegetables reported that $>2 \times 10^6$ predominantly Gram-negative CFU/g could be isolated from the outer leaves of cabbage, in contrast to 4×10^3 bacterial CFU/g that were isolated from inner leaves [39]. The results of numerous studies have been in agreement with these findings, with viable aerobic bacteria ranging between 32 CFU/g on inner leaves of lettuce to 10^7 CFU/g on spinach or peas in warm, humid conditions [38]. Washing vegetables in water usually will decrease the number of bacteria present only marginally (2.5- to 3-fold), reflecting the relatively tight attachment of bacteria to the surface. Viable bacteria have been detected in the interior tissue of cucumbers and tomatoes, locations that had been assumed to be sterile [38]. Thus, bacteria interact with plants by mechanisms that probably involve significant movement of bacteria on tissues, and subsequent attachment, and in some cases entry into tissues (possibly endophytic). These events are in contrast to the infiltration of bacteria into cut or damaged plant tissue, or into natural openings in fruits and vegetables like pores, stems, and calyx during food processing [40]. The internalization of human pathogens in produce is of great concern, regardless of when it occurs (pre/postharvest), because it would decrease the effectiveness of any disinfection steps to minimize contamination [40].

Many microorganisms are present in the plant rhizosphere and phyllosphere, including bacteria, fungi, protozoa, nematodes, and viruses. In addition, many different bacterial genera have been isolated from fruits and vegetables (Table 2.1) [17]. Although most of these organisms probably interact by specific and different mechanisms with plants, little is known about the interactions involved. As noted, most of the fundamental and definitive studies of plant–microbe interactions involve Gram-negative plant epiphytic or pathogenic bacteria. A review of the literature related to this subject provides a context for assessing observations obtained from very preliminary studies of human pathogens in similar environments. In addition, it is probable that some of the human pathogens use similar and/or modified mechanisms of attachment. Thus, attention to any similarities in human pathogens and plant bacteria location on plants, biochemistry of bacterial outer surface structures, and possible adhesins, could provide clues for more fundamental studies.

2.4 ATTACHMENT BY PLANT NITROGEN FIXING, EPIPHYTIC, AND PATHOGENIC BACTERIA TO PLANTS

2.4.1 *RHIZOBIUM* SPP. (*RH* SP)

The process whereby bacteria communicate with leguminous plants (e.g., soybean, bean, pea, peanut, lentil, chickpea, alfalfa, clover) and become

TABLE 2.1
Putative Attachment Factors Described for Epiphytic Bacteria and Plant Pathogens

Type of bacteria	Attachment factor(s)[a]	Ref.
Nitrogen-fixing		
Rhizobium japonicum	EPS	49
Bradyrhizobium japonicum	BJ38 lectin (Gal/Lac)[b]	52
Rhizobium trifolii	CPS	25
Rhizobium leguminosarum	Rhicadhesin	21, 26
Rhizobium leguminosarum bv. trifolii	LPS	50
Rhizobium leguminosarum bv. trifolii	RapA1[c]	53
Rhizobium meliloti (*Sinorhizobium*)	Nex18	187
Plant pathogens		
Agrobacterium tumefaciens	LPS	64
Agrobacterium tumefaciens	T-pilus, F-conjugation factor	188
Agrobacterium tumefaciens	Rhicadhesin	21
Agrobacterium tumefaciens	att-encoded proteins	60
Agrobacterium tumefaciens	CPS	61
Erwinia carotovora	Type 1 fimbriae, Hrp	88, 188
Erwinia chrysanthemi	Hrp proteins	188, 189
Erwinia chrysanthemi	HecA (FHA)	85
Klebsiella aerogenes	Type 3 fimbriae	111, 188
Pantoea stewartii	Hrp	188
Pseudomonas aeruginosa[d]	Type IV pili	104
Pseudomonas aeruginosa	PA-IIL lectin[e]	190
Pseudomonas aeruginosa	Type II pseudopilus	81
Pseudomonas fluorescens	Type III?	95
Pseudomonas fluorescens	Type IVB pili, Hrp	188
Pseudomonas syringae pathovars	Type IVB pili, Hrp	188, 191, 192
Ralstonia solanacearum	Type IVB pili, Hrp	79, 188
Ralstonia solanacearum	FHA homologs[f]	80
Ralstonia solanacearum	RSL and RS-IIL lectins[g]	83
Xanthomonas campestris	Type IVB pili, Hrp	105, 188
Xanthomonas campestris pathovars	FHA homologs	106
Xylella axonopodis	Flagellum, type IV pili	193
Xylella fastidiosa	Type IVB pili?	188
Xylella fastidiosa	FHA homologs	194
Epiphytic/biocontrol bacteria		
Azospirillum spp.	EPS and CPS	108
Azospirillum brasilense	Flagellum (polar)	78, 109
Azospirillum brasilense	MOMP	110
Pseudomonas fluorescens	Fimbriae	195
Pseudomonas fluorescens	Flagella	97
Pseudomonas putida	Agglutinin?	99

[a] EPS, exopolysaccharide; CPS, capsular polysaccharide; LPS, lipopolysaccharide; Hrp, hypersensitive response and pathogenicity; HecA, homologous to FHA; FHA, filamentous hemagglutinin.
[b] Gal, galactose; Lac, lactose.
[c] Secreted from bacteria and binds to carbohydrate on bacteria.
[d] *P. aeruginosa* can be both a human and a facultative plant pathogen.
[e] PA-IIL has high affinity for L-fucose.
[f] FHA, filamentous hemagglutinin; based on gene homologs in different pathovars.
[g] RSL has affinity for L-fucose > L-galactose > L-arabinose/D-fructose/D-mannose; RS-IIL has high affinity for fructose and mannose.

endosymbionts involved in nodule formation and nitrogen fixation on roots provides the most advanced model of plant–microbe interactions. Research continues on understanding the fundamental steps involved [3]. It involves specificity between Gram-negative Rhsp and the host [41], including the site on roots where it is initiated [42]. The early steps of the process do not involve attachment; chemical signals are released from the plant, inducing bacterial genes that encode the release of corresponding signals to the plant that induce nodule development on roots. Rhsp then adhere tightly to the surface of the curling tips of root hair cells [21,42].

2.4.1.1　Two-Step Model of Attachment

The general consensus model that describes Rhsp root attachment proceeds in two steps (Figure 2.1) [21]. Chemicals released by the plant (e.g., flavonoids) induce movement of bacteria by chemotaxis towards chemicals exuding from the root (Figure 2.1D) [3]; this results in close contact between the roots and the bacteria and initiation of attachment [21,43]. The first step in attachment involves a bacterial Ca^{2+}-binding protein called rhicadhesin ($\sim$14,000 Da) which is responsible for attachment of mostly single (not aggregated, Figure 2.1C) bacterial cells directly to the root hair (Table 2.1) [43]. Growth of the bacteria under low Ca^{2+} conditions decreases direct attachment considerably [44], possibly due to the release of rhicadhesin under low Ca^{2+} [21]. The second step in attachment involves bacterial synthesis of cellulose fibrils that bind rhicadhesin leading to auto-aggregation, and/or firm binding of other bacteria at the site of infection [45]. Under carbon-limitation, *R. leguminosarum* bv *viciae* cells form aggregates on root hair tips by attaching to other rhizobia cells (Figure 2.1I and J).

2.4.1.2　Attachment Factors

Rhicadhesin inhibits the attachment of many *Rhizobiaceae* spp. to pea root hair tips, including *R. leguminosarum* biovars, *R. meliloti*, *R. lupini*, *Bradyrhizobium japonicum,* as well as *Agrobacterium tumefaciens* (*Agt*) and *A. rhizogenes*, indicating that rhicadhesin or rhicadhesin homologs are part of a common mechanism of attachment to root hairs [43]. When Mn^{2+} concentration is limiting, the bacterial cells attach and form aggregates also, but apparently this process is accelerated by a pea plant lectin that binds to a carbohydrate receptor/ligand on the bacteria (Figure 2.1J) [46]. Transgenic alfalfa plants transformed with pea lectin, bound *B. japonicum*, and *R. leguminosarum* better than did untransformed lines [47]. Expression of rhizobium exopolysaccharide (EPS) has been reported to be essential for infection thread entry into the root hairs, possibly due to lectin–carbohydrate interactions [47]. Earlier studies reported that EPS, and not lipopolysaccharide (LPS), was the probable bacterial receptor responsible for specific interactions between *R. japonicum* cells and soybean root hairs by means of a soybean lectin (Figure 2.1J-2) [48,49].

LPS and lipooligosaccharides (LOS) are potential attachment factors as receptors for lectins expressed by the plant host, since they are prominent cell-surface glycoconjugates in Gram-negative bacteria. Plant–microbe studies indicate that EPS and LPS/LOS both are possible receptors for plant lectins, and can be highly variable within a population of cells depending on the environment and mechanism of gene expression. In two studies, LPS of *R. leguminosarum* was abundant on cells during their attachment to the rhizoplane of *Zea mays* compared to cells present in the root cortex [50]; a 38,000 M_r cell surface lectin in *B. japonicum* (BJ38), inhibitable by lactose and galactose, also was identified [51,52]. Thus, these separate results define a candidate bacterial carbohydrate receptor and a bacterial lectin (possibly pili) that are involved putatively in attachment to plant lectins (Table 2.1) [50], or plant carbohydrates (Table 2.1 and Figure 2.1J), respectively [51,52].

In an attempt to clone the rhicadhesin gene of *R. leguminosarum* bv. trifolii, a unipolar cell surface protein, RapA1, was identified that bound to cognate carbohydrates on the bacteria [53]. The RapA1 protein was proposed to be a bacterial lectin. The unipolar location of the lectin and activity in agglutination/aggregation of bacteria indicated it is similar to the *B. japonicum* BJ38 lectin described above (Table 2.1). In an earlier study, polar attachment of *R. trifolii* by the clover root hair lectin trifolin A to bacterial extracellular microfibrils composed of capsular polysaccharide (CPS) was described [25]. Thus, it is speculated that these bacterial lectins are capable of recognizing carbohydrate structures present both on the bacteria and plant root hairs [53], a binding activity that could result in both aggregation of bacteria and attachment of bacteria (possibly singly or as aggregates) to root hairs (Figure 2.1I and J). The expression of RapA1 was stable at any growth phase, but the expression of the bacterial receptors was highest during exponential phase of growth, corresponding to higher agglutination [53]. These results reflect the dynamic state of the rhizobial cell surface, which changes due to growth phase and contact with plants, and affects attachment.

2.4.2 *AGROBACTERIUM TUMEFACIENS* (AGT)

Agt is a Gram-negative bacterium that when inoculated on wounded dicotyledonous plant tissue causes crown gall tumors by transferring a portion (T-DNA) of a resident plasmid (Ti-plasmid) into the plant [54]. The essential nature of *Agt* attachment to plant wound tissue for *Agt* root transformation in a pinto bean leaf model was first reported by Lippincott and Lippincott [55]. Subsequent studies have identified multiple *Agt* mutants or strains defective in attachment to different plant tissues [56–63], although for many of the mutants the functions of the predicted proteins have not been identified or characterized.

Whatley *et al.* reported that both LPS on *Agt* cells and purified LPS inhibited specifically *Agt* tumorigenic activity on pinto bean leaves by >50% [64]. Although specific genes or gene products were not identified, the authors suggested that LPS interacts with the sites of attachment on the leaves.

Five Agt Tn5 transposon mutant strains unable to attach to carrot cells in suspension (10^7 bacteria/10^5 cells), and non-tumorigenic on carrot disks and wounded bean leaves, were identified with an associated loss of 33, 34, and 38 kDa proteins [65]. Revertants of the nonattaching mutants were isolated and shown to have regained virulence and ability to attach, confirming the involvement of the proteins in attachment. In addition, LPS purified from the parent and each of the mutant strains, inhibited by 30 to 60% the attachment of Agt to carrot cells, supporting the hypothesis that LPS plays a role also in attachment [65]. Agt biovar 3 ($A.$ $vitis$), which is predominantly isolated from grapes and causes root decay, produces a polygalacturonase, that appears to function by modifying specifically grape root cells in a manner that increases attachment of the biovar 3 Agt [58].

In attachment (att) mutants characterized by Matthysse et $al.$ [60], open reading frames (ORFs) were identified that have homology to genes encoding the membrane-spanning proteins of periplasmic binding protein-dependent (ABC) transporter systems and ATP-binding proteins of Gram-negative bacteria, and to an ORF in an operon of $Campylobacter$ $jejuni$ associated with attachment. These results do not identify a specific attachment factor; rather they suggest other mechanisms involved in attachment, including secretion in or out of cells of a substance required to condition the medium for bacterial attachment, or ATP-transporter-dependent transfer of plant signals into bacteria with induction of a substance important for attachment. One of the attachment mutants was mutated in a gene, $attR$, homologous to bacterial transacetylase genes [61]. The $attR$ mutant strain lacked an acetylated CPS present in the parent strain Agt C58, and consequently did not attach to wound sites and was avirulent for legumes and nonlegumes [63]. The $attR$ mutant strain also did not attach to root hairs and root epidermis of nonlegumes, but did attach to these areas on legumes (alfalfa, bean, pea). These results suggested that $attR$ plays a role in binding of Agt to, and in colonization of, root hairs on nonlegume plants, but that $attR$ has no role in colonization of root hairs on legume plants [63]. Thus, two systems for Agt attachment and colonization are available and may function depending on the plant species.

A polysaccharide purified from the water-soluble fraction of a phenol–water extraction of Agt strain C58 cells inhibited the attachment of Agt to carrot cells. The extracted polysaccharide was acidic, acetylated, and composed of glucose, glucosamine, and an unidentified deoxy-sugar [61]. Interestingly, the ligand in carrot cells that binds the Agt polysaccharide may be a homolog of vitronectin (S protein), a serum-spreading factor in animals and part of the extracellular matrix [66]. The vitronectin-like protein was detected immunochemically as present on the surface of carrot cells [66], and was detected previously on tomatoes, soybeans, and broad beans [67]. If Agt is bound to plant cell vitronectin, and vitronectin is linked by integrin to actin in the cytoskeletal network as it is in animal cells, this would provide an intimate contact for initiation of the crucial step of transport of Agt T-DNA and proteins to the nucleus of plant cells [66].

2.4.2.1 *Agt* and Rhicadhesin

Rhicadhesin, noted above as a proteinaceous attachment factor for *R. leguminosarum* (Table 2.1), was reported to be important in attachment of *Agt* to pea root hair tips [59]. However, attachment by rhicadhesin in this system was dependent upon sufficient expression in *Agt* of cyclic β-1,2-glucan, an osmoregulating molecule synthesized by the *chvB* encoded protein [59].

2.4.2.2 Pili

Although there are several *Agt* virulence proteins suspected of interacting with proteins on different plants [68], a strong candidate for having a role in attachment of *Agt* to plants is a pilus, the structure of which in *Agt* is composed predominantly of multiple copies of the VirB2 protein (propilin) [69,70]. The mature pilus is expressed as $\sim$10 nm diameter filaments on the cell surface and is required for transformation, presumably, by interacting with plant cell wall or membrane molecules [68]. This conjugative T-pilus has significant sequence homology to the conjugative F-pilus of *E. coli* [70]. Mutants produced in *Arabidopsis thaliana* by T-DNA insertion revealed plant lines resistant to *Agt* transformation (*rat*) and modified in *Agt* attachment to root hairs: *rat1*, which encodes an arabinogalactan-related enzyme, and *rat3*, which encodes putatively a plant cell wall protein [68]. Therefore, both carbohydrate and proteins of the plant are implicated as receptors for *Agt* pili.

2.4.2.3 Cellulose

Previous to these descriptions of the mechanisms of attachment of *Agt* to plant cells and their role in transformation, Matthysse *et al.* described the synthesis of cellulose fibrils by *Agt* induced by the attachment of the bacteria to carrot cells [71]. Although the cellulose fibrils appeared not to be necessary for initial attachment, they were shown to be important in anchoring *Agt* and associated bacteria to the plant cell surface, and enmeshing *Agt* in aggregates associated with tumor formation [71,72]. An 11 kb region (*celABCDE*) was identified containing two operons involved in cellulose synthesis [73]. Thus, cellulose is important in establishing a stable and perhaps more complex interaction of *Agt* with plant tissue subsequent to attachment.

2.4.3 *RALSTONIA (PSEUDOMONAS) SOLANACEARUM (RS)*

2.4.3.1 EPS and LPS

Rs is a Gram-negative soilborne plant pathogen (PP) that infects more than 200 species of plants including fresh produce-related plants like tomato, potato, eggplant, banana, and papaya, and causes bacterial wilt disease [74]. *Rs* enters the root tissue and invades the plant through the xylem, then moves through the vascular system into the aerial parts of the plant (Figure 2.1K). EPS and LPS were identified as major cell surface molecules associated with virulence of *Rs*, perhaps functioning by blocking the xylem vessel and preventing water movement [75]. Sugars identified in composition analyses of EPS include different proportions of N-acetylgalactosamine, glucose, rhamnose, basillosamine, and uronic acids [75]. At least one *Rs* LPS O-antigen was characterized chemically and reported to contain rhamnose, N-acetylglucosamine, and xylose [76]. Mutations in *Rs* EPS genes (*ops* gene cluster) modified unexpectedly the synthesis of both EPS and LPS, and corresponded to a significant decrease in the ability of *Rs* to attach to (presumably), and to infect, two-week-old axenic eggplant seedlings (inoculated in cotyledon) and three-week-old eggplant plants (inoculated in leaf stem). Although five of seven complemented *ops* mutants had nearly their full virulence restored, no association of LPS or EPS with attachment to the plant tissue was defined [75].

2.4.3.2 Type III Secretion System (T3SS)

Many of the Gram-negative bacterial plant and animal pathogens described in this chapter produce a T3SS, which has been shown in multiple systems to be crucial to the delivery of multiple virulence factors into the extracellular milieu, but more importantly, directly into plant and/or animal cells. The T3SS was discovered by characterization of gene clusters present in pathogenicity islands and in large plasmids with similarities to flagellar assembly genes [77]. The common genes and functions of the T3SS in plants and details regarding the assembly of pilin and avirulence (*avr*) genes are provided in an excellent review [77].

The T3SS in plant pathogenic bacteria involve the *hrp* (hypersensitive reaction and pathogenicity) genes [77]. *hrp*-related genes that are relevant to attachment of bacteria to plants include those involved in the synthesis of the novel Hrp pilus. The potential role of the Hrp pilus in delivery of bacterial proteins into plant cells by direct interaction suggests that it also is an attachment factor for other plant pathogens and human pathogens with T3SS (e.g., *Pseudomonas, Erwinia, Xanthomonas, Ralstonia, Salmonella, Shigella, Yersinia*). However, experiments with *Rs* in a tobacco plant cell co-culture model indicated that T3SS-encoded pili, composed mainly of HrpY protein, had no role in attachment [78]. An interesting finding was the observation of the HrpY pili and fimbriae concentrated at the same end of the *Rs* cells, perhaps indicating that a unipolar location was important biologically, possibly in attachment with plants other than tobacco. However, in another study, *Rs* mutants lacking Hrp pili retained twitching motility and, by electron microscopy, a different polarly located pili structure was observed [79]. This *Rs* pilus is a 17 kDa protein encoded by *pilA*, and 46% identical to *P. aeruginosa* type IV pilin. *Rs* type IV pili were shown in this study to have a role in autoaggregation, biofilm formation on plastic surfaces, and transformation. However, a PilA$^-$ mutant retained capability to bind to tobacco cells and to tomato roots, but in a nonpolar fashion, indicating that

PilA has a qualitative role in attachment [79]. The lack of a quantitative effect on attachment of Rs lacking either Hrp pili (type III) or type IV pili to tobacco or tomato plant cells/roots indicates that their major role may be more relevant in natural plant environments (e.g., nutrient acquisition, genetic exchange, and movement and biofilm formation in xylem).

2.4.3.3 Type II Secretion System (T2SS)

Multiple T2SS loci have been identified also in the genome sequence of Rs strain GMI1000 [80]. In $P.$ $aeruginosa$, the T2SS produces bundled fibrils called type II pseudopilins that increase adherence of the bacteria to plastic surfaces and are involved in production of biofilms [81]. The Rs and other species T2SS-encoded fibrils may have a role in attachment to plants.

2.4.3.4 Rs Lectins, Fimbriae, FHA

Two Rs protein lectins with potential roles in attachment have been characterized recently: RSL (9.9 kDa subunit) with activity/specificity for L-fucose > L-galactose > D-arabinose; and RS-IIL (11.6 kDa subunit) with activity/specificity for D-fructose and D-mannose [82,83]. The RS-IIL is similar, but not identical, to the PA-IIL lectin described for $P.$ $aeruginosa$. The activity of the RS-IIL lectin for sugars also prominent in plant cell walls has stimulated studies of the role of Rs lectins in attachment to more relevant and complex plant glycoconjugates [83].

Finally, multiple ORFs identified in the Rs genome strain GMI1000 are similar to nonfimbrial adhesins or hemagglutinin (e.g., FHA) molecules, some of which promote strong adhesion to mammalian cells [80,84]. Future work is necessary to determine whether these proteins have any role in attachment of Rs to plants in a complex soil environment.

2.4.4 $ERWINIA$ SPP.

The soft-rot pathogen $Erwinia$ $chrysanthemi$ ($Echr$) expresses an adhesin (HecA) that has homology to filamentous hemagglutinins (FHA) expressed in both plant and animal pathogens (Table 2.1) [85,86]. A $hecA$ mutant of $Echr$ had decreased virulence in seedlings of a particular tobacco cultivar, but not other cultivars or plants, indicating a relatively specific attachment [85]. Observation of green fluorescent protein (GFP)-labeled mutant and wild-type strains by confocal microscopy illustrated that the mutant cells did not aggregate by end-to-end attachment, nor attach to seedling roots, nearly as well as the wild-type cells. Attachment of the mutant strain to the leaf surface was decreased dramatically and no $Echr$ aggregates on leaves were observed. Thus, at least with the specific tobacco cultivar described, HecA appears to function as an important $Echr$ adhesin [85].

Cell suspension cultures of $Gypsophila$ $paniculata$ ("baby's breath") leaf segments with a pathogenic strain of $Erwinia$ $herbicola$ pv. gypsophilae (Ehg) resulted in a greater than five-fold increase in plant cell aggregation compared to a nonpathogenic *Ehg* strain, indicating attachment had occurred; electron microscopy revealed intimate attachment by a possible "bridge" between *Ehg* and the plant surface [87]. However, no attachment molecules related to this interaction were identified.

Both *Echr* and *E. carotovora* (*Ecar*) possess *hrp* genes encoding a T3SS, but their role in virulence of *Ecar* had been unclear [88]. Recently, mutations in Hrp system structural genes confirmed that T3SS proteins are required for full virulence of *Ecar* for potatoes [89]. *Erwinia amylovora*, the cause of fire blight in many plants, was observed by scanning electron microscopy (SEM) at regions on and in plant leaves, on the epidermis around detachment sites of leaf hairs, and on stems and roots of apple seedlings [90]. Although intimate interactions between the T3SS Hrp pili and Hrp virulence proteins of *E. amylovora* have been observed by transmission electron microscopy (TEM) [91], no evidence for direct attachment of the pilus to host cells has been described. Finally, the *in vitro* specificity of a presumed *Erwinia rhapontici* (pathogenic for rhubarb) lectin for *N*-acetyllactosamine (galactose-β1-4N-acetylglucosamine) would be intriguing if it were related to an attachment factor for plants, but it is apparently nonfimbrial, and no attachment factor has been described [92].

2.4.5 *PSEUDOMONAS* SPP.

Pseudomonas syringae (*Ps*) causes disease in more than 80 plant species, including many important produce-related plants [32]. *Ps* has been studied extensively as both a pathogen and an epiphyte on plant leaves. Attempts to remove epiphytic populations of bacteria on leaves, including *Ps*, by vigorous washing and sonication revealed that: (1) some bacteria were bound more strongly than others, (2) the phylloplane was heterogeneous relative to optimal attachment/colonization, and (3) pili-minus *Ps* strains were washed more easily from leaves compared to wild-type *Ps* [32,93,94].

P. fluorescens (*Pf*) strains colonize roots and are important competitors for biocontrol of plant pathogens [95], and also possibly of human pathogens [96]. Flagellin was shown to be important in motility of *Pf* for chemotaxis and colonization of potato roots [97]. Piliated/fimbriated strains of *Pf* were shown to bind to roots of corn seedlings better than a nonpiliated variant strain [98]. Fimbriae/pili (34 kDa) purified from *Pf* strains also bound to the roots. The fibrillar nature of the pili suggests that they may be T2SS pili (see below). *Pf* hemagglutination activity was inhibited by all sugars representative of those present in plant root exudates, thus indicating carbohydrate-specific binding activity.

Pseudomonas putida (*Pp*) is common in soil and acts as a plant growth promoter and suppressor of fungal pathogens. A kidney bean root surface glycoprotein was described that agglutinated *Pp* cells; agglutination-negative *Pp* mutants were reported to be 20- to 30-fold less effective in attaching to root surfaces of seedlings [99]. Motility was shown to be associated with efficient *Pp* attachment to sterile wheat roots in a simplified model system, implicating flagellin as a potential attachment factor [100]. However, no definitive attachment factor was identified for Pp in these two studies.

$P.\ aeruginosa$ (Pa) provides a good example of a species that can be a nonpathogenic or pathogenic organism in plants and animals [101,102]. This broad pathogenesis for different hosts reflects the conservation of virulence mechanisms for attacking quite different hosts, including possibly mechanisms of attachment [103]. Pa is very relevant also to fundamental studies of the biology of enteric human pathogens in produce; it may provide clues about mechanisms of attachment, communication, and invasion, important for development of methods for minimizing both human and plant pathogens on plants.

The type IV pilus has been described as the most important "virulence-associated adhesin" of Pa [104]. However, this is due to the emphasis on characterization of attachment of Pa to specific glycosphingolipids of mammalian epithelial cells. In plant models, differences in Pa attachment to leaves of different arabidopsis ecotypes have been reported [102]. Pa cells were observed attached perpendicularly to, and degrading regions of, the surfaces of leaves; in other regions, cells bound to trichomes in multiple layers probably as biofilms. The perpendicular orientation noted for Pa cells on arabidopsis surfaces is reminiscent of the unipolar location of rhicadhesin, of rhizobium, and of Hrp pili and fimbriae of Rs (described above), and suggests that an attachment factor may be localized similarly. The movement of Pa cells (possibly by type IV pili twitching motility) towards stomatal openings, entry into them, then attachment to host cell walls, reflects pathogenesis similar to erwinia and ralstonia pathogens [102].

2.4.6 XANTHOMONAS CAMPESTRIS (Xc)

Type IV-encoded bundle-forming fimbriae have been characterized in Xc pv. $vesicatoria$ (Xcv) [105]. The fimbriae are composed mainly of a protein subunit of 15.5 kDa (FimA). The use of a FimA-mutant strain indicated that fimbriae had no role in colonization of tomato leaves. However, major differences were noted between the wild-type and mutant strains in the amount of cell–cell aggregation in laboratory cultures, on tomato leaf surfaces, and on trichomes, with the wild-type always more prevalent [105]. These results suggest that if the fimbriae assist attachment of Xcv to tomato surfaces, they may be specific for selected regions of the plant, such as trichomes. Putative FHA proteins suspected of being involved in attachment in other species also are present in Xc [106].

2.4.7 AZOSPIRILLUM SPP.

$Azospirillum$ spp. have been investigated because of their nitrogen-fixing capability while in close contact with grass roots [107]. Two different polysaccharide structures were identified in strains of $A.\ brasilense$ (Abr) and $A.\ lipoferum$ ($Alip$): a CPS tightly associated with the cell surface, and an EPS appearing to be less dense and extending from the cell [108]. A wheat lectin, wheat germ agglutinin, bound to *Abr* and *Alip* cells, and the binding was inhibited by *N*-acetylglucosamine. Thus, these results provided evidence of carbohydrate surface structures that are candidate receptors for specific plant lectins [108].

Abr strains express a polar flagellum [109]. Attachment of a nonmotile *Abr* flagellar mutant to wheat roots was reduced dramatically, whereas purified flagella bound directly to wheat roots. The major outer membrane protein of *Abr* was reported to be an adhesin responsible for *Abr–Abr* aggregation, and the variable attachment of *Abr* to root extracts of wheat > corn > sorghum > bean >> chickpea > tomato [110]. The degree of aggregation of *Abr* cells is related possibly to the amount and composition of the EPS and/ or CPS [110].

2.4.8 *KLEBSIELLA* SPP.

Klebsiella spp. are enteric bacteria that can be soilborne, saprophytic, and cause serious human illness. An associative nitrogen-fixing strain of *K. aerogenes* expressing type 3 fimbriae was characterized for attachment [111]. The fimbriae of 23.5 kDa were associated with hemagglutination of human O erythrocytes, and the adhesion of bacteria to plant seedling roots. In addition, the purified fimbriae also bound directly to root tissue. Other strains of klebsiella also were shown to bind to roots by the type 3 fimbriae [111]. Subsequently, type 1 fimbriae were reported to mediate adherence of *K. pneumoniae* and *Enterobacter agglomerans* (*Pantoea agglomerans*) strains to plant roots, with binding inhibitable by α-methyl-D-mannoside [112]. Thus, different types of fimbriae appeared to mediate adherence of different enteric species to plant roots.

2.5 FUNGI AND VIRUSES AND PLANTS

The invasion of plant tissues by phytopathogenic fungi is likely initiated by attachment steps [113]. There have been few reports describing this attachment process [114], but degradation of the cuticle and fungal hyphae germ tube formation is important [113]. Studies of the survival of gastrointestinal viruses in water, soil, fruits, and vegetables have been reported [115], but there is no definitive information about the mechanisms of virus attachment to plants. A recent report on the specific attachment of cucumber necrosis virus (CNV) to fungal zoospores and invasion of cucumbers may be relevant to the introduction of foodborne gastrointestinal viruses in produce by similar mechanisms [116].

2.6 POTENTIAL ATTACHMENT FACTORS OF ENTERIC BACTERIAL PATHOGENS FOR PLANTS

The increasing amount of genetic and molecular data obtained for a variety of microorganisms, including those that interact with, and/or are pathogenic

TABLE 2.2
Potential Attachment Factors in Human Pathogens That are Similar to Those Described for Plant Bacteria

Enteric bacteria/human pathogen	Attachment factor(s)	Ref.
Escherichia coli	G-fimbriae	196
Escherichia coli	Flagellin	197
Enterohemorrhagic *E. coli*	Fimbriae	198
Enteropathogenic *E. coli*	Type IV bundle forming pili	199
Enteropathogenic *E. coli*	Fimbriae	200
Enterotoxigenic *E. coli*	Fimbriae	201
Enterotoxigenic *E. coli*	Pili	202
Uropathogenic *E. coli*	Type 1, P, S pili	203
Enterobacter (*Pantoea*) *agglomerans*	Type 1 and 3 fimbriae	112, 188
Klebsiella pneumoniae	Type 1 and 3 fimbriae	112, 188
Listeria monocytogenes	Flagellin	171, 204
Salmonella enterica	Type 1 fimbriae	205
Salmonella enterica	Type IVB pili	206
Salmonella enterica	Thin aggregative fimbriae (curli)	207
Shigella flexneri	Pili	208
Shigella flexneri	Type 1 fimbriae	209
Vibrio cholerae	Fimbriae	210

in, plants and humans, has revealed a remarkable amount of conservation in the mechanisms available to them for survival and sometimes pathogenicity [103,117–119]. Table 2.2 lists some of the species of enteric bacteria that are potential foodborne pathogens, and the corresponding proteins that are putative factors for attachment to their animal hosts. Clearly, multiple types of fimbriae (G, types 1 and 3, IVB (curli, bundle forming)), pili (type 1, P, S), and flagella represent the major known attachment factors in enteric bacteria. In addition, all of the Gram-negative enteric bacteria shown in Table 2.2 express LPS, and some express CPS; both are major surface glycoconjugates that could serve as receptors for plant lectins, similar to the mechanisms described previously for plant nitrogen-fixing, epiphytic, and pathogenic bacteria (Table 2.1). *Klebsiella* spp., *Enterobacter* spp., and *Pseudomonas* spp. bridge the environments of plants and animals by their capability to colonize both hosts, presumably with Type 1 and 3 fimbriae, pili, and lectins involved in attachment (Table 2.2) [111,112,120]. Although certain species of these three bacterial genera have been recognized to have biologically relevant interactions with plants (e.g., nitrogen fixation, pathogenesis, competition), there is no evidence that the major foodborne enteric human pathogens (Table 2.3), many of which have been associated with produce outbreaks, are either pathogenic or beneficial for the plant. However, the apparent interactions of human pathogens with plant tissues, and their ability to survive and grow on plants under certain conditions (e.g., temperature, Table 2.3),

TABLE 2.3
Summary of Some Human Pathogen–Plant Models and Observations Related to Attachment

Plant–human pathogen model	Observations and conclusions	Ref.
Lettuce–*E.coli* O157:H7 (pre- and postharvest)	Attachment of human pathogen to edges/grooves of seed coat and root hairs Attachment to stomates and trichomes Cells concentrated at leaf epidermal cell junctions Aggregates on roots and leaves Strain differences in adherence Internalization (45 μm below surface)	121, 127, 133, 134, 136, 211
Tomatoes–*S. enterica*	Attachment to roots, stems, leaves, flowers Attachment to stem scar > intact fruit skin Strain differences in survival and growth (adherence?) Viable cells isolated from stem scars up to 49 days after inoculation of fruit and at least 9 days after inoculation of roots Internalization; protection from sanitization	142–144, 212
Apples–*S. enterica*	Attachment to stem, calyx, broken skin > intact skin Sanitization less effective when attached to stem, calyx, broken skin	145
Sprouts–*S. enterica*	Human pathogen and aerobic bacteria concentrate on damaged seeds Attachment to root hairs and edges of seed coats Tight attachment; hard to sanitize Human pathogen attached better than 2 of 3 plant epiphytic species Internalization through emerging root hairs (endophytic?)	13, 27, 120, 148
Sprouts–*E.coli* O157:H7	Minimal attachment to sprout tissue Nonpathogenic *E. coli* isolated from cabbage attached to sprouts effectively Grows well with nutrients in sprout irrigation water	13, 27, 148

Cilantro leaf–*S. enterica*	Grew on leaves at 30°C > 22°C (3 days) Tolerated low humidity (60%) Attachment to leaf veins, senescent and damaged regions Aggregates with plant pathogen and epiphytic bacteria Leaf extract compounds bind *Se*	128
Cut radish–*L. monocytogenes*	Attaches well to cut radish Attachment dependent on temperature Attachment of flagellar, export, and sugar phosphotransferase system mutants decreased	171
Cantaloupe–*S. enterica*, *E. coli*, and *L. monocytogenes*	*S. enterica* attaches more strongly compared to *E. coli* and *L. monocytogenes* *S. enterica* strains variable in surface hydrophobicity and electrostatic charge Attachment correlated with bacterial cell surface hydrophobicity and charge	125
A. thaliana–E.coli O157:H7 and *S. enterica*	Attachment to root hairs, stems, leaves, flowers Internalization at emerging root hairs Attached tightly; washing and sanitization ineffective Competitive epiphytic bacteria identified	175
Spinach/radish–*C. jejuni*	Minimal attachment to leaves and roots Attachment to soil components Survives best at 10–16°C	129

indicate that human pathogens interact with plants/produce by more than simply physical and nonspecific ways.

2.7 ATTACHMENT OF HUMAN ENTERIC PATHOGENS TO PLANTS AND OTHER INTERACTIONS

Recent studies of human pathogens in produce models have suggested that foodborne pathogens, many of which are Gram-negative, may interact with plants by mechanisms evolutionarily conserved, and at least somewhat similar to those described above for plant bacteria (Table 2.3). Studies of human pathogens and produce have involved, generally, assessments of the attachment and survival of human pathogens on postharvest, retail market products [121–126]. Other studies have used human pathogen–plant models to investigate attachment on seeds and seedlings or young plants that are contaminated in the laboratory and grown in chambers under various conditions (e.g., humidity, light, temperature, competitors) [27,127–129]. The biology of attachment of human pathogens in preharvest (soil, other microbes, temperature, UV, extended exposure time) and postharvest (rinse water, shorter exposure time, temperature) environments may be quite different. Samples of human pathogen-contaminated produce/plants have been examined mostly by conventional culture methods, polymerase chain reaction (PCR), fluorescence microscopy, or other methods. These studies have provided no definitive information regarding the molecular interactions that may be involved in attachment of human pathogens to plant tissues. However, it is anticipated that future work in this area will provide more fundamental biochemical or genetic data related to attachment. A few examples of studies pertinent to the concepts of plant–human pathogen attachment are presented below and summarized in Table 2.3.

2.7.1 LETTUCE AND *E. COLI* O157:H7

In the last decade there have been more than 15 foodborne outbreaks linked to contaminated lettuce or salad [10,130,131]. This provided the impetus for initiation of studies of attachment of human pathogens in both pre- and postharvest lettuce model systems [127,132–137]. Recent studies of *E. coli* O157:H7 (*Ec*O157) on store-bought lettuce indicated that cells attached in a relatively short time period, and that not all cells could be removed by vigorous washing or treatment with chlorine (Table 2.3) [133,134,138]. *Ec*O157, and other human pathogens and microbes, often are most concentrated at cut surfaces since there are vast nutrient resources released that can be metabolized by human pathogens. Cut lettuce leaves immersed in a suspension of a strain of *Ec*O157 (up to 10^8 CFU/ml) were exposed to fluorescent anti-*Ec*O157 antibody and observed by confocal microscopy. *Ec*O157 attached predominantly to the cut edges of leaves; fewer cells attached to the intact cuticle of leaves, but were observed attached near stomates, on trichomes [133], and concentrated on vein areas of the leaf [133]. Strains of EcO157, pseudomonas, salmonella, and $L.$ $monocytogenes$ (Lm) attached to different regions of cut lettuce leaves, indicating different and specific mechanisms of attachment for different species or strains [134]. Recent studies of EcO157 and postharvest lettuce have addressed the general nature and force of the plant–EcO157 interactions by measuring the effect of surfactants and other treatments. More hydrophobic surfactants were the most effective in detaching EcO157 from the leaf cuticle, but cells at cut edges remained attached [137]. Attachment of EcO157 to the lettuce leaf surface was 0.8 log_{10} higher after treatment with $CaCl_2$; treatment with NaCl had no significant effect [137]. Neither $CaCl_2$ nor NaCl, however, had any significant effect on attachment to the cut edges.

Interestingly, it was reported that the medium in which EcO157 cells were grown affected attachment. Cells grown in tryptic soy broth were more hydrophilic, produced more CPS, and attached better to edges of lettuce (0.4 log_{10}) and to the surface of both lettuce and apple (0.8 to 1.0 log_{10}) than those grown in nutrient broth, suggesting that CPS may be involved directly in attachment [139]. These studies suggest that EcO157 has different mechanisms for attaching to different regions of lettuce leaves, possibly involving hydrophobic interactions, surface carbohydrates (CPS/LPS), neutralization of ionic charge, or bridging of anionic moieties by divalent cations.

In sprout models of EcO157–lettuce attachment, strains of EcO157 implicated in produce outbreaks attached to lettuce roots approximately one log_{10} better than did two of five nonpathogenic $E.$ $coli$ strains, indicating variability in attachment among strains [127]. Attachment of EcO157 strains was highest to seed coats and roots compared to the shoots.

GFP-labeled EcO157 was observed under a fluorescence stereomicroscope to bind in aggregates to the grooves and edges of the seed coats, and to small root hairs of sprouted seedlings [127]. High concentrations of EcO157 added to soil prior to growth of lettuce seedlings resulted in pathogen bound to all parts of the plant. Aggregates of EcO157 cells were observed on cotyledon and root tissue of lettuce seedlings grown for 5 days in spiked soil [127]. These studies illustrate the potential for $E.$ $coli$ to colonize both pre- and postharvest lettuce.

In similar studies, an EcO157-GFP strain spiked in manure-contaminated soil (10^4 to 10^8 CFU/g) was monitored by confocal microscopy for presence on lettuce plants grown in the soil and then treated with chlorine and $HgCl_2$ [136]. EcO157-GFP remained attached to the edible portion of treated lettuce seedlings grown in soil spiked with the highest concentration of EcO157 (10^8 CFU/g). In addition, cells were observed as aggregates on three-day-old leaf surfaces, with some cells present 45 µm below the outer leaf surface. There are multiple potential routes of entry for human pathogens in plants (lateral roots, stomates, pores, cuts, lesions, "invasion"), but whether specific mechanisms of attachment are involved during internalization on preharvest produce is not known (Figure 2.1 and Figure 2.2).

2.7.2 TOMATOES AND APPLES AND *SALMONELLA ENTERICA* (*SE*)

Outbreaks of salmonella illness associated with raw tomatoes have been reported [140,141]. Tomatoes inoculated with high doses of *S.* Montevideo (*Se*M) and stored for 3 days retained viable cells on skin and stem scars [142]. However, *Se*M survived at 2 to 4 $\log_{10}$ higher concentrations in scars and cracks compared to skin, both after washes in water and 100 µg/ml aqueous chlorine. Tomato plants inoculated at stems or flowers with a combination of five different *S. enterica* serovars including *Se*M were analyzed to determine the incidence and length of time salmonella survived in fruit [143]. SeM was isolated from stem scar tissue up to 49 days after inoculation, but *S.* Poona, *S.* Michigan, and *S.* Enteritidis also were isolated at 22 to 39 days from pulp and stem scar tissue. Further studies with the five strain combination with hydroponically grown tomato plants reported the uptake and survival of salmonella for at least nine days on hypocotyls, cotyledons, stems, and leaves of plants inoculated at intact or cut roots [144]. These studies confirmed the capability of *Se* to survive and grow on and in tomato plants and fruit, and indicated the possibilities of strain differences in attachment, and multiple types of attachment involved in the interaction of *Se* with a variety of plant tissues.

Apple fruit provides a surface and environment for human pathogens similar to that of tomato fruit. The intact skin is composed of a waxy cuticle less conducive to attachment by human pathogens than other regions of the fruit [145]. Apple fruit immersed in 10^8 CFU/ml of a strain of *S.* Chester and dried for 10 minutes retained human pathogen on broken skin, and the calyx and stem, at 20°C better than at lower temperatures; also, more human pathogen cells in these regions survived chemical sanitization than those on intact skin.

2.7.3 SPROUTS AND *E. COLI* O157 (*Ec*O157) AND *SE*

Numerous outbreaks of *Ec*O157 and *Se* associated with contaminated sprouts have occurred since 1995, and a number of the outbreaks have been traced to seeds contaminated with relatively low levels of pathogen [146]. Although calcium hypochlorite at 20 mg/ml has been recommended for sanitizing seeds [147], it does not remove the entire natural microbial flora on the seed, suggesting that bacteria are attached in sites inaccessible to chemical treatments [13]. A comparison of the growth of multiple *Ec*O157 and *Se* strains on alfalfa sprouts revealed major differences in attachment among the strains [27]. Six strains of *Ec*O157 grew an average of 1.5 $\log_{10}$ less on sprouts compared to five strains of *Se* (Table 2.3). An *Ec*O157-GFP strain attached poorly to sprout roots and shoots, whereas individual cells and aggregates of *Se* Newport-GFP were observed adhering to sprout seed coat edges and root hairs (Figure 2.3D). The 10- to 1000-fold difference in attachment to sprout tissues by *Ec*O157 compared to *Se* strains was confirmed

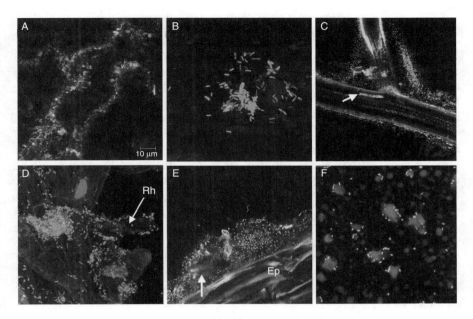

FIGURE 2.3 (Color insert follows page 594) Confocal micrographs of bacteria on plant leaf, stem, and root tissues, and bacteria bound to material extracted from leaves. (A) Natural microorganisms, mostly bacteria, bound to junction of epidermal cells on a lettuce leaf. The bacteria were stained with LIVE *Bac*Light Gram stain (Molecular Probes, OR). (B) GFP-labeled *S. enterica* and dsRed-labeled *P. agglomerans* cells bound singly and in aggregates after their inoculation and incubation on the leaves of cilantro plants. Natural epiphytic bacteria were stained with SYTO® 62 (Molecular Probes) and were detected in the close vicinity of the inoculated strains. The SYTO® 62 signal was assigned the pseudocolor blue. (C) GFP-labeled *Ec*O157:H7 bound in the region of a lateral root emerging from an *Arabidopsis thaliana* plant. The arrow points to a region where the *Ec*O157 cells have become internalized. (D) GFP-labeled *S. enterica* bound to the root hairs (Rh) of an alfalfa sprout. (E) A thick biofilm of natural microorganisms colonizing the root of an alfalfa sprout and stained with LIVE *Bac*Light Gram stain. (F) GFP-labeled *S. enterica* cells attached to a dried compound extracted from cilantro leaves and identified as stigmasterol. (Brandl and Mandrell, unpublished data.).

in subsequent studies [148]. Interestingly, a *Se* Newport strain appeared to be as fit in sprouts up to at least three days as three epiphytic strains isolated from sprouts. In addition, four nonpathogenic *E. coli* strains isolated from field-grown cabbage (Table 2.3) attached as well as the epiphytic strains, but less than *Se* Newport. The results of these and other studies of *Se* on sprouts [149] indicated that specific interactions occur between human pathogens and plants, and suggested that enteric bacteria and human pathogens isolated recently from plant surfaces may retain fitness and attachment capability for plants.

2.7.4 CILANTRO AND Se THOMPSON (SeT)

The investigation of an outbreak of SeT associated with cilantro suggested that the cause of the outbreak was a result of preharvest contamination of imported cilantro [150]. The outbreak strain of SeT increased on cilantro leaves $\sim$1.0 $\log_{10}$ and 2.0 $\log_{10}$ at 24 and 30°C, respectively, 18 hours after inoculation [128]. Observation by confocal microscopy of SeT-GFP incubated on the leaves of cilantro plants revealed that SeT cells localized to the vein area of leaves. Small microcolonies of cells were observed on the leaf veins, but larger concentrations of both individual and aggregated cells were observed on senescent portions of the leaf and in lesions, suggesting that the release of nutrients from leaky or damaged plant cells enhanced growth of the pathogen (Figure 2.2L) [128]. Co-inoculation of cilantro with SeT-GFP and *P. agglomerans*, a plant epiphyte isolated from cilantro and containing a red fluorescent protein, revealed that SeT cells (Figure 2.3B, green cells) were attached to the leaf in aggregates with *P. agglomerans* (red/pink cells) and other natural epiphytic bacteria (purple cells) [128]. These results suggest that SeT interacted with the plant and native bacteria after prolonged exposure to the leaf surface.

In an attempt to assess the mechanism of attachment of SeT to cilantro leaves, a chloroform–methanol (2:1) extract of cilantro leaf surfaces was prepared and fractions obtained by separation by thin-layer chromatography (TLC). Multiple TLC-purified bands were applied to glass slides, exposed to a suspension of SeT-GFP cells and the slides were incubated. Unbound cells were washed from the slide and the slide was observed by confocal microscopy. Figure 2.3F is representative of the results observed with one of the samples that bound SeT cells most effectively. This sample was analyzed by mass spectrometry and shown to be composed of >90% stigmasterol, a sterol compound that is present in the cuticle and has been detected in other leaf extracts [151]. Interaction with cuticular waxes or sterols in regions of the leaf where nutrients are more available is a reasonable strategy for bacteria. A recent study reported the growth of epiphytic *Pseudomonas* spp. on apple cuticle membranes without disrupting the membrane, and the release of a variety of bacterial proteins (e.g., flagellin, porin, ABC transporter binding component) through the membrane [31].

2.7.5 PRODUCE SAMPLES AND
L. MONOCYTOGENES (Lm)

Lm is a Gram-positive, facultative intracellular pathogen acquired most often through the consumption of contaminated food. Listeriosis is a serious illness that can cause a variety of symptoms including septicemia, liver failure, meningitis, and spontaneous abortion and death [152,153].

Lm can survive as a saprophyte on decaying plants and grows at a wide range of temperatures [154]. Outbreaks have occurred due to produce contaminated with Lm [155]. Lm has been isolated from market produce such as cabbage, corn, lettuce, peppers, sprouts, radishes, potatoes, cucumbers, grains, parsley, and watercress [154–162]. It has been reported to grow on asparagus, broccoli, cantaloupe, cauliflower, and leafy vegetables [163–166], and attach to cut potato tissue [167].

A 60-minute exposure of whole cucumber to 10^8 CFU/ml of Lm followed by washing resulted in 10^3 to 10^4 CFU/g of Lm remaining attached [168]. Lm attached to unwaxed cucumber better than to waxed cucumber, in contrast to the decrease in attachment of Se typhimurium and $Staphylococcus\ aureus$ to unwaxed cucumber, indicating either that the Lm cell surface is relatively more hydrophilic [168] or that openings on the cucumber surface (stomates, pores, cracks) are sealed by wax. Similarly, 80% of the Lm cells added to whole cantaloupes at approximately 10^3 CFU/cm^2 remained attached; however, higher concentrations of Se and EcO157 attached initially to the fruit surface compared to Lm [125].

In a study comparing the attachment of multiple foodborne human pathogens to cut lettuce with that of $P.\ fluorescens$ (Pf) by observations under a confocal scanning laser microscope, Lm and EcO157 attached preferentially to the cut edge of lettuce, Pf attached to the uncut surface, and Se Typhimurium attached to both locations [134]. The nutrient-rich and hydrophilic nature of a plant cut surface compared to the hydrophobic waxy cuticle, is consistent with Lm, but not Pf, attaching and concentrating in this location [134,169].

There have been numerous studies of Lm attachment to postharvest produce, but few studies of Lm attachment to preharvest plants/produce. In one study, a 100- to 1000-fold difference was reported in the attachment and colonization of different strains of Lm to alfalfa sprouts grown from inoculated seeds [170]. No association of attachment with any known Lm surface characteristic (serotype) or genotype could be discerned. The same investigators reported minimal differences in the attachment of seven different strains of Lm to radish tissue after 2 hours (4.76 to 5.39 CFU/g tissue), indicating that the mechanism of attachment in this system was relatively conserved among strains. A screen of a library of Tn917-LTV3 Lm mutants of one of the strains with fresh-cut radish tissue (4 hours, 30°C) resulted in identification of three attachment-defective mutants [171]. Two mutations were in genes of unknown function within an operon-encoding flagellar biosynthesis; only one of the mutants lacked flagella and was nonmotile. A third mutant carried an insertion in an operon necessary for the transport of arabitol. All three mutants attached at least 10-fold less compared to the parent strain, which bound to the radish tissue at levels as high as 5 log$_{10}$ CFU/g at 30°C. However, none of the mutants attached less than the parent strain when the samples were incubated at 37°C. Incubation temperatures of 10 and 20°C affected the attachment of the single motility mutant negatively, whereas the arabitol transport mutant was decreased in attachment at 10 and 30°C. Changes in the Lm cell surface at low temperatures (e.g., 10 versus 37°C) have been shown previously to occur, including decreased chain lengths and branching of membrane fatty acids [172], and up-regulation of three genes predicted to encode cell surface proteins: *fbp* (putative fibronectin binding protein), *flaA* (flagellin), and *psr* (putative penicillin binding protein) [173]. These variable results associated with temperature suggest that *Lm* might express different attachment factors in different environments (e.g., temperature) [171].

2.7.6 CANTALOUPE AND *SE*, *EcO157*, AND *LM*

Ukuku and Fett studied the type and strength of attachment of multiple strains of *Se*, *E. coli*, and *Lm*, including outbreak- or food-associated strains, to the surface of cantaloupes [125]. The bacterial cell surface charge and hydrophobicity of each of the strains were determined and compared to the strength of the interaction, as measured by the number of cells retained on the cantaloupe surface after immersing whole melons in water. Attachment was measured both on melons spiked with individual strains and mixtures of strains. *Se* had the highest and most variable surface hydrophobicity, and the highest negative and positive surface charge; *E. coli*, *EcO157*, and *Lm* strains were similar in hydrophobicity, but *Lm* had a much higher negative surface charge compared to *E. coli*. Although more *E.coli* cells attached initially to the melon surface compared to *Se* and *Lm*, *Se* attached more strongly than either *E. coli* or *Lm* after storage at 4°C up to 7 days, regardless of whether strains were added individually or as mixtures [125]. The strength of attachment of each of the species was correlated significantly with the hydrophobicity and the negative and positive surface charge of the strains, indicating that all of these parameters were important in attachment.

2.7.7 *ARABIDOPSIS THALIANA* AND *EcO157* AND *SE*

Many genetic tools are available for studying *A. thaliana* (thale cress) [174]. Thus, it provides an opportunity to gain insight into the response of a plant to human pathogens. Single strains of *EcO157* and *Se* Newport were assessed in an *A. thaliana* model for attachment and growth characteristics [175]. In initial experiments, the human pathogens, applied to sterile roots under ideal humidity, remained attached at high concentrations (10^9 CFU/g tissue) with eventual migration to the stems/shoots (2×10^7 CFU/g). Examination of the roots by confocal microscopy revealed that *EcO157*-GFP and *Se* Newport-GFP strains appeared to have "invaded" the plant interior specifically at locations where lateral roots emerge (Figure 2.3C, Figure 2.1H). A similar result was obtained recently with a *Se* typhimurium strain in an alfalfa seedling model using relatively low numbers of cells ($\sim 10^2$ CFU) [120]. Single cells and cell aggregates of *EcO157* and *Se* Newport were observed also on shoots and flowers [175]; surprisingly, *EcO157* was isolated also from seed and chaff harvested from contaminated plants, and from plants grown from contaminated seed (unpublished results). The interaction of these two important human pathogens with multiple plant tissues suggests that multiple attachment mechanisms are involved [175].

2.7.8 PLANT–MICROBE BIOFILMS

Intact plant surfaces, especially those of leaves, are relatively inhospitable environments for microorganisms, providing limited sites for attachment, surface retention of water, and nutrients. Nevertheless, many microorganisms have developed mechanisms to attach, survive, or grow in microniches on different plants. The micrograph shown in Figure 2.3A demonstrates the localization and high density of epiphytic bacteria on a lettuce leaf. Both Gram-positive and Gram-negative bacteria are interacting in aggregates and possibly competing for the limited nutrients available in the microniche at the junction of epidermal cells where cuticular waxes are less dense, water accumulates, and nutrients are more available than in other sites. Although biofilms with classic structures described in recent studies are rarely found on plants, thick three-dimensional biofilms have been observed on sprouts sampled from a commercial sprout facility (Figure 2.3E). The image reveals the potential for complex interactions to occur between Gram-positive and Gram-negative resident bacteria under ideal conditions of plentiful water, exuded nutrients, and warm temperatures during food production or processing. A thick mat of mostly aggregated bacterial cells was detected on the root hairs of the sprouts (Figure 2.3E, "Ep"). Although plant tissue was likely present within the biofilm (Figure 2.3E, arrow), it appears that multiple layers of cells compose the biofilm, and that the presence of EPS at the surface of, or within, the biofilm is possible. Similar aggregates of bacteria have been observed using SEM on roots of alfalfa, broccoli, clover, sunflower, and mung bean sprouts [176,177]. Following attachment of bacteria as individual cells on leaf surfaces, aggregation is crucial as a strategy to ensure survival under environmental stresses such as water or nutrient depletion, UV irradiation, unfavorable temperatures, or predation [178,179]. In many enteric bacteria, fimbriae composed of curli protein interact with a cellulose polysaccharide resulting in aggregation and either pellicle formation or biofilms (Table 2.2) [180]. T3SS-encoded proteins in other bacteria (Table 2.1 and Table 2.2) are analogous to curli. In a recent report, T3SS-encoded protein in $Echr$ was shown to interact with β-glucanlike (noncellulose) carbohydrates, and this interaction was crucial for pellicle and biofilm formation $in\ vitro$ [181]. Thus, T3SS proteins, and possibly type 1 pili, conjugative pili, and curli (fimbriae), are important in aggregation leading to biofilm formation.

Biofilm formation is thought to be a major reason for the persistence of microorganisms, including pathogens, for long periods in food processing environments [182]. Bacteria, filamentous fungi, yeasts, and even viruses may be represented within biofilms on a plant surface. Therefore, the mechanisms of initiating bacterial autoaggregation and mixed-species aggregation, and the attachment of bacteria singly or as aggregates to plant surfaces or to microorganisms/EPS in preexisting biofilms on plant surfaces, could involve attachment factors such as those described in this review (Table 2.1 and Table 2.2). Understanding the mechanisms could yield intervention strategies for decontamination of produce.

2.8 CONCLUSIONS

The many years of difficult and labor-intensive studies on plant–microbe interactions involved in plant symbiosis and disease have begun to yield fundamental molecular information regarding bacterial attachment to plants (Table 2.1). The attachment factors designated in Table 2.1 can be grouped essentially into five categories: polysaccharides (EPS, CPS, LPS), outer membrane proteins, flagella, pili, and fimbriae. In some systems, bacterial protein factors have been identified that bind to plant carbohydrates (e.g., rhicadhesin), and in others, a bacterial polysaccharide is bound by a plant lectin (e.g., Rhsp EPS/CPS/LPS). It is probable that attachment for some bacteria will involve both strategies ("dual bridge") simultaneously, or with different hosts and/or in different environments. The attachment factors identified in human pathogens mostly relate to studies with animal cell lines or animal models (Table 2.2). However, it is very likely that flagella, pili, and fimbriae might have roles as attachment factors for human pathogens on plants, considering their prominent outer surface location and length. Absent from Table 2.2 are EPS (e.g., colanic acid), CPS (e.g., K-antigens), and LPS (e.g., O-antigens), all very important complex carbohydrate-containing molecules synthesized by human pathogens; these molecules are surface-expressed and often regulated by environmental cues [183–185]. Surface complex carbohydrates are excellent candidates for possible interactions with plant lectins of the appropriate specificity [186]; a precedent is the well-defined rhizobiaceae EPS interaction with pea plant lectin (Table 2.1).

The model studies of enteric human pathogens with plants/produce indicate the general fitness of human pathogens in these environments. Similar to plant bacteria, human pathogens appear to possess multiple specific mechanisms of attachment and growth (Table 2.3). The interactions of human pathogens with host plants probably will involve many unique factors depending upon the plant and the pathogen. It is probable that events occur preceding the direct interaction of a human pathogen with a plant that are important for attachment. For example, the environment in which the pathogen has remained viable (water, manure, soil, eukaryotic micro-organisms, insects, animals) will dictate what surface molecules are expressed and the metabolic state of the human pathogen prior to interaction with the plant host. Also, the human pathogen may be associated with other microorganisms in aggregates or in a detached biofilm. The plant may release chemicals that are signals and/or chemotaxis factors for some human pathogens. The availability of different types of plant receptors (specific and nonspecific) will determine the efficiency of attachment. After the human pathogen cell or cells make direct contact with the potential host plant, the human pathogen attaches either specifically or nonspecifically by weak or strong interactions depending upon the site of attachment. Flagellated cells may move (e.g., twitching motility) along a surface until an optimal attachment site is recognized. Initial attachment likely occurs by biochemical forces or by human pathogen proteins extended from the surface (pili/fimbriae, flagella), with tighter attachment established later by other surface molecules. Based on other plant–microbe interactions (Table 2.1), a possible strategy for attachment may combine human pathogen protein–plant receptor (e.g., carbohydrate) and plant lectin–human pathogen polysaccharide (e.g., EPS, CPS, LPS) interactions. The human pathogen may then be further secured by human pathogen cell–cell aggregation (possibly involving T3SS) or human pathogen–plant microbe aggregation, both of which likely require expression of different attachment factors.

The presence of putative attachment factors in enteric human pathogens that are similar to those of plant-related bacteria, point to obvious approaches for identifying fundamental mechanisms of human pathogen attachment to produce. Fimbriae, pili, flagella, polysaccharides, and porin proteins are all candidates for direct attachment to and aggregation of human pathogens on plant tissue. Recent advances by researchers in studies of how native microbes attach and interact with the rhizoplane and phylloplane provides inspiration and guidance for researchers studying the biology of human pathogens in similar environments.

ACKNOWLEDGMENTS

The authors thank Dr. Jeri Barak for providing information prior to publication and Dr. Amy Charkowski for an image of GFP-labeled *S. enterica* on sprouts. This work was supported by the U.S. Department of Agriculture, Agricultural Research Service CRIS project 5325-42000-040.

REFERENCES

1. Rovira, A.D., Plant root excretions in relation to the rhizosphere effect. I. The nature of root exudate from oats and peas, *Plant Soil,* 7, 178, 1956.
2. Rovira, A.D., Root excretions in relation to the rhizosphere effect. IV. Influence of plant species, age of plant, light, temperature, and calcium nutrition on exudation, *Plant Soil,* 11, 53, 1959.
3. Long, S.R., Rhizobium symbiosis: nod factors in perspective, *Plant Cell,* 8, 1885, 1996.
4. Van Wees, S.C. *et al.*, Differential induction of systemic resistance in Arabidopsis by biocontrol bacteria, *Mol. Plant. Microbe Interact.,* 10, 716, 1997.
5. Glick, B.R. and Bashan, Y., Genetic manipulation of plant growth-promoting bacteria to enhance biocontrol of phytopathogens, *Biotechnol. Adv.,* 15, 353, 1997.
6. Simons, M. *et al.*, Gnotobiotic system for studying rhizosphere colonization by plant growth-promoting *Pseudomonas* bacteria, *Mol. Plant. Microbe Interact.,* 9, 600, 1996.
7. Preston, G.M., Plant perceptions of plant growth-promoting *Pseudomonas*, *Philos. Trans. R. Soc. Lond. B. Biol. Sci.,* 359, 907, 2004.

8. Sessitsch, A., Reiter, B., and Berg, G., Endophytic bacterial communities of field-grown potato plants and their plant-growth-promoting and antagonistic abilities, *Can. J. Microbiol.*, 50, 239, 2004.

9. CDC, Centers for Disease Control and Prevention, U.S. Foodborne Disease Outbreak Line Listings, 1990–2002, 2004, 2004, http://www.cdc.gov/foodborneoutbreaks/us_outb.htm.

10. CSPI, Outbreak alert: closing the gaps in our federal food safety net, *Report from Center for Science in the Public Interest*, 58, 2004, http://www.cspinet.org/reports/index.html.

11. Katznelson, H. and Sutton, M.D., A rapid phage plaque count method for the detection of bacteria as applied to the demonstration of internally borne bacterial infections of seed, *J. Bacteriol.*, 61, 689, 1951.

12. Leben, C., Soybean flower-to-seed movement of epiphytic bacteria, *Can. J. Microbiol.*, 22, 429, 1976.

13. Charkowski, A.O., Sarreal, C.Z., and Mandrell, R.E., Wrinkled alfalfa seeds harbor more aerobic bacteria and are more difficult to sanitize than smooth seeds, *J. Food Prot.*, 64, 1292, 2001.

14. Torsvik, V., Goksoyr, J., and Daae, F.L., High diversity in DNA of soil bacteria, *Appl. Environ. Microbiol.*, 56, 782, 1990.

15. Geldreich, E.E., Kenner, B.A., and Kabler, P.W., Occurrence of coliforms, fecal coliforms, and streptococci on vegetation and insects, *Appl. Microbiol.*, 12, 63, 1964.

16. Geldreich, E.E. and Bordner, R.H., Fecal contamination of fruits and vegetables during cultivation and processing for market. A review, *J. Milk Food Technol.*, 34, 1971.

17. Gould, W.A., Micro-contamination of horticultural products, *HortScience*, 8, 12, 1973.

18. Ercolani, G.L., Bacteriological quality assessment of fresh marketed lettuce and fennel, *Appl. Environ. Microbiol.*, 31, 847, 1976.

19. Sivapalasingam, S., Friedman, C.R., and Tauxe, C.R.V., Fresh produce: a growing cause of outbreaks of foodborne illness in the United States, 1973 through 1997, *J. Food Prot.*, 67, 2342, 2004.

20. Collmer, A. and Bauer, D.W., *Erwinia chrysanthemi* and *Pseudomonas syringae*: plant pathogens trafficking in extracellular virulence proteins, *Curr. Top. Microbiol. Immunol.*, 192, 43, 1994.

21. Smit, G. *et al.*, Molecular mechanisms of attachment of *Rhizobium* bacteria to plant roots, *Mol. Microbiol.*, 6, 2897, 1992.

22. Guttman, D.S., Plants as models for the study of human pathogenesis, *Biotechnol. Adv.*, 22, 363, 2004.

23. Curl, E.A. and Truelove, B., The structure and function of roots, in *The Rhizosphere*, Yaron, B., Ed., Springer-Verlag, Berlin, 1986, p. 9.

24. Curl, E.A. and Truelove, B., Root exudates, in *The Rhizosphere*, Yaron, B., Ed., Springer-Verlag, Berlin, 1986, p. 55.

25. Dazzo, F.B. *et al.*, Specific phases of root hair attachment in the *Rhizobium trifolii*–clover symbiosis, *Appl. Environ. Microbiol.*, 48, 1140, 1984.

26. Smit, G. *et al.*, Purification and partial characterization of the *Rhizobium leguminosarum* biovar viciae Ca^{2+}-dependent adhesin, which mediates the first step in attachment of cells of the family Rhizobiaceae to plant root hair tips, *J. Bacteriol.*, 171, 4054, 1989.

27. Charkowski, A.O. *et al.*, Differences in growth of *Salmonella enterica* and *Escherichia coli* O157:H7 on alfalfa sprouts, *Appl. Environ. Microbiol.*, 68, 3114, 2002.

28. Jeffree, C.E., Structure and ontogeny of plant cuticles, in *Plant Cuticles*, Kerstiens, G., Ed., Bios, Oxford, 1996, p. 33.

29. Beattie, G.A., Leaf surface waxes and the process of leaf colonization by microorganisms, in *Phyllosphere Microbiology*, Lindow, S.E., Hecht-Poinar, E.I., and Elliott, V.J., Eds., APS Press, St. Paul, MN, 2002, p. 3.

30. Purdy, R.E. and Kolattukudy, P.E., Hydrolysis of plant cuticle by plant pathogens. Properties of cutinase I, cutinase II, and a nonspecific esterase isolated from *Fusarium solani pisi*, *Biochemistry (Mosc)*. 14, 2832, 1975.

31. Singh, P. *et al.*, Investigation on epiphytic lining *Pseudomonas* species from *Malus domestica* with an antagonistic effect to *Venturia inaequalis* on isolated plant cuticle membranes, *Environ. Microbiol.*, 6, 1149, 2004.

32. Hirano, S.S. and Upper, C.D., Bacteria in the leaf ecosystem with emphasis on *Pseudomonas syringae*: a pathogen, ice nucleus, and epiphyte, *Microbiol. Mol. Biol. Rev.*, 64, 624, 2000.

33. Handelsman, J. and Stabb, E.V., Biocontrol of soilborne plant pathogens, *Plant Cell*, 8, 1855, 1996.

34. Lindow, S.E. and Brandl, M.T., Microbiology of the phyllosphere, *Appl. Environ. Microbiol.*, 69, 1875, 2003.

35. de Weert, S. *et al.*, Flagella-driven chemotaxis towards exudate components is an important trait for tomato root colonization by *Pseudomonas fluorescens*, *Mol. Plant. Microbe Interact.*, 15, 1173, 2002.

36. Kinkel, L.L., Wilson, M., and Lindow, S.E., Plant species and plant incubation conditions influence variability in epiphytic bacterial population size, *Microb. Ecol.*, 39, 1, 2000.

37. Hirano, S.S. *et al.*, Lognormal distribution of epiphytic bacterial populations on leaf surfaces, *Appl. Environ. Microbiol.*, 44, 695, 1982.

38. Lund, B.M., Ecosystems in vegetable foods, *J. Appl. Bacteriol. Symp. Suppl.*, 1992.

39. Keipper, C.H. and Fred, E.B., The microorganisms of cabbage and their relation to sauerkraut production, *J. Bacteriol.*, 19, 53, 1930.

40. Sapers, G.M., Efficacy of washing and sanitizing methods for disinfection of fresh fruit and vegetable products, *Food Technol. Biotechnol.*, 39, 305, 2001.

41. Dazzo, F.B., Napoli, C.A., and Hubbell, D.H., Adsorption of bacteria to roots as related to host specificity in the *Rhizobium*–clover symbiosis, *Appl. Environ. Microbiol.*, 32, 166, 1976.

42. Higashi, S. and Mikiko, A., Scanning electron microscopy of *Rhizobium trifolii* infection sites on root hairs of white clover, *Appl. Environ. Microbiol.*, 40, 1094, 1980.

43. Smit, G., Kijne, J.W., and Lugtenberg, B.J., Roles of flagella, lipopolysaccharide, and a Ca^{2+}-dependent cell surface protein in attachment of *Rhizobium leguminosarum* biovar viciae to pea root hair tips, *J. Bacteriol.*, 171, 569, 1989.

44. Smit, G., Kijne, J.W., and Lugtenberg, B.J., Involvement of both cellulose fibrils and a Ca^{2+}-dependent adhesin in the attachment of *Rhizobium leguminosarum* to pea root hair tips, *J. Bacteriol.*, 169, 4294, 1987.

45. Ausmees, N. *et al.*, Structural and putative regulatory genes involved in cellulose synthesis in *Rhizobium leguminosarum* bv. *trifolii*, *Microbiology*, 145, 1253, 1999.

46. Kijne, J.W. *et al.*, Lectin-enhanced accumulation of manganese-limited *Rhizobium leguminosarum* cells on pea root hair tips, *J. Bacteriol.*, 170, 2994, 1988.

47. van Rhijn, P. *et al.*, Sugar-binding activity of pea lectin enhances heterologous infection of transgenic alfalfa plants by *Rhizobium leguminosarum* biovar *viciae*, *Plant Physiol.*, 126, 133, 2001.

48. Bohlool, B.B. and Schmidt, E.L., Lectins: a possible basis for specificity in the *Rhizobium*–legume root nodule symbiosis, *Science*, 185, 269, 1974.

49. Tsien, H.C. and Schmidt, E.L., Localization and partial characterization of soybean lectin-binding polysaccharide of *Rhizobium japonicum*, *J. Bacteriol.*, 145, 1063, 1981.

50. Schloter, M. *et al.*, Root colonization of different plants by plant-growth-promoting *Rhizobium leguminosarum* bv. trifolii R39 studied with monospecific polyclonal antisera, *Appl. Environ. Microbiol.*, 63, 2038, 1997.

51. Ho, S.C., Wang, J.L., and Schindler, M., Carbohydrate binding activities of *Bradyrhizobium japonicum*. I. Saccharide-specific inhibition of homotypic and heterotypic adhesion, *J. Cell Biol.*, 111, 1631, 1990.

52. Loh, J.T. *et al.*, Carbohydrate binding activities of *Bradyrhizobium japonicum*: unipolar localization of the lectin BJ38 on the bacterial cell surface, *Proc. Natl. Acad. Sci. USA*, 90, 3033, 1993.

53. Ausmees, N., Jacobsson, K., and Lindberg, M., A unipolarly located, cell-surface-associated agglutinin, RapA, belongs to a family of Rhizobium-adhering proteins (Rap) in *Rhizobium leguminosarum* bv. *trifolii*, *Microbiology*, 147, 549, 2001.

54. Gelvin, S.B., *Agrobacterium* and plant genes involved in T-DNA transfer and integration, *Annu. Rev. Plant Physiol. Plant Mol. Biol.*, 51, 223, 2000.

55. Lippincott, B.B. and Lippincott, J.A., Bacterial attachment to a specific wound site as an essential stage in tumor initiation by *Agrobacterium tumefaciens*, *J. Bacteriol.*, 97, 620, 1969.

56. Douglas, C.J., Halperin, W., and Nester, E.W., *Agrobacterium tumefaciens* mutants affected in attachment to plant cells, *J. Bacteriol.*, 152, 1265, 1982.

57. Crews, J.L., Colby, S., and Matthysse, A.G., *Agrobacterium rhizogenes* mutants that fail to bind to plant cells, *J. Bacteriol.*, 172, 6182, 1990.

58. Brisset, M. *et al.*, Attachment, chemotaxis, and multiplication of *Agrobacterium tumefaciens* biovar 1 and biovars 3 on grapevine and pea, *Appl. Environ. Microbiol.*, 57, 3178, 1991.

59. Swart, S. *et al.*, Rhicadhesin-mediated attachment and virulence of an *Agrobacterium tumefaciens chvB* mutant can be restored by growth in a highly osmotic medium, *J. Bacteriol.*, 176, 3816, 1994.

60. Matthysse, A.G., Yarnall, H.A., and Young, N., Requirement for genes with homology to ABC transport systems for attachment and virulence of *Agrobacterium tumefaciens*, *J. Bacteriol.*, 178, 5302, 1996.

61. Reuhs, B.L., Kim, J.S., and Matthysse, A.G., Attachment of *Agrobacterium tumefaciens* to carrot cells and *Arabidopsis* wound sites is correlated with the presence of a cell-associated, acidic polysaccharide, *J. Bacteriol.*, 179, 5372, 1997.

62. Matthysse, A.G. and McMahan, S., Root colonization by *Agrobacterium tumefaciens* is reduced in cel, attB, attD, and attR mutants, *Appl. Environ. Microbiol.*, 64, 2341, 1998.

63. Matthysse, A.G. and McMahan, S., The effect of the *Agrobacterium tumefaciens attR* mutation on attachment and root colonization differs between legumes and other dicots, *Appl. Environ. Microbiol.*, 67, 1070, 2001.

64. Whatley, M.H. *et al.*, Role of *Agrobacterium* cell envelope lipopolysaccharide in infection site attachment, *Infect. Immun.*, 13, 1080, 1976.

65. Matthysse, A.G., Characterization of nonattaching mutants of *Agrobacterium tumefaciens*, *J. Bacteriol.*, 169, 313, 1987.

66. Wagner, V.T. and Matthysse, A.G., Involvement of a vitronectin-like protein in attachment of *Agrobacterium tumefaciens* to carrot suspension culture cells, *J. Bacteriol.*, 174, 5999, 1992.

67. Sanders, L.C. *et al.*, A homolog of the substrate adhesion molecule vitronectin occurs in four species of flowering plants, *Plant Cell*, 3, 629, 1991.

68. Gelvin, S.B., *Agrobacterium*-mediated plant transformation: the biology behind the "gene-jockeying" tool, *Microbiol. Mol. Biol. Rev.*, 67, 16, 2003.

69. Belanger, C. *et al.*, Genetic analysis of nonpathogenic *Agrobacterium tumefaciens* mutants arising in crown gall tumors, *J. Bacteriol.*, 177, 3752, 1995.

70. Lai, E.M. and Kado, C.I., Processed VirB2 is the major subunit of the promiscuous pilus of *Agrobacterium tumefaciens*, *J. Bacteriol.*, 180, 2711, 1998.

71. Matthysse, A.G., Holmes, K.V., and Gurlitz, R.H., Elaboration of cellulose fibrils by *Agrobacterium tumefaciens* during attachment to carrot cells, *J. Bacteriol.*, 145, 583, 1981.

72. Matthysse, A.G., Role of bacterial cellulose fibrils in *Agrobacterium tumefaciens* infection, *J. Bacteriol.*, 154, 906, 1983.

73. Matthysse, A.G., White, S., and Lightfoot, R., Genes required for cellulose synthesis in *Agrobacterium tumefaciens*, *J. Bacteriol.*, 177, 1069, 1995.

74. Hayward, A.C., Biology and epidemiology of bacterial wilt caused by *Pseudomonas solanacearum*, *Annu. Rev. Phytopathol.*, 29, 65, 1991.

75. Kao, C.C. and Sequeira, L., A gene cluster required for coordinated biosynthesis of lipopolysaccharide and extracellular polysaccharide also affects virulence of *Pseudomonas solanacearum*, *J. Bacteriol.*, 173, 7841, 1991.

76. Baker, J.M. *et al.*, Chemical characterization of th lipopolysaccharide of *Ralstonia solanacearum*, *Appl. Environ. Microbiol.*, 47, 1096, 1984.

77. He, S.Y., Type III protein secretion systems in plant and animal pathogenic bacteria, *Annu. Rev. Phytopathol.*, 36, 363, 1998.

78. Van Gijsegem, F. *et al.*, *Ralstonia solanacearum* produces hrp-dependent pili that are required for PopA secretion but not for attachment of bacteria to plant cells, *Mol. Microbiol.*, 36, 249, 2000.

79. Kang, Y. *et al.*, *Ralstonia solanacearum* requires type 4 pili to adhere to multiple surfaces and for natural transformation and virulence, *Mol. Microbiol.*, 46, 427, 2002.

80. Salanoubat, M. *et al.*, Genome sequence of the plant pathogen *Ralstonia solanacearum*, *Nature*, 415, 497, 2002.

81. Durand, E. *et al.*, Type II protein secretion in *Pseudomonas aeruginosa*: the pseudopilus is a multifibrillar and adhesive structure, *J. Bacteriol.*, 185, 2749, 2003.

82. Sudakevitz, D., Imberty, A., and Gilboa-Garber, N., Production, properties and specificity of a new bacterial L-fucose- and D-arabinose-binding lectin of the plant aggressive pathogen *Ralstonia solanacearum*, and its comparison to related plant and microbial lectins, *J. Biochem. (Tokyo)*, 132, 353, 2002.

83. Sudakevitz, D. *et al.*, A new *Ralstonia solanacearum* high-affinity mannose-binding lectin RS-IIL structurally resembling the *Pseudomonas aeruginosa* fucose-specific lectin PA-IIL, *Mol. Microbiol.*, 52, 691, 2004.

84. Genin, S. and Boucher, C., Lessons learned from the genome analysis of *Ralstonia solanacearum*, *Annu. Rev. Phytopathol.*, 42, 107, 2004.

85. Rojas, C.M. *et al.*, HecA, a member of a class of adhesins produced by diverse pathogenic bacteria, contributes to the attachment, aggregation, epidermal cell killing, and virulence phenotypes of *Erwinia chrysanthemi* EC16 on *Nicotiana clevelandii* seedlings, *Proc. Natl. Acad. Sci. USA*, 99, 13142, 2002.

86. Clantin, B. *et al.*, The crystal structure of filamentous hemagglutinin secretion domain and its implications for the two-partner secretion pathway, *Proc. Natl. Acad. Sci. USA*, 101, 6194, 2004.

87. Salman, M.N., Establishment of callus and cell suspension cultures from *Gypsophila paniculata* leaf segments and study of the attachment of host cells by *Erwinia herbicola* pv. *gypsophilae*, *Plant Cell Tissue Organ Culture*, 69, 189, 2002.

88. Rantakari, A. *et al.*, Type III secretion contributes to the pathogenesis of the soft-rot pathogen *Erwinia carotovora*: partial characterization of the hrp gene cluster, *Mol. Plant. Microbe Interact.*, 14, 962, 2001.

89. Holeva, M.C. *et al.*, Use of a pooled transposon mutation grid to demonstrate roles in disease development for *Erwinia carotovora* subsp. *atroseptica* putative type III secreted effector (DspE/A) and helper (HrpN) proteins, *Mol. Plant. Microbe Interact.*, 17, 943, 2004.

90. Bogs, J. *et al.*, Colonization of host plants by the fire blight pathogen *Erwinia amylovora* marked with genes for bioluminescence and fluorescence, *Phytopathology*, 88, 416, 1998.

91. Jin, Q. *et al.*, Visualization of secreted Hrp and Avr proteins along the Hrp pilus during type III secretion in *Erwinia amylovora* and *Pseudomonas syringae*, *Mol. Microbiol.*, 40, 1129, 2001.

92. Korhonen, T.K. *et al.*, A N-acetyllactosamine-specific cell-binding activity in a plant pathogen, *Erwinia rhapontici*, *FEBS Lett.*, 236, 163, 1988.

93. Romantschuk, M. and Bamford, D.H., The causal agent of halo blight in bean, *Pseudomonas syringae* pv. phaseolicola, attaches to stomata via its pili, *Microb. Pathog.*, 1, 139, 1986.

94. Hirano, S.S., Baker, L.S., and Upper, C.D., Raindrop momentum triggers growth of leaf-associated populations of *Pseudomonas syringae* on filed-grown snap bean plants, *Appl. Environ. Microbiol.*, 62, 2560, 1996.

95. Rainey, P.B., Adaptation of *Pseudomonas fluorescens* to the plant rhizosphere, *Environ. Microbiol.*, 1, 243, 1999.

96. Liao, C.H. and Fett, W.F., Analysis of native microflora and selection of strains antagonistic to human pathogens on fresh produce, *J. Food Prot.*, 64, 1110, 2001.

97. De Weger, L.A. *et al.*, Flagella of a plant-growth-stimulating *Pseudomonas fluorescens* strain are required for colonization of potato roots, *J. Bacteriol.*, 169, 2769, 1987.

98. Vesper, S.J., Production of pili (fimbriae) by *Pseudomonas fluorescens* and correlation with attachment to corn roots, *Appl. Environ. Microbiol.*, 53, 1397, 1987.

99. Anderson, A.J., Habibzadegah-Tan, P., and Tepper, C.S., Molecular studies on the role of a root surface agglutinin in adherence and colonization by *Pseudomonas putida*, *Appl. Environ. Microbiol.*, 54, 375, 1988.

100. Turnbull, G.A. *et al.*, The role of motility in the in vitro attachment of *Pseudomonas putida* PaW8 to wheat roots, *FEMS Microbiol. Ecol.*, 35, 57, 2001.

101. Elrod, R.P. and Braun, A.C., *Pseudomonas aeruginosa*: its role as a plant pathogen, *J. Bacteriol.*, 44, 633, 1942.

102. Plotnikova, J.M., Rahme, L.G., and Ausubel, F.M., Pathogenesis of the human opportunistic pathogen *Pseudomonas aeruginosa* PA14 in Arabidopsis, *Plant Physiol.*, 124, 1766, 2000.

103. Rahme, L.G. *et al.*, Plants and animals share functionally common bacterial virulence factors, *Proc. Natl. Acad. Sci. USA*, 97, 8815, 2000.

104. Hahn, H.P., The type-4 pilus is the major virulence-associated adhesin of *Pseudomonas aeruginosa*: a review, *Gene*, 192, 99, 1997.

105. Ojanen-Reuhs, T. *et al.*, Characterization of the *fimA* gene encoding bundle-forming fimbriae of the plant pathogen *Xanthomonas campestris* pv. vesicatoria, *J. Bacteriol.*, 179, 1280, 1997.

106. Van Sluys, M.A. *et al.*, Comparative genomic analysis of plant-associated bacteria, *Annu. Rev. Phytopathol.*, 40, 169, 2002.

107. Burdman, S., Okon, Y., and Jurkevitch, E., Surface characteristics of *Azospirillum brasilense* in relation to cell aggregation and attachment to plant roots, *Crit. Rev. Microbiol.*, 26, 91, 2000.

108. Del Gallo, M., Negi, M., and Neyra, C.A., Calcofluor- and lectin-binding exocellular polysaccharides of *Azospirillum brasilense* and *Azospirillum lipoferum*, *J. Bacteriol.*, 171, 3504, 1989.

109. Croes, C.L. *et al.*, The polar flagellum mediates *Azospirillum brasilense* adsorption to wheat roots, *J. Gen. Microbiol.*, 139, 2261, 1993.

110. Burdman, S. *et al.*, Purification of the major outer membrane protein of *Azospirillum brasilense*, its affinity to plant roots, and its involvement in cell aggregation, *Mol. Plant. Microbe Interact.*, 14, 555, 2001.

111. Korhonen, T.K. *et al.*, Type 3 fimbriae of *Klebsiella* sp.: molecular characterization and role in bacterial adhesion to plant roots, *J. Bacteriol.*, 155, 860, 1983.

112. Haahtela, K., Tarkka, E., and Korhonen, T.K., Type 1 fimbria-mediated adhesion of enteric bacteria to grass roots, *Appl. Environ. Microbiol.*, 49, 1182, 1985.

113. Mendgen, K., Hahn, M., and Deising, H., Morphogenesis and mechanisms of penetration by plant pathogenic fungi, *Annu. Rev. Phytopathol.*, 34, 367, 1996.

114. Doss, R.P. *et al.*, Adhesion of germlings of *Botrytis cinerea*, *Appl. Environ. Microbiol.*, 61, 260, 1995.

115. Seymour, I.J. and Appleton, H., Foodborne viruses and fresh produce, *J. Appl. Microbiol.*, 91, 759, 2001.

116. Kakani, K., Robbins, M., and Rochon, D., Evidence that binding of cucumber necrosis virus to vector zoospores involves recognition of oligosaccharides, *J. Virol.*, 77, 3922, 2003.

117. Galan, J.E. and Collmer, A., Type III secretion machines: bacterial devices for protein delivery into host cells, *Science,* 284, 1322, 1999.

118. Staskawicz, B.J. *et al.,* Common and contrasting themes of plant and animal diseases, *Science,* 292, 2285, 2001.

119. Cao, H., Baldini, R.L., and Rahme, L.G., Common mechanisms for pathogens of plants and animals, *Annu. Rev. Phytopathol.,* 39, 259, 2001.

120. Dong, Y. *et al.,* Kinetics and strain specificity of rhizosphere and endophytic colonization by enteric bacteria on seedlings of *Medicago sativa* and *Medicago truncatula, Appl. Environ. Microbiol.,* 69, 1783, 2003.

121. Takeuchi, K. and Frank, J.F., Penetration of *Escherichia coli* O157:H7 into lettuce tissues as affected by inoculum size and temperature and the effect of chlorine treatment on cell viability, *J. Food Prot.,* 63, 434, 2000.

122. Takeuchi, K. and Frank, J.F., Direct microscopic observation of lettuce leaf decontamination with a prototype fruit and vegetable washing solution and 1% NaCl–NaHCO₃, *J. Food Prot.,* 64, 1235, 2001.

123. Takeuchi, K. and Frank, J.F., Quantitative determination of the role of lettuce leaf structures in protecting *Escherichia coli* O157:H7 from chlorine disinfection, *J. Food Prot.,* 64, 147, 2001.

124. Burnett, S.L., Chen, J., and Beuchat, L.R., Attachment of *Escherichia coli* O157:H7 to the surfaces and internal structures of apples as detected by confocal scanning laser microscopy, *Appl. Environ. Microbiol.,* 66, 4679, 2000.

125. Ukuku, D.O. and Fett, W.F., Relationship of cell surface charge and hydrophobicity to strength of attachment of bacteria to cantaloupe rind, *J. Food Prot.,* 65, 1093, 2002.

126. Richards, G.M. and Beuchat, L.R., Attachment of *Salmonella* Poona to cantaloupe rind and stem scar tissues as affected by temperature of fruit and inoculum, *J. Food Prot.,* 67, 1359, 2004.

127. Wachtel, M.R., Whitehand, L.C., and Mandrell, R.E., Association of *Escherichia coli* O157:H7 with preharvest leaf lettuce upon exposure to contaminated irrigation water, *J. Food Prot.,* 65, 18, 2002.

128. Brandl, M.T. and Mandrell, R.E., Fitness of *Salmonella enterica* serovar Thompson in the cilantro phyllosphere, *Appl. Environ. Microbiol.,* 68, 3614, 2002.

129. Brandl, M.T. *et al.,* Comparison of survival of *Campylobacter jejuni* in the phyllosphere with that in the rhizosphere of spinach and radish plants, *Appl. Environ. Microbiol.,* 70, 1182, 2004.

130. Hilborn, E.D. *et al.,* A multistate outbreak of *Escherichia coli* O157:H7 infections associated with consumption of mesclun lettuce, *Arch. Intern. Med.,* 159, 1758, 1999.

131. Ackers, M.L. *et al.,* An outbreak of *Escherichia coli* O157:H7 infections associated with leaf lettuce consumption, *J. Infect. Dis.,* 177, 1588, 1998.

132. Beuchat, L.R., Survival of enterohemorrhagic *Escherichia coli* O157:H7 in bovine feces applied to lettuce and the effectiveness of chlorinated water as a disinfectant, *J. Food Prot.,* 62, 845, 1999.

133. Seo, K.H. and Frank, J.F., Attachment of *Escherichia coli* O157:H7 to lettuce leaf surface and bacterial viability in response to chlorine treatment as demonstrated by using confocal scanning laser microscopy, *J. Food Prot.,* 62, 3, 1999.

134. Takeuchi, K. *et al.*, Comparison of the attachment of *Escherichia coli* O157:H7, *Listeria monocytogenes*, *Salmonella typhimurium*, and *Pseudomonas fluorescens* to lettuce leaves, *J. Food Prot.*, 63, 1433, 2000.

135. Li, Y. *et al.*, Survival and growth of *Escherichia coli* O157:H7 inoculated onto cut lettuce before or after heating in chlorinated water, followed by storage at 5 or 15 degrees C, *J. Food Prot.*, 64, 305, 2001.

136. Solomon, E.B., Yaron, S., and Matthews, K.R., Transmission of *Escherichia coli* O157:H7 from contaminated manure and irrigation water to lettuce plant tissue and its subsequent internalization, *Appl. Environ. Microbiol.*, 68, 397, 2002.

137. Hassan, A.N. and Frank, J.F., Influence of surfactant hydrophobicity on the detachment of *Escherichia coli* O157:H7 from lettuce, *Int. J. Food Microbiol.*, 87, 145, 2003.

138. Wachtel, M. and Charkowski, A., Cross-contamination of lettuce with *Escherichia coli* O157:H7, *J. Food Prot.*, 65, 465, 2002.

139. Hassan, A.N. and Frank, J.F., Attachment of *Escherichia coli* O157:H7 grown in tryptic soy broth and nutrient broth to apple and lettuce surfaces as related to cell hydrophobicity, surface charge, and capsule production, *Int. J. Food Microbiol.*, 96, 103, 2004.

140. Hedberg, C.W., MacDonald, K.L., and Osterholm, M.T., Changing epidemiology of food-borne disease: a Minnesota perspective, *Clin. Infect. Dis.*, 18, 671, 1994.

141. Cummings, K. *et al.*, A multistate outbreak of *Salmonella enterica* serotype Baildon associated with domestic raw tomatoes, *Emerg. Infect. Dis.*, 7, 1046, 2001.

142. Wei, C.I. *et al.*, Growth and survival of *Salmonella montevideo* on tomatoes and disinfection with chlorinated water, *J. Food Prot.*, 58, 829, 1995.

143. Guo, X. *et al.*, Survival of salmonellae on and in tomato plants from the time of inoculation at flowering and early stages of fruit development through fruit ripening, *Appl. Environ. Microbiol.*, 67, 4760, 2001.

144. Guo, X. *et al.*, Evidence of association of salmonellae with tomato plants grown hydroponically in inoculated nutrient solution, *Appl. Environ. Microbiol.*, 68, 3639, 2002.

145. Liao, C.H. and Sapers, G.M., Attachment and growth of *Salmonella* Chester on apple fruits and in vivo response of attached bacteria to sanitizer treatments, *J. Food Prot.*, 63, 876, 2000.

146. Mohle-Boetani, J.C. *et al.*, *Escherichia coli* O157 and *Salmonella* infections associated with sprouts in California, 1996-1998, *Ann. Intern. Med.*, 135, 239, 2001.

147. Anonymous, Guidance for industry: reducing microbial food safety hazards for sprouted seeds and guidance for industry: sampling and microbial testing of spent irrigation water during sprout production, *Fed. Regist.*, 64, 57893, 1999.

148. Barak, J.D., Whitehand, L.C., and Charkowski, A.O., Differences in attachment of *Salmonella enterica* serovars and *Escherichia coli* O157:H7 to alfalfa sprouts, *Appl. Environ. Microbiol.*, 68, 4758, 2002.

149. Gandhi, M. *et al.*, Use of green fluorescent protein expressing *Salmonella* Stanley to investigate survival, spatial location, and control on alfalfa sprouts, *J. Food Prot.*, 64, 1891, 2001.

150. Campbell, J.V. *et al.*, An outbreak of *Salmonella* serotype Thompson associated with fresh cilantro, *J. Infect. Dis.*, 183, 984, 2001.

151. Esmelindro, A.A. *et al.*, Influence of agronomic variables on the composition of mate tea leaves (*Ilex paraguariensis*) extracts obtained from CO_2 extraction at 30 degrees C and 175 bar, *J. Agric. Food Chem.*, 52, 1990, 2004.

152. Schlech, W.F., III, Epidemiology and clinical manifestations of *Listeria monocytogenes* infection, in *Gram-Positive Pathogens*, Fishetti, V.A. *et al.*, Eds., ASM Press, Washington D.C., 2000, p. 473.

153. Dorozynski, A., Seven die in French listeria outbreak, *BMJ*, 320, 601, 2000.

154. Weis, J. and Seeliger, H.P.R., Incidence of *Listeria monocytogenes* in nature, *Appl. Microbiol.*, 30, 29, 1975.

155. Aureli, P. *et al.*, An outbreak of febrile gastroenteritis associated with corn contaminated by *Listeria monocytogenes*, *N. Engl. J. Med.*, 342, 1236, 2000.

156. Brackett, R., Incidence and behavior of *Listeria monocytogenes* in products of plant origin, in *Listeria, Listeriosis, and Food Safety*, Ryser, E. and Marth, E., Eds., Marcel Dekker, New York, 1999, p. 631.

157. Ho, J.L. *et al.*, An outbreak of type 4b *Listeria monocytogenes* infection involving patients from eight Boston hospitals, *Arch. Intern. Med.*, 146, 520, 1986.

158. Farber, J.M. *et al.*, Listeriosis traced to the consumption of alfalfa tablets and soft cheese, *N. Engl. J. Med.*, 322, 338, 1990.

159. Heisick, J.E. *et al.*, *Listeria* spp. found on fresh market produce, *Appl. Environ. Microbiol.*, 55, 1925, 1989.

160. Porto, E. and Eiroa, M.N.U., Occurrence of *Listeria monocytogenes* in vegetables, *Dairy Food Environ. Sanit.*, 21, 282, 2001.

161. Prazak, A.M. *et al.*, Prevalence of *Listeria monocytogenes* during production and postharvest processing of cabbage, *J. Food Prot.*, 65, 1728, 2002.

162. FDA Enforcement Report Index, 2002, U.S. Food and Drug Administration, 2002, http://www.fda.gov/opacom/Enforce.html.

163. Berrang, M.E., Brackett, R.E., and Beuchat, L.R., Growth of *Listeria monocytogenes* on fresh vegetables stored under controlled atmosphere, *J. Food Prot.*, 52, 702, 1989.

164. Farber, J.M. *et al.*, Changes in populations of *Listeria monocytogenes* inoculated on packaged fresh-cut vegetables, *J. Food Prot.*, 61, 192, 1998.

165. Ukuku, D.O. and Fett, W., Behavior of *Listeria monocytogenes* inoculated on cantaloupe surfaces and efficacy of washing treatments to reduce transfer from rind to fresh-cut pieces, *J. Food Prot.*, 65, 924, 2002.

166. Steinbruegge, E.G., Maxcy, R.B., and Liewen, M.B., Fate of *Listeria monocytogenes* on ready to serve lettuce, *J. Food Prot.*, 51, 596, 1988.

167. Garrood, M.J., Wilson, P.D., and Brocklehurst, T.F., Modeling the rate of attachment of *Listeria monocytogenes*, *Pantoea agglomerans*, and *Pseudomonas fluorescens* to, and the probability of their detachment from, potato tissue at 10 degrees C, *Appl. Environ. Microbiol.*, 70, 3558, 2004.

168. Reina, L.D., Fleming, H.P., and Breidt, F., Jr., Bacterial contamination of cucumber fruit through adhesion, *J. Food Prot.*, 65, 1881, 2002.

169. Taiz, L. and Zeiger, E., *Plant Physiology*, 2nd ed., Sinauer Associates, Sunderland, MA, 1998.

170. Gorski, L., Palumbo, J.D., and Nguyen, K.D., Strain-specific differences in the attachment of *Listeria monocytogenes* to alfalfa sprouts, *J. Food Prot.*, 67, 2488, 2004.

171. Gorski, L., Palumbo, J.D., and Mandrell, R.E., Attachment of *Listeria monocytogenes* to radish tissue is dependent upon temperature and flagellar motility, *Appl. Environ. Microbiol.,* 69, 258, 2003.

172. Annous, B.A. *et al.,* Critical role of anteiso-C15:0 fatty acid in the growth of *Listeria monocytogenes* at low temperatures, *Appl. Environ. Microbiol.,* 63, 3887, 1997.

173. Liu, S. *et al.,* Identification of *Listeria monocytogenes* genes expressed in response to growth at low temperature, *Appl. Environ. Microbiol.,* 68, 1697, 2002.

174. Ghassemian, M. *et al.,* An integrated *Arabidopsis* annotation database for Affymetrix Genechip data analysis, and tools for regulatory motif searches, *Trends Plant Sci.,* 6, 448, 2001.

175. Cooley, M., Miller, W., and Mandrell, R., Colonization of *Arabidopsis thaliana* with *Salmonella enterica* or enterohemorrhagic *Escherichia coli* O157:H7 and competition by an *Enterobacter asburiae, Appl. Environ. Microbiol.,* 69, 4915, 2003.

176. Fett, W.F., Naturally occurring biofilms on alfalfa and other types of sprouts, *J. Food Prot.,* 63, 625, 2000.

177. Fett, W.F. and Cooke, P.H., Scanning electron microscopy of native biofilms on mung bean sprouts, *Can. J. Microbiol.,* 49, 45, 2003.

178. Monier, J.M. and Lindow, S.E., Differential survival of solitary and aggregated bacterial cells promotes aggregate formation on leaf surfaces, *Proc. Natl. Acad. Sci. USA,* 100, 15977, 2003.

179. Morris, C.E. and Monier, J.M., The ecological significance of biofilm formation by plant-associated bacteria, *Annu. Rev. Phytopathol.,* 41, 429, 2003.

180. Zogaj, X. *et al.,* Production of cellulose and curli fimbriae by members of the family *Enterobacteriaceae* isolated from the human gastrointestinal tract, *Infect. Immun.,* 71, 4151, 2003.

181. Yap, M.N. *et al.,* The *Erwinia chrysanthemi* type III secretion system is required for multicellular behavior, *J. Bacteriol.,*187, 639, 2005.

182. Zottola, E.A., Microbial attachment and biofilm formation: a new problem for the food industry?, *Food Technol.,* 48, 107, 1996.

183. Orskov, I. *et al.,* Serology, chemistry, and genetics of O and K antigens of *Escherichia coli, Bacteriol. Rev.,* 41, 667, 1977.

184. Caroff, M. and Karibian, D., Structure of bacterial lipopolysaccharides, *Carbohydr. Res.,* 338, 2431, 2003.

185. Whitfield, C. and Paiment, A., Biosynthesis and assembly of Group 1 capsular polysaccharides in *Escherichia coli* and related extracellular polysaccharides in other bacteria, *Carbohydr. Res.,* 338, 2491, 2003.

186. Varki, A. *et al.,* Plant lectins, in *Essentials of Glycobiology,* Cold Spring Harbor Laboratory Press, Cold Spring Harbor, NY, 1999, p. 653.

187. Oke, V. and Long, S.R., Bacterial genes induced within the nodule during the *Rhizobium*–legume symbiosis, *Mol. Microbiol.,* 32, 837, 1999.

188. Romantschuk, M., Bacterial attachment to leaves, in *Encyclopedia of Plant and Crop Science,* Goodman, R.M., Ed., Marcel Dekker, New York, 2004, p. 75.

189. Collmer, A. *et al., Pseudomonas syringae* Hrp type III secretion system and effector proteins, *Proc. Natl. Acad. Sci. USA,* 97, 8770, 2000.

190. Gilboa-Garber, N., *Pseudomonas aeruginosa* lectins, *Methods Enzymol.,* 83, 378, 1982.

191. Roine, E. *et al.*, Hrp pilus: an hrp-dependent bacterial surface appendage produced by *Pseudomonas syringae* pv. tomato DC3000, *Proc. Natl. Acad. Sci. USA*, 94, 3459, 1997.

192. Roine, E. *et al.*, Characterization of type IV pilus genes in *Pseudomonas syringae* pv. tomato DC3000, *Mol. Plant. Microbe Interact.*, 11, 1048, 1998.

193. Moreira, L.M. *et al.*, Comparative genomics analyses of citrus-associated bacteria, *Annu. Rev. Phytopathol.*, 42, 163, 2004.

194. Simpson, A.J. *et al.*, The genome sequence of the plant pathogen *Xylella fastidiosa*. The *Xylella fastidiosa* Consortium of the Organization for Nucleotide Sequencing and Analysis, *Nature*, 406, 151, 2000.

195. Vesper, S.J. and Bauer, W.D., Role of pili (fimbriae) in attachment of *Bradyrhizobium japonicum* to soybean roots, *Appl. Environ. Microbiol.*, 52, 134, 1986.

196. Kukkonen, M. *et al.*, Identification of two laminin-binding fimbriae, the type 1 fimbria of *Salmonella enterica* serovar typhimurium and the G fimbria of *Escherichia coli*, as plasminogen receptors, *Infect. Immun.*, 66, 4965, 1998.

197. Wang, L. *et al.*, Species-wide variation in the *Escherichia coli* flagellin (H-antigen) gene, *J. Bacteriol.*, 185, 2936, 2003.

198. Doughty, S. *et al.*, Identification of a novel fimbrial gene cluster related to long polar fimbriae in locus of enterocyte effacement-negative strains of enterohemorrhagic *Escherichia coli*, *Infect. Immun.*, 70, 6761, 2002.

199. Cleary, J. *et al.*, Enteropathogenic *Escherichia coli* (EPEC) adhesion to intestinal epithelial cells: role of bundle-forming pili (BFP), EspA filaments and intimin, *Microbiology*, 150, 527, 2004.

200. Giron, J.A., Ho, A.S., and Schoolnik, G.K., Characterization of fimbriae produced by enteropathogenic *Escherichia coli*, *J. Bacteriol.*, 175, 7391, 1993.

201. Mooi, F.R. and de Graaf, F.K., Molecular biology of fimbriae of enterotoxigenic *Escherichia coli*, *Curr. Top. Microbiol. Immunol.*, 118, 119, 1985.

202. Giron, J.A. *et al.*, Longus pilus of enterotoxigenic *Escherichia coli* and its relatedness to other type-4 pili: a mini review, *Gene*, 192, 39, 1997.

203. Mulvey, M.A., Adhesion and entry of uropathogenic *Escherichia coli*, *Cell Microbiol.*, 4, 257, 2002.

204. Dons, L., Rasmussen, O.F., and Olsen, J.E., Cloning and characterization of a gene encoding flagellin of *Listeria monocytogenes*, *Mol. Microbiol.*, 6, 2919, 1992.

205. Muller, K.H. *et al.*, Type 1 fimbriae of *Salmonella enteritidis*, *J. Bacteriol.*, 173, 4765, 1991.

206. Zhang, X.L. *et al.*, *Salmonella enterica* serovar typhi uses type IVB pili to enter human intestinal epithelial cells, *Infect. Immun.*, 68, 3067, 2000.

207. White, A.P. *et al.*, Extracellular polysaccharides associated with thin aggregative fimbriae of *Salmonella enterica* serovar enteritidis, *J. Bacteriol.*, 185, 5398, 2003.

208. Utsunomiya, A. *et al.*, Studies on novel pili from *Shigella flexneri*. I. Detection of pili and hemagglutination activity, *Microbiol. Immunol.*, 36, 803, 1992.

209. Snellings, N.J., Tall, B.D., and Venkatesan, M.M., Characterization of *Shigella* type 1 fimbriae: expression, FimA sequence, and phase variation, *Infect. Immun.*, 65, 2462, 1997.

210. Hall, R.H. *et al.*, Morphological studies on fimbriae expressed by *Vibrio cholerae* 01, *Microb. Pathog.*, 4, 257, 1988.

211. Warriner, K. *et al.*, Interaction of *Escherichia coli* with growing salad spinach plants, *J. Food Prot.*, 66, 1790, 2003.
212. Zhuang, R.Y., Beuchat, L.R., and Angulo, F.J., Fate of *Salmonella* montevideo on and in raw tomatoes as affected by temperature and treatment with chlorine, *Appl. Environ. Microbiol.*, 61, 2127, 1995.

3 Internalization and Infiltration

Jerry A. Bartz

CONTENTS

3.1 OVERVIEW OF INTERNALIZED MICROORGANISMS

Microorganisms embedded in plant tissues may be defined as "internalized," derived from "internal," meaning located inside the plant surface. Functionally, internalized microbes cannot be washed off the plant, they are protected from environmental stresses, and they cannot be inactivated by contact biocides or other surface disinfectants. Inside the plant, most microorganisms are located in spaces between cells called intercellular spaces, whereas plant viruses and certain other pathogens are inside host cells. Microbes in the intercellular spaces are bathed in nearly saturated relative humidity with a gas composition that enables aerobic metabolic activities [1]. The main threat to the survival of internalized microorganisms appears to be mechanisms that protect the plant against microbial attack [1]. As such, the microbe must either evade, counteract, or not induce its host's defenses. Plant pathogenic microorganisms, which by nature harm plant tissues, have developed ways to cope with host defense reactions. In contrast, nonplant pathogens usually do not harm living tissues and, as a consequence, appear unlikely to stimulate plant defenses. Moreover, the absence of tissue damage reduces the likelihood that nonpathogens will be exposed to preformed antimicrobial chemicals, which would be compartmentalized in the cytoplasm or specialized cells.

Microorganisms that are resistant to washing, surface disinfectants, or environmental stresses are not always internalized. Romantschuk et al. [2] noted that washing leaves, with or without sonication, does not remove all bacteria that live entirely on the plant surface, perhaps because portions of this population may embed in surface biofilms or other attached aggregates (see Chapter 2). Additionally, bacteria have been observed partially buried in surface waxes [3] and in cracks in the cuticle [4]. Microorganisms embedded in aggregates, biofilms, surface waxes or ruptures in the cuticle are somewhat protected against environmental stresses [5] and surface treatments. However, truly internalized organisms, which are located beneath layers of plant cells, would have much greater protection.

Proof that microorganisms exist inside healthy, unblemished fruits and vegetables was provided by Samish et al. [6]. Using special surface sterilization procedures, her group isolated Gram-negative, motile, and rod-shaped bacteria frequently from tomatoes, cucumbers, English peas, and green beans sampled from farm fields. Populations were found less frequently in melons and bananas, whereas successful isolations were infrequent in grapes, citrus fruits, olives, and peaches. Internal populations of microorganisms would likely be highest in root tissues [7] and lowest in the acidic environment within certain fruit tissues [6].

Internalized microorganisms are part of a complex microbial ecosystem associated with plants [8,9]. Epiphytic microorganisms survive and multiply on the plant surface, whereas endophytes colonize the interior of plants without causing noticeable damage [9]. Those that grow on or in plants and cause damage are plant pathogens [10]. Epiphytes, endophytes, and plant pathogens may be considered resident microorganisms because they compose the plant-associated microbial ecosystem. Individual species that fail to establish a presence in this ecosystem despite one or more introductions are called casual microorganisms [11]. Casual microbes are usually ill-suited to survive on the plant surface. Once inside the plant, however, casuals can survive for prolonged periods of time depending on their ability to adapt to an environment that is high in humidity but low in available nutrition. However, under certain conditions, internalized casuals multiply. For example, Dong et al. [12] observed endophytic growth of Escherichia coli and Salmonella enterica (strains of Cubana, Typhimurium, and Infantis serovars) in alfalfa and barrel medic seedlings grown in test tubes. Young [13] noted that water congestion of leaf tissues enabled a wide range of bacteria to multiply. King and Bolin [14] reported that severe tissue water congestion caused plant cell membranes to leak minerals and metabolites, which supported the growth of saprophytes. Furthermore, the development of large populations of microorganisms on fresh-cut vegetables could produce nonspecific spoilage, likely because plant defense mechanisms were compromised by anoxia.

3.2 LOCATION OF INTERNALIZED ORGANISMS IN PLANTS

Plants are covered by a protective layer made up of cutin polymers embedded in waxes [1,3,15–17]. This layer is relatively impervious to water penetration or loss, gas exchange, and penetration by particulates. Various structures in the plant surface enable the gas exchange required for vital metabolic and photosynthetic processes occurring in the underlying cells. Therefore, to internalize, a microorganism must either directly penetrate the surface layer or enter through a surface opening (aperture) or wound.

The surface coating of plants and structures beneath it may be categorized as either symplast or apoplast [16,17]. The symplast or living matter includes the cytoplasm of cells, whereas the apoplast includes the surface layer, cell walls, air spaces between cells and in the cell wall matrix, and the primary water-conducting tissues (xylem) [17]. Sieve tube elements, which are primarily devoted to movement of the products of photosynthesis and other cellular processes, accompany the xylem vessels [16]. However, the sieve tube elements are filled with a cytoplasm-like material. As such, whether they should be included in the apoplast is unclear. In certain parts of the plant, sieve tubes transport water, whereas in other parts, xylem vessels carry sugars, etc.

The xylem, composed of specialized vessels, tracheids, and associated parenchyma, connects the water-absorbing tissues in the root system with the rest of the plant [18]. Vessels and tracheids are filled with water containing dissolved minerals and occasionally organic solutes. The general structure of these water-conducting elements tends to exclude microorganisms such that only a few specialized types are able to enter and move through the system. Individual vessel cells connect through perforation plates that would appear to allow passage of suspended particulates such as bacteria [18]. However, Pao et al. [19] observed multiple, helical perforations in the walls and ends of the vessels in the stem scar of orange fruits that blocked movement of bacteria. Whether these were xylem vessels or tracheid cells, which do not possess perforation plates [18], is unclear. Both types of water-conducting cells attach to adjacent cells through pits in their secondary walls. The pits are paired with those in walls of an adjacent parenchyma or vessel cell. The base of each pit pair contains a membrane composed of the initial primary cell wall of the adjacent cells and the middle lamella. Pit membranes contain pores that are slightly larger than plasmodesmata. At a reported $0.3\,\mu m$ in diameter [1], such pores would not allow passage of bacteria. However, the pits are freely permeable to water and solutes. Microbes that can enzymatically digest the pits, such as the wilt pathogens, inhabit xylem vessels [1]. Moreover, microbes that are able to weaken pit membranes that interface with adjacent parenchyma could egress from the vessel as well as obtain nutrition from the parenchyma cells.

A large portion of the apoplast of most plants consists of interconnected intercellular air spaces, which are linked with openings in the plant surface [18]. Less than 1% of the volume of potato tubers is devoted to intercellular spaces, whereas up to 66% of certain leaves is air space [20].

3.3   STRUCTURES THAT ENABLE INTERNALIZATION

Naturally occurring surface apertures and wounds are keys to the internalization of microbes. Two apertures, stomata and lenticels, function in gas exchange, whereas hydathodes provide relief of excessive internal water pressures. Stomata occur in the epidermis of all above-ground parts of plants. Specialized stomata function as nectaries (secrete nectar) in certain types of flowers [1]. Stomata are apertures in the plant's epidermis that are created by two specialized cells called guard cells. The turgor of the guard cells changes with exposure to sunlight, darkness, or moisture stress [21]. The guard cells swell during daylight opening the pore, and shrink during darkness or with water stress, closing the pore. Epidermal cells adjacent to the guard cells may grow under the stoma forming a substomatal chamber [18]. Schönherr and Bukovac [22] suggested that stomata be viewed as narrow capillaries having inclined walls.

Lenticels are specialized portions of a periderm, which is an impervious secondary surface layer that replaces the epidermis or forms on the surfaces of wounds [18]. The periderm is composed of a phellogen (cambium), phellum (corky cells), and phelloderm (resembles parenchyma cells formed inside the phellogen). A lenticel is similar in organization to the surrounding periderm, except that the lenticel phellogen is more active and contains intercellular spaces [1,18]. It produces a phellum that is loosely organized with many intercellular spaces. Thus, gases readily diffuse through lenticels into the underlying tissues of the plant organ. Phellum cells in lenticels may or may not be suberized (cell walls infiltrated with and coated by a polymeric organic chemical complex that is a barrier to moisture diffusion) [15], whereas the phellum of the regular periderm is nearly always suberized.

Lenticel-like structures may form on certain types of fruit [18]. In certain types of apple fruit, a periderm-like structure forms under stomata but a phellogen is usually absent. Certain types of melons crack as they approach maturity. Living cells beneath the crack develop into a phellogen that produces the characteristic net common to cantaloupes and certain other fruit. The net resembles a lenticel in structure. Certain lenticels respond to changes in the environment around them. For example, cells in lenticels on potato tubers proliferate when the soil becomes moist [23,24]. These proliferated cells are thin-walled, surrounded by large intercellular spaces, and highly susceptible to microbial attack.

Hydathodes, apparently designed to release excessive water pressure in the plant, vary in complexity among different plant species but all provide a connection between the water-conducting elements and the external environment [18]. Certain ones resemble stomata except for not closing during darkness. Others are specialized for water release and may be better termed "water glands." Gas exchange could occur through hydathodes that are not water congested. Water congestion develops in above-ground tissues of plants when the roots absorb water more rapidly than above-ground parts lose it to evapotranspiration [25]. The excess water can pool under the epidermis causing edemas or, more often, water moves from the ends of the vascular strands through the leaf mesophyll and then into and out of hydathodes in a process called guttation [21]. Guttation droplets, which are derived from xylem sap, appear on the edges of leaves and are often confused with dew. However, guttation may occur at any time of the day, particularly if the soil is moist, plants are growing rapidly, and evapotranspiration is low [26]. Lawn grasses and corn have been observed to excrete water in bright sunlight. Guttation may be part of a natural detoxification method in certain plants, particularly when rainfall, fog, or dew cause the droplets to fall from the plant surface [17].

Fruit attachment structures on certain plants contain natural openings involved with gas exchange. Most of the gas exchange required by the internal cells of tomato fruit occurs through the stem scar [27]. If the stem scar is covered with wax, the carbon dioxide levels in the intercellular spaces increase two to four times above normal, evidence that the wax layer blocks equilibration of respiratory CO_2 with the external environment. If the rest of the fruit is waxed and the stem scar is not, the CO_2 level in the fruit remains similar to that in nonwaxed fruit. Air injected into a tomato fruit submerged in water bubbles from cracks in the edges of the stem scar [28]. Only rarely are any bubbles observed at the blossom end of the fruit. If the stem is still attached, the air bubbles from the area between the stem and fruit.

Wounds also connect a plant's intercellular air-space network with the surrounding environment. Wounds can arise from various biotic and abiotic factors including insects, storms, wind-blown particles, harvest crews, etc. Excessive water uptake or even normal growth may produce cracks in the surface of plant organs. Trichomes, defined as outgrowths of the epidermis [18], are easily damaged and are a frequent site for infection by plant pathogenic bacteria and growth of epiphytes [9]. Whether broken trichomes enable the internalization of resident or casual microbes is unclear. The porosity of wounds to gases and moisture often changes over time due to healing processes involving the formation of closing layers such as a periderm, or suberization and lignification of cell layers [15]. These changes usually quickly restore the wound to an imperviousness to water loss and penetration by particulate matter similar that of the intact surface layers [18].

3.4 PROCESS OF INTERNALIZATION

The internalization process whereby microorganisms enter the plant apoplast is either active or passive. During active internalization, microbes grow through the plant surface into intercellular spaces, which is consistent with the activity of various plant pathogens [10]. During active internalization, plant pathogens penetrate directly through the cuticle or indirectly through stomata, lenticels, hydathodes, or wounds. Passive internalization implies that microbes are carried into the apoplast due to contact with an object causing injury or by a penetration of apertures by water, aerosol, or particulate that contains microbes. Plant viruses may internalize in plant tissues that are being fed upon by insects [10]. Aerosols may enter open stomata during a mass flow of gases into leaves [29]. Aqueous suspensions of microorganisms may infiltrate surface apertures or wounds either spontaneously [22] or because of pressure differentials between the apoplast and the external environment [30–32]. Suspensions also may diffuse or be drawn into plants through water channels, which are a direct liquid connection between a plant's intercellular spaces and its exterior environment [25].

Most surface apertures of plants are large enough to allow passage of bacteria and smaller particulates, whereas fungal spores would likely be excluded. The stem scar of tomato fruit may allow the passage of spores of the sour rot fungus, *Geotrichum candidum* [33], although the evidence was not conclusive. Lesions of *Rhizopus stolonifer* and *Geotrichum candidum* developed around and beneath the stem scar of tomato fruit that had been previously treated to cause an internalization of the spores of these fungi [28]. Vigneault *et al.* [34] reported that tomatoes cooled with water containing spores of *R. stolonifer* usually decayed during subsequent storage. However, whether the spores in these examples internalized through the stem scar is unclear. In contrast, wounds involving tissues with large intercellular spaces appear likely to internalize fungal spores.

3.5 INTERNAL STRUCTURES OF THE PLANT
INVOLVED IN INTERNALIZATION

The morphology of the surface pores and interconnected intercellular spaces has a direct influence on how readily particulate matter moves through the plant surface as well as the size of particles admitted. Intercellular spaces are delimited by the walls of the surrounding cells. The spaces form when cells dissolve (lysigenous), tear (rhexigenous), or separate (schizogenous) [18]. Cell walls are primarily composed of cellulose existing as microfibers bound to hemicellulose, specialized structural proteins, and pectins [1,16–18]. The walls of adjacent cells are initially cemented together by pectic compounds that compose a middle lamella. As these cells mature, they assume a more rounded as compared with an initial square or rectangular shape. The rounding splits the middle lamella apart at cell-to-cell contact points leaving a pectic sheath on the exposed walls [1]. Micropores, sometimes called micro-capillaries, exist in the lattice of microfibers and associated carbohydrates. These pores may be partially filled with pectic compounds or other wall material. Additionally, the microcapillaries contain water in a pectin gel or as free water, such that the relative humidity in the intercellular spaces ranges from 98 to 100%. Sakurai [16] suggested that a plant's symplast is surrounded by a liquid medium.

The precise environment within apertures and intercellular spaces is unclear. Internalized, nonplant pathogenic microbes are not likely to be in direct contact with plant cell membranes due to the thickness and structure of the plant cell walls, which was referred to as a matrix by Sattelmacher *et al.* [17]. Apoplastic fluid containing an array of solutes exists in the wall matrix. However, the fluid's solute concentration and pH is not likely to equal those reported to make plant tissues a favorable nutritional environment for growth of bacteria [35]. The pH of the apoplastic fluid, which varies with the location in the plant, the nutrition of the plant, and even the time of day [17], would seldom be as low as that reported for macerated plant tissues, where cell vacuoles have been ruptured. Xylem sap generally has a pH between 5 and 7, whereas the average pH of all apoplastic fluid ranges from 4.5 to 7.0. In ripening fruit, the apoplast pH is reduced due to leakage of organic acids through the plasmalemma and exposure of carboxyl groups from the hydrolysis of pectin [16]. However, the contents of cell vacuoles normally have a much lower pH than does the xylem sap or apoplastic fluid [21]. The pH of vacuoles in lemon fruit was measured down to 2.4, whereas a pH of 0.9 was reported for fluid in the cell vacuoles of a species of begonia.

Certain reports conclude that the cell walls bounding intercellular spaces have a coating of water, whereas others have suggested the exposed wall is actually hydrophobic due to an incrustation of cutin [17]. The plant cuticle has been observed to cover the guard cells and pore of a stoma and to extend partially into the substomatal chamber [1,3,15]. Schönherr and Bukovac [22] noted that the chemical characteristics of the surfaces of the cuticle on the plant surface were similar to those within the substomatal chamber. The cuticular complex, however, contains both polar carbohydrates and relatively nonpolar cuticular components [15]. Cutin has been described as a polyester with polar properties and an affinity for water [1]. The thickness of cutin on the plant surface increases with light intensity and exposure to moisture stress, which seem related to a restriction in water loss [3]. The thickness of an internal cuticle in the succulent tissues of fruits and vegetables could be quite different from that in leaf tissues where water loss through transpiration can be a major stress on the plant. Thus, an incrustation of cutin might not make cell walls hydrophobic, particularly in fruits and vegetables.

A combination of waxes and epidermal hairs help keep stomata from being clogged with water as a result of dew formation or rainfall [2]. Such a plug of water might substantially impair gas exchange [20]. The waxes on the stomata surfaces repel water, whereas the stomatal pore contains a bubble of air [1]. Thus, during dew formation, water droplets would bead up over the surface waxes and air bubble associated with the aperture. Goodman *et al.* [1] suggest, however, that changes in temperature or leaf movement could create pressure differentials that would draw surface water into stomata. By contrast, a wind and rainstorm during daylight hours would produce substantial water soaking of leaves through open stomata [36].

3.6 TYPES OF INTERNALIZATION

3.6.1 AEROSOLS

A mass flow of air through open stomata on leaves [29] could internalize floating aerosol-sized particles including bacteria and viruses. Such aerosols can disperse long distances from sources. Fattal *et al.* [37] detected aerosolized enteric bacteria and viruses as far as 730 m downwind of wastewater sprinkler irrigated field plots (note that the authors did not attempt detection at greater than 730 m). Gottwald *et al.* [38] concluded that in the spread of citrus canker, the pathogenic bacteria could be dispersed as an aerosol, in leaf debris, or wind-driven rainfall more than 5 miles by a single severe rainstorm.

3.6.2 WATER CHANNELS AND WATER CONGESTION

Free water in surface apertures such as stomata constitutes a "water channel" that connects a plant's apoplast with its external environment. Microorganisms can internalize through water channels in various ways. Additionally, persistent congestion of the apoplast by water may restrict oxygen availability, which could compromise the resistance of the cells to microbial attack [1,13,14]. Burton [20] noted that cells in respiring plant tissues become anaerobic if water congestion blocked them from direct contact with air in intercellular spaces. Tissues in a potato tuber covered with a film of water and stored at 20°C become anaerobic within 2.5 hours [39]. Wet tubers are susceptible to bacterial soft rot [40]. The loss of natural resistance to the disease associated with tissue anaerobiosis occurs relatively quickly. Bartz and Kelman [41] reported that freshly harvested and then washed tubers developed soft rot during subsequent storage at 20°C if their surfaces remained wet for 20 hours, whereas if the tuber surfaces dried within 16 hours the disease did not develop.

Water channels in leaf tissues have been associated with a large-scale internalization of plant pathogenic bacteria. Massive wildfire and blackfire lesions developed on field-grown tobacco only if leaf tissues were water congested at the time the plants were exposed to inoculum [36,42]. In the absence of water congestion, lesions tended to be small and of little consequence. By contrast, in the absence of inoculum (disease absent from the field), water-congested leaves recovered from a water-soaked appearance without evidence of necrosis or other damage.

Experimentally, water-soaked areas on leaf surfaces were correlated with rapid internalization of bacteria [36,42,43–46]. Leaves of various plant species were water-soaked by applying water under pressure to the root system or cut surface of petioles [42,43] or as a water stream from a syringe or sprayer [36,45,46]. Bacteria misted or poured on such surfaces were rapidly internalized as were the carbon particles in India ink, solutions of water-soluble dye, and suspensions of plant viruses. In the absence of water congestion, a similar application of aqueous cell suspensions or India ink led to little or no evidence of internalization. Cocci of *Staphylococcus aureus* penetrated rapidly into water-congested leaf tissues providing clear evidence that internalized bacteria need not be motile [46] or from a plant-associated ecological niche. Johnson [43] concluded that bacterial suspensions were pulled into water-congested tissues by capillary forces, which is inconsistent with the concept that intercellular spaces are bounded with hydrophobic surfaces [17]. Even with established water channels, however, water does not totally flood intercellular spaces on submerged or partially submerged leaves. Partially flooded intercellular spaces should function like closed capillary tubes. Water would enter until pressure on trapped air balanced the capillary forces. Thus, aqueous suspensions or solutions could penetrate quickly through water channels but would move only a few cell layers due to a developing back pressure.

The guttation of water through hydathodes creates water channels where bacteria and similarly sized microbes can passively internalize in plant leaves [26]. Under normal conditions, guttation disappears when leaves begin to transpire. Curtis [26] concluded that most guttation droplets are sucked back into the leaf at this time. The drying of guttation moisture may concentrate solutes such that certain ones may damage the leaf surface. Mechanical movement of guttation moisture back into hydathodes could passively internalize bacteria and any other particulates that are small enough to pass through the pore.

Mild water congestion of leaf tissues, which would not be visible as water soaking also appears to enhance microbial internalisation [25]. For example, preinoculation incubation of plants under high humidity leads to more disease than postinoculation incubation [45,47]. Citrus leaves become infected by *Xanthomonas citri* only if the substomatal chambers are filled with water [48]. This level of water congestion would not be visible to the unaided eye. A bacterial disease of cucumber, angular leaf spot, progressed most rapidly when the soil was warm and contained high moisture despite daytime air temperatures that inhibited pathogen development [50]. It is precisely this type of environment that favors guttation.

3.6.3 Internalization in Wounds

Fresh wounds feature an immediate release of fluid from ruptured vacuoles and plasmalemma. This "cell sap" congests the intercellular spaces in and beneath the damaged cells creating instant fluid channels [25]. Within seconds of contact, particulate matter or aqueous suspensions may be transported up to 1 cm laterally from a puncture wound in a leaf [43]. This concept of rapid internalization in wounds is supported by tests on the disinfection of wounds on tomato fruit. Bartz *et al.* [50] observed that within 5 seconds of application of an aqueous cell suspension of *E. carotovora* subsp. carotovora to the flat surface of a fresh wound on a tomato fruit, a portion of that population could not be completely eliminated when the fruit was washed for 2 minutes in 100 ppm free chlorine at pH 7.0 in a scale model flume. Gently rubbing the submerged wound surface with a soft bristle brush or with a gloved finger did not improve disinfection efficacy. In contrast, 10 ppm free chlorine present over similar wounds on fruit in the same flume prevented inoculation by a similar suspension, whereas just 5 ppm prevented most wounds from becoming inoculated. A water-soluble dye could be completely rinsed from these wound surfaces if the fruit was rinsed under running tap water within 6 seconds of dye application. If the wash was delayed more than 6 seconds a portion of the dye could be observed embedded in intercellular spaces beneath the wound.

3.6.4 Infiltration of the Plant Surface by Aqueous Cell Suspensions

Water or aqueous cell suspensions of bacteria may infiltrate apertures as well as wounds on fresh fruits and vegetables during harvest and handling [30,34, 51–53]. This infiltration can directly internalize microbes and can be either spontaneous or pressure driven. Schönherr and Bukovac [22] observed spontaneous penetration of stomata on leaves by water that was amended with a surfactant. A biosurfactant produced by *Pseudomonas fluorescens* alters the wax crystals on the surface of broccoli florets and may aid in spontaneous penetration of that structure by plant pathogenic strains of this bacterium [3]. Fresh wounds also appear susceptible to spontaneous infiltration by surface moisture [43,50].

Pressure-driven infiltration of fruits and vegetables means that pressure on water covering plant surfaces forces water into surface apertures despite air bubbles and the waxy nature of the pore surfaces. The cooling of fruits and vegetables leads to a reduction of gas pressures in the apoplast [54], particularly if the surface apertures are clogged with liquid. This pressure differential would persist until internal temperatures and gas pressures equilibrate with the external environment. Tomatoes allowed to cool while submerged in water may increase in weight due to water uptake [30,51]. Hydrocooled tomatoes increased in weight as they cooled [34] as did hydrocooled strawberries [52]. If the water contained cells of *Erwinia carotovora* subsp. carotovora or spores of *Botrytis cinerea*, water uptake correlated with a rapid development of internal lesions when the tomatoes or strawberries, respectively, were subsequently stored. When submerged in an aqueous cell suspension of *E. coli* at 2°C, fruit of four different apple cultivars initially at 22°C internalized the bacterium in the outer core region of the fruit during a 20-minute exposure [53]. A water-soluble dye was observed to internalize in similar treatments. However, evidence for the penetration of the skin, likely through open lenticels, appeared to be limited to injuries to the surface. Kenney *et al.* [55] observed *E. coli* cells up to 24 µm deep in open lenticels on "Delicious" apple fruit that had been cooled in an ice bath. Bruising the surface increased the number of internalized bacteria, particularly with respect to those embedded in cracks in the surface waxes. However, washing the apples in distilled water prior to examination led to an apparent reduction in penetration to depths no greater than 6 µm. Cooling hot water-treated mango fruits (46°C) in water (22°C) for 10 minutes led to infiltration of the fruit by a dye solution or by a suspension of *Salmonella enterica* [56]. The dye and bacteria primarily entered through the stem scar.

Direct injury to plant tissues may be caused by an infiltration by water, likely because the congestive water is absorbed by the cells causing them to swell. Tomato fruits that absorbed water equal to 3% or more of their original weight developed visible cracks, usually near the shoulders [30]. Studer and Kader [51] reported that a high percentage of freshly harvested tomatoes submerged for 15 to 120 minutes in water of various temperatures developed splits (breaks in the surface), whereas those stored overnight before the water treatment did not. Warming the water to above the fruit temperatures reduced but did not prevent the splits.

Hydrostatic pressure also can force water into apertures on fruits and vegetables [32]. Fruit or vegetables at the bottom of containers of submerged products would be exposed to a hydrostatic pressure on product surfaces equal to the total depth of submersion. Hydrostatic pressures would be additive to pressure differentials associated with cooling but counteract those associated with warming. However, water depth pressures would be exerted more rapidly than those associated with temperature changes. Hydrostatic pressures not only directly force water into surface apertures but also tend to squeeze submerged products and may cause air to bubble out of openings. When the hydrostatic pressure is removed, the product is likely to expand to its original volume leading to an internal pressure differential that will draw water into the product.

An abrupt impact with water can cause microbial internalization and water channels in fruits and vegetables. Water impact forces occur when field containers of freshly harvested fruits or vegetables are emptied into water or when a pile of a product is dispersed into a packinghouse flume by a heavy stream of water. Pressure washing systems in packinghouses also are likely to produce water congestion in surface apertures.

3.6.5 EVENTS IN PLANT DEVELOPMENT

In the field, plant root systems and hydathodes appear most likely to internalize microorganisms. Plant roots are likely to internalize soil microbes because wounds form during root growth. The development of lateral roots in plants usually begins at the pericycle, which underlies the endodermis [18]. The endodermis is a tightly packed cell structure located several cell layers below the root surface. As the root tip forms and then emerges, it breaks through the endodermis and cortex creating an open wound, which is a frequent site for colonization by soilborne bacteria [1,12]. Even casual bacteria can grow in wounds created by the emergence of lateral roots. *Escherichia coli* O157:H7 internalized in lettuce apparently through the root system when the plants were fertilized with contaminated manure or irrigated with contaminated water [57]. In controlled studies with seedlings of several plant species grown in test tubes, strains of *Salmonella enterica* were able to colonize the lateral root emergence wounds and then colonize intercellular spaces in the interior of the root [12]. Certain bacterial types applied to these plantlets were observed in xylem vessels, whereas others were rarely found in such cells. All applied bacteria were observed in the cortex of the root. Populations of a known endophytic bacterium, *Klebsiella pneumoniae*, exceeding log 8.0 CFU/g fresh weight were found in the root system of seedling rice plants grown in test tubes [7]. Endophytic populations were successfully initiated by the inoculation of seedlings with as few as 1 CFU per seedling [12]. Whether populations multiplied on the rhizoplane prior to entering the plant or found sufficient nutrition to multiply totally inside plants could not be determined; however, endophytic populations were correlated with those on the rhizoplane.

The soilborne, bacterial wilt pathogen *Ralstonia solanacearum* (*Pseudomonas solanacearum*) was observed to penetrate tobacco roots through epidermal cells that were damaged by lateral root emergence [58]. After penetration, the bacteria moved intercellularly in the cortex. The entrance of bacteria into xylem vessels appeared to occur where the endodermis was not fully developed or as a consequence of hypertrophy of xylem parenchyma cells, which appeared to disrupt young xylem vessels.

Growth cracks in plant surfaces during maturation processes could, at least temporarily, provide microbe internalization sites. Wide temperature changes, rainfall, the planting of crack-susceptible cultivars, and fertilization programs featuring high nitrogen and low potash have been associated with growth cracks in tomatoes [59]. Growth cracks in tomato fruit surfaces are a frequent site of microbial attack and predispose the fruit to pre- and postharvest decay [60]. Such cracks could enable internalization of a wide range of microorganisms. Many other crops have cultivars designated as crack resistant. In any fresh fruit or vegetable, the development of cracks or punctures in surfaces leading up to harvest, at harvest, or after harvest could enable various microbes to internalize. While most plant organs with growth cracks are culled during the packing process, items with minor cracks or punctures could be shunted to fresh-cut processing and lead to a contaminated product. Alternatively, microbes could internalize in plant organs that naturally crack during development such as cantaloupes [18].

3.7 IMPLICATIONS AND CONTROL

Many of the examples of the internalization of microorganisms by fruits and vegetables cited above involve situations occurring during crop production or harvest that cannot be controlled. Many internalization hazards can be controlled. The results of internalization can range from poor shelf life due to decay to unwholesomeness due to contamination by hazardous microorganisms. The list of human pathogens that can be internalized by fruits and vegetables is extensive [4]. For crops intended for consumption as raw products, such contamination is, at present, irreversible. With the globalization of agriculture and the consumer demand for fresh crop items all year round [61], there are nearly endless opportunities for microorganisms originating in the fields and surface waters of underdeveloped countries to end up in salad or fresh fruit items served in homes and restaurants in developed countries.

The inability of even the strongest surface disinfectants to eradicate completely human pathogens from contaminated fresh fruits and vegetables and yet be compatible with a product appearance that meets marketing requirements is well documented [4]. The failure of chlorinated water treatments, even at concentrations exceeding 5000 ppm, to eradicate plant pathogens from inoculated wounds has been known since 1945 [62]. Much conjecture has been focused on the inability of chlorine, a strong oxidizer, to disinfect contaminated wounds. Often authors suggest that active chlorine reacts with wounded tissues such that pathogen structures are not exposed to a critical dose. However, with contaminated wounds on tomato fruit, increasing doses and mechanical scrubbing of the wound surface have not led to significant increases in efficacy [63]. With contaminated cantaloupes, however, Ukuku and Fett [64] observed an increase in efficacy of 200 ppm chlorine (pH 6.4) or 5% H_2O_2 if the fruits were rubbed during the 2-minute immersion treatment, although not all contamination was eradicated. Based on Johnson's [43] theories on capillary movement of suspensions into leaf tissues and observations on dye movement and suspensions of soft rot bacteria into wounds on tomato fruit, Bartz et al. [50] suggested that solutions of active chlorine applied to inoculated wound surfaces on tomatoes displaced the pathogen cells further into the underlying intercellular spaces.

Internalization risks can be minimized through use of HACCP-type (hazard analysis critical control point) analyses and practices in production systems [65]. The ultimate goals of such a program are to minimize water penetration of plant tissues, crop contact with hazardous microorganisms, open wounds on plant surfaces, and situations likely to cause fluid penetration of plant surface apertures. Particularly, crops intended for raw consumption should never be treated, irrigated, washed, or cooled with poor-quality water [66]. Improperly composted manures should never be used in fields intended for production of fresh fruits and vegetables. In a recent survey of fruit and vegetable producers in Minnesota, one grower spread untreated manure throughout the growing season and 90% of the fruit and vegetable samples from that farm were positive for *E. coli* [67]. Fields should be fenced to keep out domestic or wild animals and should be located at least 5 miles from the nearest feed lots or other concentrations of animals [65]. Field workers should not be allowed to work with or harvest a crop if they are ill or have recently been ill. Working with water-congested plants creates a special hazard and should be avoided. Cultivars selected for production should resist the development of growth cracks or other characteristics that enable penetration by microorganisms.

Certain handling steps after harvest can reduce the internalization hazard. For example, the porosity of tomato stem scars to water is greatest immediately after harvest and then decreases over time [32]. Leaving a stem attached until just before water treatment only slightly reduces this characteristic. Studer and Kader [51] observed that tomato fruits split readily (from water uptake) if they were submerged in water immediately after harvest but did not if stored overnight before treatment. Additionally, warm fruit is more likely to absorb water than cool fruit during exposure to hydrostatic pressure as well as during exposure to water cooler than the fruit [30,31,51]. Thus, allowing tomatoes to cool overnight before packing them should decrease the likelihood of water infiltration during the unloading and washing processes at packing-houses. Although this prepacking storage would allow pathogen growth on damaged fruit (which otherwise would have been culled), small wounds would begin to heal and the stem scar would dry, thereby reducing the number of water channels. Additionally, the loss of a small amount of water from each fruit should decrease the likelihood for handling injuries to the tomato surface. With citrus, Eckert [62] noted that a standard practice in California was to "wilt" the fruit before washing and packing to reduce susceptibility to surface injuries.

The water used to handle or wash fruits and vegetables must be continually sanitized during the workday, particularly if the water is recycled. Moreover, the sanitizer must be present where the unwashed product enters the water system to minimize the chances for an internalization of hazardous microorganisms at the initial contact point. Highly reactive chemicals such as ozone [68] may be too unstable for maintenance of adequate residuals. Currently, hypochlorous acid from solutions of sodium hypochlorite, liquefied elemental chlorine, or solid powder or pellets of calcium hypochlorite best combines efficacy, speed of action, and stability for minimizing internalization hazards at packinghouses. Moreover, residues from the chlorinated water treatment either quickly dissipate from treated products or are harmless salts. Unfortunately, water chlorination cannot make badly contaminated surface waters safe to use for handling and washing produce as it is not effective against the resting stages (cysts, oocysts) of certain human parasites [68]. Additionally, where high chlorine demand exists, such as with shredded vegetables or with certain root crops, maintenance of adequate residuals is difficult.

Whether water chlorination eliminates the need to suppress completely water infiltration during postharvest handling is unclear. The infiltration of tomatoes with chlorinated water failed to prevent the development of postharvest decay when submerged fruits were treated with hydrostatic pressure at room temperature [69], but did prevent such decays when fruits were hydrocooled [34]. The presence of chlorine in the water appeared to increase the porosity of tomato stem scars [69]. The chlorination of the water used to hydrocool strawberries led to a significant reduction in botrytis fruit rot [52]. As noted above, however, chlorinated water treatments have consistently failed to eradicate completely microorganisms from fruit or vegetables likely to have internalized a portion of the contamination. For example, the washing of contaminated wounds on tomato fruit with over 500 ppm free chlorine at pH 7.0 reduced the subsequent development of soft rot by 50% in one test and had no effect in two [70]. In two separate reports on tomatoes that had been contaminated in the laboratory, washing wounds or stem scars with 100 ppm or more of free chlorine failed to eliminate *Salmonella* Montevideo [71,72].

When fruits or vegetables are unloaded into or washed by water, infiltration of natural apertures due to a temperature related pressure differential may be controlled by maintaining water temperatures above those of the incoming fruits and vegetables [30]. Current recommendations for water handling steps with tomato fruit are to keep water temperatures about $5°C$ ($10°F$) above those of incoming fruit and to limit fruit contact with water to 2 minutes [73]. This handling recommendation also includes provision for maintaining 100 to 150 ppm free chlorine in the water. The pH of chlorinated water should be in the range of pH 6.5 to 7.5 to ensure ample concentrations of the killing agent, HOCl, and minimal corrosion [65]. Warming the water increases chlorine's efficacy and decreases its stability [68]. In cooler weather, use of warm water to handle tomatoes has been associated with a reduction of surface injuries [74].

Selection of crop cultivars may also help reduce internalization hazards. The relative tendency of a tomato stem scar to absorb water appears to be a varietal characteristic [75]. Certain varieties consistently absorbed more water than others over different harvests of the same crop or the same cultivars in different fields and seasons. Heggestad [76] reported that leaves of certain tobacco cultivars were less likely to develop water congestion than others. In naturally occurring outbreaks of wildfire disease, lines that were less prone to water congestion had less disease. McLean and Lee [48] noted that structural differences in stomata were responsible for the resistance of mandarin orange to a bacterial disease, citrus canker. Therefore, the tendency of plant tissues to resist water intrusion and microbial internalization might be enhanced by selection and breeding.

Encouraging tissue respiration has been suggested as an internalization reduction treatment during preparation of fresh-cut lettuce. Takeuchi and Frank [77] reported that a high respiration rate in minimally processed lettuce produced a "counterforce" that reduced the internalization of cells of

E. coli. Thus, warming lettuce to encourage respiration during sensitive stages of fresh-cut lettuce preparation might decrease the potential for internalization of bacteria from wash water. Subsequently, a group of authors noted that reducing the O_2 over the lettuce to 2.7% reduced the internalization associated with low-temperature incubation [78]. Ostensibly, the counterforce was CO_2 released from respiration. The methodology used in these reports, however, raised questions about the validity of the authors' conclusions [79]. The lettuce was purchased from local stores and stored at 4°C. Tissue sections were prepared and submerged in water or an aqueous cell suspension of *E. coli* for 24 hours at 4, 10, 22, or 37°C [77,78]. The authors did not indicate if the tissues and fluids were equilibrated to these temperatures prior to the incubation. In fact, Takeuchi *et al.* [78] commented in their discussion that "... subsequent infiltration of the bacteria into the lettuce as it cooled during the inoculation period." If the lettuce tissues cooled during the incubation, then a pressure differential would occur in the intercellular spaces, as discussed in Section 3.6.4. This would lead to an infiltration of the bacterial suspension into the cut edges. Conversely, if the tissue sections warmed, internal gases would expand and tend to prevent infiltration. Without knowledge of tissue and fluid temperatures at the beginning of the test, it is impossible to interpret the results. Additionally, the relatively high cold-water solubility of CO_2 as compared with O_2 may be involved. Due to respiration, O_2 in intercellular spaces would be absorbed by the lettuce cells. If CO_2 production matches O_2 absorption, gas pressures in the intercellular spaces should not change. However, at low temperatures, a significant portion of the CO_2 released by mitochondria is likely to remain dissolved in cell sap. As such, the uptake of O_2 could contribute to a reduction in internal gas pressure. (Note that this would be much like the standard laboratory exercise on measuring plant tissue respiration with a Warburg apparatus where the CO_2 produced is scrubbed by an alkali solution and oxygen uptake is measured with a manometer.) In the absence of significant respiration, changes in the partial pressures of O_2 and CO_2 should not be factors in pressure differentials developing within the tissues.

REFERENCES

1. Goodman, R.N., Király, Z., and Zaitlin, M, *The Biochemistry and Physiology of Infectious Plant Disease*, Van Nostrand, Princeton, NJ, 1967.
2. Romantschuk, M. *et al.*, The role of pili and flagella in leaf colonization by Pseudomonas syringae, in *Phyllosphere Microbiology*, Lindow, S.E., Hecht-Poinar, E.I., and Elliot, V.R., Eds., APS Press, St Paul, MN, 2002, chap. 7.
3. Beattie, G.A., Leaf surface waxes and the process of leaf colonization by microorganisms, in *Phyllosphere Microbiology*, Lindow, S.E., Hecht-Poinar, E.I., and Elliot, V.R., Eds., APS Press, St Paul, MN, 2002, chap. 1.
4. Burnett, S.L. and Beuchat, L.R., Human pathogens associated with raw produce and unpasteurized juices, and difficulties in decontamination, *J. Ind. Micro. Biotech.*, 25, 281, 2000.

5. Morris, C.E., Barnes, M.B., and McLean, R.J.C., Biofilms on leaf surfaces: implications for the biology, ecology and management of populations of epiphytic bacteria, in *Phyllosphere Microbiology*, Lindow, S.E., Hecht-Poinar, E.I., and Elliot, V.R., Eds., APS Press, St Paul, MN, 2002, chap. 10.

6. Samish, Z., Etinger-Tulczynska, R., and Bick, M., The microflora within the tissue of fruits and vegetables, *J. Food Sci.*, 28, 259, 1963.

7. Dong, Y., Iniguez, A.L., and Triplett, E.W. Quantitative assessments of the host range and strain specificity of endophytic colonization by *Klebsiella pneumoniae* 342, *Plant and Soil*, 257, 49, 2003.

8. Upper, C.D. *et al.*, *The Ecology of Plant-Associated Microorganisms*, National Academy Press, Washington D.C., 1989.

9. Manceau, C.R. and Kasempour, M.N., Endophytic versus epiphytic colonization of plants: what comes first?, in *Phyllosphere Microbiology*, Lindow, S.E., Hecht-Poinar, E.I., and Elliot, V.R., Eds., APS Press, St Paul, MN, 2002, chap. 8.

10. Agrios, G.A., *Plant Pathology*, 4rth ed., Academic Press, San Diego, CA, 1997.

11. Leben, C., Microorganisms on cucumber seedlings, *Phytopathology*, 51, 553, 1961.

12. Dong, Y. *et al.*, Kinetics and strain specificity of rhizosphere and endophytic colonization by enteric bacteria on seedlings of *Medicago sativa* and *Medicago truncatula*, *Appl. Environ. Microbiol.*, 69, 1783, 2003.

13. Young, J.M., Effect of water on bacterial multiplication in plant tissue, *NZ J. Agric. Res.*, 17, 115, 1974.

14. King, A.D. and Bolin, H.R., Physiological and microbiological storage stability of minimally processed fruits and vegetables, *Food Technol.*, 43, 132, 1989.

15. Kolattukudy, P.E., Biochemistry and function of cutin and suberin, *Can. J. Bot.*, 62, 2918, 1984.

16. Sakurai, N., Dynamic function and regulation of apoplast in the plant body, *J. Plant Res.*, 111, 133, 1998.

17. Sattelmacher, B., Mühling, K.-H., and Pennewiß, K., The apoplast: its significance for the nutrition of higher plants, *Z. Pflanzenernähr. Bodenk.*, 161, 485, 1998.

18. Esau, K., *Anatomy of Seed Plants*, John Wiley, New York, 1960.

19. Pao, S., Davis, C.L., and Parish, M.E., Microscopic observation and processing validation of fruit sanitizing treatments for the enhanced microbiological safety of fresh orange juice, *J. Food Prot.*, 64, 310, 2001.

20. Burton, W.G., Some biophysical principles underlying the controlled atmosphere storage of plant material, *Ann. Appl. Biol.*, 78, 149, 1974.

21. Meyer, B.S., Anderson, D.B., and Böhning, R.H., *Introduction to Plant Physiology*, Van Nostrand, Princeton, NJ, 1960.

22. Schönherr, J. and Bukovac, M.J., Penetration of stomata by liquids, *Plant Physiol.*, 49, 813, 1972.

23. Adams, M.J., Potato tuber lenticels: susceptibility to infection by *Erwinia carotovora* var. *atroseptica* and *Phytophthora infestans*, *Ann Appl. Biol.*, 79, 275, 1975.

24. Smith, E.F., Bacteria in Relation to Plant Diseases: History, General Considerations, Vascular Diseases, Carnegie Institution of Washington, Publication No. 27, Vol. 2, 174, 1911.

25. Johnson, J., Water-congestion in plants in relation to disease, *Univ. Wis. Res. Bull.*, 160, 1947.

26. Curtis, L.C., Deleterious effects of guttated fluids on foliage, *Am. J. Bot.*, 30, 778, 1943.

27. Brooks, C., Some effects of waxing tomatoes, *Proc. Am. Soc. Hort. Sci.*, 35, 720, 1937.

28. Bartz, J.A., unpublished, 1998.

29. Shive, J.B., Jr., Leaf gas exchange: does bulk flow occur?, *What's New in Plant Phys.*, 11, 1, 1980.

30. Bartz, J.A. and Showalter, R.K., Infiltration of tomatoes by aqueous bacterial suspensions, *Phytopathology*, 71, 515, 1981.

31. Bartz, J.A., Ingress of suspensions of *Erwinia carotovora* subsp. *carotovora* into tomato fruit, Proc. 5th Int. Conf. Plant Path. Bact., Cali, Colombia, 1981, p. 452.

32. Bartz, J.A., Infiltration of tomatoes immersed at different temperatures to different depths in suspensions of *Erwinia carotovora* susbp. *carotovora*, *Plant Dis.*, 66, 302, 1982.

33. Pritchard, F.J. and Porte, W.S., Watery-rot of tomato fruits, *J. Agric. Res.*, 24, 895, 1923.

34. Vigneault, C., Bartz, J.A., and Sargent, S.A., Postharvest decay risk associated with hydrocooling tomatoes, *Plant Dis.*, 84, 1314, 2000.

35. Skovgaard, N., Vegetables as an ecological environment for microbes, in *Microbial Associations and Interactions: Proceedings of the 12th International IUMS-ICFMH Symposium*, Kiss, I., Deak, T., and Incze, K., Eds., D. Reidel, Budapest, 1984, p. 27.

36. Clayton, E.E., Water soaking of leaves in relation to development of the wildfire disease of tobacco, *J. Agric. Res.*, 52, 239, 1936.

37. Fattal, B. *et al.*, Prospective epidemiological study of health risks associated with wastewater utilization in agriculture, *Wat. Sci. Tech.*, 18, 199, 1986.

38. Gottwald, T.R., Graham, J.H., and Schubert, T.S., Citrus canker in urban Miami: an analysis of spread and prognosis for the future, *Citrus Industry*, Aug. 5, 1997.

39. Burton, W.G., and Wigginton, M.J., The effect of a film of water upon the oxygen status of a potato tuber, *Potato Res.*, 13, 180, 1970.

40. Lund, B.M. and Kelman, A., Determination of the potential for development of bacterial soft rot of potatoes, *Am. Potato J.*, 54, 211, 1977.

41. Bartz, J.A., and Kelman, A., Infiltration of lenticels of potato tubers by *Erwinia carotovora* pv. *carotovora* under hydrostatic pressure in relation to bacterial soft rot, *Plant Dis.*, 69, 69, 1985.

42. Johnson, J., Relation of water-soaked tissues to infection by *Bacterium angulatum* and *Bact. tabacum* and other organisms, *J. Agric. Res.*, 55, 599, 1937.

43. Johnson, J., Infection experiments with detached water-congested leaves, *Phytopathology*, 35, 1017, 1945.

44. Diachun, S., Relation of stomata to infection of tobacco leaves by *Bacterium tabacum, Phytopathology*, 30, 268, 1940.

45. Diachun, S., Valleau, W.D., and Johnson, E.M., Relation of moisture to invasion of tobacco leaves by *Bacterium tabacum* and *Bacterium angulatum*, *Phytopathology*, 32, 379, 1942.

46. Diachun, S., Valleau, W.D., and Johnson, E.M., Invasion of water-soaked tobacco leaves by bacteria, solutions, and tobacco-mosaic virus, *Phytopathology*, 34, 250, 1944.

47. Davis, D. and Halmos, S., The effect of air moisture on the predisposition of tomato to bacterial spot, *Plant Dis. Reptr.*, 42, 110, 1958.
48. McLean, F.T. and Lee, H.A., Pressures required to cause stomatal infections with the citrus-canker organism, *Philippine J. Sci.*, 20, 309, 1922.
49. Wiles, A.B. and Walker, J.C., Epidemiology and control of angular leaf spot of cucumber, *Phytopathology*, 42, 105, 1952.
50. Bartz, J.A. *et al.*, Internalization of microorganisms into tomato fruit through water congested tissues, Abstracts, 8th International Congress of Plant Pathology, 2003, p. 316.
51. Studer, H.E., and Kader, A.A., Handling tomatoes in water, *Annual Report 1976–77 California Fresh Market Tomato Research Program*, Fresh Market Tomato Advisory Board, Bakersfield, CA, 1977.
52. Ferriera, M.D. *et al.*, An assessment of the decay hazard associated with hydrocooling strawberries, *Plant Dis.*, 80, 1117, 1996.
53. Buchanan, R.L. *et al.*, Contamination of intact apples after immersion in an aqueous environment containing *Escherichia coli* O157:H7, *J. Food Prot.*, 62, 444, 1999.
54. Corey, K.A. and Tan, Z.-Y., Induction of changes in internal gas pressure of bulky plant organs by temperature gradients, *J. Am. Soc. Hort. Sci.*, 115, 308, 1990.
55. Kenney, S.J., Burnett, S.L., and Beuchat, L.R., Location of *Escherichia coli* O157:H7 on and in apples as affected by bruising, washing, and rubbing, *J. Food Prot.*, 64, 1328, 2001.
56. Penteado, A.L., Eblen, B.S., and Miller, A.J., Evidence of salmonella internalization into fresh mangos during simulated postharvest insect disinfection procedures, *J. Food Prot.*, 181, 2004.
57. Solomon, E.B., Yaron, S., and Matthews, K.R., Transmission of *Escherichia coli* O157:H7 from contaminated manure and irrigation water to lettuce plant tissue and its subsequent internalization, *Appl. Environ. Microbiol.*, 68, 397, 2002.
58. Quimo, A., Penetration of Tobacco Roots By, and Nature of Resistance To, *Pseudomonas solanacearum*, Ph.D. Thesis, North Carolina State University, Raleigh, University Microfilms, Ann Arbor, MI, 1971.
59. Scott, J.W., Growth cracks, in *Compendium of Tomato Diseases*, Jones, J.B., Jones, J.P., Stall, R.E., and Zitter, T.A. Eds., APS Press, St. Paul, MN, 1991, p. 56.
60. Bartz, J.A., Predisposition to postharvest diseases, in *Compendium of Tomato Diseases*, Jones, J.B., Jones, J.P., Stall, R.E., and Zitter, T.A. Eds., APS Press. St. Paul, MN, 1991, p. 47.
61. Bartz, J.A. and Brecht, J.K., Introduction, in *Postharvest Physiology and Pathology of Vegetables,* Bartz, J.A. and Brecht, J.K., Eds., Marcel Dekker, New York, 2003, chap. 1.
62. Eckert, J.W., Control of postharvest diseases, in *Antifungal Compounds,* Vol. 1, Siegel, M.R. and Sisler, H.D., Eds., Marcel Dekker, New York, 1977, chap. 9.
63. Bartz, J.A., Mahovic, M., and Concelmo, D., Rapid movement of inoculum into wounds on tomato fruit (abstr.), *Phytopathology*, 91, S6, 2001.
64. Ukuku, D.O. and Fett, W.F., Method of applying sanitizers and sample preparation affects recovery of native microflora and *Salmonella* on whole cantaloupe surfaces, *J. Food Prot.*, 67, 999, 2004.

65. Bartz, J.A., and Tamplin, M.L., Sales of vegetables for the fresh market: the requirement for hazard analysis and critical control points (HACCP) and sanitation, in *Postharvest Physiology and Pathology of Vegetables,* Bartz, J.A. and Brecht, J.K., Eds., Marcel Dekker, New York, 2003, chap. 23.

66. Suslow, T., Production practices affecting the potential for persistent contamination of plants by microbial foodborne pathogens, in *Phyllosphere Microbiology,* Lindow, S.E., Hecht-Poinar, E.I., and Elliot, V.R., Eds., APS Press, St Paul, MN, 2002, chap. 16.

67. Mukherjee, A. *et al.*, Preharvest evaluation of coliforms, *Escherichia coli, Salmonella,* and *Escherichia coli* O157:H7 in organic and conventional produce grown by Minnesota farmers, *J. Food Prot.*, 67, 894, 2004.

68. White, G.C., *Handbook of Chlorination and Alternative Disinfectants*, 4th ed., John Wiley, New York, 1998.

69. Bartz, J.A., Potential for postharvest disease in tomato fruit infiltrated with chlorinated water, *Plant Dis.*, 72, 9, 1988.

70. Bartz, J.A. *et al.*, Chlorine concentration and the inoculation of tomato fruit in packinghouse dump tanks, *Plant Dis.*, 85, 885, 2001.

71. Wei, C.I. *et al.*, Growth and survival of *Salmonella montevideo* on tomatoes and disinfection with chlorinated water, *J. Food Prot.*, 58, 829, 1995.

72. Zhuang, R.-Y., Beuchat, L.R., and Angulo, F.J., Fate of *Salmonella montevideo* on and in raw tomatoes as affected by temperature and treatment with chlorine, *Appl. Environ. Microbiol.*, 61, 2127, 1995.

73. Sherman, M. *et al.*, Tomato packinghouse dump tank sanitation, Vegetable Crops Fact Sheet, VC-31, Florida Coop. Ext. Serv., University of Florida Institute of Food and Agricultural Science, Gainesville, FL, 1981.

74. Kasmire, R.F., Hot water treatments for tomatoes, *Fruit Veg. Perishables Handling*, 29, 3, 1971.

75. Bartz, J.A., Relation between resistance of tomato fruit to infiltration by *Erwinia carotovora* subsp. *carotovora* and bacterial soft rot, *Plant Dis.*, 75, 152, 1991.

76. Heggestad, H.E., Varietal variation and inheritance studies on natural water-soaking in tobacco, *Phytopathology*, 35, 754, 1945.

77. Takeuchi, K. and Frank, J.F., Penetration of *Escherichia coli* O157:H7 into lettuce tissue as affected by inoculum size and temperature and the effect of chlorine treatment on cell viability, *J. Food Prot.*, 63, 434, 2000.

78. Takeuchi, K., Hassan, A.N., and Frank, J.F., Penetration of *Escherichia coli* O157:H7 into lettuce as influenced by modified atmosphere and temperature, *J. Food Prot.*, 64, 1820, 2001.

79. Gorny, J.R., Letter to the editor: Penetration of Escherichia coli O157:H7 into lettuce as influenced by modified atmosphere and temperature, a comment on *J. Food Prot.*, 64(11):1820–1823 (2001), *J. Food Prot.*, 65, 739, 2002.

4 Microbial Stress Adaptation and Safety of Produce

Luis A. Rodriguez-Romo and Ahmed E. Yousef

CONTENTS

4.1 MICROBIAL STRESS ADAPTATION PHENOMENON

4.1.1 STRESS

Microbial stress can be defined as any deleterious physical, chemical, or biological factor that induces modifications in the physiology of microorganisms (i.e., changes in the genome or proteome) that adversely affect microbial growth or survival [1–3]. The application of this broad definition of stress in food processing implies that many preservation treatments (e.g., heat, cold, and acid) are considered stresses, and, as a result, these may

95

significantly influence the behavior of foodborne pathogenic and spoilage microorganisms. Depending on the severity, stresses affect a microbial population in a number of ways. Exposing microorganisms to a sublethal stress (simply, this will be referred to as "stress") affects their metabolic activities unfavorably, leads to cell injury, and consequently retards or temporarily arrests their growth. When a microorganism is exposed to a severe adverse condition (i.e., lethal stress), this causes irreversible cell damage and, consequently, a decrease in population viability.

4.1.2 STRESS RESPONSE

The microbial cell has the means to sense stresses such as those leading to ribosomal disruption (e.g., heat stress) or modification in cell membrane fluidity (e.g., cold shock). Response to these stresses is presumed beneficial to the cell, but it occasionally has detrimental consequences. Protective responses require physiological adaptations to compensate for stress damage and permit the cell to continue its growth and ensure its survival. Similarly, the bacterial cell responds to stress induced by inherent physiological change. Entry of a cell population into the stationary phase, for example, triggers a general stress response, which results in microbial resistance to multiple stresses. Adaptive stress response involves the induction of a number of genetic and physiological mechanisms, as well as morphological events, which include: (1) synthesis of protective proteins that participate in damage repair, cell maintenance, or suppression of stress agents, (2) temporary increase in resistance to lethal factors, (3) transformation of cells to a latent state, e.g., spore formation or induction of viable-but-not-culturable state, (4) evasion of the host's defense mechanisms, and (5) adaptive mutations [1,2,4–7].

Environmental or physiological conditions may hinder a cell's ability to respond to stress. Chilled or metabolically exhausted cells, for example, may not respond to radiation stress. Similarly, when dormant bacterial spores are exposed to an injurious stress they are incapable of responding until conditions are favorable for germination and outgrowth. Lack of response to a stress may sensitize a microbial cell to subsequent stresses that are otherwise innocuous. Response to a stress also may exhaust a cell's ability to cope with subsequent stresses, causing a stress-sensitizing effect.

4.1.3 STRESS ADAPTATION AND THE GENERAL STRESS RESPONSE

Exposure of a microorganism to stress triggers a series of metabolic responses that may adapt the cell to subsequent lethal levels of the same type of stress or to multiple lethal stresses. The cell's adaptive response is generally referred to as stress adaptation. Stress, ensuing adaptive response, and the manifestation of this phenomenon in food preservation are collectively described as stress hardening. Food microbiota are regularly subjected to stress hardening. Therefore, the stress adaptation phenomenon is of paramount importance when evaluating the efficacy of intervention strategies to achieve food safety and to preserve food quality. Although stress adaptation is usually associated with the undesirable acquired resistance of foodborne pathogens to processing, this phenomenon also plays a key role in the survival of beneficial microorganisms used as probiotics or fermentation starters.

The microbial mechanisms to survive adverse environmental conditions can be divided into two classes, consisting of limited and multiple adaptive responses [8]. A limited or specific adaptive response results when microorganisms are exposed to a sublethal dose of a physical, chemical, or biological stress, which protects cells against subsequent lethal treatment with the same stress [8,9]. A multiple adaptive response, also known as cross protection, occurs when microbial cells adapt to an inherent physiological condition or to an environmental factor, which results in protection against subsequent lethal treatments, including stresses to which the microorganism had not been previously exposed [8,10–12]. This cross-protective response requires the induction of the general stress response, and it is triggered by stresses relevant to produce, both preharvest and postharvest, including cell starvation, exposure to high or low temperatures, high osmolarity, and low pH [12,13]. The activation of the general stress response is characterized by reduced growth rate or induced entry into stationary phase. The regulation of the general stress response has been well characterized in several microorganisms. This regulation is under the control of the alternative sigma factors, which bind to core RNA polymerase, mediating cellular responses through redirection of transcription initiation. Sigma S (σ^S, also known as RpoS) and σ^B regulate the general stress response in *Escherichia coli* and other Gram-negative bacteria, and in *Bacillus subtilis* and other Gram-positive bacteria, respectively [13,14].

4.1.4 REGULATION OF THE GENERAL STRESS RESPONSE

The general stress response is regulated by the *rpoS*, a gene that encodes the σ^S in *E. coli* and other bacteria such as *Shigella flexneri* and *Salmonella enterica* serovar Typhimurium [15–17]. Although the regulation of the general stress response has been studied in a variety of microorganisms, the regulation mechanisms covered in this section will refer to *E. coli*, an organism in which these mechanisms have been well characterized. During rapid growth, microbial cells, not exposed to any particular stress, have hardly detectable levels of σ^S. Exposure of these cells to stress (e.g., entry into stationary phase, high osmolarity, high or low temperature) results in rapid σ^S accumulation to high levels, and subsequent expression of more than 50 genes involved in stress adaptation [16]. The regulation of *rpoS*, which determines the cellular concentration of σ^S, occurs at multiple levels, including transcription, translation, and post-translational modifications (i.e., σ^S proteolysis), with the level of control being dependent on the type of stress affecting the cells [15,16,18]. In general, sudden exposure of bacteria to lethal stresses, which requires a rapid response (i.e., a shocking stress), involves σ^S

proteolysis-mediated regulation, while gradual exposure to stress usually requires stimulation of $rpoS$ expression at the transcription or translation level [16,19].

Enhanced cellular accumulation of σ^S occurs during microbial growth in rich media, while cells are transitioning from late exponential phase to stationary phase [15,19]. At the transcriptional level, the two-component system, cAMP and its receptor protein, the catabolite regulatory protein (CRP), act as negative regulators of $rpoS$. Conversely, small molecules such as guanosine-$3',5'$-bispyrophosphate (ppGpp), homoserine lactone, and polyphosphate may enhance $rpoS$ transcription [16,18]. Translational control involves a series of complex mechanisms in which stress conditions such as high osmolarity, low temperature, or entry into late exponential phase stimulate the translation of $rpoS$ mRNA [13]. It has been suggested that these stresses can play an important role in stabilizing the mRNA secondary structure, allowing its accessibility to ribosomes, and therefore enhancing its translation [16]. Activation of $rpoS$ mRNA translation requires the presence of Hfq, a small mRNA binding-protein that stabilizes the secondary structure of the polynucleotide. Translation of $rpoS$ can also be enhanced by the stabilization of the mRNA with a small RNA fragment (DsrA RNA) in cells stressed by temperature downshifts [13]. Control at the post-translational level involves regulation of the sigma-factor proteolysis rate. In cells growing exponentially, the levels of σ^S are very low because of its continuous proteolysis. Sudden stresses, including carbon starvation, shift to low pH, high temperature, and high osmolarity, prevent σ^S proteolysis and permit its accumulation in the cells to trigger the general stress response. Proteolysis of σ^S requires ClpXP protease, which is regulated by the RssB protein. The level of phosphorylation or dephosphorylation of RssB, influenced by the stresses already mentioned, determines its affinity for σ^S and the subsequent recognition of the σ^S–RssB complex by the ClpXP protease [13,16].

The activation of the general stress response, mediated by σ^S, results in the expression of stress-adaptive genes, including $bolA$ (involved in controlling cell morphology), cfa (involved in cyclopropane fatty acid synthesis), $uspB$ (involved in ethanol resistance), and $katE$ and $katG$ (encoding catalases), among many others [15,20]. Sensitivity of bacteria, defective in the $rpoS$ gene, to a series of stresses such as heat shock, oxidative environment, starvation, acid, ethanol, and ultraviolet radiation provides additional, and indisputable, evidence of the role of σ^S in the control of the general stress response [21,22].

4.2 PRODUCE MICROBIOTA AS INFLUENCED BY STRESS HISTORY

4.2.1 PREHARVEST STRESS

Fruits and vegetables can be contaminated with bacterial pathogens and spoilage microorganisms by contact with feces, soil, irrigation water, improperly composted manure, air-carried dust, wild and domestic animals, and human handling [23–28]. Survival and potential proliferation of contaminants on produce depends on the type of microorganism, the type and condition of produce, and the environment (e.g., temperature, humidity). Examples of environmental stresses that may affect microorganisms on fresh produce include nutrient restrictions, temperature and pH fluctuations, water availability limitations, exposure to ultraviolet radiation, presence of organic compounds (e.g., pesticides) and metal contaminants, inhibitory plant tissue reactions, and microbiota competition, among many others.

Many foodborne pathogens have enteric origin, which could limit their ability to survive in other environments, colonize plant tissues, and compete with plant-associated microorganisms [23]. However, it is known that *Salmonella* can survive, adapt, and proliferate in soil, a nonhost environment characterized by its thermal variability, high osmolarity, pH fluctuations, and variable nutrient availability [28]. Brandl and Mandrell [29] reported that *Salmonella* Thompson was able to colonize the surface of cilantro leaves and proliferate when plants were incubated at warm temperature (30°C). In addition, it was observed that the microorganism tolerated plant dry conditions (60% relative humidity) at least as well as usual bacterial plant colonizers (e.g., *Pantoea agglomerans* and *Pseudomonas chlororaphis*). There is evidence that secretions of plant seeds can induce microbial stress. Miché *et al.* [30] studied the response of *E. coli*, containing *luxCDABE* reporter genes, to germinating rice seed exudates, and reported that these secretions enhanced the expression of microbial genes involved in general stress, heat shock, and oxidative stress responses.

4.2.1.1 Temperature Fluctuation

Temperature variations at different stages of preharvest can affect the behavior of microbial populations present on produce. Besides influencing microbial growth, sudden fluctuations of temperature can cause heat or cold shock, and consequently may enhance the tolerance of foodborne pathogens to subsequent stresses. In this section, the effect of heat-induced stress will be discussed; cold stress will be addressed in another section of the chapter. Sublethal heat stress refers to the stress resulting from exposing a microbial population to temperatures higher than the maximum for growth and lower than that causing considerable cell death. Response to heat stress is most obvious when this stress causes minimal (less than one log) reduction in cell population.

Sublethal heat stress causes damage in the macromolecular structure of bacterial cells (e.g., protein denaturation), causing disruption of metabolic activities, which consequently affects microbial growth [31]. Microbial cells react against heat by inducing a universal protective response, generally known as the heat-shock response. This response involves the transient overexpression of heat-shock proteins that protect the cells against heat damage and other stresses. Heat-shock proteins include molecular chaperones (e.g., DnaK and GroEL) which repair cell injury by refolding the denatured proteins [2]. Other heat-shock proteins have protease activity (e.g., ClpP), dependant on ATP, and are involved in the degradation of heat-damaged proteins [32]. Additionally, microorganisms may adapt to mild heat by modifying the fluidity of their cell membranes; this is accomplished by increasing the length or level of saturation of the membrane's fatty acids [11].

The transcription of the majority of the heat-shock proteins in $E. coli$ is controlled by the alternative sigma factor, σ^{32} [33]. Additionally, σ^E is involved in the regulation of heat-induced genes of this bacterium [34,35]. Induction of the heat-shock response in $B. subtilis$ requires several regulatory groups including the HrcA-CIRCE system, which controls the major chaperone genes. The general stress response, controlled by σ^B, and the genes encoding Clp protease system are also involved in regulating the heat-shock response of this bacterium [3,33].

In addition to heat, several other stresses may trigger the synthesis of heat-shock proteins, and, as a result, induce multiple stress-protective responses. These stresses include changes in pH or osmolarity, ultraviolet irradiation, and the presence of substances such as ethanol, antibiotics, aromatic compounds, and heavy metals [36]. Synthesis of heat-shock proteins after exposure to other stresses may be attributed to the presence of a common stress sensing mechanism in the cells, which detects accumulated abnormal proteins in the cytoplasm [36–37].

There is substantial evidence confirming that exposure to sublethal heat increases the resistance of microorganisms to single or multiple lethal stresses. Seyer $et\ al.$ [38] observed that $E. coli$, heated at 55°C for 105 minutes and permitted to recover, had enhanced tolerance to subsequent lethal, thermal treatments (60°C for 50 minutes). In addition, the investigators reported that internal cell concentration of the DnaK chaperone played a key role in microbial recovery and stress tolerance. Lou and Yousef [39] indicated that stressing $Listeria\ monocytogenes$ with heat (45°C for 60 minutes) protected the cells to subsequent exposure to lethal concentrations of ethanol, hydrogen peroxide, and sodium chloride. Lin and Chou [40] observed a similar behavior in the same microorganism under comparable sublethal stress conditions. These researchers also indicated that thermal stress at a higher temperature and shorter time (48°C for 10 minutes) than those used by Lou and Yousef [39], protected $L. monocytogenes$ against sodium chloride but decreased the resistance of the pathogen to lethal concentrations of hydrogen peroxide.

4.2.1.2 Ultraviolet Radiation

Microorganisms are exposed to ultraviolet (UV) radiation from sunlight while present on the surface of fruits and vegetables. The main fraction of solar UV radiation that reaches the Earth's surface consists of long-wavelength UV (320–400 nm), which is usually designated as ultraviolet A (UVA). This type of radiation affects the microbial cell membrane and causes oxidation of unsaturated fatty acids. In addition, UVA participates in an oxygen-dependent reaction that involves the photosensitization of pigments, which results in the generation of reactive oxygen species (e.g., O_2^-) with antimicrobial activity [41]. In *E. coli*, UVA induces lethal and sublethal stress that may cause temporary growth inhibition, loss in phage sensitivity, and inhibition of tryptophanase induction [42,43]. There is evidence that when *E. coli* is treated with sublethal UVA radiation while in the stationary phase the microorganism recovers rapidly and acquires resistance to subsequent lethal irradiation treatments; however, this tolerance is not associated with the general stress response involving *rpoS* [41,42].

Short-wave UV radiation (200–280 nm), designated as ultraviolet C (UVC), causes damage to microbial DNA and RNA by inducing formation of pyrimidine–base dimers and DNA–protein crosslinks, which results in cell growth cessation, decreased viability, or cell death [44]. This microbicidal UVC radiation, particularly at 254 nm, has been implemented as a preservation treatment in a variety of foods; however, its application as an intervention strategy to be used alone is not recommended [11,45]. Sublethal doses of UVC may induce mutations and render cells tolerant to lethal irradiation and other stresses [46,47]. Hartke *et al.* [48] reported that irradiation of *Lactococcus lactis* with sublethal UVC (at 254 nm) induces the production of numerous protective proteins and enhances the tolerance of the microorganism to subsequent lethal treatments with heat, acid, and hydrogen peroxide. Bacterial cells can recognize damage caused to DNA as a consequence of UVC exposure and trigger a series of mechanisms to repair deleterious nucleic acid modifications. These processes require the participation of enzymes, which can be induced in the absence or presence of visible light, and these are named dark-repair and photoreactivation mechanisms, respectively [44]. Damage caused to microbial DNA can be counterbalanced by induction of the SOS response, which regulates the expression of genes involved in DNA repair [49].

4.2.1.3 Osmotic Stress

Although osmotic stress of microorganism on produce surfaces is not common at the preharvest stage, the following scenarios are likely to occur. Microorganisms may experience osmotic stress when they are exposed to saps released from bruises and wounds on produce surfaces. Dryness of microorganisms on produce surfaces also may result in osmotic stress. Exposure of surface microbiota to salts may occur with some commodities if brine flotation is used in conveying, sorting, or sizing operations. In order to survive osmotic stress, microorganisms must keep a balance between the water inside the cells and the concentration of solutes in the environment. Generally, bacterial cells use two protective mechanisms to survive hyperosmotic stress: (1) discharging the excess of solutes within the cells to the outside and (2) accumulating compatible solutes or osmolytes. In addition, microorganisms adapt to this stress by modifying their cell membranes, e.g., by increasing the ratio of *trans* to *cis* unsaturated fatty acids [2,4]. Microbial accumulation of compatible solutes is a mechanism that has been well characterized. Compatible solutes are small, polar, organic molecules that remain water soluble at relatively high concentrations without affecting intracellular structures or metabolic activities. These solutes include compounds such as carnitine, trehalose, glycerol, sucrose, proline, mannitol, glycine-betaine, and small peptides, among others [4,15]. The accumulation of compatible solutes, as a result of osmotic stress, requires the expression of proteins involved in the synthesis of the osmoprotectants or their transport systems [2,50]. The synthesis of several proteins that participate in osmolyte accumulation is under the control of the general stress response sigma factors σ^S and σ^B in *E. coli* and *B. subtilis*, respectively, which regulate the expression of chaperones and proteases [3,15]. Therefore, adaptation of microorganisms to osmotic stress may render them resistant to subsequent stresses of different types. Pretreatment of *B. cereus* with NaCl (1%) caused microbial resistance to subsequent lethal treatments with heat, ethanol, hydrogen peroxide, and acid. However, stress adaptation failed to protect the microorganism against a medium containing 12% NaCl [51,52]. Periago *et al.* [53] observed that pre-exposure of the same microorganism to osmotic stress, with 2.5% NaCl for 30 minutes, induced cell tolerance to lethal heating at 50°C. Osmotic stress of *L. monocytogenes*, by previous exposure to 3.5% NaCl for 2 hours, increased microbial tolerance to acid (pH 3.5), but the adaptation was strain-specific [54].

4.2.2 POSTHARVEST STRESS

Many sources of microbial contamination of produce at the postharvest stage have been identified. These include humans (i.e., workers and consumers), wild and domestic animals, insects, improperly sanitized harvesting equipment, transportation vehicles and containers, air-carried dust, wash, rinse, and cooling water, ice, processing and packaging equipment, and storage facilities, among many others [23,24,26,55–57]. Foodborne pathogens can survive on the intact outer surface of fresh fruits and vegetables, but they may not proliferate due to restriction of nutrients and water, or as a result of their inability to synthesize degradative enzymes against protective barriers covering produce. Survival and proliferation of pathogens on produce are enhanced by physical damage (e.g., punctures and bruises) of the protective epidermal barrier or the infection of the produce with pests and microorganisms [26]. Microbial stress adaptation may occur at various postharvest stages, and can involve transportation conditions, use of wash and rinse water at variable temperature, application of intervention strategies (e.g., use of sanitizers), pH fluctuations, and storage and packaging conditions.

4.2.2.1 Cold Stress

Microorganisms respond to cold stress by undergoing an adaptive response known as the cold-shock response. Adaptation to cold stress involves the expression and accumulation of cold-shock proteins, which could protect the cells to subsequent freezing or against other lethal stresses [53,58,59]. Broadbent and Lin [58] observed that cold shocking *L. lactis* at 10°C for 2 hours increased its resistance to freezing (−60°C for 24 hours) and lyophilization. Bollman *et al.* [60] reported that stress-adapted *E. coli* O157:H7, previously cold-shocked at 10°C for 1.5 hours, had enhanced survival in several foods including milk, whole egg, and sausage when compared to the nonadapted bacterium. A previous study indicated that cold shock of *B. cereus* (7°C for 2 hours) increased the survival of the microorganism to subsequent lethal thermal treatment [53]. In a different study, cold shocking *Clostridium perfringens* at 15°C for 30 minutes increased the thermotolerance of the bacterium at 55°C [59].

The cold-shock response involves a number of physiological adjustments, which include modifications in the cell membrane fluidity via increasing the unsaturation of membrane lipids or decreasing the chain length of its fatty acids, synthesis of protective proteins that bind to DNA and RNA, and importation of compatible solutes [4]. The cytoplasmic membrane, nucleic acids, and ribosomes participate in sensing temperature variations in microbial cells, and temperature downshifts induce the synthesis of up to 50 different cold protection-associated proteins [61,62]. Microbial response to cold stress involves the overexpression of two types of proteins, the cold-shock proteins (Csps) and the cold-acclimation proteins (Caps). A sudden drop in temperature induces the rapid, and transient, synthesis of Csps. Conversely, Caps are synthesized for extended time periods under continuous microbial growth at low temperatures; the expression of both protein types, however, can overlap during stress adaptation [63,64].

The cold-shock response has been well characterized in *E. coli*, and its Csps fall into two classes, I and II. Class I Csps are expressed at very low levels at 37°C, and are induced and overexpressed after a temperature downshift to 15°C. These class I proteins include the major cold-shock protein, CspA (a RNA- and DNA-binding chaperone), ribosomal binding factors (e.g., RbfA, CsdA), and the transcriptional termination and antitermination factors (e.g., NusA) [4,61,65]. Class II Csps are present in cells at 37°C, and are induced at moderate levels (< 10-fold) after the cold shock. Among the induced Csps, there are recombination factors (e.g., RecA), a subunit of DNA gyrase (GyrA), and energy-generating enzymes (e.g., dihydrolipoamide transferase and pyruvate dehydrogenase) [61,64,66].

In spite of the evidence of the protective effects of cold shock against multiple stresses, other researchers indicated that previous exposure to low temperatures sensitized *L. monocytogenes* [67,68] and *Vibrio parahaemolyticus* [69] to subsequent thermal treatments. As discussed earlier in this chapter, exposing microorganisms to a stress may lead to their adaptation or sensitization to more severe stresses. This variable behavior of pathogens in response to cold stress should be considered when treating produce that has been previously refrigerated to antimicrobial processes such as surface pasteurizing.

4.2.2.2 Acid Stress

Foodborne bacteria usually encounter drastic pH variations in the environment, and are exposed to acidic conditions while present in foods, during processing, and when they invade the gastrointestinal tract of animals and humans [70]. Acidification is a common food preservation method, in which organic acids (e.g., acetic, propionic, and lactic) are produced during fermentation or added as preservatives to foods. These weak acids, in their nondissociated form, are capable of diffusing into microbial cells; once inside the cytoplasm, they dissociate and decrease the intracellular pH, which results in disruption of metabolic activities. Acid stress of foodborne microorganisms results from the combination of the biological effect of low pH and the direct effect of weak acids [15].

Microorganisms have developed strategies to respond to acid stress by inducing a protective response known as the acid-tolerance response (ATR) [71]. Microbial cells develop an ATR when exposed to a moderately low pH (e.g., 4.5 to 5.5), and this results in the induction of proteins that protect the cells against lethal acid conditions (e.g., pH < 4). In addition, cells respond to acid environments by modifying their membrane composition, increasing proton efflux and amino acid catabolism, and by synthesizing enzymes involved in DNA repair [3,4].

In *Salmonella* Typhimurium, two different acid adaptation systems are recognized: these are the log-phase and the stationary-phase ATR [71]. Log-phase ATR is triggered when cells are grown under moderately acid conditions, and involves the synthesis of acid-shock proteins under the control of σ^S, the signaling protein, PhoP, and the iron regulator, Fur [15,70]. The stationary-phase ATR consists of σ^S-independent and σ^S-dependent mechanisms. The response independent of σ^S requires acid induction, and involves the participation of the response regulator, OmpR, to control the synthesis of acid-shock proteins. The induction of the ATR dependent on σ^S does not require previous exposure of the microorganism to acid, and it is triggered by entry of the cells into stationary phase [3,4,71]. Therefore, the latter ATR involves the induction of the general stress response, which is associated with multiple stress adaptation. Wong *et al.* [72] reported that *V. parahaemolyticus*, pretreated in acid medium (pH 5.0 to 5.8), showed increased resistance to treatments with low salinity and heat (45°C). In a different study, Rowe and Kirk [73] indicated that exposing pathogenic *E. coli* to acid shock (pH 4 for 1 hour) enhanced microbial tolerance against subsequent lethal treatments with osmotic stress (20% NaCl) or heat at 56°C.

Microorganisms grown under mild acid conditions are more resistant to lethal acid environments, as well as to other lethal stresses, than those grown at neutral pH [71]. Tosun and Gönül [74] indicated that *Salmonella* Typhimurium, grown at pH 5.8, developed tolerance to lethal doses of heat, salt, and organic acids, but not to cold shock. Ryu and Beuchat [75] observed that acid-adapted *E. coli* O157:H7, grown under gradual pH reduction in a medium containing 1% glucose, showed enhanced tolerance to thermal treatments (52°C) in apple cider and orange juice. In a different study, Bacon et $al.$ [76] reported that stress-adaptation of $Salmonella$ spp., grown under gradually increasing acid conditions (i.e., in a medium containing 1% glucose), caused cross-protection against lethal heat treatments. $Listeria$ $monocytogenes$, growing under similar gradually increasing acid conditions, or previously treated at pH 5.0 to 5.5 for 90 minutes, showed enhanced survival to a lethal acid medium (pH 3.5). Nonetheless, exposure to other stress conditions including high osmolarity, heat, and cold was unable to protect the microorganism against acid [77]. These results may have implications in the washing of fruit since many of the commonly used washing agents are acidic in nature. Application of these agents in a manner that sensitizes, rather than hardens, the pathogens to other stresses would improve the safety of produce.

4.2.2.3 Oxidative Stress

Foodborne microorganisms are exposed to oxidative stress, which may be induced endogenously as a result of microbial metabolism or exogenously due to treatments that increase the levels of reactive oxygen species, i.e., hydrogen peroxide (H_2O_2), superoxide anion (O_2^-), hydroxyl radical ($HO^\bullet$), and singlet oxygen ($^1O_2$). Similarly, microbial oxidative stress can be triggered by conditions that lead to depletion of protective antioxidant molecules or enzymes. Reactive oxygen species can be generated during processing as a result of radiation, presence of heavy metals, or treatments with oxidizing sanitizers. Reactive oxygen species are deleterious to microorganisms, and can cause extensive damage to their cellular components such as lipids, proteins, and nucleic acids; this negatively affects cell functionality and reduces its viability [78–80]. Microorganisms respond to oxidative stress by synthesizing (1) protective proteins (e.g., glutathione reductase, thioredoxin 2) and other organic molecules (e.g., methylerythrol, cyclopyrophosphate) with antioxidant capacity or (2) proteins that participate in repairing oxidative damage (e.g., exonuclease III and endonuclease IV), specifically repairing deleterious modifications affecting nucleic acids [2,81,82].

In $E.$ $coli$, response to oxidative stress caused by H_2O_2 and O_2^- is under the control of $oxyR$ and $soxRS$ regulons, respectively [79,81,83]. Genes controlled by $oxyR$ include those encoding the hydroperoxidase I (HPI), glutaredoxin, glutathione reductase, NADPH-dependent alkyl hydroperoxide reductase, and a protective DNA-binding protein (Dps) [83,84]. The regulon $soxRS$ controls the expression of genes encoding Mn-superoxide dismutase (Mn-SOD), endonuclease IV, glucose-6-phosphate dehydrogenase, fumarase, aconitase, and ferredoxin reductase, among others [80,83]. In unstressed cells, both proteins OxyR and SoxR are present in an inactive form. During oxidative damage, e.g., by exposure of cells to H_2O_2, OxyR senses the stress and is activated by the formation of intramolecular disulfide bonds [81,84]. There is evidence that the colanic acid polysaccharide produced by many strains of $E.$ $coli$ O157:H7 protects the microorganism against oxidative stress and other environmental conditions such as acid, heat, and osmotic stresses [85]. Van der Straaten *et al.* [86] reported that RamA, a protein synthesized in response to oxidative stress in *Salmonella* Typhimurium, could be involved in antibiotic resistance and virulence.

Produce microbiota are often exposed to oxidative stress. Sanitizers that may be used in washing produce (e.g., chlorine, chlorine dioxide, and ozone) undoubtedly lead to oxidative stress, which may trigger stress adaptation among these microorganisms. Metal ions in washing water and oxygen in package headspace are additional factors that may contribute to the oxidative stress adaptation of microorganism on produce.

4.2.2.4 Minimal Processing

Recently there has been an increase in consumer demand for high-quality and safe foods with fresh-like attributes. Minimally processed fruits and vegetables can be defined as products that are processed with methods (e.g., low-level irradiation and active packaging) that achieve food preservation and safety while causing minimal quality modifications or alteration of the fresh characteristics compared to produce treated by conventional food preservation treatments [87]. Applying minimal processing involves using preservation factors singly or in combination. Therefore, minimal processing may be considered an implementation of the "hurdle concept" which refers to the application of mild preservation factors (i.e., hurdles) in combinations, either in sequence or simultaneously, to enhance microbial inactivation by additive or synergistic effects [88,89].

Combination of sublethal stresses, although potentially acting synergistically to inactivate microorganisms in foods, could lead occasionally to stress adaptation and cross-protective responses [11,90]. Examples of cross protection were reported by Lou and Yousef [39] who observed that stressing *L. monocytogenes* with heat (45°C for 60 minutes) protected the cells to subsequent exposure to lethal concentrations of ethanol, hydrogen peroxide, and sodium chloride. During food processing, microorganisms are treated with sublethal stresses sequentially rather than simultaneously. Therefore, microbial exposure to a sublethal stress could harden the microorganisms and protect them against subsequent treatment factors or hurdles. Consequently, stress hardening could pose limitations to the possible benefits of the hurdle concept. However, careful application of minimal processing could alleviate the consequences of stress adaptation of microbiota in produce.

4.3 MICROBIAL STRESS ADAPTATION ON PRODUCE

Information in published literature regarding microbial stress adaptation on produce is scarce. However, greater processing resistance of natural microbial contaminants on produce surfaces compared with those inoculated onto these products may support the hypothesis that most produce microbiota are adapted to stresses encountered in the field and throughout the production chain. Readers are cautioned that apparent processing resistance of produce microbiota also could be attributed to their inaccessibility to treatments, or the inability of the analyst to recover these microorganisms using common sample preparation and processing techniques. Association of microbial contaminants with pores, stem scars, wounds, and other surface irregularities could protect microorganisms and make them appear resistant to processing.

Attachment of microorganisms to fruit and vegetable surfaces could initiate stress adaptive response against physical and chemical treatments. Gawande and Bhagwat [92] reported that *Salmonella* spp. attached to apple, tomato, or cucumber had enhanced surface contact-mediated acid tolerance and increased survival, by 4 to 5 log, to acid stress induced by exposure to sodium citrate (50 mM, pH 3) for 2 hours when compared to cell suspensions treated under the same conditions. When these investigators inoculated *Salmonella* Typhimurium on the surface of fresh-cut apples, and stored them at 4°C for 2 hours, the tolerance of the pathogen to acid stress increased. Han *et al.* [93] treated green pepper, contaminated on the surface with *E. coli* O157:H7, using chlorine dioxide gas (0.2 to 1.2 mg/l), and reported that attachment of the microorganism to injured pepper surfaces protected the cell against the gas when compared to cells attached to uninjured surfaces.

Francis and O'Beirne [94] inoculated acid-adapted *L. monocytogenes*, previously exposed to pH 5.5 for 60 minutes, on lettuce, swedes, dry coleslaw mix, and bean-sprouts, which were packed under modified atmosphere and subsequently stored at 8°C for 14 days. The researchers observed that the stress-adapted microorganism had enhanced survival compared to unstressed controls, under relatively high (25 to 30%) carbon dioxide atmospheres. Hsin-Yi and Chou [95] stressed *E. coli* O157:H7 in acidified medium (pH 5 for 4 hours) and inoculated the microorganism in acidic mango or asparagus juice with subsequent storage at 7°C. The investigators reported that acid adaptation and low temperature increased microbial survival in both fruit juices.

4.4 ASSESSING STRESS ADAPTATION AND ASSOCIATED RISKS

Safety of food may be achieved using treatment factors that halt the growth of pathogens. The optimization of these treatments requires an understanding of the limits between conditions that support growth and those in which growth is not possible, also known as the growth–no-growth interface [96]. The growth–no-growth interface can be defined as the boundary at which the microbial growth rate is zero and the lag phase is infinite [97]. The behavior of foodborne pathogens at the growth–no-growth interface has been assessed using models that take into account combinations of temperature, pH, a_w, and concentrations of chemical compounds [77,98–100]. These and other predictive and risk assessment models should consider the contribution of stress adaptation to the survivability and behavior of pathogens in food. The majority of models available to evaluate survival or inactivation of chemically stressed bacteria are based on primary models, which describe the fate of microbial populations as a function of time [96]. To develop reliable microbial inactivation models, researchers should consider the physiological state of the organism and the potential induction of stress-tolerance responses [97]. However, including stress adaptation in these models depends greatly on researchers' ability to monitor accurately and quantify the stress adaptation phenomenon experimentally.

Rapid and quantitative assessment of microbial adaptive response to a predefined stress remains a great challenge. Advances in this area would improve our understanding of how the microbial cell responds to multiple stresses, or its ability to exhibit multiple responses to a single stress, leading to cross protection. These techniques would also enable researchers to measure the response of microbial cells to a complex battery of stresses. Advances in genomic and proteomic research may bring the scientific community closer to this goal [20,33]. Although genome-wide microarray analysis enables researchers to identify genes expressed in response to stress, the technique does not distinguish between expressions leading to adaptation and those that are not directly related to this phenomenon. Fluorescence staining is a promising technique for rapidly assessing stress response. Instrumentation advances may enable researchers to monitor the effect of stress on membranes in real time with the use of fluorescent dyes. Reliable, quantitative measures of stress adaptation should facilitate the efforts to develop mathematical predictive models of stress-, adaptive-, and cross-protective responses. Change in these responses as a function of stress type and intensity, for example, would help predict the behavior of pathogens during food processing and storage, and their virulence in infected individuals.

4.5 SUMMARY

Adaptation of foodborne pathogens to environmental and processing stresses is a potential risk that may greatly compromise the safety of food. Only anecdotal evidence is available, supporting the notion that microbiota on produce owe their processing resistance to the stress adaptation phenomenon. Repeated demonstration of stress adaptation under laboratory conditions provides indirect proof that the phenomenon is of paramount importance to the safety of food. Modern and efficient techniques are needed to assess and monitor accurately the adaptive response in foodborne pathogens.

REFERENCES

1. Aertsen, A. and Michiels, C.W., Stress and how bacteria cope with death and survival, *Crit. Rev. Microbiol.*, 30, 263, 2004.

2. Vorob'eva, L.I., Stressors, stress reactions, and survival of bacteria: a review, *Appl. Biochem. Microbiol.*, 40, 261, 2004.

3. Yousef, A.E. and Courtney, P.D., Basics of stress adaptation and implications in new-generation foods, in *Microbial Stress Adaptation and Food Safety*, Yousef, A.E. and Juneja, V.K., Eds., CRC Press, Boca Raton, FL, 2003, chap. 1.

4. Beales, N., Adaptation of microorganisms to cold temperatures, weak acid-preservatives, low pH, and osmotic stress: a review, *Comp. Rev. Food Sci. Food Saf.*, 3, 1, 2004.

5. Lombardo, M-J., Aponyi, I., and Rosenberg, S., General stress response regulator RpoS in adaptive mutation and amplification in *Escherichia coli*, *Genetics*, 166, 669, 2004.

6. Sonenshein, A.L., Bacterial sporulation: a response to environmental signals, in *Bacterial Stress Responses*, Storz, G. and Hengge-Aronis, R., Eds., American Society for Microbiology, Washington D.C., 2000, p. 199.

7. Voyich, J.M. *et al.*, Genome-wide protective response used by group A *Streptococcus* to evade destruction by human polymorphonuclear leukocytes, *PNAS*, 100, 1996, 2003.

8. De Angelis, M, and Gobbetti, M., Environmental stress responses in *Lactobacillus*: a review, *Proteomics*, 4, 106, 2004.

9. Sanders, J.W., Venema, G., and Kok, J., Environmental stress responses in *Lactococcus lactis*, *FEMS Microbiol. Rev.*, 23, 483, 1999.

10. Hecker, M., Schumann, W., and Völker, U., Heat-shock and general stress response in *Bacillus subtilis*, *Mol. Microbiol.*, 19, 417, 1996.

11. Juneja, V.K. and Novak, J.S., Adaptation of foodborne pathogens to stress from exposure to physical intervention strategies, in *Microbial Stress Adaptation and Food Safety*, Yousef, A.E. and Juneja, V.K., Eds., CRC Press, Boca Raton, FL, 2003, chap. 2.

12. Pichereau, V., Hartke, A., and Auffray, Y., Starvation and osmotic stress induced multiresistances, influence of extracellular compounds, *Int. J. Food Microbiol.*, 55, 19, 2000.

13. Hengge-Aronis, R., Interplay of global regulators and cell physiology in the general stress response of *Escherichia coli*, *Curr. Opin. Microbiol.*, 2, 148, 1999.

14. Price, C.W., Protective function and regulation of the general stress response in *Bacillus subtilis* and related gram-positive bacteria, in *Bacterial Stress Responses*, Storz, G. and Hengge-Aronis, R., Eds., American Society for Microbiology, Washington D.C., 2000, p. 179.

15. Abee, T. and Wouters, J.A., Microbial stress response in minimal processing, *Int. J. Food Microbiol.*, 50, 65, 1999.

16. Hengge-Aronis, R., The general stress response in *Escherichia coli*, in *Bacterial Stress Responses*, Storz, G., and Hengge-Aronis, R., Eds., American Society for Microbiology, Washington D.C., 2000, p. 161.

17. Komitopoulou, E., Bainton, N.J., and Adams, M.R., Oxidation-reduction potential regulates RpoS levels in *Salmonella*, *J. Appl. Microbiol.*, 96, 271, 2004.

18. Venturi, V., Control of *rpoS* transcription in *Escherichia coli* and *Pseudomonas*: why so different?, *Mol. Microbiol.*, 49, 1, 2003.

19. Ihssen, J. and Egli, T., Specific growth rate and not cell density controls the general stress response in *Escherichia coli*, *Microbiology*, 150, 1637, 2004.

20. Schweder, T. and Hecker, M., Monitoring of stress responses, *Adv. Biochem. Eng. Biotechnol.*, 89, 47, 2004.

21. Farewell, A., Kvint, K., and Nyström, T., *uspB*, a new sigma S-regulated gene in *Escherichia coli* which is required for stationary phase resistance to ethanol, *J. Bacteriol.*, 180, 6140, 1998.

22. Hengge-Aronis, R., Regulation of gene expression during entry into stationary phase, in *Escherichia coli and Salmonella*, Neidhardt, F.C., Ed., American Society for Microbiology, Washington D.C., 1996, p. 1497.

23. Beuchat, L.R., Ecological factors influencing survival and growth of human pathogens on raw fruits and vegetables, *Micr. Infect.*, 4, 413, 2002.

24. Beuchat, L.R. and Ryu, J.-H., Produce handling and processing practices, *Emerg. Infect. Dis.*, 3, 459, 1997.

25. Guan, T.Y. *et al.*, Fate of foodborne bacterial pathogens in pesticidal products, *J. Sci. Food Agric.*, 81, 503, 2001.

26. Harris, L.J. *et al.*, Outbreaks associated with fresh produce: incidence, growth, and survival of pathogens in fresh and fresh-cut produce, *Comp. Rev. Food Sci. Food Saf.*, 2, 78, 2003.

27. Islam, M. *et al.*, Fate of *Escherichia coli* O157:H7 in manure compost-amended soil and on carrots and onions grown in an environmentally controlled growth chamber, *J. Food Prot.*, 67, 574, 2004.

28. Winfield, M.D. and Groisman, E.A., Role of nonhost environments in the lifestyles of *Salmonella* and *Escherichia coli*, *Appl. Environ. Microbiol.*, 69, 3687, 2003.

29. Brandl, M.T. and Mandrell, R.E., Fitness of *Salmonella enterica* serovar Thompson in the cilantro phyllosphere, *Appl. Environ. Microbiol.*, 68, 3614, 2002.

30. Miché, L. *et al.*, Rice seedling whole exudates and extracted alkylresorcinols induce stress-response in *Escherichia coli* biosensors, *Environ. Microbiol.*, 5, 403, 2003.

31. Russell, A.D., Lethal effects of heat on bacterial physiology and structure, *Sci. Prog.*, 86, 115, 2003.

32. Krüger, E.D. *et al.*, Clp-mediated proteolysis in gram-positive bacteria is autoregulated by the stability of a repressor, *EMBO J.*, 20, 852, 2001.

33. Rosen, R. and Ron, E.Z., Proteome analysis in the study of the bacterial heat-shock response, *Mass Spect. Rev.*, 21, 244, 2002.

34. Alba, B.M. and Gross, C.A., Regulation of the *Escherichia coli* σ^E-dependent envelope stress response, *Mol. Microbiol.*, 52, 613, 2004.

35. Raivio, T.L. and Silhavy, T.J., Sensing and responding to envelope stress, in *Bacterial Stress Responses*, Storz, G. and Hengge-Aronis, R., Eds., American Society for Microbiology, Washington D.C., 2000, p. 19.

36. Ramos, J.L. *et al.*, Responses of Gram-negative bacteria to certain environmental stressors, *Curr. Opin. Microbiol.*, 4, 166, 2001.

37. Wawrzynów, A.B. *et al.*, ATP hydrolysis is required for the DnaJ-dependent activation of DnaK chaperone for binding to both native and denatured protein substrates, *J. Biol. Chem.*, 270, 19307, 1995.

38. Seyer, K.M. *et al.*, *Escherichia coli* heat shock protein DnaK: production and consequences in terms of monitoring cooking, *Appl. Environ. Microbiol.*, 69, 3231, 2003.

39. Lou, Y. and Yousef, A.E., Adaptation of sublethal environmental stresses protects *Listeria monocytogenes* against lethal preservation factors, *Appl. Environ. Microbiol.*, 63, 1252, 1997.

40. Lin, Y. and Chou, C., Effect of heat shock on thermal tolerance and susceptibility of *Listeria monocytogenes* to other environmental stresses, *Food Microbiol.*, 21, 605, 2004.

41. Bintsis, T., Litopoulu-Tzanetaki, E., and Robinson, R.K., Existing and potential applications of ultraviolet light in the food industry: a critical review, *J. Sci. Food Agric.*, 80, 637, 2000.

42. Dantur, K.I. and Pizarro, R.A., Effect of growth phase on the *Escherichia coli* response to ultraviolet-A radiation: influence of conditioned media, hydrogen peroxide and acetate, *J. Photochem. Photobiol. B: Biol.*, 75, 33, 2004.

43. Favre, A. *et al.*, Mutagenesis and growth delay induced in *Escherichia coli* by near ultraviolet radiation, *Biochimie*, 67, 335 1985.

44. Blatchley, E.R., III and Peel, M.M., Disinfection by ultraviolet irradiation, in *Disinfection, Sterilization and Preservation*, Block, S.S., Ed., Lippincott Williams and Wilkins, Philadelphia, 2001, p. 823.

45. Rodriguez-Romo, L.A. and Yousef, A.E., Inactivation of *Salmonella enterica* serovar Enteritidis on shell eggs by ozone and ultraviolet radiation, *J. Food Prot.*, 68, 711, 2005.

46. Lado, B.H. and Yousef, A.E., Alternative food-preservation technologies: efficacy and mechanisms, *Micr. Infect.*, 4, 433, 2002.

47. Rowbury, R.J., UV radiation-induced enterobacterial responses, other processes that influence UV tolerance and likely environmental significance, *Sci. Prog.*, 86, 313, 2003.

48. Hartke, A. *et al.*, UV-inducible proteins and UV-induced cross-protection against acid, ethanol, H_2O_2 or heat treatments in *Lactococcus lactis* subsp. *lactis*, *Arch. Microbiol.*, 163, 329, 1995.

49. Walker, G.C., Smith, B.T., and Sutton, M.D., The SOS response to DNA damage, in *Bacterial Stress Responses*, Storz, G. and Hengge-Aronis, R., Eds., American Society for Microbiology, Washington D.C., 2000, p. 131.

50. Van de Guchte, M. *et al.*, Stress responses in lactic acid bacteria, *Antonie van Leeuwenhoek*, 82, 187, 2002.

51. Browne, N. and Dowds, B.C.A., Heat and salt stress in the food pathogen *Bacillus cereus*, *J. Appl. Microbiol.*, 91, 1085, 2001.

52. Browne, N. and Dowds, B.C.A., Acid stress in the food pathogen *Bacillus cereus*, *J. Appl. Microbiol.*, 92, 404, 2002.

53. Periago, P.M. *et al.*, Identification of proteins involved in the heat stress response of *Bacillus cereus* ATCC 14579, *Appl. Environ. Microbiol.*, 68, 3486, 2002.

54. Faleiro, M.L., Andrew, P.W., and Power, D., Stress response of *Listeria monocytogenes* isolated from cheese and other foods, *Int. J. Food Microbiol.*, 84, 207, 2003.

55. Janisiewicz, W.J. *et al.*, Fate of *Escherichia coli* O157:H7 on fresh-cut apple tissue and its potential for transmission by fruit flies, *Appl. Environ. Microbiol.*, 65, 1, 1999.

56. Michaels, B. *et al.*, Prevention of food worker transmission of foodborne pathogens: risk assessment and evaluation of effective hygiene intervention strategies, *Food Serv. Technol.*, 4, 31, 2004.

57. Zagory, D., Effects of post-processing handling and packaging on microbial populations, *Postharvest Bio. Technol.*, 15, 313, 1999.

58. Broadbent, J.R. and Lin, C., Effect of heat shock or cold shock treatment on the resistance of *Lactococcus lactis* to freezing and lyophilization, *Cryobiology*, 39, 88, 1999.

59. Garcia, S., Limon, J.C., and Heredia, N.L., Cross protection by heat and cold shock to lethal temperatures in *Clostridium perfringens, Braz. J. Microbiol.*, 32, 110, 2001.

60. Bollman, J., Ismond, A., and Blank, G., Survival of *Escherichia coli* O157:H7 in frozen foods: impact of the cold shock response, *Int. J. Food Microbiol.*, 64, 127, 2001.

61. Phadtare, S., Yamanaka, K., and Inouye, M., The cold shock response, in *Bacterial Stress Responses,* Storz, G. and Hengge-Aronis, R., Eds., American Society for Microbiology, Washington D.C., 2000, p. 33.

62. Russell, N.J., Bacterial membranes: the effects of chill storage and food processing. An overview, *Int. J. Food Microbiol.*, 79, 27, 2002.

63. Graumann, P. and Marahiel, M.A., Some like it cold: response of microorganisms to cold shock, *Arch. Microbiol.*, 166, 293, 1996.

64. Panoff, J.-M. *et al.*, Cold stress responses in mesophilic bacteria, *Cryobiology,* 36, 75, 1998.

65. Inouye, M. and Phadtare, S., Cold shock response and adaptation at near-freezing temperature in microorganisms, *Science's STKE* [serial online]. Available at: http://www.stke.org/cgl/content/full/sigtrans;2004/237/pe26.

66. Ermolenko, D.N. and Makhatadze, G.I., Bacterial cold-shock proteins, *Cell. Mol. Life Sci.*, 59, 1902, 2002.

67. Bayles, D.O. *et al.*, Cold shock and its effect on ribosomes and thermal tolerance in *Listeria monocytogenes, Appl. Environ. Microbiol.*, 66, 4351, 2000.

68. Miller, A.J., Bayles, D.O., and Eblen, B.S., Cold shock induction of thermal sensitivity in *Listeria monocytogenes, Appl. Environ. Microbiol.*, 66, 4345, 2000.

69. Lin, C., Yu, R.-C., and Chou, C.-C., Susceptibility of *Vibrio parahaemolyticus* to various environmental stresses after cold shock treatment, *Int. J. Food Microbiol.*, 92, 207, 2004.

70. Sharma, M., Taormina, P.J., and Beuchat, L.R., Habituation of foodborne pathogens exposed to extreme pH conditions: genetic basis and implications in foods and food processing environments, *Food Sci. Technol. Res.*, 9, 115, 2003.

71. Foster, J.W., Microbial responses to acid stress, in *Bacterial Stress Responses,* Storz, G. and Hengge-Aronis, R., Eds., American Society for Microbiology, Washington D.C., 2000, p. 99.

72. Wong, H.-C. *et al.*, Effect of mild acid treatment on the survival, enteropathogenicity, and protein production in *Vibrio parahaemolyticus, Infect. Immun.*, 66, 3066, 1998.

73. Rowe, M.T. and Kirk, R.B., Cross-protection phenomenon in *Escherichia coli* strains harbouring cytotoxic necrotizing factors and cytolethal distending toxins, *Lett. Appl. Microbiol.,* 32, 67, 2001.

74. Tosun, H. and Gönül, A., Acid adaptation protects *Salmonella typhimurium* from environmental stresses, *Turk. J. Biol.*, 27, 31, 2003.

75. Ryu, J.-H. and Beuchat, L.R., Influence of acid tolerance responses on survival, growth, and thermal cross-protection of *Escherichia coli* O157:H7 in acidified media and fruit juices, *Int. J. Food Microbiol.*, 45, 185, 1998.

76. Bacon, R.T. *et al.*, Thermal inactivation of susceptible and multiantimicrobial-resistant *Salmonella* strains grown in the absence or presence of glucose, *Appl. Environ. Microbiol.*, 69, 4123, 2003.

77. Koutsoumanis, K.P., Kendall, P.A., and Sofos, J.N., Effect of food processing-related stresses on acid tolerance of *Listeria monocytogenes*, *Appl. Environ. Microbiol.*, 69, 7514, 2003.

78. De Spiegeleer, P. *et al.*, Source of tryptone in growth medium affects oxidative stress resistance in *Escherichia coli*, *J. Appl. Microbiol.*, 97, 124, 2004.

79. Lu, C., Bentley, W.E. and Rao, G., Comparisons of oxidative stress response genes in aerobic *Escherichia coli* fermentations, *Biotech. Bioeng.*, 83, 864, 2003.

80. Storz, G. and Zheng, M., Oxidative stress, in *Bacterial Stress Responses*, Storz, G. and Hengge-Aronis, R., Eds., American Society for Microbiology, Washington D.C., 2000, p. 47.

81. Lushchack, V.I., Oxidative stress and mechanisms of protection against it in bacteria, *Biochemistry*, 66, 476, 2001.

82. Ritz, D. *et al.*, Thioredoxin 2 is involved in the oxidative stress response in *Escherichia coli*, *J. Biol. Chem.*, 275, 2505, 2000.

83. Cabiscol, E., Tamarit, J., and Ros, J., Oxidative stress in bacteria and protein damage by reactive oxygen species, *Int. Microbiol.*, 3, 3, 2000.

84. Tkachenko, A.G. and Nesterova, Yu.L., Polyamines as modulators of gene expression under oxidative stress in *Escherichia coli*, *Biochemistry*, 68, 850, 2003.

85. Chen, J., Lee, S.M., and Mao, Y., Protective effect of exopolysaccharide colanic acid of *Escherichia coli* O157:H7 to osmotic and oxidative stress, *Int. J. Food Microbiol.*, 93, 281, 2004.

86. Van der Straaten, T. *et al.*, *Salmonella enterica* serovar Typhimurium RamA, intracellular oxidative stress response, and bacterial virulence, *Infect. Immun.*, 72, 996, 2004.

87. Alzamora, S.M., López-Malo, A., and Tapia, M.S., Overview, in *Minimally Processed Fruits and Vegetables*, Alzamora, S.M., Tapia, M.S., and López-Malo, A., Eds., Aspen Publishers, Gaithersburg, MD, 2000, p. 1.

88. Leistner, L. and Gould, G., *Hurdle Technologies, Combination Treatments for Food Stability, Safety and Quality*, Kluwer Academic/Plenum, New York, 2002.

89. Scott, V.N., Interaction of factors to control microbial spoilage of refrigerated foods, *J. Food Prot.*, 52, 431, 1989.

90. Lou, Y. and Yousef, A.E., Resistance of *Listeria monocytogenes* to heat after adaptation to environmental stresses, *J. Food Prot.*, 59, 465, 1996.

91. Gawande, P.V. and Bhagwat, A.A., Protective effects of cold temperature and surface-contact on acid tolerance of *Salmonella* spp., *J. Appl. Microbiol.*, 93, 689, 2002.

92. Gawande, P.V. and Bhagwat, A.A., Inoculation onto solid surfaces protects *Salmonella* spp. during acid challenge: a model study using polyetherosulfone membranes, *Appl. Environ. Microbiol.*, 68, 86, 2002.

93. Han, Y. *et al.*, Inactivation of *Escherichia coli* O157:H7 on surface-uninjured and -injured green pepper (*Capsicum annuum* L.) by chlorine dioxide gas as demonstrated by confocal laser scanning microscopy, *Food Microbiol.*, 17, 643, 2000.

94. Francis, G.A. and O'Beirne, D., Effects of acid adaptation on the survival of *Listeria monocytogenes* on modified atmosphere packaged vegetables, *Int. J. Food Sci. Technol.*, 36, 477, 2000.

95. Hsin-Yi, C. and Chou, C.-C., Acid adaptation and temperature effect on the survival of *E. coli* O157:H7 in acidic fruit juice and lactic fermented milk product, *Int. J. Food Microbiol.*, 70, 189, 2001.

96. Devlieghere, F. *et al.*, Effect of chemicals on the microbial evolution in foods, *J. Food Prot.*, 67, 1977, 2004.
97. McMeekin, T.A. *et al.*, Predictive microbiology: towards the interface and beyond, *Int. J. Food Microbiol.*, 73, 395, 2002.
98. Lanciotti, R. *et al.*, Growth/no growth interfaces of *Bacillus cereus*, *Staphylococcus aureus* and *Salmonella enteritidis* in model systems based on water activity, pH, temperature and ethanol concentration, *Food Microbiol.*, 18, 659, 2001.
99. Presser, K.A., Ross, T., and Ratkowsky, D.A., Modeling the growth limits (growth/no growth interface) of *Escherichia coli* as a function of temperature, pH, lactic acid concentration, and water activity, *Appl. Environ. Microbiol.*, 64, 1773, 1998.
100. Stewart, C.M. *et al.*, *Staphylococcus aureus* growth boundaries: moving towards mechanistic predictive models based on solute-specific effects, *Appl. Environ. Microbiol.*, 68, 1864, 2002.

Section II

Microbial Spoilage of Fruits and Vegetables

5 Bacterial Soft Rot

Ching-Hsing Liao

CONTENTS

5.1 INTRODUCTION

Global production and international trade of fresh fruits and vegetables have increased very sharply during the past two decades, mainly because of consumers' awareness of the health benefit expected from this popular diet [1]. More than 300 fresh and fresh-cut produce items are available for sale at supermarkets throughout the U.S. [2], largely due to the advanced postharvest technologies, improved crop varieties, and efficient distribution systems. To meet the market demand, new strategies are required to increase the production of fresh produce in farms and to reduce the postharvest losses caused by biotic and abiotic factors. It has been estimated that between 10 and 30% of fresh fruits and vegetables produced in the U.S. are wasted, mainly due to three factors: mechanical injuries, physiological decays, and microbial spoilage [3].

Microbial spoilage accounts for a substantial proportion of postharvest losses of fresh produce, which can be caused by a wide variety of microorganisms including bacteria, fungi, or yeasts [4,5]. In general, the spoilage of acidic fruits such as apple, orange, and berries is caused by molds, lactic acid bacteria, or yeasts. The spoilage of fresh produce with neutral pH such as salad vegetables and edible roots or tubers is caused by bacteria capable of producing pectolytic enzymes required for degradation of plant cell walls. Bacterial spoilage of fresh produce is usually found in the form of soft rot, which is characterized by water-soaking and total disintegration of plant tissues [6,7]. As reported in the literature, bacterial soft rot has been identified as the leading cause of disorders in many types of produce, including potato [8], lettuce [9,10], bell pepper [11], cucumber [12], and tomato [13]. This disorder can cost the fresh produce industry and consumers hundreds of millions of dollars annually [3].

In addition to its economic impact, soft-rotted plant tissue may serve as a carrier or reservoir for foodborne human pathogens and pose a potential threat to the safe supply of fresh produce. Wells and Butterfield [14,15] reported that the rotted plant tissues were more likely to harbor salmonella than the apparently healthy counterparts. They found a 5- to 10-fold increase in the population of *Salmonella* Typhimurium in potato slices co-inoculated with soft-rot bacteria [14]. Therefore, an integrated approach to control the proliferation of both soft-rot bacteria and foodborne human pathogens on fresh produce is required.

In this chapter, the diversity of soft-rot bacteria associated with postharvest losses of horticultural commodities and the factors affecting their survival in nature are discussed. In addition, the enzyme and molecular genetic mechanism by which soft-rot bacteria (especially fluorescent pseudomonas) cause maceration of plant tissues is reviewed. Furthermore, the synergistic and antagonistic interactions between spoilage microorganisms and human pathogens on fresh produce are discussed. Farm practices that are useful for controlling the dissemination and proliferation of both soft-rot and human pathogens on fruit and vegetable crops are presented.

5.2 DIVERSITY OF SOFT-ROT BACTERIA

Soft rot of fresh produce can be caused by diverse groups of bacteria including erwinia, pseudomonas, xanthomonas, clostridium, bacillus, and cytophaga [4].

The characteristics of these bacteria and their association with the spoilage of fresh produce under different conditions have been briefly described [7]. As strict anaerobes, *Clostridium* spp. cause soft rot of potatoes under oxygen-depleted conditions, especially when a more aggressive plant pathogen such as erwinia is present [16,17]. Pectolytic *Clostridium* spp. also play a role in the spoilage of fresh-cut produce that is packaged using an impermeable film [4]. Pectolytic bacillus including *Bacillus polymyxa* and *B. subtilis* have also been shown to be associated with soft rot in a wide variety of crops including potatoes, tomatoes, carrot, onion, and cucumber grown at elevated temperatures from ambient to 37°C [18]. Like clostridium and bacillus, pectolytic cytophaga [19] and xanthomonas [20] are generally considered the secondary pathogens which invade plants following the attack of a more aggressive pathogen such as erwinia or pseudomonas. Based on a series of studies previously conducted in our laboratory [21], erwinia and pseudomonas combined account for over 90% of soft rot of fresh produce while in storage or at markets. Less than 10% of soft rot of fresh produce found at the markets could be caused by xanthomonas, cytophaga, bacillus, or other unidentified genera [4,21].

5.2.1 PECTOLYTIC *ERWINIA* SPP.

The soft-rot erwinia group, consisting of three species or subspecies, *E. carotovora* subsp. carotovora (Ecc), *E. carotovora* subsp. atroseptica (Eca), and *E. chrysanthemi* (Ech), is the major single cause of microbial spoilage of vegetables. The losses due to soft-rot erwinia cost tens or hundreds of millions of dollars yearly [3,7]. Ecc has the broadest host range causing diseases in almost every species of vegetable crops grown in temperate and subtropical regions [7]. Eca is present at cooler regions and is more often associated with black leg of potatoes [22] in the field than with soft rot of fresh produce after harvesting. In contrast, Ech causes diseases of crops grown in subtropical or tropical regions [23,24]. Both Ecc and Ech grow poorly and fail to induce soft rot of fresh produce at 10°C or below. At 20°C or higher, Ecc is considered the most destructive soft-rotting pathogen of fruits and vegetables. The soft-rot erwinia group is widespread in nature and very closely associated with plant vegetation and can be readily isolated from weed, plant debris, rhizosphere soil, and lenticels of potato tubers [25,26]. However, soft-rot erwinia is rarely detected on the surfaces of plant leaves or true seeds [22]. In addition to erwinia, other enteric bacteria including enterobacter, klebsiella, and serratia are commonly present on the surfaces of many different types of vegetable crops [4,5]. A vast majority of enteric bacteria are nonpectolytic and not expected to cause the spoilage of fresh produce. However, they may play a critical role in maintaining the quality and safety of fresh produce by enhancing or suppressing the growth of spoilage and pathogenic microorganisms on the surfaces of plants.

5.2.2 PECTOLYTIC FLUORESCENT (PF) PSEUDOMONADS

For fruits and vegetables that are stored at refrigeration temperatures, pectolytic fluorescent (PF) pseudomonads are responsible for a substantial proportion of soft-rot disorder observed in markets. PF pseudomonads as a group are physiologically and taxonomically heterogeneous, mainly consisting of *P. viridiflava* and five biovars of *P. fluorescens* [21,27–29]. The latter was often designated as *P. marginalis* in the plant pathology literatures [21,28,29]. These pseudomonads are widespread in nature and can be isolated from diverse ecological niches including soils, irrigation water, rhizosphere, and surfaces of fruits and vegetables [27–29]. PF pseudomonads account for over 40% of total bacterial rot found at retail and wholesale produce markets [21]. They are especially abundant on the surfaces of leafy or salad vegetables including spinach [30], lettuce [31–33], cabbage [34], potato lenticels [25], tomatoes [35,36], and bell pepper [37]. On salad vegetables including lettuce, cabbage, and spinach, PF pseudomonads account for over 30% of total native bacteria recovered. Because of their prevalence in nature, PF pseudomonads are expected to play an important role in maintaining the safety and quality of refrigerated or ready-to-eat vegetables or fruits. They could be readily isolated from very diverse ecological niches including soil [38], rhizosphere [39], surfaces of fresh vegetables [30–37], and wash water from produce processing plants [38].

The importance of PF pseudomonads as the leading cause of spoilage of refrigerated fresh produce is primarily due to their psychrotrophic nature, nutritional versatility, and predominant presence on the surfaces of fresh produce. PF pseudomonads are responsible for a very large proportion of decay of fresh fruits and vegetables stored at low temperatures [21]. In addition, some *P. marginalis* and *P. viridiflava* strains can also cause soft-rot disease of horticultural crops in the field, e.g., the "pink eye" of potato tubers [39]. A few reports also showed that other fluorescent pseudomonads including *P. aeruginosa* [40], *P. tolasii* [7], and *P. chorii* [9] were involved in postharvest spoilage of vegetables or mushrooms.

5.3 FACTORS AFFECTING THE SURVIVAL OF SOFT-ROT BACTERIA IN NATURE

5.3.1 PLANT VEGETATION

Soft-rot erwinia and pseudomonas are widespread in nature and can be readily isolated from decayed tissue, plant debris, rhizosphere soil, and weeds [25,26,38,41,42]. De Boer [41] reported that Eca was isolated more often from soil in which potatoes had been grown in previous years than from soils in which other types of crops had been grown. Thus, survival and over-wintering of erwinia in soil can be greatly affected by the type of crop grown in the previous season. Soft-rot erwinia is rarely found on the surfaces of leafy vegetables and true seeds and survives poorly in sterilized soil [22]. Plant vegetation appears to be important for long-term survival of erwinia and pseudomonas in soil [26,28,41]. Because of the widespread distribution in plant and nonplant environments, it is impossible to eliminate completely soft-rot erwinia and pseudomonas from propagation materials, irrigation water, or soils in the field.

5.3.2 TEMPERATURE AND ATMOSPHERIC CONDITIONS

Refrigeration is the most convenient and effective means to maintain the organoleptic properties, to reduce the spoilage, and to extend the shelf life of fresh produce. The International Fresh-Cut Produce Association (IFPA) [42] recommends minimally processed produce be stored at 1 to 4°C to maintain the quality and safety. Refrigeration of fresh produce at between 4 and 10°C is commonly used by the industry to extend the shelf life and to prevent the soft rot caused by bacteria (such as Ecc and Ech) and fungi. At this temperature range, the development of soft rot by Eca and PF pseudomonads will occur. The minimum temperature for growth of Eca has been estimated to be between 3 and 6°C [7] and the minimal temperature for growth of PF pseudomonads estimated to be 4°C or below [21].

The seven genera of soft-rot bacteria mentioned above require somewhat different optimal atmospheric conditions for growth and induction of spoilage. For instance, *Clostridium* spp. are strictly anaerobic and PF pseudomonads (with the exception of nitrate-denitrifying strains) are strictly aerobic. Induction of soft rot in potatoes by erwinia and clostridium is greatly enhanced by the depletion of oxygen [16]. Reduction in oxygen concentration or increase in carbon dioxide concentration in the atmosphere reduced the growth of PF pseudomonads [43] and their ability to induce soft rot on fresh produce [44].

5.3.3 LATENT INFECTION AND INTERNALIZATION

Although the internal parts of plant organs are generally considered sterile [6], many different types of bacteria including soft-rot erwinia, pseudomonas, and serratia can be detected within apparently healthy tomatoes [45] and cucumber fruits [46]. These bacteria presumably exist in a commensalistic or quiescent state, which can be activated only when the stressed conditions in fruits are removed. A large proportion of storage rot of fruits is due to external contamination by soft-rotting microorganisms and a small proportion of them may be caused by the activation of latent bacteria inside the fruits. The route by which soft-rot erwinia penetrates into the internal parts of apparently healthy tomato fruits is unclear but possibly may be through the connective tissue at the stem end of the fruits [47]. Bartz and Kelman [48] also reported that the bacterial soft-rot potential in potato tubers was affected by difference in temperature between tubers and suspensions of erwinia at the time of inoculation by immersion. A series of laboratory experiments have conclusively demonstrated that human pathogenic bacteria including *E. coli* O157:H7 and salmonella can be infiltrated into apple [49,50], orange [51], tomatoes [51], and lettuce [52–54] if the temperature of bacterial suspension is lower than that of fruits. Surface cleaning and sanitization treatments are not expected to eliminate completely the undesirable bacteria that become internalized [55] and those attached to the surfaces of intact or injured fruits [56,57].

5.4   ENZYMATIC AND MOLECULAR MECHANISM OF TISSUE MACERATION BY SOFT-ROT BACTERIA

5.4.1   BIOCHEMICAL CHARACTERIZATION OF PECTATE LYASE (PL)

5.4.1.1   Analysis of PL Isozymes

Soft-rot erwinia including Ech, Ecc, and Eca are characterized by their ability to produce an array of pectolytic enzymes including pectin methylesterase (PME), polygalacturonase (PG), pectin lyase (PNL), and pectate lyase (PL). These enzymes can be readily detected in filtrates of bacterial cultures and assayed by the standard biochemical procedures [58]. The PLs produced by Ech, Ecc, and Eca are usually present in multiple (three to five) isozymic forms in culture filtrates, which can be readily identified by isoelectric focusing (IEF) gel electrophoresis and overlay enzyme staining techniques [59]. In Ech, a second set of PL isozymes, inducible only in the presence of plant constituents, have been identified using molecular genetic and enzyme analyses [60]. The biological and pathological function of each pectic enzyme produced by soft-rot erwinia has not been fully determined. It is also unclear if production of certain pectic enzymes is restricted to specific tissues or organs or limited to specific stages of plant development. With the aid of molecular genetic technologies, experimental results [61–63] have shown that no single pectic enzyme produced by soft-rot erwinia is absolutely required for the pathogen to initiate disease development. However, the PL isozymes, especially alkaline PLe, usually display the highest degree of tissue macerating ability *in vitro* [64] and are assumed to be the principal enzymes required for development of soft rot by erwinia *in vivo* [62].

Because of their complex pectic enzyme system, it is difficult to purify a single PL isozyme from culture filtrates of *Erwinia* spp. However, due to the simplicity of the pectic enzyme system in other spoilage bacteria, including *P. fluorescens*, *P. viridiflava*, and *Xanthomonas campestris*, it is relatively easy to purify the PLs from their culture filtrates [65]. Normally, following two simple steps (ammonium sulfate precipitation and anion exchange chromatography), the PL can be purified from culture filtrates of these two PF pseudomonads to near homogeneity [65,66]. Enzymological properties of PLs purified from culture filtrates of *P. fluorescens* and *P. viridiflava* have been characterized, and a minute amount of purified enzyme was capable of causing total maceration of potato tuber tissue even in the absence of live bacteria [67].

5.4.1.2   Production of Other Pectic Enzymes

In addition to PLs, soft-rot erwinia produces an array of other pectic enzymes including pectin methyesterase (PME), polygalacturonase (PG) and pectin lyase (PNL). Production of PME, PG, and PNL by soft-rot pathogens does not seem to play a significant role in initiating the maceration of plant tissues. However, they may be required for interactions with host plants or coping with adverse environments [68]. It has been reported that purified PG, but not purified PME or PNL by itself, is sufficient to induce soft rot of potato tuber slices. Production of PNL by soft-rot *Erwinia* spp. [69] and *Pseudomonas* spp. [70] is inducible only after exposing the bacteria to DNA-damaging agents such as ultraviolet radiation and mitomycin C. The ecological and pathological significance of producing PNL by erwinia and pseudomonas remains obscure. The role of PME in soft-rot pathogenesis is minimal and probably not required. However, it has been suggested that a coordinated action between PME and PL may be necessary for complete degradation of native pectins in plant cell walls.More information about the enzymatic mechanism of soft-rot pathogenesis by *Erwinia* spp. can be found in earlier reviews [61–63,68].

5.4.2 PL as the Principal Tissue-Macerating Factor

5.4.2.1 Transposon Mutagenesis

The notion that PL is the principal or sole pathogenicity factor of soft-rotting pseudomonads can be supported by a series of molecular genetic studies. By using transposon (*Tn5*)-mediated mutagenesis, Liao *et al.* [66] isolated several types of *P. viridiflava* mutants that became defective in production or secretion of PL. When assayed on plants, nonpectolytic *P. viridiflava* mutants were unable to induce soft rot on potato tuber slices. The loss in the ability to produce or secrete PL is accompanied by the loss in the ability to induce soft rot. This result provides the first unequivocal evidence that PL is the sole enzyme required for the induction of soft rot by *P. viridiflava* [66].

5.4.2.2 Cloning and Analysis of PL Genes

The gene encoding PL has been cloned from the genomes of *P. fluorescens* [71], *P. viridiflava* [72], and *Xanthomonas campestris* [73]. When cloned PL gene was mobilized into nonpectolytic mutants of *P. viridiflava* or *P. fluorescens*, the PL-producing and soft-rotting ability of nonpectolytic mutants was restored [74–76]. These results provide direct genetic evidence that the gene coding for PL is the principal or sole pathogenicity or virulence determinant of soft-rotting *P. viridiflava* or *P. fluorescens*.

5.4.3 Control of PL Production and Pseudomonas Rot

5.4.3.1 Two-Component Regulatory Gene System

The enzymatic and molecular genetic mechanism of soft-rot pathogenesis caused by erwinia has been extensively investigated and reviewed [61–63,68]. However, very little is known about the mechanism by which PF pseudomonads regulate the production of PL and induction of tissue maceration in plants. Pleotropic mutants of *P. fluorescens* and *P. viridiflava* showing simultaneous loss of production of both pectolytic and proteolytic enzymes have been identified by transposon mutagenesis [74–76]. Results from Southern Blot analysis revealed that mutants were derived from the insertion of Tn5 into one of two distinct genomic fragments. Two genes regulating the production of pectolytic enzyme and induction of soft rot, designated as $gacS$ (=$repA$ or $lemA$) and $gacA$ (=$repB$), have been identified in these two fragments and subsequently cloned and confirmed by complementation studies [74–76].

Based on the nucleotide sequence analyses, the $gacS$ and $gacA$ genes were respectively predicted to encode a sensory and a regulator protein in the two-component regulatory protein family [74,76]. The $gacS/gacA$ genes were predicted to act in pairs to mediate the production of an array of extracellular compounds including PL, protease (PRT), exopolysaccharide (EPS), and ion-chelating siderophores [74–76], possibly in response to environmental signals. The $gacS/gacA$ genes in biological control strains of $P.\ fluorescens$ have also been shown to regulate the production of phospholipase C [77], lipase [78], and antibiotics [79–81]. Proper function of the $gacS/gacA$ gene system is also required for the formation of disease lesions on snap beans by $Pseudomonas$ $syringae$ pv. syringae [82]. This two-component $gacS/gacA$ gene system can also interact with the stationary-phase factor δ^s (encoded on $rpoS$) in a biological control strain of $P.\ fluorescens$ to control the responses of this strain to environmental stimuli [83]. In $P.\ aeruginosa$, the activator GacA will interact with two quorum sensing proteins (LuxR, LuxI) to regulate the production of an autoinducer (butylhomoserine lactone) [84]. It has not yet been investigated, however, if RpoS, LuxR, and LuxI would act in concert to regulate the production of PL and other extracellular compounds in soft-rotting $P.$ $fluorescens$ and $P.\ viridiflava$.

A group of $P.\ viridiflava$ mutants failing to excrete PL and Prt across the outer membrane have also been generated by transposon mutagenesis during the isolation of nonpectolytic mutants [66]. These secretion-defective mutants (designated Out⁻) were assumed to result from the insertion of Tn5 into a gene belonging to the Type II secretory gene family [85,86]. Out⁻ mutants were also unable to induce soft rot on potato tuber slices and bell pepper fruits [66]. This indicates that the synthesis and the secretion of PL are two critical steps required for induction of soft rot.

5.4.3.2 Role of Calcium Ions

Production of PL in certain strains of $P.\ fluorescens$ is inducible by pectic substrates [87,88] or plant tissue extracts [89–91]. However, in other $P.\ fluorescens$ strains production of PL is not affected by the type of carbon source included in the medium [91]. Recently, we investigated the mode of PL production in 24 strains of $P.\ fluorescens$ and found that production of PL in certain $P.\ fluorescens$ strains (4 out of 24) was not induced by pectic substrates but by Ca^{2+} [92]. These four strains produce ten times more PL in medium containing 1 mM $CaCl_2$ than in one containing no $CaCl_2$ supplement. Supplement of $CaCl_2$ in the medium not only affects the amount but also the final destination of PL. Over 86% of total PL produced by strain CY091 in $CaCl_2$-supplemented medium was excreted into the culture fluid. By comparison, only 13% of total PL produced by this strain in $CaCl_2$-deficient medium was detected in the extracellular fraction. The effect of Ca^{2+} on PL production is concentration-dependent and can be replaced by Sr^{2+}, but not by Zn^{2+}, Fe^{2+}, Mn^{2+}, Mg^{2+}, or Ba^{2+} [92].

5.4.3.3 Use of Ion-Chelating Agents for Control of Pseudomonas Rot

Because of the indispensable role of Ca^{2+} in the production, secretion, and catalytic activity of PLs, the potential of using ion-chelating agents such as EDTA for control of pseudomonas rot has been investigated [92]. Application of ion-chelating agents such as EDTA to limit the availability of Ca^{2+} to *P. fluorescens* infecting the plants thus offers a potential strategy for control of soft rot caused by pseudomonads. We have demonstrated that application of $0.05\,\mu M$ (or 40 ppm) of EDTA, alone [92] or in combination with a bacteriocin (nisin) [93], suppresses the induction of soft rot by *P. fluorescens*. Zucker and Hankin [94] also reported that EDTA treatments reduced the soft rot potential of potato tubers.

It should be noted, however, that the EDTA treatment would not be effective for control of soft rot caused by erwinia, because *Erwinia* spp. produce not only Ca^{2+}-dependent PL but also Ca^{2+}-independent PG. However, infiltration of potato tubers or apple fruits with $CaCl_2$ can enhance their resistance to attack by Ecc or Eca [95] or *Penicillium expansum* [96]. Changes in calcium fertilization in potato fields could also affect the susceptibility of potato tubers to bacterial soft rot [97]. Infiltration of potato tubers and fruits with Ca^{2+} was thought to strengthen the cell walls and consequently increase their resistance to postharvest rot pathogens [98]. None of the above control strategies have been applied on a large scale for commercial operations.

5.5 INTERACTIONS BETWEEN SOFT-ROT AND HUMAN PATHOGENS ON FRESH PRODUCE

Despite the lack of a known mechanism for attacking plants, the gastrointestinal human pathogens including salmonella, *E. coli* O157:H7, and *L. monocytogenes* can survive and even grow on fruits or vegetables over a long period of time [99]. Their survival and growth can be affected by the indigenous microflora and by the storage conditions [4]. The dynamics of the interactions between native microflora, spoilage bacteria, and human pathogens, especially under modified atmospheres, have been investigated [5]. The results obtained thus far indicate that the effect of spoilage or saprophytic microorganisms on the proliferation of human pathogens on fresh produce could be either synergistic or antagonistic, largely depending on the type of pathogens, fresh produce, and storage conditions examined.

5.5.1 SYNERGISTIC INTERACTIONS

The interactions between soft-rot and human pathogens on fresh produce began to catch the attention of public health officials when Wells and Butterfield [15] reported that the rotted plant tissue more often harbored salmonella than the healthy counterpart. They also demonstrated that the population of salmonella increased by 5- to 10-fold on potato or carrot slices that were co-inoculated with soft-rotting *E. carotovora* or *P. viridiflava*. Carlin *et al.* [100] later showed a positive correlation between the number of *L. monocytogenes* and the extent of soft rot observed with endive leaves. These studies indicate that the rotted plant tissues may provide extra nutrients to enhance the growth of human pathogens. Contaminated plant tissues can then serve as a reservoir or vehicle for the dissemination of clinically important pathogens in farms or food processing facilities.

Surveys of salmonella contamination on rotted fruits and vegetables induced by molds or fungi have also been reported [16]. The incidence of salmonella contamination on rotted tissue induced by molds or fungi was about one tenth of that induced by soft-rotting bacteria. Nevertheless, the fungi-induced rotted tissues are three times more likely to contain salmonella than the healthy counterpart. In spite of this, the investigators [16] concluded that rotted tissues pose little or no greater safety risk than the healthy tissues. Gastrointestinal pathogens including salmonella and *L. monocytogenes* usually do not grow, or grow very poorly, on acidic fruits (pH ≤ 4.0) such as apple and orange [101]. Conway *et al.* [102] demonstrated that these bacterial pathogens were able to multiply in rotted tissues induced by specific groups of fungal pathogens such as *Glomerella cingulata* but not in rotted tissues induced by other groups of fungal pathogens such as *Penicillium expansum*. They found that the population of *L. monocytogenes* increased in apple fruits infected with *G. cingulata* but declined in fruits infected with *P. expansum*. Conway *et al.* [102] revealed that the pH in *P. expansum*-induced rotted apples decreased from 4.7 to 3.7 as opposed to the increase in pH from 4.6 to 7.7 in *G. cingulata*-induced rotted fruits. Riordan *et al.* [103] also showed that the population of *E. coli* O157:H7 increased 1 to 3 logs on apple infected with *G. cingulata* but continued to decrease in rotted tissues infected with *P. expansum*. The pH change in rotted tissues induced by different groups of fungi thus plays a critical role in the fate of human pathogens on fresh produce.

5.5.2 ANTAGONISTIC INTERACTIONS

In contrast to the synergistic effect, a number of studies have shown that the growth of human pathogens on fresh produce could be suppressed by the presence of postharvest rot pathogens. For examples, the growth of *L. monocytogenes* on potato slices [104], spinach [30], and endive [105] could be markedly reduced by diverse strains of fluorescent pseudomonads. The inhibition was thought to be caused by the production of iron-chelating fluorescent siderophores or antimicrobials by the pseudomonads [106]. Carlin *et al.* [105]

reported that more growth of *L. monocytogenes* was detected on endive leaves that were rinsed with disinfectants than those rinsed with water. Two pseudomonad antagonists possibly responsible for inhibiting the growth of *L. monocytogenes* on endive leaves have been identified [105]. Additional strains of fluorescent pseudomonads capable of inhibiting the growth of *L. monocytogenes* or *L. innocua* on different types of produce including carrot, lettuce, bell pepper, and sprouting seeds have been isolated [106,107].

L. monocytogenes is in general more susceptible than salmonella or *E. coli* O157:H7 to the antagonists naturally present on the surfaces of fresh produce [106]. In addition to the saprophytic antagonists, postharvest rot pathogens including *P. fluorescens* and *P. expansum* can also inhibit the growth of human pathogens such as *E. coli* O157:H7 and *L. monocytogenes* [102–104]. It has been suggested that elimination of native microflora (including bacterial and fungal rot pathogens) from fruits and vegetables may create a less competitive environment for the proliferation of human pathogens on fresh produce [4].

5.6 SELECTED FARM PRACTICES FOR CONTROL OF BOTH SOFT-ROT AND HUMAN PATHOGENS

To minimize the dissemination and proliferation of the harmful microorganisms on growing plants, it is necessary to take preventive measures to intervene in the introduction of contamination sources in the field. Sources of soft-rotting erwinia and pseudomonas [22,28,38] and foodborne human pathogens that may contaminate or infect growing plants in the field have been previously reviewed [99,108]. Preventive measures and good agricultural practices (GAPs) for reducing the contamination of field crops with foodborne pathogens have been suggested in several guidance references including one published by the U.S. Food and Drug Administration [109] and another one by the IFPA [110]. Preventive control strategies for bacterial soft rot have also been reviewed by Eckert and Ogawa [111] and by Lund [6]. A few practices useful for control of both soft-rot bacteria and foodborne human pathogens in the fields are indicated as follows:

- Use seeds and propagation materials that are free of soft-rot bacteria and human pathogens for planting. Although soft-rot erwinia are generally not considered seedborne [22], long-term survival of salmonella and soft-rot pseudomonas in water [112] and on alfalfa seeds destined for sprouting has been reported [113].
- Properly dispose of the decayed plant materials in the field, which can become the inoculum source of soft-rot bacteria [26] and serve as a fertile ground for the proliferation of foodborne human pathogens such as salmonella [14,15].
- Avoid the use of improperly treated manure or compost in the field. Long-term survival of salmonella and *E. coli* O157:H7 in feces and in partially composted manure or biosolids has been documented [99].

Application of improperly treated compost possibly containing decayed materials may also serve as the inoculum source of soft-rot bacteria.

- Monitor and ensure that water to be used for irrigation, washing, and preparation of protective chemicals is devoid of harmful microorganisms. Both soft-rot bacteria and human pathogens have been known to survive in water for several years [112].
- Harvest the crop at the optimal stage of maturity and with the minimal mechanical injury. It has been reported that the mature crop exhibits a higher level of resistance to attack by soft-rot bacteria [22] and to the colonization by human pathogens [99]. Injured plant surfaces can serve as the points of entry for soft-rot bacteria and as the sites for attachment by human pathogens [56,57].
- Maintain sanitary conditions and enforce good worker hygiene in the field to prevent the contamination of growing or harvested crops with pathogens carried by farm workers. Outbreaks of foodborne illness due to the contamination of fresh produce with foodborne pathogens originating from farm workers have been previously reported [108].
- Use clean and sanitary vehicles for transporting produce from farms to processing plants.
- Keep the orchards and vegetable farms away from domestic and wild animals and far away from poultry and dairy farms. Feces and animal wastes are believed to be the two most important carriers or reservoirs of foodborne human pathogens [109].
- Remove weeds grown in the field, which may become alternative inoculum sources for soft-rot erwinia [42] and human pathogens [100].

REFERENCES

1. NACMCF (National Advisory Committee on Microbiological Criteria for Foods), Microbiological safety evaluations and recommendations on fresh produce, *Food Control*, 10, 117, 1999.
2. Kaufman, P.R. *et al.*, Understanding the dynamics of produce markets: consumption and consolidation grow, USDA, Economic Research Service, Agriculture Information Bulletin No. 758, http://www.ers.usda.gov/publications/aib758, accessed Aug. 24, 2001.
3. Harvey, J.M., Reduction of losses in fresh market fruits and vegetables, *Ann. Rev. Phytopathol.*, 16, 321, 1978.
4. Nguyen-The, C. and Carlin, F., The microbiology of processed fresh fruits and vegetables, *Crit. Rev.Food Sci. Nutr.*, 34, 37, 1994.
5. Nguyen-The, C. and Carlin, F., Fresh and processed vegetables, in *The Microbiological Safety and Quality of Food*, Vol. 1, Lund, B., Baird-Parker, T.C., and Gould, G.W. Eds., Aspen, Gaithersburg, MD, 2000, chap. 25.
6. Lund, B.M., The effect of bacteria on post-harvest quality of vegetables and fruits, with particular reference to spoilage, in *Bacteria and Plants*, Rhodes-Roberts, M. and Skinner, F.A., Eds., Academic Press, New York, 1982, p. 135.

7. Lund, B.M., Bacterial spoilage, in *Post-harvest Pathology of Fruits and Vegetables*, Dennis, C., Ed., Academic Press, London, 1983, p. 219.
8. Cappellini, R.A. *et al.*, Disorders in potato shipments to the New York market, 1972–1980, *Plant Dis.*, 68, 1018, 1984.
9. Ceponis, M.J., Diseases of California head lettuce on the New York market during the spring and summer months. *Plant Dis.*, 54, 964, 1970.
10. Ceponis, M.J., Kaufman, J., and Butterfield, J.E., Relative importance of gray mold rot and bacterial soft rot of western lettuce on the New York market. *Plant Dis.*, 54, 263, 1970.
11. Ceponis, M.J. and Butterfield, J.E. Causes of cullage of Florida bell peppers in New York wholesale and retail markets, *Plant Dis.*, 58, 367, 1974.
12. Ceponis, M.J. and Butterfield, J.E., Market losses in Florida cucumbers and bell peppers in metropolitan New York, *Plant Dis.*, 58, 558, 1974.
13. Ceponis, M.J., Cappellini, R.A., and Lightner, G.W., Disorders in tomato shipments to the New York market, 1972–1984, *Plant Dis.*, 70, 261, 1986.
14. Wells, J.M. and Butterfield, J.E., *Salmonella* contamination associated with bacterial soft rot of fresh fruits and vegetables in the marketplace, *Plant Dis.*, 81, 867, 1997.
15. Wells, J.M., and Butterfield, J.E., Incidence of *Salmonella* on fresh fruits and vegetables affected by fungal rots or physical injury, *Plant Dis.*, 83, 722, 1999.
16. Pérombelon, M.C.M., Gullings-Handley, J., and Kelman, A., Population dynamics of *Erwinia carotovora* and pectolytic *Clostridium* spp. in relation to decay of potatoes, *Phytopathology*, 69, 167, 1978.
17. Lund, B.M., Isolation of pectolytic clostridia from potatoes, *J. Appl. Bacteriol.*, 35, 609, 1972.
18. Dowson, W.J., Spore-forming bacteria in potatoes, *Nature* (London), 152, 331, 1943.
19. Liao, C.-H. and Wells, J.M., Properties of *Cytophaga johnsonae* strains causing spoilage of fresh produce at food markets, *Appl. Environ. Microbiol.*, 52, 1261, 1986.
20. Liao, C.-H. and Wells, J.M., Association of pectolytic strains of *Xanthomonas campestris* with soft rots of fruits and vegetables at retail markets, *Phytopathology*, 77, 418, 1987.
21. Liao, C.-H. and Wells, J.M., Diversity of pectolytic, fluorescent pseudomonads causing soft rot of fresh vegetables at produce markets, *Phytopathology*, 77, 673, 1987.
22. Pérombelon, M.C.M. and Kelman, A., Ecology of the soft rot *Erwinia*, *Ann. Rev. Phytopathol.*, 18, 361, 1980.
23. Farrar, J.J., Nunez, J.J., and Davis, R.M., Influence of soil saturation and temperature on *Erwinia chrysanthemi* soft rot of carrot, *Plant Dis.*, 84, 665, 2000.
24. Hsu, S.-T. and Tzeng, K.-C., Species of *Erwinia* associated with soft rot disease of plant in Taiwan, in *Proceedings 5th International Conference Plant Pathology Bacteria*, Lozano, J.C., Ed., CIAT, Cali, Columbia, 1981, p. 9.
25. Cuppels, D. and Kelman, A., Evaluation of selective media for isolation of soft-rot bacteria from soil and plant tissue, *Phytopathology*, 64, 469, 1974.
26. Burr, T.J. and Schroth, M.N., Occurrence of soft-rot *Erwinia* spp. in soil and plant material, *Phytopathology*, 67, 1382, 1977.
27. Lelliott, R.A., Billing, E., and Hayward, A.C., A determinative scheme for the fluorescent plant pathogenic pseudomonads, *J. Appl. Bacteriol.*, 29, 470, 1966.

28. Sands, D.C. and Hankin, L., Ecology and physiology of fluorescent pectolytic pseudomonads, *Phytopathology*, 65, 921, 1975.

29. Fahy, P.C. and Lloyd, A.B., *Pseudomonas*: the fluorescent pseudomonads, in *Plant Bacterial Diseases: A Diagnostic Guide*, Fahy, P.C. and Persley, G.J., Eds., Academic Press, Australia, 1983, chap 8.

30. Babic, I., *et al.*, Changes in microbial populations of fresh cut spinach, *Int. J. Food Microbiol.*, 31, 107, 1996.

31. Magnuson, J.A., King, A.D., Jr., and Török, T., Microflora of partially processed lettuce, *Appl. Environ. Microbiol.*, 56, 3851, 1990.

32. Bolin, H.R. *et al.*, Factors affecting the storage stability of shredded lettuce, *J. Food Sci.*, 42, 1319, 1977.

33. King, A.D., Jr. *et al.*, Microbial flora and storage quality of partially processed lettuce, *J. Food Sci.*, 56, 459, 1991.

34. Chesson, A., The fungal and bacterial flora of stored white cabbage, *J. Appl. Bacteriol.*, 46, 189, 1979.

35. Garg, N., Churey, J.J., and Splittstoesser, D.F., Effect of processing conditions on the microflora of fresh-cut vegetables, *J. Food Prot.*, 53, 701, 1990.

36. Bartz, J.A., Causes of postharvest losses in a Florida tomato shipment, *Plant Dis.*, 64, 934, 1980.

37. Coplin, D.L., *Erwinia carotovora* var. *carotovora* on bell peppers in Ohio, *Plant Dis.*, 64, 191, 1980.

38. Cuppels, D.A. and Kelman, A., Isolation of pectolytic fluorescent pseudomonads from soil and potatoes, *Phytopathology*, 70, 1110, 1980.

39. Hagar, S.S. and McIntyre, G.A., Pectic enzymes produced by *Pseudomonas fluorescens*, an organism associated with "pink eye" disease of potato tubers, *Can. J. Botany*, 50, 2479, 1972.

40. Cother, E.J., Darbyshire, B., and Brewer, J., *Pseudomonas aeruginosa*: cause of internal brown rot of onion, *Phytopathology*, 66, 828, 1976.

41. De Boer, S.H., Allan, E., and Kelman, A., Survival of *Erwinia carotovora* in Wisconsin soils, *Am. Potato J.*, 56, 243, 1979.

42. McCarter-Zorner, N.J. *et al.*, Soft rot *Erwinia* bacteria in the rhizosphere of weeds and crop plants in Colorado, United States and Scotland, *J. Appl. Bacteriol.*, 59, 357, 1985.

43. Gill, C.O. and Tan, K.H., Effect of carbon dioxide on growth of *Pseudomonas fluorescens*, *Appl. Environ. Microbiol.*, 38, 237, 1979.

44. Wells, J.M., Growth of *Erwinia carotovora*, *E. atroseptica*, and *Pseudomonas fluorescens* in low oxygen and high carbon dioxide atmosphere, *Phytopathology*, 64, 1012, 1974.

45. Samish, Z. and Etinger-Tulczynska, R., Distribution of bacteria within the tissue of healthy tomatoes, *Appl. Microbiol.*, 11, 7, 1963.

46. Meneley, J.C. and Stanghellini, M.E., Establishment of an inactive population of *Erwinia carotovora* in healthy cucumber fruit, *Phytopathlogy*, 65, 670, 1975.

47. Bartz, J.A. and Showalter, R.K., Infiltration of tomatoes by aqueous bacterial suspensions, *Phytopathology*, 71, 515, 1981.

48. Bartz, J.A. and Kelman, A., Bacterial soft rot potential in washed potato tubers in relation to temperatures of tubers and water during simulated commercial handling practices, *Am. Potato J.*, 61, 485, 1984.

49. Buchanan, R.L. et al., Contamination of intact apples after immersion in an aqueous environment containing Escherichia coli O157:H7, J. Food Prot., 62, 444, 1999.

50. Burnett, S.L., Chen, J., and Beuchat, L.R., Attachment of Escherichia coli O157:H7 to the surfaces and internal structures of apples as detected by confocal scanning laser microscopy, Appl. Environ. Microbiol., 66, 4679, 2000.

51. FDA (U.S. Food and Drug Administration), Preliminary studies on the potential for infiltration, growth and survival of Salmonella enterica serovar Hartford and Escherichia coli O157:H7 within oranges, http://vm.cfsan.fad.gov/%7Ecomm/juicsstud.html, accessed Jan. 27, 2000.

52. Takeuchi, K., Hassan, A.N., and Frank, J.F., Penetration of Escherichia coli O157:H7 into lettuce as influenced by modified atmosphere and temperature, J. Food Prot., 64, 1820, 2001.

53. Wachtel, M.R, Whitehand, L.C., and Mandrell, R.E., Association of Escherichia coli O157:H7 with preharvest leaf lettuce upon exposure to contaminated irrigation water, J. Food Prot., 65, 18, 2002.

54. Solomon, E.B., Yaron, S., and Matthews, K.R., Transmission of Escherichia coli O157:H7 from contaminated manure and irrigation water to lettuce plant tissue and its subsequent internalization, Appl. Environ. Microbiol., 68, 397, 2002.

55. Zhuang, R.-Y., Beuchat, L.R., and Angulo, F.J., Fate of Salmonella Montevideo on and in raw tomatoes as affected by temperature and treatment with chlorine, Appl. Environ. Microbiol., 61, 2127, 1995.

56. Liao, C.-H. and Sapers, G.M., Attachment and growth of Salmonella Chester on apple fruits and in vivo response of attached bacteria to sanitizer treatments, J Food Prot., 63, 876, 2000.

57. Liao, C.-H. and Cooke, P.H., Response to trisodium phosphate treatment of Salmonella Chester attached to fresh-cut pepper slices, Can. J. Microbiol., 47, 25, 2001.

58. Collmer, A., Ried, J.L., and Mount, M.S., Assay methods for pectic enzymes, Methods Enzymol., 161, 329, 1988.

59. Ried, J.L. and Collmer, A., Activity stain for rapid characterization of pectic enzymes in isoelectric focusing and sodium dodecyl sulfate-polyacrylamide gels, Appl. Environ., Microbiol., 50, 615, 1985.

60. Kelemu, S. and Collmer, A., Erwinia chrysanthemi EC16 produces a second set of plant-inducible pectate lyase isoenzymes, Appl. Environ. Microbiol., 59, 1756, 1993.

61. Kotoujansky, A., Molecular genetics of pathogenesis by soft-rot erwinias, Ann. Rev. Phytopathol., 25, 405, 1987.

62. Barras, F., van Gijsegem, F., and Chatterjee, A.K., Extracellular enzymes and pathogenesis of soft-rot Erwinia, Ann. Rev. Phytopathol., 32, 201, 1994.

63. Py, B. et al., Extracellular enzymes and their role in Erwinia virulence, Methods Microbiol., 27, 157, 1998.

64. Payne, J.H. et al., Multiplication and virulence in plant tissue of Escherichia coli clones producing pectate lyase isozymes PLb and PLe at high levels and an Erwinia chrysanthemi mutant deficient in PLe, Appl Environ. Microbiol., 53, 2315, 1987.

65. Liao, C.-H., Analysis of pectate lyase produced by soft rot bacteria associated with spoilage of vegetables, Appl. Environ. Microbiol., 55, 1677, 1989.

66. Liao, C.-H., Hung, H.Y., and Chatterjee, A.K., An extracellular pectate lyase is the pathogenicity factor of the soft-rotting bacterium *Pseudomonas viridiflava*, *Mol. Plant Microbe Interact.*, 1, 199, 1988.

67. Liao, C.-H. *et al.*, Biochemical characterization of pectate lyases produced by fluorescent pseudomonads associated with spoilage of fresh fruits and vegetables, *J. Appl. Microbiol.*, 83, 10, 1997.

68. Collmer, A., and Keen, N.T., The role of pectic enzymes in plant pathogenesis, *Ann. Rev. Phytopathol.*, 24, 383, 1986.

69. McEvoy, J.L., Murata, H., and Chatterjee, A.K., Molecular cloning and characterization of an *Erwinia carotovora* subsp. *carotovora* pectin lyase gene that respond to DNA-damaging agents, *J. Bacteriol.*, 166, 172, 1984.

70. Sone, H. *et al.*, Production and properties of pectin lyase in *Pseudomonas marginalis* induced by mitomycin C, *Agric. Biol. Chem.*, 52, 3205, 1988.

71. Liao, C.-H., Cloning of pectate lyase gene *pel* from *Pseudomonas fluorescens* and detection of sequences homologous to *pel* in *Pseudomonas viridiflava* and *Pseudomonas putida*, *J. Bacteriol.*, 173, 4386, 1991.

72. Liao, C.-H. *et al.*, Cloning and characterization of a pectate lyase gene from the soft-rotting bacterium *Pseudomonas viridiflava*, *Mol. Plant Microbe Interact.*, 5, 301, 1992.

73. Liao, C.-H. *et al.*, Cloning of a pectate lyase gene from *Xanthomonas campestris* pv. *malvacearum* and comparison of its sequence relationship with *pel* genes of soft-rot *Erwinia* and *Pseudomonas*, *Mol. Plant Microbe Interact.*, 9, 14, 1996.

74. Liao, C.-H., McCallus, D.E., and Fett, W.F., Molecular characterization of two gene loci required for production of the key pathogenicity factor pectate lyase in *Pseudomonas viridiflava*, *Mol. Plant Microbe Interact.*, 7, 391, 1994.

75. Liao, C.-H. *et al.*, Identification of gene loci controlling pectate lyase production and soft-rot pathogenicity in *Pseudomonas marginalis*, *Can. J. Microbiol.*, 43, 425, 1997.

76. Liao, C.-H. *et al.*, The *repB* gene required for production of extracellular enzymes and fluorescent siderophores in *Pseudomonas viridiflava* is an analog of the *gacA* gene of *Pseudomonas syringae*, *Can. J. Microbiol.*, 42, 177, 1996.

77. Sacherer, P., Défago, G., and Haas, D., Extracellular protease and phosphlipase C are controlled by the global regulatory gene *gacA* in the biocontrol strain *Pseudomonas fluorescens* CHA0, *FEMS Microbiol. Lett.*, 116, 155, 1994.

78. Woods, R.G. *et al.*, The *aprX-lipA* operon of *Pseudomonas fluorescens* B52: a molecular analysis of metalloprotease and lipase production, *Microbiology* 147, 345, 2001.

79. Laville, J. *et al.*, Global control in *Pseudomonas fluorescens* mediating antibiotic synthesis and suppression of black root rot of tobacco, *Proc. Natl. Acad. Sci. USA*, 89, 1562, 1992.

80. Gaffney, T.D., *et al.*, Global regulation of expression of anti-fungal factors by a *Pseudomonas fluorescens* biological control strain, *Mol. Plant Microbe Interact.*, 7, 455, 1994.

81. Corbell, N. and Loper, J.E., A global regulator of secondary metabolite production in *Pseudomonas fluorescens* Pf-5, *J. Bacteriol.*, 177, 6230, 1995.

82. Hrabak, E.M. and Willis, D.K., The *lemA* gene required for pathogenicity of *Pseudomonas syringae* pv. syringae on bean is a member of a family of two-component regulators, *J. Bacteriol.*, 174, 3011, 1992.

83. Whistler, C.A. *et al.*, The two-component regulators GacS and GacA influence accumulation of stationary-phase sigma factor δ^s and the stress response in *Pseudomonas fluorescens* Pf-5, *J. Bacteriol*, 180, 6635, 1998.

84. Reimmann, C. *et al.*, The global activator GacA of *Pseudomonas aeruginosa* PAO positively controls the production of the autoinducer *N*-butyryl-homoserine lactone and the formation of the virulence factors pyocyanin, cyanide, and lipase, *Mol. Microbiol.*, 24, 309, 1997.

85. Sandkvist, M., Biology of type II secretion, *Mol. Microbiol.*, 40, 271, 2001.

86. Koster, M., Bitter, W., and Tommassen, J., Protein secretion mechanisms in Gram-negative bacteria, *Int. J. Med. Microbiol.*, 290, 325, 2000.

87. Nasuno, S. and Starr, M.P., Pectic enzymes of *Pseudomonas marginalis*, *Phytopathology*, 56, 1414, 1966.

88. Fuchs, A., The *trans*-eliminative breakdown of Na-polygalacturonate by *Pseudomonas fluorescens*, *Antonie van Leeuwenhoek J. Microbiol. Serol.*, 31, 323, 1965.

89. Zucker, M. and Hankin, L., Regulation of pectate lyase synthesis in *Pseudomonas fluorescens* and *Erwinia carotovora*, *J. Bacteriol.*, 104, 13, 1970.

90. Zucker, M. and Hankin, L., Inducible pectate lyase synthesis and phytopathogenicity of *Pseudomonas fluorescens*, *Can. J. Microbiol.*, 17, 1313, 1971.

91. Zucker, M., Hankin, L., and Sands, D., Factors governing pectate lyase synthesis in soft rot and non-soft rot bacteria, *Physiol. Plant Pathol.*, 2, 59, 1972.

92. Liao, C.-H., McCallus, D.E., and Wells, J.M., Calcium-dependent pectate lyase production in the soft-rotting bacterium *Pseudomonas fluorescens*, *Phytopathology*, 83, 813, 1993.

93. Wells, J.M., Liao, C.-H., and Hotchkiss, A.T., *In vitro* inhibition of soft-rotting bacteria by EDTA and nisin and *in vivo* response on inoculated fresh cut carrots, *Plant Dis.*, 82, 491, 1998.

94. Zucker, M. and Hankin, L., Effectiveness of ehtylenediaminetetraacetic acid (EDTA) in controlling soft rot potatoes, *Plant Dis. Reptr.*, 54, 863, 1970.

95. McGuire, R.G. and Kelman, A., Reduced severity of *Erwinia* soft rot in potato tubers with increased calcium content, *Phytopathology*, 74, 1250, 1984.

96. Conway, W.S., and Sams, C.E., Calcium infiltration of Golden Delicious Apples and its effect on decay, *Phytopathology*, 73, 1068, 1983.

97. Bartz, J.A., Locascio, S.J., and Weingartner, D.P., Calcium and potassium fertilization of potatoes grown in north Florida. II. Effect on the bacterial soft rot potential in the tubers, *Am. Potato J.*, 69, 39, 1992.

98. Conway, W.S. *et al.*, Calcium treatment of apples and potatoes to reduce postharvest decay, *Plant Dis.*, 76, 329, 1992.

99. FDA (U.S. Food and Drug Administration), Analysis and evaluation of preventive control measures for the control and reduction/elimination of microbial hazards on fresh and fresh-cut produce, http://www.cfsan.fda.gov/~comm/ift3-1.html, accessed Jan. 15, 2002.

100. Carlin, F., Nguyen-The, C., and Abreu da Silva, A., Factors affecting the growth of *Listeria monocytogenes* on minimally processed fresh endive, *J. Appl. Bacteriol.*, 78, 636, 1995.

101. Lund, B.M. and Snowdon, A.L., Fresh and processed fruits, in *The Microbiological Safety and Quality of Food*, Vol. I, Lund, B.M., Baird-Parker, T.C., and Gould, G.W., Eds., Aspen, Gaithersbrug, MD, 2000, chap. 27.

102. Conway W.S. *et al.*, Survival and growth of *Listeria monocytogenes* on fresh-cut apples slices and its interaction with *Glomerella cingulata* and *Penicillium expansum*, *Plant Dis.*, 84, 177, 2000.

103. Riordan, D.C.R., Sapers, G.M., and Annous, B.A., The survival of *Escherichia coli* O157:H7 in the presence of *Penicillium expansum* and *Glomerella cingulata* in wounds on apple surfaces, *J. Food Prot.*, 63, 1637, 2000.

104. Liao, C.-H. and Sapers, G.M., Influence of soft rot bacteria on growth of *Listeria monocytogenes* on potato tuber slices, *J. Food Prot.*, 62, 343, 1999.

105. Carlin, F., Nguyen-The, C., and Morris, C.E., Influence of background microflora on *Listeria monocytogenes* on minimally processed fresh broad-leaved endive (*Cichorium endivia* var. *latifolia*), *J. Food Prot.*, 59, 698, 1996.

106. Liao, C.-H. and Fett, W.F., Analysis of native microflora and selection of strains antagonistic to human pathogens on fresh produce, *J. Food Prot.*, 64, 1110, 2001.

107. Francis, G.A. and O'Beirne, D., Effects of the indigenous microflora of minimally processed lettuce on the survival and growth of *Listeria innocua*, *Int. J. Food Sci. Technol.*, 33, 477, 1998.

108. Beuchat, L.R., Pathogenic microorganisms associated with fresh produce, *J. Food Prot.*, 59, 204, 1996.

109. FDA (U.S. Food and Drug Administration), Guidance for industry: Guide to minimize microbial food safety hazards for fresh fruits and vegetables, www.foodsafety.gov/~dms/prodguid.htm, accessed Aug. 30, 2000.

110. IFPA, *Food Safety Guidelines for the Fresh-Cut Produce Industry*, 3rd ed., International Fresh-Cut Produce Association, Alexandria, VA, 1996, p. 125.

111. Eckert, J.W., and Ogawa, J.M., The chemical control of postharvest diseases: deciduous fruits, berries, vegetables and root/tuber crops, *Ann. Rev., Phytopathol.*, 26, 433, 1988.

112. Liao, C.-H., and Shollenberger, L.M., Survivability and long-term preservation of bacteria in water and in phosphate-buffered saline, *Lett. Appl. Microbiol.*, 37, 45, 2003.

113. Liao, C.-H., and Fett, W.F., Isolation of *Salmonella* from naturally contaminated alfalfa seeds and demonstration of impaired recovery of heat-injured cells in alfalfa seed homogenates, *Int. J. Food Microbiol.*, 82, 245, 2003.

6 Microbial Spoilage of Fresh Mushrooms

Naveen Chikthimmah and Robert B. Beelman

CONTENTS

6.1 FRESH MUSHROOMS

6.1.1 INTRODUCTION

Based on phylum classification, fungi are classified as Ascomycota, Basidiomycota, Chytridiomycota, Deuteromycota, and Zygomycota [1]. While edible fungi such as truffles and morels belong to the phylum Ascomycota, most commercially cultivated edible fungal genera including agaricus, lentinula, and pleurotus belong to the phylum Basidiomycota.

Mushrooms, the common name for a large group of edible fungi, are a common and popular food product. The reproductive portion or the fruiting body of the mushroom usually lies above the growing substrate. It is the portion that is commonly used for consumption. Because of their unique earthy aroma and taste, many wild mushroom species have been traditionally consumed. However only a few mushroom species have been extensively cultivated on a commercial basis [2].

Agaricus bisporus (J. Lge) Imbach (button mushroom) is the most widely cultivated species of edible mushroom, representing approximately 32% of world production in 1997 [3]. China, the U.S., and the Netherlands are the top three producers of *A. bisporus* in the world [4]. *Lentinula edodes* (Berk.) (shiitake) and *Pleurotus ostreatus* (Jacq.:Fr) Kumm. (oyster mushroom), the second and third most cultivated edible mushrooms, account for approximately 25 and 14% of world production, respectively [3,4]. Commercial mushroom production makes a significant contribution to the total agricultural output of the U.S. In 2002–2003 the U.S. mushroom crop totaled 844 million pounds, valued at $889 million. White and off-white *A. bisporus* mushrooms still have by far the largest market share, in particular in the western hemisphere, accounting for about 80% [5].

Since agaricus is the major genera of cultivated mushrooms around the world, this chapter mainly describes the microbiology and microbial spoilage of the cultivated button mushroom, *A. bisporus*.

6.1.2 COMMERCIAL GROWING PRACTICES

The agaricus mushroom growing process is unique in that it requires decomposed organic matter as both a substrate for growing and as a source of essential nutrients. A typical growth substrate contains straw-bedded horse or chicken manure, hay, corn cob, brewer's grain, cotton seed, cocoa seed hull, and water. The substrate mixture is aerobically fermented under semicontrolled conditions [6,7], a process known as Phase I composting. Ingredients are mixed and placed in aerated bunkers or formed into long rows that are periodically turned, watered, and reformed. Rapid microbial growth over a 15- to 25-day period causes the substrate (compost) temperatures to reach as high as 175°F (80°C). During the Phase I process substrate nutrients are converted into forms efficiently assimilated by the mushrooms.

Phase II composting begins when the finished substrate is transferred in bulk into controlled atmosphere tunnels, or in trays into controlled atmosphere rooms where further microbial activity and nutrient conversion occur. Phase II includes a controlled pasteurization step designed to eliminate mushroom and human pathogens [8], weeds, and insect pests. A successful crop requires that the compost temperature reach 130 to 140°F (60°C) for at least 2 hours [9].

Agaricus mycelial starter cultures grown on cereal grains, commonly known as mushroom spawn, are then mixed into the substrate and allowed to

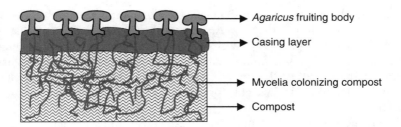

Agaricus fruiting body

Casing layer

Mycelia colonizing compost

Compost

FIGURE 6.1 Schematic diagram of the mushroom substrate (compost), the peat-based casing layer, and fruiting bodies.

grow throughout the compost for 14 days. Following complete colonization of the substrate by *A. bisporus* mycelia, a two-inch casing layer (consisting of peat soil amended with calcium carbonate and water) is applied on top of the colonized substrate bed (Figure 6.1). The casing soil enhances retention of irrigation water on the growing beds, and promotes mushroom fruit body formation.

Mycelial growth occurs throughout the substrate and into the casing layer. After 14 to 21 days, mushroom primordia are formed at the fruiting stage known as pinning. The primordia develop into mature fruiting bodies over a one-week period. During the development process, the growing beds are irrigated to maintain substrate moisture, prevent disease, and maintain postharvest mushroom quality [2,10,11]. At maturity, the mushrooms are harvested, stipe-trimmed, packaged, and moved into cold storage.

6.1.3 GENERAL COMPOSITION

Edible mushrooms, especially *A. bisporus* (button), tend to be high in moisture. Mattila *et al.* [12] reported that the dry matter (percent solids) content of *A. bisporus* grown in Finland was 7.7%. These values for *A. bisporus* mushrooms are similar to those normally experienced in North America, but moisture can be as high as 95% when mushrooms are excessively irrigated [13].

Mushrooms contain large amounts of carbohydrates including polysaccharides (such as glucans and glycogen), monosaccharides, and disaccharides (such as trehalose), sugar alcohols (such as mannitol), and chitin. Mattila *et al.* [12] reported that *A. bisporus* contained 4.5% (fresh weight) total carbohydrates. Most of the polysaccharides are structural components of the cell walls — chitin and glucans — and are indigestible by humans and can be considered as dietary fiber. The *A. bisporus* mushroom species is also known to contain significant amounts (20 to 30%, dry weight) of the sugar alcohol mannitol, and 1 to 3% of the disaccharide trehalose [14].

While mushrooms contain only low levels of crude fat (0.31 to 0.35%, fresh weight) [15,16], they contain a significant amount of protein, vitamins, and minerals. Mattila *et al.* [12] found that *A. bisporus* mushrooms contained about 2.0% net protein (fresh weight). These mushrooms are also known to be high in the B-complex vitamins: niacin, folate, pantothenic acid, and ribo-flavin [17]. It was found that *A. bisporus* mushrooms contained almost 0.4% riboflavin (fresh weight) [12]. With respect to minerals and trace elements, *A. bisporus* mushrooms contain relatively high concentrations of potassium (0.36% fresh weight) [12], copper (0.22% fresh weight) [18], and selenium (3.2 and 1.4 mg/kg, dry weight for brown and white *A. bisporus* strains, respectively) [12].

From the standpoint of nutrients, fresh mushrooms are capable of supporting growth of microorganisms. Agaricus mushrooms have a neutral pH value, and fall in the category of foods with a water activity of 0.98 or higher. These factors favor the growth of microorganisms, leading to the microbial-induced quality degradation and spoilage of fresh mushrooms.

6.2 MICROBIOLOGY OF FRESH MUSHROOMS

Doores *et al* [19] demonstrated that normal healthy mushrooms have high bacterial populations. Total bacterial numbers ranged from 6.3 to 7.2 log CFU/g of fresh mushroom tissue. The majority (54.0%) of bacteria isolated from the mushrooms were identified as fluorescent pseudomonads with flavobacteria comprising the second largest group (10.0%). Recent experiments in our laboratory have confirmed this pattern, but we have also been able to isolate the chryseobacterium genus (5.5 log CFU/g) and the coryneform bacterial genus (5.6 log CFU/g) from freshly harvested mushrooms. Halami *et al.* [20] isolated lactic acid bacteria belonging to the *Lactobacillus* sp. and *Pediococcus* sp. from fresh mushrooms by incubating agaricus mushrooms in deMan Rogosa and Sharpe (MRS) broth for enrichment of resident lactic acid bacteria. However, the bacterial counts were not enumerated in their study.

Mushrooms also contain significant levels of yeasts and molds. Studies in our laboratory have shown that freshly harvested mushrooms harbor approximately 3 log CFU of molds and 6 log CFU of native yeast per gram of fresh tissue (Figure 6.2).

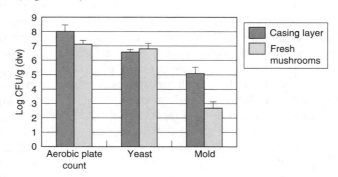

FIGURE 6.2 Microbiology of the mushroom casing layer and fresh mushrooms (dw, dry weight). Error bars represent standard deviation of the mean.

6.3 SPOILAGE OF FRESH MUSHROOMS

Quality is the single most important factor affecting retail mushroom sales [21]. Whiteness, cleanliness, and brown blotches on fresh mushrooms are the principal factors determining mushroom quality. Consumers prefer to purchase mushrooms that are bright white, free of casing material or other unwanted particulate contaminants clinging to the mushroom surface, and free of brown blotches. The brown blotch discoloration of mushrooms is perceived as a symptom of decreased freshness or microbiological deterioration (spoilage).

Enzymatic browning catalyzed by the enzyme tyrosinase (polyphenol oxidase) [22] is the most important factor involved in quality deterioration of fresh mushrooms. The browning reactions are initiated by tissue breakdown due to either mechanical damage or bacterial activity [23]. It has been suggested that the role of tyrosinase in mushrooms is to function as a stress metabolite [24]. Tyrosinase naturally occurs at high levels in the mushroom surface tissue, and is normally found in a latent form [5]. When activated during senescence [22] the enzyme oxidizes mushroom phenolic compounds into brown melanins [25–27] resulting in brown discoloration. In fresh mushrooms, tyrosinase and its substrates have been hypothesized to be located in separate subcellular compartments [22]. When mushrooms are mishandled or bruised the cellular membrane is damaged, and rapid browning of the mushroom cap is observed. It has been hypothesized that the loss of membrane integrity provides greater access of tyrosinase to its substrates, resulting in formation of brown compounds [22,28] and associated brown discoloration of fresh mushrooms.

The presence of high bacterial populations in fresh mushrooms is a major factor that significantly diminishes quality by causing a brown, blotchy appearance [23] (Figure 6.3). The rate of postharvest deterioration of fresh mushrooms has been directly related to the initial microbial load [23]. Doores *et al.* [19] found that bacterial populations during postharvest storage at 13°C increased from an initial load of 7 log CFU/g to almost 11 log CFU/g over a 10-day storage period. The authors also reported that deterioration of mushroom quality as indicated by maturity and color measurement appeared to be concomitant with increase in bacterial numbers. *Pseudomonas* spp. and *Flavobacterium* spp. were the two main groups that predominated during agaricus mushroom postharvest storage. Similarly we have observed that bacterial populations tend to increase from 7.3 to 8.4 log CFU/g during a 6-day storage period at 4°C (Figure 6.4). Populations of yeast increased from 6.9 to 8.0 log CFU/g during the storage period. Populations of molds remained constant (3 log CFU/g) during the storage period [29,30].

A majority of mushrooms of good quality and color, harvested and marketed, develop blotches at retail or in consumer homes, even while kept at refrigeration temperatures. Symptoms of brown blotch disease are sunken, dark, and brown spots [31] on the mushroom fruit body surface. Pseudomonas is the major spoilage genus associated with blotch

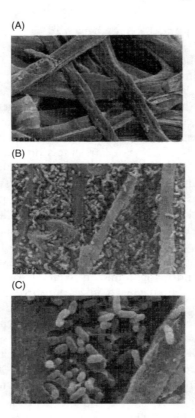

FIGURE 6.3 Scanning electron micrographs of mushroom cap surfaces: (A) healthy tissue (×3000); and blotched tissue showing invading bacteria (B) (×3000) and (C) (×10,000).

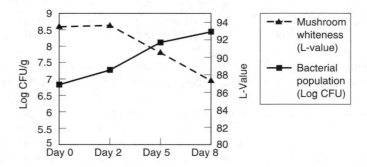

FIGURE 6.4 Increase in aerobic bacterial populations and a concomitant decrease in the whiteness (measured by L-value) of fresh *Agaricus bisporus* mushrooms during postharvest storage at 12°C. The solid line represents aerobic bacterial populations (log CFU/g fresh mushroom tissue). The broken line represents the L-value of the mushroom cap during postharvest storage. Data are the average of four independent samplings.

formation of fresh mushrooms [32–34]. Paine [35] attributed *Pseudomonas tolaasii* as the causative organism of the classic bacterial blotch disease of cultivated mushrooms. Application of *P. tolaasii* cells as low as $20\,CFU/cm^2$ of growing beds resulted in blotch formation in mushrooms [36]. Symptoms of mushroom blotch became visible when $5.4 \times 10^6\,CFU/cm^2$ were detectable in the mushrooms [36]. When *P. tolaasii* was placed directly onto caps, $6 \times 10^7\,CFU/cm^2$ were necessary to produce a blotch lesion (though only $3.5 \times 10^6\,CFU$ could be recovered). The researchers of the study [36] concluded that the number of cells of *P. tolaasii* present in the early primordial stages of mushroom growth controls the extent of blotch disease seen at harvesting. It has also been shown that tyrosinase is activated during infection by the bacterium *P. tolaasii* or exposure to its toxin, tolaasin, causing brown blotch disease symptoms of fresh mushrooms [37]. Wells *et al.* [38], by isolating and reinoculating the bacteria on freshly harvested healthy mushrooms, confirmed that postharvest blotch formation and associated discoloration was caused by three phenotypic groups (pathotypes) of fluorescent pseudomonads. Severe infections with darkened or yellowed lesions were caused by strains of pathotypes A or B, respectively. Mild infections with superficial discoloration were caused by the pathotype C. Based on cellular fatty acid analysis, the authors concluded that each pathotype corresponded to one or several mushroom-related pseudomonads reported in the literature as follows: pathotype A = *Pseudomonas tolaasii*, pathotype B = *Pseudomonas "gingeri"*, and pathotype C = *Pseudomonas "reactans"*. Isolates from mushroom casing material yielded all three pathotypes.

Fluorescent pseudomonads also produce exopolysaccharides (EPSs) associated with the sliminess accompanying spoilage of mushrooms. Fett *et al.* [39] isolated, partially purified, and characterized acidic EPSs from 63 strains of mushroom-associated fluorescent pseudomonads. The strains were originally isolated from discolored lesions on mushroom caps, or from commercial lots of mushroom casing soil. An acidic galacto-glucan named marginalan was produced by mucoid strains of the saprophyte *Pseudomonas putida* and the majority of mucoid strains of saprophytic *P. fluorescens* isolated from casing medium. Other strains produced EPSs that included alginate, and unique EPSs containing neutral and amino sugars and glucuronic acid.

There has been a long and complex association between the fungal genus trichoderma and mushroom cultivation since Beach [40] first reported disease symptoms on caps of agaricus mushrooms. In a study by Sharma *et al.* [41] colonization assessments confirmed that *Trichoderma harzianum* biotypes Th1, Th2a, Th2b, and Th3 inoculated into the mushroom substrate became established in the mushroom substrate. The extension rate of two Th2 isolates in the substrate was over 1000 times that of Th1 and Th3. Results confirmed that while Th1 and Th3 did not significantly affect yield, Th2 could reduce mushroom quality and productivity by as much as 80%. *In vitro* studies by Mumpuni *et al.* [42] suggested that the growth of *T. harzianum* biotypes could be related to the release of metabolites by *A. bisporus* into the compost substrate. Dilute aqueous solutions of *n*-butanol extracts of *A. bisporus* culture filtrates and fruit bodies inhibited Th1 and Th3 but stimulated Th2 isolates, suggesting that the active compound(s) may be constitutive components of the *A. bisporus* species.

6.3.1 SOURCES OF MICROORGANISMS CAUSING SPOILAGE

It has been demonstrated that the casing microflora have a vital role in the sporophore (fruit body) formation of mushrooms from the mycelia stage [43–45]. The requirement for biotic agents in the initiation of fruit body formation [45] excludes the possibility of mushroom cultivation on a commercial scale under axenic conditions. This factor, combined with the intensity of production within a confined area, results in the introduction of microorganisms on fresh mushrooms that contribute to spoilage during postharvest storage.

The casing layer on which the mushroom fruiting bodies develop is a significant reservoir for the microflora of fresh mushrooms [19]. Doores *et al.* [19] found that aerobic bacterial populations from casing material ranged between 8.2 and 8.5 log CFU/g. In a study conducted by Wong and Preece [46], the primary sources of *Pseudomonas tolaasii* on a mushroom farm were the peat and limestone used in the casing process. This mushroom pathogen could not be detected in the farm soil, water supply, the mushroom spawn used, or in compost after spawning, but was isolated from the casing (peat/limestone mixture) layer of symptom-free mushroom beds and both the casing layer and compost of beds bearing blotched mushrooms. Secondary sources were numerous once the pathogen was present in mushroom beds. These included symptomless and blotched mushrooms, the fingers and shoes of people handling the crop, their baskets, knives, and ladders. *P. tolaasii* was also isolated from dust in the air of infected houses. While spores of infected mushrooms may transport the bacterium, sciarid flies can act as vectors contributing to bacterial transfer.

6.3.2 CULTURAL (GROWING) PRACTICES FAVORING SPOILAGE

The extent of irrigation significantly affects the bacterial populations and the quality of the mushroom crop. Wong and Preece [36] concluded that very frequent irrigation, resulting in over-watering, increased blotch symptoms on mushrooms during growing.

6.3.3 CULTURAL PRACTICES TO SUPPRESS SPOILAGE OF FRESH MUSHROOMS

Significant efforts have been directed to improve mushroom quality by adding calcium salts or antimicrobial treatments to irrigation water during cultivation. Barden *et al.* [47] demonstrated that the postharvest shelf life of fresh mushrooms increased by 2 days when mushrooms were irrigated with 0.5% calcium chloride. The increase in shelf life was mainly due to a decreased rate of postharvest bacterial growth and a concomitant reduction of surface browning. Solomon *et al.* [48] demonstrated a significant improvement in quality and shelf life when mushroom crops were irrigated with tap water containing 50 ppm stabilized chlorine dioxide and 0.25% calcium chloride. Initial and postharvest bacterial counts and degree of browning were lower in these mushrooms as compared to mushrooms irrigated with water without chlorine dioxide or calcium chloride. Irrigation treatments involving the addition of calcium salts to irrigation water to reduce bacterial populations and improve initial and postharvest mushroom quality have been extensively studied [10,24,48,49], and are now a common commercial growing practice.

Kukura *et al.* [11] conducted a study to examine the influence of 0.3% $CaCl_2$ added to irrigation water on mushroom tyrosinase activity and postharvest browning. With the addition of $CaCl_2$ to the irrigation water, the calcium content of mushrooms significantly increased, accompanied by reduced postharvest browning. Irrigation with $CaCl_2$ had no effect on inherent tyrosinase activity. The $CaCl_2$ irrigation treatment had even more pronounced improvement on mushroom shelf life following a standard bruising treatment, as indicated by reduced browning. Based on transmission electron micrographs, the authors speculated that increased levels of calcium in mushrooms irrigated with $CaCl_2$ may have decreased browning by increasing vacuolar membrane integrity, thereby reducing the opportunity for tyrosinase to react with its phenolic substrates.

In other studies in our laboratory [29] we evaluated irrigation with modified acidic electrolyzed oxidizing (EO) water in combination with 0.3% calcium chloride on the reduction in bacterial populations of fresh mushrooms. Crops were grown using standard growing practices except for the experimental additions to the irrigation water of acidic EO water (diluted with 2 parts of regular irrigation water) and/or 0.3% calcium chloride. Compared to the control, all treatments reduced bacterial populations on the fresh mushrooms. While no significant differences in color were observed between the treatments on the day of harvest, irrigation with modified acidic EO water and/ or calcium chloride resulted in enhanced whiteness, point-of-sale appearance, and quality after a 7-day holding period of the fresh mushrooms. Recently we investigated the effect of irrigation with water containing 0.75% hydrogen peroxide on reduction in bacterial populations on fresh mushrooms. Irrigation with 0.75% hydrogen peroxide in combination with 0.3% calcium chloride added to the irrigation water consistently reduced the bacterial populations on fresh mushrooms by 85% (compared to bacterial populations on mushrooms irrigated with water without hydrogen peroxide and calcium chloride). This irrigation combination treatment shows promise as an effective preharvest method to enhance the quality of fresh mushrooms.

Research has been conducted to investigate the effect of natural antimicrobial secondary metabolites added into the irrigation water. In a study by Geels [50], a 1% aqueous solution of kasugamycin, an antibiotic produced by *Streptomyces kasugaensis*, was evaluated for reducing bacterial blotch after artificial infection of the mushroom crop with *P. tolaasii*. An artificial infection was established in the first flush (harvest) by inoculating the button-sized mushrooms with a suspension of *P. tolaasii*. A 1% aqueous solution of kasugamycin supplied through irrigation water on the second-flush mushrooms drastically reduced bacterial blotch symptoms on these mushrooms at picking stage. Disease incidence in the second flush in the control treatment (inoculated with *P. tolaasii*) was composed of 18% lightly, 29% moderately, and 10% heavily affected mushrooms, which totaled to 57% affected. The 1% kasugamycin treatment significantly reduced total disease incidence to only 9% (lightly) affected. In the same study, a sodium hypochlorite-based irrigation treatment showed no beneficial results.

Studies with canned products processed from mushrooms grown under experimental cultural conditions indicated that canned product spoilage was reduced significantly by employing peat versus soils as the casing material [51]. While this study has no implication on the spoilage of fresh mushrooms, it does indicate that casing type may have an effect on the microbiology and microbial spoilage of fresh mushrooms.

Aerated steam treatment is sometimes employed to treat thermally (pasteurize) the casing layer. Though steam treatment of casing material is not a common cultural practice, some commercial growers employ pasteurization ($60°C$, $140°F$) of the casing layer to control diseases associated with some materials they employ. However, most growers do not heat-treat their casing material because of the additional cost involved and anecdotal evidence that crop yield will be reduced.

It has been speculated that reducing the microbial load in the casing layer may result in reduced bacterial populations associated with the mushrooms and improve postharvest quality [23]. Hence, we conducted an experiment to evaluate casing pasteurization on reduction in bacterial populations in fresh mushrooms and its effect on crop yield and quality. The crop was grown at the Mushroom Test Demonstration Facility (MTDF) on the Penn State University campus using standard growing practices used at the MTDF except for the pasteurization treatment to the mushroom casing. Unpasteurized casing served as the control. For pasteurization, the casing material was held in a steam vault designed for direct steam injection. Steam was generated on-site. Pasteurization of the casing was conducted by forcing a mixture of air and steam into the vault to increase the temperature of the casing material to $60°C$ ($140°F$). The casing material was held at $60°C$ for at least 2 hours. Following the application of the pasteurized and untreated (control) casing layers to the colonized compost the rest of the growing, irrigation, and harvesting procedures were conducted as per standard MTDF practices. Pasteurization of the casing layer (Figure 6.5) resulted in a 2.9 log CFU/g reduction in total bacterial populations (reducing the total population in the pasteurized casing from 5.9 to 3 log CFU per gram of casing layer material). However, bacterial numbers of the pasteurized casing

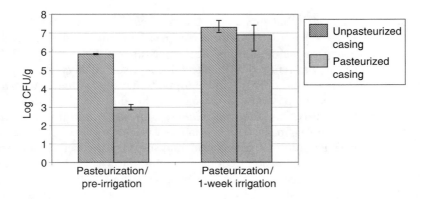

FIGURE 6.5 Effect of pasteurization at 60°C followed by irrigation on total bacterial populations in the mushroom casing soil.

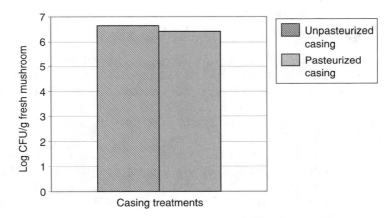

FIGURE 6.6 Aerobic bacterial populations on fresh *Agaricus bisporus* mushrooms grown using either pasteurized or unpasteurized casing soil.

increased by 3.9 log CFU/g (from 3 log to 6.9 log) following 1 week of irrigation. At the same time the bacterial numbers increased by 1.4 log CFU/g in the unpasteurized casing (from 5.9 log to 7.3 log). Interestingly, there was no significant difference in bacterial numbers in mushrooms grown using unpasteurized or pasteurized casing (Figure 6.6). Mushrooms grown on steam-treated casing material showed improved postharvest shelf life. However the crop yield decreased by 10% when the pasteurized casing was used.

From a food safety perspective, a recommendation to steam-treat mushroom casing soils to reduce pathogenic bacterial populations should be delayed, since steam treatment may negatively affect a hurdle (beneficial soil microflora) to inhibit foodborne pathogens introduced into the soil (via irrigation water or cross contamination). Preliminary research in our laboratory has indicated that survival of *Listeria monocytogenes* is enhanced in pasteurized casing soil (60°C, 2 hours), compared to untreated soil. Under mushroom growing casing conditions (80% moisture, 22°C), 6.8 log CFU/g of *L. monocytogenes* was reduced to undetectable levels in 10 days in untreated casing soil. During this time period, populations of *L. monocytogenes* remained unchanged in pasteurized casing soil. So far, we have been able preliminarily to identify that the *Penicillium* sp. present naturally in casing soils may play a vital role in the destruction of *L. monocytogenes*. It is possible that thermal pasteurization of casing soil may destroy the penicillium and other beneficial microbial populations, thereby allowing survival of *L. monocytogenes* in the casing soil. Hence practical nonthermal methods are urgently required to destroy selectively the foodborne pathogens without significantly affecting the beneficial microbial populations in casing soils. Interestingly, *L. monocytogenes* demonstrated enhanced survival in casing soils colonized with the agaricus mycelia than in soils without the mycelia present in it. This situation warrants research on casing soil handling and disinfecting crop irrigation procedures to achieve preharvest food safety and quality goals.

Biocontrol has been evaluated as an alternative cultural practice to reduce bacterial populations and subsequently enhance quality and postharvest shelf life. Nair and Fahy [52] reported the isolation of three bacteria antagonistic to *P. tolaasii* from soil and peat. These were a nonfluorescent *Pseudomonas* species from soil, and strains of *P. fluorescens* and *Enterobacter aerogenes* from peat. When the antagonists and the pathogen (*Pseudomonas tolaasii*) were added in the ratio of 7.9:6 log CFU/ml to unsterilized peat and applied to mushroom trays, infection of mushroom sporophores by the pathogen was effectively controlled. *In vitro* studies failed to show lysis or growth inhibition of *P. tolaasii* by the antagonists. While biocontrol-based products have been introduced into the market in the recent past for controlling bacterial brown blotch of mushrooms, they have not been a significant commercial success.

6.3.4 POSTHARVEST CONDITIONS FAVORING SPOILAGE OF FRESH MUSHROOMS

Postharvest storage conditions significantly contribute to mushroom quality and shelf life. Pai [53] evaluated the effect of storage temperature (5, 10, and 15°C), and relative humidity (RH) (91, 94, 97, and 99%) on weight loss, whiteness change, and microbial activity of *A. bisporus* mushrooms. Weight loss of tested samples was correlated highly with storage time at each RH level. Increasing storage temperature and decreasing RH significantly enhanced ($p < 0.05$) the rate of weight loss. Mushroom whiteness values were not affected ($p > 0.05$) by changes in RH. Microbial growth increased with increasing storage temperatures. It was concluded that the use of clean mushrooms with low initial microbial counts, an environment of high RH, and minimal condensation in packages are important factors for maximizing the shelf life of mushrooms under refrigerated storage.

Temperature abuse during storage is an important factor contributing to the spoilage of fresh mushrooms. Tano et al. [54] evaluated the effects of temperature fluctuation on the atmosphere inside modified atmosphere containers and their impact on the quality of fresh mushrooms within the containers. Mushrooms were packaged in 4-liter modified atmosphere (MA) containers, and an atmosphere of 5% O_2 and 10% CO_2 was maintained at 4°C. Temperature was fluctuated from 4 to 20°C during a 12-day storage period in cycles of 2 days at 4°C followed by 2 days at 20°C. The severity of bacterial blotch on mushrooms was assessed using a rating of 1 to 4, with 1 = no bacterial blotch and 4 = above 25% of the mushrooms cap area with symptoms of blotch disease. Temperature increase during fluctuations caused anoxic atmospheres both in O_2 (1.5%) and CO_2 (22 to 10%). The quality of mushrooms stored under temperature fluctuating regime was severely affected as indicated by extensive browning, loss of firmness, and a high level of ethanol in the tissue compared to mushrooms stored at constant temperature. For the control group, the bacterial blotch index was negligible over a 6-day storage period, whereas with mushrooms stored under temperature abuse conditions, the index increased rapidly from 2.6 to 3.6 after 4 days. This study clearly demonstrated that temperature abuse and temperature fluctuation seriously compromise the benefits of MA packaging of fresh mushrooms.

Condensation of water in packages can severely affect the quality of fresh packaged mushrooms. Apart from making the appearance of the mushroom packs unattractive, condensation is not desirable since a water layer on mushroom caps supports the growth of Pseudomonas tolaasii [55]. Gormley and MacCanna [56] studied the effect of overwrapping mushrooms with different types of perforated and nonperforated films on changes in mushroom quality during storage. They found that water condensation occurred on the underside of the nonperforated film. At the same time excessive water loss through the perforated films caused wrinkling and brown patches on the mushroom caps [56]. Hence it is important to select packaging material taking into consideration the high respiration rate of mushrooms and the potential fluctuating storage temperatures during warehouse storage and retail display.

6.3.5 POSTHARVEST PRACTICES TO SUPPRESS SPOILAGE OF FRESH MUSHROOMS

Various postharvest treatments have been investigated in order to impede browning and reduce rate of spoilage of fresh mushrooms. While proper cold storage is a primary requirement during postharvest storage, new or novel packaging techniques, washing treatments, and irradiation of mushrooms can further contribute to spoilage suppression [2].

6.3.5.1 Packaging

Overwrapping mushrooms with plastic film improves their quality as observed by rate of cap opening, color, and weight loss [56–58]. Since mushrooms respire heavily (500 mg CO_2/kg fresh weight/hour at ambient temperature) [59], it is important to ensure proper ventilation of the packages to maintain a high O_2 environment within the packages. Freshly harvested mushrooms were found to induce a near anaerobic environment ($<2\%$ O_2) in unventilated, PVC-overwrapped packages within 2 to 6 hours when incubated at 20 to 30°C [60]. To prevent in-package atmospheres from turning anaerobic which can increase risk of *Clostridium botulinum* growth, conventional mushroom packages are also perforated at the top with 2 mm holes in accordance with a U.S. Food and Drug Administration (FDA) recommendation [61].

New technologies such as modified atmosphere packaging (MAP) have been developed in order to delay quality loss and to extend storage life of mushrooms [62–64]. The MAP method changes the mixture of gases surrounding a respiring product to a composition other than that of air. The gas composition of a storage atmosphere may reduce both microbial and physiological spoilage of fresh mushrooms [65] Lopez-Briones *et al.* [66] demonstrated that while up to 2.5% CO_2 seems to benefit mushroom whiteness, CO_2 concentrations higher than 5% enhanced mushroom discoloration during storage. The authors suggested that a desirable modified atmosphere for mushrooms storage should contain 2.5 to 5.0% CO_2 and 5 to 10% O_2.

Water persisting on mushroom caps after irrigation supports the growth of *Pseudomonas tolaasii* [55] and subsequent appearance of blotch. Roy *et al.* [67,68] evaluated sorbitol as a moisture absorber in mushroom packages at 12°C. Surface moisture content of mushrooms decreased in the presence of a sorbitol pouch. Mushrooms packaged with 10 g sorbitol pouches had constant surface moisture content and those packaged with 15 g sorbitol pouches had the best overall color. Lowering the in-package relative humidity did not affect the maturation rate of mushrooms but reduced bacterial growth, suggesting that improvement in color was probably due to reduced bacterial activity.

Martin and Beelman [60] evaluated the potential of *Staphylococcus aureus* to grow and produce staphylococcal enterotoxin in ventilated and unventilated fresh mushroom packages when stored at 25 to 35°C. Mushrooms were inoculated with an enterotoxigenic strain of *S. aureus* and incubated in overwrapped trays at different temperatures. *S. aureus* grew and produced staphylococcal enterotoxin (SE) in unventilated PVC-overwrapped mushroom packages when inoculated at levels of 3, 4, and 5 log CFU/g of mushroom after 4 days of incubation at 30°C. Growth of *S. aureus* was observed at all levels of inoculation at 25°C, but no SE was detected after 7 days of incubation. When mushroom packages were ventilated, *S. aureus* growth was suppressed and no SE was detected after 7 days at 25°C and 4 days at 30°C. However, *S. aureus* growth in ventilated packs exceeded growth in unventilated packages when the incubation temperature was increased to 35°C; SE was detected within 18 hours of incubation at this temperature, even in mushrooms inoculated at a low level (2 log CFU/g). These results show the extreme importance of proper sanitation and worker hygiene during mushroom harvesting and packaging, ventilation of fresh mushroom packages, and use of proper storage temperatures for fresh mushrooms at all points of the food chain since SE is extremely thermotolerant and can even survive the rigorous thermal process used in canning mushrooms [69].

6.3.5.2 Washing Treatments

Washing mushrooms has recently gained commercial popularity as a means of removing casing soil particles and for the application of browning and microbial inhibitors. Prior to 1986, aqueous solutions of sulfite, particularly sodium metabisulfite, were used to wash mushrooms for the purpose of removing unwanted particulate matter and to enhance mushroom whiteness. While sulfite treatment yielded mushrooms of excellent initial whiteness and overall quality, it did not inhibit the growth of spoilage bacteria. Therefore, the quality improvement brought about by sulfite use was transitory. After 3 days of refrigerated storage, bacterial decay of sulfited mushrooms becomes evident. In 1986 the FDA banned the application of sulfite compounds to fresh mushrooms due to severe allergic reactions to sulfites among certain asthmatics. Following the ban on sulfite compounds for washing fresh mushrooms, there have been several efforts to develop wash solutions for use as a suitable replacement for sulfites.

McConnell [70] conducted a review of potential wash additives for mushrooms including sodium hypochlorite, hydrogen peroxide, potassium sorbate, and sodium salts of benzoate, EDTA, and phosphoric acids. The researcher concluded that effective antioxidants, in addition to antimicrobial compounds, were required to enhance shelf life of fresh mushrooms by washing. A fresh mushroom wash solution containing 10,000 ppm hydrogen peroxide and 1000 ppm calcium disodium EDTA was developed. Hydrogen peroxide present in the wash solution acts as a bactericide. Copper is a functional cofactor of the mushroom browning enzyme tyrosinase. EDTA in the wash solution binds copper more readily than tyrosinase, thereby sequestering copper and reducing tyrosinase activity and associated enzymatic browning of mushroom tissue.

Beelman and Duncan [71] developed a mushroom wash process (U.S. Patent 5,919,507). The method employed a first-stage high pH (pH of 9.0 or above) antibacterial wash followed by a neutralizing wash containing browning inhibitors. The neutralizing wash contained a buffered solution of erythorbic acid and sodium erythorbate. Other browning inhibitors such as ascorbates, EDTA, or calcium chloride were identified as suitable ingredients for addition to the neutralizing solution. The process also helped remove debris and delayed microbial spoilage of fresh mushrooms.

Sapers *et al.* [72] developed a two-stage mushroom wash process employing 10,000 ppm (1%) hydrogen peroxide in the first stage aqueous solution, and 2.25 to 4.5% sodium erythorbate, 0.2% cysteine-HCl, and 500 ppm to 1000 ppm EDTA in aqueous solution in the second stage. The two-stage washing typically yielded mushrooms nearly as white as sulfited mushrooms initially, and whiteness surpassed that of sulfited mushrooms after 1 to 2 days of storage at 12°C [73,74]. The treatment was effective in reducing bacterial populations in wash water and on mushroom surfaces [75] and had minimal effects on mushroom structure and composition [76]. The process was further modified and optimized [72] to include a prewash step using 0.5% (5000 ppm) to 1% (10,000 ppm) hydrogen peroxide. Mushrooms washed by this process were free of adhering soil, less subject to brown blotch than conventionally washed mushrooms, and at least as resistant to enzymatic browning as unwashed mushrooms during storage at 4°C. However, storage at 10°C accelerated development of brown blotch and browning.

6.3.5.3 Irradiation

In 1986 the FDA approved gamma irradiation doses up to 1 kGy on fruits and vegetables for the purpose of insect and/or growth and maturation control. Low-dose gamma irradiation has been reported to be a very effective method of controlling deterioration and improving quality and shelf life of fresh mushrooms [77–79]. Radiation, usually from a cobalt-60 source, is most effective when applied to the mushrooms shortly after harvest. A dose of 1 kGy, an FDA-approved dose, greatly reduced bacterial counts and slowed the rate of senescence [78]. A dose of 0.25 kGy was ineffective in controlling senescence, while 2 kGy showed no significant improvement over 1 kGy in terms of postharvest quality [78]. Cap opening, stipe elongation, surface darkening, and tissue softening were either delayed or prevented by the application of irradiation [78]. Sensory data comparing irradiated mushrooms with unirradiated controls showed that the former had equal or superior flavor and texture scores for both raw and cooked samples [78]. In another study, Ajlouni *et al.* [14] concluded that low-dose gamma irradiation (1 kGy) was an effective method for improving quality and extending the shelf life of mushrooms under commercial retail conditions, but it would need to be coupled with refrigerated storage to be most effective. Commercial application of irradiation for enhancing the quality of mushrooms has not yet been used in the U.S. However, cultivated mushrooms appear to be a good candidate for irradiation because of their high market value and short shelf life.

Recently, electron-beam irradiation was evaluated for its application to fresh sliced mushrooms [80]. The effects of electron-beam irradiation on microbial counts, color, texture, and enzyme activity of mushroom slices were evaluated at dose levels of 0.5, 1, 3.1, and 5.2 kGy. Irradiation levels above 0.5 kGy reduced total plate counts, yeast and mold, and psychrotrophic bacteria counts to below detectable levels, and prevented microbial-induced browning. Firmness of all samples was similar during storage except for the 5.2 kGy sample. Color was not affected by the irradiation treatments. Electron-beam irradiation at the levels tested did not affect the polyphenol oxidase activity. Irradiation at 1 kGy was most effective in extending shelf life of mushroom slices [80].

6.3.5.4 Pulsed Ultraviolet Light Treatment

Ultraviolet (UV) light is a portion of electromagnetic spectrum ranging from 100 to 400 nm wavelengths. UV light in the wavelength range 100 to 280 nm has germicidal properties due to DNA damage in microorganisms. Several researchers have demonstrated that the UV light can be used for the inactivation of foodborne pathogens without adversely affecting the quality of food. UV light treatment of foods can be accomplished using a pulsed UV system, whereby the energy is stored in a high-power capacitor and is released periodically in short pulses (often in nanoseconds). The pulsed UV light system reduces the temperature buildup as compared to that obtained with a continuous UV light, due to short pulse durations and cooling periods between pulses. Thus, the pulsed UV light process may be considered a nonthermal process.

Beelman *et al.* [81] conducted an experiment to evaluate the pulsed UV light sterilization system to reduce bacterial populations in/on fresh mushrooms. Pulsed UV light treatment was carried out with a laboratory scale, batch, pulsed light sterilization system (SteriPulse®-XL 3000, Xenon Corporation, Woburn, MA). The system generated $5.6 \, \text{J/cm}^2$ per pulse on the strobe surface for an input voltage of 3800 V and with 3 pulses per second. The output from the pulsed UV light system followed a sinusoidal wave pattern, with $5.6 \, \text{J/cm}^2$ per pulse being the peak value of the pulse. The pulse width (duration of pulse) was 360 μs. Packed mushrooms were placed in the pulsed UV light sterilization chamber and treated with pulsed light. The first study used a 30-second treatment at a distance of 8 cm from the UV strobe. The control samples did not undergo any pulsed UV treatment. In the second study, treatments with varying treatment time (2 or 4 seconds) and distance from UV strobe (8 or 13 cm) combinations were evaluated. Treated mushrooms were analyzed for total aerobic bacteria, yeast/mold, and coliform populations.

The microbiological results from the first experiment are shown in Table 6.1. The 30-second pulsed UV treatment at 8 cm distance demonstrated a greater than 1 log (90%) reduction for yeast and mold and aerobic bacterial populations. The UV treatment did not significantly affect coliform populations. On visual analysis, the color of the mushrooms as a result of the pulsed UV treatment was negatively impacted due to surface browning.

The microbiological results of the second experiment are depicted in Table 6.2. In general, the UV treatments of 2 or 4 s duration and 8 or 13 cm distance from the UV strobe resulted in 0.9 to 1.6 log reduction in total

TABLE 6.1
Microbiological Populations of Fresh *Agaricus bisporus* Mushrooms Treated with Pulsed UV Light (30 Second Application at a Distance of 8 cm from the UV Strobe)

Microbiological test	Control	Pulsed UV
Coliforms	6×10^2	3×10^2
Yeast and mold	$6 \times 10^3/2 \times 10^3$	$<1 \times 10^2$
Aerobic plate count	1×10^7	4.5×10^5

TABLE 6.2
Microbiological Populations of Fresh *Agaricus bisporus* Mushrooms Treated with Pulsed UV Light at Varying Treatment Times and Distances from the UV Strobe

Microbiological test	Control	2 s/8 cm	4 s/8 cm	2 s/13 cm	4 s/13 cm
Coliforms	5×10^2	4×10^2	$<1 \times 10^2$	$<1 \times 10^2$	$<1 \times 10^2$
Yeast and mold	1.6×10^4	2.1×10^4	3.4×10^3	1.3×10^3	1.2×10^3
Aerobic plate count	1.6×10^6	4×10^4	1.7×10^5	8×10^4	4×10^4

aerobic populations. Also, increasing treatment time improved reduction in microbial populations. However, all pulsed UV treatments had a negative impact on the color of the mushrooms due to surface browning.

The results from the pulsed UV study indicate little potential use for pulsed UV treatments with white strains but could be useful with crimini or portabella mushrooms, since the surface discoloration resulting from the treatments would most likely not be observable by consumers due to the inherent brown color associated with those types of mushrooms. Also, treatment with UV light could be useful to increase the vitamin D_2 content of mushrooms [15].

6.4 CONCLUSIONS

This chapter mainly describes the microbiology and microbial spoilage of the white button mushroom *Agaricus bisporus*. Cultural and postharvest practices to enhance the quality of fresh white button mushrooms have also been reviewed. Since the casing layer largely influences the microbiology of fresh mushrooms, it is possible that the microbiology of mushrooms grown using casing from the similar sources is largely similar. Cultural and postharvest practices that enhance agaricus quality may also be applicable to other commercial varieties such as crimini, portabella, shiitake, oyster, maaitake, and other exotic mushrooms varieties commonly seen in retail outlets.

While we have noted significant increases in yeast populations during postharvest storage of fresh mushrooms, the role played by yeast in the microbial spoilage of fresh mushrooms is largely unknown. Hence, as a starting point, the predominant yeast varieties in fresh mushrooms need to be characterized.

While it is our understanding that mushroom growers strive to maintain refrigeration temperatures during storage prior to shipping, temperature fluctuations and abuse can be commonly encountered during transportation and retailing. This seriously compromises the quality of fresh mushrooms. Hence it becomes essential for food transportation companies and retailers to understand the implications of postharvest storage conditions on the quality and shelf life of fresh mushrooms.

HACCP (hazard analysis critical control point) is increasingly being adopted by mushroom growers as a system to enhance the safety of fresh mushrooms. Studies at Penn State have been conducted to validate critical control points to ensure the safety of irrigation water [29,30] and the mushroom compost substrate [8]. Since heat pasteurization is not a practical method to disinfect the casing layer, research is needed to understand the microbial ecology of this material and thereby identify and validate other practical casing or mushroom disinfection procedures.

REFERENCES

1. Anonymous, Classification of fungi, The online msn-Encarta Encyclopedia, retrieved from http://encarta.msn.com.
2. Mau, J.-L., Miklus, M.B., and Beelman, R.B., The shelf life of *Agaricus* mushrooms, in *Shelf Life Studies of Foods and Beverages: Chemical, Biological, Physical and Nutritional Aspects*, Charalambous, G., Ed., Elsevier Science, 1993, pp. 255–288.
3. Chang, S.T., World production of edible and medicinal mushrooms in 1997 with emphasis on *Lentinus edodes* (Berk.) Sing. in China., *Int. J. Medicinal Mushrooms*, 1, 291–301, 1999.
4. Van Griensven, L.J.L.D., The edible and medicinal button mushroom (*Agaricus bisporus* (J. Lge) Imbach) and its relatives: present status, use, and future in commerce and research, *Int. J. Medicinal Mushrooms*, 3, 311–331, 2001.
5. Soler-Rivas, C., Arpin, N., Olivier, J.M., and Wichers, H.J., Discoloration and tyrosinase activity in *Agaricus bisporus* fruit bodies infected with various pathogens, *Mycol. Res.*, 104, 351–356, 2000.
6. Schisler, L.C., Biochemical and mycological aspects of mushroom composting, in *Penn State Mushroom Shortcourse Manual*, Pennsylvania State University, University Park, PA, 1982, pp. 3–10.
7. Beyer, D.M., *Basic Procedures for Agaricus Mushroom Growing*, Pennsylvania State University, University Park, PA, 2003.

8. Weil, J., *The Effect of Phase II Pasteurization on Populations of Select Human Pathogenic Bacteria in Mushroom Compost*, Pennsylvania State University, University Park, PA, 2004.

9. Wuest, P.J. and Bengston, G.D., *Penn State Handbook for Commercial Mushroom Growers*, Pennsylvania State University, University Park, PA, 1982.

10. Beelman, R.B., Effect of type of peat casing layer and the addition of calcium chloride to watering treatments on quality and shelf-life of fresh mushrooms, in *Developments in Crop Science: Cultivating Edible Fungi*, Wuest, P.J., Royse, D.J., and Beelman, R.B., Eds., Pennsylvania State University, University Park, PA, 1987, pp. 271–282.

11. Kukura, J.L., Beelman, R.B., Peiffer, M., and Walsh, R., Calcium chloride added to irrigation water of mushrooms (*Agaricus bisporus*) reduces postharvest browning, *J. Food Sci.*, 63, 454–457, 1998.

12. Mattila, P., Salo-Vaananen, P., Konko, K., Aro, H., and Jalava, Basic composition and amino acid contents of mushrooms cultivated in Finland, *J. Agri. Food Chem.*, 50, 6419–6422, 2002.

13. Frankhuizen, R. and Boekestein, A., Non-destructive determination of moisture content of fresh mushrooms by near infrared (NIR) spectroscopy, in *Mushroom Science XIV: Science and Cultivation of Edible Fungi, 2*, Elliott, T.J., Ed., A.A. Balkema, Rotterdam, 1995, pp. 755–763.

14. Ajlouni, S.O., Beelman, R.B., and Thompson, D.B., Influence of gamma radiation on quality characteristics, sugar content, and respiration rate of mushrooms during postharvest storage, in *Food Flavors, Ingredients and Composition*, Charalambous, G.E., Ed., Elsevier Science, 1993, pp. 103–121.

15. Mattila, P.A.-M.L., Ronkainen, R., Toivo, J., and Piironen, V., Sterol and vitamin D_2 contents in some wild and cultivated mushrooms, *Food Chem.*, 76, 293–298, 2002.

16. Beelman, R.B. and Edwards, C.G., Variability in the composition and nutritional value of the cultivated mushroom, *Agaricus bisporus*, *Mushroom News*, 37, 20–26, 1989.

17. USDA, USDA Nutrient Database for Standard Reference, Release 14, Nutrient Data Laboratory, 2001.

18. Spaulding, T. and Beelman, R.B., Survey evaluation of selenium and other minerals in *Agaricus* mushrooms commercially grown in the United States, *Mushroom News*, 51, 6–9, 2003.

19. Doores, S., Kramer, M., and Beelman, R., Evaluation and bacterial populations associated with fresh mushrooms (*Agaricus bisporus*), in *Proceeding of the International Symposium on Technical Aspects of Cultivating Edible Fungi,.* Wuest, P.J., Royse, D.J., and Beelman, R.B. Eds., Pennsylvania State University, University Park, PA, 1986, pp. 283–294.

20. Halami, P.M., Chandrashekar, A., and Joseph, R., Characterization of bacteriocinogenic strains of lactic acid bacteria in fowl and fish intestines and mushroom, *Food Biotech.*, 13, 121–136, 1999.

21. Anonymous, 1994.

22. Burton, K.S., Quality investigations into mushroom browning, *Mushroom J.*, 158, 68–70, 1986.

23. Beelman, R.B., Guthrie, B.D., and Royse, D.J., Influence of bacterial populations on postharvest deterioration of fresh mushrooms, *Mushroom Sci.*, 12, 655–665, 1989.

24. Beelman, R.B. and Simons, S.S., Influence of selected cultural factors on relative tyrosinase activity in cultivated mushrooms, in *Proceedings of the Second International Conference, Mushroom Biology and Mushroom Products*, Royse, D.E., Ed., Pennsylvania State University, University Park, PA, 1996, pp. 543–551.

25. Boekelheide, K., Graham, D.G., Mize, P.D., Anderson, C.W., and Jeffs, P.W., Synthesis of gamma-L-glutaminyl-[3,5-3H] 4-hydroxybenzene and the study of reactions catalyzed by the tyrosinase of *Agaricus bisporus.*, *J. Biol. Chem.*, 254, 12185–12191, 1979.

26. Soulier, L., Foret, V., and Arpin, N., Occurrence of agaritine and gamma-glutaminyl-4-hydroxybenzene (GHB) in the fructifying mycelium of *Agaricus bisporus*, *Mycol. Res.*, 97, 529–532, 1993.

27. Jolivet, S., Voiland, A., Pellon, G., and Arpin, N., Main factors involved in the browning of *Agaricus bisporus*, *Mushroom Sci.*, 14, 695–702, 1995.

28. Atkey, P.T. and Nichols, R., Surface structure of *Agaricus bisporus* by scanning electron microscopy, *Mushroom J.*, 129, 334–335, 1983.

29. Chikthimmah, N., McMillen, J., LaBorde, L.F., Demirci, A., and Beelman, R.B., Irrigation with electrolyzed oxidizing water to reduce bacterial populations on fresh mushrooms, in *Book of Abstracts*, Institute of Food Technologists Annual Meeting, Chicago, IL, 2003.

30. Chikthimmah, N., LaBorde, L.F., and Beelman, R., Irrigation with hydrogen peroxide and calcium chloride as a strategy to reduce bacterial populations on fresh mushrooms, in *Book of Abstracts*, Institute of Food Technologists Annual Meeting, Las Vegas, NV, 2004.

31. Olivier, J.M., Guillaumes, J., and Martin, D., Study of a bacterial disease of mushroom caps, in *4th International Conference in Plant Pathology and Bacteriology*, INRA, Angers, 1978, pp. 903–916.

32. Geels, F.P., Hesen, L.P.W., and van Griensven, L.J.L.D., Brown discolouration of mushrooms caused by *Pseudomonas agarici*, *J. Phytopath.*, 140, 249–259, 1994.

33. Wong, W.C., Fletcher, J.T., Unsworth, B.A., and Preece, T.F., A note on ginger blotch, a new bacterial disease of the cultivated mushrooms, *Agaricus bisporus*, *J. Appl. Bacteriol.*, 52, 43–48, 1982.

34. Rainey, P.B., Brodey, C.L., and Johnstone, K., Biology of *Pseudomonas tolaassii*, cause of brown blotch disease of the cultivated mushroom, in *Advances in Plant Pathology*, Academic Press, 1992, pp. 95–117.

35. Paine, S.G., Studies in bacteriosis. II. A brown blotch disease of cultivated mushrooms, *Ann. Appl. Biol.*, 5, 206–219, 1919.

36. Wong, W.C. and Preece, T.F., *Pseudomonas tolaasii* in cultivated mushroom (*Agaricus bisporus*) crops: numbers of the bacterium and symptom development on mushrooms grown in various environments after artificial inoculation, *J. Appl. Bacteriol.*, 53, 87–96, 1982.

37. Soler-Rivas, C., Arpin, N., Olivier, J.M., and Wichers, H.J., Activation of tyrosinase in *Agaricus bisporus* strains following infection by *Pseudomonas tolaasii* or treatment with a tolaasin-containing preparation, *Mycol. Res.*, 97, 529–532, 1997.

38. Wells, J.M., Sapers, G.M., Fett, W.F., Butterfield, J.E., Jones, J.B., Bouzar, H., and Miller, F.C., Postharvest discoloration of the cultivated mushroom *Agaricus bisporus* caused by *Pseudomonas tolaasii*, *P. 'reactans'*, and *P. 'gingeri'*, *Postharvest Pathol. Mycotoxins*, 86, 1098–1104, 1996.

39. Fett, W.F., Wells, J.M., Cescutti, O., and Wijey, C., Identification of exopolysaccharides produced by fluorescent pseudomonads associated with commercial mushroom (*Agaricus bisporus*) production, *Appl. Environ. Microbiol.*, 61, 513–517, 1995.

40. Beach, W.S., Control of mushroom disease and weed fungi, *Penn State College Agric. Bull.*, 351, 1–32, 1937.

41. Sharma, H.S.S., Kilpatrick, M., and Ward, F., Colonisation of phase II compost by biotypes of *Trichoderma harzianum* and their effect on mushroom yield and quality, *Appl. Microbiol. Biotech.*, 51, 572–578, 1999.

42. Mumpuni, A., Sharma, H.S.S., and Brown, A.E., Effect of metabolites produced by *Trichoderma harzianum* biotypes and *Agaricus bisporus* on their respective growth radii in culture, *Appl. Environ. Microbiol.*, 64, 5053–5056, 1998.

43. Visscher, H.R., Fructification of *Agaricus bisporus* (Lge.) Imb in relation to the relevant microflora in the casing soil, *Mushroom Sci.*, 10, 641–655, 1978.

44. Reddy, M.S. and Patrick, Z.A., Effect of bacteria associated with mushroom compost and casing materials on basidomata formation in *Agaricus bisporus*, *Can. J. Plant Path.*, 12, 236–242, 1990.

45. Hayes, W.A., Randle, P.E., and Last, F.T., The nature of the microbial stimulus affecting sporophore formation in *Agaricus bisporus* (Lange) Sing., *Ann. Appl. Biol.*, 64, 177–187, 1969.

46. Wong, W.C. and Preece, T., *Pseudomonas tolaasii* in mushroom crops: a note on primary and secondary sources of the bacterium on a commercial farm in England, *J. Appl. Bacteriol.*, 49, 305–314, 1980.

47. Barden, C.L., Beelman, R.B., Bartley, C.E., and Schisler, L.C., The effect of calcium chloride added to the irrigation water on quality and shelf life of harvested mushrooms, *J. Food Prot.*, 53, 759–762, 1990.

48. Solomon, J.M., Beelman, R.B., and Bartley, C.E., Addition of calcium chloride and stabilized chlorine dioxide to irrigation water to improve quality and shelf-life of *Agaricus bisporus*, in *Science and Cultivation of Edible Fungi*, Maher Balkema, Rotterdam, 1991, pp. 695–701.

49. Miklus, M.B. and Beelman, R.B., $CaCl_2$ treated irrigation water applied to mushroom crops (*Agaricus bisporus*) increases Ca concentration and improves postharvest quality and shelf life, *Mycologia*, 88, 403–409, 1996.

50. Geels, F.P., *Pseudomonas tolaasii* control by kasugamycin in cultivated mushrooms, *J. Appl. Bacteriol.*, 79, 38–42, 1995.

51. McArdle, F.J., Beelman, R.B., Gavin, A., Abdollahi, A., and Wuest, P.J., Production and processing factors influencing potential thermophilic spoilage of canned mushrooms, in *Mushroom Science X (Part II)*, Proceedings of the 10th International Congress on the Science and Cultivation of Edible Fungi, France, 1978.

52. Nair, N. and Fahy, G., Bacteria antagonistic to *Pseudomonas tolaasii* and their control of brown blotch of the cultivated mushroom *Agaricus bisporus*, *J. Appl. Bacteriol.*, 35, 439–442, 1972.

53. Pai, T., Effects of storage environmental conditions on weight loss, whiteness change, and microbial activity of mushrooms (*Agaricus bisporus*), *Agric. Chem. Biotech.*, 43, 161–164, 2000.

54. Tano, K., Arul, J., Doyon, G., and Castaigne, F., Atmospheric composition and quality of fresh mushrooms in modified atmosphere packages as affected by storage temperature abuse, *J. Food Sci.*, 64, 1073–1077, 1999.

55. Barber, W.H. and Summerfield, M.R.D., Environmental control of bacterial blotch on Pennsylvania shelf farms, *Mushroom News*, 38, 8–17, 1990.
56. Gormley, T.R. and MacCanna, C., Prepackaging and shelf life of mushrooms, *Irish J. Agric. Res.*, 6, 255–265, 1967.
57. Nichols, R. and Hammond, J.B.W., Storage of mushrooms in pre-packs: the effect of changes in carbon dioxide and oxygen on quality., *J. Sci. Food Agric.*, 24, 1371–1381, 1973.
58. Nichols, R. and Hammond, J.B.W., The relationship between respiration, atmosphere and quality in intact and perforated mushroom pre-packs, *J. Food Technol.*, 10, 427–435, 1975.
59. Burton, K.S. and Twyning, R.V., Extending mushroom storage-life by combining modified atmosphere packaging and cooling, *Acta Horticulturae*, 258, 565–571, 1989.
60. Martin, S.T. and Beelman, R.B., Growth and enterotoxin production of *Staphylococcus aureus* in fresh packaged mushrooms (*Agaricus bisporus*), *J. Food Prot.*, 59, 819–826, 1996.
61. Kautter, D.A., Lilly, T.J., and Lynt, R., Evaluation of the botulism hazard in fresh mushrooms wrapped in commercial poly vinyl chloride film, *J. Food Prot.*, 41, 120–121, 1978.
62. Burton, K.S., The quality and storage life of *Agaricus bisporus*, in *Mushroom Science XII*, Proceedings of the 12th International Congress on the Science and Cultivation of Edible Fungi, Braunschwey, Germany, 1989, pp. 683–688.
63. Burton, K.S. and Twyning, R.V., Extending mushroom storage-life by combining modified atmosphere packaging and cooling, *Acta Horticulturae*, 258, 565–571, 1989.
64. Hotchkiss, J.M. and Banco, M.J., Influence of new packaging technologies on the growth of microorganisms in produce, *J. Food Prot.*, 55, 815–820, 1992.
65. Lopez-Briones, G., Varoquaux, P., Bureau, G., and Pascat, B., Modified atmosphere packaging of common mushrooms, *Int. J. Food Sci. Technol.*, 28, 57–68, 1993.
66. Lopez-Briones, G., Varoquaux, P., Yves, B.J., Bureau, G., and Pascat, B., Storage of common mushroom under controlled atmospheres, *Int. J. Food Sci. Technol.*, 28, 57–68, 1992.
67. Roy, S., Anantheswaran, R.C., and Beelman, R.B., Sorbitol increases shelf life of fresh mushrooms stored in conventional packages, *J. Food Sci.*, 60, 1254–1259, 1995.
68. Roy, S., Anantheswaran, R.C., and Beelman, R.B., Modified atmosphere and modified humidity packaging of fresh mushrooms, *J. Food Sci.*, 61, 391–397, 1996.
69. Anderson, J.E., Beelman, R.B., and Doores, S., Persistence of serological and biological activities of Staphylococcus enterotoxin A in canned mushrooms, *J. Food Prot.*, 59, 1292–1299, 1996.
70. McConnell, A.L., Evaluation of Wash Treatments for the Improvement of Quality and Shelf-Life of Fresh Mushroom (*Agaricus bisporus*). M.S. thesis, Pennsylvania State University, University Park, PA, 1991.
71. Beelman, R.B. and Duncan, E.M., Preservation Composition and Methods for Mushrooms, U.S. Patent 5919507, 1999.
72. Sapers, G.M., Miller, R.L., Pilizota, V., and Kamp, F., Shelf life extension of fresh mushrooms (*Agaricus bisporus*) by application of hydrogen peroxide and browning inhibitors, *J. Food Sci.*, 66, 362–366, 2001.

73. Sapers, G.M., Miller, R.L., Miller, F.C., Cooke, P.H., and Choi, S.W., Enzymatic browning control in minimally processed mushrooms, *J. Food Sci.*, 59, 1042–1047, 1994.

74. Sapers, G.M., Miller, R.L., and Choi, S.-W., Mushroom discoloration: new process for improving shelf life and appearance, *Mushroom News*, 43, 7–13, 1995.

75. Sapers, G.M. and Simmons, G.F., Hydrogen peroxide disinfections of minimally processed fruits and vegetables, *Food Technol.*, 52, 48–52, 1998.

76. Sapers, G.M., Miller, R.L., Choi, S.W., and Cooke, P.H., Structure and composition of mushrooms as affected by hydrogen peroxide wash, *J. Food Sci.*, 59, 889–892, 1999.

77. Wozna, J., Effect of ionizing radiation on the microflora and storage life of fresh mushrooms (*Psalliota campestris*), *Roczniki Technologii i Chemii Zywnosci.*, 19, 89–101, 1970.

78. Kramer, M.E., Doores, S., and Beelman, R.B., The effect of radiation processing on mushroom (*Agaricus bisporus*) shelf life at two storage temperatures, in *Changing Food Technology, Institute of Food Technologists 4th Eastern Food Science and Technology Symposium*, Kroger, M. and Shapiro, R. Eds., Technomic, Lancaster, PA, 1987.

79. Beaulieu, M., Lacroix, M., Charbonneau, R., Laberge, I., and Gagnon, M., Effects of gamma irradiation dose rate on microbiological and physical quality of mushrooms (*Agaricus bisporus*), *Sciences des Aliments*, 12, 289–303, 1992.

80. Koorapati, A., Foley, D., Pilling, R., and Prakash, A., Electron beam irradiation preserves the quality of white button mushroom (*Agaricus bisporus*) slices, *J. Food Sci.*, 69, 25–29, 2004.

81. Beelman, R.B., Demirci, A., and Weil, D.A., unpublished data, 2003.

7 Spoilage of Juices and Beverages by *Alicyclobacillus* spp.

Mickey E. Parish

CONTENTS

7.1 INTRODUCTION

The versatility and diversity of the microbial world often lead to unique and valuable discoveries that expand our knowledge and yield advancements for humankind. Antibiotics, fermented foods, health and beauty aids, and other commonly utilized items in our everyday lives were discovered or produced based upon unusual physiological and phenotypical characteristics of microorganisms. In the past five decades the study of microbial extremophiles in geothermal sites with high temperatures and high acidity has advanced our understanding of spore-forming bacteria and led to the establishment of the new genera alicyclobacillus and sulfobacillus.

Until the mid-1980s, the presence of bacterial spore-formers in low pH foods was thought to be insignificant. The reigning dogma of the time declared that Gram-positive, sporogenous bacteria could not outgrow to any great extent at pH levels below 4.5. Therefore, the first report of spoilage in shelf-stable, low pH fruit juices by Gram-positive, spore-forming bacteria [1] was met with some skepticism. However, by the mid-1990s spoilage of acidic juice products by members of the recently named genus alicyclobacillus [2] was well established and the impact of this situation began to clarify. Fruit juice and juice-containing beverages, bottled tea, isotonic drinks, and other low pH, shelf-stable products were at risk of spoilage by a widespread thermotolerant-to-thermophilic organism that could survive pasteurization and hot-fill treatments, and was, surprisingly, acidophilic in nature. At present, roughly 20 years after the initial reports of spoilage in fruit juice, concerns by food processors about these thermoacidophilic, spore-forming bacteria remain strong, economic losses continue, and effective commercial protocols to address the situation are limited.

7.2 TAXONOMIC HISTORY

Ecological studies of extreme environments, such as geothermal hot springs, during the latter half of the 20th century amplified scientific awareness of unique, spore-forming, acidophilic bacteria with the ability to survive and reproduce at high temperatures (40 to 100°C). This awareness, coupled with characterization studies of various isolates, led to the discovery of one such group, now recognized as the genus alicyclobacillus. The alicyclobacilli have optimal growth conditions in warm to hot, acidic, low-nutrient environments and have been isolated from a variety of sources. These bacteria are rod-shaped, approximately 2 to 4 μm in length and <1 μm in width. Cells produce swollen, terminal to subterminal sporangia with refractile endospores (Figure 7.1) that are significantly heat resistant and capable of surviving typical pasteurization and thermal concentration conditions of juice/beverage manufacturing. On agar, colonies are usually a white to cream color, slightly raised, with smooth to irregular margins (Figure 7.2). Older, larger colonies take on a translucent character and may have slightly raised edges.

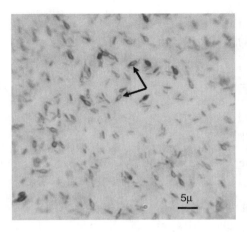

FIGURE 7.1 (Color insert follows page 594) Gram stain of *Alicyclobacillus acidoterrestris*. Note swollen sporangia at arrows. (Magnification ×1000.)

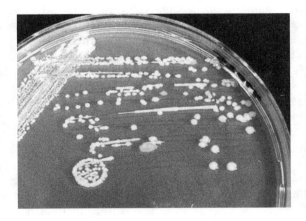

FIGURE 7.2 (Color insert follows page 594) Colonies of *Alicyclobacillus acidoterrestris* ATCC 49025 on Ali agar after 24 hours at 45°C. Key characteristics: white/cream color, smooth to irregular edges. Older, larger colonies may develop translucent quality with slightly raised margins.

In 1967 Uchino and Doi [3] reported the isolation of thermoacidophilic bacteria from geothermal hot springs in Japan. Similar organisms were subsequently isolated from other geographically distinct geothermal sources [4,5]. Brock and Darland [4] isolated thermophilic bacteria in about 300 hot springs of various pH levels in the western U.S., New Zealand, Japan, and Iceland. Microbial populations were found in "virtually every spring in the neutral and alkaline pH range" despite extreme temperatures up to 100°C. Bacterial isolates in acidic hot springs were temperature dependent and were not apparent at 90°C in springs with pH < 4.0 or at 70°C in springs with pH < 2.0. Darland and Brock [5] named the species *Bacillus acidocaldarius* to represent the unusual thermoacidophilic nature of these bacteria. Hippchen *et al.* [6] isolated thermoacidophilic bacilli from common soil samples, and Cerny *et al.* [1] reported the first account of a thermoacidophile, later identified as *B. acidoterrestris*, isolated from spoiled fruit juice. De Lucca *et al.* [7] reported the first isolation of *B. acidocaldarius* from another agricultural source, sugar refineries. Other research has established the presence of these organisms in high-fructose corn syrup (HFCS) commonly used as a sweetener in beverages (R. Worobo, personal communication).

De Rosa *et al.* [8,9] reported the presence of several forms of ω-cyclohexane fatty acids in the membranes of *B. acidocaldarius*. Poralla *et al.* [10] suggested a "cholesterol-like" function for hopanoids in the membranes of *B. acidocaldarius*. Poralla and König [11] isolated strains with ω-cycloheptane fatty acids that were later designated *B. cycloheptanicus* by Deinhard *et al.* [12]. Deinhard *et al.* [13] also reported on the new species, *B. acidoterrestris*, the thermoacidophile reportedly most involved in fruit juice spoilage.

Wisotzkey *et al.* [2] suggested reclassification of *B. acidocaldarius*, *B. acidoterrestris*, and *B. cycloheptanicus,* as species of alicyclobacillus based upon their thermoacidophilic phenotype and alicyclic fatty acids in the cellular membranes. In a review of classification schemes for endospore-forming bacteria, Berkeley and Ali [14] reported high homology (98.8%) between DNA from *A. acidocaldarius* and *A. acidoterrestris* and suggested that these might belong to one species rather than two. At the present time, the species remain separate due largely to their differences in optimum growth temperature and range.

Species of alicyclobacillus (Table 7.1) have been identified in a variety of samples from six continents, including Antarctica [15]. Differentiation of isolated strains from known species has been based largely upon genotypic and phenotypic characterization with phylogenetic and chemotaxonomic analyses. Albuquerque *et al.* [16] isolated thermoacidophilic strains from volcanic soil in the Azores Islands and subsequently designated one as *A. hesperidum*. Matsubara *et al.* [17] isolated a new species, *A. acidiphilus*, from acidic beverages based upon phylogenetic analysis of the 16S rRNA gene sequence and phenotypic differences related to spore morphology, growth temperatures, and acid production from carbon sources.

Goto *et al.* [18] isolated the species, *A. herbarius*, from hibiscus-flavored herbal tea. This strain contained ω-cycloheptane fatty acid as the major membrane lipid component and could be distinguished from other species by phylogenetic analysis of the 16S rDNA sequence. In 2003 Goto *et al.* [19] isolated a new species from fruit juice and named it *A. pomorum*. This novel species does not contain alicyclic fatty acids but clusters among the alicyclobacilli based upon phylogenetic analysis of the 16S rDNA sequence with a level of similarity between 92.5 and 95.5%. Tsuruoka *et al.* [20] isolated a collagenase positive strain that was closely related to species of the alicyclobacilli based on 16S rDNA sequence analysis but had less than 33%

TABLE 7.1
Current Species of Alicyclobacillus and Corresponding pH and Temperature Characteristics

Species	Source	pH		Temperature (°C)		Ref.
		Range	Optimum	Range	Optimum	
Alicyclobacillus acidocaldarius subsp. acidocaldarius	Acidic hot spring, U.S.	2.0–6.0	3.0–4.0	45–70	60–65	3, 13, 14
A. acidocaldarius subsp. rittmannii	Geothermal soil, Antarctica	2.5–5.0	4.0	45–70	63	10
A. acidiphilus	Acidic beverages, Japan	2.5–5.5	3.0	20–55	50	9
A. acidoterrestris	Garden soil and apple juice, Europe	2.5–5.5	3.5–4.0	42–53	45–50	2, 4, 8, 14
A. cycloheptanicus	Soil, Europe	2.5–5.5	3.5–4.0	42–53	45–50	5
A. herbarius	Hibiscus herbal tea, Japan	3.5–6.0	4.5–5.0	35–65	55–60	6
A. hesperidum	Solfataric soil, Azores	2.5–5.5	3.5–4.5	40–55	50	1
A. pomorum	Fruit juice, Japan	3.0–6.0	4.0–4.5	30–60	45–50	7
A. sendaiensis	Soil, Japan	2.5–6.5	5.5	40–65	55	12
A. vulcanalis	Geothermal pool, U.S.	2.0–6.0	4.0	35–65	55	11

DNA-to-DNA reassociation homology with known alicyclobacillus-type strains. This organism was deemed a new species and designated *A. sendaiensis* in recognition of the Japanese city where it was isolated. Simbahan *et al.* [21] reported a new species, *A. vulcanalis*, isolated from a geothermal pool at Coso Hot Springs in the Mojave Desert, California, U.S.

Rodgers *et al.* [22] described growth enhancement of *Acidiphilium cryptum* by two "alicyclobacillus-like" strains isolated from liquor (pH 2.1) of uranium leaching operations. These strains had 94 to 96% homology to the alicyclobacillus 16S rDNA sequence but were differentiated from other alicyclobacilli in their ability to oxidize iron. These strains might represent new alicyclobacillus species.

Taxonomy of alicyclobacillus and other thermoacidophilic genera/species remains fluid and continued additions and revisions to species and genus names should be expected for the foreseeable future. As more species-specific hypervariable DNA regions are discovered in bacterial genomes, it is likely that changes in taxonomy will continue.

7.3 PHYSIOLOGICAL AND PHENOTYPIC CHARACTERISTICS

7.3.1 DISTINGUISHING FEATURES

The alicyclobacilli are Gram-positive, spore-forming bacteria having thermoacidophilic characteristics. Key diagnostic traits of the alicyclobacillus phenotype include the presence of ω-alicyclic fatty acids in cell membranes, growth and endospore production under aerobic to facultative conditions at 45°C and pH 3.0, limited, if any, spore production under anaerobic conditions, and, to differentiate from sulfobacillus, no utilization of ferrous iron, sulfide, or sulfur as energy sources under any conditions [14].

It should be noted that some publications refer to the alicyclobacilli as Gram variable. This is due to their propensity to destain very rapidly during the Gram stain procedure, which yields a visual appearance that can be interpreted as indeterminate.

7.3.2 THERMOACIDOPHILIC GROWTH

The acidophilic and thermophilic nature of this genus is generally well characterized although specific growth conditions appear to be species and strain dependent [5,12,13]. Farrand *et al.* [23] conducted response surface analyses to investigate growth of *B. acidocaldarius* over a range of temperatures and pH, thereby establishing the extremes at which growth can occur, and confirming their thermoacidophilic character. Growth optima and upper/lower limits for temperature and pH are shown in Table 7.1. In general, *A. acidocaldarius* is the most tolerant of very high temperatures with optimum growth at 60 to 65°C. *A. herbarius* has a slightly lower optimum growth temperature (55 to 60°C) while all other species have temperature optima generally in the 45 to 50°C range. Most species have similar pH optima for growth (pH 3.0 to 4.5) although *A. herbarius* and *A. sendaiensis* optima are higher at 4.5 to 5.0 and 5.5, respectively.

Darland and Brock [5] described the strain-dependent nature for growth requirements for 14 strains of *A. acidocaldarius*. Two of the 14 strains had a minimum pH for growth of 3.0 while the remaining 12 were capable of growth down to pH 2.0. The lower temperature limit for all 14 strains was 45°C while the upper limit was 65 and 70°C for eight and six strains, respectively. Sinigaglia *et al.* [24] modeled the effects of temperature, water activity, and pH on germination of *A. acidoterrestris* spores. Their results support the thermoacidophilic characteristics of these organisms. Confirmation of their model by other laboratories is needed.

7.3.3 Alicyclic Fatty Acids in Membrane

In all alicyclobacillus species except one, alicyclic fatty acids are the major lipid components in the cell membrane. While this is a key distinguishing feature, it should be noted that all species of sulfobacillus, as well as *Curtobacterium pusillum* and *Propionibacterium cyclohexanicum*, also contain significant quantities of these fatty acids [25–27]. Additionally, one recently named species, *A. pomorum*, does not contain significant quantities of ω-alicyclic fatty acids [19].

Since there is no consistent correlation between the thermophilic/acidophilic phenotype and the presence of ω-alicyclic fatty acids, the purpose of these lipids in cell membranes is not clearly understood and requires further study.

7.4 THERMAL RESISTANCE CHARACTERISTICS

7.4.1 *D*- and *z*-values

Although juice spoilage from heat-resistant molds has been known for decades, the thermal resistance of spore-forming bacteria in low pH fruit juice and beverages was of little concern to fruit juice manufacturers prior to the 1990s. Despite the isolation of thermoacidophilic spore-formers from apple juice in 1982 [1], the importance of this genus to juice stability and consumer acceptance received little attention until reports of spoilage surfaced in the early 1990s from Europe and the U.S. Since that time, considerable research has been published illustrating the abundance of heat-resistant alicyclobacilli in low pH juices and beverages.

Kinetic parameters of *A. acidoterrestris* have been elucidated by several research teams using various techniques with different controlled conditions and heating menstrua. Early research by Splittstoesser *et al.* [28,29] produced *D*- and *z*-values that supported previous empirical observations of spore survival in thermally treated juices. *D*-values (the time necessary at a specific temperature to reduce the overall microbial population by 90%)

reported by these researchers in apple and grape juices ranged from almost 60 minutes at 85°C to between 2 and 3 minutes at 95°C (Table 7.2). Since typical commercial thermal process conditions are in the range 85 to 100°C for 10 to 30 seconds, Splittstoesser's results demonstrated conclusively that spores of the alicyclobacilli could survive traditional pasteurization and hot-fill processes to cause spoilage in shelf-stable products.

Similar kinetic results for *A. acidoterrestris* have been reported in other juices, beverages, model broth systems, and distilled water by various laboratories (Table 7.2). Reported D-values range from 81 minutes at 88°C to about 1 second at 125°C. Although specific D- and z-values from the various studies differ, there are general similarities in magnitude. Average D-values from Table 7.2 are 47 minutes (81 to 85°C), 24 minutes (86 to 90°C), 17 minutes (91 to 95°C), 7 minutes (96 to 100°C), 3.8 minutes (at 110°C), and 0.025 minutes (at 125°C). This is illustrated in Figure 7.3, which shows an overall thermal death time curve for data points in Table 7.2.

A z-value is the temperature increase needed to reduce by 1-log cycle the time necessary to produce a 90% reduction in cell populations. This is a valuable tool when attempting to alter commercial processing conditions to either decrease the time needed to achieve product safety and stability, or decrease the temperature to enhance product quality. In essence, when process time is decreased, the z-value is used to determine the new target processing temperature. Likewise, if a lower temperature is desired to improve product flavor, the z-value provides the increased time needed to achieve the same product safety and stability as with the previous process conditions.

Except for one study that will be discussed below, z-values in Table 7.2 for *A. acidoterrestris* in juices, beverages, model systems, and water are relatively similar with an average value of $8.3 \pm 1.9°C$. This means that on average the time needed to inactivate a specific population of spores will decrease by a factor of 10 if the pasteurization temperature is increased by 8.3°C.

7.4.2 FACTORS AFFECTING THERMAL RESISTANCE

It is important to remember that the inactivation kinetics of wild-type alicyclobacillus strains may differ from those obtained from laboratory strains that have been subjected to long-term cultivation. Many factors affecting inactivation kinetics for other microorganisms have been studied and may provide insights into thermal inactivation of the alicyclobacilli. Factors include, among others, culture and inoculum incubation temperatures, sporulation temperature, nutrient composition and pH of the growth medium, nutrient composition and pH of the heating menstruum (e.g., test juice), presence or absence of divalent cations, storage temperature of inoculum stock, osmolarity of test juice matrix, or presence of antimicrobial compounds. It is also well documented that specific strains within a species can vary considerably in D- and z-values.

While it might be assumed that various factors would affect thermal resistance of the alicyclobacilli in a similar manner, studies on this topic

TABLE 7.2
D- and z-Values Reported for *Alicyclobacillus acidoterrestris*

Strain	D (minutes)	z (°C)	Heating menstruum	Ref.
Alicyclobacillus acidoterrestris	$D_{85°C}$ 56 $D_{90°C}$ 23 $D_{95°C}$ 2.8	7.7	Apple juice (11.4°B, pH 3.5)	28
Alicyclobacillus acidoterrestris	$D_{85°C}$ 57 $D_{90°C}$ 16 $D_{95°C}$ 2.4	7.2	Grape juice (15.8°B, pH 3.3)	28
Alicyclobacillus acidoterrestris	$D_{88°C}$ 11 $D_{91°C}$ 3.8 $D_{95°C}$ 1.0	7.2	Berry juice	39
Alicyclobacillus acidoterrestris *Alicyclobacillus acidoterrestris*	$D_{95°C}$ 5.3 $D_{95°C}$ 2.2–3.3	Not reported 6.4–7.5	Orange juice drink (5.3°B, pH 4.1) Clear apple juice	43 63
Alicyclobacillus acidoterrestris strain VF	$D_{91°C}$ 31.3 $D_{97°C}$ 7.9	10.0	Malic acid (0.4%) model broth (12°B, pH 3.1)	33
Alicyclobacillus acidoterrestris strain VF	$D_{88°C}$ 81.2 $D_{100°C}$ 0.8	5.9	Malic acid (0.4%) model broth (12°B, pH 3.4)	33
Alicyclobacillus acidoterrestris strain VF	$D_{91°C}$ 54.3 $D_{97°C}$ 8.8	7.7	Malic acid (0.4%) model broth (12°B, pH 3.7)	33
Alicyclobacillus acidoterrestris strain VF	$D_{91°C}$ 46.1 $D_{97°C}$ 8.2	8.5	Citric acid (0.58%) model broth (12°B, pH 3.1)	33
Alicyclobacillus acidoterrestris strain VF	$D_{91°C}$ 57.9 $D_{97°C}$ 10.8	8.2	Citric acid (0.58%) model broth (12°B, pH 3.7)	33
Alicyclobacillus acidoterrestris strain VF	$D_{91°C}$ 49.1 $D_{97°C}$ 8.4	7.8	Tartaric acid (0.45%) model broth (12°B, pH 3.1)	33

(Continued)

TABLE 7.2
Continued

Strain	D (minutes)	z (°C)	Heating menstruum	Ref.
Alicyclobacillus acidoterrestris strain VF	$D_{91°C}$ 69.5 $D_{97°C}$ 10.0	7.1	Tartaric acid (0.45%) model broth (12°B, pH 3.7)	33
Alicyclobacillus acidoterrestris	$D_{88°C}$ 24 $D_{95°C}$ 2.7	7.1	McIlvaine buffer, pH 3.0	64
Alicyclobacillus acidoterrestris	$D_{88°C}$ 29 $D_{95°C}$ 2.7	6.6	McIlvaine buffer, pH 5.0	64
Alicyclobacillus acidoterrestris	$D_{88°C}$ 26 $D_{95°C}$ 2.3	6.4	McIlvaine buffer, pH 8.0	64
Alicyclobacillus acidoterrestris	$D_{85°C}$ 50 $D_{90°C}$ 17 $D_{95°C}$ 2.7	7.9	Orange juice	65
Alicyclobacillus acidoterrestris	$D_{80°C}$ 41 $D_{90°C}$ 7.4 $D_{95°C}$ 2.3	12.2	Apple juice (pH 3.5)	34
Alicyclobacillus acidoterrestris	$D_{80°C}$ 24 $D_{90°C}$ 4.6 $D_{95°C}$ 2.0	13.8	Apple juice (pH 3.5) with 50 IU nisin/ml	34
Alicyclobacillus acidoterrestris	$D_{80°C}$ 37.9 $D_{90°C}$ 6.0 $D_{95°C}$ 1.8	11.6	Grapefruit juice (pH 3.4)	34
Alicyclobacillus acidoterrestris	$D_{80°C}$ 54.3 $D_{90°C}$ 10.3 $D_{95°C}$ 3.6	12.9	Orange juice (pH 3.9)	34

Alicyclobacillus acidoterrestris NCIMB 13137	$D_{85°C}$ 17.5 $D_{97°C}$ 0.6	9.0	Cupuaçu extract (11.3°B, pH 3.6)	66
Alicyclobacillus acidoterrestris NCIMB 13137	$D_{85°C}$ 65.6 $D_{91°C}$ 11.9	7.8	Orange juice (11.7°B, pH 3.5)	66
Alicyclobacillus acidoterrestris NCIMB 13137	$D_{91°C}$ 3.8 $D_{91°C}$ 24.1	Not reported	Blackcurrant concentrate (26.1°B, pH 2.5) Blackcurrant concentrate (58.5°B, pH 2.5)	66
Alicyclobacillus acidoterrestris STCC 5137	$D_{110°C}$ 3.9 $D_{125°C}$ 0.03	7	Orange juice	35
Alicyclobacillus acidoterrestris STCC 5137	$D_{110°C}$ 3.7 $D_{125°C}$ 0.02	7	Distilled water	35
Alicyclobacillus acidoterrestris	$D_{90°C}$ 20.8 $D_{90°C}$ 19.3 $D_{90°C}$ 15.5 $D_{90°C}$ 14.8	Not reported	Clear apple drink (no nisin added) With 50 IU nisin/ml With 100 IU nisin/ml With 200 IU nisin/ml	35
Alicyclobacillus acidoterrestris NCIMB 13137	$D_{95°C}$ 5.3 $D_{95°C}$ 3.8	7.8 (no storage) 29 (spores stored at −18°C)	Cupuaçu nectar (18°B, pH 3.2)	32

Note: *D*-value is the time at a specific temperature necessary to reduce the microbial population 1 log cycle (90%); *z*-value is the temperature increase necessary to reduce by 1 log cycle the time needed to achieve a 1 log reduction.

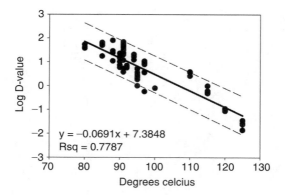

FIGURE 7.3 Compiled literature values for $\log D$ vs. temperature, $n = 98$. Dashed lines represent the 95% mean prediction confidence boundary.

are limited and generally not comprehensive in scope. However, there are interesting results to consider. Two studies [30,31] determined that the presence of divalent cations (Ca^{2+}, Mg^{2+}, Ba^{2+}, Mn^{2+}, and Sr^{2+}) in the sporulation medium did not affect the heat resistance of $A.$ $acidoterrestris$. They also noted that techniques to demineralize and remineralize spores with calcium had no effect on their calcium content or thermal inactivation even though these techniques were capable of decreasing the thermal resistance of $Bacillus$ $subtilis$ spores. Vieira et $al.$ [32] concluded that extended cold storage of spores at $-18°C$ substantially increased z-values. Freshly prepared spores had a z-value of $7.8°C$ while spores stored at $-18°C$ for 4 and 8 months had z-values of 22 and $29°C$, respectively. Since juice concentrates are typically stored at -5 to $-10°C$ in large cold-stored tanks (380,000 L or more per tank) or $-18°C$ in 55 gal drums, establishment of time/temperature processing parameters may require knowledge of the concentrate storage history prior to use by juice and beverage manufacturers.

Of three $A.$ $acidoterrestris$ strains tested by Pontius et $al.$ [33], one was less heat resistant than the other two at both 91 and $97°C$ in a model juice system. However, the type of organic acid (citric, malic, or tartaric) in the model system had no significant effect ($P > 0.05$) on results. These researchers determined that results were significantly ($P > 0.05$) influenced by pH at $91°C$ but not at $97°C$. Their results were similar to those of Komitopoulou et $al.$ [34] who showed a decrease in D-values when the pH was changed from 4 to 3 in grapefruit juice. This effect was more apparent at $80°C$ (32 minutes at pH 3.0 and 52 minutes at pH 4.0) than at $95°C$ (1.5 minutes at pH 3.0 and 1.7 minutes at pH 4.0).

Under test conditions used by Palop et $al.$ [35], D-values at 110 to $125°C$ were not different for spores suspended in orange juice, distilled water, or citrate–phosphate buffers at pH 4 and 7. These researchers noted that a sporulation temperature of $65°C$ yielded spores with increased heat resistance compared to those produced at $45°C$. A linear correlation was noted between sporulation temperature and $\log D_{110°C}$. No differences in z-values were observed in this study among spores in the juice, water, and buffer test matrices.

Further research into processes and relevant parameters for control of the alicyclobacilli is warranted. While it is obvious that the quality of most juice and beverage products would not tolerate typical thermal conditions needed to control these organisms, alternative processes or combined processes may provide a solution.

7.4.3 OTHER CONTROL MEASURES

Processors of 100% juices often choose to rely on thermal treatments as the only hurdle for the production of a safe and stable juice. For those processors selling refrigerated, pasteurized products, there is little cause for concern since the alicyclobacilli do not grow under common refrigeration conditions of 4 to 8°C. These organisms cause problems in shelf-stable, low pH products. In some of these products, sodium benzoate, sorbate or some other preservative may be added to inhibit growth of potential contaminants, but their effects on the alicyclobacilli are not clearly understood. While sorbates have been shown to inhibit germination of spores in some food products [36], this preservative has not been the subject of public, refereed research in relation to *Alicyclobacillus* spp.

The addition of the antimicrobial peptide nisin to juice was found to enhance the lethality of thermal treatment [34,37]. These references report use of a commercial nisin compound containing 2.5% active nisin with 10^6 international units (IU) per gram. Growth of *A. acidoterrestris* (Z CRA 7182) at 25°C was controlled in apple, grapefruit, and orange juices by as little as 5 IU/ml nisin [34]. In the same study, $D_{80°C}$ of spores in apple juice decreased from 41 minutes without nisin to 24 minutes with 50 IU nisin/ml. As with pH effects, the effect on D-values was more obvious at the lowest (80°C) process temperature tested than at the highest (95°C). Although citrus juices are known to be susceptible to spoilage by the alicyclobacilli, it is interesting to note a greater degree of stability for citrus juice in this study when compared to apple juice. Natural oils in citrus juices are known to have antimicrobial activity, which might partially explain this observation.

When stored at 44°C, growth was observed in apple juice containing 50 IU nisin/ml but was not seen at the 100 IU/ml level over a 6-day period [34]. In contrast, Yamazaki *et al.* [37] reported substantial growth of *A. acidoterrestris* (AB-5) over a 12-day storage trial in a clear apple drink with up to 600 IU nisin/ml. In that study, $D_{90°C}$ values were reduced from 20.8 to 14.8 with the addition of 200 IU nisin/ml of apple drink. Yamazaki *et al.* [30] reported that lysozyme increased thermal sensitivity of *A. acidoterrestris* spores in a citrate buffer (pH 4.0).

Alternative processes other than thermal treatment could provide another avenue of research to investigate alicyclobacillus inactivation. Lee *et al.* [38] reported reductions greater than 5 log in spore populations from the application of high-pressure processing (HPP) with thermal treatments. Although HPP potentially reduces the heat resistance of bacterial spores, high pressure alone would not provide adequate kill to prevent spoilage. Other alternative processes, such as pulsed electric field or ultraviolet light treatment, are generally not very effective against bacterial spores. At present, there are no adequate alternatives to stringent sanitation operating procedures coupled with current good manufacturing practices and appropriate raw material specifications for control of alicyclobacilli in beverages. This makes the establishment of specifications difficult. Many customers of juice/beverage ingredients desire to set a zero tolerance for the presence of the alicyclobacilli in concentrates and sweeteners; however, this standard may be unachievable with current processing and sanitizing technology.

7.5 INDUSTRIAL IMPORTANCE

7.5.1 EMERGENCE AS SPOILAGE ORGANISMS

Food manufacturers continually investigate innovations in product development, processing aids, and processing equipment to meet changing consumer demands in order to maintain adequate profit margins. Changes in products and manufacturing procedures frequently result in unique quality and stability challenges. For example, in the 1980s and 1990s the use of oxygen barrier films in gable-top cartons allowed orange juice packers to increase refrigerated shelf life of orange juice from 35 to 60+ days. This increased shelf life provided enough time for fungal propagules within the paperboard matrix of the carton to germinate and outgrow into the product thereby increasing consumer complaints of mold spoilage and leading to use of a different package design.

Between the mid-1980s and 1990s juice and beverage manufacturers became aware of spoilage in low pH, shelf-stable products by spore-forming bacteria. Processors of shelf-stable apple juices noticed occasional development of strong off-aromas in finished product after a few weeks or months of storage. Also, a few citrus juice processors reported the slow growth of a spore-forming bacterium in aseptic juice samples thermally abused at warm temperatures during quality control testing. Since the final product was marketed as a refrigerated product, these spore-formers did not result in spoilage of the finished product.

7.5.2 TYPES OF SPOILAGE

The earliest documented fruit juice spoilage event related to alicyclobacillus was reported by Cerny *et al.* [1] and involved an antiseptic off-aroma in apple juice. The causative organism was described as "related to" *Bacillus acidocaldarius* but would likely be classified today as *A. acidoterrestris*. Splittstoesser *et al.* [28] reported growth characteristics of two thermoaciduric spore-forming isolates that resulted in strong off-aromas in finished apple juice products. Other reports also documented the isolation of thermoacido-philes from juices, other low pH beverages, soil, fruit surfaces, and recycled water [39–46]. Based on reports such as these, investigations were conducted into the type of spoilage observed and the necessary conditions for such spoilage to become apparent.

Prior to the recognition of alicyclobacillus as a spoilage agent of juices and beverages, microbial stability problems were limited to yeasts, lactic acid bacteria, acetic acid bacteria, or heat-resistant molds. Typical juice spoilage involves production of CO_2 and off-flavors by fermentative organisms resulting in bulging or exploding containers. Spoilage events caused by the alicyclobacilli are different. Since there is no CO_2 to indicate microbial growth, the powerful antiseptic and medicinal aromas may not be detected until consumers open the package. The only visual clue to the presence of this organism is the possible presence of a slight haze in clear liquids such as clarified apple juice. Therefore, food manufacturers must implement new quality assurance measures to detect the presence of these spoilage organisms.

The alicyclobacilli produce at least three odiferous phenolic compounds with low detection thresholds. It was established in early research that guaiacol is the principal off-aroma produced by the alicyclobacilli although 2,6-dibromophenol (DBP) and 2,6-dichlorophenol (DCP) were also isolated from products with large populations of *Alicyclobacillus* spp. Pettipher *et al.* [42] found detectable levels of guaiacol in fruit juices, including orange and apple juices. Baumgart *et al.* [43] and Borlinghaus and Engel [47] reported the presence of DBP in tea and juice products. Orr *et al.* [48] reported that guaiacol content in apple juice was not correlated with numbers of cells, and that the best estimate threshold for guaiacol added to apple juice was 2.23 ppb. Jensen and Whitfield [49] described the production of DCP by the alicyclobacilli. Göçmen *et al.* [50] reported production of guaiacol, DBP, and DCP by five strains of alicyclobacilli in orange juice.

7.5.3 SANITATION

For some manufacturers it is not feasible to consider the addition of preservatives, such as nisin discussed above, to juice and beverage products as a microbial control measure due to regulatory reasons. In these cases, processors must reduce alicyclobacillus in the product through intensive sanitation of the processing environment, and through strict specifications and current good manufacturing practices for the product ingredients. Greater emphasis on appropriate use of cleaners and sanitizers to control these organisms on food contact surfaces should reduce the potential for product contamination.

It is also important for processors to examine carefully all potential sources of contamination since these organisms are widespread in nature and are common inhabitants of soil. As thermoacidophiles, alicyclobacilli would be expected to proliferate under warm acidic conditions in a manufacturing facility. These conditions can occur in juice/beverage processing plants, especially during the warm summer months. (Empirical evidence of this exists in the large amount of spoilage observed in Europe during the unusually hot summers of 1994 and 1995.) Some facilities that produce concentrated fruit juices reclaim water that is released from the juice during the evaporation process, and utilize it for cleaning purposes within the plant. Since the recovered water is warm to hot and slightly acidic, it provides an appropriate environment for proliferation of the alicyclobacilli.

Wisse and Parish [44] and Eguchi *et al.* [45] reported the presence of significant alicyclobacillus populations in "condensate water" recovered for use in citrus concentrate facilities. Not only was the water used to wash equipment and incoming fruit, in some cases pulp extracted from the juice was washed with this water to recover sugar solids. This wash water was then diverted to the evaporator and mixed with juice just prior to the concentration process thereby guaranteeing a constant source of contamination in juice concentrates produced by those facilities. Since that discovery, efforts to address the alicyclobacillus issue at citrus processing facilities have emphasized the cleanliness of the condensate water recovery system and treatment of water used in the facility.

Considering the substantial economic losses sustained due to growth of the alicyclobacilli in juices and other low pH beverages, there are few refereed publications discussing cleaning and sanitation requirements to control these organisms. Orr and Beuchat [51] exposed five strains of *A. acidoterrestris* spores to sodium hypochlorite, acidified sodium chlorite, trisodium phosphate, hydrogen peroxide, and Tsunami® sanitizer (Ecolab Inc., St. Paul, MN), for 10 minutes at 23°C. A 5 log reduction in spore population was observed after treatment with 1000 ppm sodium hypochlorite or 4% hydrogen peroxide. At lower concentrations, significant ($P \leq 0.05$) reductions of 2 log, 0.4 log, and 0.1 log were observed with 200 ppm hypochlorite, 500 ppm acidified sodium chlorite, and 0.2% hydrogen peroxide, respectively. Based on these results, these researchers continued with practical experiments to determine chemical effectiveness against *A. acidoterrestris* on apple surfaces. Reductions in spore populations after 1 minute exposure to 500 ppm hypochlorite or 1200 ppm acidified sodium chlorite were statistically significant ($P \leq 0.05$) but did not inactivate spores more than 1 log as compared to 5 and 2.5 log in the earlier direct challenge experiments.

7.6 DETECTION AND IDENTIFICATION

7.6.1 CONTROVERSY

Methods for accurate and sensitive detection, isolation, identification, and quantification of the alicyclobacilli in foods have developed slowly and remain somewhat controversial. In general, detection of these organisms in juices and beverages has relied upon their thermoacidophilic character and their ability to produce odiferous phenolic compounds. No standard method detection/recovery protocols have yet been developed that are universally accepted.

Research has been conducted on the use of DNA and PCR-based technologies to detect the alicyclobacilli in food samples. Specific primers for detection have been developed [52] and a real-time PCR-based detection method has been developed [53]. Rapid test kits for detection or identification of *Alicyclobacillus* spp. are commercially available from Vermicon AG, MicroBio Corporation, and BioSys. As with any rapid test methods, customers should verify and validate these products for their usefulness in the specific commodity or environmental sample of interest. Customers should further question the manufacturers for information on false negative and false positive tests in order to make an informed decision on the usefulness of such products for a particular application.

7.6.2 MEDIA

There are two general types of media commonly used for the isolation or detection of the alicyclobacilli. The first type contains four specific mineral salts (ammonium sulfate, magnesium sulfate, calcium chloride, potassium phosphate) with minimal amounts of carbon and/or nitrogen sources (Table 7.3). Most of these media are similar in composition and are based upon early reports by Uchino and Doi [3] and Darland and Brock [5]. Modifications of these media have been published by Farrand *et al.* [23], Deinhard *et al.* [13], Yamazaki *et al.* [40], Wisse and Parish [44], and the Internationale Fruchtsaft-Union (IFU) [54]. Some of these media require addition of a trace mineral solution although the need for the full complement of minerals in a recovery medium is unclear.

The second class of media does not contain a complement of minerals and may be based on more traditional nutrient media with reduced pH. Examples are acidified versions of potato dextrose agar, orange serum agar, and plate count agar. YSG agar, recently advanced by Japanese organizations as part of a universal method for detection of the alicyclobacilli in juices, contains only yeast extract, soluble starch, and glucose. One particular nonmineral-containing medium, K agar, was developed specifically for the isolation of alicyclobacillus [55,56]. While nonmineral media may provide adequate recovery of alicyclobacilli in certain situations, some research suggests that mineral-containing media are more effective for enumeration or for situations where detection of small cell populations via enrichment is required. A recent study of more than 1500 environmental samples indicates that a minimal mineral medium, Ali agar [44], recovered significantly ($\alpha = 0.05$) more alicyclobacillus strains than two nonmineral media, acidified potato dextrose agar and K agar, under the conditions of that study (Parish, unpublished data). Continued research to compare isolation media or investigate alternative media is warranted.

The official method of the IFU for detection of alicyclobacillus in fruit juices [54] depends upon the organisms' thermoacidophilic trait. A juice sample

TABLE 7.3
Composition of Minimal Mineral Media for Isolation of the Alicyclobacilli[a]

Ingredient	Ref. 3	Ref. 5	Ref. 1	Refs. 13, 23	Ref. 40	Ref. 44	Refs. 45, 54
$CaCl_2 \cdot 2H_2O$	0.1 g	0.25 g	0.25 g	0.25 g	0.25 g	0.25 g	0.25 g
$FeSO_4 \cdot 7H_2O$	0.025 g	—	0.28 mg	—	—	—	—
KH_2PO_4	1.0 g	3.0 g	3.0 g	3.0 g	0.6 g	3.0 g	3.0 g
$MgSO_4 \cdot 7H_2O$	0.5 g	0.5 g	0.5 g	0.5 g	0.5 g	0.5 g	0.5 g
$MnCl_2 \cdot 4H_2O$	—	—	1.25 mg	—	—	—	—
$MnSO_4 \cdot 4H_2O$	0.025 g	—	—	—	—	—	—
NaCl	1.0 g	—	—	—	—	—	—
$Na_2HPO_4 \cdot 12H_2O$	2.5 g	—	—	—	—	—	—
$(NH_4)_2SO_4$	—	0.2 g	0.2 g	0.2 g	0.2 g	0.2 g	0.2 g
NH_4Cl	1.0 g	—	—	—	—	—	—
$ZnSO_4 \cdot 7H_2O$	—	—	0.4 mg	—	—	—	—
Glucose	5.0 g	0.1–1.0 g	1.0 g	5.0 g	1.0 g	1.0 g	5.0 g
Glycerol	—	0.1–1.0 g	—	—	—	—	—
Peptone	—	—	—	—	—	—	—
Ribose	—	0.1–1.0 g	—	—	—	—	—
Soluble starch	—	—	2.0 g	—	—	—	—
Tween 80	—	—	—	—	—	2.0 g	—
Yeast extract	—	0.1 g	2.0 g	1.0 g	1.0 g	2.0 g	2.0 g
Deionized water	1000 ml	1000 ml	1000 ml	1000 ml	1000 ml	1000 ml	1000 ml
Trace elements[b]	—	—	—	1 ml	—	—	1 ml
Biotin	10 μg	—	—	—	—	—	—
Agar	—	—	—	15 g	—	15–20 g	15–20 g

[a] Media are typically acidified with organic or mineral acids to pH 3.5 to 4.0 after autoclaving.
[b] Solution of 0.66 g/l $CaCl_2 \cdot 2H_2O$, 0.30 g/l $Na_2MoO_4 \cdot 2H_2O$, 0.18 g/l $ZnSO_4 \cdot 7H_2O$, 0.18 g/l $CoCl_2 \cdot 6H_2O$, 0.16 g/l $CuSO_4 \cdot 5H_2O$, 0.15 g/l $MnSO_4 \cdot 4H_2O$, 0.10 g/l H_3BO_3.

is diluted, heat shocked for 10 minutes at 80°C, incubated at 45°C for 7 days, and plated onto an acidic, low-nutrient medium (BAT agar reported by Eguchi *et al.* [45]; see Table 7.3). Growth at 45°C after 5 days is microscopically examined to exclude yeasts, followed by restreaking onto BAT and PCA. After incubation at 45°C, strains that grow on BAT but not PCA are considered probable alicyclobacillus although "further bio- and genotyping of suspect alicyclobacillus" is encouraged to ensure the identification.

The most recent edition of the *Compendium of Methods for the Microbiological Examination of Foods* describes a direct plating and enrichment procedure [57]. While the protocols are fundamentally sound, the use of a mineral-containing medium in addition to, or in place of, K agar is suggested. Additionally, direct plating of concentrate is not recommended since higher Brix levels may inhibit colony formation. In the enrichment protocol, the heat shock at 90°C is too severe and should be reduced to a range of 70 to 80°C.

7.6.3 HEAT SHOCK CONDITIONS

There are two purposes for using a heat shock during recovery protocols. First, it eliminates vegetative cells (including vegetative alicyclobacillus) to allow germination of spores without competition from other organisms. Second, it activates spores to germinate and outgrow although the actual increase in spore recovery is not well established. Unpublished results by Parish show as much as a 400% increase in recovery of alicyclobacillus from orange juice by use of a mild heat shock. However, heat shock may not be necessary in finished, shelf-stable, low pH products that do not contain competitive microflora. Further research is warranted to determine the effect of heat shocks on spore viability.

Heat shock regimes should be of appropriate duration to eliminate competing microflora but not so stringent as to inactivate spores. Baumgart *et al.* [43] suggest 20 minutes at 70°C as an effective heat shock regime. Parish and Goodrich [58] investigated various times and temperatures from 60°C for 30 minutes to 90°C for 5 minutes, but ultimately recommended 75°C for 10 minutes to recover alicyclobacilli from diluted orange juice. They reported optimal recovery over a range of times at specific temperatures: 10 to 30 minutes at 60 and 65°C; 5 to 25 minutes at 70 and 75°C; and up to 5 minutes at 80 and 85°C. Recovery at 90°C was inadequate compared to results at the other temperatures. Additionally, the heating menstruum may affect recovery. Several studies show that percent recovery increases as samples are diluted to lower sugar content.

7.6.4 ENUMERATION

Enumeration of the cell population may provide important information related to spoiled packaged products having obvious medicinal off-aroma. Basic plating techniques for enumeration are spread plating, including spiral plates and pour plating [59,60]. In the case of clear liquids, a filtration technique may be used to concentrate cells from a large sample volume, which increases test sensitivity. For liquids containing particulate matter, such as cloudy juices, a most probable number (MPN) technique is appropriate [60]. Enumeration by the MPN technique would be conducted using an appropriate broth medium such as filter-sterilized juice, or a low pH minimal nutrient broth such as Ali Broth [44] or BAT Broth [45,54].

7.6.5 DETECTION BY ENRICHMENT

Enrichment of samples is of considerable importance for the recovery of small numbers of alicyclobacillus from concentrated juices, purees, and nectars. Although the general concept of enrichment is consistent from one study to the next, conditions for assaying different samples vary and are not applicable to all situations. Protocols typically involve incubation of a sample that is diluted in either a minimal broth medium that contains minerals with nutrients, or plain water. The level of dilution varies from single-strength to as little as $2°$ Brix. Sample sizes typically range from 1 to 100 ml of concentrate, puree, or nectar.

After dilution, samples are heat shocked followed by incubation at 40 to $55°C$ for 3 to 7 days. A yeast/mold inhibitor may be added in cases where a heat shock is not used or is inadequate. At the end of incubation, samples are streaked onto appropriate media and plates are incubated aerobically at 45 to $50°C$ for several days. Development of colonies suggests the possible presence of alicyclobacillus although confirmation requires further analysis.

7.6.6 IDENTIFICATION AND CONFIRMATION

Colonies produced by enrichment or direct plating on low pH media must be subjected to appropriate analytical techniques to confirm their identification as alicyclobacillus. Deciding which analytical techniques are considered appropriate is the basis of controversy that has yet to be adequately addressed. A microscopic examination is necessary to confirm that the isolate is not fermentative yeast or another type of acidophilic organism, and that it is a Gram-positive spore-forming bacterium. The sporogenous nature of the isolate may require further plating and incubation on other media followed by spore staining and microscopic observations.

Confirmation activities include growth on low pH media at elevated temperatures (45 to $55°C$) with concurrent lack of growth on media of neutral pH, such as traditional plate count agar or nonacidified minimal media, and lack of growth at reduced temperatures. One possible protocol is to streak suspect colonies onto duplicate plates of low pH and neutral pH media. One plate of each is incubated at 45 to $50°C$ is incubated and the remaining plates at 20 to $25°C$ for appropriate time periods. Observation of growth at $45°C$ and little or no growth at $25°C$ on low pH agar with no growth on the remaining plates at either temperature is consistent with the genus alicyclobacillus. Further testing is suggested for final confirmation.

The most solid proof of identity is obtained from sequence analysis of species-specific DNA or RNA segments, such as for the 16S ribosomal subunit, coupled with observations of basic phenotypic characteristics (thermoacidophilic, Gram-positive bacterial spore-former). Ribotyping has been investigated by the National Food Processors Association for identification of the alicyclobacilli with limited success. A protocol for use of randomly amplified polymorphic DNA was developed by Yamazaki *et al.* [61] to identify *A. acidoterrestris*. Further research is required by other laboratories to verify the use of these technologies.

7.7 SIGNIFICANCE OF DETECTION/ISOLATION FROM FOODS

Information regarding the storage conditions and number of alicyclobacilli necessary to cause spoilage is sparse. Since these organisms can be routinely isolated from unspoiled products, the question of their significance has been raised. A few researchers suggest that 10^3 cells/ml of *A. acidoterrestris* can produce enough phenolic compounds to cause an antiseptic/medicinal off-aroma within a few days to weeks. On the other hand, a publication by ABECitrus [62] states "Spore counts in the average range of 10^2 to 10^3 CFU/ml in the concentrated juice may be at an acceptable level which does not compromise further utilization and processing of juice given that adequate processing practices are employed, particularly after heat treatment or pasteurization." Further, Eguchi *et al.* [45] reported that only 2 of 13 wild-type strains produced spoilage when inoculated into orange juice. The differences of opinion among researchers regarding the presence of alicyclobacillus in foods indicate that questions regarding conditions that lead to product spoilage are unresolved and require additional investigation.

7.8 FUTURE DIRECTION

Due to their ubiquitous nature in the environment, the alicyclobacilli can be routinely isolated from liquid sugars, such as HFCS, concentrated juices, purees, and other agricultural products. Although there is no doubt that the alicyclobacilli have been involved in spoilage of juice and beverage products, the significance of detecting them in an unspoiled food sample remains in question. It is unlikely that the detection of a single spore in 100 g of concentrated juice by quality control testing would correlate with the potential for spoilage in the final product.

A substantial amount of research is needed to ascertain the significance of finding these organisms in food products. It is understandable that beverage-manufacturing companies wish to obtain ingredients (juice, HFCS, granulated sugar, etc.) that do not contain alicyclobacillus. This should be tempered with the knowledge that the significance of finding small numbers of these organisms in raw ingredients is not fully understood, and that current technology does not allow HFCS or concentrate/puree facilities to completely eliminate the alicyclobacilli from their products. Research to determine the probability of spoilage in products that contain specific numbers of spores and are stored under various conditions is lacking. Additionally, it would be helpful to investigate the existence of processing and storage parameters that might prevent outgrowth of alicyclobacilli in packaged products. Further discoveries may provide the answers needed to establish acceptable standards for alicyclobacillus in foods.

ACKNOWLEDGMENT

This document was approved by the Florida Agricultural Experiment Station for publication as Journal Series No. N-02531.

REFERENCES

1. Cerny, G., Hennlich, W., and Poralla, K., Fruchtsaftverderb durch Bacillen: Isolierung und Charakterisierung des Verderbserregers, *Z. Lebensm. Unters. Forsch.*, 179, 224, 1984.
2. Wisotzkey, J. *et al.*, Comparative sequence analyses on the 16S rRNA (rDNA) of *Bacillus acidocaldarius, Bacillus acidoterrestris*, and *Bacillus cycloheptanicus* and proposal for creation of a new genus, *Alicyclobacillus* gen. nov., *Int. J. Syst. Bacteriol.*, 42, 263, 1992.
3. Uchino, F. and Doi, S., Acido-thermophilic bacteria from thermal waters, *Agric. Biol. Chem.*, 31, 817, 1967.
4. Brock, T. and Darland, G., Limits of microbial existence: temperature and pH, *Science*, 169, 1316, 1970.
5. Darland, G. and Brock, T., *Bacillus acidocaldarius* sp. nov., an acidophilic thermophilic spore-forming bacterium, *J. Gen. Microbiol.*, 67, 9, 1971.
6. Hippchen, B., Röll, A., and Poralla, K., Occurrence in soil of thermoacidophilic bacilli possessing ω-cyclohexane fatty acids and hopanoids, *Arch. Microbiol.*, 129, 53, 1981.
7. De Lucca, A.J., II *et al.*, Mesophilic and thermophilic bacteria in a cane sugar refinery, *Zuckerind.*, 117, 237, 1992.
8. De Rosa, M. *et al.*, Cyclohexane fatty acids from a thermophilic bacterium, *J. Chem. Soc., Chem. Commun.*, 1971, 1334, 1971.
9. De Rosa, M. *et al.*, Isoprenoids of *Bacillus acidocaldarius*, *Phytochemistry*, 12, 1117, 1973.
10. Poralla, K., Kannenberg, E., and Blume, A., A glycolipid containing hopane isolated from the acidophilic, thermophilic *Bacillus acidocaldarius*, has a cholesterol-like function in membranes, *FEBS Lett.*, 113, 107, 1980.
11. Poralla, K. and König, W., The occurrence of ω-cycloheptane fatty acids in a thermo-acidophilic bacillus, *FEMS Microbiol. Lett.*, 16, 303, 1983.
12. Deinhard, G. *et al.*, *Bacillus cycloheptanicus* sp. nov., a new thermotolerant acidophile isolated from different soils, *Syst. Appl. Microbiol.*, 10, 68, 1987.

13. Deinhard, G. *et al.*, *Bacillus acidoterrestris* sp. nov., a new thermotolerant acidophile isolated from different soils, *Syst. Appl. Microbiol.*, 10, 47, 1987.

14. Berkeley, R.C.W. and Ali, N., Classification and identification of endospore-forming bacteria, *J. Appl. Bacteriol. Symp. Suppl.*, 76, 1S, 1994.

15. Nicolaus, B. *et al.*, Alicyclobacilli from an unexplored geothermal soil in Antarctica: Mount Rittmann, *Polar Biol.*, 19, 133, 1998.

16. Albuquerque, L. *et al.*, *Alicyclobacillus hesperidum* sp. nov. and a related genomic species from solfataric soils of São Miguel in the Azores, *Int. J. Syst. Evol. Microbiol.*, 50, 451, 2000.

17. Matsubara, H. *et al.*, *Alicyclobacillus acidiphilus* sp. nov., a novel thermo-acidophilic, ω-alicyclic fatty acid-containing bacterium isolated from acidic beverages, *Int. J. Syst. Evol. Microbiol.*, 52, 1681, 2002.

18. Goto, K. *et al.*, *Alicyclobacillus herbarius* sp. nov., a novel bacterium containing ω-cycloheptane fatty acids, isolated from herbal tea, *Int. J. Syst. Evol. Microbiol.*, 52, 109, 2002.

19. Goto, K. *et al.*, *Alicyclobacillus pomorum* sp. nov., a novel thermo-acidophilic, endospore-forming bacterium that does not possess T-alicyclic fatty acids, and emended description of the genus *Alicyclobacillus*, *Int. J. Syst. Evol. Microbiol.*, 53, 1537, 2003.

20. Tsuruoka, H. *et al.*, *Alicyclobacillus sendaiensis* sp. nov., a novel acidophilic, slightly thermophilic species isolated from soil in Sendai, Japan, *Int. J. Syst. Evol. Microbiol.*, 53, 1081, 2003.

21. Simbahan, J., Drijber, R., and Blum, P., *Alicyclobacillus vulcanalis* sp. nov., a thermophilic, acidophilic bacterium isolated from Coso Hot Springs, California, USA., *Int. J. Syst. Evol. Microbiol.*, paper in press at http://www.sgm.ac.uk/IJSEM/PiP/ijsem03012.pdf, accessed May 28, 2004.

22. Rodgers, L., Holden, P., and Foster, L., Culture of *Acidiphilium cryptum* BV1 with halotolerant *Alicyclobacillus*-like spp.: effects on cell growth and iron oxidation, *Biotechnol. Lett.*, 24, 1519, 2002.

23. Farrand, S. *et al.*, The use of response surface analysis to study growth of *Bacillus acidocaldarius* throughout the growth range of temperature and pH, *Arch. Microbiol.*, 135, 272, 1983.

24. Sinigaglia, M. *et al.*, Combined effects of temperature, water activity, and pH on *Alicyclobacillus acidoterrestris* spores, *J. Food Prot.*, 66, 2216, 2003.

25. Suzuki, K. *et al.*, Occurrence of ω-cyclohexyl fatty acids in *Curtobacterium pusillum* strains, *J. Gen. Appl. Microbiol.*, 27, 261, 1981.

26. Dufresne, S. *et al.*, *Sulfobacillus disulfidooxidans* sp. nov., a new acidophilic, disulfide-oxidizing, gram-positive, spore-forming bacterium, *Int. J. Syst. Bacteriol.*, 46, 1056, 1996.

27. Kusano, K. *et al.*, *Propionibacterium cyclohexanicum* sp. nov., a new acid-tolerant ω-cyclohexyl fatty acid-containing propionibacterium isolated from spoiled orange juice, *Int. J. Syst. Bacteriol.*, 47, 825, 1997.

28. Splittstoesser, D., Churey, J., and Lee, C., Growth characteristics of aciduric sporeforming bacteria isolated from fruit juices, *J. Food Prot.*, 57, 1080, 1994.

29. Splittstoesser, D., Lee, C., and Churey, J., Control of *Alicyclobacillus* in the juice industry, *Dairy Food Environ. San.*, 18, 585, 1998.

30. Yamazaki, I. *et al.*, Thermal resistance and prevention of spoilage bacterium, *Alicyclobacillus acidoterrestris*, in acidic beverages, *Nippon Shokuhin Kagaku Kaishi*, 44, 905, 1997.

31. Yamazaki, K. *et al.*, Influence of sporulation medium and divalent ions on the heat resistance of *Alicyclobacillus acidoterrestris* spores, *Lett. Appl. Microbiol.*, 25, 153, 1997.

32. Vieira, M.C. *et al.*, *Alicyclobacillus acidoterrestris* spores as a target for Cupuaçu (*Theobroma grandiflorum*) nectar thermal processing: kinetic parameters and experimental methods, *Int. J. Food Microbiol.*, 77, 71, 2002.

33. Pontius, A., Rushing, J., and Foegeding, P., Heat resistance of *Alicyclobacillus acidoterrestris* spores as affected by various pH values and organic acids, *J. Food Prot.*, 61, 41, 1998.

34. Komitopoulou, E. *et al.*, *Alicyclobacillus acidoterrestris* in fruit juices and its control by nisin, *Int J. Food Sci. Technol.*, 34, 81, 1999.

35. Palop, A. *et al.*, Heat resistance of *Alicyclobacillus acidocaldarius* in water, various buffers, and orange juice, *J. Food Prot.*, 63, 1377, 2000.

36. Sofos, J., Busta, F., and Allen, C., Sodium nitrite and sorbic acid effects on *Clostridium botulinum* spore germination and total microbial growth in chicken frankfurter emulsions during temperature abuse, *Appl. Environ. Microbiol.*, 37, 1103, 1979.

37. Yamazaki, K. *et al.*, Use of nisin for inhibition of *Alicyclobacillus acidoterrestris* in acidic drinks, *Food Microbiol.*, 17, 315, 2000.

38. Lee, S., Dougherty, R., and Kang, D., Inhibitory effects of high pressure and heat on *Alicyclobacillus acidoterrestris* spores in apple juice, *Appl. Environ. Microbiol.*, 68, 4158, 2002.

39. McIntyre, S. *et al.*, Characteristics of an acidophilic *Bacillus* strain isolated from shelf-stable juices, *J. Food Prot.*, 58, 319, 1995.

40. Yamazaki, K., Teduka, H., and Shinano, H., Isolation and identification of *Alicyclobacillus acidoterrestris* from acidic beverages, *Biosci. Biotech. Biochem.*, 60, 543, 1996.

41. Pinhatti, M. *et al.*, Detection of acidothermophilic bacilli in industrialized fruit juices, *Fruit Processing*, 7, 350, 1997.

42. Pettipher, G., Osmundson, M., and Murphy, J., Methods for the detection and enumeration of *Alicyclobacillus acidoterrestris* and investigation of growth and production of taint in fruit juice and fruit juice-containing drinks, *Lett. Appl. Microbiol.*, 24, 185, 1997.

43. Baumgart, J., Husemann, M., and Schmidt, C., *Alicyclobacillus acidoterrestris*: Vorkommen, Bedeutung und Nachweis in Getränken und Getränkegrundstoffen, *Flüssiges Obst.*, 64, 178, 1997.

44. Wisse, C. and Parish, M., Isolation and enumeration of sporeforming, thermoacidophilic, rod-shaped bacteria from citrus processing environments, *Dairy Food Environ. San.*, 18, 504, 1998.

45. Eguchi, S. *et al.*, An ecological study of acidothermophilic sporulating bacteria (*Alicyclobacillus*) in the citrus industry, *Ann. of the 23rd IFU Symposium*, Havana, 2000, p. 257.

46. Jensen, N., *Alicyclobacillus* in Australia, *Food Australia*, 52, 282, 2000.

47. Borlinghaus, A. and Engel, R., *Alicyclobacillus* incidence in commercial apple juice concentrate (AJC) supplies: Method development and validation, *Fruit Processing*, 7, 262, 1997.

48. Orr, R. *et al.*, Detection of guaiacol produced by *Alicyclobacillus acidoterrestris* in apple juice by sensory and chromatographic analyses, and comparison with spore and vegetative cell populations, *J. Food Prot.*, 63, 1517, 2000.

49. Jensen, N. and Whitfield, F., Role of *Alicyclobacillus acidoterrestris* in the development of a disinfectant taint in shelf-stable fruit juice, *Lett. Appl. Microbiol.*, 36, 9, 2003.

50. Göçmen, D. *et al.*, Identification of off-flavors generated by *Alicyclobacillus* species in orange juice using GC-Olfactometry and GC-MS, *Lett. Appl. Microbiol.*, 40, 172, 2004.

51. Orr, R. and Beuchat, L., Efficacy of disinfectants in killing spores of *Alicyclobacillus acidoterrestris* and performance of media for supporting colony development by survivors, *J. Food Prot.*, 63, 1117, 2000.

52. Yamazaki, K. *et al.*, Specific primers for detection of *Alicyclobacillus acidoterrestris* by RT-PCR, *Lett. Appl. Microbiol.*, 23, 350, 1996.

53. Luo, H., Yousef, A.E., and Wang, H.H., A real-time polymerase chain reaction-based method for rapid and specific detection of spoilage *Alicyclobacillus* spp. in apple juice, *Lett. Appl. Microbiol.*, 39, 376, 2004.

54. IFU, Method 12: First standard IFU method on the detection of *Alicyclobacillus* in fruit juices, in *Microbiological Methods*, Internationale Fruchtsaft-Union Microbiology Working Group, Bischofszell, Switzerland, 2004.

55. Walls, I. and Chuyate, R., *Alicyclobacillus*: historical perspective and preliminary characterization study, *Dairy Food Environ. San.*, 18, 499, 1998.

56. Walls, I. and Chuyate, R., Isolation of *Alicyclobacillus acidoterrestris* from fruit juices, *J. AOAC Int.*, 83, 1115, 2000.

57. Evancho, G. and Walls, I., Aciduric flat sour sporeformers, in *Compendium of Methods for the Microbiological Examination of Foods*, American Public Health Association, Washington D.C., 2001.

58. Parish, M. and Goodrich, R., Detection and Enumeration of Presumptive *Alicyclobacillus* Species and Other Spore-Forming Thermotolerant Acidophilic Rod-Shaped Bacteria in FCOJ, Final report to the Florida Department of Citrus, August 31, 2000.

59. Morton, R.D., Aerobic plate count, in *Compendium of Methods for the Microbiological Examination of Foods*, American Public Health Association, Washington D.C., 2001.

60. Swanson, K., Petran, R., and Hanlin, J., Culture method for enumeration of microorganisms, in *Compendium of Methods for the Microbiological Examination of Foods*, American Public Health Association, Washington D.C., 2001.

61. Yamazaki, K. *et al.*, Randomly amplified polymorphic DNA (RAPD) for rapid identification of the spoilage bacterium *Alicyclobacillus acidoterrestris*, *Biosci. Biotech. Biochem.*, 61, 1016, 1997.

62. Anon., Acidothermophilic sporeforming bacteria (ATSB) in orange juices: detection methods, ecology, and involvement in the deterioration of fruit juices, *Fruit Processing*, 11, 95, 2001.

63. Previdi, M. *et al.*, Thermoresistenza di spora di *Alicyclobacillus* in succhi di frutta, *Industria Conserve*, 72, 353, 1997.

64. Murakami, M., Tedzuka, H., and Yamazaki, K., Thermal resistance of *Alicyclobacillus acidoterrestris* spores in different buffers and pH, *Food Microbiol.*, 15, 577, 1998.

65. Eiroa, M., Junqueira, V., and Schmidt, F., *Alicyclobacillus* in orange juice: occurrence and heat resistance of spores, *J. Food Prot.*, 62, 883, 1999.

66. Silva, F. *et al.*, Thermal inactivation of *Alicyclobacillus acidoterrestris* spores under different temperature, soluble solids and pH conditions for the design of fruit processes, *Int. J. Food Microbiol.*, 51, 95, 1999.

Section III

Food Safety Issues

8 Interventions to Ensure the Microbial Safety of Sprouts

William F. Fett

CONTENTS

8.1 INTRODUCTION

Sprouts are considered a natural healthy food by many consumers in the U.S. and elsewhere. The North American sprouting industry has grown rapidly from only a very few commercial growers in 1970 to approximately 300 growers today with a total product market value of approximately $250,000,000 [1]. Over 20 seed types are used for sprouting in commercial operations and in the home [2]. Commercial sprouting operations are indoor facilities and in the U.S. are usually small in size with less than 10 employees [3]. Distribution of sprouts to retail outlets is local or regional.

Sprouts can be classified as either green sprouts or bean sprouts. Green sprouts such as alfalfa, clover, broccoli, radish, and sunflower have been subjected to light at some point in the growing process to allow for chlorophyll development. Bean (mung bean and soybean) sprouts are propagated under continuous dark and thus do not produce chlorophyll. Mung bean sprouts make up the major portion of the market for sprouts in the U.S. Green sprouts

---

Mention of trade names or commercial products in this chapter is solely for the purpose of providing specific information and does not imply recommendation or endorsement by the U.S. Department of Agriculture.

are consumed raw while bean sprouts are most often, but not always, served after at least light cooking.

Unfortunately, since 1995, both in the U.S. and in other countries, there have been numerous outbreaks of foodborne illness due to the consumption of sprouts contaminated with the bacterial pathogens salmonella and *Escherichia coli* O157 [4,5]. Raw sprouts were identified as a special food safety problem due to the potential for bacterial human pathogens to multiply from low levels on contaminated seed to high levels on sprouts due to favorable conditions of moisture, temperature, and nutrient availability during the sprouting process [4]. The U.S. Food and Drug Administration (FDA) has released a number of consumer advisories informing the consuming public about the risks associated with eating raw sprouts, the latest occurring in November 2003 [6], and raw sprouts are considered a "potentially hazardous food" in the FDA Food Code [7]. The consumer advisory states: "Those persons who wish to reduce the risk of foodborne illness from sprouts are advised not to eat raw sprouts." Particularly vulnerable to foodborne illness are the young, the elderly, and the immunocompromised.

This chapter provides an overview of the incidence and causes of sprout-related foodborne illness, interventions that have been tested for eliminating human pathogens from seeds and sprouts, means for reducing the risk of future outbreaks, and finally, further research needs.

8.2 FOODBORNE ILLNESS ASSOCIATED WITH SPROUTS

Several foodborne human pathogens have been isolated from sprouts and consumption of contaminated sprouts has been associated with numerous outbreaks of foodborne illness in the U.S. (Table 8.1). Some of these outbreaks have been international in scope due to the international distribution of sprout seed [10,12,24,25]. In addition to those in the U.S., sprout-related outbreaks of foodborne illness have been reported in several other countries including Canada, Japan, Sweden, Denmark, Holland, Finland, and the U.K. [4,5]. The earliest documented outbreak in the U.S. occurred in 1973 and was associated with consumption of raw sprouts grown with home sprouting kits containing soybean, cress, and mustard seed contaminated with enterotoxigenic *Bacillus cereus* [8]. There were no additional sprout-related outbreaks of foodborne illness recorded in the U.S. until 1990. Since 1995 there have been many outbreaks due to contamination of alfalfa and clover sprouts with various serovars of salmonella or *E. coli* O157. The first foodborne outbreak due to mung bean sprouts in the U.S. occurred in 2000 due to contamination with salmonella [9]. Previously, the only documented mung bean-associated outbreak of salmonellosis took place in England and Sweden in 1988 [26]. The number of culture confirmed cases in the U.S. has ranged from less than 10 to over 400 per outbreak. The actual number of cases was most likely much higher due to the significant underreporting normally encountered for

TABLE 8.1
Incidence of Foodborne Illness Due to Contaminated Sprouts in the U.S.

Year	Bacterium	Location	Sprout type	No. of culture confirmed cases	Ref.
1973	*Bacillus cereus*	TX	Soybean, cress, mustard	4	8
1990	*Salmonella* Anatum	WA	Alfalfa	15	9
1995	*Salmonella* Stanley	17 states/Finland	Alfalfa	242	10
1995	*Salmonella* Newport	OR	Alfalfa	69	9
1995–1996	*Salmonella* Newport	7 states/Canada/ Denmark	Alfalfa	>133	5, 11, 12
1996	*Salmonella* Stanley	VA	Alfalfa	30	9
1996	*Salmonella* Montevideo/ Meleagridis	CA/NV	Alfalfa/clover	492	13
1997	*Salmonella* Infantis/Anatum	KS/MO	Alfalfa	109	5
1997	*Escherichia coli* O157:H7	Multistate	Alfalfa	85	9, 14
1997–1998	*Salmonella* Senftenberg	CA/NV	Alfalfa/clover	60	13
1998	*Salmonella* Havana/Cubana	Multistate	Alfalfa	40	13, 15
1998	*Escherichia coli* O157:NM	CA/NV	Clover/alfalfa	8	13
1999	*Salmonella* Mbandaka	Multistate	Alfalfa	87	9, 16
1999	*Salmonella* spp.	MI	Alfalfa	34	9
1999	*Salmonella* Typhimurium	CO, CT	Alfalfa/clover	119	9, 17
1999	*Salmonella* Saint Paul	CA	Clover	36	9
1999	*Salmonella* Muenchen	Multistate	Alfalfa	~157	18
2000	*Salmonella* Enteritidis PT33	Multistate	Mung bean	75	9, 19
2001	*Salmonella* Kottbus	Multistate	Alfalfa	31	20
2001	*Salmonella* Enteritidis PT1	HI	Mung bean	26	21
2001	*Salmonella* Enteritidis PT913	FL	Mung bean	35	9
2002	*Escherichia coli* O157:H7	CA/NV	Alfalfa	5	18
2002	*Salmonella* Enteritidis	ME	Mung bean	16	22
2003	*Salmonella* Saint Paul	OR/WA	Alfalfa	8	22
2003	*Escherichia coli* O157:NM	CO/WY	Alfalfa	13	22
2003	*Escherichia coli* O157:H7	MN	Alfalfa	5	23
2003	*Salmonella* Chester	OR	Alfalfa	24	22
Salmonella	Bovismorbificans	Multistate	Alfalfa	28	22
Escherichia coli	O157:NM	GA	Alfalfa	5	22

foodborne illnesses [27]. The first recognized sprout-related outbreak due to *E. coli* O157:H7 occurred in Japan in 1996 and was associated with contaminated Daikon radish sprouts. To date this is the largest recorded foodborne outbreak due to contaminated sprouts worldwide with well over 7000 confirmed cases [28,29]. The first recorded sprout-related outbreak of foodborne illness in the U.S. due to contamination with *E. coli* O157:H7 was in 1997 [14]. Contaminated sprout seed is thought to be the primary source of the pathogens responsible for most sprout-related outbreaks of foodborne illness [4,5]. This conclusion is based on direct isolation of pathogens from seed of implicated lots and/or epidemiological evidence.

Several studies have indicated that salmonella and *E. coli* O157:H7 present initially on artificially as well as naturally contaminated seed have the potential to increase up to 10,000-fold on sprouts propagated at 20 to 30°C. The majority of growth of salmonella and *E. coli* O157:H7 on sprouting seed occurs during the first 48 hours. For sprouts grown from artificially inoculated seed, maximum populations of salmonella and *E. coli* O157:H7 ranging from 5 to 8 $\log_{10}$ colony-forming units (CFU)/g have been reported [30–39]. The maximum pathogen population obtained was not dependent on the initial inoculum level present on the seed [36]. For comparison, populations of total aerobes reported for sprouts typically range from 7 to 9 $\log_{10}$ CFU/g [30,40–42]. For salmonella on alfalfa, the doubling time was estimated at 47 minutes during the initial rapid growth phase and growth was not dependent on pathogen serovar, isolation source, or virulence [33]. Populations of salmonella and *E. coli* O157:H7 were stable from 48 hours to harvest at 3 to 5 days and then declined only slightly during subsequent storage of contaminated alfalfa sprouts at 5 to 9°C for 6 to 10 days [33,34,37]. Populations of *B. cereus* on sprouts grown from naturally contaminated alfalfa and mung bean seed reached approximately 4 $\log_{10}$ CFU/g [43]. The maximum pathogen populations attained during germination and growth of naturally contaminated seed under commercial practice may be several $\log_{10}$ units less than that for artificially inoculated seed [44]. Maximum populations of salmonella attained on alfalfa sprouts grown from two different lots of naturally contaminated seed were only 2 to 4 $\log_{10}$ MPN/g for salmonella. The reduced growth may be due to several factors. The first is the much lower overall contamination levels on naturally contaminated seed when compared to even the lowest initial pathogen populations utilized for laboratory studies. Second, pathogen populations on naturally contaminated seed may contain a higher percentage of injured cells. Third, differing methods of irrigation and increased irrigation frequency employed in commercial operations may affect the final pathogen populations attained. Interestingly, salmonella serovars attach more tightly to surfaces of alfalfa sprouts than do strains of *E. coli* O157:H7 and the difference in strength of attachment was proposed to explain, at least in part, the greater number of outbreaks of foodborne illness associated with contaminated sprouts due to salmonella [39].

Studies in several independent laboratories have indicated that bacterial human pathogens can be internalized in sprouts. By use of immunofluorescence and scanning immunoelectron microscopy, *E. coli* O157:H7 was located in stomata and the vascular system of radish sprouts grown from inoculated seed [45]. Bioluminescent *Salmonella* Montevideo and various salmonella serovars expressing the autofluorescent green-fluorescent protein were also located in the internal tissues of mung bean and alfalfa sprouts, respectively, after inoculation of seed or roots [38,45,46]. The mode of entry of bacterial human pathogens into plants remains unknown, but it is likely due to passive uptake at the site of injury where lateral roots emerge [46,48], as salmonella and *E. coli* O157:H7 have not been reported to excrete cell-wall-degrading enzymes (e.g., pectinases or cellulases) that might facilitate active entry. Pathogens may form biofilms on sprout surfaces and/or become part of biofilms produced by native microorganisms [49,50] (Figure 8.1).

FIGURE 8.1 Biofilm consisting of native bacteria on the surface of an alfalfa sprout hypocotyl.

8.3 INTERVENTIONS: SEEDS

8.3.1 CHEMICAL AND PHYSICAL

Sanitizing sprout seed presents a unique challenge in the arena of produce safety in that even a low residual pathogen population remaining on contaminated seed after treatment appears capable of growing to very high levels (up to $8 \log_{10}$ CFU/g) due to favorable conditions of moisture, relative humidity, temperature, and nutrient availability during seed germination and subsequent sprout growth [51,52]. In addition, after a sanitizing procedure seed germination as well as sprout yield and quality need to be maintained at commercially acceptable levels. In 1999, based on research available at the time, the FDA published guidance documents recommending that commercial sprout growers treat sprout seed with one or more antimicrobial treatments such as 20,000 ppm of $Ca(OCl)_2$ that have been approved for reduction of pathogens on seeds or sprouts, with at least one approved antimicrobial treatment applied immediately before sprouting [53]. Also, in 2000 the FDA and the California Department of Health Services, Food and Drug Branch jointly released a food safety training video [54] for use by commercial sprout growers. The video, based on the FDA guidance documents, contains a recommendation to treat sprout seed with 20,000 ppm available chlorine from $Ca(OCl)_2$ for 15 minutes (continuous mixing) with potable water rinses both before and after seed treatment. Since sprout seed is considered a raw agricultural product, chemical seed treatments are subject to approval by the U.S. Environmental Protection Agency and not the FDA.

Population reductions reported after treatment of alfalfa seed artificially inoculated with salmonella or *E. coli* O157:H7 using 16,000 to 20,000 ppm of available chlorine has varied considerably among different laboratories, but usually are in the range of 2 to 4 $\log_{10}$ (Table 8.2). Lesser reductions were achieved after treatments with lower amounts of chlorine. A number of factors likely contribute to the variability in results. Such factors include the percentage of treated inoculated seed with broken, cracked, or wrinkled seed coats [81], differences in the initial pathogen population on the seed, the extent of mixing of sanitizer during treatment, the initial organic load on the seed, and the use of rinse steps before and after seed treatment. Some studies have been done with relatively low initial pathogen populations on the seed allowing for maximum population reductions of 2 to 3 $\log_{10}$. One consistent finding among the various laboratories is that the two pathogens when artificially inoculated onto sprout seed are not eliminated even by treatment with 16,000 to 20,000 ppm of available chlorine for 10 to 15 minutes.

The findings for similar studies with naturally contaminated seed are not consistent among laboratories (see below).

Investigations of recent foodborne outbreaks of salmonellosis due to contaminated sprouts indicates that treatment of sprout seed with high levels of chlorine by commercial growers reduces, but may not always eliminate, the risk of human illness [16–18,20]. The inability of seed treatments with high levels of chlorine to always ensure a pathogen-free seed under commercial practice may be due to several factors including the use of differing protocols for administering seed treatments at grower locations. Also, the particular seed treated, if naturally contaminated, may differ in the level of contamination present and the location of the pathogens on the contaminated seed (e.g., deep in cracks, crevices, and/or natural openings) (Figure 8.2). The ability of bacterial human pathogens to be internalized in seed under natural conditions in the field is not known, but seeds in general can harbor internalized native bacteria [82]. If present in internal tissues of the seed, pathogens may escape contact with chemical sanitizers.

Numerous chemical treatments in addition to chlorine as well as several physical treatments have been tested individually or in combination for eliminating pathogens from artificially inoculated sprout seed. To date there are few reports of stand alone chemical or physical interventions capable of eliminating pathogens from artificially inoculated sprout seed or consistently achieving the recommended 5 $\log_{10}$ reductions [4] without significant adverse affects on seed germination and/or sprout yield (Table 8.2). Most of the interventions included in Table 8.2 have been tested using more than the single set of conditions listed. Additional chemicals tested in the references cited, but not included in Table 8.2, are aqueous acetic acid, calcinated calcium, carvacrol, cinnamic aldehyde, citric acid, Citricidal® (NutriTeam, Inc., Reston, VT), CitroBio™ (= Pangermex) (CitroBio, Inc., Sarasota, FL), Environnè Fruit and Vegetable Wash™ (Consumer Health Research, Inc., Brandon, OR), ethanol, eugenol, linalool, methyl jasmonate, sodium carbonate, sodium hypochlorite, thymol, *trans*-anethole, trisodium phosphate, Tsunami 200® (Ecolab, Mendota Heights,

TABLE 8.2
Chemical and Physical Interventions for Reducing Pathogens on Inoculated Sprouting Seeds

Treatment	Conditions	Time	Seed type	Bacterium	Log reduction (CFU/g)	Seed germination	Ref.
Acetic acid, vapor	242 µl/l air, 45°C	12 h	Mung bean	Salmonella	>5, no survivors	No effect	55
Acetic acid, vapor	242 µl/l air, 45°C	12 h	Mung bean	E. coli O157:H7	>6, no survivors	No effect	55
Acetic acid, vapor	242 µl/l air, 45°C	12 h	Mung bean	L. monocytogenes	4.0	No effect	55
Acetic acid, vapor	300 mg/l air, 50°C	24 h	Alfalfa	Salmonella	0.8	No effect	56
Acidic EO water	1081 mV, 84 ppm chlorine	10 min	Alfalfa	Salmonella	1.5	No effect	57
Acidic EO water	1150 mV, 50 ppm chlorine	64 min	Alfalfa	E. coli O157:H7	1.6	Significant reduction	58
Acidic EO water	1079 mV, 70 ppm chlorine	15 min	Alfalfa	Salmonella	2.0	No effect	59
Allyl isothiocyanate	50 µl/950 cm³ jar, 47°C	24 h	Alfalfa	E. coli O157:H7	>2.0, survivors present	Slight reduction	60
Ammonia, gas	300 mg/l	22 h	Alfalfa	Salmonella	2.0	No effect	61
Ammonia, gas	300 mg/l	22 h	Mung bean	Salmonella	5.0	No effect	61
Ammonia, gas	300 mg/l	22 h	Alfalfa	E. coli O157:H7	3.0	No effect	61
Ammonia, gas	300 mg/l	22 h	Mung bean	E. coli O157:H7	6.0	No effect	61
Ca(OH)₂	1%	10 min	Alfalfa	E. coli O157:H7	3.2	No effect	62
Ca(OH)₂	1%	10 min	Alfalfa	Salmonella	2.8–3.8	No effect	62, 63
Ca(OCl)₂	20,000 ppm	3 min	Alfalfa	E. coli O157:H7	>2.3, survivors present	Reduced rate	64
Ca(OCl)₂	20,000 ppm	10 min	Alfalfa	Salmonella	2.0	Slight reduction	63
Ca(OCl)₂	18,000 ppm	10 min	Alfalfa	Salmonella	3.9	No effect	65
Ca(OCl)₂	18,000 ppm	10 min	Alfalfa	E. coli O157:H7	4.5	No effect	65
Ca(OCl)₂	16,000 ppm	10 min	Mung bean	Salmonella	5.0	No effect	66
Ca(OCl)₂	16,000 ppm	10 min	Mung bean	E. coli O157:H7	3.9	No effect	66
Chlorine dioxide, acidified	500 ppm	10 min	Alfalfa	E. coli O157:H7	>2.4, survivors present	Significant reduction	64

(continued)

TABLE 8.2
Continued

Treatment	Conditions	Time	Seed type	Bacterium	Log reduction (CFU/g)	Seed germination	Ref.
Citrex™	20,000 ppm	10 min	Alfalfa	Salmonella	3.6	No effect	67
Citrex™	20,000 ppm	10 min	Alfalfa	E. coli O157:H7	3.4	No effect	67
Dry heat	50°C	60 min	Alfalfa	E. coli O157:H7	1.7	No effect	68
Dry heat	70°C	3 h	Alfalfa	Salmonella	3.0	Slight reduction	56
Fit™	According to label	15 min	Alfalfa	Salmonella	2.3	No effect	69
Fit™	According to label	15 min	Alfalfa	E. coli O157:H7	>5.4	No effect	69
H_2O_2	8%	3 min	Alfalfa	E. coli O157:H7	>2.9, survivors present	No effect	64
H_2O_2	8%	10 min	Alfalfa	Salmonella	3.2	No effect	63
Hydrostatic pressure	300 mPa	15 min	Garden cress	Salmonella	5.8	Reduced rate	70
Hydrostatic pressure	300 mPa	15 min	Garden cress	Shigella flexneri	4.5	Reduced rate	70
Lactic acid	5%, 42°C	10 min	Alfalfa	E. coli O157:H7	3.0	No effect	52
Radiation, gamma	Various		Alfalfa	Salmonella	D-value of 0.97 kGy	Dosage dependent	71
Radiation, gamma	Various		Alfalfa	E. coli O157:H7	D-value of 0.60 kGy	Dosage dependent	71
Radiation, gamma	Various		Broccoli	Salmonella	D-value of 1.10 kGy	Dosage dependent	72
Radiation, gamma	Various		Broccoli	E. coli O157:H7	D-value of 1.11 kGy	Dosage dependent	72
Sodium chlorite, acidified	1200 ppm, 55°C	3 min	Alfalfa	E. coli O157:H7	>1.9, survivors present	Slight reduction	64
Sulfuric acid	2 N	20 min	Alfalfa	E. coli O157:H7	5.0	No effect	73
Ozone, aqueous	21 ppm, w/sparging	64 min	Alfalfa	E. coli O157:H7	2.2	No effect	74
Ozone, aqueous	21.3 ppm, w/sparging	20 min	Alfalfa	L. monocytogenes	1.5	No effect	75
Pulsed UV light	5.6 J/cm², 270 pulses	90 sec	Alfalfa	E. coli O157:H7	4.9	Significant reduction	76
Dielectric heating, radio frequency	39 MHz, 1.6 kV/cm	26 sec	Alfalfa	Salmonella	1.7	No effect	77
Supercritical CO_2	4000 psi, 50°C	60 min	Alfalfa	E. coli, generic	1.0	No effect	78
Water, hot	3-stage: 25 to 50 to 85°C	30 min, 9 sec, 9 sec	Alfalfa	E. coli, generic	>4, no survivors	No effect	79
Water, hot	54°C	5 min	Alfalfa	Salmonella	2.5	No effect	34
Water, hot	80°C	2 min	Mung bean	Salmonella	>6	No effect	80

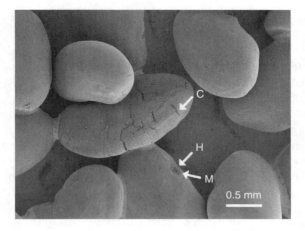

FIGURE 8.2 Scanning electron micrograph of alfalfa seeds showing extensive cracking of a seed coat and natural openings: C, crack in the seed coat; H, hilum; M, micropyle.

MN), Tween 80, Vegi-clean^{TM} (Microcide, Inc., Detroit, MI), and Vortex® (Ecolab). Treating with aqueous chemicals at elevated temperatures can lead to greater reductions of pathogen populations on seed, but is often detrimental to seed germination [83]. Addition of high levels of the surfactant Tween 80 (1%, w/v) to 1% Ca(OH)$_2$ led to only an additional 1 log$_{10}$ reduction or less in the population of salmonella on alfalfa seed [62,63]. Sonication of seed during treatment with aqueous antimicrobial compounds also did not have a significant effect, only slightly increasing the log$_{10}$ kill obtained [68,83].

Treatment with gaseous acetic acid was reported to eliminate both salmonella and *E. coli* O157:H7, but not *Listeria monocytogenes*, from artificially inoculated mung bean seed without reducing seed germination [55]. Similar treatments of inoculated alfalfa seed led to either unacceptable reductions of seed germination [84] or were not effective [56]. Hot water treatments of alfalfa seed inoculated with generic *E. coli* were reported to eliminate the bacterium [79], but results both with alfalfa seed artificially inoculated with human pathogens as well as naturally contaminated seed have not been as promising due to lowered effectiveness and/or detrimental effects on seed germination [34,85]. Under commercial practice, the ability of hot water treatments to ensure consistent elimination of bacterial human pathogens from alfalfa seed was put into question by a recent multistate outbreak of salmonellosis due to contaminated alfalfa sprouts grown from seed treated with hot water followed by a soak in low levels (2000 ppm) of chlorine [20]. However, a recent laboratory study indicates that treatment of mung bean seed with hot water may be an effective seed-sanitizing step. Treatment of seed inoculated with salmonella at 55°C for 20 minutes, 60°C for 10 minutes, or 70°C for 5 minutes led to an approximate 5 log$_{10}$ reduction [80]. Treating seed at 80°C for 2 minutes was even more effective resulting in an over 6 log$_{10}$ reduction. None of these temperature/time treatments led to a decrease in germination of the treated seed.

There have been a very limited number of studies on seed sanitization using naturally contaminated rather than artificially inoculated seed and these studies have evaluated the efficacy of hot water and chlorine treatments only [44,65, 85]. The use of naturally contaminated seed rather than artificially inoculated seed may give a more accurate prediction of the efficacy of seed treatments for eliminating bacterial human pathogens in commercial practice. This may be due to differences in bacterial populations per gram of seed (normally much lower on naturally contaminated seed than on artificially contaminated seed used for laboratory studies), possible differences in the location and physiological status of the pathogens and the potential presence of pathogens in biofilms. In contrast to studies with artificially inoculated seed treated with high levels of chlorine, research conducted independently in two laboratories using alfalfa seed lots naturally contaminated with salmonella indicated that treatment with chlorine (unbuffered and buffered to neutral pH, from 2,000 to 20,000 ppm) completely eliminated the pathogen [65,85]. However, a third laboratory published contrasting results using 20,000 ppm of unbuffered active chlorine also using seed naturally contaminated with salmonella [44]. The reasons for the differing results between laboratories may include differences in the degree of mixing during seed treatment as well as differences in the population and location of the pathogen on the particular naturally contaminated seed tested even if originating from the same seed lot.

Several physical treatments have also been tested for sanitizing sprout seed (Table 8.2). In 2000 the FDA approved exposure of sprout seed to ionizing radiation at doses up to 8 kGy [86]. Treatment with ionizing radiation can significantly reduce bacterial pathogens on sprout seed. Exposure of inoculated alfalfa seed to a 2 kGy dose of gamma irradiation led to a 3.3 and 2.0 $\log_{10}$ reduction in *E. coli* O157:H7 and salmonella populations, respectively, while still maintaining commercially acceptable yields as well as nutritive values of sprouts grown from the treated seed [71,87,88]. Higher dosages led to unacceptable reductions in yields. For alfalfa seed naturally contaminated with salmonella and treated with gamma radiation, Thayer *et al.* [89] reported a D-value of 0.81 kGy. An absorbed dose of 4 kGy was required to eliminate the pathogen, a dosage that results in significant reductions in yield. A required dosage of 4 kGy for pathogen elimination along with a D-value of 0.81 kGy indicates that individual naturally contaminated seeds may harbor pathogen populations in excess of 4 $\log_{10}$ CFU. Electron beam radiation or use of so-called soft electrons (low-energy electron beam, energies ≤ 300 kV) may also be useful for reducing pathogen populations on the surface of seed [90], but both have lowered penetration ability compared to gamma radiation.

Various treatment combinations (hurdle concept) for reducing contaminants on sprout seed have also been tested. Bari *et al.* [68] reported that the combination of dry heat (50°C, 1 hour) followed by treatment with hot acidic electrolyzed oxidizing (EO) water and sonication was able to reduce populations of *E. coli* O157:H7 on artificially inoculated mung bean seed by 4.6 $\log_{10}$, but the combination treatment was less effective when tested against inoculated radish and alfalfa seed. Seed germination and subsequent sprout growth were not adversely affected. In the same study, a dry heat (50°C, 1 hour) seed treatment in combination with exposure to 2 to 2.5 kGy of gamma radiation led to the elimination of the pathogen on mung bean, radish, and alfalfa seed, but resulted in decreases in yield, most significantly for mung bean and radish. Lang et $al.$ [52] found that successive treatments of alfalfa seed artificially inoculated with $E.$ $coli$ O157:H7 with lactic acid and chlorine (2000 ppm) were slightly more effective than lactic acid treatments alone, but were less effective than high levels of chlorine (20,000 ppm). Sharma et $al.$ [91] found that treating alfalfa seed inoculated with $E.$ $coli$ O157:H7 first with ozone (continuous sparging in water) followed by a dry heat treatment (60°C, 3 hours) led to a greater than 4 $\log_{10}$ reduction of the pathogen population, but survivors were detected by enrichment. A sequential washing treatment with thyme oil (5 ml/l) followed by ozonated water (14.3 mg/l) and aqueous ClO_2 (25 mg/l) led to a 3.3 $\log_{10}$ reduction of $E.$ $coli$ O157:H7 on inoculated alfalfa seed [92].

The large body of research reported subsequent to the release of the FDA guidance documents [53] indicates that several alternative chemical and physical treatments may be similar or greater in efficacy to high levels of chlorine for reducing pathogen populations on sprout seed. For sanitizing alfalfa seed such treatments include seed soaks in 1% $Ca(OH)_2$, 1% calcinated calcium, FIT®, 8% H_2O_2, or 2% CITREX™ [62,63,67,69,83,93]. For sanitizing mung bean seed exposure to gaseous acetic acid or soaking seed in hot water appear especially promising [55,80]. The efficacy of these alternative chemical and physical treatments needs to be confirmed by other researchers ideally using naturally contaminated seed. In contrast to high levels of chlorine, several of these alternative methods of sanitizing seed may be acceptable for use by organic growers as well as conventional growers pending any required regulatory approvals. The cost of some of these alternative methods to the commercial grower may be prohibitive, however. Cost may not be as much of an issue for home growers.

8.3.2 BIOLOGICAL

In contrast to the voluminous literature concerning biological control of plant pathogens [94] as well as numerous studies on the biological control (competitive exclusion) of pathogens in poultry, meat, and dairy products [95,96], there is little published information on the use of antagonistic microorganisms to control human pathogens on produce. The ideal biocontrol product for use on sprout seed and sprouts would contain a nonpathogenic microorganism(s) that is genetically stable, easily cultured and formulated using low-cost substrates and materials, has a long shelf life, is easily applied to seeds and/or sprouts, is highly effective on a variety of sprout types and against several human pathogens, and is affordable for the grower. For control of pathogens in poultry, meat, and dairy products, single microbial strains or defined or undefined consortia of microbes have been tested as antagonists.

Most of the studies on biological control of bacterial human pathogens on produce have examined the use of lactic acid bacteria (LAB) as antagonists [95]. LAB are attractive candidates for commercial biological control agents due to their common occurrence on sprout surfaces [41,97], their ability to produce multiple antimicrobial agents including bacteriocins, hydrogen peroxide, and organic acids *in vitro*, their extensive use in the food industry for fermentation, and their lack of known pathogenicity [95]. A strain of *Lactococcus lactis* inhibitory *in vitro* against *Listeria monocytogenes* due to acid production was tested for control of the pathogen when the two bacteria were co-inoculated onto alfalfa seed before sprouting [98]. Results indicated that the strain was much less inhibitory towards the pathogen *in situ* than *in vitro*, reducing pathogen populations on the sprouts by only 1 $\log_{10}$. In a second study on LAB, Wilderdyke *et al.* [99] found that of 58 isolates of LAB isolated from alfalfa seeds and sprouts, 32 were inhibitory towards the three pathogens salmonella, *E. coli* O157:H7, and *Listeria monocytogenes* in agar spot tests. One strain of *Lactococcus lactis* subsp. *lactis* was particularly inhibitory towards all three pathogens on agar media and in broth culture. The same group reported a significant reduction in populations of *Listeria monocytogenes* on alfalfa sprouts after application of a strain of LAB in the seed soak solution [100]. A commercial product containing a lactic acid bacterium is available in Japan for controlling *E. coli* O157:H7 on Daikon radish sprouts [101]. This product is to be sprayed onto seeds and sprouts several times during the sprouting process. In our laboratory, we have tested hundreds of plant-associated bacteria, primarily isolated from sprout surfaces, for their ability to inhibit growth of salmonella inoculated to alfalfa seed in small-scale laboratory bioassays [102]. Of these, a few isolates (none are LAB) have been identified that consistently reduce growth by several $\log_{10}$ units in small-scale laboratory bioassays. Currently, the effectiveness of these antagonists is being evaluated in larger scale experiments and studies on their mode of action are also underway.

More than a single antagonist may be required for controlling pathogens on germinating seeds of various sprout types due to compositional differences in the native microflora [103]. Treating artificially contaminated alfalfa seed with a novel purified bacteriocin, colicin HU194, led to reductions ranging from 3 $\log_{10}$ CFU/g to complete elimination of *E. coli* O157:H7. Efficacy was dependent on the particular strain of *E. coli* O157:H7 used for seed inoculation [104]. Bacteriophages are also being researched as a possible antimicrobial intervention for application to sprout seed [105]. Biological control agents may also be useful for reducing spoilage caused by soft-rotting bacteria [106].

8.4 INTERVENTIONS: SPROUTS

A variety of antimicrobial chemicals have been tested as additives to sprout irrigation water for the purpose of preventing or reducing the growth of native microflora and bacterial human pathogens. A study in our laboratory indicated that addition of H_2O_2, Tsunami®, acidified $NaClO_2$, Aquatize™ (Bioxy, Raleigh, NC), EDTA, $NaPO_4$, and NaOCl at varying concentrations to the irrigation water did not reduce the populations of the native micro-flora on alfalfa sprouts grown in a commercial-scale tray system by more than approximately 1 $\log_{10}$ without evidence of phytotoxicity [107]. Piernas and Guiraud [108] reported that spray irrigation of tray-grown rice sprouts with chlorinated water (100 mg/l) every 6 hours was not effective in reducing populations of total aerobic bacteria, *B. cereus*, or *L. innocua*. Daily spraying of alfalfa sprouts grown from artificially inoculated seed with chlorine (100 mg/l) led to reduction of less than 2 $\log_{10}$ in the population of salmonella at day 4 of sprouting [51]. Daily irrigation with ClO_2 (100 mg/l) did not reduce the population of total aerobic bacteria on alfalfa sprouts grown in trays, but did reduce populations of *V. cholera* up to 2 $\log_{10}$ when sprouts were grown from seed inoculated with the pathogen [32]. A reduction of 4 $\log_{10}$ for total coli-forms was obtained for mung bean sprouts that were subject to irrigation with 0.2 ppm gaseous ozone and 0.3 to 0.5 mg/l of ozonated water at days 4 to 7 of sprouting [105]. Rinsing of inoculated alfalfa seed growing in plastic jars with aqueous ClO_2 (25 mg/l) or ozonated water (9.27 mg/l) after 48 or 72 hours of sprouting was ineffective in reducing populations of *E. coli* O157:H7 [92]. However, rinsing with thyme oil (5.0 mg/l) alone or in sequence with ClO_2 and ozonated water led to reductions of up to 2 $\log_{10}$ in pathogen populations when carried out at 24 and 48 hours into the sprouting process. None of the rinsing treatments were effective at 72 hours, however. Rinsing with water was ineffective at all time points. Taormina and Beuchat [110] tested a variety of aqueous antimicrobial chemicals as spray treatments for reducing or elimi-nating *E. coli* O157:H7 from the surface of growing alfalfa sprouts. None of the chemicals were effective for reducing pathogen populations and only acidified $NaOCl_2$ (1200 ppm) controlled the growth of the pathogen. A complication of addition of antibacterial compounds to the irrigation water is that any patho-gens present in the spent irrigation water may be killed, but viable pathogen populations may remain on the sprouts rendering the testing of spent irrigation water for viable pathogens meaningless [4].

Several postharvest treatments for reducing the populations of native microbes and pathogens have been examined. Water rinses are not highly effective in reducing microbes on sprouts with resultant population reductions of 1 $\log_{10}$ or less [31,39,110,111]. A 2-minute treatment with aqueous ozone (23 ppm) did not reduce the population of aerobic microorganisms on alfalfa sprouts [75]. Dipping in hot water (60°C) for 30 seconds led to a reduction of 2 $\log_{10}$ in the population of total microbes on soybean sprouts [112] and a similar treatment for 5 minutes led to a reduction of 5 $\log_{10}$ in aerobic plate counts on rice sprouts [113]. Blanching in hot water (90°C, 1 minute) was reported to reduce microbial counts by 5 $\log_{10}$ units for mung bean sprouts [111]. Rinsing of mung bean sprouts in 1 and 2% lactic or acetic acid reduced the native microflora by less than 2 $\log_{10}$ [111]. Treatment of rice sprouts with chlorine (100 mg/l) for up to 10 minutes decreased aerobic plate counts by only 1.5 $\log_{10}$ [111]. Treatment (10 minutes) of inoculated mung bean sprouts with chlorous acid ($HClO_2$; 268 ppm), NaOCl (200 ppm), or lactic acid (2%) resulted in a maximum reduction of 1 $\log_{10}$ for total aerobes [114].

Blanching in hot water (100°C, 30 seconds) did not eliminate $E.$ $coli$ O157:H7 from alfalfa sprouts [115]. Treatment (10 minutes) of alfalfa sprouts with EO water (84 mg/l of available chlorine) in conjunction with sonication led to a reduction of 1.5 $\log_{10}$ in the population of salmonella [116]. Treatment (64 minutes) with EO water (50 mg/l of available chlorine) resulted in a reduction of 3 $\log_{10}$ of $E.$ $coli$ O157:H7 on alfalfa sprouts without any reported changes in appearance [117]. Aqueous ozone treatments (maximum concentration of 20 to 23 ppm, treatment time of 20 to 64 minutes) of alfalfa sprouts led to a maximum population reduction of approximately 1 to 2 $\log_{10}$ for $L.$ $monocytogenes$ and $E.$ $coli$ O157:H7, respectively [75,118]. The greatest log reductions reported for a postharvest aqueous chemical treatment were for $HClO_2$. Treatment (10 minutes) of inoculated mung bean sprouts with $HClO_2$ (268 ppm) resulted in a reduction of approximately 5 $\log_{10}$ of salmonella and $L.$ $monocytogenes$. Lactic acid (2%) was also tested in this study, but was less effective [114]. Exposure of inoculated alfalfa sprouts to gaseous acetic acid or allyl isothiocyanate vapor led to significant reductions in the population of salmonella, but also led to undesirable changes in sensory quality [56].

Most likely the only postharvest treatment able to inactivate pathogens that have been internalized into sprouts during the growing process is irradiation. A postharvest treatment with gamma radiation at 2 kGy extended the shelf life of alfalfa and broccoli sprouts by 10 days due to significant decreases in the native microflora [72,88]. Doses up to 2.6 kGy did not significantly change the appearance or nutrient quality of alfalfa sprouts [118]. Salmonella was eliminated from alfalfa sprouts grown from naturally contaminated seed when exposed to gamma radiation at a minimum dose of 0.5 kGy [120]. Irradiation of inoculated alfalfa sprouts with 3.3 kGy of beta radiation (electron beam) eliminated $L.$ $monocytogenes$ without an adverse effect on quality [121].

8.5 REDUCING THE RISK OF FUTURE OUTBREAKS

Several steps can be taken to minimize the risk of future sprout-related outbreaks of foodborne illness including the use of good agricultural practices (GAPs) during the production of sprouting seed as detailed in several recent government, university, and produce organization publications [122–124]. Sprout seed is obtained from plants grown in the open field and thus subject to potential contamination by nonpotable irrigation water, manure, domestic and wild animals, birds, farm machinery, and farm workers. To the author's knowledge, there are no fields in the U.S. or elsewhere designated solely for the production of seed destined for use by sprout growers. Settings on harvesting machinery should be such as to minimize damage to the seed. Cross contamination between clean and contaminated lots of harvested seed can occur in seed cleaning (conditioning) facilities and also when lots of seed are mixed before packaging and distribution. Several salmonella serovars were detected in the waste streams of a seed-cleaning machine in a U.S. alfalfa seed-cleaning facility indicating the presence of salmonella in the local alfalfa fields where the seed originated [125]. Seed-cleaning machines should be thoroughly cleaned and sanitized before and between lots of seed destined for sprouting. Seed scarification has been used historically to increase the germinability of seed lots that contain a significant amount of hard seed. Scarification involves the mechanical abrasion of the seed coat to allow for entry of water facilitating germination. Damage to the seed coat may make elimination of bacterial pathogens by treatment with chemical sanitizers more difficult [62,81] and probably should be avoided if possible. There is also the potential for contamination during transit and storage of seed as well as during seed germination, growth, and harvest.

Commercial sprout growers need to follow good manufacturing practices (GMPs) and have written standard sanitation operating procedures (SSOPs) and a hazard analysis and critical control point (HACCP) plan in place [126]. Growers should be thoroughly familiar with the recommendations contained in the FDA guidance documents which include detailed methods for testing of spent irrigation water for salmonella and *E. coli* O157:H7 [53]. Seed should be of high quality and all bags of seed should be inspected for evidence of rodent activity (gnawed holes and presence of urine stains using a blacklight). Thorough testing of all lots of sprout seed for bacterial pathogens is desirable and should reduce the risk of sprout-related outbreaks of foodborne disease. A sampling and testing protocol for use with sprout seed for human pathogens has been proposed [127]. However, due to the sporadic and low level of contamination with human pathogens often encountered, a negative sample test cannot guarantee that the entire lot is pathogen free. Thus, an effective, approved seed-sanitizing step should be applied by the grower, and the spent irrigation water or sprouts should be tested for the presence of pathogens. Irrigation water needs to be of high quality and the use of well water also requires regular testing for adequate levels of residual chlorine. Postharvest contamination of sprouts can occur during transit, storage, display, and by cross contamination in restaurant or home kitchens and adequate precautions need to be taken.

8.6 RESEARCH NEEDS

Despite considerable research efforts towards the development of sprout seed-sanitizing methods there is still a need for highly effective, low-cost, easily implemented, and environmentally benign seed-sanitizing strategies that can be used by organic and conventional sprout growers. The use of 20,000 ppm $Ca(OCl)_2$ presents worker and environmental safety concerns, may not always be effective in eliminating human pathogens from contaminated seed lots under commercial practice, and can be highly detrimental to the germination capacity of some seed types [65]. The potential for internalization of bacterial human pathogens into sprouts during germination and growth from contaminated sprouting seed has been demonstrated, but the location of pathogens on naturally contaminated seed is still not known. Are the pathogens solely surfaceborne, sometimes entering into cracks and natural openings such as the hilum and micropyle in the seed coat, or are they also present internally in the seed coat as are some seedborne plant pathogenic bacteria [128]? The optimization and commercialization of biological control agents for use on sprouting seed as an alternative to chemical sanitizers is highly desirable.

The ecology of human pathogens on sprouts is not well defined and several questions remain unanswered. Are pathogens capable of forming biofilms on sprout surfaces or can they become part of biofilms formed by the native microflora making their eradication more problematic? What microbial cell surface components (e.g., curli, fimbriae, flagella, and extracellular poly-saccharides such as colanic acid and cellulose) are important for the initial attachment to plant surfaces and subsequent biofilm formation? Does the plant react in any way to the presence of pathogens on surfaces or in internal tissues? Could sprout seed cultivars be developed that release high levels of antibac-terial compounds upon germination that might inhibit growth and survival of pathogens?

Further research in the areas mentioned above should assist in the development of improved strategies for reducing the risk of future foodborne outbreaks allowing for greater consumer confidence in the microbiological safety of sprouts and ensuring the survival of a strong sprout industry worldwide. Intervention strategies developed for seeds and sprouts may also be applicable to ensuring the microbiological safety of other types of produce.

ACKNOWLEDGMENTS

Sincere thanks are due to Drs. Francisco Diez-Gonzalez and Mindy Brashears for supplying articles in press, to Drs. Jeff Farrar and Glenn Henderson for supplying information concerning sprout-related outbreaks, and to Drs. Jeff Farrar, James Smith, and Mary Lou Tortorello for critically reviewing the manuscript before publication.

REFERENCES

1. International Sprout Growers Association, personal communication, 2003.
2. Meyerowitz, S., *Sprouts, The Miracle Food: The Complete Guide to Sprouting*, Sproutman Publications, Great Barrington, MA, 1998, p. 57.
3. Thomas, J.L. *et al.*, Industry practices and compliance with the U.S. Food and Drug Administration Guidelines among California sprout firms, *J. Food Prot.*, 66, 1253, 2003.
4. National Advisory Committee on Microbiological Criteria for Foods, Micro-biological safety evaluations and recommendations on sprouted seeds, *Int. J. Food Microbiol.*, 52, 123, 1999.
5. Taormina, P.J., Beuchat, L.R., and Slutsker, L., Infections associated with eating seed sprouts: an international concern, *Emerg. Infect. Dis.*, 5, 626, 1999.
6. U.S. Food and Drug Administration, Consumers in Oregon area advised of risks associated with raw sprouts, www.cfsan.fda.gov/~lrd/fpsprout.html (accessed March 1, 2004).

7. U.S. Food and Drug Administration, 2001 Food Code, http://www.cfsan.fda.gov/~dms/fc01-1.html (accessed Feb. 17, 2004).

8. Portnoy, B.L., Goepfert, J.M., and Harmon, S.M., An outbreak of *Bacillus cereus* food poisoning resulting from contaminated vegetable sprouts, *Am. J. Epidemiol.*, 103, 589, 1976.

9. Centers for Disease Control and Prevention. U.S. Foodborne Disease Outbreaks, www.cdc.gov/foodborneoutbreaks/us_outb.htm (accessed Dec. 29, 2003).

10. Mahon, B.E. *et al.*, An international outbreak of *Salmonella* infections caused by alfalfa sprouts grown from contaminated seeds, *J. Infect. Dis.*, 175, 876, 1997.

11. Abbo, S. and Baggesen, D.L., Growth of *Salmonella* Newport in naturally contaminated alfalfa sprouts and estimation of infectious dose in a Danish *Salmonella* Newport outbreak due to alfalfa sprouts, *Salmonella and Salmonellosis '97*, 425, 1997.

12. Van Beneden, C.A. *et al.*, A health food fights back: an international outbreak of *Salmonella newport* infections due to alfalfa sprouts, in *Proc. 36th Interscience Conference on Antimicrobial Agents and Chemotherapy*, American Society for Microbiology, 1996.

13. Mohle-Boetani, J.C. *et al.*, *Escherichia coli* O157 and *Salmonella* infections associated with sprouts in California, 1996–1998, *Ann. Int. Med.*, 135, 240, 2001.

14. Breuer, T. *et al.*, A multistate outbreak of *Escherichia coli* O157:H7 infections linked to alfalfa sprouts grown from contaminated seeds, *Emerg. Infect. Dis.*, 7, 977, 2001.

15. Backer, H.D. *et al.*, High incidence of extra-intestinal infections in a *Salmonella havana* outbreak associated with alfalfa sprouts, *Public Health Rep.*, 115, 339, 2000.

16. Gill, C.J. *et al.*, Alfalfa seed decontamination in a *Salmonella* outbreak, *Emerg. Intfect. Dis.*, 9, 474, 2003.

17. Brooks, J.T. *et al.*, *Salmonella* Typhimurium infections transmitted by chlorine-pretreated clover sprout seeds, *Am. J. Epidemiol.*, 154, 1020, 2001.

18. Proctor, M.E. *et al.*, Multistate outbreak of *Salmonella* serovar Muenchen infections associated with alfalfa sprouts grown from seeds pretreated with calcium hypochlorite, *J. Clin. Microbiol.*, 39, 3461, 2001.

19. Farrar, J., California Department of Health Services, personal communication, 2003.

20. Winthrop, K.L. *et al.*, Alfalfa sprouts and *Salmonella* Kottbus infection: a multistate outbreak following inadequate seed disinfection with heat and chlorine, *J. Food Prot.*, 66, 13, 2003.

21. Centers for Disease Control and Prevention, Update on *Salmonella* serotype Enteritidis infections, outbreaks, and the importance for traceback and timely reporting of outbreaks, http://www.cdc.gov/ncidod/dbmd/diseaseinfo/files/2001SECSTE.pdf (accessed Jan. 5, 2004).

22. Henderson, G., U.S. Food and Drug Administration, personal communication, 2004.

23. Smith, K., Minnesota Department of Health, personal communication, 2003.

24. Fleming, D., Multinational outbreak of *Salmonella enterica* serotype Newport infections due to contaminated alfalfa sprouts, *JAMA*, 281, 158, 1999.

25. Puohiniemi, R., Heiskanen, T., and Siitonen, A., Molecular epidemiology of two international sprout-borne *Salmonella* outbreaks, *J. Clin. Microbiol.*, 35, 2487, 1997.

26. O'Mahony, M. *et al.*, An outbreak of *Salmonella saint-paul* infection associated with bean sprouts, *Epidemiol. Infect.*, 104, 229, 1990.

27. Mead, P.S. *et al.*, Food-related illness and death in the United States, *Emerg. Infect. Dis.*, 5, 607, 1999.

28. Michino, H. *et al.*, Massive outbreak of *Escherichia coli* O157:H7 infection in school children in Sakai City, Japan, associated with consumption of white radish sprouts, *Am. J. Epidemiol.*, 150, 787, 1999.

29. Watanabe, Y. *et al.*, Factory outbreak of *Escherichia coli* O157:H7 infection in Japan, *Emerg. Infect. Dis.*, 5, 424, 1999.

30. Andrews, W.H., *et al.*, Microbial hazards associated with bean sprouting, *J. Assoc. Off. Anal. Chem.*, 65, 241, 1982.

31. Castro-Rosas, J. and Escartin, E.F., Survival and growth of *Vibrio cholerae* O1, *Salmonella typhi*, and *Escherichia coli* O157:H7 in alfalfa sprouts, *J. Food Sci.*, 65, 162, 2000.

32. Castro-Rosas, J. and Escartin, E.F. Incidence and germicide sensitivity of *Salmonella typhi* and *Vibrio cholerae* O1 in alfalfa sprouts, *J. Food Saf.*, 19, 137, 1999.

33. Howard, M.B. and Hutcheson, S.W., Growth dynamics of *Salmonella enterica* strains on alfalfa sprouts and in waste seed irrigation water, *Appl. Environ. Microbiol.*, 69, 548, 2003.

34. Jaquette, C.B., Beuchat, L.R., and Mahon, B.E., Efficacy of chlorine and heat treatment in killing *Salmonella stanley* inoculated onto alfalfa seeds and growth and survival of the pathogen during sprouting and storage, *Appl. Environ. Microbiol.*, 62, 2212, 1996.

35. Hara-Kudo, Y. *et al.*, Potential hazard of radish sprouts as a vehicle of *Escherichia coli* O157:H7, *J. Food Prot.*, 60, 1125, 1997.

36. Stewart, D. *et al.*, Growth of *Escherichia coli* O157:H7 during sprouting of alfalfa seeds, *Lett. Appl. Microbiol.*, 33, 95, 2001.

37. Taormina, P.J. and Beuchat, L.R., Behavior of enterohemorrhagic *Escherichia coli* O157:H7 on alfalfa sprouts during the sprouting process as influenced by treatments with various chemicals, *J. Food Prot.*, 62, 850, 1999.

38. Warriner, K. *et al.*, Internalization of bioluminescent *Escherichia coli* and *Salmonella* Montevideo in growing bean sprouts, *J. Appl. Microbiol.*, 95, 719, 2003.

39. Barak, J.D., Whitehand, L.C., and Charkowski, A.O., Differences in attachment of *Salmonella enterica* serovars and *Escherichia coli* O157:H7 to alfalfa sprouts, *Appl. Environ. Microbiol.*, 68, 4758, 2002.

40. Prokopowich, D. and Blank, G., Microbiological evaluation of vegetable sprouts and seeds, *J. Food Prot.*, 4, 560, 1991.

41. Patterson, J.E. and Woodburn, M.J., *Klebsiella* and other bacteria on alfalfa and bean sprouts at the retail level, *J. Food Sci.*, 45, 492, 1980.

42. Thunberg, R.L. *et al.*, Microbial evaluation of selected fresh produce obtained at retail markets, *J. Food Prot.*, 65, 677, 2002.

43. Harmon, S.M., Kautter, D.A., and Solomon, H.M., *Bacillus cereus* contamination of seeds and vegetable sprouts grown in a home sprouting kit, *J. Food Prot.*, 50, 62, 1987.

44. Stewart, D.S. *et al.*, Growth of *Salmonella* during sprouting of alfalfa seeds associated with salmonellosis outbreaks, *J. Food Prot.*, 64, 618, 2001.

45. Itoh, Y. *et al.*, Enterohemorrhagic *Escherichia coli* O157:H7 present in radish sprouts. *Appl. Environ. Microbiol.*, 64, 1532, 1998.

46. Dong, Y. *et al.*, Kinetics and strain specificity of rhizosphere and endophytic colonization by enteric bacteria on seedlings of *Medicago sativa* and *Medicago trunculata*, *Appl. Environ. Microbiol.*, 69, 1783, 2003.

47. Gandhi, M. *et al.*, Use of green fluorescent protein expressing *Salmonella* Stanley to investigate survival, spatial location, and control on alfalfa sprouts, *J. Food Prot.*, 64, 1891, 2001.

48. Hallman, J. *et al.*, Bacterial endophytes in agricultural crops, *Can. J. Microbiol.*, 43, 895, 1997.

49. Fett, W.F., Naturally occurring biofilms on alfalfa and other types of sprouts, *J. Food Prot.*, 63, 625, 2000.

50. Fett, W.F. and Cooke, P., Scanning electron microscopy of native biofilms on mung bean sprouts, *Can. J. Microbiol.*, 49, 45, 2003.

51. Gandhi, M., and Matthews, K.R., Efficacy of chlorine and calcinated calcium treatment of alfalfa seeds and sprouts to eliminate *Salmonella*, *Int. J. Food Microbiol.*, 87, 301, 2003.

52. Lang, M.M., Ingham, B.H., and Ingham, S.C., Efficacy of novel organic acid and hypochlorite treatments for eliminating *Escherichia coli* O157:H7 from alfalfa seeds prior to sprouting, *Int. J. Food Microbiol.*, 58, 73, 2000.

53. U.S. Food and Drug Administration, Guidance for industry: reducing microbial food safety hazards for sprouted seeds and guidance for industry: sampling and microbial testing of spent irrigation water during sprout production, *Fed. Registr.*, 64, 57893, 1999.

54. U.S. Food and Drug Administration and the California Department of Health Services, Food and Drug Branch, Safer Processing of Sprouts. Information available at: http://www.cfsan.fda.gov/~dms/sprouvid.html, 2000 (accessed Dec. 29, 2003).

55. Delaquis, P.J., Sholberg, P.L., and Stanich, K., Disinfection of mung bean seed with gaseous acetic acid, *J. Food Prot.*, 62, 953, 1999.

56. Weissinger, W.R., McWatters, K.H., and Beuchat, L.R., Evaluation of volatile chemical treatments for lethality to *Salmonella* on alfalfa seeds and sprouts, *J. Food Prot.*, 64, 442, 2001.

57. Kim, C. *et al.*, Efficacy of electrolyzed oxidizing water in inactivating *Salmonella* on alfalfa seeds and sprouts, *J. Food Prot.*, 66, 208, 2003.

58. Sharma, R.R. and Demirci, A., Treatment of *Escherichia coli* O157:H7 inoculated alfalfa seeds and sprouts with electrolyzed water, *Int. J. Food Microbiol.*, 86, 231, 2003.

59. Stan, S.D. and Daeschel, M.A., Reduction of *Salmonella enterica* on alfalfa seeds with acidic electrolyzed oxidizing water and enhanced uptake of acidic electrolyzed oxidizing water into seeds by gas exchange, *J. Food Prot.*, 66, 2017, 2003.

60. Park, C.M., Taormina, P.J., and Beuchat, L.R., Efficacy of allyl isothiocyanate in killing enterohemorrhagic *Escherichia coli* O157:H7 on alfalfa seeds, *Int. J. Food Microbiol.*, 56, 13, 2000.

61. Himathongkham, S. *et al.*, Reduction of *Escherichia coli* O157:H7 and *Salmonella* Typhimurium in artificially contaminated alfalfa seeds and mung beans by fumigation with ammonia, *J. Food Prot.*, 64, 1817, 2001.

62. Holliday, S.L., Scouten, A.J., and Beuchat, L.R., Efficacy of chemical treatments in eliminating *Salmonella* and *Escherichia coli* O157:H7 on scarified and polished alfalfa seeds, *J. Food Prot.*, 64, 1489, 2001.

63. Weissinger, W.R. and Beuchat, L.R., Comparison of aqueous chemical treatments to eliminate *Salmonella* on alfalfa seeds, *J. Food Prot.*, 63, 1475, 2000.

64. Taormina, P.J. and Beuchat, L.R., Comparison of chemical treatments to eliminate enterohemorrhagic *Escherichia coli* O157:H7 on alfalfa seeds, *J. Food Prot.*, 62, 318, 1999.

65. Fett, W.F., Factors affecting the efficacy of chlorine against *Escherichia coli* O157:H7 and *Salmonella* on alfalfa seed, *Food Microbiol.*, 19, 135, 2002.

66. Fett, W.F., Reduction of *Escherichia coli* O157:H7 and *Salmonella* spp. on laboratory-inoculated mung bean seed by chlorine treatment, *J. Food Prot.*, 65, 848, 2002.

67. Fett, W.F. and Cooke, P.H., Reduction of *Escherichia coli* O157:H7 and *Salmonella* on laboratory-inoculated alfalfa seed with commercial citrus-related products, *J. Food Prot.*, 66, 1158, 2003.

68. Bari, M.L. *et al.*, Chemical and irradiation treatments for killing *Escherichia coli* O157:H7 on alfalfa, radish, and mung bean seeds, *J. Food Prot.*, 66, 767, 2003.

69. Beuchat, L.R., Ward, T.E., and Pettigrew, C.A., Comparison of chlorine and a prototype produce wash product for effectiveness in killing *Salmonella* and *Escherichia coli* O157:H7 on alfalfa seeds, *J. Food Prot.*, 64, 152, 2001.

70. Wuytack, E.Y. *et al.*, Decontamination of seeds for sprout production by high hydrostatic pressure, *J. Food Prot.*, 66, 918, 2003.

71. Thayer, D.W. *et al.*, Inactivation of *Escherichia coli* O157 and *Salmonella* by gamma irradiation of alfalfa seed intended for production of food sprouts, *J. Food Prot.*, 66, 175, 2003.

72. Rajkowski, K.T., Boyd, G., and Thayer, D.W., Irradiation D-values for *Escherichia coli* O157:H7 and *Salmonella* sp. on inoculated broccoli seeds and effects of irradiation on broccoli sprout keeping quality and seed viability, *J. Food Microbiol.*, 66, 760, 2003.

73. Pandrangi, S. *et al.*, Efficacy of sulfuric acid scarification and disinfectant treatments in eliminating *Escherichia coli* O157:H7 from alfalfa seeds prior to sprouting, *J. Food Sci.*, 68, 613, 2003.

74. Sharma, R.R. *et al.*, Inactivation of *Escherichia coli* O157:H7 on inoculated alfalfa seeds with ozonated water and heat treatment, *J. Food Prot.*, 65, 447, 2002.

75. Wade, W.N. *et al.*, Efficacy of ozone in killing *Listeria monocytogenes* on alfalfa seeds and sprouts and effects on sensory quality of sprouts, *J. Food Prot.*, 66, 44, 2003.

76. Sharma, R.R. and Demirci, A., Inactivation of *Escherichia coli* O157:H7 on inoculated alfalfa seeds with pulsed ultraviolet light and response surface modeling, *J. Food Sci.*, 68, 1448, 2003.

77. Nelson, S.O. *et al.*, Radio-frequency heating of alfalfa seed for reducing human pathogens, *Trans. Am. Soc. Agric. Eng.*, 45, 1937, 2002.

78. Mazzoni, A.M. *et al.*, Supercritical carbon dioxide treatment to inactivate aerobic microorganisms on alfalfa seeds, *J. Food Saf.*, 21, 215, 2001.

79. Enomoto, K. *et al.*, Hot-water treatments for disinfecting alfalfa seeds inoculated with *Escherichia coli* ATCC 25922, *Food Sci. Technol. Res.*, 8, 247, 2002.

80. Weiss, A. and Hammes, W.P., Thermal seed treatment to improve the food safety status of sprouts, *J. Appl. Bot.*, 77, 152, 2003.

81. Charkowski, A.O., Sarreal, C.Z., and Mandrell, R.E. Wrinkled alfalfa seeds harbor more aerobic bacteria and are more difficult to sanitize than smooth seeds, *J. Food Prot.*, 64, 1292, 2001.

82. Mundt, J.O., and Hinkle, N.F., Bacteria within ovules and seeds, *Appl. Environ. Microbiol.*, 32, 694, 1976.

83. Scouten, A.J. and Beuchat, L.R., Combined effects of chemical, heat and ultrasound treatments to kill *Salmonella* and *Escherichia coli* O157:H7 on alfalfa seeds, *J. Appl. Microbiol.*, 92, 668, 2002.

84. Delaquis, P., Agriculture and Agri-Food Canada, personal communication, 2004.

85. Suslow, T.V. *et al.*, Detection and elimination of *Salmonella* Mbandaka from naturally contaminated alfalfa seed by treatment with heat or calcium hypochlorite, *J. Food Prot.*, 65, 452, 2002.

86. U.S. Food and Drug Administration, Irradiation in the production, processing and handling of food, *Fed. Registr.*, 65, 64605, 2000.

87. Rajkowski, K.T. and Thayer, D.W., Alfalfa seed germination and yield ratio and alfalfa sprout microbial keeping quality following irradiation of seeds and sprouts, *J. Food Prot.*, 64, 1988, 2001.

88. Fan, X., Thayer, D., and Sokorai, K.J.B., Changes in growth and antioxidant status of alfalfa sprouts during sprouting as affected by gamma irradiation of seeds, *J. Food Prot.*, 67, 561, 2004.

89. Thayer, D.W., Boyd, G., and Fett, W.F., γ-radiation decontamination of alfalfa seeds naturally contaminated with *Salmonella* Mbandaka., *J. Food Sci.*, 68, 1777, 2003.

90. Kikuchi, O.K. *et al.*, Efficacy of soft-electron (low-energy electron beam) for soybean decontamination in comparison with gamma-rays, *J. Food Sci.*, 68, 649, 2003.

91. Sharma, R.R. *et al.*, Inactivation of *Escherichia coli* O157:H7 on inoculated alfalfa seeds with ozonated water and heat treatment, *J. Food Prot.*, 65, 447, 2002.

92. Singh, N., Singh, R.K., and Bhunia, A.K., Sequential disinfection of *Escherichia coli* O157:H7 inoculated alfalfa seeds before and during sprouting using aqueous chlorine dioxide, ozonated water, and thyme essential oil, *Lebensm.-Wiss. U.-Technol.*, 36, 235, 2002.

93. Beuchat, L.R. and Scouten, A.J., Combined effects of water activity, temperature and chemical treatments on the survival of *Salmonella* and *Escherichia coli* O157:H7 on alfalfa seeds, *J. Appl. Microbiol.*, 92, 382, 2002.

94. Campbell, R.C., *Biological Control of Microbial Plant Pathogens*, Cambridge University Press, Cambridge, 2003.

95. Breidt, F. and Fleming, H.P., Using lactic acid bacteria to improve the safety of minimally processed fruits and vegetables, *Food Technol.*, 51, 44, 1997.

96. Nisbet, D., Defined competitive exclusion cultures in the prevention of enteropathogenic colonization in poultry and swine, *Antonie Leeuwenhoek*, 81, 481, 2002.

97. Cai, Y., Ng, L.-K., and Farber, J.M., Isolation and characterization of nisin-producing *Lactococcus lactis* subsp. *lactis* from bean-sprouts, *J. Appl. Microbiol.*, 83, 499, 1997.

98. Palmai, M. and Buchanan, R.L., Growth of *Listeria monocytogenes* during germination of alfalfa sprouts, *Food Microbiol.*, 19, 195, 2002.

99. Wilderdyke, M.R., Smith, D.A., and Brashears, M.M., Isolation, identification, and selection of lactic acid bacteria from alfalfa sprouts for competitive inhibition of food-borne pathogens. *J. Food Prot.*, 67, 947, 2004.

100. Harris, M.R. and Brashears, M.M., Evaluation of the use of lactic acid bacteria to control pathogens on alfalfa sprouts, Annual Meeting of the International Association for Food Protection, June 30–July 3, 2002, *Program and Abstract Book*, abstract no. T32, p. 130.

101. Okada, S., Tokyo University of Agriculture, personal communication, 2000.

102. Matos, A. and Fett, W.F., unpublished data, 2003.

103. Matos, A., Garland, J.L. and Fett, W.F., Composition and physiological profiling of sprout-associated microbial communities, *J. Food Prot.*, 65, 1903, 2002.

104. Nandiwada, L.S., Schamberger, G.P., Schafer, H.W., and Diez-Gonzalez, F., Characterization of a novel E2-type colicin and its application to treat alfalfa seeds to reduce *Escherichia coli* O157:H7. *Int. J. Food Microbiol.*, 93, 267, 2004.

105. Kostrzynska, M. *et al.*, Application of bacteriophages to control food-borne pathogens on contaminated alfalfa sprouts. Program 9th International Symposium on Microbial Ecology, Amsterdam, 2001, abstr. no. P.22.025.

106. Enomoto, K., Use of bean sprout *Enterobacteriaceae* isolates as biological control agents of *Pseudomonas fluorescens*, *J. Food Sci.*, 69, 17, 2004.

107. Fett, W.F., Reduction of the native microflora on alfalfa sprouts during propagation by addition of antimicrobial compounds to the irrigation water, *Int. J. Food Microbiol.*, 72, 13, 2002.

108. Piernas, V. and Guiraud, J.P., Control of microbial growth on rice sprouts, *Int. J. Food Sci. Technol.*, 33, 297, 1998.

109. Naito, S. and Shiga, I., Effect of ozone treatment on elongation of hypocotyls and microbial counts of bean sprouts, *J. Jpn. Soc. Food Sci. Technol.*, 36, 181, 1989.

110. Taormina, P.J. and Beuchat, L.R., Behavior of enterohemorrhagic *Escherichia coli* O157:H7 on alfalfa sprouts during the sprouting process as influenced by treatments with various chemicals, *J. Food Prot.*, 62, 850, 1999.

111. Becker, B. and Holzapfel, W.H., Microbiological risk of prepacked sprouts and measures to reduce total counts. *Arch. Lebensmittelhygiene* 48, 81, 1997.

112. Park, W.P., Cho, S.H., and Lee, D.S., Effect of minimal processing operations on the quality of garlic, green onion, soybean sprouts and watercress, *J. Sci. Food Agric.*, 77, 282, 1998.

113. Piernas, V. and Guiraud, J.P., Microbial hazards related to rice sprouting, *Int. J. Food Sci. Technol.*, 32, 33, 1997.

114. Lee, S-Y. *et al.*, Inhibition of *Salmonella* Typhimurium and *Listeria monocytogenes* in mung bean sprouts by chemical treatment, *J. Food Prot.*, 65, 1088, 2002.

115. Fratamico, P.M. and Bagi, L.K., Comparison of an immunochromatographic method and the Taqman® *E. coli* O157:H7 assay for detection of *Escherichia coli* O157:H7 in alfalfa sprout spent irrigation water and in sprouts after blanching, *J. Ind. Microbiol. Biotechnol.*, 27, 129, 2001.

116. Kim, C. *et al.*, Efficacy of electrolyzed oxidizing water in inactivating *Salmonella* on alfalfa seeds and sprouts, *J. Food Prot.*, 66, 208, 2003.

117. Sharma, R.R. and Demirci, A., Treatment of *Escherichia coli* O157:H7 inoculated alfalfa seeds and sprouts with electrolyzed oxidizing water. *Int. J. Food Microbiol.*, 86, 231, 2003.

118. Sharma, R.R. *et al.*, Application of ozone for inactivation of *Escherichia coli* O157:H7 on inoculated alfalfa sprouts, *J. Food Proc. Preserv.*, 27, 51, 2003.

119. Fan, X. and Thayer, D.W., Quality of irradiated alfalfa sprouts, *J. Food Prot.*, 64, 1574, 2001.
120. Rajkowski, K.T. and Thayer, D.W., Reduction of *Salmonella* spp. and strains of *Escherichia coli* O157:H7 by gamma radiation of inoculated sprouts, *J. Food Prot.*, 63, 871, 2000.
121. Schoeller, N.P., Ingham, S.C., and Ingham, B.H., Assessment of the potential for *Listeria monocytogenes* survival and growth during alfalfa sprout production and use of ionizing radiation as a potential intervention treatment, *J. Food Prot.*, 65, 1259, 2002.
122. U.S. Food and Drug Administration, Department of Agriculture and Centers for Disease Control and Prevention. Guide to Minimize Food Safety Hazards for Fresh Fruits and Vegetables, 1998, http://www.foodsafety.gov/~dms/prodguid.html (accessed Feb. 6, 2004).
123. Rangarajan, A., *et al.*, Food Safety Begins on the Farm: A Grower's Guide, Cornell University, 2001, http://www.gaps.cornell.edu/pubs/Farm_Boo.pdf (accessed Feb. 13, 2004).
124. International Fresh-Cut Produce Association and the Western Growers Association, Voluntary Food Safety Guidelines for Fresh Produce, 1997.
125. Fett, W.F. and Sapers, G., USDA, ARS, Eastern Regional Research Center, unpublished data, 1998.
126. Stevenson, K.E. and Bernard, D. T., *HACCP: A Systematic Approach to Food Safety*, Food Processors Institute, Washington D.C., 1999.
127. International Specialty Supply, http://www.sproutnet.com/sprouting_seed_safety.htm and http://www.sproutnet.com/Research/seed_sampling_and_testing.htm (accessed March 1, 2004).
128. Maude, R.B., *Seedborne Diseases and Their Control: Principles and Practice*, CAB International, Wallingford, U.K., 1996, p. 28.

9 Microbiological Safety of Fresh Citrus and Apple Juices

Susanne E. Keller and Arthur J. Miller

CONTENTS

9.1 INTRODUCTION

Historically, citrus juices and apple juice or cider were not considered to be beverages associated with a high risk for causing foodborne illness. These products were not typically thought of as being exposed to pathogens that were animal derived, such as salmonella. Secondly, the pH and organic acid content of these foods was presumed to be too adverse for the survival or growth of bacterial foodborne pathogens. Nonetheless, incidents of foodborne illness associated with citrus juice and apple cider occurred as far back as 1922 [1].

211

Documented evidence of pathogen survival in juice has also existed for some time, along with proposed mechanisms for acid resistance [2–12]. Therefore, survival of foodborne pathogens, and the occurrence of serious foodborne illness outbreaks, including fatalities, have led to new regulation requiring the implementation of hazard analysis critical control point (HACCP) programs by juice manufacturers [13]. The regulation requires implementation of a process capable of reducing the pertinent pathogen by 100,000-fold (5 log units).

This chapter briefly describes production of citrus and apple juices, their physical characteristics, and typical microflora. The emphasis is on pathogens that have been associated with fresh juice and on recent regulations related to the prevention of foodborne illness outbreaks. Sources of contamination and intervention methods are also discussed.

9.2 JUICE PRODUCTION

The U.S. Food and Drug Administration (FDA) defines juice as "the aqueous liquid expressed or extracted from one or more fruits or vegetables, purees of the edible portions of one or more fruits and vegetables, or any concentrates of such liquid or puree" [13]. Produce production is beyond the scope of this chapter, but it is critical to recognize that fruit and vegetables used for juice manufacture should be produced, harvested, and transported using good agricultural practices (GAPs). Only high-quality produce should be used. Juice processing begins with the reception of the raw produce at the processing facility. Raw produce is inspected and culled according to established good manufacturing practices (GMPs). Removal of defective raw material is critical to the production of a quality juice product. This is discussed in greater detail in Chapter 16. Sound raw produce is then cleaned and sanitized prior to extraction or maceration. For some products, such as apples and oranges, a mechanical means of washing may be employed, such as a brusher-washer. Such methods can efficiently remove soil and extraneous materials, but may be too harsh for more fragile fruits, such as berries. Sanitizing follows cleaning, which generally results in some reduction of microbial load at the surface of the fruit. Both cleaning and sanitizing are described in more detail in Chapter 17.

After appropriate culling, cleaning, and sanitizing, most noncitrus produce is macerated. Generally, produce is mechanically conveyed to size reduction equipment such as a hammer mill, crusher, or a grater for processing into a mash or pulp-like material from which the juice may be extracted. Rice hull may be added to the mash to improve juice yield during extraction. Any such added ingredients must also be approved and used according to established GMPs. Following maceration, some product types, such as tomatoes or grapes, may be given a mild heat treatment to set color, inactivate enzymes, and/or improve yield. In general, this treatment is not an effective means to reduce microbial load. Extraction of the juice follows mild heat treatment, if it is employed.

The most common juice extraction method from a mash or pulp is batch hydraulic pressing. The whole or chopped raw fruit or vegetable is placed into bags that are stacked alternately with plastic separator grid interleaves and then subjected to hydraulic pressure. Pulpers, with tapered screws or paddles, that squeeze juice and puree through a cylindrical screen, while carrying the pomace to one end for discharge, are also common.

Juices may be marketed in both clarified and unclarified styles. For clarified juices, additional processing aids, such as approved pectinolytic enzymes, are added to facilitate removal of particles and cloud.

For citrus products an entirely different extraction procedure is commonly employed by large-scale manufacturers. Citrus fruits are generally not macerated; rather, the juice is extracted while largely maintaining peel integrity. This results in limited contact between the juice and the peel. Two types of equipment are in common use: a mechanical reamer and a pin-point extractor [14]. Mechanical reamers first cut the fruit in half. The halves are then held against rotating burrs to extract the juice and pulp. A pin-point extractor contains a small hollow tube that punctures the peel at one point. Then, intermeshing mechanical fingers squeeze the fruit surface to force the juice and pulp out through the hollow extraction tube. Seeds and pulp are separated from the juice using cylindrical pulpers and finishers.

9.3 PHYSICOCHEMICAL PROPERTIES AND ENDOGENOUS JUICE MICROFLORA

9.3.1 CITRUS JUICE

Citrus are nonclimacteric fruits that are allowed to mature on the tree, since postharvest maturation will not occur. The indices of fruit maturity include °Brix, acid content, and the Brix/acid ratio. Major U.S. citrus producing regions, such as California, Florida, and Texas, all have legal maturity standards. Typically, oranges range from 7 to 14, grapefruit from 10 to 12, and tangerines from 16 to 17 °Brix [15]. Citric acid is the major acid present in citrus fruit. At maturity, concentrations of total acid for oranges, grapefruit, and tangerines are 0.5 to 1.5%, 1.0 to 2.0%, and 0.6 to 2.3%, respectively. The U.S. Department of Agriculture (USDA) has published standards for grades of orange juice. Brix standards for pasteurized juice are 11° for grade A and 10.5° for grade B. An acid concentration or pH standard is not established. However, Brix/acid minimum and maximum ratios are given for grade A and grade B juice. The typical pH range for most citrus juice is from 3.0 to 4.0 and cannot be legally altered by added acidulants [16].

Typical aerobic microbial load on citrus fruit is approximately $4.0 \log$ CFU/cm^2 [17–19]. Yeast and mold populations measured alone seem to show greater variability than total aerobic populations, but have been reported as nearly as high as total aerobic microbial load. In citrus juices, acidic conditions, coupled with higher sugar content, result in a microbiological population made up primarily of acidolactic bacteria, yeasts, and molds. Lactic acid bacteria are reported to be the major spoilage organisms [16,20,21]. Populations found in fresh citrus juices are reduced compared to populations found on fruit when appropriate sanitation and extraction methods are used [22].

9.3.2 APPLE CIDER

Apples are climacteric fruit, with respiration increasing as the fruit matures. The ripening process will continue postharvest. Indices of maturation are variety specific; they include hardness, °Brix, and/or color. The composition of apple juice is dependent on fruit variety. A study by Mattick and Moyer examined 15 varieties of apple from 8 geographic locations over 3 years [23]. Although 15 varieties were examined, only composite data were reported. Composite results over this 3-year period showed a mean °Brix of 12.74, with minimum and maximum values of 9.8 and 16.9, respectively. The mean estimate of composite pH for this same period was 3.69, with a range of 3.23 to 6.54. Total acid calculated as % malic acid was 0.42% with minimum and maximum values of 0.15 and 0.91%, respectively. Malic acid represents $>85\%$ of the acid present in apples [24]. In an earlier study by Goverd et al., pH values and acid levels (% malic) were listed for six different apple varieties [25]. Acid content and pH values varied from 0.17 % with a pH of 4.03 for variety Sweet Coppin to 1.43 % with a pH of 2.92 for variety Bramley's Seedling. Brix was not reported.

In a study initiated by the FDA in 1998, seven apple varieties were examined for pH, % acid, and °Brix prior to fresh juice/cider manufacture [26]. Brix, pH, and % titratable acidity (TA, calculated as malic acid) averages for fresh juice/cider made from both fresh and stored apples that were harvested from trees and from the ground are shown in Table 9.1. All three parameters were significantly influenced by apple variety ($p < 0.0001$).

TABLE 9.1
Influence of Apple Variety on pH, °Brix, and % TA Content of Fresh Apple Juice/Cider

Apple variety	pH[a]	°Brix[a]	% Acid (malic)[a]
Fuji	3.88 ± 0.06	16.8 ± 1.2	0.40 ± 0.08
Gala	3.94 ± 0.09	14.6 ± 0.4	0.26 ± 0.04
Golden Delicious	3.71 ± 0.10	13.0 ± 0.4	0.35 ± 0.04
Granny Smith	3.46 ± 0.05	13.1 ± 0.9	0.57 ± 0.06
McIntosh	3.48 ± 0.05	12.0 ± 0.7	0.55 ± 0.07
Red Delicious	4.06 ± 0.09	13.6 ± 0.9	0.20 ± 0.03
Red Rome	3.58 ± 0.10	13.4 ± 1.0	0.43 ± 0.06

[a] Values are means and standard deviations of all juices made from fresh and stored, tree- and ground-harvested apples during the 1999 harvest season in northern California.
Adapted from Keller, S.E., Chirtel, S.J., Merker, R.I., Taylor, K.T., Tan, H.L., and Miller, A.J., J. Food Prot., 67, 2240, 2004.

TABLE 9.2
Influence of Apple Variety on the Total Aerobic Microbial Populations Found in Fresh Apple Juice/Cider

Apple variety	Aerobic microflora in fresh cider/juice from tree-harvested unsorted fruit	
	Total aerobic plate count (log CFU/ml)[a]	Yeast and mold count (log CFU/ml)[a]
Fuji	3.93 ± 0.36	3.66 ± 0.13
Gala	2.65 ± 0.04	3.31 ± 0.69
Golden Delicious	3.57 ± 0.57	3.27 ± 0.56
Granny Smith	3.32 ± 0.56	2.87 ± 0.66
McIntosh	2.47 ± 0.52	1.97 ± 0.83
Red Delicious	3.77 ± 0.60	3.16 ± 0.40
Red Rome	3.85 ± 0.51	3.88 ± 0.13

[a] Values are means $\pm$ standard deviation of 6 replicate composite samples.
Adapted from Keller, S.E., Chirtel, S.J., Merker, R.I., Taylor, K.T., Tan, H.L., and Miller, A.J., *J. Food Prot.*, 67, 2240, 2004.

Although cider/juice physiochemical parameters such as °Brix, pH, and % TA are influenced by variety, there is large overlap in the range of values obtained for each apple variety. In addition, the range of any given parameter can be large. For example, the pH of cider/juice from over 32 batches of Fuji apples averaged 4.02 ± 0.30 (FDA, unreported data) [26]. However, individual batches from Fuji apples ranged from a low pH of 3.80 to a high pH of 4.65. Damaged and dropped fruit also have higher pH values. In a study by Dingman, fresh, undamaged tree-picked variety Red Delicious apples had a mean pH value of 3.98 ± 0.05, whereas bruised tree-picked, undamaged dropped fruit, and bruised dropped fruit had mean pH values of 4.57 ± 0.11, 4.15 ± 0.07, and 4.90 ± 0.09, respectively [27]. The range of physiochemical parameters such as pH of different apple varieties, as well as the influence of unsound fruit can be significant when compared to minimum levels required for growth of many foodborne pathogens.

Natural microflora found in fresh apple juice/cider also varies with apple variety and is significantly influenced by pH, % TA, and Brix (Table 9.2) [26]. Tree-harvested fruit that were culled to remove damaged fruit had total aerobic microbial populations in juice that ranged from 1.90 to 3.40 log CFU/ml. Yeast and mold populations in the same juice/cider ranged from 1.99 to 3.32 log CFU/ml. Total aerobic microbial and yeast and mold populations were also measured in poorer quality, ground harvest fruit. For this group, juice/cider microbial populations were substantially higher, ranging from 4.19 to 5.43 log CFU/ml for total aerobic populations and from 3.84 to 5.23 log CFU/ml for yeast and mold populations. Other studies reported similar population density ranges, particularly when ground harvested apples were included in cider production [28–30].

The types of organisms normally associated with fresh apple juice/cider are typically aciduric microorganisms, due to the pH, acid, and sugar content normally associated with this product. As with orange juice, human pathogenic microorganisms such as salmonella and enterohemorrhagic *Escherichia coli* are not considered endogenous microflora of the fruit or juice. Rather they are environmental contaminants originating from animal sources. However, unlike fresh citrus juice, populations of microorganisms in fresh apple juice/cider are generally higher than populations found on apples [28,29,31,32].

9.4 PATHOGENS ASSOCIATED WITH FRESH JUICE AND THEIR ENVIRONMENTAL SOURCES

9.4.1 ENTEROHEMORRHAGIC *ESCHERICHIA COLI*

Escherichia coli is one of the most studied bacteria. It is part of the normal bacterial flora resident in the intestines of many animals, including humans, and is commonly used as a nonpathogenic indicator of recent fecal contamination and of fecally associated pathogenic organisms such as salmonella [33]. However, numerous strains of *E. coli* exist which are not commensal. Pathogenic *E. coli* produces toxins of various types and toxicities that cause various diseases. These toxins have been described previously [34]. Diarrheagenic *E. coli* are subdivided into six classes based on the symptoms they produce and virulence factors they possess [35]. Of these groups, the enterohemorrhagic (EHEC) class is of most concern, due to its low infectious dose and its association with hemorrhagic colitis (HC), hemolytic uremic syndrome (HUS), and thrombotic thrombocytopenic purpura (TTP). HUS occurs primarily in children under 10 years of age and has a mortality of 3 to 5% [35]. Children's susceptibility to HUS led the FDA to issue a warning in November 2001 to the public concerning the health risk of consuming untreated juices by children [36].

Although there are several serotypes of EHEC known, the most common serotype, particularly in the U.S., Canada, Great Britain, and parts of Europe, is *E. coli* O157:H7 [35]. In the years from 1998 to 2000 the Center for Disease Control (CDC) recorded 86 outbreaks attributed to *E. coli*. Of these, 68 were identified as outbreaks caused by *E. coli* O157:H7 [37]. The great majority of these outbreaks were either from meat products or had an unknown source. Despite the fact that the majority of EHEC outbreaks are not associated with fresh fruit or juice made from fresh fruit, outbreaks associated with fresh fruit or fresh juice are of concern, since these products are associated with a healthy lifestyle and are generally consumed raw.

EHEC strains of *E. coli* are not normal endogenous microflora of fresh juice or of the fruit used to produce fresh juice. Their presence on fruit and in fruit juice is believed to be the consequence of some form of fecal contamination prior to consumption. Cattle have been implicated as a major reservoir of this organism [38–41]. Wild animals such as deer may be an additional source of the organism [42]. Wild birds have also been implicated as vectors for contamination, particularly those living near landfills [43]. Presumably, birds become infected at landfills and then may carry infection to farm fields and/or cattle. In addition to birds, transfer of *E. coli* O157:H7 by fruit flies has been demonstrated [44].

From epidemiological data it is clear that *E. coli* O157:H7 can survive well enough in low pH juice to result in serious illness. Of particular note was the Western states outbreak during October 1996 from contaminated apple cider that resulted in 66 cases of illness and one death [45,46]. Although the pH of most apple and orange juice is low enough to either significantly slow or inhibit growth of *E. coli*, EHEC strains have tolerance to high levels of acid allowing for extended survival time [3,47]. Tolerance to high acid levels is a complex induced response involving three distinct mechanisms and is enhanced in stationary phase cells [3,7,8,47].

9.4.2 SALMONELLA SPECIES

There are over 2000 serotypes of the genus salmonella that cause human disease [48]. According to the CDC, there are an estimated 1.4 million cases annually, with an estimated 500 fatalities [48]. Approximately half of all cases are caused by serotypes Enteritidis or Typhimurium [48].

Salmonella infections are more commonly associated with animal-derived foods, such as meat, seafood, dairy, and egg products, rather than juices. However, outbreaks associated with fresh juice have occurred as far back as 1922 [1]. Early outbreaks resulting in typhoid fever were associated with poor hygiene by asymptomatic *S.* Typhi shedding food handlers. As disinfection of water, sanitation procedures, and hygiene practices have improved, outbreaks of typhoid fever have become far less common in developed countries. Nonetheless, given the dramatic increase of fresh fruit imported from developing countries, typhoid fever outbreaks associated with these commodities remain a concern [49]. More recent outbreaks of nontyphoidal salmonellosis in fresh juice have been attributed to fecal-associated contamination of fruit or poor processing practices [50–52].

Both *E. coli* O157:H7 and salmonella are tolerant to extreme acid environments. As with *E. coli* O157:H7, tolerance in salmonella is inducible and increases when cells have been adapted either to acid conditions or are in stationary phase [4,7]. For *S.* Typhimurium, two major acid tolerance systems were identified, one associated with log phase and one associated with stationary phase [2]. Not surprisingly, survival in juice for extended periods has been observed. Goverd *et al.* reported survival of *S.* Typhimurium in apple cider [25]. Survival in juice above pH 3.6 held at 22°C was reported as greater than 30 days. Survival was decreased by lower pH and lower temperature. Survival in orange juice by various salmonella serovars was studied by Parish *et al.* [53]. Salmonella serovars Gaminara, Hartford, Rubislaw, and Typhimurium were inoculated at log 6 CFU/ml into orange juice at pH 3.5, 3.8, 4.1, and 4.4. Survival (to below levels of detection) at pH 3.5 ranged from a low of 14.3 ± 0.9 days for $S.$ Typhimurium to a high of 26.7 ± 4.0 days for $S.$ Hartford.

9.4.3 CRYPTOSPORIDIUM PARVUM

$Cryptosporidium$ $parvum$ is a highly infectious protozoan parasite causing persistent diarrhea. Common reservoirs are ruminants including cattle, deer, and sheep [54,55]. Infection with cryptosporidium does not always result in severe disease symptoms and the organism is far more dangerous for the immunocompromised [56]. Cryptosporidium is more commonly associated with contaminated water. The largest waterborne outbreak in U.S. history occurred in Milwaukee, WI, in 1993 and affected an estimated 403,000 people [56]. Cryptosporidium cannot replicate in the environment; however, the oocysts are thick-walled, resistant to chlorine, and persist in the environment. Presumably, the thick wall also confers some acid resistance, as outbreaks of cryptosporidiosis have also occurred from fresh-pressed cider [54,55]. Apple cider-associated outbreaks were reported in 1993, 1996, and 2003.

9.4.4 LISTERIA MONOCYTOGENES

Although not implicated in a foodborne outbreak associated with fresh juices, other foodborne pathogens do exist that also exhibit a tolerance/resistance to high levels of acid. Given an opportunity to contaminate fresh juice, these acid-resistant organisms could result in foodborne outbreaks. Chief among these possible pathogens is $Listeria$ $monocytogenes$. $L.$ $monocytogenes$ is ubiquitous within the environment, carried by animals, and frequently found on fruits and vegetables [57–59].

The minimum pH for growth of $L.$ $monocytogenes$ is dependant on the acidulant. For malic acid, the primary acid found in apple cider/juice, the lowest pH value for growth of $L.$ $monocytogenes$ is from 4.4 to 4.6 depending on the strain [60]. Although this pH is somewhat higher than typical fresh apple cider/juice, some apple cider/juice may fall within a range that will allow $L.$ $monocytogenes$ growth, particularly if unsound fruit is used in production. Not all apple cider/juice may have a pH low enough to prohibit growth of $L.$ $monocytogenes$. In addition, although $L.$ $monocytogenes$ may not grow at lower pH values, survival at lower pH similar to $E.$ $coli$ O157:H7 and salmonella is possible [60,61]. $L.$ $monocytogenes$ has been isolated from unpasteurized apple juice [62]. The recently completed $L.$ $monocytogenes$ risk assessment indicated that consumption of fresh fruit has a low risk for listeriosis [63]. However, two risk factors need to be considered concerning juice-associated listeriosis. First, comingling fruit to make juice or cider spreads the risk over a much larger exposed population, when compared to a single or limited serving size typically associated with the fruit itself. Second, fresh juice is frequently consumed by subpopulations at risk for listeriosis, e.g., children and adults with compromised immune systems. Consequently, it is reasonable to consider as somewhat likely outbreaks or sporadic cases of listeriosis associated with fresh juice.

9.5 JUICE HACCP RULE

In response to a series of juice-associated outbreaks, the FDA published its final rule on January 19, 2001 requiring the application of HACCP to juice production [13]. The rule became effective January 22, 2002 for large businesses. For small and very small businesses effective dates were January 21, 2003 and January 20, 2004, respectively. As part of the rule, the FDA issued a performance standard that requires all juice receive a treatment wherein the pertinent pathogen is reduced in concentration by 100,000-fold (5-log units). The "5-log reduction standard" was established based on recommendations by the National Advisory Committee on Microbiological Criteria for Foods (NACMCF). NACMCF considered worst-case scenarios, such as might occur if apples were contaminated directly with bovine feces. The committee included a 100-fold safety factor in their recommendation for a 5-log reduction process to ensure the safety of juice. The FDA also considered regulatory precedence when setting the 5-log pathogen reduction performance standard. This same standard is also required for *E. coli* O157:H7 reduction in fermented sausage, and FDA has advised that a 5-log process for salmonella should be used for in-shell pasteurization of eggs [13].

9.6 IMPORTANCE OF SSOPs TO HACCP

HACCP, as applied to juice production, requires that sanitation standard operating procedures (SSOPs) be developed and consistently applied. For juice manufacture the SSOP must address eight points: (1) safety of the water that comes into contact with the product; (2) conditions and cleanliness of food contact surfaces; (3) prevention of cross contamination; (4) maintenance of hand washing, sanitizing, and toilet facilities; (5) protection of food, food contact surfaces, and packaging material from adulteration; (6) proper labeling, storage, and use of any toxic compounds; (7) control of employee health that could result in microbial contamination of the food; and (8) exclusion of pests from the food processing facilities [13].

The juice HACCP rule sparked considerable public comment, especially the requirement of HACCP for juice processing in lieu of better enforcement of GMPs and SSOPs. The FDA conducted a survey of cider production facilities in 1997 [30]. In that survey, 67% of firms had good sanitation, 27% were marginal, and only 4% were categorized as poor. Microbiological trends indicated that the cider plant GMPs and SSOPs implemented during processing did not substantially reduce the level of microbiological contamination between incoming raw ingredients and finished juice.

Senkel *et al.* followed the performance of 11 Maryland cider producers before and after HACCP training [29]. Although a significant decrease in

E. coli positive juice was reported, not all cider processed was negative for *E. coli* after the implementation of more stringent processing control. The efficacy of GMPs and SSOPs in controlling contamination in an experimental commercial cider plant was examined by Keller *et al.* [32]. As expected, the lack of appropriate sanitation controls resulted in significant increases in *E. coli* K-12 in juice, when inoculated apples were used for cider processing. However, re-implementation of appropriate GMPs and SSOPs during cider manufacture failed to yield reduced *E. coli* levels in juice to below detectable limits, suggesting that GMPs and SSOPs alone are incapable of ensuring safety of fresh cider. In addition, considerable cross contamination was observed between batches of cider when intentionally contaminated and uncontaminated apple batches were alternately processed. Cross contamination occurred even under stringent application of SSOPs. The key findings of this study were as follows: (1) preharvest prevention of contamination of apples is essential; (2) if contaminated apples enter the processing environment, juice safety will be compromised both in the contaminated and subsequent juice batches; (3) SSOPs and GMPs are critical to the maintenance of a sanitary establishment, but are insufficient controls to reduce hazard levels in juice once a contaminated batch has been processed; (4) for apple cider production, a terminal 5-log treatment is essential to ensure product safety. In addition, any process needs to be validated and verified through a HACCP program.

Processing apple juice/cider is a considerably different process than is used for most citrus products. During apple juice/cider processing, the whole fruit including the skin is macerated and juice expressed with pressure applied on the fruit mash. For citrus processing, juice is commonly extracted using an automated extractor that separates juice from the fruit peel, seeds, and large pieces of pulp simultaneously [14]. In theory, only a small hole (approximately 1 inch) is cut into the fruit to extract the juice, and the juice has limited exposure to peel surface. Pao and Davis studied the transfer of microorganisms from fruit surface to juice during extraction [64]. Both natural and artificially inoculated microorganisms were followed during this study. Results indicated that significant transfer could occur (1.7%) for both naturally occurring and artificially inoculated organisms. Several different strains of *E. coli* were included among the artificially inoculated fruit. Significant cross contamination was also observed by Martinez-Gonzales *et al.* during the preparation of fresh orange juice [65]. Initial counts for inoculated orange surfaces were 2.3 log CFU/cm^2 for *S.* Typhimurium, 3.6 log CFU/cm^2 for *E. coli* O157:H7, and 4.4 log CFU/cm^2 for *L. monocytogenes*. Transfer from the fruit to the juice during mechanical extraction resulted in 1.0 log CFU/ml of *S.* Typhimurium, 2.3 log CFU/ml *E. coli* O157:H7, and 2.7 log CFU/ml *L. monocytogenes* in final orange juice. The authors concluded that strict sanitation programs and decontamination treatments for fruit might be effective control measures to prevent cross contamination and to reduce risk of foodborne illness.

In summary, studies with both apple and orange juice clearly show the importance of SSOPs. For both types of product, transfer of both pathogens and indigenous microflora into product can be reduced by strict adherence to

GMPs and SSOPs. However, despite reductions in microbial levels, results also clearly indicate that such procedures will be insufficient to ensure elimination of all pathogens present. Consequently, HACCP, GMPs, and SSOPs are all required to ensure juice safety.

9.7 APPLICATION OF THE 5-LOG STANDARD

The 5 log pathogen reduction standard must be applied to the most resistant pertinent pathogen present on the fruit or in the juice. For most juices, this requires that the standard be applied after the juice is expressed. The surface skin of many fruits and vegetables is an imperfect barrier to microorganisms [66]. Internalization is particularly problematic for fruit such as apples. Temperature differentials between wash water and fruit clearly exacerbate pathogen internalization [67]. However, even in undamaged apples without the assistance of a temperature differential, pathogens have been shown to internalize through the floral tube and other structures or defects. Surface interventions such as washing or treating with chlorine or hydrogen peroxide solutions do not typically result in complete destruction of pathogens [68–73]. Ozone (Table 9.3) was similarly unsuccessful in destroying *E. coli* O157:H7 inoculated onto apples.

Surface treatments such as with ethanol or heat can result in greater than 5 log reductions in pathogens when the pathogens are spot inoculated onto the surface of apples (Table 9.3). However, when apples are inoculated by immersion, without any temperature differential between the fruit and the inoculum menstruum, surface treatments fail. Surface heat treatment applied to apples inoculated by immersion into an inoculum menstruum achieved

---

TABLE 9.3
Reduction in *E. coli* O157:H7 in Apples Using Different Surface Treatments and Inoculation Methods

Inoculation method	Surface treatment method	Length of treatment (min)	Reduction (log CFU/ml)
Immersion in inoculum menstruum	Water immersion at 40°C	1.5	1.10
Inoculum spot dried on surface	Water immersion at 40°C	0.5	0.00
Immersion in inoculum menstruum	Water immersion at 95°C	0.5	2.40
Inoculum spot dried on surface	Water immersion at 95°C	0.5	6.20
Inoculum spot dried on surface	Immersion in ozone saturated water (1.7–1.1 ppm)	10.0	1.16
Inoculum spot dried on surface	Immersion in 50% ethanol	5.0	6.49

no greater than a 3 log reduction [74]. Penetration of heat into the fruit was measured during the study. Subsurface fruit temperatures remained significantly lower than treatment temperatures even after 60 seconds of exposure. Consequently the lower level of pathogen reduction achieved using an immersion method of inoculation implies penetration of the pathogen into the fruit.

With apples, pathogen internalization into the tissue can occur. Typical processing conditions used ensure a reasonable likelihood of its occurrence. Consequently, to ensure that all pathogens in noncitrus fruit juice received an intervention treatment sufficient to result in the prescribed 5 log reduction, the FDA ruled that the intervention must occur after the juice is expressed. For fruit such as citrus, internalization potential or likelihood was not as clearly defined. In December 1999 the FDA asked NACMCF to consider the potential for internalization of microorganisms into citrus fruit [75]. Data were considered that demonstrated the internalization and survival potential of artificially inoculated foodborne pathogens in seemingly intact oranges [76,77]. NACMCF concluded that although laboratory evidence indicated potential internalization in sound, intact citrus fruit, there was no demonstrated evidence that such internalization was likely under current industry practices. As a result of NACMCF conclusions, the FDA determined that the 5 log pathogen intervention treatment in citrus juice can be applied prior to extraction, providing only tree-picked, sound, intact fruits are used. The fruit must be cleaned and culled prior to the 5 log intervention treatment.

9.8 INTERVENTION TREATMENTS

The juice HACCP rule required a 5 log reduction (100,000-fold decrease) in the pertinent pathogen, but did not specify the means by which the reduction was to be achieved. This approach facilitates development and implementation of treatment alternatives to thermal pasteurization. Many of these alternative treatments are discussed elsewhere in this book. Since most are still under development, the primary means of juice pathogen reduction remains thermal pasteurization.

Thermal pasteurization is a well-studied method of pathogen reduction. D- and z-values for various pathogens are available in the literature [78–81]. To apply thermal pasteurization to juice, processors must first determine the pertinent pathogen for their product. In most cases the pertinent pathogen will be either $E.\ coli$ O157:H7, salmonella, or $C.\ parvum$. Physicochemical factors influencing thermal death time in juice include pH, viscosity, particulates, and Brix. Since actual time and temperature parameters may vary depending on heating menstrua and the equipment used, each processor should verify their systems once a process has been validated.

An alternative intervention treatment that is currently in use for apple cider processing is ultraviolet (UV) light irradiation. UV light irradiation was shown to destroy pathogens in apple cider [82–86]. The efficacy of UV light treatment on any liquid is strongly and negatively affected by turbidity and the sizes of any particles present [87–91]. For this reason, using UV treatment requires turbulent flow to expose all portions of the juice to the light treatment [92]. It is important to note that the majority of studies examining the efficacy of UV light irradiation in apple cider have been undertaken using systems that do not achieve turbulent flow. Although limited efficacy with nonturbulent flow devices has been demonstrated, these units would not conform to current FDA regulations, which require turbulent flow. In the application of any processing technology, juice processors have the responsibility of ensuring that all appropriate government regulations are met.

Juice manufacturers also have the option of combining treatments to achieve the 5 log pathogen reduction standard. Comes and Beelman demonstrated 5 log reductions in populations of *E. coli* O157:H7 in apple cider using a combination of fumaric acid, sodium benzoate, and a 25°C holding time prior to refrigeration [93]. In an earlier study, Uljas and Ingham achieved 5 log reduction through a combination of freeze–thaw cycles and preservatives [94]. Ingham and Schoeller went on to test the acceptability to consumers of a multistep intervention method capable of a 5 log reduction [95]. In this study, despite treatment similar to their first study, some juices did not show the expected 5 log reduction. The cause of the failure to achieve a 5 log reduction in these juices was unknown, indicating more research was necessary before this system could be commercially applied. In addition, consumers typically rated the multistep-treated juice lower than untreated juice.

Citrus processors may utilize surface treatments to achieve the required 5 log reduction in the pathogen that is most resistant to the intervention treatment applied. Interventions applied to the surface of citrus fruit must be applied after the fruit has been cleaned and culled. Interventions aimed at the surface of citrus fruits are often similar in type to that applied to the extracted juice. One of the most effective interventions remains thermal treatment. Pao *et al.* demonstrated greater than 5 log reductions in pathogen levels when oranges were submerged at 80°C for 1 or 2 minutes [96].

When surface interventions, such as a thermal treatment, are used in the production of citrus juice, additional microbiological process verification is mandated by 21 CFR 120.25. These requirements specify the number and volume of juice samples that must be tested for generic *E. coli*. Testing must be performed according to established standard methods [13]. A "moving window" scheme is used for the testing protocol. If two of seven samples tested are positive for *E. coli*, then the intervention measures in place are considered inadequate and corrective measure must be taken.

9.9 OTHER JUICE HACCP CONSIDERATIONS

Although the presence of pathogens is a primary concern in the production of fresh juice, it may not be the only hazard present. HACCP principles require that an initial hazard analysis be conducted prior to the establishment of controls for any identified hazards. Other hazards may include mycotoxins, such as patulin, and are discussed elsewhere in this book. Depending on processing and/or packaging methods employed, physical hazards may also exist. All of these issues should be addressed when developing a process for the production of juice.

9.10 LABELING

As a last consideration, the juice processor must be aware that interventions directed at the whole juice, such as thermal pasteurization or UV irradiation, will prevent the use of the term "fresh" on the juice label. Only citrus juices produced using a surface intervention, where the juice is expressed after treatment, may be labeled as "fresh" juice products.

9.11 CONCLUSION

Unpasteurized juices have been part of the American culture since colonial times. Current trends in food and beverage consumption indicate a growing interest in foods and beverages with a more healthful image. Fresh foods and beverages, as well as more foods that are minimally processed, are viewed as part of a healthy diet. Consequently, Americans are demanding more foods that are minimally processed. To keep such foods and beverages safe and to meet consumer needs, processors are being challenged to develop multiple prevention and intervention systems that begin on the farm or orchard and are carried through to the point of consumption. Research has demonstrated that GAPs, GMPs, SSOPs, and HACCP are all essential measures to ensure the food safety of juices.

REFERENCES

1. Parish, M.E., Public health and nonpasteurized fruit juices, *Crit. Rev. Microbiol.*, 23, 109–119, 1997.
2. Bang, I.S., Kim, B.H., Foster, J.W., and Park, Y.K., OmpR regulates the stationary-phase acid tolerance response of *Salmonella enterica* serovar Typhimurium, *J. Bacteriol.*, 182, 2245–2252, 2000.
3. Benjamin, M.M. and Datta, A.R., Acid tolerance of enterohemorrhagic *Escherichia coli*, *Appl. Environ. Microbiol.*, 61, 1669–1672, 1995.
4. Foster, J.W. and Hall, H.K., Adaptive acidification tolerance response of *Salmonella typhimurium*, *J. Bacteriol.*, 172 (2), 771–778, 1990.
5. Garren, D.M., Harrison, M.A., and Russell, S.M., Acid tolerance and acid shock response of *Escherichia coli* O157:H7 and non-O157:H7 isolates provide cross protection to sodium lactate and sodium chloride, *J. Food Prot.*, 61, 158–161, 1998.
6. Leyer, G.J., Wang, L., and Johnson, E.A., Acid adaptation of *Escherichia coli* O157:H7 increases survival in acidic foods, *Appl. Environ. Microbiol.*, 61, 3752–3755, 1996.

7. Lin, J., Lee, I.S., Frey, J., Slonczewski, J.L., and Foster, J.W., Comparative analysis of extreme acid survival in *Salmonella typhimurium, Shigella flexneri,* and *Escherichia coli, J. Bacteriol.,* 177, 4097–4104, 1995.

8. Lin, J., Smith, M.P., Chapin, K.C., Baik, H.S., Bennett, G.N., and Foster, J.W., Mechanisms of acid resistance in Enterohemorrhagic *Escherichia coli, Appl. Environ. Microbiol.,* 62, 3094–3100, 1996.

9. McKellar, R.C. and Knight, K.P., Growth and survival of various strains of enterohemorrhagic *Escherichia coli* in hydrochloric and acetic acid, *J. Food Prot.,* 62, 1466–1469, 1999.

10. Miller, L.G. and Kaspar, C.W., *Escherichia coli* O157:H7 acid tolerance and survival in apple cider, *J. Food Prot.,* 57, 460–464, 1994.

11. Mitscherlich, E. and Marth, E.H., *Microbial Survival in the Environment,* Springer-Verlag, New York, 1984.

12. Waterman, S.R. and Small, P.L.C., Acid-sensitive enteric pathogens are protected from killing under extremely acidic conditions of pH 2.5 when they are inoculated onto certain solid food sources, *Appl. Environ. Microbiol.,* 64, 3882–3886, 1998.

13. U.S. Food and Drug Administration, 21 CFR Part 120. Hazard analysis and critical control point (HACCP); Procedures for the safe and sanitary processing and importing of juice, Final rule, *Fed. Regist.,* 66, 6137–6202, 2001.

14. Kimball, D.A., *Citrus Processing: A Complete Guide,* 2nd ed., Aspen Publishers, Gaithersburg, MD, 1999.

15. Ringer, M., Citrus fruits: Citrus spp., www.uga.edu/fruit/citrus.htm, 2002.

16. Parish, M.E., Microbiological concerns in citrus juice processing, *Food Technol.,* 45, 128–133, 1991.

17. Chun, D. and McDonald, R.E., Seasonal trends in the population dynamics of fungi, yeasts, and bacteria on the fruit surfaces of grapefruit in Florida, *Proc. Fla. State Hortic. Soc.,* 100, 23–25, 1987.

18. Pao, S. and Brown, G.E., Reduction of microorganisms on citrus fruit surfaces during packinghouse processing, *J. Food Prot.,* 61, 903–906, 1998.

19. Pao, S. and Davis, C., Enhancing microbiological safety of fresh orange juice by fruit immersion in hot water and chemical sanitizers, *J. Food Prot.,* 62, 756–760, 1999.

20. Parish, M.E. and Higgins, D., Isolation and identification of lactic acid bacteria from samples of citrus molasses and unpasteurized orange juice, *J. Food Sci.,* 53, 645–646, 1988.

21. Parish, M.E., Microbiological aspects of fresh squeezed citrus juice, in *Ready to Serve Citrus Juices and Juice Added Beverages, Proceedings of the Food Industry Short Course,* University of Florida, Gainesville, FL, 1988, pp. 79–87.

22. Pao, S. and Davis, C.L., Maximizing microbiological quality of fresh orange juice by processing sanitation and fruit surface treatments, *Dairy Food Environ. Sanit.,* 21, 287–291, 2001.

23. Mattick, L.R. and Moyer, J.C., Composition of apple juice, *J. AOAC,* 66, 1251–1255, 1983.

24. Anon., Determination of Organic Acids in Fruit Juices, application note 143, Dionex Corporation, Sunnyvale, CA, 2003, pp. 1–5.

25. Goverd, K.A., Beech, F.W., Hobbs, R.P., and Shannon, R., The occurrence and survival of colifoms and salmonellas in apple juice and cider, *J. Appl. Bacteriol.,* 46, 521–530, 1979.

26. Keller, S.E., Chirtel, S.J., Merker, R.I., Taylor, K.T., Tan, H.L., and Miller, A.J., Influence of fruit variety, harvest technique, quality sorting, and storage on the native microflora of unpasteurized apple cider, *J. Food Prot.*, 67, 2240, 2004.

27. Dingman, D.W., Growth of Escherichia coli O157:H7 in bruised apple (Malus domestica) tissue as influences by cultivar, date of harvest, and source, *Appl. Environ. Microbiol.*, 66, 1077–1083, 2000.

28. Cummins, A., Reitmeier, C., Wilson, L., and Glatz, B., A survey of apple cider production practices and microbial loads in cider in the state of Iowa, *Dairy Food Environ. Sanit.*, 22, 745–751, 2002.

29. Senkel, I.A.J., Henderson, R.A., Jobitado, B., and Meng, J., Use of hazard analysis critical control point and alternative treatments in the production of apple cider, *J. Food Prot.*, 62, 778–785, 1999.

30. U.S. Food and Drug Administration, Report of 1997 inspections of fresh, unpasteurized apple cider manufacturers, http://vm.cfsan.fda.gov/~dms/ciderrpt.html, 1999.

31. Dingman, D.W., Prevalence of *Escherichia coli* in apple cider manufactured in Connecticut, *J. Food Prot.*, 62, 567–573, 1999.

32. Keller, S.E., Merker, R.I., Taylor, K.T., Tan, H.L., Melvin, C.D., Chirtel, S.J., and Miller, A.J., Efficacy of sanitation and cleaning methods in a small apple cider mill, *J. Food Prot.*, 65, 911–917, 2002.

33. Edberg, S.C., Rice, E.W., Karlin, R.J., and Allen, M.J., *Escherichia coli*: the best biological drinking water indicator for public health protection, *J. Appl. Microbiol.*, 88, 106S–116S, 2000.

34. Chart, H., Toxigenic *Escherichia coli*, *J. Appl. Microbiol.*, 84 (Symp. Suppl.), 77S–86S, 1998.

35. Buchanan, R.L. and Doyle, M.E., Foodborne disease significance of *Escherichia coli* O157:H7 and other enterohemorrhagic *E. coli*, *Food Technol.*, 51, 69–76, 1997.

36. U.S. Food and Drug Administration, Untreated Juice May Pose Health Risk to Children, http://www.cfsan.fda.gov/~dms/juicsaf2.html, 2001.

37. Centers for Disease Control, U.S. Foodborne Disease Outbreaks, annual listing 1990–2000, http://www.cdc.gov/foodborneoutbreaks/report_pub.htm, 2003.

38. Whipp, S.C., Rasmussen, M.A., and Cray, W.C., Public veterinary medicine: food safety and handling. Animals as a source of *Escherichia coli* pathogenic for human beings, *J. Am. Vet. Med. Assoc.*, 204, 1168–1175, 1994.

39. Zhao, T., Doyle, M.E., Shere, J., and Garber, L., Prevalence of enterohemorrhagic *Escherichia coli* O157:H7 in a survey of dairy herds, *Appl. Environ. Microbiol.*, 61, 1290–1293, 1995.

40. Hancock, D.D., Rice, D.H., Herriott, D.E., Besser, T.E., Ebel, E.D., and Carpenter, L.V., Effects of farm manure-handling practices on *Escherichia coli* O157:H7 prevalence in cattle, *J. Food Prot.*, 60, 363–366, 1997.

41. Tauxe, R.V., Emerging foodborne diseases: an evolving public health challenge, *Emerg. Infect. Dis.*, 3, 425–434, 1997.

42. Rice, D.H. and Besser, T.E., Verotoxigenic E. coli O157 colonisation of wild deer and range cattle, *Vet. Rec.*, 137, 524, 1995.

43. Wallace, J.S., Cheasty, T., and Jones, K., Isolation of vero cytotoxin-producing *Escherichia coli* O157:H7 from wild birds, *J. Appl. Microbiol.*, 82, 399–404, 1997.

44. Janisiewicz, W.J., Conway, W.S., Brown, M.W., Sapers, G.M., Fratamico, P.M., and Buchanan, R.L., Fate of *Escherichia coli* O157:H7 on fresh-cut apple tissue and its potential for transmission by fruit flies, *Appl. Environ. Microbiol.*, 65, 1–5, 1999.

45. Centers for Disease Control, Outbreaks of *Escherichia coli* O157:H7 infections associated with drinking unpasteurized commercial apple juice: October 1996, *Morbid. Mortal. Weekly Rep.*, 45, 975, 1996.

46. Deliganis, C.V., Death by apple juice: the problem of foodborne illness, the regulatory response, and further suggestions for reform, *Food Drug Law J.*, 53, 681–728, 1998.

47. Buchanan, R.L. and Edelson, S.G., Culturing enterohemorrhagic *Escherichia coli* in the presence and absence of glucose as a simple means of evaluating the acid tolerance of stationary-phase cells, *Appl. Environ. Microbiol.*, 62, 4009–4013, 1996.

48. Centers for Disease Control, Salmonellosis, http://www.cdc.gov/ncidod/dbmd/diseaseinfo/salmonellosis_t.htm, 2002.

49. Anon., FDA Warns Consumers About Frozen Mamey, U.S. Food and Drug Administration, http://vm.cfsan.fda.gov/~lrd/tpmamey.html, 1999.

50. Centers for Disease Control, Outbreak of *Salmonella* serotype Muenchen infections associated with unpasteurized orange juice: United States and Canada June 1999, *Morbid. Mortal. Weekly Rep.*, 48, 582–585, 1999.

51. Cook, K.A., Dobbs, T.E., Hlady, W.G., Wells, J.G., Bearrett, T.J., Puhr, N.D., Lancette, G.A., Bodager, D.W., Toth, B.L., Genese, C.A., Hoighsmith, A.K., Pilot, K.E., Finelli, L., and Swerdlow, D.L., Outbreak of *Salmonella* serotype Hartford infections associated with unpastuerized orange juice, *JAMA*, 280, 1504–1509, 1998.

52. Krause, G., Terzagian, R., and Hammond, R., Outbreak of *Salmonella* serotype Anatum infection associated with unpasteurized orange juice, *South. Med. J.*, 94, 1168–1172, 2001.

53. Parish, M.E., Narcisco, J.A., and Fredrich, L.M., Survival of Salmonellae in orange juice, *J. Food Saf.*, 17, 273–281, 1997.

54. Centers for Disease Control, Outbreaks of *Escherichia coli* O157:H7 infection and cryptosporidiosis associated with drinking unpasteurized apple cider: Connecticut and New York, October 1996, *Morbid. Mortal. Weekly Rep.*, 46, 4–8, 1997.

55. Millard, P.S., Gensheimer, K.F., Addiss, D.G., Sosin, D.M., Beckett, G.A., Houck-Jankoski, A., and Hudson, A., An outbreak of Cryptosporidiosis from fresh-pressed apple cider, *JAMA*, 272, 1592–1596, 1994.

56. Guerrant, R.L., Cryptosporidiosis: an emerging highly infectious threat, *Emerg. Infect. Dis.*, 3, 51–57, 1997.

57. Cox, L.J., Kleiss, T., Cordier, J.L., Cordellana, C., Konel, P., Pedrazzini, C., Beumer, R., and Siebenga, A., Listeria spp. in food processing, non-food and domestic environments, *Food Microbiol.*, 6, 49–61, 1989.

58. Fenlon, D.R., Wilson, J., and Donachie, W., The incidence and level of *Listeria monocytogenes* contamination of food sources at primary production and initial processing, *J. Appl. Bacteriol.*, 81, 641–650, 1996.

59. Beuchat, L.R., Pathogenic microorganisms associated with fresh produce, *J. Food Prot.*, 59, 204–216, 1995.

60. Sorrells, K.M., Enigl, D.C., and Hatfield, J.R., Effect of pH, acidulant, time, and temperature on the growth and survival of *Listeria monocytogenes*, *J. Food Prot.*, 52, 571–573, 1989.

61. Beuchat, L.R. and Brackett, R.E., Behavior of *Listeria monocytogenes* inoculated into raw tomatoes and processed tomato products, *Appl. Environ. Microbiol.*, 57, 1367–1371, 1991.

62. Sado, P.N., Jinneman, K.C., Husby, G.J., Sorg, S.M., and Omiecinski, C.J., Identification of *Listeria monocytogenes* from unpasteurized apple juice using rapid test kits, *J. Food Prot.*, 61, 1199–1202, 1998.

63. U.S. Food and Drug Administration, Quantitative Assessment of Relative Risk to Public Health from Foodborne Listeria Monocytogenes Among Selected Categories of Ready-to-Eat Foods, http://www.cfsan.fda.gov/~dms/lmr2-toc.html, 2003.

64. Pao, S. and Davis, C.L., Transfer of natural and artificially inoculated microorganisms from orange fruit to fresh juice during extraction, *J. Food Sci. Technol. India*, 34, 113–117, 2001.

65. Martinez-Gonzales, N.E., Hernandez-Herrera, A., Martinez-Chavez, L., Rodriguez-Garcia, M.O., Torres-Vitela, M.R., Mota de la Garza, L., and Castillo, A., Spread of bacterial pathogens during preparation of freshly squeezed orange juice, *J. Food Prot.*, 66, 1490–1494, 2003.

66. Burnett, S.L. and Beuchat, L.R., Human pathogens associated with raw produce and unpasteurized juices, and difficulties in decontamination, *J. Ind. Microbiol. Biotechnol.*, 27, 104–110, 2001.

67. Buchanan, R.L., Edelson, S.G., Miller, R.L., and Sapers, G.M., Contamination of intact apples after immersion in an aqueous environment containing *Escherichia coli* O157:H7, *J. Food Prot.*, 62, 444–450, 1999.

68. Wright, J.R., Sumner, S.S., Hackney, C.R., Pierson, M.D., and Zoecklein, B.W., Reduction of *Escherichia coli* O157:H7 on apples using wash and chemical sanitizer treatments, *Dairy Food Environ. Sanit.*, 20, 120–126, 2000.

69. Wisniewsky, M.A., Glatz, B.A., Gleason, M.L., and Reitmeier, C.A., Reduction of *Escherichia coli* O157:H7 counts on whole fresh apples by treatment with sanitizers, *J. Food Prot.*, 63, 703–708, 2000.

70. Ukuku, D.O., Pilizota, V., and Sapers, G.M., Influence of washing treatment on native microflora and *Escherichia coli* population of inoculated cantaloupes, *J. Food Saf.*, 21, 31–47, 2001.

71. Sapers, G.M. and Simmons, G.F., Hydrogen peroxide disinfection of minimally processed fruits and vegetables, *Food Technol.*, 52, 48–52, 1998.

72. Kenney, S.J. and Beuchat, L.R., Survival of *Escherichia coli* O157:H7 and *Salmonella* Muenchen on apples as affected by application of commercial fruit waxes, *Int. J. Food Microbiol.*, 77, 223–231, 2001.

73. Kenney, S.J. and Beuchat, L.R., Comparison of aqueous commercial cleaners for effectiveness in removing Escherichia coli O157:H7 and Salmonella muenchen from the surface of apples, *Int. J. Food Microbiol.*, 74, 47–55, 2002.

74. Fleischman, G.J., Bator, C., Merker, R., and Keller, S.E., Hot water immersion to eliminate *Escherichia coli* O157:H7 on the surface of whole apples: thermal effects and efficacy, *J. Food Prot.*, 64, 451–455, 2001.

75. Anon., National Advisory Committee on Microbiological Criteria for Foods: Meeting on Fresh Citrus Juice: Transcript of Proceedings. U.S. Food and Drug Administration, http://www.cfsan.fda.gov/~comm/tr991208.html, 1999.

76. Eblen, B.S., Walderhaug, M.O., Edelson-Mammel, S., Chirtel, S.J., De Jesus, A., Merker, R.I., Buchanan, R.L., and Miller, A.J., Potential for internalization,

growth and survival of *Salmonella* spp. and *Escherichia coli* O157:H7 in oranges, *J. Food Prot.*, 67, 1578–1584, 2004.

77. Walderhaug, M.O., Edelson-Mammel, S.G., DeJesus, A.J., Eblen, B.S., Miller, A.J., and Buchanan, R.L., Preliminary studies on the potential for infiltration, growth and survival of *Salmonella enterica* serovar Hartford and *Escherichia coli* O157:H7 within oranges, U.S. Food and Drug Administration, http://www.cfsan.fda.gov/~comm/juicstud.html, 1999.

78. Mak, P.P., Ingham, B.H., and Ingham, S.C., Validation of apple cider pasteurization treatments against *Escherichia coli* O157:H7, *Salmonella*, and *Listeria monocytogenes*, *J. Food Prot.*, 64, 1679–1689, 2001.

79. Splittstoesser, D.F., McLellan, M.R., and Churey, J.J., Heat resistance of *Escherichia coli* O157:H7 in apple juice, *J. Food Prot.*, 59, 226–229, 1995.

80. Deng, M.Q. and Cliver, D.O., Inactivation of *Cryptosporidium parvum* oocysts in cider by flash pasteurization, *J. Food Prot.*, 64, 523–527, 2001.

81. Mazzotta, A.S., Thermal inactivation of stationary-phase and acid adapted *Escherichia coli* O157:H7, *Salmonella*, and *Listeria monocytogenes* in fruit juices, *J. Food Prot.*, 64, 315–320, 2001.

82. Chang, J.C.H., Ossoff, S.F., Lobe, D.C., Dorfman, M.H., Dumais, C.M., Qualls, R.G., and Johnson, J.D., UV inactivation of pathogenic and indicator microorganisms, *Appl. Environ. Microbiol.*, 49, 1361–1365, 1985.

83. Hanes, D.E., Orlandi, P.A., Burr, D.H., Miliotis, M.D., Robi, M.G., Bier, J.W., Jackson, G.J., Arrowood, M.J., Churey, J.J., and Worobo, R.W., Inactivation of *Crytosporidium parvum oocysts* in fresh apple cider using ultraviolet irradiation, *Appl. Environ. Microbiol.*, 68, 4168–4172, 2002.

84. Duffy, S., Churey, J., Worobo, R., and Schaffner, D.W., Analysis and modeling of the variability associated with UV inactivation of *Escherichia coli* in apple cider, *J. Food Prot.*, 63, 1587–1590, 2000.

85. Harrington, W.O. and Hills, C.H., Reduction of the microbial population of apple cider by ultraviolet irradiation, *Food Technol.*, 22, 117–120, 1968.

86. Wright, J.R., Sumner, S.S., Hackney, C.R., Pierson, M.D., and Zoecklein, B.W., Efficacy of ultraviolet light for reducing *Escherichia coli* O157:H7 in unpasteurized apple cider, *J. Food Prot.*, 63, 563–567, 2000.

87. Anon., Kinetics of Microbial Inactivation for Alternative Food Processing Technologies, Institute of Food Technologists, http://vm.cfsan.fda.gov/~comm/ift-pref.html, 2000.

88. Liltved, H. and Cripps, S.J., Removal of particle-associated bacteria by prefiltration and ultraviolet irradiation, *Aquacult. Res.*, 30, 445–450, 1999.

89. Qualls, R.G., Flynn, M.P., and Johnson, J.D., The role of suspended particles in ultraviolet disinfection, *J. WPCF*, 55, 1280–1285, 1983.

90. Shama, G., Ultraviolet irradiation apparatus for disinfecting liquids of high ultraviolet absorptivities, *Lett. Appl. Microbiol.*, 15, 69–72, 1992.

91. Whitby, G.E. and Palmateer, G., The effect of UV transmission, suspended solids and photoreactivation on microorganisms in wastewater treated with UV light, *Water Sci. Technol.*, 27, 379–386, 1993.

92. U.S. Food and Drug Administration, 21 CFR Part 179, Irradiation in the production, processing and handling of food, *Fed. Regist.*, 65, 71056–71058, 2000.

93. Comes, J.E. and Beelman, R.B., Addition of fumaric acid and sodium benzoate as an alternative method to achieve a 5-log reduction of *Escherichia coli* O157:H7 populations in apple cider, *J. Food Prot.*, 65, 476–483, 2002.

94. Uljas, H.E. and Ingham, S.C., Combinations of intervention treatments resulting in a 5-$\log_{10}$-unit reductions in numbers of *Escherichia coli* O157:H7 and *Salmonella typhimurium* DT104 organisms in apple cider, *Appl. Environ. Microbiol.*, 65, 1924–1929, 1999.
95. Ingham, S.C. and Schoeller, N.P., Acceptability of a multi-step intervention system to improve apple cider safety, *Food Res. Int.*, 35, 611–618, 2002.
96. Pao, S., Davis, C.L., and Parish, M.E., Microscopic observation and processing validation of fruit sanitizing treatments for the enhanced microbiological safety of fresh orange juice, *J. Food Prot.*, 64, 310–314, 2001.

10 Microbiological Safety Issues of Fresh Melons

Dike O. Ukuku and Gerald M. Sapers

CONTENTS

10.1 INTRODUCTION

In the U.S., melons are widely available year round and represent an important dietary component. In 2001 annual per capita consumption was estimated to be

231

14.9, 11.2, and 2.1 pounds for watermelon, cantaloupe, and honeydew melons respectively [1]. The value of these commodities in 2003 was reported to be $346,022,000, $372,965,000, and $93,241,000, respectively [2]. In recent years fresh-cut melons have become increasingly popular with consumers and now account for a large and growing proportion of melon consumption.

For most consumers, melons represent a refreshing and healthy dessert or snack. However, for a small number of consumers, the situation is quite different; melon consumption has been a source of foodborne illness. At least 17 melon-related outbreaks involving hundreds of cases have been reported since 1990 [3–5]. Additional outbreaks ascribed to "multiple fruit" or "fresh-cut fruit" also may have been due to contamination of an unspecified melon component. While the largest melon-related outbreaks have been attributed to various salmonella serotypes, other human pathogens including *Escherichia coli* O157:H7, *Campylobacter jejuni*, and Norwalk-like virus also have been implicated [4].

Survival and growth of human pathogens including salmonella, *E. coli* O157:H7, and *Listeria monocytogenes* in melon flesh has been demonstrated [6–8]. Annous *et al.* [9] reported growth of *S.* Poona on cantaloupe rind at 20°C.

Salmonella outbreaks in 2000–2002 were traced by the U.S. Food and Drug Administration (FDA) to melons imported from Mexico [10]. On-farm investigations in Mexico conducted by the FDA concluded that "measures were not in place to minimize microbial contamination in growing, harvesting, packaging, and cooling of cantaloupe." Detection of *L. monocytogenes* in cut melons resulted in a recent product recall [11]. FDA surveys of imported and domestic produce have documented the presence of salmonella and shigella in cantaloupe [12,13]. The incidence of salmonella on imported cantaloupe (from Mexico, Costa Rica, and Guatemala) was 5.3% and on domestic cantaloupes was 2.6%. Shigella also was detected on these samples, an incidence of 2% on the imports and 0.9% on domestic melons. On October 28, 2002 the FDA issued an import alert on cantaloupes from Mexico, halting all such shipments. Subsequently, export of Mexican cantaloupes to the U.S. by a small number of grower/packers who met FDA safety criteria was resumed [10].

In this chapter some of the production and postharvest handling conditions that may contribute to microbial contamination of melons are examined. Studies of the efficacy of conventional washing practices in reducing the microbial load on melons are reviewed. Finally, current research results pointing to means of improving the efficacy of melon disinfection are examined.

10.2 MICROFLORA OF MELONS

Melons, especially cantaloupe, present a variety of surfaces to which microorganisms may bind. In cantaloupe the epidermal cell surface is ruptured with a meshwork of raised tissue (the net). This net consists of lenticels and phellum (cork) cells. These cells have hydrophobic suberized walls to reduce water loss and protect against pathogen ingress. Also imparting a hydrophobic nature to the outer surface of cantaloupe is the cuticle composed of waxes and cutin that cover the epidermal cells [14].

The ability of pathogenic and spoilage-causing bacteria to adhere to surfaces of melons represents a food safety problem of great concern as well as a source of economic loss to the produce and fresh-cut industry. The mechanism of attachment of bacterial cells to plant surfaces has been studied most extensively for plant pathogens and symbionts [15,16]. The predominant class of organisms on cantaloupe and honeydew melon were aerobic mesophilic bacteria followed by lactic acid bacteria, Gram-negative bacteria, yeasts and molds, and *Pseudomonas* spp. [17]. The populations of each of the categories of microorganisms were found to be higher on cantaloupe than on honeydew, both for whole and fresh-cut melon. Differences in the populations of the native microflora on honeydew and cantaloupe melons are most likely due to the rougher surface of the cantaloupe compared to the relatively smooth surface of honeydew melon. The extensive raised netting on the surface of cantaloupe melon no doubt provides more microbial attachments sites and helps to protect attached microbes from being washed from the surface, and possibly from environmental stresses such as UV radiation and desiccation. In unwrapped and wrapped sliced watermelon, *Pseudomonas* spp., *E. coli*, *Enterobacter* spp., and micrococci comprised the predominant microflora [18].

10.2.1 SPOILAGE ORGANISMS

The primary causative agents for microbial spoilage of melons are mostly yeasts and molds and, to a lesser extent, bacteria. Several studies have demonstrated the presence of enteric bacteria, including Enterobacteriaceae and Pseudomonadaceae, on whole and fresh-cut melons [17]. Microorganisms responsible for postharvest diseases are not necessarily dominant on the surface of sound fruits; they are abundant in the environment and can easily contaminate the melon surfaces. In a study conducted at the Eastern Regional Research Center, it was found that the spoilage organisms in fresh-cut melon were mostly yeasts and molds, *Pseudomonas* spp., and *Erwinia* spp. [19]. The level of these organisms in freshly prepared cut melons was very low but gradually increased during storage at 5 or 20°C.

10.2.2 HUMAN BACTERIAL PATHOGENS

The ability of human bacterial pathogens to attach to melon surfaces [20] and their virulence characteristics must both be considered. Results of a study examining attachment of bacteria from a mixed cocktail containing multiple suspensions of individual strains of each genus (salmonella, *E. coli*, and *L. monocytogenes*) on the surface of cantaloupes stored at 4°C for up to 7 days showed that salmonella has the strongest attachment to the cantaloupe surface followed by *L. monocytogenes* and *E. coli*, either as individual strains or as a mixed cocktail [20]. The strength of attachment increased slightly for *E. coli* over the 7 days of storage, but decreased for *L. monocytogenes*. Efficacy of sanitizer treatments applied to inoculated cantaloupes at 7 days postinoculation was greatly reduced for *L. monocytogenes* and *E. coli* but not for salmonella. Surface irregularities such as roughness, crevices, and pits have been shown to increase bacterial adherence by increasing cell attachment and reducing the ability to remove cells [21].

Salmonella is among the most frequently reported causes of foodborne outbreaks of gastroenteritis in the U.S. [22]. Salmonellosis has been steadily increasing as a public health problem in the U.S. since reporting began in 1943 [23]. Five multistate outbreaks of salmonellosis have been associated epidemiologically with cantaloupes. The first in 1990 involved *S.* Chester, which affected 245 individuals (two deaths) in 30 states [22]. The second in 1991 involved more than 400 laboratory-confirmed *S.* Poona infections and occurred in 23 states and Canada [22]. A 1997 outbreak associated with *S.* Saphra was reported (www.cdc.gov/mmwr/preview/mmwrhtml/mm5146a2. htm). The most recent outbreaks (2000, 2001, and 2002) were due to *S.* Poona [5]. Other melons including watermelon have been associated with outbreaks of foodborne illness [5,24–26]. The implication of these outbreaks is that improvements are needed at the farm level to limit or minimize contact of melons with sources of human pathogens, and at the packinghouse level in sanitizing and processing conditions.

Other human pathogens including *E. coli* O157:H7 and shigella are capable of growth on melon flesh [6,7]. A 1993 outbreak of foodborne illness was attributed to cantaloupe contaminated with *E. coli* O157:H7 (M. Diermayer, Oregon Health Division, Portland, OR, personal communication).

10.3　FACTORS CONTRIBUTING TO MELON CONTAMINATION

10.3.1　PREHARVEST AND HARVEST CONDITIONS

Relatively little definitive information on sources of human pathogen contamination of melons is available. The FDA suggested that preharvest contamination of Mexican melons with human pathogens may have resulted from use of sewage-contaminated irrigation water [10]. Irrigation water, transported over long distances and distributed to farms through open and unprotected aqueducts and channels, may become contaminated by animal or human activity (Table 10.1). Other potential sources may be from feces of birds [28,29], reptiles [5], or other wildlife in fields, or exposure to airborne contamination. The latter scenario was demonstrated by Annous *et al.* [30] in studies conducted in an apple orchard in close proximity to a pasture. Animal production activity was observed by one of the authors within several miles of melon production locations in California and Mexico. However, the limits of airborne distribution and survival of human pathogens attached to aerosols has not been reported. Suslow [31] was unable to recover salmonella from more than 900 individual field-collected melons produced in different regions of California during 1999–2001. It may be that contamination events in some

TABLE 10.1
Potential Sources of Melon Contamination

Preharvest
  Direct fecal contamination — human, birds, reptiles, insects, other wildlife
  Indirect fecal contamination — irrigation water, dust from animal production
During harvest
  Poor worker hygiene
Packing plant
  Contaminated process water
  Poor plant sanitation
  Ineffective washing
  Cross contamination during washing
  Poor worker hygiene

locations are highly sporadic and localized, e.g., to individual melons with adhering avian feces or insect damage, a melon defect observed by one of the authors in a California packing shed. Duffy *et al.* [32] reported that salmonella isolates obtained from washed cantaloupes in Texas were most closely related to isolates obtained from equipment and irrigation water, but DNA fingerprinting did not conclusively establish relationships between contamination sources. Contamination of melons could occur during harvest if worker hygiene was deficient [10].

Research is needed to identify specific sources of preharvest contamination of melons and to develop guidelines and good agricultural practices (GAPs) that reduce the risk of contamination. Appropriate training of farm workers in personal hygiene and avoidance of behaviors that result in melon contamination is essential.

10.3.2 POSTHARVEST CONDITIONS

Gagliardi *et al.* [33] reported in most cases little change or an increase of indicator microorganisms (total and fecal coliforms and enterococci) on melons during washing in samples obtained at packing facilities in the Rio Grande River Valley of Texas. They attributed contamination to the management of primary wash tanks or hydrocoolers, e.g., use of contaminated river water, buildup of soil in tanks, and depletion of chlorine. The contamination of cantaloupes in Mexico may have been due to cooling and washing with contaminated water [10]. The potential for such contamination also exists in the U.S. One of the authors has observed melon processing operations in which cantaloupes were tightly packed in tanks containing chlorinated water, with minimal opportunities for agitation of the melons or mixing of the water, prior to fresh-cut processing. Under such conditions, rapid depletion of chlorine at the melon surface and survival of attached bacteria on contaminated melons might be expected with the possibility of cross contamination of other melons in the tank.

Other potential sources of postharvest sources of contamination include poor personal hygiene or work practices by workers (one of the authors observed the failure of packinghouse employees to wear gloves or hairnets while handling melons; another worker used his foot to move cantaloupes down a ramp from a receiving platform to a conveyor) and inadequate plant sanitation. Accumulation of debris from incoming melons was visible on the aforementioned ramp and conveyors. Conveyors and processing equipment must be cleaned and sanitized on a regular schedule with sufficient frequency so as not to allow debris to accumulate and microbial populations to build up on food contact surfaces.

Such deficiencies can be addressed by development and implementation of a hazard analysis critical control point (HACCP) plan and an effective cleaning and sanitation program, and adherence to good manufacturing practices (GMPs). Of equal importance is employee training in food safety. Such training should be appropriate to the employee's job and in the employee's native language.

10.3.3 MODE OF MICROBIAL ATTACHMENT TO MELONS

The external surface of cantaloupe melons is characterized by the presence of a net comprising porous lenticellar tissue on the epidermis [14]. Such tissue provides numerous attachment sites for microorganisms and also may shield attached cells from contact with cleaning or antimicrobial agents (Figure 10.1). Microbial attachment and the possibility of internalization may occur in the stem scar region. In contrast, honeydew melon and watermelon have a smooth surface that should be less favorable for attachment and protection of

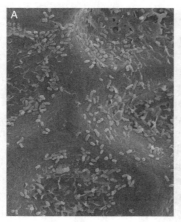
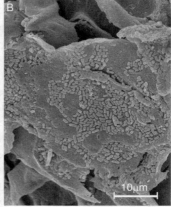

FIGURE 10.1 Scanning electron microscopy image showing bacteria on cantaloupe rind surface (A) and in lenticel (B).

microorganisms. Park and Beuchat [34] reported that greater numbers of $E.\ coli$ O157:H7 and salmonella cells were inactivated or detached from inoculated honeydew melon than from cantaloupe when the melons were washed with sanitizer solutions. Similarly, the population of aerobic microorganisms on honeydew melon could be reduced to lower levels than the population on cantaloupes by washing with 200–2000 ppm chlorine solutions [35]. Similar results were reported by Ukuku and Fett [17].

10.4 EFFICACY OF CONVENTIONAL WASHING

10.4.1 WASHING IN THE PACKINGHOUSE

Field-packed melons are not generally washed because of the difficult logistics of supplying adequate water to mobile washing equipment. Melons transported to packing plants may be washed by spraying over rollers in flat-bed brush washers or by immersion in a wash tank [33]. However, these investigators found little or no reduction and in some cases an elevation in microbial populations on cantaloupes and honeydew melons washed with commercial equipment in packing plants in the Rio Grande Valley of Texas. This may have resulted from contamination of the wash water and/or depletion of chlorine by reaction with organic material. It also may have been due to the limited efficacy of brush washers in detaching microbial contaminants from melon surfaces. Annous $et\ al.$ [36] demonstrated the inability of a flat-bed brush washer to reduce the population of $E.\ coli$ on inoculated apples. In contrast, Materon [37] reported reductions of 3.2 logs in the populations of aerobic microorganisms on cantaloupes washed by unspecified means in four commercial packinghouses, also located in the Rio Grande Valley of Texas.

10.4.2 LABORATORY-SCALE WASHING STUDIES

Laboratory washing studies in which the melons are fully immersed in a sanitizing solution with scrubbing or agitation have demonstrated that significant reductions in microbial populations can be achieved. Ayhan $et\ al.$ [35] reported reductions of 1 and 2 logs for the aerobic plate count on whole honeydew melons and cantaloupes, respectively, after dipping in 200 ppm chlorine (as sodium hypochlorite) solutions; reductions exceeding 3 logs were obtained on cantaloupes dipped in 1000 ppm chlorine. Park and Beuchat [34] compared 200 or 2000 ppm chlorine, 850 or 1200 ppm acidified sodium chlorite, 0.2 or 1.0% hydrogen peroxide, and 40 or 80 ppm peroxyacetic acid (Tsunami^{TM}) as sanitizers for cantaloupes inoculated with human pathogens. Population reductions of $E.\ coli$ O157:H7 and salmonella cocktails approached or exceeded 3 logs for all of these treatments except hydrogen peroxide, which was less effective. Population reductions of total aerobic microorganisms were substantially smaller than reductions of human pathogen populations. Similar results were reported for honeydew melons.

TABLE 10.2
Effect of Postinoculation Storage at 5°C on Efficacy of Chlorine Wash in Inactivating Salmonella and Listeria and *E. coli* on Inoculated Cantaloupes

Bacteria	Days postinoculation	Survivors ($\log_{10}$ CFU/cm^2)^a		
		Treatment^b		
		Control	H$_2$O	Cl$_2$ (1000 ppm)
Salmonella^c	0	4.6 ± 0.2	4.6 ± 0.1	1.5 ± 0.2
	3	4.7 ± 0.1	4.6 ± 0.2	2.0 ± 0.1
	5	4.6 ± 0.1	4.6 ± 0.1	2.4 ± 0.2
L. monocytogenes^d	0	3.6 ± 0.2	3.1 ± 0.1	ND
	3	3.5 ± 0.2	3.3 ± 0.2	ND
	5	3.5 ± 0.2	3.3 ± 0.2	ND
E. coli 25922^e	0	5.0 ± 0.1	4.5 ± 0.1	0.3 ± 0.1
	3	4.5 ± 0.2	4.0 ± 0.1	2.0 ± 0.1
	5	2.0 ± 0.1	2.2 ± 0.1	2.2 ± 0.1

Note: ND = not detected by plating.
^{a} Values are means ± standard deviation of three experiments with duplicate determinations per experiment.
^{b} Treatments applied for 3 min.
^{c} Cocktail of *Salmonella* spp. containing *S.* Stanley H0558, *S.* Poona RM2350, and *S.* Saphra 97A3312. (Data from Ukuku and Fett [74].)
^{d} Cocktail of *L. monocytogenes* containing strains Scott A., ATCC 15313, LM-4, and H7778. (Data from Ukuku and Fett [8].)
^{e} Data from Ukuku *et al.* [41].

Sapers *et al.* [38] reported reductions in the aerobic plate count on cantaloupe surface of less than 1 log when rind plugs were washed by immersion in 1000 ppm chlorine, 1% APL KLEEN 246 (an acidic detergent formulation supplied by Cerexagri; www.cerexagri.com), or 4% trisodium phosphate. Immersion of whole cantaloupes freshly inoculated with *Salmonella* spp. in 1000 ppm chlorine solution for 5 minutes resulted in population reductions of 3 logs for salmonella [39,40]; however, the reduction was only 2 logs when the treatment was applied 5 days after inoculation (Table 10.2). With a nonpathogenic *E. coli* (ATCC 25922), the reduction was greater than 4 logs with freshly inoculated melons but less than 1.5 logs when the treatment was applied 72 hours after inoculation [41]. However, with *L. monocytogenes*, the time interval between inoculation and treatment had no effect on treatment efficacy [8].

Barak *et al.* [42] obtained a 1 log reduction in the population of *Pantoea agglomerans* (a surrogate for *S.* Poona) on inoculated cantaloupe by immersion in 150 ppm sodium hypochlorite for 20 seconds, followed by a 2-minute cold water rinse. In studies with cantaloupes inoculated with *E. coli* O157:H7, Materon [37] reported reductions generally exceeding 5 logs from washing by immersing the melons for 1 or 10 minutes in solutions containing 200 ppm chlorine, 1.5% lactic acid, or 1.5% lactic acid + 1.5% hydrogen peroxide at 25 or 35°C. In view of the efficacy data obtained by other investigators, these extraordinary results are difficult to explain. It is possible that the recovery of attached bacteria from the melon surface by rubbing with a sponge was substantially less efficient than predicted by the investigator's validation procedure. Alternatively, the presence of residual lactic acid or hydrogen peroxide in the lowest dilutions plated may have been inhibitory to *E. coli* O157:H7 on Petrifilm^{TM}.

The FDA advises consumers to wash melons with cool tap water with scrubbing but without use of soap or detergents immediately before eating. Consumers are also advised to wash cutting boards, utensils, and counter tops often using hot soapy water followed by diluted bleach as a sanitizer. Avoidance of cross contamination with meat, poultry, or fish is essential (www.fda.gov/bbs/topics/ANSWERS/2002/ANS01167.html/). Fresh-cut processing studies conducted by one of the authors clearly demonstrated the need to develop and rigorously adhere to a strict protocol for sanitizing knives, cutting boards, and other food contact surfaces and equipment to avoid cross contamination and achieve an acceptable product shelf life. Attention to detail was found to be critical [38].

While the literature on efficacy of washing melons is limited and contradictory, the overall trend suggests that microbial populations attached to melon surfaces can be reduced by several logs if sanitizers are applied by immersion of melons in the solution with scrubbing and/or agitation. Treatment efficacy may be reduced if the time interval between contamination and washing is greater than one day, a likely situation with preharvest contamination. Since human pathogens transferred from the rind to the flesh are capable of growth on the flesh surface, the presence of even small numbers of survivors following a sanitizing wash represents a significant risk to consumers. Consequently, there is a great need for better methods of disinfecting melons so that this risk is minimized.

10.5 NOVEL DISINFECTION TREATMENTS

10.5.1 HYDROGEN PEROXIDE

Hydrogen peroxide is classified as generally recognized as safe (GRAS) for use in food products [43]. It is used as a bleaching agent, oxidizing and reducing agent, and antimicrobial agent. The FDA specifies approved food uses of hydrogen peroxide such as treatment of milk used for cheese, preparation of modified whey, and production of thermophile-free starch. However, the FDA requires that the residual hydrogen peroxide be removed by physical or chemical means during processing. Hydrogen peroxide has not yet been approved by the FDA for washing fruits and vegetables. Antimicrobial activity of hydrogen peroxide as a preservative for fruits and vegetables [44], salad vegetables, berries, and fresh-cut melons [45] has been reported. Also it has been used to control postharvest decay in table grapes [46]. When used as a

TABLE 10.3
Population of *Salmonella* spp. on Cantaloupe Rind and Recovered from Fresh-Cut Pieces Before or After Washing Treatments and Fresh-Cut Preparation

Melon	Treatment	Salmonella population[a]			
		Log CFU/cm^2 whole melon	Log reduction	Log CFU/g fresh-cut pieces	Log reduction
Cantaloupe	Control	4.4 ± 0.1	—	2.1 ± 0.1	—
	Water	4.3 ± 0.2	0.1	2.1 ± 0.1	0.0
	H_2O_2 (2.5%)	1.9 ± 0.0	2.5	0.4 ± 0.1	1.7
	H_2O_2 (5%)	2.1 ± 0.1	2.3	0.3 ± 0.1	1.8
Honeydew	Control	3.1 ± 0.1		1.3 ± 0.1	—
	Water	2.7 ± 0.2	0.4	1.2 ± 0.1	0.1
	H_2O_2 (2.5%)	ND	~3.0	ND	~1.3
	H_2O_2 (5%)	ND	~3.0	ND	~1.3

Note: Cocktail of *Salmonella* spp. containing *S.* Stanley H0558, *S.* Poona RM2350, and *S.* Newport H1275 in the inoculum. Melons were completely submerged in bacterial inoculum ($\sim 20°C$) for 10 min. ND = not detected by plating.
[a] Values are mean ± standard deviation of duplicate determinations from three experiments.
From Ukuku D.O., *Int. J. Food Microbiol.*, 95, 137, 2004.

sanitizer for whole melon surfaces at a concentration in the range 2.5 to 5% H_2O_2, there were significant ($p \leq 0.05$) reductions in the populations of inoculated *E. coli* and indigenous microflora [41] and approximately 2.3 to 2.6 and 3.0 log CFU/cm^2 reductions of salmonella on cantaloupe and honeydew melon, respectively (Table 10.3) [40,47]. Treatment of cantaloupes with 5% hydrogen peroxide at 70°C for 1 minute resulted in a 5.0 log reduction of total mesophilic aerobes, a 3 log reduction of yeasts and molds, and a 3.8 log reduction of inoculated salmonella [48]. When the initial level of salmonella on the melons was 1.9 log CFU/cm^2, no survivors were detected after treatment with 5% hydrogen peroxide at 70°C, even with enrichment. However, when the initial population on melon surfaces was at 3.5 log CFU/cm^2, the treated samples were negative for salmonella by plating but were positive upon enrichment.

10.5.2 HOT WATER

Hot water decontamination of whole cantaloupes designated for fresh-cut processing was found to have major advantages over the use of sanitizers, including a significant reduction of microbiological populations on melon surfaces [48]. The major advantage was that it reduced the probability of potential transfer of pathogenic bacteria from the rind to the interior tissue during cutting. In experiments carried out in our laboratory, treatment of cantaloupes, inoculated with *S.* Poona, with hot water for 1 minute resulted in a 2.1 log reduction at 70°C and a 3.6 log reduction at 97°C (Table 10.4) [48].

TABLE 10.4
Inactivation of *E. coli* ATCC 25922 and *S.* Poona on Inoculated Cantaloupe by Surface Pasteurization With Hot Water for 2 min and Reduction of Transfer to Fresh-Cut Flesh

Experiment	Target organism	Treatment		Surviving population	
		Time (min)	Temp. ($°C$)	On melons ($\log_{10}$ CFU/cm^2)	On fresh-cut ($\log_{10}$ CFU/g)
A	*E. coli*	2	Control[a]	4.0 ± 0.6	—
			76	0.6 ± 0.6	—
			86	ND	—
			97	ND	—
B	*S.* Poona	1	Control	4.7 ± 0.1	2.9 ± 0.1
			70	2.6 ± 0.1	0.7 ± 0.1
			97	1.1 ± 0.2	—[b]

Note: ND = Not detected by plating.
[a] Untreated.
[b] Detectable by enrichment.
Experiment A data from Pilizota, V. and Sapers, G.M., Unpublished data, 2000; Experiment B data from Ukuku, D.O., Pilizota, V., and Sapers, G.M., *J. Food Prot.*, 67, 432, 2004.

Surviving *S.* Poona could not be detected by plating on fresh-cut pieces prepared from cantaloupes treated at 97°C but could be detected after enrichment, evidence that a small number of survivors were transferred during fresh-cut preparation. When the initial level of salmonella on the melons was 1.9 log CFU/cm^2, no survivors were detected after this treatment, even with enrichment, but with an initial population of 3.5 log CFU/cm^2, the treated samples were negative for salmonella by plating but were positive upon enrichment. Similar reductions in the population of salmonella occurred when treatments were applied to cantaloupes stored at 5°C for 5 days as for 3 days. In experiments with *E. coli*, the efficacy of hot water treatments at lower temperatures was compared with that at 96°C (Table 10.4). Surviving *E. coli* could be detected on inoculated cantaloupe by plating following treatment at 76°C; no survivors were detected at 86 or 97°C. These hot water treatments, which approach population reductions of 4 log CFU/cm^2, represent a substantial improvement over chlorinated water (1000 ppm) or hydrogen peroxide at ~20°C which yielded reductions of only 2 to 3.0 logs.

Additional information concerning hot water treatment of melons can be found in Chapter 21.

10.5.3 STEAM

The use of steam to treat fruits is somewhat difficult to control due to time and exact temperature needed to maintain the desired texture. The application of steam on whole cantaloupe surface for reduction of microbial population would be appropriate since melon has a thick rind that may protect the interior flesh from deleterious effects of the steam. In a preliminary study in our laboratory, steam pasteurization of melon surface was not promising compared to hot water treatment. The inability of the steam to reduce effectively total microbial populations on whole melon surfaces can be attributed to the surface roughness where the netting, cracks, and possible openings due to detached trichome can provide protection to the attached organisms.

10.5.4 OTHER

The application of an effective antibacterial agent to the surface of whole melons may be desirable. There are several reports that nisin, used in combination with a chelating agent, exhibits a bactericidal effect towards both Gram-positive and Gram-negative bacteria [49–53]. Treatment of whole and fresh-cut cantaloupe and honeydew melon with nisin-EDTA significantly reduced the natural microflora and extended the shelf life [17]. We also found that sodium lactate was inhibitory to the native microflora on melons [19]. The antimicrobial activity of lactic acid is due both to a lowering of pH and to disruption of the outer membrane of Gram-negative bacteria [54]. Application of lactic acid (2%) as an antimicrobial spray applied to animal carcasses to reduce surface populations of $E.\ coli$ O157:H7 and salmonella has been reported [55]. Sorbic acid (pK_a of 4.76) and its potassium salt are widely used in foods at a concentration of 0.02 to 0.3% to inhibit yeasts and molds, but they also have antibacterial activity [56]. However, washing inoculated whole melons with sodium lactate (2%), potassium sorbate (0.02%), EDTA (0.2 M), or nisin (50 μg/ml), when tested individually, did not cause significant ($p > 0.05$) reductions in salmonella populations. Treatment of whole cantaloupe with nisin-EDTA may lead to both increased shelf life and a reduced risk of foodborne illness due to contamination with salmonella or other pathogens [17].

10.6 ISSUES WITH FRESH-CUT MELONS

The visual symptoms of deterioration of fresh-cut produce are flaccidity due to loss of water, changes in color resulting from oxidative browning at the cut surfaces, and microbial contamination [57]. Minimally processed fresh fruits and vegetables provide a good substrate for microbial growth [58,59]. Such substrate may allow proliferation of human pathogenic organisms like salmonella, $L.\ monocytogenes$, and enterotoxigenic $E.\ coli$ that contaminate food when proper sanitation is not employed. Microbial spoilage of fresh-cut melons will depend on storage conditions and the initial microbial population of the melon. Honeydew melon generally has a lower initial microbial population than cantaloupe and also has been found to have a longer refrigerated shelf life [17,47]. Similar results were reported for minimally processed honeydew and cantaloupe melon stored at 4°C, and the authors concluded that both the length of shelf life and type of spoilage were related to the type of fruit [60].

10.6.1 TRANSFER OF BACTERIA FROM RIND TO FLESH

Fresh-cut pieces prepared from whole cantaloupe or honeydew melons showed the presence of mesophilic aerobic bacteria, Gram-negative bacteria, lactic acid bacteria, *Pseudomonas* spp., and yeasts and molds [17,47]. The predominant categories of microorganisms on fresh-cut cantaloupe immediately after fresh-cut preparation from unwashed whole melons were mesophilic aerobic bacteria and lactic acid bacteria. For fresh-cut honeydew, mesophilic aerobic bacteria predominated immediately after fresh-cut preparation. As days of refrigerated storage increased, other categories of microbes were detected in all samples, irrespective of initial treatment before fresh-cut preparation. The fact that the same categories of microorganisms were detected on fresh-cut pieces during storage as on the whole melon surface indicates that the microbes were transferred from the rind to the flesh during fresh-cut preparation. Transfer occurred during cutting and removal of melon rinds.

Salmonella inoculated on whole melon surfaces was recovered in fresh-cut pieces prepared from inoculated melons [39]. Similarly, Ukuku and Fett [8] reported survival and transfer of *L. monocytogenes* population from whole cantaloupe to fresh-cut pieces. The population on fresh-cut pieces also survived and increased during storage at an abusive temperature.

Ukuku *et al.* [48] reported that fresh-cut pieces prepared from cantaloupes inoculated with initial salmonella populations of 1.9, 3.5, or 4.6 log and treated with 97°C water or 5% hydrogen peroxide at 70°C were negative for salmonella by dilution plating, although positive by enrichment (Table 10.4). However, the populations of salmonella and all classes of native microflora in fresh-cut pieces prepared from sanitized melons were low compared to populations in fresh-cut pieces from untreated whole melon.

10.6.2 OUTGROWTH ON FLESH

Populations of all groups of native microorganisms increased in fresh-cut samples as storage time increased, regardless of the treatment. The population of salmonella transferred from the untreated melons to the flesh during cutting averaged 2 log CFU/g for cantaloupe and 1.3 log CFU/g for honeydew. The population of salmonella on fresh-cut cantaloupe inoculated with 2.56 log CFU/g increased as storage time increased, especially at an abusive temperature [19,39] (Figure 10.1). Golden *et al.* [6] reported growth of salmonella inoculated directly onto fresh-cut cantaloupe, watermelon, and honeydew melons during storage at 23°C. Ukuku and Sapers [39] reported growth of *S.* Stanley on fresh-cut cantaloupe during storage at 8 and 20°C. Other investigators have reported that interior watermelon tissues support the growth of *Salmonella* spp. [7,61]. All melon-related foodborne outbreaks noted so far involved melons that were precut and held at unknown temperatures for some period of time at restaurants and retail food stores prior to being purchased and consumed. The inner flesh of melons comprises mainly parenchyma cells containing sugars, organic acids, and other substances that may be released upon plant cell injury and support microbial growth. Tamplin [3] suggested that attention should be directed to cleaning the melons at the time of cutting, using clean and sanitized utensils and surfaces to minimize contamination of the edible portion, and immediately consuming or holding cut melon pieces at cold temperatures.

10.6.3 SUPPRESSION OF OUTGROWTH

The application of effective antibacterial agents to the surface of fresh-cut melons may suppress outgrowth of the native microflora and any human pathogens. Studies showing antilisterial activity of nisin in TSB or PBS62, and demonstrating its activity against native microflora on whole and minimally processed cantaloupe have been reported [17]. However, total elimination of salmonella on the surface of whole or fresh-cut melon could not be achieved, probably due to surface irregularities and internalization which reduced the ability of antimicrobial treatments to contact or remove bacterial cells. However, treatment with the combinations sodium lactate–potassium sorbate or nisin–sodium lactate may lead to an increased shelf life and a reduced risk of foodborne illness from salmonella or other human pathogens; such treatments also appeared acceptable from a quality standpoint [17,63]. The use of nisin for treating fresh-cut melon may reduce the risk of *L. monocytogenes* outgrowth [64].

Bacteriophage was used to control growth and reduce population of *S.* Enteritidis on fresh-cut melons [65]. In our most recent study, we found that the native microflora of cantaloupe and honeydew melon was inhibitory to *L. monocytogenes* [66]. Lactic acid bacteria were used to improve microbial safety of minimally processed fruits and vegetables [67]. Other researchers have used antagonistic microorganisms isolated from the field to control postharvest pathogens and colonization of apple surfaces [68].

10.7 METHODOLOGY FOR MICROBIOLOGICAL EVALUATION OF MELONS

Accurate assessment of the microbiological quality and safety of melons requires use of suitable sampling, recovery, and detection methods that take into account the mode of attachment of microorganisms to the melon surface. This is especially important with cantaloupes because of their complex surface morphology characterized by netting and the presence of fissures, both of which are absent on honeydew melons [14,69]. The cantaloupe surface morphology provides numerous microbial attachment sites and opportunities for inaccessibility not present on other non-netted melons. However, all melons will show variations in surface features that could affect microbial attachment and growth, especially in the stem scar and ground spot regions.

Beuchat and Scouten [70] conducted a detailed study of survival and recovery of S. Poona on spot- and dip-inoculated cantaloupes sampled at three sites: the intact rind, a wound, or the stem scar. Recovery was accomplished by stomaching excised rind in a wash solution containing 0.1% peptone, with or without added Tween 80, or by rubbing melons in the same wash solution within a plastic bag. They demonstrated the equivalence of a number of combinations of preenrichment broth, enrichment broth, and selective agar medium in detection of S. Poona recovered from the rind surface. They reported no difference in recovery of S. Poona from the three sites compared to when the inoculum was suspended in water or an organic matrix (horse serum); growth occurred in both spot- and dip-inoculated wounds over 24 hours at 21 and 37°C but not at 4°C. Addition of up to 1.0% Tween 80 to peptone may have enhanced detachment of S. Poona, recovered by the washing procedure. The stomaching and wash solution procedures appeared to give equivalent results.

Annous *et al.* [71] examined recovery and survival of *E. coli* NRRL B-766 on spot- and dip-inoculated cantaloupe rind. Less than 1% of the inoculum applied by spot inoculation to the rind surface could be recovered by excising plugs containing the inoculation sites and blending. *E. coli* survival on inoculated cantaloupe after treatment with 300 ppm chlorine or water at 60°C was greater if applied by dip inoculation of the melon surface compared to spot inoculation. The investigators compared two sampling methods for recovering bacteria from the melon rind surface: (1) excision and blending of 20 replicate plugs containing inoculation sites for spot inoculation or taken at random locations for dip inoculation, and (2) removal of the entire spot- or dip-inoculated rind with an electric peeler. With both methods, the rind samples were homogenized with peptone water, serially diluted, and plated. The methods were applied to melons inoculated with *E. coli* B-766 or S. Poona. A method was developed for calculating the melon surface area from measurements of the polar and equatorial diameters, based on an assumption that the cantaloupe was a sphere, oblate spheroid, or prolate spheroid. When expressed on an area basis, the population estimates for the two methods were the same with both test organisms (Table 10.5). Expression of the population estimate on a weight basis would be invalid, however, because of poor correlation between the rind weight and external surface area. The whole rind method is less time-consuming and requires less handling than the rind plug method.

Barak *et al.* [42] compared two elution methods with peeling and blending for recovery of S. Poona from inoculated cantaloupes. They reported better recovery with Butterfield's buffer containing Tween 80 as the eluant than with phosphate-buffered saline, similar recovery when agitation was provided by shaking or rolling, and better recovery by the elution methods than by peeling and blending. The last result was attributed to the release of inhibitory substances during blending.

TABLE 10.5
Comparison of Rind Plug and Whole Rind Sampling Methods for Recovery of *Salmonella* Poona RM 2350 from the Surface of Dip-Inoculated Cantaloupes

Storage of inoculated melon at 20°C (h)	*S.* Poona population[a] ($\log_{10}$ CFU/cm^2)	
	Plug method[b]	Whole rind method[c]
2	4.7	4.3
24	6.3	6.8
48	6.7	7.0
72	6.9	7.0

Note: Inoculum in water; population was 8.7 log10 CFU/ml. XLT-4 agar medium used to enumerate *S.* Poona cell densities.
[a] Mean for 3 melons per trial; no significant difference between plug and whole rind methods.
[b] Based on total cross-sectional area of 20 rind plugs, each with 20 mm diameter.
[c] Based on calculated surface area for spheroid or sphere.

Hammack *et al.* [72] compared methods for the recovery of salmonella from cantaloupes spot inoculated at levels to provide fractionally positive results. They obtained better recoveries by soaking in preenrichment broth as compared to rinsing with the broth, and by detecting the salmonella using a culture procedure. Such methods would be useful in evaluating melons subjected to antimicrobial treatments such as surface pasteurization in which surviving populations are very small or not detectable by ordinary plating.

10.8 RESEARCH NEEDS

While extensive research has been conducted in a number of areas relating to the microbiological safety and quality of melons, a number of gaps exist that impede further progress. One deficiency is the relatively small amount of information concerning melons other than cantaloupe. Another area requiring more attention is the nature of microbial attachment to melons, especially conditions favoring biofilm formation and internalization in the netting of cantaloupes and stem scar of melons. A better understanding of salmonella adhesion to cantaloupe is needed for the development of more effective washing treatments to control this organism on melon surfaces and fresh-cut pieces. With regard to sanitation methods for melons, the promising results obtained with hot water surface pasteurization should be extended to additional melons besides cantaloupe, and the possibility of adverse effects on quality and shelf life should be given further study. As a back-up strategy, research should be conducted on lower temperature surface treatments used in combination with other treatments that may be synergistic. Finally, because of the possibility of low-level survival of pathogens on melon surfaces following such treatments and transfer to the flesh during fresh-cut processing, better means of suppressing outgrowth of survivors by treatment of fresh-cut melon with preservatives, irradiation, or other means should be investigated.

REFERENCES

1. Agricultural Statistics 2003, Vegetables and Melons, USDA National Agricultural Statistics Service (http://www.usda.gov/nass/pubs/agstatistics.htm).
2. Vegetable Annual Summary, USDA National Agricultural Statistics Service, Jan. 29, 2004 (http://usda.mannlib.cornell.edu/reports/nassr/fruit/pvg-bban).
3. Tamplin, M., Salmonella and cantaloupes, *Dairy Food Environ. Sanit.*, 17, 284, 1997.
4. Dewaal, C.S. and Barlow, K., Outbreak Alert! Closing the Gaps in our Federal Food Safety Net, Center for Science in the Public Interest, Washington D.C., September 2002 (http://cspinet.org/foodsafety/reports.html).
5. CDC, Multistate outbreaks of Salmonella serotype Poona infections associated with eating cantaloupe from Mexico: United States and Canada, 2000–2002, *MMWR*, 51, 1044, 2002.
6. Golden, D.A., Rhodehamel, E.J., and Kautter, D.A, Growth of *Salmonella* spp. in cantaloupe, watermelon and honeydew melons, *J. Food Prot.*, 56, 194, 1993.
7. Del Rosario, B.A. and Beuchat, L.R., Survival and growth of enterohemorrhagic *Escherichia coli* O157:H7 in cantaloupe and watermelon, *J. Food Prot.* 58, 105, 1995.
8. Ukuku, D.O. and Fett, W., Behavior of *Listeria monocytogenes* inoculated on cantaloupe surfaces and efficacy of washing treatments to reduce transfer from rind to fresh-cut pieces, *J. Food Prot.*, 65, 24, 2002.
9. Annous, B.A., Burke, A., and Sites, J.M., Surface pasteurization of cantaloupe surfaces inoculated with *Salmonella* Poona RM 2350 or *Escherichia coli* ATCC 25922, *J. Food Prot.*, 67, 1876, 2004.
10. FDA, Detention Without Physical Examination of Cantaloupes from Mexico, Import Alert IA2201 Attachment, 2003 (www.fda.gov.ora/fiars/ora_import_ia2201.html).
11. FDA, Duck Delivery Produce Recalls Cut Honeydew and Cut Cantaloupe Melon for Possible Health Risk, Recall: firm press release, 2003 (www.fda.gov/oc/po/firmrecalls/duck07_03.html).
12. FDA, Survey of Imported Fresh Produce, FYT 2000 Field Assignment, U.S. Food and Drug Administration, Center for Food Safety and Applied Nutrition, Office of Plant and Dairy Foods and Beverages, Jan. 30, 2001 (http://vm.cfsan.fda.gov/~dms/).
13. FDA, Survey of Domestic Fresh Produce: Interim Results, U.S. Food and Drug Administration, Center for Food Safety and Applied Nutrition, July 31, 2001 (http://www.cfsan.fda.gov/~dms/).
14. Webster, B.D. and Craig, M.E, Net morphogenesis and characteristics of the surface of muskmelon fruit, *J. Am. Soc. Hort. Sci.*, 101, 412, 1976.
15. Romantschuk, M., Attachment of plant pathogenic bacteria to plant surfaces, *Ann. Rev. Phytopathol.*, 30, 225, 1992.

16. Romantschuk, M. *et al.*, Microbial attachment to plant aerial surfaces, in *Aerial Plant Surface Microbiology*, Morris, C.E., Nicot, P.C., and Nguyen-The, C., Eds., Plenum Press, New York, 1996, p. 43.

17. Ukuku, D.O. and Fett, W., Effectiveness of chlorine and nisin-EDTA treatments of whole melons and fresh-cut pieces for reducing native microflora and extending shelf-life, *J. Food Saf.*, 22, 231, 2002.

18. Abbey, S.D. *et al.*, Microbiological and sensory quality changes in unwrapped and wrapped sliced watermelon, *J. Food Prot.*, 51, 531, 1988.

19. Ukuku, D.O., unpublished data, 2003.

20. Ukuku D.O. and Fett, W.F., Relationship of cell surface charge and hydrophobicity to strength of attachment of bacteria to cantaloupe rind, *J. Food Prot.*, 65, 1093, 2002.

21. Frank, J.F. and Koffi, R.A., Surface adherent growth of *Listeria monocytogenes* is associated with increased resistance to surfactant sanitizer and heat, *J. Food Prot.*, 53, 550, 1990.

22. CDC, Multistate outbreak of Salmonella poona infections: United States and Canada, *MMWR*, 40, 549, 1991.

23. Rise, A.A. *et al.*, A multistate outbreak of *Salmonella chester* linked to imported cantaloupe, in *Program and Abstracts of the 30th Interscience Conference on Antimicrobial Agents and Chemotherapy*, American Society for Microbiology, Washington D.C., 1990, abstr. 915.

24. Dewaal, C.S., Alderton, L., and Jacobson, M.F., Outbreak Alert! Closing the Gaps in Our Federal Food Safety Net, Center for Science in the Public Interest, Washington D.C., 2000.

25. Gayler, G.E. *et al.*, An outbreak of salmonellosis traced to watermelon, *Public Health Rep.*, 70, 311, 1955.

26. NACMCF, Microbiological safety evaluations and recommendations on fresh produce. National Advisory Committee on Microbiological Criteria for Foods, *Food Control*, 10, 117, 1999.

27. Castillo, A. and Escartin, E.F., Survival of *Campylobacter jejuni* on sliced watermelon and papaya, *J. Food Prot.*, 57, 166, 1994.

28. Kullas, H. *et al.*, Prevalence of *Escherichia coli* serogroups and human virulence factors in faeces of urban Canada geese (*Branta Canadensis*), *Int. J. Environ. Health Res.*, 12, 153, 2002.

29. Clark, L., and McLean, R.G., A review of pathogens of agricultural and human health interest found in blackbirds, in *Management of North American Blackbirds*, Linz, G.M., Ed., National Wildlife Research Center, Fort Collins, CO, 2003, p. 103.

30. Annous, B.A., Unpublished data, 2001.

31. Suslow, T.V., Key Points of Control and Management of Microbial Food Safety for Melon Producers, Handlers snd Processors, University of California: Good Agricultural Practices, 2004 (http://ucgaps.ucdavis.edu/Key_Points_Melons/, accessed April 26, 2004).

32. Duffy, E.A. *et al.*, Genetic diversity and antibiotic resistance profiling of *Salmonella* isolated from irrigation water, packing shed equipment, and fresh produce in Texas, IAFP 90th Annual Meeting, August 10–13, 2003, New Orleans, LA, abstr. P251.

33. Gagliardi, J.V. *et al.*, On-farm and postharvest processing sources of bacterial contamination to melon rinds, *J. Food Prot.*, 66, 82, 2003.

34. Park, C.M. and Beuchat, L.R., Evaluation of sanitizers for killing *Escherichia coli* O157:H7, *Salmonella*, and naturally occurring microorganisms on cantaloupes, honeydew melons, and asparagus, *Dairy, Food and Environ. Sanit.*, 19, 842, 1999.

35. Ayhan, Z., Chism, G.W., and Richter, E.R., The shelf-life of minimally processed fresh cut melons, *J. Food Qual.*, 21, 29, 1998.

36. Annous, B.A. *et al.*, Efficacy of washing with a commercial flat-bed brush washer, using conventional and experimental washing agents, in reducing populations of *Escherichia coli* on artificially inoculated apples, *J. Food Prot.*, 64, 159, 2001.

37. Materon, L.A., Survival of *Escherichia coli* O157:H7 applied to cantaloupes and the effectiveness of chlorinated water and lactic acid as disinfectants, *World J. Microbiol. Biotechnol.*, 19, 867, 2003.

38. Sapers, G.M. *et al.*, Anti-microbial treatments for minimally processed cantaloupe melon, *J. Food Sci.*, 66, 345, 2001.

39. Ukuku, D.O. and Sapers, G.M., Effect of sanitizer treatments on *Salmonella* Stanley attached to the surface of cantaloupe and cell transfer to fresh-cut tissues during cutting practices, *J. Food Prot.*, 64, 1286, 2001.

40. Ukuku, D.O. and Fett, W.F., Research note: method of applying sanitizers and sample preparation affects recovery of native microflora and *Salmonella* on whole cantaloupe surfaces, *J. Food Prot.*, 67, 999, 2004.

41. Ukuku, D.O., Pilizota, V., and Sapers, G.M., Influence of washing treatment on native microflora and *Escherichia coli* population of inoculated cantaloupes, *J. Food Saf.*, 21, 31, 2001.

42. Barak, J.D., Chue, B., and Mills, D.C., Recovery of surface bacteria from and surface sanitization of cantaloupes, *J. Food Prot.*, 66, 1805, 2003.

43. CFR, Hydrogen Peroxide, Code of Fed. Reg. 21, Parts 170–199, Section 184.1366, April 1, U.S. Government Printing Office, Washington D.C., 1994.

44. Honnay, R., Process for Improving the Preservation of Fresh Vegetables and Fruits, European Patent 0 255 814, 1988.

45. Sapers, G.M. and Simmons, G., Hydrogen peroxide disinfection of minimally processed fruits and vegetables, *Food Technol.*, 52, 48, 1998.

46. Forney, C.F. *et al.*, Vapor phase hydrogen peroxide inhibits postharvest decay of table grapes, *HortScience*, 26, 1512, 1991.

47. Ukuku D.O., Effect of hydrogen peroxide treatment on microbial quality and appearance of whole and fresh-cut melons contaminated with *Salmonella* spp., *Int. J. Food Microbiol.*, 95, 137, 2004.

48. Ukuku, D.O., Pilizota, V., and Sapers, G.M., Effect of hot water and hydrogen peroxide treatments on survival of *Salmonella* and microbial quality whole and fresh-cut cantaloupes, *J. Food Prot.*, 67, 432, 2004.

49. Blackburn, P. *et al.*, Nisin Compositions for Use as Enhanced, Broad Range Bacteriocins, International patent application PCT/US89/02625; international publication WO 89/12399, Applied Microbiology, New York, 1989.

50. Cutter, C.N. and Siragusa, G.R., Population reductions of gram negative pathogens following treatments with nisin and chelators under various conditions, *J. Food Prot.*, 58, 977, 1995.

51. Stevens, K.A. *et al.*, Nisin treatment for inactivation of *Salmonella* species and other gram-negative bacteria, *Appl. Environ. Microbiol.*, 57, 3613, 1991.

52. Stevens, K.A. *et al.*, Antimicrobial action of nisin against *Salmonella typhimurium* lipopolysaccharide mutants, *Appl. Environ. Microbiol.*, 58, 1786, 1992.
53. Stevens, K.A. *et al.*, Effect of treatment conditions on nisin inactivation of gram-negative bacteria, *J. Food Prot.*, 55, 763, 1992.
54. Alakomi, H.-L. *et al.*, Lactic acid permeabilizes gram-negative bacteria by disrupting the outer membrane, *App. Environ. Microbiol.*, 66, 2001, 2001.
55. Castillo, A. *et al.*, Lactic acid sprays reduce bacterial pathogens on cold beef carcass surfaces and in subsequently produced ground beef, *J. Food Prot.*, 64, 58, 2001.
56. Sofos, J.N. and Busta, F.F., Sorbic acid and sorbates, in *Antimicrobials in Foods*, 2nd ed., Davidson, P.M. and Branen, A.L., Eds., Marcel Dekker, New York, 1993.
57. King, A.D., Jr. and Bolin, H.R., Physiology and microbiological storage stability of minimally processed fruits and vegetables, *Food Technol.*, 2, 132, 1989.
58. Marston, E.V., Fresh-cut fruits: maximizing quality, *Cutting Edge*, 9, 3, 1995.
59. Nguyen-the, C. and Carlin, F., The microbiology of minimally processed fresh fruits and vegetables, *Crit. Rev. Food Sci. Nutr.*, 34, 371, 1994.
60. O'Connor-Shaw, R.E. *et al.*, Shelf-life of minimally processed honeydew, kiwifruit, papaya, pineapple and cantaloupe, *J. Food Sci.*, 59, 1202, 1994.
61. Escartin, E F., Ayala, A.C., and Lozano, J.S., Survival and growth of *Salmonella* and *Shigella* on sliced fresh fruits, *J. Food Prot.*, 52, 471, 1989.
62. Ukuku, D.O. and Shelef, L.A., Sensitivity of six strains of *Listeria monocytogenes* to nisin, *J. Food Prot.*, 60, 867, 1997.
63. Ukuku, D.O. and Fett, W.F., Effect of nisin in combination with EDTA, sodium lactate and potassium sorbate for reducing *Salmonella* on whole and fresh-cut cantaloupe, *J. Food Prot.*, 67, 2143, 2004.
64. Leverentz, B. *et al.*, Biocontrol of *Listeria monocytogenes* on fresh-cut produce by treatment with lytic bacteriophages and bacteriocin, *Appl. Environ. Microbiol.*, 69, 4519, 2003.
65. Leverentz, B. *et al.*, Examination of bacteriophage as a biocontrol method for *Salmonella* on fresh-cut fruits: a model study, *J. Food Prot.*, 64, 116, 2001.
66. Ukuku, D.O., Fett, W., and Sapers, G.M., Inhibition of *Listeria monocytogenes* by native microflora of cantaloupes, *J. Food Saf.*, 24, 129, 2004.
67. Breidt, F. and Fleming, H.P., Using lactic acid bacteria to improve the safety of minimally processed fruits and vegetables, *Food Technol.* 51, 44, 1997.
68. Leibinger, W. *et al.*, Control of postharvest pathogens and colonization of the apple surface by antagonistic microorganisms in the field, *Phytopathology*, 87, 1103, 1997.
69. Lester, G., Comparisons of "Honey Dew" and netted muskmelon fruit tissues in relation to storage life, *HortScience*, 23, 180, 1988.
70. Beuchat, L.R. and Scouten, A.J., Factors affecting survival, growth, and retrieval of Salmonella Poona on intact and wounded cantaloupe rind and stem scar tissue, *Food Microbiol.*, 21, 683, 2004.
71. Annous, B.A. *et al.*, Improved recovery procedure for evaluation of sanitizer efficacy in disinfecting contaminated cantaloupes, *J. Food Sci.*, 70, 242–247, 2005.

72. Hammack, T.S. *et al.*, Relative effectiveness of the Bacteriological Analytical Manual method for the recovery of Salmonella from whole cantaloupes and cantaloupe rinses with selected pre-enrichment media and rapid methods, *J. Food Prot.*, 67, 870, 2004.

73. Pilizota, V. and Sapers, G.M., Unpublished data, 2000.

74. Ukuku, D.O. and Fett, F.W., Unpublished data, 2002.

11 Fresh-Cut Vegetables

Pascal Delaquis

CONTENTS

11.1 INTRODUCTION

Natural biological processes require that all plants eventually undergo senescence, death, and decomposition by microorganisms. Physical stresses, environmental factors, or disease can hasten this end in otherwise healthy plants, including species that are cultivated by humans for consumption as fresh vegetables. In this light, the trauma inflicted by harvest may be viewed as the first in a series of events that ultimately lead to decomposition. Harvest provokes physiological alterations associated with attempts to maintain homeostasis, repair injury, and prevent infection by opportunistic microorganisms. Postharvest technological interventions are applied in distribution systems for whole, fresh vegetables to delay quality changes that arise from these reactions. Unit operations applied in fresh-cut processing invariably contribute further stresses, particularly where tissues are cut or sliced. The latter operations are of critical importance, as cutting irrevocably alters metabolic processes and provides ample opportunity for invasion of tissues by microorganisms. Additional measures are therefore necessary to preserve the eating quality of fresh-cut vegetables.

Vegetables destined for fresh-cut processing carry complex microbial populations that may include saprophytic species living in mutually beneficial, symbiotic relationships with the healthy plant, potential phytopathogens, or accidental contaminants derived from environmental sources. Microorganisms derived from the field are described in detail in Chapter 1. Additional species may be acquired during subsequent handling and processing. Any or all of these microorganisms can exploit opportunities for growth provided by access to the rich source of nutrients contained within plant tissues. Selective pressures derived from agronomic factors (source, field conditions), post-harvest treatments, the intrinsic properties of raw materials (physical structure, pH, availability of growth substrates, antimicrobial factors), and processing (washing, application of antimicrobials, storage atmospheres, temperature) influence the success of individual species and the composition of microbial populations in products derived from individual vegetables.

A considerable body of scientific literature indicates that spoilage associations in fresh-cut vegetables are product-specific, and a complete description of microbiological phenomena in all such products is beyond the scope of this work. Instead, three commodities have been selected to illustrate the influence of inherent and processing variables on the development of spoilage associations and the fate of undesirable microorganisms in fresh-cut vegetables.

11.2 FRESH-CUT CARROTS

Fresh-cut carrots are distributed as slices, sticks, shreds (grated carrots), or bite-size pared products referred to as "baby" carrots. These are normally derived from the mature root, although specialty items prepared with imma-ture plants grown for the purpose are available in the marketplace. The appearance of "white blush" on cut surfaces is a serious quality problem with all fresh-cut carrot products. The phenomenon is believed to result mainly from dehydration of the cut surface but enzymatic reactions leading to lignification may contribute to the phenomenon. Some processors dip carrot products in citric acid to delay enzymatic whitening, and most package the product without drying to avoid rapid dehydration of the surface. The use of proprietary humectants that prevent dehydration and preserve the appearance of freshly sliced carrots is also practiced.

Freshly harvested carrots carry microbial populations dominated by species derived from soil. In-field washing in chlorinated water to remove excess soil is a common practice, and the harvested crop is often stored prior to distribution and processing. In some jurisdictions, fungicides such as thiabendazole and/or iprodione (Rovral) and bactericides such as chlorane are applied for the control of storage disease [1]. Under ideal conditions storage temperature is held close to $0°C$, and relative humidity is maintained above 90% to reduce respiration rates, limit weight loss, and discourage growth of microorganisms responsible for postharvest diseases. There is little doubt that postharvest treatments such as these alter the microflora of raw carrots destined for processing, although their influence on microorganisms of significance in fresh-cut products has not been examined in detail. Similarly, the processing plant is also expected to contribute spoilage microorganisms to the finished product. Unfortunately, the microbial ecology of the processing environment is poorly understood.

Microbiological examination of carrot sticks at various stages of processing has shown that peel is a major source of microbial contaminants on the finished product. A survey of processing establishments by Garg et al. [2] revealed that peeling reduced mean total aerobic populations from 6.5×10^6 to 3.6×10^4 CFU/g. It should be noted that the samples analyzed for this work were prepared by blending cut pieces. Hence the actual population density on the peel of raw carrots may have been several orders of magnitude greater. The data derived from this study clearly showed that peeling results in the transfer of microorganisms from the outer epidermis of the raw material to surfaces exposed by cutting. It is also interesting to note that populations remained unchanged during slicing or dipping in a chlorinated water ice bath. Mean total aerobic populations on the finished product were in the range 10^4 to 10^5 CFU/g, in close agreement with a separate study by Odumeru et al. [3]. Garg et al. [2] found large populations of Gram-negative psychrotrophic bacteria, particularly Pseudomonas spp., lactic acid bacteria, and fungi on freshly prepared carrot sticks. The fate of these microorganisms during refrigerated storage was subject to the influence of preparation method, composition of the atmosphere in the package, and storage temperature. Izumi et al. [4] compared various quality attributes in carrot slices, sticks, and shreds stored in air or under controlled atmosphere (0.5% O_2, 10% CO_2). Although differences were not always statistically significant, total aerobic microbial populations were highest in shreds, followed by sticks and slices. Hence there appears to be a relationship between the degree of damage to plant tissues, available surface area for colonization, and the extent of microbial growth.

Rapid quality changes in commercially packaged, shredded carrots and the suspected involvement of microorganisms led to detailed microbiological examination of these products by Carlin et al. [5–7]. Most are distributed in impermeable films or rigid containers. Rapid depletion of oxygen to less than 2% and accumulation of carbon dioxide to more than 30% due to accelerated respiration in plant tissues leads to the establishment of atmospheres conducive to the growth of microaerophilic and facultatively anaerobic species. Lactic acid bacteria, particularly the heterofermentative species Leuconostoc mesenteroides, and yeasts quickly become the predominant groups, and both lactic acid and ethanol accumulate in the product [5]. Controlled atmospheres containing 15 to 20% CO_2 and 5% O_2 can delay these changes [6–7]. Kakiomenou et al. [8] also examined the influence of temperature and various modified atmospheres on these events. Lactic acid bacteria were always the dominant group, particularly at higher temperatures ($10°C$).

11.3 FRESH-CUT CABBAGE

Shredded cabbage is usually distributed in polymeric film bags. Washing in chlorinated water followed by dewatering by centrifugation are common in the industry. Citric acid-containing dips are also employed by some manufacturers, purportedly to delay both physiological disorders and microbial growth, although there is little experimental evidence that such treatments are effective. Quality defects due to the development of off-flavors, odors, and changes in color are not uncommon in distribution systems.

Whole cabbage is frequently stored for extended periods of time before processing. Temperatures are maintained near $0°C$ and relative humidity above 90%. Stored cabbage is at risk of spoilage by fungi and the use of fungicides (such as benomyl, thiabendazole) is practiced in some countries [9]. Microbiological examination of whole cabbage heads by King et $al.$ [10] showed that outer leaves carry much higher microbial populations than tissues close to the core. These ranged from a density of $1.4 \times 10^6 \, CFU/g$ on the outer leaves to $3.8 \times 10^2 \, CFU/g$ in core samples of stored cabbage sampled over a period of several months. Psychrotrophic bacteria belonging to the genera brevibacterium, chromobacterium, citrobacter, pseudomonas, and xanthomonas were the largest group of microorganisms present, although yeast and mold were also recovered in large numbers. Geeson [9] also found that the outer leaves of stored cabbage carry large populations of fluorescent pseudomonads, pectolytic bacteria, yeasts, and molds. In the latter study, postharvest application of benomyl and thiabendazole by drenching was correlated with higher bacterial populations in cabbage stored for a period of several months at various relative humidities.

Unfortunately, the impact of unit operations on the microbiology of shredded cabbage has not been examined in detail. However, commercial experience has shown that removal of outer leaves prior to slicing improves the microbiological stability of shredded products. Shredding likely distributes remaining contaminants throughout the product and provides extensive opportunities for growth on cut surfaces. Hao et $al.$ [11] examined the microbiology of shredded cabbage stored in both low- and high-oxygen transmission films at 4, 13, and $21°C$. Carbon dioxide levels rose and oxygen levels declined faster in the headspace above product packaged in less permeable films. Surprisingly, the observed spoilage patterns were consistent for all treatments, and populations of aerobic bacteria tended to dominate the spoilage association. However, lactic acid bacteria populations increased to high densities regardless of packaging treatment or storage temperature. The development of sourness and blowing of packages can occur in these products and has been associated with extensive growth of these bacteria. Growth of this relatively minor component of the raw cabbage microflora can be stimulated by the addition of salt for the manufacture of fermented products. Reasons for their rapid growth in shredded cabbage despite the presence of active, large populations of Gram-negative, psychrotrophic bacteria remain unclear.

11.4 FRESH-CUT LETTUCE

Iceberg, leaf, and mixed lettuces remain the most common fresh-cut products in the marketplace. Unlike carrot or cabbage products, raw materials for their manufacture are not stored for extensive periods of time prior to processing.

Field lettuce carries populations of microorganisms derived from soil [12]. Gram-negative bacteria, particularly fluorescent pseudomonads, and smaller numbers of yeast make up the bulk of the initial microbial flora, which ranges in density between 10^4 and 10^6 CFU/g [2,13–15]. Densities of microorganisms also vary with location in the plant, and outer leaves tend to carry higher microbial populations.

The need to control browning reactions catalyzed by oxygen in packaged cut lettuce has led to extensive use of technologies designed to maintain anoxic atmospheres during storage. Control over browning can be achieved by vacuum packaging, flushing with nitrogen, or the incorporation of atmospheres containing various levels of oxygen, nitrogen, and carbon dioxide. Selective pressures exerted by gas composition of the atmosphere are expected to influence the spoilage association. However, observations drawn from numerous studies on the microbiology of lettuce packed under oxygen-reduced atmospheres suggest that storage temperature exerts more influence on the development of the spoilage association than gaseous composition. Mesophilic and psychrotrophic populations tend to be similar in stored shredded lettuce, while yeast and mold and lactic acid bacteria populations generally remain low [2,13,14]. Spoilage microflora are always dominated by species of psychrotrophic *Pseudomonas* spp. [14–16]. Evidence of microbial spoilage appears earlier in products with high initial microbial loads and at higher storage temperatures.

The effects of unit operations on the microbiology of fresh-cut lettuce are not well characterized. Bolin *et al.* [17] examined the influence of cutting method, washing, centrifugation, initial microbial load, packaging, and temperature on the storage stability of iceberg lettuce. Unfortunately, microbiological assessments were not carried out for all experimental treatments. Deleterious quality effects were associated with tissue damage induced by cutting with dull blades, inadequate water removal, and storage at high temperatures. There is little doubt that some of these factors could influence the development of microbial populations in the package. High initial microbial loads were also correlated with reduced keeping quality.

11.5 OCCURRENCE AND BEHAVIOR OF HUMAN PATHOGENS IN FRESH-CUT VEGETABLES

Any food product delivered to consumers without the application of treatments that completely eliminate microbial contaminants may serve as a vehicle for the transmission of microorganisms capable of causing disease. Hence, concerns about the potential for transmission of human pathogens through fresh-cut vegetables are warranted because raw horticultural products are naturally contaminated with large numbers of microorganisms, and they are not subjected to lethal processing treatments prior to distribution. Microbial populations on field vegetables may occasionally include potential pathogens acquired in the production environment, either from natural sources

TABLE 11.1
Some Outbreaks of Foodborne Illness Linked to Fresh-Cut Cabbage and Lettuce Products

Etiological agent	Commodity	Ref.
Listeria monocytogenes	Cabbage coleslaw	44
Clostridium botulinum (A)	Cabbage salad	45
Shigella sonnei	Iceberg lettuce	46
E. coli O157:H7	Mesclun lettuce	47
Hepatitis A	Iceberg lettuce	48
E. coli O157:H7	Shredded carrots	49

or as a result of human activity (see Chapter 1). Foodborne pathogens have been isolated from market fresh-cut vegetable products in various parts of the world [18–20]. In addition, processing plants frequently receive raw materials from a variety of sources and the potential exists for cross contamination of product. Despite these apparent vulnerabilities, relatively few documented outbreaks of foodborne illness have been conclusively linked to fresh-cut vegetables. Table 11.1 lists some outbreaks that have implicated fresh-cut carrots, cabbage, and lettuce, and these examples serve to illustrate the potential for transmission of various infectious agents through fresh-cut vegetables. It should be stressed that epidemiological investigations of outbreaks involving such products are fraught with difficulties. Considerable time delays between definitive association and sampling of suspected products may reduce the likelihood of detection. In addition, microbiological analyses often lack the sensitivity required to detect small populations of target microorganisms in environmental samples containing complex background microflora in high populations. This problem is acute for some pathogens, particularly species of shigella, due to a lack of effective selective enrichment protocols. Hence, currently held assumptions about the behavior of human pathogens in fresh-cut vegetable products are largely derived from research.

The behavior of pathogenic microorganisms has been examined in fresh-cut cabbage under laboratory conditions. For example, Kallander et al. [21] were unable to detect Listeria monocytogenes (detection limit $10^2\,CFU/g$) in shredded cabbage stored under air or under 70% carbon dioxide and 30% nitrogen after 6 days at 25°C. A reduction in pH related to the growth of lactic acid bacteria was evident at this temperature. In contrast, the species was capable of growth in cabbage stored at 5°C, irrespective of atmosphere composition, and little or no growth of lactic acid bacteria was detected. Omary et al. [23] inoculated shredded cabbage with Listeria innocua, a surrogate species for L. monocytogenes. Packaging films with oxygen transmission rates ranging from 5.6 to $6000\,cm^3\,O_2/m^2/24$ hours were employed for storage of samples at 11°C. Listeria innocua populations initially declined but eventually increased in samples subjected to all treatments. Hence, the results of studies carried out in the laboratory may lead to different individual conclusions. Nevertheless, the sum of this research suggests that *Listeria* spp. can grow in packaged fresh-cut cabbage and that temperature is a critical determinant for the fate of this species.

The behavior of *L. monocytogenes* in cut lettuce has also been examined and variable results are reported from individual studies. A gradual decline in populations was reported by Francis and O'Beirne [22] at 3°C and by Kakiomenou et al. [8] at 4°C. Beuchat and Brackett [24] found there was no change in cut lettuce stored at 5°C, but Steinbruegge et al. [25] observed exponential growth at higher temperatures. The effect of modified atmospheres on the fate of *L. monocytogenes* is also unpredictable. Francis and O'Beirne [22] found that growth was enhanced in shredded lettuce stored under 100% N_2 instead of an aerobic packaging system. Jacxsens et al. [26] recorded similar observations for shredded iceberg lettuce stored at 7°C under a 2 to 3% O_2, 2 to 3% CO_2, 94 to 96% N_2 atmosphere. However, contradictory conclusions were drawn from the work of Beuchat and Brackett [24] where no difference was found between samples stored at 10°C in air or in a 3% O_2/97% N_2 gas mixture. Hence, the effects of modified atmospheres on the fate of *L. monocytogenes* in cut lettuce remain uncertain.

Comparatively little is known about the behavior of human pathogens in fresh-cut carrots, likely due to the limited number of foodborne infections conclusively associated with this commodity. Nevertheless, some interesting observations have been derived from research with inoculated products. Bagamboula et al. [27] found that *Shigella* spp. populations gradually fell in grated carrots stored at either 7 or 12°C. The spoilage association in the product was dominated by lactic acid bacteria, and it was concluded that a gradual decline in pH was largely responsible for a gradual die-off of the inoculum. Viable cells were recovered from all products after 7 days.

Analysis of published research indicates that temperature has a major influence on the fate of pathogens in fresh-cut vegetable products. The maintenance of low temperatures evidently provides the best means to prevent growth, particularly for members of the family Enterobacteriaceae. *Listeria monocytogenes* poses a unique challenge given the ability of the species to grow at refrigeration temperatures. However, the plurality of experimental outcomes at higher temperatures or where modified atmospheres are applied to packaging systems suggests that additional factors influence the behavior of this pathogen in fresh-cut vegetable products.

11.6 INTERACTIONS BETWEEN MICROORGANISMS AND PLANT TISSUES

The shelf life of fresh-cut vegetables including lettuce, cabbage, and carrots is limited mainly by the appearance of enzymatically induced discolorations. Control over these quality defects currently relies on the maintenance of low-oxygen atmospheres or the application of dips containing enzyme inhibitors, but additional means to alleviate the problem are under investigation.

The application of mild heat treatments or heat shocks before packaging shows promise for this purpose [28–31]. Heat shocks applied between 47 and 50°C for 90 to 180 seconds delay browning in iceberg lettuce by several days in refrigerated product. In addition, these treatments improve disinfection of the product, and reductions of up to 3 log CFU/g are readily achieved. Unfortunately, the advantages of heat treatments are negated by accelerated microbial growth during subsequent refrigerated storage. Faster growth of spoilage bacteria has been observed, and inoculation with *L. monocytogenes* and *E. coli* O157:H7 confirmed that growth is enhanced by prior heat treatment of the lettuce [28,32]. A similar effect occurs in shredded cabbage subjected to heat treatments. The example provided in Figure 11.1 shows total aerobic microbial populations in shredded cabbage stored under refrigeration in low oxygen permeability films after treatment in 100 µg/ml NaOCl solutions heated to various temperatures (Delaquis, unpublished data). Evidently, washing in a cold chlorinated water solution had little effect on the development of the spoilage association. Application of heated chlorinated water reduced the initial population by 2 log CFU/g or more. However, the rate of population growth and the ultimate size of the spoilage population were greater in cabbage subjected to the heat treatments.

Removal of a substantial part of the native microflora could provide a competitive advantage to survivors of heat treatments or to contaminants acquired postprocess. There is evidence that additional factors are responsible for these effects, however. The classic plant pathology literature describes numerous constitutive and induced antimicrobial systems in plants [33]. The economic impact of species that are pathogenic to crops in the field or those responsible for losses during prolonged storage has stimulated considerable

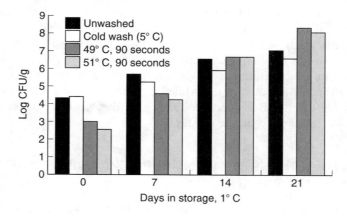

FIGURE 11.1 Total aerobic populations in shredded cabbage stored at 1°C in oxygen-permeable film after treatment in a 100 µg/m; NaOCl solution heated to various temperatures. The cabbage was dried in a centrifuge prior to packaging. (P.J. Delaquis, 2004. Unpublished data.)

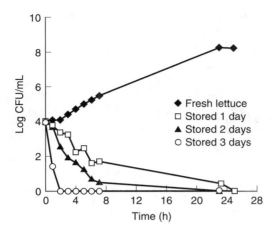

FIGURE 11.2 Fate of *Listeria monocytogenes* in aqueous extracts prepared from packaged, shredded cut iceberg lettuce stored aerobically for 0, 1, 2, and 3 days at 15°C. (P.J. Delaquis, 2004. Unpublished data.)

research in this area. In contrast, comparatively little is known about the influence of intrinsic plant defense mechanisms on the fate of significant spoilage microorganisms or foodborne pathogens in fresh-cut vegetable or fruit products. Reports of antimicrobial activity in vegetable extracts provide evidence that some plant constituents may have a role in the microbial ecology of these products. Conner *et al.* [34] found that an unidentified cabbage juice extract inhibits the growth of *Listeria monocytogenes*. Carrot extracts have also been widely reported to inhibit fungi [35], foodborne bacteria and yeast [36], and *Listeria monocytogenes* [36–40].

Accelerated development of microbial populations in products subjected to heat shocks hints that intrinsic barriers to growth normally present in physiologically intact plant tissues may be disrupted by processing. This hypothesis was tested in our laboratory by inoculation of *Listeria monocytogenes* in iceberg lettuce tissue extracts prepared from tissues stored aerobically for up to three days, as shown in Figure 11.2. The results of these experiments provided evidence that an antilisterial factor or factors is elaborated by cut lettuce tissues stored under aerobic conditions. Application of heat treatments before storage and preparation of the extracts reduces this effect, however, as shown in Figure 11.3. The chemical nature of the inhibitor(s) responsible for this effect in iceberg lettuce is not yet known. There is little doubt that heat shocks and unit operations applied in processing and preservation have a major impact on the physiology of plant tissues. For example, heat shocks [41] and modified atmospheres [42] inhibit the activity of phenylalanine ammonia lyase (PAL), a key enzyme in the development of discolorations in cut lettuce tissues. The enzyme catalyzes the first step in a series of complex reactions that leads to the accumulation of phenylpropanoid intermediates, including phenolic compounds such as caffeic, ferulic, and chlorogenic acids. Several of these

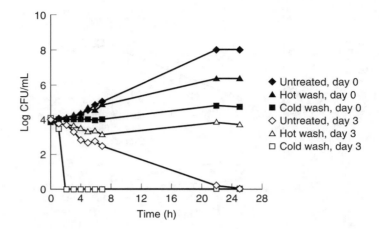

FIGURE 11.3 Fate of *Listeria monocytogenes* in aqueous extracts prepared from packaged, shredded iceberg lettuce stored aerobically for 0 and 3 days at 15°C following a three-minute wash in cold (4°C) or warm (47°C) water. (P.J. Delaquis, 2004. Unpublished data.)

compounds exhibit antimicrobial activity *in vitro* [43], but it remains unknown whether they influence the fate of microorganisms in packaged cut lettuce.

Clearly, much remains to be learned about the interaction between microorganisms and plant tissues. Improved characterization of intrinsic factors, which affect the fate of microorganisms in fresh-cut vegetables, will undoubtedly enhance our understanding of fundamental interactions in these complex microbial ecosystems. Furthermore, a more complete understanding of these interactions could lead to significant practical outcomes. In the future, it may be possible to exploit intrinsic barriers to restrict microbial growth for the development of novel preservation processes for fresh-cut vegetables.

REFERENCES

1. Afek, U., Orenstein, J., and Nuriel, E., Steam treatment to prevent carrot decay during storage, *Crop Prot.*, 18, 639, 1999.
2. Garg, N., Churey, J.J., and Splittstoesser, D.F., Effect of processing conditions on the microflora of fresh-cut vegetables, *J. Food Prot.*, 53, 701, 1990.
3. Odumeru, J., Mitchell, S.J., Alves, D.M., Lynch, J.A., Yee, A.J., Wang, S.L., Styliadis, S., and Farber, J., Assessment of the microbiological quality of ready-to-use vegetables for health-care food services, *J. Food Prot.*, 60, 954, 1997.
4. Izumi, H., Watada, A.E., Nathanee, P.K., and Douglas, W., Controlled atmosphere storage of carrot slices, sticks and shreds, *Postharv. Biol. Technol.*, 9, 165, 1996.
5. Carlin, F., Nguyen-The, C., Cudennec, P., and Reich, M., Microbiological spoilage of fresh, "ready-to-use" grated carrots, *Science des Aliments*, 9, 371, 1989.

6. Carlin, F., Nguyen-The, C., Hilbert, G., and Chambroy, Y., Modified atmosphere packaging of fresh, "ready-to-use" grated carrots in polymeric films, *J. Food Sci.*, 55, 1033, 1990.

7. Carlin, F., Nguyen-The, C., Chambroy, Y., and Reich, M., Effect of controlled atmospheres on microbial spoilage, electrolyte leakage and sugar content of fresh "ready-to-use" carrots, *Int. J. Food Sci. Technol.*, 25, 110, 1990.

8. Kakiomenou, K., Tassou, C., and Nychas, G.-J., Survival of *Salmonella enteritidis* and *Listeria monocytogenes* on salad vegetables, *World. J. Microbiol. Biotechnol.*, 14, 381, 1998.

9. Geeson, H., The fungal and bacterial flora of stored white cabbage, *J. Appl. Bacteriol.*, 46, 189, 1979.

10. King, A.D., Michener, H.D., Bayne, H.G., and Mihara, K.L., Microbial studies on shelf life of cabbage and coleslaw, *Appl. Environ. Microbiol.*, 31, 404, 1976.

11. Hao, Y.Y., Brackett, R.E., Beuchat, L.R., and Doyle, M.P., Microbiological quality and the inability of proteolytic *Clostridium botulinum* to produce toxin in film-packaged fresh-cut cabbage and lettuce, *J. Food Prot.*, 61, 1148, 1998.

12. Boyette, M.D., Ritchie, D.F., Carballo, S.J., Blankenship, S.M., and Sanders, D.C., Chlorination and postharvest disease control, *HortTechnol.*, 3, 395, 1993.

13. Barriga, M.I., Trachy, G., Willemot, C., and Simard, R.E., Microbial changes in shredded iceberg lettuce stored under controlled atmospheres, *J. Food Sci.*, 56, 1586, 1991.

14. Marchetti, R., Casadei, M.A., and Guerzoni, M.E., Microbial population dynamics in ready-to-use vegetable salads, *Ital. J. Food Sci.*, 2, 97, 1992.

15. King, A.D., Magnuson, J.A., Torok, T., and Goodman, N., Microbial flora and storage quality of partially processed lettuce, *J. Food Sci.*, 56, 459, 1991.

16. Delaquis, P.J., Stewart, S., Toivonen, P.M.A.. and Moyls., A.L., Effect of warm, chlorinated water on the microbial flora of shredded iceberg lettuce, *Food Res. Int.*, 32, 7, 1999.

17. Bolin, H.R., Stafford, A.E., King, A.D., Jr., and Huxsoll, C.C., Factors affecting the storage stability of shredded lettuce, *J. Food Sci.*, 42, 1319, 1977.

18. Farber, J.M., Sanders, G.W., and Johnston, M.A., A survey of various foods for the presence of *Listeria* species, *J. Food Prot.*, 52, 456, 1989.

19. Hessick, J.E., Wagner, D.E., Nierman, M.L., and Peeler, J.T., *Listeria* spp. found on fresh market produce, *Appl. Environ. Microbiol.*, 55, 1925, 1989.

20. Francis, G. A., Thomas, C., and O'Beirne, D., The microbiological safety of minimally processed vegetables, *Int. J. Food Sci. Technol.*, 34, 1, 1999.

21. Kallander, K.D., Hitchins, A.D., Lancette, G.A., Schmieg, J.A., Garcia, G.R., Solomon, H.M., and Sofos, J.N., Fate of *Listeria monocytogenes* at 5 and 25°C under a modified atmosphere, *J. Food Prot.*, 54, 302, 1991.

22. Francis, G.A. and O'Beirne, D., Effects of gas atmosphere, antimicrobial dip and temperature on the fate of *Listeria innocua* and *Listeria monocytogenes* on minimally processed lettuce, *Int. J. Food Sci. Technol.*, 32, 141, 1997.

23. Omary, M.B., Testin, R.F., Barefoot, S.F., and Rushing, J.W., Packaging effects on the growth of *Listeria innocua* in shredded cabbage, *J. Food Sci.*, 58, 623, 1993.

24. Beuchat, L.R. and Brackett, R.E., Survival and growth of *Listeria monocytogenes* on lettuce as influenced by shredding, chlorine treatment, modified atmosphere packaging and temperature, *J. Food Sci.*, 55, 755, 1990.

25. Steinbruegge, E.G., Maxcy, R.B., and Liewen, M.B., Fate of *Listeria monocytogenes* on ready to serve lettuce, *J. Food Prot.*, 51, 596, 1988.

26. Jacxsens, L., Devlieghere, F., Falcato, P., and Debevre, J., Behavior of *Listeria monocytogenes* and *Aeromonas* spp. on fresh-cut produce packaged under equilibrium-modified atmosphere, *J. Food Prot.*, 62,1128, 1999.

27. Bagamboula, C.F., Uyttendaele, M., and Debevere, J., Growth and survival of *Shigella sonnei* and *S. flexneri* in minimally processed vegetables packed under equilibrium modified atmosphere and stored at 7°C and 12°C, *Food Microbiol.*, 19, 536, 2002.

28. Delaquis, P.J., Stewart, S., Cazaux, S., and Toivonen, P., Survival and growth of *Listeria monocytogenes* and *Escherichia coli* O157:H7 in ready-to-eat iceberg lettuce washed in warm, chlorinated water, *J. Food Prot.*, 65, 495, 2002.

29. Loaize-Velarde, J.G., Tomas-Barberan, F.A., and Saltveit, M.E., Effect of intensity and duration of heat-shock treatments on wound-induced phenolic metabolism in iceberg lettuce, *J. Am. Soc. Hort. Sci.*, 122, 873, 1997.

30. Loaize-Velarde, J.G. and Saltveit, M.E., Heat shocks applied either before or after wounding reduce browning of lettuce leaf tissue, *J. Am. Soc. Hort. Sci.*, 126, 227, 2001.

31. Saltveit, M.E., Wound induced changes in phenolic metabolism and tissue browning are altered by heat shock, *Postharv. Biol. Technol.*, 21, 61, 2000.

32. Li, R., Brackett, R.E., Chen, J., and Beuchat, L.R., Mild heat treatment of lettuce enhances growth of *Listeria monocytogenes* during subsequent storage at 5°C or 15°C, *J. Appl. Microbiol.*, 92, 269, 2002

33. Walker, J.R.L. Antimicrobial compounds in food plants, in *Natural Antimicrobial Systems and Food Preservation*, Dillon, V.M. and Board, R.G., Eds., CAB International, Wallingford, UK, 1994, p. 181.

34. Conner, D.E., Brackett, R.E., and Beuchat, L.R., Effect of temperature, sodium chloride and pH on growth of *Listeria monocytogenes* in cabbage juice, *Appl. Environ. Microbiol.*, 52, 59, 1996.

35. Batt, C., Solberg, M., and Ceponis, M., Inhibition of aflatoxin production by carrot root extract, *J. Food Sci.*, 45, 1210, 1980.

36. Babic, I., Nguyen-The, C., Amiot, M.J., and Aubert, S., Antimicrobial activity of shredded carrot extracts on food-borne bacteria and yeast, *J. Appl. Bacteriol.*, 76, 135, 1994.

37. Beuchat, L.R. and Brackett, R.E., Inhibitory effect of raw carrots on *Listeria monocytogenes*, *Appl. Environ. Microbiol.*, 56, 1734, 1990.

38. Nguyen-the, C. and Lund, B.M., An investigation of the antibacterial effect of carrot on *L. monocytogenes*, *J. Appl. Bacteriol.*, 73, 23, 1992.

39. Beuchat, L.R., Brackett, R.E., and Doyle, M.P., Lethality of carrot juice to *Listeria monocytogenes* as affected by pH, sodium chloride and temperature, *J. Food Prot.*, 57, 470, 1994.

40. Beuchat, L.R. and Doyle, M.P., Survival and growth of *Listeria monocytogenes* in foods treated or supplemented with carrot juice, *Food Microbiol.*, 12, 73, 1995.

41. Fukumoto, L.R., Toivonen, P.M.A., and Delaquis, P.J., Effect of wash water temperature and chlorination on phenolic metabolism and browning of stored iceberg lettuce photosynthetic and vascular tissues, *J. Agric. Food Chem.*, 50, 4503, 2002.

42. Lopez-Galvez, G., Saltveit, M., and Cantwell, B., The visual quality of minimally processed lettuces stored in air or controlled atmospheres with emphasis on romaine and iceberg types, *Postharv. Biol. Technol.*, 8, 179, 1996

43. Wen, A., Delaquis, P.J., and Stanich, K., Antilisterial activity of selected phenolic acids, *Food Microbiol.*, 20, 305, 2003.
44. Schlech, W.F., Epidemic listeriosis: evidence for transmission by feed, *N. Engl. J. Med.*, 308, 203, 1983.
45. Solomon, H.M., Kautter, D.A., Lilly, T., and Rhodehamel, E.J., Outgrowth of *Clostridium botulinum* in shredded cabbage at room temperature under modified atmosphere, *J. Food Prot.*, 53, 831, 1990.
46. Davis, H.J., Taylor, P., Perdue, J.N., Stelma, G.N., Humphreys, J.M., Rowntree, R., and Green, K.D., A shigellosis outbreak linked to commercially distributed lettuce, *Am. J. Epidemiol.*, 128, 1312, 1988.
47. Hillborn, E.D., Mermin, J.H., Mshar, P.A., Hadler, J.L., Voetsch, A., Wojtkunski, C., Swartz, M., Mshar, R., Lambert-Fair, M.A., and Farrar, J.A., A multistate outbreak of *Escherichia coli* 0157:H7 infections associated with consumption of mesclun lettuce, *J. Am. Med. Assoc.*, 159, 1758, 1999.
48. Rosenblum, L.S., Mirkin, I.R., Allen, D.T., Safford, S., and Hadler, S.C., A multifocal outbreak of hepatitis A traced to commercially distributed lettuce, *AJPH*, 80, 1075, 1990.
49. CDC, Foodborne outbreaks of enterotoxigenic *Escherichia coli*: Rhode Island and New Hampshire, 1993, *MMWR*, 43, 81, 1994.

12 Outbreaks Associated with *Cyclospora* and *Cryptosporidium*

Ynes R. Ortega and Charles R. Sterling

CONTENTS

12.1 Introduction 267
12.2 Overview of the Parasites 267
12.3 Sources of Contamination 269
12.4 Description of the Foodborne Outbreaks for Both Parasites 270
12.5 Detection and Enumeration Methodologies 271
12.6 Interventions for Decontamination 273
12.7 Conclusions 274
References 275

12.1 INTRODUCTION

Parasites have frequently been implicated in food and waterborne outbreaks in the U.S. and elsewhere. This may be related to the development of more sensitive and specific assays for parasite identification and changes in diet and food consumption habits. To satisfy the demands of consumers, more fresh produce is being imported. Food management and production practices in developing countries are not necessarily similar to those in the developed world. This chapter covers the parasites cryptosporidium and cyclospora, as they have been implicated in several food and waterborne outbreaks.

12.2 OVERVIEW OF THE PARASITES

Cryptosporidium is a protozoan parasite that belongs to the subphylum Apicomplexa. Because of large and significant waterborne outbreaks associated with it, much has been learned about the epidemiology, immunology, and biology of this parasite. *Cryptosporidium parvum* was considered to be the only species of public health significance; however, studies involving pediatric populations, the immunocompromised, travelers, and endemic populations have demonstrated that humans can also be infected with *C. meleagridis* (of turkeys), *C. canis* (of dogs), *C. felis* (of cats), and *C. muris*

267

(of cattle and rodents). Molecular studies have also demonstrated that the previously described *C. parvum* is actually at least two different species: *C. parvum* (zoonotic) and *C. hominis* (considered anthroponotic, i.e., human-to-human). Both are morphologically similar [1,2].

The infective stage of cryptosporidium is the oocyst, which is excreted in the feces of infected hosts. These oocysts are immediately infectious to a susceptible host and contain four fully differentiated and infectious sporozoites. Once ingested, oocysts excyst in the gastrointestinal tract, releasing infective sporozoites. Sporozoites infect the epithelial cells of the ileum preferentially and may continue colonization of all the small intestine and bile ducts if the host is immunocompromised, thus making eradication of the parasite more difficult. Once parasites have entered the epithelial cells they are compartmentalized in a parasitic vacuole which is extracytoplasmatic, but intracellular. This site localization is unique, and the parasite depends upon an elaboration of membrane surface at the cytoplasmic interface called the feeder organelle which allows for the transport of select nutrients to the parasite. Once established, the parasite multiplies asexually, producing type I and II meronts containing 8 and 4 merozoites, respectively. The first asexual meront generation can multiply indefinitely in the absence of immunity while those that have progressed to the second generation eventually differentiate and produce gamonts (macro- and microgametocytes). A microgametocyte fertilizes the macrogamont and leads to the formation of a zygote. When mature, the zygote will become either a thin- or thick-walled oocyst. The former constitutes approximately 20% of oocysts produced and is autoinfective, while the latter is environmentally resistant and fully infectious when excreted to the environment [3].

Cryptosporidiosis is characterized by abundant and persistent diarrhea, fever, and abdominal pain. There is no effective therapy but new drugs are being evaluated with promising results [4–6]. The immune system of the infected individual plays a crucial role in eradicating the parasite. In immunocompromised patients cryptosporidiosis can be fatal [7].

Cyclospora cayetanensis is also a parasite of the subphylum Apicomplexa. Fourteen different species of cyclospora have been described in rodents and insectivores [8]. *Cyclospora cayetanensis* is considered to be exclusively anthroponotic [9]. In 1999 three other species infecting nonhuman primates were described. These cyclospora species are morphologically similar to *C. cayetanensis*, but, based on their host specificity and 18S DNA sequence homology, are different [10,11].

The life cycle of cyclospora begins when oocysts are excreted in the feces of an infected individual. These oocysts are unsporulated and take approximately two weeks under optimal environmental conditions to sporulate fully and become infectious. Differentiated oocysts contain two sporocysts, each containing two sporozoites. When ingested and upon passage through the gastrointestinal tract, oocysts rupture and sporocysts are released. Enzymes and bile salts induce the release (excystation) of the sporozoites, which in turn invade the epithelial cells of the small intestine, forming an intracellular parasitic vacuole. Based on histological observations of biopsies of infected individuals, type I and II meronts are produced. Gametocytes have also been observed, suggesting that cyclospora has a life cycle similar to that of cryptosporidium and other coccidia. After zygote formation, oocysts are formed and excreted to the environment [12].

There are nonspecific fingerprinting tools for traceback studies; however, characterization of the internal transcribed spacers 1 (ITS1) sequences may be used for these purposes. The ITS1 sequences of clones of all five raspberry-associated isolates were identical, consistent with their origin from a single source. One of the two Guatemala isolates and two Peruvian isolates contained multiple ITS1 sequences [13,14].

12.3 SOURCES OF CONTAMINATION

Cryptosporidium can be acquired by ingestion of contaminated water and foods. The oocysts are already infectious when excreted; therefore, cryptosporidiosis can also be transmitted via the fecal–oral route involving person-to-person or animal-to-person transmission. Recreational [15–18] and drinking waters [19,20] have been responsible for various cryptosporidium waterborne outbreaks. Chlorine concentrations suitable for drinking water have proven insufficient to inactivate cryptosporidium oocysts. Prepared foods and apple cider have also been described as sources for cryptosporidium contamination [21–24]. Inappropriate manipulations of foods by food handlers [25] or fruits contaminated with cattle feces have also been sources of cryptosporidium contamination. Zoonotic transmission has also been important, particularly for hikers who drink river or lake water without disinfecting it. Cryptosporidium can infect a diverse variety of animals, particularly cattle. Zoonotic contaminations have been described in veterinary students and animal caretakers [26,27]. Outbreaks in day care centers have also been reported. Children are highly susceptible to infection, and transmission of this parasite can be high if proper hygiene practices are not in place, either at home or day care centers [28–33]. The same conditions are favorable for cryptosporidium outbreaks in hospitals [34, 35].

Investigations of cyclospora outbreaks suggest that it can be acquired when ingesting water [36–38] and food [39–44] that contain the parasite's oocysts. Cyclospora has been isolated from fresh produce in the U.S. and elsewhere [36,45,46], suggesting that foods play an important role in cyclospora transmission. The mechanisms and dynamics of transmission of cyclospora are more complicated than those of cryptosporidium since oocysts are not fully sporulated and therefore infective when passed from an infected human.

Contaminated water used for irrigation [47] or pesticide spraying support oocyst survival [48]. Aerosols, insects [49–51], and contaminated water courses and streams used in crop irrigation may factor in the introduction of viable oocysts into fresh produce.

12.4 DESCRIPTION OF THE FOODBORNE OUTBREAKS FOR BOTH PARASITES

Cryptosporidium can be found worldwide and infects a variety of hosts, including humans. The incidence of cryptosporidium can vary depending upon the population and location from 0.6 to 20%. In the U.S., cryptosporidiosis is associated with 0.4 to 1% of cases of diarrhea [52]. The cryptosporidium oocyst is resistant to normal environmental conditions, but desiccation can render oocysts noninfectious. Foodborne outbreaks of cryptosporidium have been associated with foods prepared in homes, suggesting direct contamination by food handlers. In Maine, in1993, an outbreak was linked to unpasteurized apple cider. A farm with livestock used dropped apples for the preparation of the apple cider [24]. In 1995, in Minnesota, chicken salad was implicated in 15 cases of cryptosporidiosis. Another outbreak involving apple cider occurred in 1996 in New York [53]. Sixty-six persons developed cryptosporidiosis and one died. In 1997, 54 cases of cryptosporidiosis were probably connected to the consumption of green onions. Two of the 14 food preparers were positive for cryptosporidium. In 1998 an outbreak was associated with consumption of meals in one of two cafeterias of a university in Washington D.C. Epidemiological investigation concluded that the outbreak was caused by *C. hominis* and the most probable source was an ill food handler who prepared raw produce [25].

Milk, salad, sausage, and tripe have also been suspect foods in travelers with cryptosporidiosis entering the U.S. from Mexico, the U.K., and Australia. Although no cases of cryptosporidiosis associated with shellfish have been reported, the presence of human strain of cryptosporidium has been reported in mussels and oysters retailed for human consumption [54–57].

Cyclospora is endemic in certain countries of tropical regions. Much of what we know about cyclospora has resulted from studies performed in those settings. A disease of the tropics and developing countries found its way to the developed countries when the latter started importing produce that is in demand throughout the year. If cyclospora is endemic in such exporting countries, it is possible that if good agricultural practices (GAPs) are not implemented in those particular fields, human feces can be carried to the products, either by crop manipulation with contaminated hands, or contaminated irrigation water. Cyclospora is highly resistant to environmental conditions and will attach to the surface of the produce and remain viable for longer periods of time. Most fecal contaminants may not remain viable for long periods of time, thus explaining why other foodborne outbreaks have not been reported in parallel with cyclospora outbreaks.

Four commodities have been implicated with cyclospora foodborne outbreaks: raspberries, basil, lettuce, and snow peas. Since the early 1990s sporadic cases of cyclosporiasis were reported in the U.S., but no source of contamination was identified. The first large cyclospora outbreak occurred in 1995 in Florida. Strawberries were initially implicated in the outbreak, but later epidemiological investigations suggested that raspberries were responsible. In 1996, 1465 cases of cyclosporiasis were reported in 20 states in the U.S. In 1997, 41 clusters comprising 762 cases were reported during the months of April and May in 9 states. Raspberries, basil, and lettuce were implicated in this outbreak. Also during April to June of the same year, 250 laboratory-confirmed sporadic cases were reported. In all instances, imported raspberries were associated with the outbreak. As a result of this, Guatemala voluntarily suspended the exportation of raspberries. In 1998 a few sporadic cases of cyclosporiasis were reported in the U.S., but Canada continued to import berries from Guatemala and experienced a large outbreak [58,59].

Surveys of fresh produce have described the presence of cyclospora [40,41,46]. In 2004 the U.S. Food and Drug Administration (FDA) issued an alert to consumers that two outbreak clusters of cyclosporiasis may be associated with raw basil and mesclun/spring salads served in Texas and Illinois. In February 2004 approximately 54 individuals in Wheaton, Illinois, and 38 people in Irvin, Texas, were stricken with cyclosporiasis. During June and July 2004 approximately 50 potential cases of cyclosporiasis were associated with a residential facility. Epidemiological and traceback studies linked the cases to consumption of raw Guatemalan snow peas [60]. Throughout the years, sporadic cases of cyclospora continue to occur in the U.S., suggesting that cyclospora is either being introduced to the U.S. by imported produce, or by food handlers who are carriers of this parasite. More studies are needed to determine the actual distribution of cyclospora in the U.S., both in human populations and environmental samples.

12.5 DETECTION AND ENUMERATION METHODOLOGIES

Methodologies for the identification and isolation of cryptosporidium in water have been thoroughly studied. The EPA Method 1623 [61,62] is based on the recovery of cryptosporidium oocysts and giardia cysts by filtration/IMS (immunomagnetic separation)/FA (fluorescent antibody) (EPA-821-R-99-006) of up to 10 l of water. Filtration can be performed using either the Pall Gelman HV Envirochek® capsule or the IDEXX Filta-Max™ filter. Cysts or oocysts are then captured by IMS using Dynabeads® GC-Combo (Dynal, Inc.) or Aureon CG (Aureon Biosystems) kits. Once the oocysts are recovered, they are identified using immunofluorescent assays from Merifluor® G/C (Meridian Diagnostics, Inc.), Aqua-Glo™ G/C Direct (Waterborne, Inc.), or Crypto-Glo™ (BioTechFrontier, Inc.). Recovery efficiency for *Cryptosporidium parvum* is 60 to 80%. Heat incubation of IMS-tagged oocysts resulted in recoveries of 71 and 51% and DAPI confirmation rates in reagent and river water of 93 and 73%, respectively [63]. Method 1623 has several limitations and interferences. IMS can be affected by water turbidity and the presence of silica, clay, humic acids, other organisms, etc. The presence of iron and pH will also affect oocyst recovery.

Electrochemiluminescence (ECL) technology has been used to identify cryptosporidium in environmental water samples of up to 10,000 nephelometric turbidity units [64].

It is also important to determine whether the parasites are viable and of public health relevance. Molecular assays will aid in speciation and sub-typing of the parasites. These include polymerase chain reaction (PCR), reverse transcription (RT)-PCR, nested PCR, and an isothermal amplification nucleic acid sequence-based amplification (NASBA) method [65–70]. After isolation of the parasites, extraction of the oocyst DNA is of critical importance. Oocysts can be broken by boiling, mechanical disruption with glass beads, digestive enzymes (proteinase-K, lysozyme) with 10% SDS, freeze/thaw, micro-wave, sonication, and commercial kits (DNA and RNA) or automated systems (contamination free) [71–73].

In water samples, various *Cryptosporidium parvum* mRNAs have been used as molecular targets for detection [65]. The mRNA coding of *C. parvum* for hsp70 was amplified using NASBA methodology with a detection limit of 80 fmol amplicon/test. [74].

Because mRNA denatures quickly, oocyst viability can be determined using RT-PCR for cryptosporidium using the *hsp*70 and the β-tubulin genes. An electrochemical enzyme-linked immobilized DNA-hybridization assay using the *C. parvum* hsp70 mRNA could distinguish dead from live oocysts. No cross-reactivity was observed with other bacterial and parasitic organisms, including *Cryptosporidium muris* [75].

In vitro cultivation recognizes parasites that are both viable and have the ability to penetrate and replicate within host cells. Infectivity can be deter-mined using animal models, but *C. hominis*, which is the anthroponotic species, is host specific and is not infectious in neonatal mouse models.

Some of these methods have been used in food matrices. An IMS-PCR assay was able to detect <10 *C. parvum* oocysts in milk [76].

A laser scanning cytometry method (ChemScanRDI), coupled with immunofluorescence detection with differential interference contrast (DIC) confirmation, has also been evaluated and compared with manual micro-scopic enumeration of cryptosporidium oocysts. The recovery rate was 50% at seeding levels from 30 to 230 oocysts. Laser scanning cytometry does eliminate the low sample throughput, operator subjectivity, and operator fatigue using conventional microscopy [77].

Although these methodologies have been described for environmental waters, they have not been fully validated in foods. The wide variety of produce and foods that could potentially be involved in parasite transmission makes selecting a unique method for isolation difficult. Detection using molecular diagnostic assays is also challenging because of the presence of inhibitors that could mask the presence of these parasites in foods.

Immunoassays have been developed for the use of cryptosporidium identi-fication in water samples. An indirect immunofluorescent assay has also proven to be useful in food matrices [23,78].

Most of the purification techniques that work for cryptosporidium have proven to be effective in purifying cyclospora oocysts (Ortega, personal com-munication). Sucrose and cesium chloride gradients used for cryptospori-dium can be used for cyclospora. Water filtration systems have also proven to concentrate cyclospora from water sources. To date, monoclonal antibodies to cyclospora have not been produced. This is not only because of the limited sources of oocysts, but also because the cell wall has poor antigenic properties. One useful approach has been to use magnetic beads coated with the lectin WGA [79].

Various methodologies have been described to identify cyclospora using conventional clinical assays. When environmental samples are examined, autofluorescence can prove useful, although this is not a specific assay. A PCR method, initially designed for clinical samples, has worked well with food matrices; however, other *Eimeria* spp. also had the same amplification product as cyclospora [80]. A restriction fragment length polymorphism using the Mnl 1 enzyme could differentiate between cyclospora and eimeria. The biggest challenges when using PCR for cyclospora are the methodologies used to extract DNA from the low number of oocysts likely to be encountered, and how to control for the presence of PCR inhibitors. Various methodologies, including chelating matrices and freeze/thaw cycles, FTA membranes, and DNA extraction kits, have been described [72,80]. The use of an extraction-free, filter-based protocol (FTA) to prepare DNA templates for use in PCR to identify *C. cayetanensis* and *C. parvum* oocysts and microsporidia spores has been described. As few as 10 to 30 *C. cayetanensis* oocysts per 100 g of fresh raspberries could be detected [72].

To control for PCR inhibitors, addition of BSA or milk has improved the sensitivity of the assay. The PCR assay in raspberries, basil, and mesclun lettuce could detect 40 or fewer oocysts per 100 g of raspberries or basil, but had a detection limit of around 1000 per 100 g in mesclun lettuce [72,81]. Real-time PCR can also detect DNA specifically from as few as 1 oocyst of *C. cayetanensis* per 5 µl reaction volume [82].

Determining the viability of cyclospora oocysts has proven to be very difficult. To date, there are no susceptible animal models or *in vitro* cultivation methods. Sporulation rates have been used to determine if a particular treatment has affected the oocyst viability. This, however, may not have any bearing on oocyst infectivity. Electrorotation has been used as a method to determine oocyst viability [83]. This method needs to be validated when an *in vivo* or *in vitro* system for cyclospora becomes available.

12.6 INTERVENTIONS FOR DECONTAMINATION

Various sanitizers and disinfectants have been evaluated for cryptosporidium. Oocysts will remain viable if kept in moist environments, but are very sensitive to desiccation. Moist heat treatments or pasteurization of cryptosporidium oocysts at 45°C for 5 to 20 minutes inactivate the parasite [84].

Chemical agents commonly used for disinfection of contaminated environmental surfaces and medical devices such as endoscopes have been evaluated for their effect on cryptosporidium viability. Exposure of *C. parvum* to steam, ethylene oxide, and Sterrad 100 and hydrogen peroxide at concentrations of 6 and 7.5% for 20 minutes resulted in population reductions of 3 logs or greater. Peracetic acid (0.2% for 20 minutes), sodium hypochlorite (5.25% for 10 minutes), a phenolic, a quaternary ammonium compound (10 minutes), 2% glutaraldehyde (45 minutes), and $ortho$-phthalaldehyde (20 minutes) did not completely inactivate oocysts [85].

The effect of ultraviolet radiation from low- and medium-pressure mercury arc lamps on $Cryptosporidium\ parvum$ oocysts has been evaluated. Two and three log units inactivation have been achieved at approximately 10 and $25\ mJ/cm^2/sec$, respectively [86]. Use of static mixers for dissolution of ozone in drinking water treatment plants may contribute to C. $parvum$ inactivation [87]. Flash pasteurization of cider inoculated with cryptosporidium oocysts at 70 or 71.7°C, both for 10 or 20 seconds, reduced viability by at least 4.9 logs (or 99.999%) when determined using a tissue culture assay. A 3.0 log (99.9%) and 4.8 log (99.9985) inactivation were achieved when oocysts were treated for 5 minutes at 70 or 71.7°C, respectively. Current practices of flash pasteurization in the juice industry are sufficient to inactivate contaminant oocysts [88]. An electrochemically produced mixed-oxidant solution (MIOX; LATA Inc.) was considerably more effective in inactivating $Cryptosporidium\ parvum$ oocysts than free chlorine. A 5 mg/l dose of mixed oxidants produced a >3 log ($>99.9\%$) inactivation of $Cryptosporidium\ parvum$ oocysts in 4 hours [89].

12.7 CONCLUSIONS

There is still limited knowledge of parasite transmission dynamics with respect to both cryptosporidium and cyclospora, and much research is required in the arena of inactivation strategies. Because of changes in population diets, food production, and management, and improved diagnostic assays, more cases of parasitic infections are being reported. It is also important to note that there is a change in population demographics, with more susceptible groups increasing in numbers (the elderly, children, and the immunocompromised).

Epidemiological features of cryptosporidium lead to the almost overwhelming conclusion that the incidence of foodborne cryptosporidiosis is underestimated. The low numbers of oocysts in suspected samples and the lack of more sensitive detection methods adapted for oocyst detection in food undoubtedly contribute to this under-reporting. Control of foodborne outbreaks caused by parasites such as cryptosporidium and cyclospora is directly related to methods that prevent food contamination in the first place. Removal or inactivation of oocysts of both parasites is a formidable task, since these organisms strongly attach themselves to produce surfaces. Oocysts have proven highly resistant to sanitizers and disinfectants, particularly at concentrations that would not affect the organoleptic characteristics of the fresh produce.

Possible vehicles of transmission have been suspected to be contaminated soils, fertilizers, pesticide solutions, and irrigation water containing human or animal waste. Washing hands, appropriate hygiene, and GAPs may contribute to the prevention of pathogens in ready-to-eat foods.

REFERENCES

1. Morgan-Ryan, U.M., Fall, A., Ward, L.A., Hijjawi, N., Sulaiman, I., Fayer, R., Thompson, R.C., Olson, M., Lal, A., and Xiao, L., *Cryptosporidium* hominis n. sp. (Apicomplexa: Cryptosporidiidae) from Homo sapiens, *J. Eukaryot. Microbiol.*, 49, 433, 2002.

2. Fayer, R., Morgan, U., and Upton, S.J., Epidemiology of *Cryptosporidium*: transmission, detection and identification, *Int. J. Parasitol.*, 30, 1305, 2000.

3. Marshall, M.M., Naumovitz, D., Ortega, Y., and Sterling, C.R., Waterborne protozoan pathogens, *Clin. Microbiol. Rev.*, 10, 67, 1997.

4. Allam,,A.F. and Shehab, A.Y., Efficacy of azithromycin, praziquantel and mirazid in treatment of cryptosporidiosis in school children, *J. Egypt. Soc. Parasitol.*, 32, 969, 2002.

5. Amadi, B., Mwiya, M., Musuku, J., Watuka, A., Sianongo, S., Ayoub, A., and Kelly, P., Effect of nitazoxanide on morbidity and mortality in Zambian children with cryptosporidiosis: a randomised controlled trial, *Lancet*, 360, 1375, 2002.

6. Rossignol, J.F., Ayoub, A., and Ayers, M.S., Treatment of diarrhea caused by *Cryptosporidium parvum*: a prospective randomized, double-blind, placebo-controlled study of Nitazoxanide, *J. Infect. Dis.*, 184, 103, 2001.

7. Clark, D.P., New insights into human cryptosporidiosis, *Clin. Microbiol. Rev.*, 12, 554, 1999.

8. Ortega, Y.R., Sterling, C.R., Gilman, R.H., Cama, V.A., and Diaz, F., *Cyclospora* species: a new protozoan pathogen of humans, *N. Engl. J. Med.*, 328, 1308, 1993.

9. Eberhard, M.L., Ortega, Y.R., Hanes, D.E., Nace, E.K., Do, R.Q., Robl, M.G., Won, K.Y., Gavidia, C., Sass, N.L., Mansfield, K., Gozalo, A., Griffiths, J., Gilman, R., Sterling, C.R., and Arrowood, M.J., Attempts to establish experimental *Cyclospora cayetanensis* infection in laboratory animals, *J. Parasitol.*, 86, 577, 2000.

10. Eberhard, M.L., da Silva, A.J., Lilley, B.G., and Pieniazek, N.J., Morphologic and molecular characterization of new *Cyclospora* species from Ethiopian monkeys: C. cercopitheci sp.n., C. colobi sp.n., and C. papionis sp.n, *Emerg. Infect. Dis.*, 5, 651, 1999.

11. Lopez, F.A., Manglicmot, J., Schmidt, T.M., Yeh, C., Smith, H.V., and Relman, D.A., Molecular characterization of *Cyclospora*-like organisms from baboons, *J. Infect. Dis.*, 179, 670, 1999.

12. Ortega, Y.R., Nagle, R., Gilman, R.H., Watanabe, J., Miyagui, J., Quispe, H., Kanagusuku, P., Roxas, C., and Sterling, C.R., Pathologic and clinical findings in patients with cyclosporiasis and a description of intracellular parasite life-cycle stages, *J. Infect. Dis.*, 176, 1584, 1997.

13. Adam, R.D., Ortega, Y.R., Gilman, R.H., and Sterling, C.R., Intervening transcribed spacer region 1 variability in *Cyclospora cayetanensis*, *J. Clin. Microbiol.*, 38, 2339, 2000.

14. Olivier, C. van de P.S., Lepp, P.W., Yoder, K., and Relman, D.A., Sequence variability in the first internal transcribed spacer region within and among *Cyclospora* species is consistent with polyparasitism, *Int. J. Parasitol.*, 31, 1475, 2001.

15. Centers for Disease Control and Prevention, Outbreak of cryptosporidiosis at a day camp: Florida, July–August 1995, *JAMA*, 275, 1790, 1996.

16. Outbreak of cryptosporidiosis at a day camp: Florida, July–August 1995, *MMWR*, 45, 442, 1996.

17. Kramer, M.H., Sorhage, F.E., Goldstein, S.T., Dalley, E., Wahlquist, S.P., and Herwaldt, B.L., First reported outbreak in the United States of cryptosporidiosis associated with a recreational lake, *Clin. Infect. Dis.*, 26, 27, 1998.

18. Lowery, C.J., Nugent, P., Moore, J.E., Millar, B.C., Xiru, X., and Dooley, J.S., PCR-IMS detection and molecular typing of *Cryptosporidium parvum* recovered from a recreational river source and an associated mussel (Mytilus edulis) bed in Northern Ireland, *Epidemiol. Infect.*, 127, 545, 2001.

19. Aragon, T.J., Novotny, S., Enanoria, W., Vugia, D.J., Khalakdina, A., and Katz, M.H., Endemic cryptosporidiosis and exposure to municipal tap water in persons with acquired immunodeficiency syndrome (AIDS): a case-control study, *BMC Public Health*, 3, 2, 2003.

20. Perz, J.F., Ennever, F.K., and Le Blancq, S.M., *Cryptosporidium* in tap water: comparison of predicted risks with observed levels of disease, *Am. J. Epidemiol.*, 147, 289, 1998.

21. Foodborne outbreak of cryptosporidiosis: Spokane, Washington, 1997, *MMWR*, 47, 565, 1998.

22. Foodborne outbreak of diarrheal illness associated with *Cryptosporidium parvum*: Minnesota, 1995, *MMWR*, 45, 783, 1996.

23. Deng, M.Q. and Cliver, D.O., Comparative detection of *Cryptosporidium parvum* oocysts from apple juice, *Int. J. Food Microbiol.*, 54, 155, 2000.

24. Millard, P.S., Gensheimer, K.F., Addiss, D.G., Sosin, D.M., Beckett, G.A., Houck-Jankoski, A., and Hudson, A., An outbreak of cryptosporidiosis from fresh-pressed apple cider, *JAMA*, 272, 1592, 1994.

25. Quiroz, E.S., Bern, C., MacArthur, J.R., Xiao, L., Fletcher, M., Arrowood, M.J., Shay, D.K., Levy, M.E., Glass, R.I., and Lal, A., An outbreak of cryptosporidiosis linked to a foodhandler, *J. Infect. Dis.*, 181, 695, 2000.

26. Pohjola, S., Oksanen, H., Jokipii, L., and Jokipii, A.M., Outbreak of cryptosporidiosis among veterinary students, *Scand. J. Infect. Dis.*, 18, 173, 1986.

27. Preiser, G., Preiser, L., and Madeo, L., An outbreak of cryptosporidiosis among veterinary science students who work with calves, *J. Am. Coll. Health*, 51, 213, 2003.

28. Alpert, G., Bell, L.M., Kirkpatrick, C.E., Budnick, L.D., Campos, J.M., Friedman, H.M., and Plotkin, S.A., Outbreak of cryptosporidiosis in a day-care center, *Pediatrics*, 77, 152, 1986.

29. Melo Cristino, J.A., Carvalho, M.I., and Salgado, M.J., An outbreak of cryptosporidiosis in a hospital day-care centre, *Epidemiol. Infect.*, 101, 355, 1988.

30. Nwanyanwu, O.C., Baird, J.N., and Reeve, G.R., Cryptosporidiosis in a day-care center, *Tex. Med.*, 85, 40, 1989.

31. Tangermann, R.H., Gordon, S., Wiesner, P., and Kreckman, L., An outbreak of cryptosporidiosis in a day-care center in Georgia, *Am. J. Epidemiol.*, 133, 471, 1991.

32. Taylor, J.P., Perdue, J.N., Dingley, D., Gustafson, T.L., Patterson, M., and Reed, L.A., Cryptosporidiosis outbreak in a day-care center, *Am. J. Dis. Child*, 139, 1023, 1985.

33. Walters, I.N., Miller, N.M., van den E.J., Dees, G.C., Taylor, L.A., Taynton, L.F., and Bennett, K.J., Outbreak of cryptosporidiosis among young children attending a day-care centre in Durban, *S. Afr. Med. J.*, 74, 496, 1988.

34. Gardner, C., An outbreak of hospital-acquired cryptosporidiosis, *Br. J. Nurs.*, 3, 152, 154, 158, 1994.
35. Navarrete, S., Stetler, H.C., Avila, C., Garcia Aranda, J.A., and Santos-Preciado, J.I., An outbreak of *Cryptosporidium* diarrhea in a pediatric hospital, *Pediatr. Infect. Dis. J.*, 10, 248, 1991.
36. Alakpa, G.E., Clarke, S.C., and Fagbenro-Beyioku, A.F., *Cyclospora cayetanensis* infection: vegetables and water as possible vehicles for its transmission in Lagos, Nigeria, *Br. J. Biomed. Sci.*, 60, 113, 2003.
37. Huang, P., Weber, J.T., Sosin, D.M., Griffin, P.M., Long, E.G., Murphy, J.J., Kocka, F., Peters, C., and Kallick, C., The first reported outbreak of diarrheal illness associated with *Cyclospora* in the United States, *Ann. Intern. Med.*, 123, 409, 1995.
38. Rabold, J.G., Hoge, C.W., Shlim, D.R., Kefford, C., Rajah, R., and Echeverria, P., *Cyclospora* outbreak associated with chlorinated drinking water, *Lancet*, 344, 1360, 1994.
39. Caceres, V.M., Ball, R.T., Somerfeldt, S.A., Mackey, R.L., Nichols, S.E., Mackenzie, W.R., and Herwaldt, B.L., A foodborne outbreak of cyclosporiasis caused by imported raspberries, *J. Fam. Pract.*, 47, 231, 1998.
40. Connor, B.A. and Shlim, D.R., Foodborne transmission of *Cyclospora*, *Lancet*, 346, 1634, 1995.
41. Doller, P.C., Dietrich, K., Filipp, N., Brockmann, S., Dreweck, C., Vonthein, R., Wagner-Wiening, C., and Wiedenmann, A., Cyclosporiasis outbreak in Germany associated with the consumption of salad, *Emerg. Infect. Dis.*, 8, 992, 2002.
42. Fleming, C.A., Caron, D., Gunn, J.E., and Barry, M.A., A foodborne outbreak of *Cyclospora cayetanensis* at a wedding: clinical features and risk factors for illness, *Arch. Intern. Med.*, 158, 1121, 1998.
43. Ho, A.Y., Lopez, A.S., Eberhart, M.G., Levenson, R., Finkel, B.S., da Silva, A.J., Roberts, J.M., Orlandi, P.A., Johnson, C.C., and Herwaldt, B.L., Outbreak of cyclosporiasis associated with imported raspberries, Philadelphia, Pennsylvania, 2000, *Emerg. Infect. Dis.*, 8, 783, 2002.
44. Lopez, A.S., Dodson, D.R., Arrowood, M.J., Orlandi, P.A., Jr., da Silva, A.J., Bier, J.W., Hanauer, S.D., Kuster, R.L., Oltman, S., Baldwin, M.S., Won, K.Y., Nace, E.M., Eberhard, M.L., and Herwaldt, B.L., Outbreak of cyclosporiasis associated with basil in Missouri in 1999, *Clin. Infect. Dis.*, 32, 1010, 2001.
45. Sherchand, J.B. and Cross, J.H., Emerging pathogen *Cyclospora cayetanensis* infection in Nepal, *Southeast Asian J. Trop. Med. Public Health*, 32 (Suppl. 2), 143, 2001.
46. Ortega, Y.R., Roxas, C.R., Gilman, R.H., Miller, N.J., Cabrera, L., Taquiri, C., and Sterling, C.R., Isolation of *Cryptosporidium parvum* and *Cyclospora cayetanensis* from vegetables collected in markets of an endemic region in Peru, *Am. J. Trop. Med. Hyg.*, 57, 683, 1997.
47. Bern, C., Hernandez, B., Lopez, M.B., Arrowood, M.J., de Mejia, M.A., de Merida, A.M., Hightower, A.W., Venczel, L., Herwaldt, B.L., and Klein, R.E., Epidemiologic studies of *Cyclospora cayetanensis* in Guatemala, *Emerg. Infect. Dis.*, 5, 766, 1999.
48. Sathyanarayanan, L. and Ortega, Y., Effects of pesticides on sporulation of *Cyclospora cayetanensis* and viability of *Cryptosporidium parvum*, *J. Food Prot.*, 67, 1044, 2004.

49. Follet-Dumoulin, A., Guyot, K., Duchatelle, S., Bourel, B., Guilbert, F., Dei-Cas, E., Gosset, D., and Cailliez, J.C., Involvement of insects in the dissemination of *Cryptosporidium* in the environment, *J. Eukaryot. Microbiol.*, Suppl., 36S, 2001.

50. Mathison, B.A. and Ditrich, O., The fate of *Cryptosporidium parvum* oocysts ingested by dung beetles and their possible role in the dissemination of cryptosporidiosis, *J. Parasitol.*, 85, 678, 1999.

51. Graczyk, T.K., Grimes, B.H., Knight, R., da Silva, A.J., Pieniazek, N.J., and Veal, D.A., Detection of *Cryptosporidium parvum* and *Giardia lamblia* carried by synanthropic flies by combined fluorescent in situ hybridization and a monoclonal antibody, *Am. J. Trop. Med. Hyg.*, 68, 228, 2003.

52. Rose, J.B. and Slifko, T.R., Giardia, *Cryptosporidium*, and *Cyclospora* and their impact on foods: a review, *J. Food Prot.*, 62, 1059, 1999.

53. Outbreaks of Escherichia coli O157:H7 infection and cryptosporidiosis associated with drinking unpasteurized apple cider: Connecticut and New York, October 1996, *MMWR*, 46, 4, 1997.

54. Fayer, R., Trout, J.M., Lewis, E.J., Santin, M., Zhou, L., Lal, A.A., and Xiao, L., Contamination of Atlantic coast commercial shellfish with *Cryptosporidium*, *Parasitol. Res.*, 89, 141, 2003.

55. Gomez-Bautista, M., Ortega-Mora, L.M., Tabares, E., Lopez-Rodas, V., and Costas, E., Detection of infectious *Cryptosporidium parvum* oocysts in mussels (Mytilus galloprovincialis) and cockles (Cerastoderma edule), *Appl. Environ. Microbiol.*, 66, 1866, 2000.

56. Graczyk, T.K., Farley, C.A., Fayer, R., Lewis, E.J., and Trout, J.M., Detection of *Cryptosporidium* oocysts and Giardia cysts in the tissues of eastern oysters (Crassostrea virginica) carrying principal oyster infectious diseases, *J. Parasitol.*, 84, 1039, 1998.

57. Negm, A.Y., Human pathogenic protozoa in bivalves collected from local markets in Alexandria, *J. Egypt. Soc. Parasitol.*, 33, 991, 2003.

58. Herwaldt, B.L. and Beach, M.J., The return of *Cyclospora* in 1997: another outbreak of cyclosporiasis in North America associated with imported raspberries. *Cyclospora* Working Group, *Ann. Intern. Med.*, 130, 210, 1999.

59. Herwaldt, B.L. and Ackers, M.L., An outbreak in 1996 of cyclosporiasis associated with imported raspberries. The *Cyclospora* Working Group, *N. Engl. J. Med.*, 336, 1548, 1997.

60. Outbreak of cyclosporiasis associated with snow peas: Pennsylvania, 2004, *MMWR*, 53, 876, 2004.

61. LeChevallier, M.W., Di Giovanni, G.D., Clancy, J.L., Bukhari, Z., Bukhari, S., Rosen, J.S., Sobrinho, J., and Frey, M.M., Comparison of method 1623 and cell culture-PCR for detection of *Cryptosporidium* spp. in source waters, *Appl. Environ. Microbiol.*, 69, 971, 2003.

62. McCuin, R.M. and Clancy, J.L., Modifications to United States Environmental Protection Agency methods 1622 and 1623 for detection of *Cryptosporidium* oocysts and Giardia cysts in water, *Appl. Environ. Microbiol.*, 69, 267, 2003.

63. Ware, M.W., Wymer, L., Lindquist, H.D., and Schaefer, F.W., III, Evaluation of an alternative IMS dissociation procedure for use with Method 1622: detection of *Cryptosporidium* in water, *J. Microbiol. Methods*, 55, 575, 2003.

64. Lee, Y.M., Johnson, P.W., Call, J.L., Arrowood, M.J., Furness, B.W., Pichette, S.C., Grady, K.K., Reeh, P., Mitchell, L., Bergmire-Sweat, D., Mackenzie, W.R., and Tsang, V.C., Development and application of a quantitative, specific assay for *Cryptosporidium parvum* oocyst detection in high-turbidity environmental water samples, *Am. J. Trop. Med. Hyg.*, 65, 1, 2001.

65. Cook, N., The use of NASBA for the detection of microbial pathogens in food and environmental samples, *J. Microbiol. Methods*, 53, 165, 2003.

66. Fontaine, M. and Guillot, E., Study of 18S rRNA and rDNA stability by real-time RT-PCR in heat-inactivated *Cryptosporidium parvum* oocysts, *FEMS Microbiol. Lett.*, 226, 237, 2003.

67. Hallier-Soulier, S. and Guillot, E., An immunomagnetic separation-reverse transcription polymerase chain reaction (IMS-RT-PCR) test for sensitive and rapid detection of viable waterborne *Cryptosporidium parvum*, *Environ. Microbiol.*, 5, 592, 2003.

68. Mayer, C.L. and Palmer, C.J., Evaluation of PCR, nested PCR, and fluorescent antibodies for detection of Giardia and *Cryptosporidium* species in wastewater, *Appl. Environ. Microbiol.*, 62, 2081, 1996.

69. Monis, P.T., Saint, C.P., Development of a nested-PCR assay for the detection of *Cryptosporidium parvum* in finished water, *Water Res.*, 35, 1641, 2001.

70. Sturbaum, G.D., Reed, C., Hoover, P.J., Jost, B.H., Marshall, M.M., and Sterling, C.R., Species-specific, nested PCR-restriction fragment length polymorphism detection of single *Cryptosporidium parvum* oocysts, *Appl. Environ. Microbiol.*, 67, 2665, 2001.

71. Nichols, R.A. and Smith, H.V., Optimization of DNA extraction and molecular detection of *Cryptosporidium* oocysts in natural mineral water sources, *J. Food Prot.*, 67, 524, 2004.

72. Orlandi, P.A. and Lampel, K.A., Extraction-free, filter-based template preparation for rapid and sensitive PCR detection of pathogenic parasitic protozoa, *J. Clin. Microbiol.*, 38, 2271, 2000.

73. Xiao, L., Singh, A., Limor, J., Graczyk, T.K., Gradus, S., and Lal, A., Molecular characterization of *Cryptosporidium* oocysts in samples of raw surface water and wastewater, *Appl. Environ. Microbiol.*, 67, 1097, 2001.

74. Esch, M.B., Baeumner, A.J., and Durst, R.A., Detection of *Cryptosporidium parvum* using oligonucleotide-tagged liposomes in a competitive assay format, *Anal. Chem.*, 73, 3162, 2001.

75. Aguilar, Z.P. and Fritsch, I., Immobilized enzyme-linked DNA-hybridization assay with electrochemical detection for *Cryptosporidium parvum* hsp70 mRNA, *Anal. Chem.*, 75, 3890, 2003.

76. Di Pinto, A. and Tantillo, M.G., Direct detection of *Cryptosporidium parvum* oocysts by immunomagnetic separation-polymerase chain reaction in raw milk, *J. Food Prot.*, 65, 1345, 2002.

77. de Roubin, M.R., Pharamond, J.S., Zanelli, F., Poty, F., Houdart, S., Laurent, F., Drocourt, J.L., and Van Poucke, S., Application of laser scanning cytometry followed by epifluorescent and differential interference contrast microscopy for the detection and enumeration of *Cryptosporidium* and Giardia in raw and potable waters, *J. Appl. Microbiol.*, 93, 599, 2002.

78. Freire-Santos, F., Oteiza-Lopez, A.M., Vergara-Castiblanco, C.A., Ares-Mazas, E., Alvarez-Suarez, E., and Garcia-Martin, O., Detection of *Cryptosporidium* oocysts in bivalve molluscs destined for human consumption, *J. Parasitol.*, 86, 853, 2000.

79. Robertson, L.J., Gjerde, B., and Campbell, A.T., Isolation of *Cyclospora* oocysts from fruits and vegetables using lectin-coated paramagnetic beads, *J. Food Prot.*, 63, 1410, 2000.

80. Jinneman, K.C., Wetherington, J.H., Hill, W.E., Adams, A.M., Johnson, J.M., Tenge, B.J., Dang, N.L., Manger, R.L., and Wekell, M.M., Template preparation for PCR and RFLP of amplification products for the detection and identification of *Cyclospora* sp. and Eimeria spp. oocysts directly from raspberries, *J. Food Prot.*, 61, 1497, 1998.

81. Steele, M., Unger, S., and Odumeru, J., Sensitivity of PCR detection of *Cyclospora cayetanensis* in raspberries, basil, and mesclun lettuce, *J. Microbiol. Methods*, 54, 277, 2003.

82. Varma, M., Hester, J.D., Schaefer, F.W., III, Ware, M.W., and Lindquist, H.D., Detection of *Cyclospora cayetanensis* using a quantitative real-time PCR assay, *J. Microbiol. Methods*, 53, 27, 2003.

83. Dalton, C., Goater, A.D., Pethig, R., and Smith, H.V., Viability of Giardia intestinalis cysts and viability and sporulation state of *Cyclospora cayetanensis* oocysts determined by electrorotation, *Appl. Environ. Microbiol.*, 67, 586, 2001.

84. Anderson, B.C., Moist heat inactivation of *Cryptosporidium* sp, *Am. J. Public Health*, 75, 1433, 1985.

85. Barbee, S.L., Weber, D.J., Sobsey, M.D., and Rutala, W.A., Inactivation of *Cryptosporidium parvum* oocyst infectivity by disinfection and sterilization processes, *Gastrointest. Endosc.*, 49, 605, 1999.

86. Craik, S.A., Weldon, D., Finch, G.R., Bolton, J.R., and Belosevic, M., Inactivation of *Cryptosporidium parvum* oocysts using medium- and low-pressure ultraviolet radiation, *Water Res.*, 35, 1387, 2001.

87. Craik, S.A., Smith, D.W., Chandrakanth, M., and Belosevic, M., Effect of turbulent gas-liquid contact in a static mixer on *Cryptosporidium parvum* oocyst inactivation by ozone, *Water Res.*, 37, 3622, 2003.

88. Deng, M.Q. and Cliver, D.O., Inactivation of *Cryptosporidium parvum* oocysts in cider by flash pasteurization, *J. Food Prot.*, 64, 523, 2001.

89. Venczel, L.V., Arrowood, M., Hurd, M., and Sobsey, M.D., Inactivation of *Cryptosporidium parvum* oocysts and Clostridium perfringens spores by a mixed-oxidant disinfectant and by free chlorine, *Appl. Environ. Microbiol.*, 63, 1598, 1997.

13 Patulin

Lauren Jackson and
Mary Ann Dombrink-Kurtzman

CONTENTS

13.1   INTRODUCTION

Mycotoxins are a chemically diverse group of toxic secondary metabolites produced by filamentous fungi. They are responsible for significant financial losses for the food industry, particularly any aspect of the industry that harvests, stores, processes, or uses commodities or ingredients. Mycotoxins elicit a variety of acute and chronic toxic effects in domestic animals and humans including reduced growth efficiency, vomiting, reproductive problems, cancer, and immunosuppression [1,2]. Worldwide, mycotoxins pose a threat to public health, agriculture, and economics [3].

Patulin is a mycotoxin produced by fungi belonging to several genera including penicillium, aspergillus, and byssochlamys. Although patulin can occur in many molding fruits, grain, and other foods, the major source of patulin contamination is apples with blue mold rot, and in apple cider or apple juice pressed from moldy fruit. *Penicillium expansum* is believed to be the major fungal species contributing to patulin in apple products. Mold growth occurs when the surface tissue of fruit has been damaged by improper handling, insect or storm damage, and is often followed by production of patulin. *P. expansum* and patulin contamination of fruit can occur before harvest, but they are more commonly found as contaminants of apples postharvest and during storage. Thermal processing is effective in destroying microorganisms such as bacteria, yeast, and most fungi. However, patulin is fairly heat resistant, especially in acidic environments. The stability of patulin is illustrated by the presence of the toxin in shelf-stable apple products (juices, concentrates, jellies, baby foods, etc.) [4–7]. Since the compound persists in heated juices, it has been suggested that the presence of patulin in processed apple products may be a good indicator of the quality of the fruit used in production.

Patulin has been demonstrated to be acutely toxic [8], genotoxic [9], teratogenic [8,10,11], and possibly immunotoxic [12,13] to animals. Although the toxicity of patulin in humans has not been demonstrated conclusively, there is a desire to limit its concentration in apple juice since young children and infants are major consumers of this product, and the effects of long-term exposure to patulin are not known. Many countries, including the U.S., have set regulatory limits for patulin in apple products of $50\,\mu g/l$ or less.

This chapter reviews the literature on the chemical properties of patulin, methods for monitoring the occurrence and levels of patulin in food, regulation of patulin levels, factors affecting growth of *P. expansum* and patulin

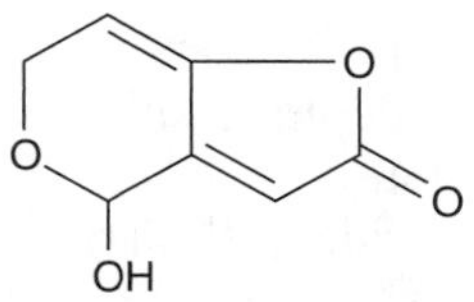

FIGURE 13.1 Chemical structure of patulin (4-hydroxy-4H-furo[3,2-c]pyran-2(6H)-one).

formation, and methods for controlling the levels of this toxin in apple products.

13.2 METHODS OF ANALYSIS

13.2.1 INTRODUCTION

Patulin, 4-hydroxy-4H-furo[3,2-c]pyran-2(6H)-one (Figure 13.1), is a low-molecular-weight (MW 154) α,β-unsaturated γ-lactone with a melting point of 110°C. Patulin is stable under acidic conditions and resistant to thermal treatments, but it is unstable at alkaline pH [14]. The toxin is soluble in water, ethyl acetate, methanol, acetonitrile, and acetone, and less soluble in diethyl ether and benzene. It reacts with sulfhydryl groups such as those in cysteine and glutathione, free amino groups, sulfur dioxide, and ascorbic acid [15–17]. Patulin is metabolized by yeast (*Saccharomyces cerevicae*) in fermenting cider into a variety of compounds including E-ascladiol and Z-ascladiol [18].

As more countries have passed regulatory limits for patulin in apple products, there have been increasing efforts to develop sensitive, selective, and rapid procedures for measuring patulin levels in food. Monitoring of patulin in apple juice, apple juice concentrates, and apple cider is performed to comply with regulatory limits set by the U.S. Food and Drug Administration (FDA) and regulatory agencies throughout the world [19]. The majority of the methods currently used are based on the Association of Official Analytical Chemists (AOAC) official methods, involving liquid–liquid extraction of patulin with ethyl acetate, followed by use of high-performance liquid chromatography (HPLC) for detection and quantification. If there is need for confirmation of the amount of patulin in a product, gas chromatography/mass spectrometry (GC/MS) is performed. Shephard and Leggott [20] published an excellent review of the chromatographic methods used to determine patulin levels in fruit products. The following is an overview of the analytical methods used for quantifying patulin in food.

13.2.2 THIN-LAYER CHROMATOGRAPHY (TLC)

The first method developed for detection of patulin and adopted as an AOAC official method (AOAC Official Method 974.18) involved the use of normal phase TLC [21]. Apple juice is extracted with ethyl acetate, and the extract partially purified on a silica gel column. Patulin is eluted, concentrated, and detected by TLC using normal phase silica gel plates which are typically developed in toluene/ethyl acetate/formic acid and then sprayed with 3-methyl-2-benzothiazolinone hydrochloride (MBTH). Patulin appears as a yellow-brown fluorescent spot under UV light at 366 nm. The method has a limit of detection of ~20 µg patulin/l apple juice. More recently, Prieta et al. [22] described an analytical method for patulin using diphasic dialysis for extraction of patulin from juice, followed by separation on TLC silica gel plates, detection with MBTH, and quantification by densitometry. The authors reported a detection limit of 50 µg patulin/l juice and extraction recovery of 65% [20]. TLC remains the method of choice for detection of patulin in many parts of the world, especially in developing countries.

13.2.3 GAS CHROMATOGRAPHY (GC)

Although not the method of choice of researchers and regulatory laboratories, GC methods have been developed for analysis of patulin. These methods generally involve the formation of trimethylsilyl ether derivatives and detection by electron capture or mass spectrometry (MS) [20]. Tarter and Scott [23] described the use of heptafluorobutyrate (HFB) derivatives of patulin with chromatographic separation on a nonpolar fused silica capillary column and electron capture detection. Application of the method to naturally contaminated apple juice gave a detection limit of 10 µg patulin/l. A recent publication by Llovera et al. [24] described the detection of underivatized patulin by MS at a detection limit of 4 µg patulin/l apple juice.

13.2.4 LIQUID CHROMATOGRAPHY (LC)

HPLC methods are the most commonly employed methods for the quantitation of patulin in fruit juices. HPLC requires a large initial cash investment, but provides good sensitivity, precision, and ease of use. In addition, a skilled and experienced staff is required to operate and maintain the HPLC equipment.

Almost all published methods involve liquid–liquid extraction of patulin into ethyl acetate, a cleanup step using a sodium carbonate solution to remove interfering phenolic compounds, and HPLC with UV detection to separate and detect patulin [25,26]. HPLC columns typically are reversed phase (C18), and the mobile phases tend to be predominately mixtures of water and acetonitrile (up to 10% v/v) or tetrahydrofuran (up to 5% v/v). Although patulin can be detected with single-wavelength UV detectors (276 nm), many laboratories use photodiode array (PDA) detectors to detect patulin and spectrally distinguish the compound from coextracted compounds such as polyphenols and hydroxymethyl furfural (HMF).

The HPLC procedure described by Brause et al. [26] was subjected to an interlaboratory study on method reproducibility and accuracy. In this collaborative study, 22 laboratories analyzed apple juice spiked with 20 to

200 µg patulin/l as well as naturally contaminated juice containing 31 µg patulin/l [25,26]. Mean recovery of patulin spiked into juice was 96%. Based on the results of this collaborative study, the method was adopted as a first action method by AOAC International (AOAC Official Method 995.10) [27]. A second AOAC official method for determination of patulin by HPLC describes the analysis of clear and cloudy apple juices and apple purees. The method (AOAC Official Method 2000.02) [28] is based on a publication by MacDonald *et al.* [29]. This method differs from that of Brause *et al.* [26] in the use of pectinase prior to extraction to remove the cloudiness present in some juice samples.

Solid phase extraction (SPE) methods recently have been developed for extracting and purifying patulin from apple juice. In the method developed by Trucksess and Tang [30], patulin was extracted from undiluted apple juice with a reversed-phase SPE (Oasis, Waters, Milford, MA) column. The column was washed to remove interfering compounds, and patulin was eluted and then detected by HPLC with a recovery of 93 to 104% [30]. Recently, Eisle and Gibson [31] modified the method of Trucksess and Tang [30] and reduced analysis time to approximately 1 hour including extraction and HPLC analysis steps.

13.2.5 MICELLAR ELECTROKINETIC CAPILLARY CHROMATOGRAPHY (MECC)

Neutral compounds, such as patulin, or mixtures of neutral and charged compounds can be analyzed by MECC [32]. There are various advantages for using MECC. Only a small amount (2 ml) of sample is used and less organic solvent is consumed compared to HPLC methods. The method is rapid, with total run time of 10 minutes, and has a low limit of detection (3.8 µg patulin/l). For samples having patulin levels < 75 µg/l, 2 ml of sample is extracted with ethyl acetate. The extract is passed through anhydrous sodium sulfate, evaporated under nitrogen, reconstituted in 0.1 ml acidic water solution (pH 4), and analyzed immediately by MECC.

13.2.6 OTHER METHODS

Although the previously described HPLC methods give accurate and precise measurements of patulin levels in fruit products, they can be laborious, and their results are not attainable for at least several hours. Efforts are being made to develop immunochemical techniques for rapidly (< 30 minutes) quantifying patulin in juice products. As opposed to other mycotoxins, no commercial enzyme-linked immunosorbent assay (ELISA) kits are available for patulin. Production of suitable antibodies for use in ELISA kits is needed. Many research groups have attempted to produce antibodies capable of detecting patulin. Unfortunately, these efforts have not met with success.

An alternative approach for detection of patulin may be to develop methods for the synthesis of molecularly imprinted polymers (MIPs), highly crosslinked polymers, capable of binding specifically to patulin. During the polymerization process the template (patulin) interacts with one or more of the functional monomers present. When the template or structurally related compound is removed from the polymer, a cavity capable of binding patulin remains. Because MIPs have the advantage of high chemical and physical stability, they have been described as "plastic antibodies" and have the potential for use in place of antibodies in applications such as affinity separation assay systems and biosensors [33]. A paper describing the synthesis of MIPs with selective binding properties for the mycotoxin ochratoxin A was published by Jodlbauer *et al.* [34]. A critical component for success in MIP synthesis is the availability of compounds that can serve as mimics of the template of interest (patulin).

13.3 TOXICOLOGICAL EFFECTS OF PATULIN

13.3.1 INTRODUCTION

Study of patulin's toxicity began over 60 years ago when the compound was first isolated from *Penicillium patulum* (now called *P. griseofulvum*) and found to possess antimicrobial properties. Patulin was later isolated from other fungal species and given the names clavacin, claviformin, expansin, mycoin, and penicidin [35]. During the 1940s research was aimed at finding pharmaceutical uses for patulin. For example, patulin was tested as a treatment for the common cold as well as an ointment for treating fungal infections [36]. However, animal studies revealed that, in addition to antibiotic properties, patulin also possessed toxic effects [37,38].

Research on the toxicological properties of patulin has shown the compound to be acutely toxic in animals and to have possible genotoxic, immunotoxic, and teratogenic effects. Patulin toxicity data have been reviewed in detail [38–40]. In addition, the FDA [19] independently reviewed the available information on patulin toxicity.

13.3.2 ACUTE TOXICITY STUDIES

In acute toxicity studies with a variety of experimental animals, the LD_{50} values for patulin, as well as the lesions observed, varied [41]. Overall, patulin produced lesions to the gastrointestinal tract, including epithelial degeneration, hemorrhaging, and ulceration of the gastric mucosa [41–43]. Other lesions included edema of the lungs and brain, visceral organ congestion, and hepatic and renal necrosis [44]. The oral LD_{50} in rats, mice, and hamsters has been reported in the range 30 to 48 mg/kg body weight [41].

13.3.3 IMMUNOTOXICITY STUDIES

Studies on the effects of patulin on the immune system have shown conflicting results. At relatively high doses, patulin has been shown to have immunosuppressive properties ranging from cytotoxicity in rat alveolar macrophages to increases in neutrophils [12,13]. However, Llewellyn *et al.* [37] found that 28 days of oral exposure to patulin at levels comparable to human exposure from apple juice did not have toxic effects to the immune system of female $B6C3F_1$ mice.

13.3.4 REPRODUCTIVE TOXICITY AND TERATOGENICITY STUDIES

Reproductive toxicity and teratogenicity studies in mice, rats, and chicken embryos indicate that patulin is a possible teratogen. Dailey *et al.* [8] reported no reproductive or teratogenic effects in mice or rats dosed with patulin at levels of up to 1.5 mg/kg body weight/day. However, maternal toxicity and an increase in the frequency of fetal resorptions were observed at higher levels, which indicate that patulin was embryotoxic. When injected into the air cell of chick eggs, patulin was reported to be embryotoxic at levels of 2.35 to 68.7 μg/egg depending on the age of the embryos, and teratogenic at levels of 1 to 2 μg/egg [45]. Roll *et al.* [10] found that patulin, when administered intraperitoneally to female mice on day 12 and 13 of pregnancy, caused an increase in the incidence of cleft palates and malformation of the kidneys of the developing fetuses.

13.3.5 GENOTOXICITY STUDIES

Although patulin failed to show mutagenicity in the Ames test and other bacteria-based assays, it has been shown to produce chromosomal damage in mammalian systems [9,10,40]. Patulin was shown to be potent inducer of chromatid-type aberrations to Chinese hamster V79E cells, but did not increase sister-chromatic exchange (SCE) frequency. In contrast, Liu *et al.* [46] reported that patulin caused a significant dose-dependent increase in SCE frequency in both Chinese hamster ovary cells and human lymphocytes. Induction of chromosome damage and micronuclei formation in mammalian cells suggest a possible clastogenic property of patulin [46]. Nucleic acid synthesis and protein synthesis have also been reported to be inhibited by patulin [47,48].

13.3.6 CARCINOGENICITY STUDIES

From the available toxicological data, it is unclear whether patulin is a carcinogen. In a study by Dickens and Jones [49], patulin, when administered subcutaneously twice a week to rats for 15 months, induced sarcomas at the injection sites. However, in two long-term studies, patulin administered orally by gavage was not carcinogenic in rats or mice [44,50]. In their review of these studies, IARC39 concluded that no evaluation could be made of the carcinogenicity of patulin to humans and that there was inadequate evidence in experimental animals [39].

13.3.7 Mechanism(s) of Toxicity

While patulin has been found to exhibit cellular toxicity in *in vivo* and *in vitro* tests, the mechanisms of cellular toxicity are not clear. Patulin alters the plasma membrane functions in cultured LLC-PK1 renal cells through an inhibition of Na^+-K^+ ATPase [51,52]. The compound also inhibits several key biosynthetic enzymes including RNA polymerase and aminoacyl-tRNA synthetases [53]. Due to its electrophilic nature, patulin reacts readily with cellular nucleophiles such as the sulfhydryl-containing compounds cysteine and glutathione [15]. The mode of action of patulin may be through oxidation of critical sulfhydryl groups in cell membranes or in enzymes [52]. Patulin adducts formed with cysteine or glutathione were less toxic than the unmodified compound in acute toxicity, teratogenicity, and mutagenicity studies [40,54]

13.4 REGULATORY ASPECTS

At present, there are no published toxicological or epidemiological data to indicate whether consumption of patulin is harmful to humans. Products containing patulin have probably been consumed for long periods of time, yet accounts of human toxicity caused by patulin exposure from food do not exist. However, there is a desire to limit patulin levels in apple products since infants and young children are major consumers of these foods and the effects of long-term exposure to patulin are not yet known.

Based on the results of reproductive toxicity and long-term toxicity studies involving animals, the Joint Food and Agriculture Organization/World Health Organization Expert Committee on Food Additives (JECFA) established a provisional maximum daily intake for patulin of 0.4 µg/kg body weight [55]. At least ten countries have established action levels of 50 µg/l for patulin in apple juice, and several have established lower limits (25 to 35 µg/l) [3,56]. The FDA has established a 50 µg/l action level for patulin in single strength and reconstituted apple juice [19].

13.5 FUNGAL SPECIES PRODUCING PATULIN IN FOODS

A variety of fungi are reported to be capable of producing patulin in defined media including *Aspergillus clavatus*, *A. giganteus*, *A. terreus*, *Byssochlamys fulva*, *B. nivea*, *Paecilomyces variotii*, *Penicillium carneum*, *P. clavigerum*, *P. concentricum*, *P. coprobium*, *P. dipodomyicola*, *P. expansum*, *P. glandicola*, *P. griseofulvum* (formerly known as *P. patulum*, *P. urticae*), *P. roqueforti*, *P. sclerotigenum*, and *P. vulpinum* [61–66]. However, *P. expansum* is considered the major producer of patulin in food, and, in particular, pome fruits such as apples. In pure culture, *P. expansum* is reported to produce over 59 secondary metabolites including patulin, citrinin, cyclopiazonic acid, chaetoglobosins A and C, roquefortine C, penicillic acid, and ochratoxin. However, only some of these metabolites (patulin, citrinin, chaetoglobosins, roquefortine) were actually detected in apples and apple products [61–66].

Reviews on the physiology and growth characteristics of *P. expansum* and other fungi that produce patulin have been published [67,68]. In addition, Doores [69] wrote a comprehensive review of the microbiology of apples and apple products. *Penicillium expansum* is a psychrophile [67]. The optimum growth temperature for this species is near 25°C, but there are reports of growth of the organism at −3°C [68]. Minimum water activities for spore germination are 0.82 to 0.83 [70]. *Penicillium expansum* has a very low requirement for oxygen; the organism was found to grow at atmospheric oxygen levels of less than 2%. Carbon dioxide concentrations of up to 15% have been found to stimulate growth of the organism [68]. The growth characteristics of the fungus help explain the finding of *P. expansum* and patulin in apples stored under modified atmosphere conditions [71].

Although patulin is typically not destroyed during pasteurization of juices, *P. expansum* and its spores typically do not survive this thermal treatment. However, other species of patulin-producing fungi (*B. fulva* and *B. nivea*) produce spores that are resistant to processing at temperatures of 90°C [72]. Consequently, there is a possibility of patulin production in stored juices if spores of these fungi germinate. At present, it is unclear if heat-resistant ascospores contribute significantly to the patulin content of apple juice products.

13.6 NATURAL OCCURRENCE OF *P. expansum* AND PATULIN IN FRUITS AND VEGETABLES

Penicillium expansum is one of the most pervasive and destructive postharvest pathogens of pome fruits such as apples and pears, but it can affect other fruits including tomatoes, strawberries, avocados, bananas, mangoes, grapes, peaches, and apricots [3,68]. The primary habitat of *P. expansum* is in fruit storage and packinghouse facilities, but it can also be found in orchard soil, seeds of various plants, and on the surface and in the core of unblemished fruit. The fungus is primarily a wound pathogen, gaining entrance through fresh mechanical injuries such as stem puncture, bruises and insect injuries, hail or weather-related damage, and fingernail scratches caused by fruit pickers [73]. There are also reports of the fungi entering apple fruits through open calyx canals, at the point of attachment of stem to fruit, and through skin lenticels [73]. The infection often occurs while apples are still on the tree, but it remains latent until the fruit is harvested and stored [66]. The appearance of the decay caused by *P. expansum* is characterized by rotten areas that are soft, watery, and light brown in color. The surface of older lesions may be covered by bluish-green spots that initially are white in color [73].

Although *P. expansum* can be isolated from the surface of a wide variety of fruits, patulin has only been detected in apples, pears, blueberries, cherries, peaches, plums, strawberries, raspberries, and mulberries [3]. The organism is rarely isolated from vegetables [67]. The mere presence of *P. expansum* does not necessarily imply that patulin will be present since mycotoxin production is influenced by many factors including environmental conditions, cultivar and nutritional status of the fruit, the microbial load on the fruit, and strain of the fungus [3,74].

Patulin is found with greater incidence and concentration in apples than in other fruit, and they contribute the vast majority of patulin in the human diet [19]. The toxin has been detected in intact fruit, juice, cider, applesauce, and apple puree. Whole apples (table fruit) are not believed to contribute significantly to human exposure since contaminated fruit is often discarded or trimmed to remove moldy areas before it is eaten. The greatest exposure to patulin comes from consumption of apple juice and cider pressed from moldy fruit [19].

Numerous surveys have been published on the incidence and concentration of patulin in apples and apple products [5–7,75–82]. Harwig *et al.* [83] surveyed 61 samples of whole apples from different orchards in Canada. *Penicillium expansum* was isolated from 42 of the samples, while patulin was found in 28 samples of expressed juice at levels up to 240 µg/l. Wilson and Nuovo [76] analyzed 100 samples of freshly pressed apple cider and detected high levels of patulin (up to 45,000 µg/l) in several samples of cider produced from organically grown fruit. The authors concluded that cider samples with the highest patulin concentrations were made from ground-harvested and rotten fruit. In contrast to these results, Malmauret *et al.* [84] and Riteni [82] reported no significant difference in patulin levels from fruit grown organically compared to conventionally grown fruit. In surveys of apple products obtained in Turkey and New South Wales, Australia, Yurdun *et al.* [79] and Burda [6] reported that >25% of juice samples contained > 50 µg patulin/l, and several samples contained 500 to 1000 µg patulin/l. Watkins *et al.* [78] analyzed apple juice purchased in Victoria, Australia, and found that >65% of samples were contaminated with patulin, and >33% had levels over 50 µg patulin/l. In contrast, Ritieni [82], Leggott and Shephard [80], and Lai *et al.* [85] reported that patulin levels in almost all tested apple products purchased in Italy, South Africa, and Taiwan, respectively, were < 50 µg/l. A survey conducted of apple juices purchased between 1994 and 2000 in the U.S. revealed that 12.6% of juices had patulin levels over 50 µg/l, and approximately 6% had levels > 100 µg/l [81]. Overall, surveys of apple products indicate that, although the incidence of patulin contamination is fairly high, levels of contamination are typically less than 50 µg patulin/l.

13.7 FACTORS AFFECTING PATULIN PRODUCTION

13.7.1 INTRODUCTION

At present, there are a number of factors known to affect production of patulin in apple products. Within a species, the mycotoxigenic potential of a fungus depends mainly on the strain of fungus. Although genetic variation may be the ultimate cause of the differences between strains with regard to fungal growth and mycotoxin production, physical and chemical properties of the food and environmental factors such as incubation temperature and time are also important factors [86]. Patulin production in fruit is believed to be affected by many factors including apple cultivar, geographical location where the fruit is grown and harvested, climate, preharvest treatments, method of harvest, surface defects on the fruit, postharvest treatments, and storage conditions. At present, it is not clear which of these factors plays the greatest role in mycotoxin production or how they can be manipulated to prevent or reduce patulin contamination of apple products. A better understanding of the aspects influencing patulin production may aid in developing effective means for controlling mycotoxin formation in food. The following is a description of some factors known to affect patulin production in apples.

13.7.2 Physical, Chemical, and Microbial Properties of Apples

Several investigators have shown that fruit cultivars differ in their susceptibilities to *P. expansum* rot and to patulin formation in the apple tissue. Jackson *et al.* [87] reported that cider pressed from ground-harvested Red Delicious apples had significantly higher levels of patulin than cider prepared from other apple cultivars (Golden Delicious, Granny Smith, Fuji, Gala, MacIntosh, Red Rome). Of the four cultivars studied by Spotts and Mielke [88], Royal Gala apples were found to be most resistant to decay from *P. expansum*, while Fuji apples were least resistant. These results, along with those published by others [65,70,76,86,89,90], indicate that apple cultivars differ substantially in susceptibility to blue mold rot. The differences in cultivars may be due to unique physical and chemical characteristics of each cultivar such as skin thickness and strength, flesh firmness, pH of flesh, sugar levels, levels of antimicrobial compounds, and other apple constituents [86,89,91]. Apple cultivars with an open calyx are at greater risk for patulin development within the apple core [92]. Since core rot is often not detected, juice or cider pressed from affected fruits may have high patulin levels.

Penicillium expansum can be isolated from the surface of unblemished fruit, although the fungus typically does not grow until it is able to make contact with the flesh of the fruit. The pathogen enters the fruit through skin breaks caused by bird, insect or weather-related damage, improper handling, and through damaged lenticels near bruised areas [73]. Several investigators reported greater patulin levels in apple juice pressed from damaged than from sound fruit [76,87,93,94]. Susceptibility to surface wounds and bruising is influenced by apple cultivar, but it also may be a function of degree of maturity of the fruit, since skin layers soften during the ripening process [3,67]. Other factors such as mineral imbalances (e.g., high nitrogen, low calcium) are believed to increase susceptibility of fruit to infection. Mineral imbalances are caused by improper fertilization, excessive or too little rain, and poor soil conditions [95].

At present, the effects of the chemical composition of fruit on patulin production are not well understood. Apples are composed of a complex mixture of sugars (primarily fructose, glucose, and sucrose), oligosaccharides, and polysaccharides, together with malic, quinic, and citric acids, polyphenols, amides and other nitrogenous compounds, soluble pectin, vitamins, minerals, water, and a variety of esters. The relative proportions of these components depend on the apple cultivar, the conditions under which the apples were grown, the state of maturity of fruit at the time of pressing, and extent of damage to the fruit. Patulin is produced over the range of pH values found in apple juice (3.2 to 3.8) and is stable at these pH values, but degrades at higher pH values [32,91,96]. McCallum et al. [36] found that the concentration of patulin formed in juice was correlated negatively with the pH value. Prusky et al. [97] reported that Penicillium spp. colonization and growth are enhanced by low pH in the host tissue. They also found that P. expansum actively reduced the pH during decay development by causing accumulation of fumaric and gluconic acid in the fruit tissue.

The mineral content of apples may influence degree of decay by postharvest pathogens. Calcium is believed to be the major mineral nutrient affecting apple quality and storage life [98,99]. The effect is thought to be partly due to the role of the mineral in preventing physiological disorders in the developing fruit [99–102]. Calcium is also believed to improve fruit firmness by forming complexes with pectic substances in the cell wall [101,103]. The influence of other minerals and antimicrobial compounds in apples (e.g., phenolic compounds) on their susceptibility to fungal rot and mycotoxin production is not known at this time. Research is needed to determine how the apple constituents affect patulin formation.

Microbiological factors can influence patulin formation. The microbiological flora on apples and other fruit differs according to geographic area, climatic conditions, pesticide or fungicide treatments, cultivar, presence of competitive microorganisms, harvest method, and postharvest treatments [69]. Spores of P. expansum are found in soil, on plant surfaces, and in air and are transferred to dump tank and flume water in packinghouses by contaminated wooden picking bins and fruit [104]. Early findings by Sommer et al. [74] indicate that the presence of a patulin-producing species does not necessarily imply patulin production in apples. Factors like incubation temperature, lesion size, and substrate, also play important roles. Substantial differences have been noted among P. expansum strains in terms of growth kinetics and patulin production [86]. McCallum et al. [86] found that highest patulin levels were those from isolated strains displaying aggressive growth and profuse mycelial development.

13.7.3 ENVIRONMENTAL FACTORS

Understanding the environmental conditions influencing mycotoxin production is important so that storage environments can be made unfavorable for fungal growth and toxin production. Temperature is one of the major factors that affect the shelf life of apple fruits and their rate of deterioration by fungi [60]. The optimum temperature for patulin production has been reported in the range 23 to 25°C [86,90]. Although patulin production tends to decrease as temperature is decreased, patulin can be produced at low temperatures (0 to 4°C). Consequently, refrigerated storage is not practical to inhibit totally patulin production [60,105]. Storage time affects the degree of decay since apples lose their natural resistance to infection with time [95].

Modified atmospheres can suppress both fungal growth and patulin formation in apples. Modified atmosphere storage has been used for over 30 years as a means for extending the storage life of fresh produce. A modified atmosphere of high carbon dioxide and low oxygen has been found to inhibit the growth and sporulation of some fungi and the production of such mycotoxins as aflatoxin, penicillic acid, and patulin [105–107]. Paster *et al.* [90] found that an atmosphere of 3% CO_2 and 2% O_2 completely inhibited patulin production by *P. expansum* at 25°C, but production occurred in atmospheres of 2% CO_2 and 10 or 20% O_2. Use of subatmospheric pressure, a type of modified atmosphere, to extend storage life of fresh produce was studied by Adams *et al.* [108] as a method for reducing growth and patulin production by *P. expansum* and *P. patulum*. This work showed that pressures as low as 160 mmHg are needed to control fungal growth and patulin production. Moodley *et al.* [109] monitored patulin formation in whole apples stored (14 days, 25°C) in polyethylene bags with different gas combinations. They found that polyethylene, the most widely used material for retail packages of apples, inhibited toxin production in apples by 99.5% and fungal growth by 68%, even in the absence of a modified atmosphere, when compared to unpackaged apples.

13.8 APPROACHES FOR CONTROLLING PATULIN LEVELS

13.8.1 INTRODUCTION

Apples and other pome fruit are major food crops, with over 40 million tons being produced worldwide [110]. Fungal diseases, and in particular blue mold rot from *P. expansum*, cause significant economic losses in the fruit growing and processing industries. The losses from this disease can be significant (up to 10% of stored fruit) but can be substantially reduced by following proper sanitation and control measures. An integrated approach, including careful handing of fruit and strict hygiene in orchard, packinghouse, and in storage, must be used for controlling *P. expansum* and hence patulin formation in fruit. Reducing patulin levels in fruit juice and other processed apple products can be achieved through the use of sound, healthy fruit, modified atmosphere storage, culling of damaged and rotted fruit, trimming of rotted tissue, filtering juice through activated carbon, and fermentation of cider with added yeast [80]. Guidelines for reducing postharvest decay of apples and other fruits have been published [73,95,111,112]. Codex Alimentarius Commission [92] published recommendations for preventing patulin contamination of apple juice.

The next sections outline preharvest, harvest, and postharvest methods for controlling patulin levels in apple products.

13.8.2 PREHARVEST

Although patulin production in fruit is believed to occur mainly postharvest, several factors pertaining to the growing conditions of fruit trees may influence fungal infection and mycotoxin production in apples. Codex Alimentarius Commission [92] outlined good agricultural practices (GAPs) that may reduce the likelihood of infection of fruit trees. Trees should be trimmed of dead and diseased wood and mummified fruits and pruned to allow proper air flow and light penetration [92]. It has been demonstrated that fruit with mineral imbalances are more susceptible to infection by *P. expansum* and other fungal pathogens. Supplementing fruit trees with foliar calcium sprays during the growing season and use of minimal amounts of nitrogen fertilizer are some methods for reducing preharvest infection of apple fruit by fungi [95,98]. Calcium is believed to reduce decay by maintaining the firmness of cell walls during ripening [98,113]. Ammonium molybdate tetrahydrate has been studied as foliar and soil treatment of several crops. When ammonium molybdate was applied as a preharvest treatment to apple trees, a significant reduction in blue mold decay was observed in the treated apples after three months' cold storage [114]. Tests *in vitro* showed that the mode of action of the chemical is by inhibiting germination of *P. expansum* spores [114].

Postharvest decay can be reduced by preharvest applications of fungicides. Studies on the effectiveness of applications of ziram fungicide showed an average reduction in decay of 25 to 50% [112,115]. Synthetic fungicides are being developed to protect produce from a number of postharvest diseases. However, problems associated with use of synthetic fungicides, such as proliferation of fungicide-resistant pathogen strains, as well as concerns about public health and environmental contamination, have increased the need for development of alternative treatments [116].

During the past five years, biological control of postharvest fungal diseases with naturally occurring antagonists (yeasts and bacteria) has become an alternative to synthetic fungicide control [116]. The commercial products Aspire™ (*Candida oleophila* strain 182, Ecogen Inc., Langhorne, PA) and Biosave™ 10 and BioSave™ 11 (*Pseudomonas syringae* strains ESC10 and ESC11, EcoScience Corp., Worcester, MA) are examples of commercial biocontrol products available in the U.S. Biocontrol agents act by colonizing the wounds of apples where decay proliferates. The organisms are believed to inhibit growth of fungal pathogens by utilizing all of the available nutrients in the wound. Although most success with biological control has been with application of the antagonists to fruit postharvest, but before storage, there has been some degree of success at preharvest treatment of apples with antagonists. Nunes *et al.* [117] reported that although preharvest treatments with *Candida sake* were less effective than postharvest treatments against *P. expansum*, about 54% control was achieved by spraying the organism on Golden Delicious apples while still on the tree. More work is needed to determine the efficacy of preharvest biocontrol of *P. expansum* and to determine if biocontrol affects postharvest formation of patulin in fruit.

13.8.3 HARVEST

The condition of produce at harvest determines the length of time the crop can be stored [112]. Stage of maturity at harvest is believed to be one of the main factors determining the susceptibility of fruit to mechanical damage and to blue mold rot during postharvest storage. Fruits become increasingly susceptible to fungal invasion during ripening as the pH of the tissue increases, soluble sugars build up, skin layers soften, and defense barriers weaken [118,119]. To reduce undesirable biochemical changes, apples should be picked when mature but not fully ripe to ensure that they can be stored for several months [112].

Studies indicate that bruising and skin punctures substantially increase the susceptibility of fruit to decay. Gentle handling of fruit by pickers during harvest and care during transport of the fruit from the orchard to the packing-house, juice processing plant, or storage may prevent injury to the fruit [73].

Rain during harvest allows for increased fungal contamination and infection [95]. Consequently, fruit should be harvested in dry weather conditions and quickly transferred to cold storage. Fallen fruit in the orchard should be discarded and not sold for the fresh market or used in processed apple products. Jackson *et al.* [87] reported significantly higher patulin levels in cider produced from ground-harvested apples than from tree-picked fruit. The process of falling from the tree may result in cuts or cracks in the apple peel that become infected from fungal spores from the soil.

One of the major methods for controlling *P. expansum* infection of apples is improved sanitary practices during harvest [95,120]. This includes reducing contamination of packing/storage bins with orchard soil by cleaning and sanitizing bins before use. Studies by Spotts and Cervantes [120] found that while steam was the most effective treatment on wood and plastic bins, chlorine compounds, sodium *o*-phenylphenate (SOPP), and quaternary ammonia compounds were also effective sanitizing agents. Sodium hypochlorite was more effective on *P. expansum* spores on plastic than on wood bins [120]. Benomyl, iprodione, and captan were generally not effective disinfectants.

13.8.4 POSTHARVEST

13.8.4.1 Introduction

Approximately 75% of the world apple crop is marketed as fresh whole apples, with the remaining 25% finding its way into processing, primarily into apple juice and cider [121]. After harvest, a portion of the apple crop is transported to packinghouses where it is packaged for the fresh (table) fruit market.

Fruit not sold for the fresh market is processed into juice and other products, or is stored at cold (0 to 4°C) temperatures with or without modified atmosphere to extend the shelf life and to provide a constant supply of raw material for the fresh market and for the processed apple industry [122]. Since the majority of patulin forms in fruit postharvest, considerable efforts have been devoted to developing strategies for reducing proliferation of fungal pathogens and contamination with patulin during storage. This section outlines some of the major postharvest controls of patulin in apple products.

13.8.4.2 Washing Treatments

Organic matter (soil, plant material, decayed apples) can act as a reservoir of fungal spores that contaminate fruit. It is important to maintain sanitary conditions in all areas where fruit is packaged, stored, and processed. Proper sanitation includes washing and sanitizing packing machinery, the walls and floors of storage rooms, and the surfaces of all processing equipment [112].

Water systems (water flumes), used to float apples from field bins to bulk tanks, minimize mechanical damage to fruit. Flume water typically contains chlorine (sodium hypochlorite) or SOPP to reduce fungal spore load [112,123]. Active chlorine levels in flume water must be maintained periodically to ensure spores are destroyed. Other chemicals that can reduce spore levels include chlorine dioxide [124] and ozone [125], although both are not commonly used disinfectants for flume water. Physical removal of fungal spores by filtration has been reported to remove >92 to 99% of $P.\ expansum$ conidia from flume water [126].

According to recent surveys of industry practices, the majority of apple packagers and processors wash apples upon receipt or immediately before chopping and pressing to remove soil, rot, pesticide residues, insects, microorganisms, and other extraneous material [127–129]. Apples are typically washed in dump tanks containing water or chlorinated water, with brusher-scrubbers, and/or with high-pressure water sprayers [122,128]. Since $P.\ expansum$ and patulin are associated with the soft rot of apples, washing may result in the removal of rotten areas of the apple and the partitioning of patulin into the cleaning water [87,93]. Jackson $et\ al.$ [87] found that washing ground-harvested apples in a dump tank before pressing reduced patulin levels in the resulting cider by 10 to 100%, depending on the initial patulin levels and type of wash solution (water vs. chlorine) used. Sydenham $et\ al.$ [94] found that patulin levels in cider decreased from 920 to 190 µg/l after Granny Smith apples were washed with water. Acar $et\ al.$ [130] reported that patulin levels were reduced by up to 54% when apples were washed with a high-pressure water spray. Total removal of patulin during the wash treatments is unlikely since patulin can diffuse up to 1 cm into healthy tissue [131]. Wash solutions other than chlorine that have had efficacy in reducing mold counts in apples include electrolyzed oxidizing water [132] and ozone [125], although their effects on the patulin content of apples are not known.

13.8.4.3 Culling, Sorting, and Trimming

Removal of decayed or damaged fruit or trimming moldy portions of apples prior to packaging or processing have been reported to reduce patulin levels in apple juice [4,93,94,131,133]. Wilson and Nuovo [76] surveyed 100 samples of fresh cider and found that samples having the highest patulin levels were produced by cider mills that did not remove decayed apples before pressing. Similarly, Sydenham *et al.* [93,94] reported that removal of rotten fruit prior to pressing significantly reduced patulin levels in cider produced from apples stored at ambient temperatures for 7 to 35 days. Jackson *et al.* [87] reported that patulin was not detected when apples were culled prior to pressing, but was found in five out of seven varieties when cider was pressed from unculled fruit. Although removing visibly decayed fruit before processing is a proven method for reducing patulin levels in apple products, there is no guarantee that culling alone can totally eliminate patulin. Apples with "invisible" sources of fungal rot (core rot) can contaminate apple juice, cider, or puree with patulin if they are not removed before processing. Apple cultivars susceptible to core rot should be cut in half and fruit with signs of decay removed before processing. In large-scale operations where this culling procedure is not practical, other methods for detecting apples with core rot are needed.

13.8.4.4 Chemical, Heat, and Biological Control, and Irradiation Treatments

Prior to storage, apples are often drenched with diphenylamine along with a fungicide (thiabendazole) to prevent superficial scald [104]. Since some strains of pathogens are developing resistance to fungicides, there has been a push to use alternative postharvest control methods [113,134]. Treatments that have shown some promise include the use of essential oils [135], organic acid fumigants [136], calcium salts [98], carbonate and bicarbonate [137,138], chitosan [139–141], 2-deoxy-D-glucose [142,143], heat, biological control, irradiation [118], and combinations of these treatments.

Spraying cinnamon oil, cinnamaldehyde, or a potassium sorbate solution on the surface of apples extended shelf life with respect to decay by *P. expansum* [135]. Complete inhibition of patulin formation in liquid culture was found with 0.2% lemon oil, and >90% inhibition was observed using 0.05% lemon oil and 0.2% orange oil [60]. It is unclear if these treatments can be used commercially to reduce fungal decay in apples or if they affect the shelf life of fruit.

Preliminary studies by Sholberg *et al.* [136] indicate that fumigation of fruit with short-chain organic acids prevents decay and could become an important alternative to liquid sterilants such as sodium hypochlorite. The number of lesions on apples caused by *P. expansum* decreased exponentially with increasing time of fumigation with vinegar or acetic acid vapors [136,144]. Drawbacks to the use of acid fumigants include the need for an airtight enclosure and the corrosiveness of the vapors to steel [136].

Fungal decay in apples was reduced by postharvest application of calcium solutions to fruit [98,113]. Direct application of calcium to fruit can be accomplished by dipping or spraying fruit with calcium solutions or with vacuum or pressure infiltration [99]. Calcium helps to maintain firmness of the apple and to decrease the incidence of physiological disorders that enable fungal pathogens to infiltrate the fruit tissue.

There has been an increased interest in the use of prestorage heat treatments to prevent fungal decay of fruit. Heat can be applied to fruit as a hot water dip, as steam, as hot dry air, and by short hot water rinses [145–148]. Leverentz et al. [134] reported that holding Golden Delicious apples at 38°C for four days reduced decay after three months of storage at 0°C without reducing fruit quality. Fallik et al. [148] reported that Golden Delicious apples treated with a 15-second hot water (55°C) rinse followed by a brushing treatment had less P. expansum decay than untreated apples or apples given a dry heat treatment (96 hours at 38°C). One explanation for the enhanced stability of the heat-treated fruit is that heated apples softened more slowly than nonheated fruit. In addition, heat treatments may recrystallize the wax layer on the surface of the apple peel or increase synthesis of wax in the peel [148,149].

A promising alternative to chemical treatments is biological control of postharvest pathogens [116,117,134,150,151]. Decay caused by P. expansum has been controlled in pome fruits by bacterial and yeast antagonists in several laboratory and pilot storage tests [134]. At least one yeast-based product and two bacteria-based products are now commercially available for treating apples after harvest. Several more are being developed for commercialization [151–153]. Although biological control agents have exhibited excellent control of fungal rot in fruit, their efficiency is sometimes lower than chemical control, and they do not always give consistent results [112,153]. Microbial antagonists have a poor ability to eradicate preexisting infections, while chemical treatments are frequently more effective at controlling established infections [153]. Use of a combination of microorganisms could improve the spectrum of activity and reduce the required concentration of biocontrol agents [116]. El Ghaouth et al. [139,140] reported enhancing the biological efficacy of the yeast Candida saitoana by combining it with either glycochitosan or with the sugar 2-deoxy-D-glucose. Both approaches increased the protective and curative activity of the yeast in controlling postharvest diseases. Droby et al. [153] found that application of 2% sodium bicarbonate in combination with Aspire^{TM} consistently enhanced its biocontrol performance against penicillium rot in apples. Similarly, McLaughlin et al. [154] demonstrated that the addition of calcium salts to yeast cell suspensions enhanced the ability of Pichia guilliermondii to control postharvest diseases of apple. Pichia guilliermondii is found as an occasional clinical isolate and therefore is of questionable safety as a biocontrol organism [155]. In another study, a combination of a heat treatment and the use of a yeast antagonist was more effective than either treatment alone [134]. Clearly more work is needed to identify a combination of treatments to control penicillium rots in apples. Additional research is also needed to determine if these treatments are able to inhibit patulin formation in fruit [153].

Aziz and Moussa [118] studied the effect of gamma irradiation on mycotoxin production in fruits stored under refrigeration conditions. After 28 days of storage, nonirradiated fruits were contaminated with higher levels of mycotoxins (including patulin) than irradiated (3.5 kGy) samples. Mycotoxin production was reported to decrease with increasing irradiation dose. Although UV light has a lethal effect on bacteria and fungi, little has been done to study the effects of UV irradiation on mold levels on apples. However, Stevens et al. [156] reported that applying a yeast antagonist to fruit after UV irradiation was the most effective treatment in reducing storage rot in peaches. Use of gamma and UV irradiation to control fungal rot in fruit deserves further study.

13.8.4.5 Storage

After harvest, apples are generally kept in cold storage at -1 to $3°C$ with or without modified atmospheres. These treatments can extend the shelf life of apples from 9 to 28 weeks, depending on apple cultivar [109]. Since apple cultivars differ in their susceptibility to postharvest diseases, cultivars with resistance to mechanical damage and infection should be chosen, especially if they will be kept in long-term storage.

Although fungal growth is dramatically reduced at temperatures $<10°C$, the growth of $P.$ $expansum$ and production of toxin were not prevented during cold storage [89,90,157]. Paster et al. [90] found that patulin levels in apples and pears inoculated with different strains of $P.$ $expansum$ generally increased with increasing storage temperature from 0 to $25°C$. Similarly, Beer and Amand [89] reported that MacIntosh apples stored at $4°C$ had substantially lower patulin levels than fruit stored at 15 or $24°C$. Apples are typically kept in cold storage; however, when suitable refrigerated storage is not available, apples are stored in the open in ambient conditions (i.e. deck storage). Sydenham et al. [94] reported that patulin levels in deck-stored apples were $2445 \mu g/l$ as opposed to $90 \mu g/l$ in comparable refrigerated stored fruit. Fungal growth and patulin formation increased with the length of storage [90,94]. Overall, the research presented here indicates that apples should be kept in refrigerated storage when possible to slow mold growth and reduce mycotoxin production.

Several researchers studied the effects of modified atmosphere conditions and found that gas composition affected $P.$ $expansum$ growth and patulin formation in fruit [90,158–160]. Lovett et al. [71] reported that juice from modified atmosphere-stored apples (3% O_2, 1 to 3% CO_2, 0 to $3.3°C$, and $>90\%$ relative humidity; 14 weeks) had $500 \mu g$ patulin/l while juice made from air-stored apples had 2000 to $3000 \mu g$ patulin/l. Stitton and Patterson [160] reported that use of high ($>3\%$) CO_2 atmospheres (greater than used for commercially stored apples) was an effective fungistatic treatment for stored apples. However, excessively high levels of CO_2 ($>8\%$) negatively affected the quality of some apple cultivars. Johnson et al. [159] reported a lower incidence of penicillium rots in apples stored at lower O_2 conditions (0.75% O_2) than under higher O_2 levels (1.0 to 1.25%). While in modified atmosphere storage, apples should be examined periodically for fungal decay [92].

13.8.4.6 Controls for Processed Apple Products

Treatments that have shown promise at reducing patulin levels in apple juice include filtration, centrifugation, use of charcoal, addition of ascorbic acid, and fermentation [16,109,161–164]. Bissessur *et al.* [164] evaluated the effectiveness of several clarification processes for the reduction of patulin in apple juice. Pressing followed by centrifugation resulted in 89% reduction in levels of the toxin. Patulin reductions using paper filtration, enzyme treatment, and fining with bentonite were 70, 73, and 77%, respectively. These data suggest that patulin tends to bind to the apple solids, which are removed from the juice during treatment. Activated carbon treatment has also shown promise as a method for reducing patulin levels in apple juice [163,165,166].

Several compounds have the ability to modify chemically patulin, rendering the toxin undetectable in some analyses. Yazici and Velioglu [167] found that adding vitamins (thiamine hydrochloride, pyridoxine hydrochloride, and calcium-d-pantothentate) to apple juice before storage at 4°C for 6 months reduced patulin levels by 55.5 to 67.7% versus controls (no vitamin addition) that had 35.8% reduction in levels of the toxin. It is unlikely that the use of these vitamins to reduce patulin levels in juice has any practical value. Adding ascorbic acid (0 to 3% w/v) to apple juice has been reported to reduce patulin levels by up to 80%, as measured by HPLC [16]. The mechanism by which patulin interacts with ascorbic acid needs to be studied in more detail.

Patulin is known to become analytically undetectable during the production of cider from contaminated apple juice [18,168]. Analysis of patulin-spiked fermentations by HPLC showed the appearance of two major metabolites of patulin, one of which appeared to be E-ascladiol [18]. More work is needed to determine the toxicity of these metabolites of patulin.

13.9 CONCLUSIONS

Contamination of apples and apple products with *P. expansum* and patulin causes considerable financial losses for apple growers and processors. Considerable efforts have been made to understand the conditions by which fungal pathogens such as *P. expansum* infect fruit and produce patulin. Fungal growth and mycotoxin production are known to result from an interaction of many factors, including the chemical and physical properties of the affected fruit crop, genetics of the fungus, environmental conditions, and preharvest, harvest, and postharvest conditions. In order to devise strategies for preventing patulin formation, more research is needed to understand how these factors separately and together can be used to prevent fungal and mycotoxin contamination. The research to date indicates that an integrated approach, including careful handling of fruit to prevent structural damage and strict hygiene in the orchard, packinghouse, storage, and processing facility, is essential for reducing *P. expansum* decay and patulin formation. Research also indicates that only sound fruit should be used for processed apple products. Fallen fruit should be discarded and not sold for the fresh market or used in the manufacture of processed apple products. Culling or removing damaged and moldy fruit before processing is an effective method for reducing patulin contamination of juice and cider. In addition, washing whole fruit before pressing and filtering juice have been successful at reducing patulin levels in juice products.

More research is needed to identify apple cultivars that are resistant to fungal decay, especially those cultivars that are stored for extended lengths of time. Although chemical treatments with synthetic fungicides traditionally have been used to control fungal pathogens in fruit, biological control has shown promise in preventing decay. As described in this chapter, postharvest treatment of whole apples with biological antagonists, heat, and calcium and other chemical treatments have been demonstrated as effective at reducing fungal rot. However, information is lacking on how these and other treatments affect patulin formation in fruit. As more countries have passed regulatory limits for patulin in juices and other apple products, there is an increasing need to develop analytical methods that can rapidly (<30 minutes) quantify patulin in food.

REFERENCES

1. Pestka, J.J. and Casale, W.L., Naturally occurring fungal toxins, in *Food Contamination from Environmental Sources*, Hriaga, J.O. and Simmons, M.S., Eds., John Wiley, New York, 1990, pp. 613–638.
2. Coulombe, R.A., Jr., Symposium: biological action of mycotoxins, *J. Dairy Sci.*, 76, 880–891, 1993.
3. Drusch, S. and Ragab, W., Mycotoxins in fruits, fruit juices, and dried fruits, *J. Food Prot.*, 66, 1514–1527, 2003.
4. Beretta, B., Gaiaschi, A., Galli, C.L., and Restani, P., Patulin in apple-based foods: occurrence and safety evaluation, *Food Addit. Contam.*, 17, 399–406, 2000.
5. Mortimer, D.N., Parker, I., Shephard, M.J., and Gilbert, J., A limited survey of retail apple and grape juices for the mycotoxin patulin, *Food Addit. Contam.*, 2, 165–170, 1985.
6. Burda, K., Incidence of patulin in apple, pear and mixed fruit products marketed in New South Wales, *J. Food Prot.*, 40, 796–798, 1992.
7. De Sylos, C.M. and Rodriguez-Amaya, D.B., Incidence of patulin in fruits and fruit juices marketed in Campinas, Brazil, *Food Addit. Contam.*, 16, 71–74, 1999.
8. Dailey, R.E., Brouwer, E., Blaschka, A.M., Reynado, E.F., Green, S., Monlux, W.S. and Ruggles, D.I., Intermediate-duration toxicity study of patulin in rats, *J. Toxicol. Environ. Health*, 2, 713–725, 1977.

9. Alves, I., Oliveira, N.G., Laires, A., Rodrigues, A.S., and Rueff, J., Induction of micronuclei and chromosomal aberrations by the mycotoxin patulin in mammalian cells: role of ascorbic acid as a modulator of patulin clastogenicity, *Mutagenesis*, 15, 229–234, 2000.

10. Roll, R., Matthiaschk, G., and Korte, A., Embryotoxicity and mutagenicity of mycotoxins, *J. Environ. Pathol. Toxicol. Oncol.*, 10, 1–7, 1990.

11. Sugiyanto, J., Inouye, M., Oda, S.-I., Takagishi, Y., and Yamamura, H., Teratogenicity of patulin, a mycotoxin in mice, *Environ. Med.* 37, 43–46, 1993.

12. Escoula, L., Thomsen, M., Bourdiol, D., Pipy, B., Peuriere, S., and Roubinet, R., Patulin immunotoxicology: effect on phagocyte activation and the cellular and humoral immune system of mice and rabbits, *Int. J. Immunopharmacol.*, 10, 983–989, 1988.

13. Sorenson, W.G., Simpson, J., and Castranova, V., Toxicity of the mycotoxin patulin for rat alveolar macrophage *in vitro*, *Environ. Res.*, 38, 407–416, 1985.

14. Lovett, J. and Peeler, J.T., Effect of pH on the thermal destruction kinetics of patulin in aqueous solution, *J. Food Sci.*, 38, 1094–1095, 1974.

15. Fliege, R. and Metzler, M., Electrophilic properties of patulin. N-acetylcysteine and glutathione adducts, *Chem. Res. Toxicol.*, 13, 373–381, 2000.

16. Brackett, R.E. and Marth, E.H., Ascorbic acid and ascorbate cause disappearance of patulin from buffer solutions and apple juice, *J. Food Prot.*, 42, 864–866, 1979.

17. Roland, J.O. and Beuchat, L.R., Biomass and patulin production by *Byssochlamys nivea* in apple juice as affected by sorbate, benzoate, SO_2 and temperature, *J. Food Sci.*, 49, 402–406, 1984.

18. Moss, M.O. and Long, M.T., Fate of patulin in the presence of the yeast *Saccharomyces cerevisiae*, *Food Addit. Contam.*, 19, 387–399, 2002.

19. FDA, Patulin in Apple Juice, Apple Juice Concentrates and Apple Juice Products, U.S. Food and Drug Administration, Center for Food Safety and Applied Nutrition Office of Plant and Dairy Foods and Beverages, September 2001, http://vm.cfsan.fda.gov/~dms/patubck2.html.

20. Shephard, G.S. and Leggott, N.L., Chromatographic determination of the mycotoxin patulin in fruit and fruit juices, *J. Chromatogr. A*, 882, 17–22, 2000.

21. AOAC International, Official Method 974.18, *Official Methods of Analysis of AOAC International*, 17th ed., AOAC International, Gaithersburg, MD, 2000.

22. Prieta, J., Moreno, M.A., Blanco, J., Suarez, G., and Dominguez, L., Determination of patulin by diphasic dialysis extraction and thin-layer chromatography, *J. Food Prot.*, 55, 1001–1002, 1992.

23. Tarter, E.J., and Scott, P.M., Determination of patulin by capillary gas chromatography of the heptafluorobutyrate derivative, *J. Chromatogr.*, 538, 441–446, 1991.

24. Llovera, M., Viladrich, R., Torres, M., and Canela, R., Analysis of underivatized patulin by GC/MS technique, *J. Food Prot.*, 62, 202–205, 1999.

25. Ware, G.M., Thorpe, C.W., and Pohland, A.E., Liquid chromatographic method for patulin in apple juice, *J. Assoc. Off. Anal. Chem.*, 57, 1111–1113, 1974.

26. Brause, A.R., Trucksess, M.W., Thomas, F.S., and Page, S.W., Determination of patulin in apple juice by liquid chromatography, *J. AOAC Int.*, 79, 451–455, 1996.

27. AOAC International, Official Method 995.10, *Official Methods of Analysis of AOAC International*, 17th ed., AOAC International, Gaithersburg, MD, 2000.

28. AOAC International, Official Method 2000.02, *Official Methods of Analysis of AOAC International*, 17th ed., AOAC International, Gaithersburg, MD, 2000.

29. MacDonald, S., Long, M., and Gilbert, J., Liquid chromatographic method for determination of patulin in clear and cloudy apple juices and apple puree: collaborative study, *JAOAC Int.*, 83, 1387–1394, 2000.

30. Trucksess, M.W. and Tang, Y., Solid-phase extraction method for patulin in apple juice and unfiltered apple juice, *J. AOAC Int.*, 82, 1109–1113, 1999.

31. Eisele, T.A. and Gibson, M.Z., Syringe-cartridge solid-phase extraction method for patulin in apple juice, *J. AOAC Int.*, 86, 1160–1163, 2003.

32. Tsao, R. and Zhou, T., Micellar electrokinetic capillary electrophoresis for rapid analysis of patulin in apple cider, *J. Agric. Food Chem.*, 48, 5231–5235, 2000.

33. Haupt, K., and Mosbach, K., Plastic antibodies: development and applications, *Trends Biotechnol.*, 16, 468–475, 1998.

34. Jodlbauer, J., Maier, N.M., and Lindner, W., Towards ochratoxin A selective moleculrarly imprinted polymers for solid-phase extraction, *J. Chromatogr. A*, 945, 45–63, 2002.

35. Bennett, J.W. and Klich, M., Mycotoxins, *Clin. Microbiol. Rev.*, 16, 497–516, 2003.

36. Raistrick, H., Birkinshaw, J.H., Bracken, A., Micael, S.E., Hopkins, W.A., and Gye, W.E., Patulin in the common cold. Collaborative research on a derivative of *Penicillium patulin* Bainier, *Lancet*, 242, 625, 1943.

37. Llewellyn, G.C., McCay, J.A., Brown, R.D., Musgrove, D.L., Butterworth, L.F., Munson, A.E., and White, K.L., Jr., Immunological evaluation of the mycotoxin patulin in female B6C3F$_1$ mice, *Food Chem. Toxicol.*, 36, 1107–1115, 1998.

38. Friedman, L., Patulin-mycotoxin or fungal metabolite? (Current state of knowledge), in *Biodeterioration Research III*, Lloewellyn, G. and O'Rear, C., Eds, Plenum Press, New York, 1990, pp. 24–51.

39. IARC (International Agency for Research on Cancer) monographs on the evaluation of carcinogenic risk of chemicals to humans: Patulin, 40, 83–98, 1986.

40. JECFA, Patulin. Safety Evaluation of Certain Food Additives And Contaminants, WHO Food Additive Series, 35, 1996, pp 377–402, http://www.inchem.org/documents/jecfa/jecmono/v35je16.htm.

41. McKinley, E.R. and Carlton, W.W., Patulin mycotoxicosis in the Syrian hamster, *Food Cosmet. Toxicol.*, 18, 173–179, 1980.

42. McKinley, E.R., Carlton, W.W., and Boon, G.D., Patulin mycotoxicosis in the rat: toxicology, pathology and clinical pathology, *Food Chem. Toxicol.*, 20, 289–300, 1982.

43. Speijers, G.J.A., Franken, M.A., and Van Leeuwen, F.X., Subacute toxicity study of patulin in the rat: effects on the kidney and the gastrointestinal tract, *Food Chem. Toxicol.*, 26, 23–30, 1988.

44. Becci, P.J., Hess, F.G., Johnson, W.D., Gallo, M.A., Babish, J.G., Dailey, R.E., and Parent, R.A., Long-term carcinogenicity and toxicity studies of patulin in the rat, *J. Appl. Toxicol.*, 1, 256–261, 1981.

45. Ciegler, A., Beckwith, A.C., and Jackson, L.K., Teratogenicity of patulin and patulin adducts formed with cysteine, *Appl. Environ. Microbiol.*, 31, 664–667, 1976.

46. Liu, B.H., Yu, F.Y., Wu, T.S., Li, S.Y., Su, M.C., Wang, M.C., and Shih, S.M., Evaluation of genotoxic risk and oxidative DNA damage in mammalian cells exposed to mycotoxins, patulin and citrinin, *Toxicol. Appl. Pharmacol.*, 191, 255–263, 2003.

47. Cooray, R., Kiessling, K.H., and Lindahl-Kiessling, K., The effects of patulin and patulin–cysteine mixtures on DNA synthesis and frequency of sister-chromatid exchanges in human lymphocytes, *Food Chem. Toxicol.*, 20, 893–898, 1982.

48. Sorenson, W.G., Gerberick, G.F., Lewis, D.M., and Castranova, V., Toxicity of mycotoxins for the rat pulmonary macrophage *in vitro*, *Env. Health Persp.*, 66, 45–53, 1986.

49. Dickens, F. and Jones, H.E.H., Carcinogenic activity of a series of reactive lactones and related substances, *Br. J. Cancer*, 15, 85–100, 1961.

50. Osswald, H., Frank, H.K., Komitowski, D., and Winter, H., Long-term testing of patulin administered orally to Sprague-Dawley rats and Swiss mice, *Food Cosmet. Toxicol.*, 16, 243–247, 1978.

51. Riley, R.T., Hinton, D.M., Showker, J.L., Rigsby, W., and Norred, W.P., Chronology of patulin-induced alterations in membrane function of cultured renal cells, LLC-PK1, *Toxicol. Appl. Pharm*, 102, 128–141, 1990.

52. Riley, R.T. and Showker, J.L., The mechanism of patulin's cytoxicity and the antioxidant activity of indole tetramic acids, *Toxicol. Appl. Pharmacol.*, 109, 108–126, 1991.

53. Arafat, W., Kern, D., and Dirheimer, G., Inhibition of aminoacyl-tRNA synthetases by the mycotoxin patulin, *Chem-Biol. Interact.*, 56, 333–349, 1985.

54. Lindroth, S. and von Wright, A., Comparison of the toxicities of patulin and patulin adducts formed with cysteine, *Appl. Environ. Microbiol.*, 35, 1003–1007, 1978.

55. World Health Organization (WHO), 44th Report of the Joint FAO/WHO Expert Committee on Food Additives, Technical Report Series 859, Geneva, Switzerland, 1995, pp. 36–38.

56. van Egmond, H.P., Current situation on regulations for mycotoxins. Overview of tolerances and status of standard methods of sampling and analysis, *Food Addit. Contam.*, 6, 139–188, 1989.

57. Scott, P.M., Collaborative study of a chromatographic method for determination of patulin apple juice, *J. Assoc. Off. Anal. Chem.*, 57, 621–625, 1974.

58. Bullerman, L.B., Significance of mycotoxins to food safety and human health, *J. Food Prot.*, 42, 65–86, 1979.

59. Palmgren, M.S. and Ciegler, A., Toxicity and carcinogenicity of fungal lactones: patulin and penicillic acid, in *Handbook of Natural Toxins*, Vol. 1, Keeler, K. and Tu, F., Eds., Marcel Dekker, New York, 1983, pp. 325–341.

60. Hasan, H., Patulin and aflatoxin in brown rot lesion of apple fruits and their regulation, *World J. Microbiol. Biotechnol.*, 16, 607–612, 2000.

61. Andersen, B., Medsgaard, J., and Frisvad, J.C., *Penicillium expansum*: consistent production of patulin, chaetoglobosins, and other secondary metabolites in culture and their natural occurrence in fruit products, *J. Agric. Food Chem.*, 52, 2421–2428, 2004.

62. Ciegler, A., Vesonder, R.F., and Jackson, L.K., Production and biological activity of patulin and citrinin from *Penicillium expansum*, *Appl. Environ. Microbiol.*, 33, 1004–1006, 1977.

63. Vinas, I., Dadon, J., and Sanchis, V., Citrin-producing capacity of *Penicillium expansum* strains from apple packing houses of Lerida (Spain), *Int. J. Food Microbiol.*, 19, 153–156, 1993.

64. Larsen, T.O., Frisvad, J.C., Ravn, G., and Skaaning, T., Mycotoxin production by *Penicillium expansum* on blackcurrant and cherry juice, *Food Addit. Contam.* 15, 671–675, 1998.

65. Martins, M.L., Gimeno, A., Martins, H.M., and Bernardo, F., Co-occurrence of patulin and citrinin in Portuguese apples with rotten spots, *Food Addit. Contam.*, 19, 568–574, 2002.

66. Pepeljnjak, S., Segvic, M., and Ozegovic, L., Citrininotoxinogenicity of *Penicillium* spp. isolated from decaying apples, *Brazilian J. Microbiol.*, 33, 134–137, 2002.

67. Pitt, J.I. and Hocking, A.D., *Fungi and Food Spoilage*, Blackie Academic and Professional, London, 1997.

68. Pitt, J.I., Biology and ecology of toxigenic Penicillium species, in *Mycotoxins and Food Safety*, DeVries, J.W., Trucksess, M.W., and Jackson, L.S., Eds., Kluwer Academic/Plenum, New York, 2002, pp. 29–41.

69. Doores, S., The microbiology of apples and apple products, *CRC Crit. Rev. Food Sci. Nutr.*, 19, 133–149, 1983.

70. Northolt, M.D., Van Egmond, H.P., and Paulsch, W.E., Patulin production by some fungal species in relation to water activity and temperature, *J. Food Prot.*, 41, 885–890, 1978.

71. Lovett, J., Thompson, R.G., Jr., and Boutin, B.K., Patulin production in apples stored in a controlled atmosphere, *J. AOAC*, 58, 912–914, 1974.

72. Rice, S.L., Beuchat, L.R., and Worthington, R.E., Patulin production by *Byssochlamys* spp. in fruit juices, *Appl. Environ. Micrbiol.*, 34, 791–796, 1977.

73. Janisiewicz, W.J., Blue Mold, *Penicillium spp.*, Fruit Disease Focus, http://www.caf.wvu.edu/kearneysville/disease_month/bluemold0199.html, 1999.

74. Sommer, N.F., Buchanan, J.R., Fortlage, R.J., and Hsieh, D.P.H., Patulin, a mycotoxin in fruit products, *Proc. IV Int. Contr. Food Sci. Technol.*, 3, 266, 1974.

75. Scott, P.M., Miles, W.F., Toft, P., and Dube, J.G., Occurrence of patulin in apple juice, *J. Agric. Food Chem.*, 20, 450–451, 1972.

76. Wilson, D.M. and Nuovo, G.J., Patulin production in apples decayed by *Penicillium expansum*, *Appl. Microbiol.*, 26, 124–125, 1973.

77. Lindroth, S.L. and Niskanen, A., Comparison of potential patulin hazard in home-made and commercial apple products, *J. Food Sci.*, 43, 446–448, 1978.

78. Watkins, K.L., Fazekas, G., and Palmer, M.V., Patulin in Australian apple juice, *Food Australia*, 42, 438–439, 1990.

79. Yurdun, T., Omurtag, G.Z., and Ersoy, O., Incidence of patulin in apple juices marketed in Turkey, *J. Food Prot.*, 64, 1851–1853, 2001.

80. Leggott, N.L. and Shephard, G.S., Patulin in South African commercial apple products, *Food Control*, 12, 73–76, 2001.
81. Roach, J.A.G., Brause, A.R., Eisele, T.A., and Rupp, H.S., HPLC detection of patulin in apple juice with GC/MS confirmation of patulin identity, in *Mycotoxins in Food Safety*, DeVries, J.W., Trucksess, M.W., and Jackson, L.S., Eds., Kluwer Academic/Plenum, New York, 2002, pp. 135–140.
82. Ritieni, A., Patulin in Italian commercial apple products, *J. Agric. Food Chem.*, 51, 6086–6090, 2003.
83. Malmauret, L., Parent-Massin, D., Hardy, J.-L., and Verger, P., Contaminants in organic and conventional foodstuffs in France, *Food Addit. Contam.*, 19, 524–532, 2002.
84. Harwig, J., Chen, Y.-K., Kennedy, B.P.C., and Scott, P.M., Occurrence of patulin and patulin-producing strains of *Penicillium expansum* in natural rots of apple in Canada, *Can. Inst. Food Sci. Technol. J.*, 6, 22–25, 1973.
85. Lai, C.-L., Fuh, Y.-M., and Shih, D.Y.-C., Detection of mycotoxin patulin in apple juice, *J. Food Drug Anal.*, 8, 85–96, 2000.
86. McCallum, J.L., Tsao, R., and Zhou, T., Factors affecting patulin production by *Penicillium expansum*, *J. Food Prot.*, 65, 1937–1942, 2002.
87. Jackson, L.S., Beacham-Bowden, T., Keller, S.E., Adhikari, C., Taylor, K.T., Chirtel, S.J., and Merker, R.I., Apple quality, storage, and washing treatments affect patulin levels in apple cider, *J. Food Prot.*, 66, 618–624, 2003.
88. Spotts, R.A. and Mielke, E.A., Variability in postharvest decay among apple cultivars, *Plant Dis.*, 83, 1051–1054, 1999.
89. Beer, S.V., and Amand, J.K., Production of the mycotoxin patulin in mature fruits of five apple cultivars infected by *Penicillium expansum*, *Proc. Am. Phytopathol. Soc.*, 1, 104–110, 1974.
90. Paster, N., Huppert, D., and Barkai-Golan, R., Production of patulin by different strains of *Penicillium expansum* in pear and apple cultivars stored at different temperatures and modified atmospheres, *Food Addit. Contam.*, 12, 51–58, 1995.
91. Damoglou, A.P., Campbell, D.S., and Button, J.E., Some factors governing the production of patulin in apples, *Food Microbiol.*, 2, 3–10, 1985.
92. Codex Alimentarius Commission, Proposed draft code of practice for the prevention of patulin contamination in apple juice and apple juice ingredients in other beverages, Joint FAO/WHO Food Standards Programme, Codex Committee on Food Additives and Contaminants, February 2002.
93. Sydenham, E.W., Vismer, H.F., Marasas, W.F.O., Brown, N., Schlecter, M., van der Westhuizen, L., and Rheeder, J.P., Reduction of patulin in apple juice samples: influence of initial processing, *Food Control*, 6, 195–200, 1995.
94. Sydenham, E.W., Vismer, H.F., Marasas, W.F.O., Brown, N.L., Schlecter, M., and Rheeder, J.P., The influence of deck storage and initial processing on patulin levels in apple juice, *Food Addit. Contam.*, 14, 429–434, 1997.
95. Agriculture and Agri-Food Canada, Postharvest Handling of Pome Fruits, Soft Fruits, and Grapes: Apple Diseases, http://res2.agr.ca/parc-crapac/pubs/phhandbook/a_disea_e.htm, 2003.
96. Eisele, T.A. and Drake, S.R., The partial compositional characteristics of apple juice from 175 apple varieties, *J. Food Comp. Anal.*, 18, 213, 2005.
97. Prusky, D., McEnvoy, J.L., Saftner, R., Conway, W.S., and Jones, R., Relationship between host acidification and virulence of *Penicillium* spp. on apple and citrus fruit, *Phytopathology*, 84, 44–51, 2004.

98. Conway, W.S., Effect of postharvest calcium treatment on decay of Golden Delicious apples, *Plant Dis.*, 74, 134–137, 1982.

99. Conway, W.S., Sams, C.E., and Hickey, K.D., Pre-and postharvest calcium treatment of apple fruit and its effect on quality, *Acta Hort.*, 594, 413–419, 2002.

100. Biggs, A.R., Effects of calcium salts on apple bitter rot caused by two *Colletotrichum* spp., *Plant Dis.*, 83, 1001–1005, 1999.

101. Conway, W.S., Tobias, R.B., and Sams, C.E., Reduction of storage decay in apples by postharvest calcium infiltration, *Acta Hort.*, 326, 115–122, 1993.

102. Sams, C.E. and Conway, W.S., Postharvest calcium infiltration improves fresh and processing qualities of apples, *Acta Hort.*, 326, 123–130, 1993.

103. Chardonnet, C.O., Sams, C.E., Conway, W.S., Draughon, F.A., and Mount, J.R., Osmotic dehydration of apple slices with $CaCl_2$ and sucrose limits decay caused by *Penicillium expansum*, *Colletotrichum acutatum*, and *Botrytis cinerea* and does not promote *Listeria monocytogenes* or total aerobic population growth, *J. Food Prot.*, 65, 172–177, 2002.

104. Sanderson, P.G. and Spotts, R.A., Postharvest decay of winter pear and apple fruit caused by species of *Penicillium*, *Phytopatholgy*, 85, 103–110, 1995.

105. Sommer, N.F., Buchanan, JR., and Fortlage, R.J., Production of patulin by *Penicillium expansum*, *J. Appl. Micrbiol.*, 28, 589–593, 1974.

106. Landers, K.E, Davis, N.D., and Diener, U.L., Influence of atmospheric gases on aflatoxin production by *Aspergillus flavus*, *Korean J. Food Sci. Technol.*, 7, 7–10, 1967.

107. Lillehoj, E.B., Milburn, M.S., and Ciegler, A., Control of *Penicillium martensii* development and penicillic acid production by atmospheric gases and temperatures, *Appl. Microbiol.*, 24, 198–201, 1972.

108. Adams, K.B., Wu, M.T., and Salunke, D.K., Effects of subatmospheric pressures on the growth and patulin production of *Penicillium expansum* and *Penicillium patulum*, *Lebensmittel-Wissenschaft und Technologie*, 9, 155, 1976.

109. Moodley, R.S., Govinden, R., and Odhav, B., The effect of modified atmospheres and packaging on patulin production in apples, *J. Food Prot.*, 65, 867–871, 2002.

110. U.S. Department of Agriculture, World Apple Situation, USDA/FAS, Horticultural and Tropical Products Division, April 2004, http://www.fas.usda.gov/htp/horticulture/Apples/World%20Apple%20Situation%202003-04.pdf.

111. Willett, M., Kupferman, G., Roberts, R., Spotts, R., Sugar, D., Apel, G., Ewart, H.W., and Bryant, B., Practices to Minimize Postharvest Decay of Apples and Pears, Washington State University, Tree Fruit Research and Extension Center, Postharvest Information Network, http://postharvest.tfrec.wsu.edu/pgDisplay.php?article=N7I3A, August 9, 2004.

112. Sholberg, P.L. and Conway, W.S., Postharvest Pathology, September 4, 2001, www.ba.ars.usda.gov/hb66/022pathology.pdf.

113. Janisiewicz, W.J., Conway, W.S., Glenn, D.M., and Sams, C.E., Integrating biological control and calcium treatment for controlling postharvest decay of apples, *Hort. Sci.*, 33, 105–109, 1998.

114. Nunes, C., Usall, J., Teixido, N., de Eribe, X.O., and Vinas, I., Control of post-harvest decay of apples by pre-harvest and post-harvest application of ammonium molybdate, *Pest Manag. Sci.*, 57, 1093–1099, 2001.

115. Sugar, D. and Spotts, R.A., Postharvest Strategies to Reduce Postharvest Decay, Washington Tree Fruit Postharvest Conference Proceedings, Washington State Horticultural Association, Wenatchee, WA, 1995.

116. Nunes, C., Usall, J., Teixido, N., Torres, R., and Vinas, I., Control of *Penicillium expansum* and *Botrytis cinerea* on apples and pears with the combination of *Candida sake* and *Pantoea agglomerans*, *J. Food Prot.*, 65, 178–184, 2002.

117. Nunes, C., Usall, J., Teixido, N., and Vinas, I., Improvement of *Candida sake* biocontrol activity against post-harvest decay by the addition of ammonium molybdate, *J. Appl. Micrbiol.*, 92, 927–935, 2002.

118. Aziz, N.H. and Moussa, L.A.A., Influence of gamma-radiation on mycotoxin producing moulds and mycotoxins in fruits, *Food Control*, 13, 281–288, 2002.

119. Torres, R., Valentines, M.C., Usall, J., Vinas, I, and Larriguadiere, C., Possible involvement of hydrogen peroxide in the development of resistance mechanisms in "Golden Delicious" apple fruit, *Postharvest Biol. Technol.*, 27, 235–242, 2003.

120. Spotts, R.A. and Cervantes, L.A., Contamination of harvest bins with pear decay fungi and evaluation of disinfectants on plastic and wood bin material, *Acta Hort. (ISHS)*, 367, 419–425, 1994.

121. Binnig, R. and Possmann, P., Apple juice, in *Fruit Juice Processing Technology*, Nagy, S., Chen, C.S., and Shaw, P.E., Eds., Agscience, Auburndale, FL, 1993, chap. 8, pp. 271–317.

122. Root, W.H., Apples and apple processing, in *Processing Fruits: Science and Technology, Volume 2. Major Processed Products*, Somogyi, L.P., Barrett, D.M., and Hui, Y.H., Eds., Technomic, Lanacaster, PA, 1996, pp. 1–35.

123. Blanpied, G.D. and Purnasiri, A., *Penicillium* and *Botrytis* rot of McIntosh apples handled in water, *Plant Dis. Rep.*, 52, 865–867, 1968.

124. Spotts, R.A. and Peters, B.B., Chlorine and chlorine dioxide for d'Anjou pear decay, *Plant Dis.*, 64, 1095–1097, 1980.

125. Spotts, R.A. and Cervantes, L.A., Effect of ozonated water on postharvest pathogens of pear in laboratory and packinghouse tests, *Plant Dis.*, 76, 256–259, 1992.

126. Spotts, R.A. and Cervantes, L.A., Use of filtration for removal of conidia of *Penicillium expansum* from water in pome fruit packinghouses, *Plant Dis.*, 77, 828–830, 1993.

127. FDA, Report of 1997 Inspections of Fresh, Unpasteurized Apple Cider Manufacturers. Summary of Results, http://vm.cfsan.fda.gov/~dms/ciderrpt.html, 1999.

128. Wright, J.R., Sumner, S.S., Hackney, C.R., Pierson, M.D., and Zoecklein, B.W., A survey of Virginia apple cider producers' practices, *Dairy Food Environ. Sanit.*, 20, 190–195, 2000.

129. Uljas, H.E. and Ingham, S.C., Survey of apple growing, harvesting, and cider manufacturing practices in Wisconsin: implications for safety, *J. Food Saf.*, 20, 85–100, 2000.

130. Acar, J., Gokmen, V., and Taydas, E.E., The effects of processing technology on the patulin content of juice during commercial apple juice concentrate production, *Zeitschrift für Lebensmitteluntersuchung und -Forschung A*, 207, 328–331, 1998.

131. Taniwaki, M.H., Hoenderboom, C.J.M., De Almeida Vitali, A. and Eiroa, M.N.U., Migration of patulin in apples, *J. Food Prot.*, 55, 902–904, 1992.

132. Okull, D.O. and LaBorde, L.F., Activity of electrolyzed oxidizing water against *Penicillium expansum* in suspension and on wounded fruit, *J. Food Sci.*, 69, 23–27, 2004.

133. Lovett, J., Thompson, R.G., Jr., and Boutin, B.K., Trimming as a means of removing patulin from fungus rotted apples, *J. AOAC*, 58, 909–911, 1975.

134. Leverentz, B., Conway, W.S., Janisiewicz, W.J., Saftner, R.A., and Camp, M.J., Effect of combining MCP treatment, heat treatment, and biocontrol on the reducing of postharvest decay of "Golden Delicious" apples, *Postharvest Biol. Technol.*, 37, 221–233, 2003.

135. Ryu, D. and Holt, D.L., Growth inhibition of *Penicillium expansum* by several commonly used food ingredients, *J. Food Prot.*, 56, 862–867, 1993.

136. Sholberg, P., Haag, P., Hocking, R., and Bedford, K., The use of vinegar vapor to reduce postharvest decay of harvested fruit, *Hort Sci.*, 35, 898–903, 2000.

137. Palmer, L.C., Horst, R.K., and Langhans, R.W., Use of bicarbonates to inhibit *in vitro* colony growth of *Botrytis cinerea*, *Plant Dis.*, 81, 1432–1438, 1997.

138. Smilanick, J.L., Margosan, D.A., Mlikota, F., Usall, J., and Michael, I.F., Control of citrus green mould by carbonate and bicarbonate salts and the influence of commercial postharvest practices on their efficacy, *Plant Dis.*, 83, 139–145, 1999.

139. El Ghaouth, A., Raul, J., Grenier, J., and Asselin, A., Antifungal activity of chitosan on two postharvest pathogens of strawberry fruits, *Phytopathology*, 82, 398–402, 1992.

140. El Ghaouth, A., Smilanick, J.L., Wisniewiski, M., and Wilson, C.A., Improved control of apple and citrus fruit decay with a combination of *Candida saitoana* and 2-deoxy-D-glucose, *Plant Dis.*, 84, 249–253, 2000.

141. El Ghaouth, A., Smilanick, J.L., Brown, G.E., Ippolito, A., Wisniewski, M., and Wilson, C.L., Application of *Candida saitoana* and glycochitosan for the control of postharvest diseases of apple and citrus fruit under semi-commercial conditions, *Plant Dis.*, 84, 243–248, 2000.

142. Abo-Dahab, N.F., Paterson, R.R.M. and Razak, A.A., Effect of fungistatic agent 2-deoxy-D-glucose on mycotoxins from *Pencillium expansum*, *Lett. Appl. Microbiol.*, 23, 171–173, 1996.

143. El Ghaouth, A., Wilson, C.L., and Wisnieski, E., Sugar analogs as potential fungicides for postharvest pathogens of apple and peach, *Plant Dis.*, 79, 254–258, 1995.

144. Sholberg, P.L., Fumigation of fruit with short-chain organic acids to reduce the potential of postharvest decay, *Plant Dis.*, 82, 689–693, 1998.

145. Lurie, S., Postharvest heat treatments of horticultural crops, *Hort. Rev.*, 22, 91–121, 1998.

146. Fallik, E., Grinsberg, S., Alkalai, S., Yekutieli, O., Wiseblum, A., Regev, R., Beres, H., and Bar-Lev, E., A unique rapid hot water treatment to improve storage quality of sweet pepper, *Postharvest Biol. and Technol.*, 15, 25–32, 1999.

147. Fallik, E., Grinsberg, S., Gambourg, M., Klein, J.D., and Lurie, S., Prestorage heat treatment reduces pathogenicity of *Penicillium expansum* in apple fruit, *Plant Pathol.*, 45, 92–97, 1995.

148. Fallik, E., Tuvia-Alkalai, S., Feng, X., and Lurie, S., Ripening characterization and decay development of stored apples after a short pre-storage hot water rinsing and brushing, *Innov. Food Sci. Emerg. Technol.*, 2, 127–132, 2001.

149. Lurie, S., Fallik, E., and Klein, J.D., The effect of heat treatment on apple epicuticular wax and calcium uptake, *Postharvest Biol. Technol.*, 8, 271–277, 1996.

150. Janisiewicz, W.J. and Bors, B., Development of microbial community of bacterial and yeasts antagonists to control wound-invading postharvest pathogens of fruits, *Appl. Environ. Microbiol.*, 61, 3261–3270, 1995.

151. Janisiewicz, W.J. and Jeffers, S.N., Efficacy of commercial formulation of two biofungicides for control of blue mold and grey mold of apples in cold storage, *Crop. Prot.*, 16, 629–633, 1997.

152. Janisiewicz, W.J. and Marchi, A., Control of storage rots on various pears with a saprophytic strain of *Pseudomonas syringae*, *Plant Dis.*, 76, 555–560, 1992.

153. Droby, S., Wisniewski, M., El Ghaouth, A., and Wilson, C., Influence of food additives on the control of postharvest rots of apple and peach and efficacy of the yeast-based biocontrol product Aspire, *Postharvest Biol. Technol.*, 27, 127–135, 2003.

154. McLaughlin, R.J., Wisniewski, M.E., Wilson, C.L., and Chalutz, E., Effect of inoculum concentration and salt solutions on biological control of postharvest diseases of apple with *Candida* sp., *Phytopathology*, 80, 456–461, 1990.

155. Kurtzman, C.P., *Pichia* E.C. Hansen emend. Kurtzman, in *The Yeasts, A Taxonomic Study*, 4th ed., Kurtzman, C.P. and Fell, J.W., Eds., Elsevier Science, Amsterdam, 1998, pp. 273–352.

156. Stevens, C., Khan, V.A., Lu, J.Y., Wilson, C.L., Pusey, P.L., Igwegbe, E.C.K., Kabwe, K., Mafolo, Y., Liu, J., Chalutz, E., and Droby, S., Integration of ultraviolet (UV-C) light with yeast treatment for control of postharvest storage rots of fruits and vegetables, *Biol. Control*, 10, 98–103, 1997.

157. Barkai-Golan, R., Postharvest disease suppression by atmospheric modification, in Food Preservation by Modified Atmospheres, Calderon, M. and Barkai-Golan, R., Eds., CRC Press, Boca Raton, FL, 1990, pp. 237–265.

158. Yackel, W.C., Nelson, A.I., Wei, L.S., and Steinberg, M.P., Effect of controlled atmosphere on growth of mold on synthetic media and fruit, *Appl. Microbiol.*, 22, 513–516, 1971.

159. Johnson, D.S., Stow, J.R., and Dover, C.J., Prospect for the control of fungal rotting in Cox's Orange Pippin apples by low oxygen and low ethylene storage, *Acta Horticult.*, 343, 334–336, 1993.

160. Sitton, J.W. and Patterson, M.E., Effect of high-carbon dioxide and low-oxygen controlled atmospheres on postharvest decay of apples, *Plant Dis.*, 76, 992–994, 1992.

161. Kadakal, C. and Nas, S., Effect of activated charcoal on patulin, fumaric acid and some other properties of apple juice, *Nahrung*, 46, 31–33, 2002.

162. Leggott, N.L., Marasas, W., Rheeder, J., Shephard, G.S., Sydenham, E., and Vismer, H., Occurrence of patulin in the commercial processing of apple juice, *S. Afr. J. Sci.*, 96, 241–243, 2000.

163. Leggott, N.L., Shephard, G.S., Stockenstrom, S., Staal, E., and van Schatkwyk, D.J., The reduction of patulin in apple juice by three different types of activated charcoal, *Food Addit. Contam.*, 18, 825–829, 2001.

164. Bissessur, J., Permaul, K., and Odhav, B., Reduction of patulin during apple juice clarification, *J. Food Prot.*, 64, 1216–1219, 2001.

165. Sands, D.C., McIntyre, J.L., and Walton, J.S., Use of activated carbon for the removal of patulin from cider, *Appl. Environ. Microbiol.*, 32, 388–391, 1976.

166. Doyle, M.P., Applebaum, R.S., Brackett, R.E., and Marth, E.H., Physical, chemical and biological degradation of mycotoxins in food and agricultural commodities, *J. Food Prot.*, 45, 964–971, 1982.
167. Yazici, S. and Velioglu, Y.S., Effect of thiamine hydrochloride, pyridoxine hydrochloride and calcium-d-pantothenate on the patulin content of apple juice concentrate, *Nahrung*, 46, 256–257, 2002.
168. Stinson, E., Osman, S.F., Huhtanen, C.N., and Bliss, D.D., Disappearance of patulin during alcoholic fermentation of apple juice, *Appl. Environ. Microbiol.*, 36, 620–622, 1978.

14 Safety of Minimally Processed, Acidified, and Fermented Vegetable Products

F. Breidt, Jr.

CONTENTS

Paper no. FSR04-21 of the Journal Series of the Department of Food Science, North Carolina State University, Raleigh, NC 27695-7624. Mention of a trademark or proprietary product does not constitute a guarantee or warranty of the product by the U.S. Department of Agriculture or North Carolina Agricultural Research Service, nor does it imply approval to the exclusion of other products that may be suitable.

14.1 INTRODUCTION

Food fermentation technology likely originated sometime between 8,000 to 12,000 years ago as plants and animals were being domesticated in the Middle East, Africa, and Asia [1–3]. The development of primitive pottery technology likely led to early fermentation experiments, either planned or unplanned. Cheese, bread, and alcoholic beverages may have resulted from the fermentation of milk, grains, fruits, and vegetables stored in ceramic jars or pots. If these "spoiled" or fermented products were found to have desirable sensory properties, they may have been developed as the first processed or fermented foods [2]. An important characteristic of fermentation was the increase in the storage lifetime during which foods could be safely eaten. The microbial nature of food fermentation or foodborne illnesses was not understood, however, until the advent of the science of microbiology in the late 19th century. The fermentation of vegetables by lactic acid bacteria (LAB) is now well understood as an effective means of preserving and ensuring the safety of foods [4,5]. LAB are being considered for use in nonfermented vegetable products as a means of ensuring safety and preventing spoilage [6–8]. Fermented and acidified vegetable products, such as sauerkraut, kimchi, olives, and cucumber pickles, not only have desirable sensory qualities, but also have an excellent safety record with no known reported cases of foodborne illness.

14.2 VEGETABLE MICROFLORA

The microflora on fresh fruits, grains, and vegetables can range from as low as 10^2 to 10^9 colony forming units (CFU) per gram [9,10]. On pickling cucumbers, for example, the aerobic microflora is typically between 10^4 to 10^6 CFU/ml for fresh fruit, with LAB less than 10^1 CFU/g [11]. In the absence of processing, degradative aerobic spoilage of plant material by mesophylic microorganisms occurs, with *Pseudomonas* spp., *Enterobacter* spp., and *Erwinia* spp. initiating the process [10]. A variety of pathogens, including *Salmonella* spp., *Shigella* spp., *Aeromonas hydrophylia*, *Yersinia enterocolitica*, *Staphylococcus aureus*, *Campylobacter*, *Listeria monocytogenes*, *Escherichia coli*, and others, may be present on fresh vegetable products [12–15]. Pathogens on fruits and vegetables may also include enteric, hepatitis, or polio viruses [16]. A variety of sources may contribute to the occurrence of pathogenic bacteria on fruit and vegetable crops, including exposure of plants to untreated manure or contaminated water, the presence of insects or birds, personal hygiene practices of farm workers, postharvest washing or hydrocooling water, and conditions of storage during distribution [12,14]. A study comparing the use of organic fertilizer (composted manure) and inorganic fertilizer from farms in Minnesota showed significantly higher coliform counts on the organically grown vegetables [17]. However, in this and related studies [18,19], pathogens, including *E. coli* O157:H7, were not detected.

Removal of pathogenic and spoilage bacteria from fruits and vegetables has proved difficult. Surface adherence of bacteria (Figure 14.1) may serve

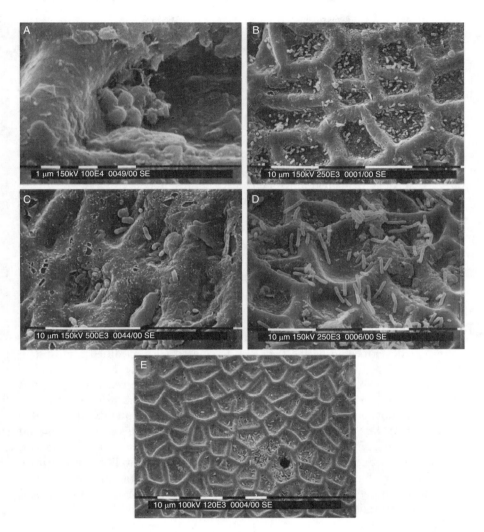

FIGURE 14.1 Attachment of pathogenic bacteria to cucumber fruit. Adhesion of bacteria to the surfaces of pickling cucumbers (Calypso variety) with wax: (A) *Staphylococcus aureus*; (B) *Lactobacillus plantarum*; (C) *Listeria monocytogenes*; (D) *Salmonella typhimurium;* (E) *Enterobacter aerogenes* ATCC 13048. Bar 5, 10 mm. (From Reina, L.D., Fleming, H.P., and Breidt, F., *J. Food Prot.*, 65, 1881–1887, 2002.)

to enhance survival of bacteria during washing or sanitizing treatments. Bacterial cell surface charge and hydrophobicity measurements have been found to correlate with the attachment of cells to surfaces of cantaloupes and cucumbers [20,21]. Dewaxing cucumber fruit led to increased adhesion of *L. monocytogenes* and decreased adhesion for other bacteria with higher relative surface hydrophobicity, including salmonella, lactobacilli, and staphylococci [20]. Biofilms of bacteria may be more resistant to sanitizing agents and organic acid treatments than free or planktonic cells [22–24]. It is likely that the vast majority of microorganisms in food processing environments occur in multispecies or multistrain biofilms on food or equipment surfaces [25,26].

14.2.1 WASHING PROCEDURES

Washing procedures with water or chemical sanitizers typically result in only a 1 to 2 $\log_{10}$ decrease in bacterial cell numbers [24]. Hydrocooling procedures used for some fruits immediately after harvest may even serve to increase internalization of bacteria due to the vacuum created as internal gases in fruits and vegetables contract with the reduction in temperature [27,28]. Bacteria may be protected in inaccessible locations on fruits and vegetables, such as the cores and calyx of apples [29]. Attachment to wounded regions or entry into the interior of fruits and vegetables through wounded regions or stomata, pores, or channels may occur [20,30–32].

The packaging and storage conditions for minimally processed vegetable products, including the use of modified atmosphere packaging, may significantly alter microbial ecology. The extended shelf life of some minimally processed vegetable products may result in an undesirable "safety index," a concept developed to define the risks associated with modified atmosphere packaged foods [33]. This safety index is defined as the ratio of spoilage to pathogenic bacteria in foods, measured as the relative cell concentrations of these organisms. It has been argued, however, that the primary effect of modified atmosphere packaging in extending the sensory quality of vegetable products may be to decrease the metabolic activity of the vegetable material [34]. In a model system, it was found that growth rates for *L. monocytogenes*, *A. hydrophilia*, and *Bacillus cereus* may be reduced by modified atmosphere conditions, but final cell density was not affected [35]. One major source of concern is that *Clostridium botulinum* spores have been isolated from a variety of vegetables, and this organism may, under the right conditions of temperature, pH, and atmosphere, grow and produce toxin in minimally processed vegetable products if the O_2 concentrations drop to 1% or lower [10].

14.2.2 BIOCONTROL IN MINIMALLY PROCESSED VEGETABLE PRODUCTS

The survival and growth of bacteria on vegetable products can depend on the competitive microflora present and the environmental conditions and processing treatments [15,36]. The use of competitive microflora to enhance the safety of minimally processed foods, including vegetable products, has been proposed by a number of authors [5,37–39]. LAB have been nominated for this role, partly because of their GRAS (generally regarded as safe) status and their common usage in food fermentations. Application of this approach for minimally processed fruit and vegetable products has led to mixed results. Vescovo and co-workers isolated LAB from salad vegetables and, subsequently, re-inoculated the vegetables with both the biocontrol cultures and selected food pathogens, including aeromonas, salmonella, staphylococcus, and listeria species [6,40]. The added LAB cultures were found to reduce or prevent the growth of microbial pathogens. Conversely, a *Lactobacillus delbruckii lactis* strain, known to inhibit *E. coli* on chicken skin due to the production of hydrogen peroxide, did not alter the survival of *E. coli* O157:H7 on fresh-cut vegetables, possibly due to the presence of catalase on the plant surfaces [8].

Competition from aerobic microflora isolated from fresh vegetables, other than LAB, including yeasts, *Bacillus* spp. and *Pseudomonas* spp., can influence the survival and growth of microbial food pathogens. *Pseudomonas* spp. have been shown to enhance [41], inhibit [42–44], or have no effect [45] on the growth of *L. monocytogenes* in fruits and vegetables. A variety of pseudomonas and aeromonas isolates from fresh vegetables were found to confer inhibitory activity against *E. coli*, salmonella, listeria, and staphylococcus strains using an agar diffusion assay [46]. Competition studies have shown iron sequestration by siderophores may influence the competition between pseudomonads and *L. monocytogenes* [42,47], although some *Listeria* spp. may be able to use exogenous siderophores as an iron source [48]. Buchanan and Bagi [49] demonstrated that the effects of salt and temperature can control the outcome of competitive growth of a *L. monocytogenes* Scott A and a *Pseudomonas fluorescens* culture that was screened for the inability to produce siderophores or bacteriocins. In a study by Del Campo *et al.* [45], competition for nutrients between a Scott A strain of *L. monocytogenes* and saprophytic bacteria from green endive was investigated. Enterobacteriaceae and pseudomonas were grown in competition with *L. monocytogenes* in minimal media and media supplemented with yeast extract. In this case, enterobacteriaceae but not pseudomonads species were effective in reducing the growth of the *L. monocytogenes* culture. Because culture filtrates from enterobacteriaceae were found to have no inhibitory effects in broth supplemented with yeast extract, the data indicated that competition for nutrients (not end product inhibition) was responsible for the inhibitory effect [45].

These studies illustrate the complexity of microbial interactions in and on fruit and vegetable products. Varying environmental conditions may include changes in the availability of nutrients, salt concentration, temperature, atmosphere, pH, and others. While further research is clearly needed, the use of protective cultures should only be considered as a supplement to good manufacturing practice, not as a substitute for the proper handling and packaging of vegetable products [5]. The use of biocontrol cultures may, therefore, be considered to enhance existing hurdle technology to prevent the growth of pathogens in foods. The hurdle concept [50] advocates the use of multiple preservative factors to prevent the growth of pathogens. In fresh fruit and vegetable products, the main factors affecting the growth of the indigenous bacterial populations are sanitation, modified atmosphere packaging, and refrigeration, as well as the competitive interactions of bacteria.

Bacteria cultures selected for use in biocontrol applications should ideally be isolated from the products for which they are intended to be used [39]. Development of successful biocontrol strategies for fresh fruit and vegetable products may include the following steps: (1) isolation of potential biocontrol LAB from the product for which they are intended to be used; (2) reduction of the total microflora in and on the vegetable product by one of a variety of procedures, including heat, washing using chemical sanitizers, irradiation, or others; (3) addition of the biocontrol culture to achieve an appropriate initial population, as determined experimentally; (4) storage of the product under refrigeration temperatures [39]. The shelf life of the product would then be dictated by the growth of the biocontrol culture, but, to be successful, the growth rates of a biocontrol culture presumably should be faster than that of the target pathogens. While rapid growth and production of inhibitory metabolites may be desirable from a safety standpoint, this may be a liability as far as the quality of the product is concerned. Breidt and Fleming [7] investigated the kinetics of acid production and inhibition of *L. monocytogenes* by *L. lactis* using a mathematical modeling approach [7]. It was observed that the growth and death of the *L. monocytogenes* culture could only be accurately predicted by the model if pH was assumed to be the limiting variable, rather than acid concentration, with cessation of growth around pH 4.6. Further studies to characterize the kinetics of bacterial competition are needed to aid in the development of biocontrol strategies.

14.3 FERMENTED VEGETABLES

Under the anaerobic conditions found with brined vegetables, rapid fermentation by LAB and yeasts occurs, resulting in the destruction of most other microflora, usually within a few days of the onset of fermentation [51]. In the U.S., cucumber pickles and sauerkraut represent the majority of fermented vegetable products. For pickles, fermentation was the primary means of preservation until the 1940s, when direct acidification and pasteurization of cucumber pickles was introduced (reviewed by Fleming *et al.* [51]). Currently, fermented cucumbers represent roughly 30% of commercial production of pickles, mostly for institutional markets (hamburger dill slices), with the majority of the retail market being nonfermented acidified pickles which are pasteurized to destroy vegetative microflora.

Vegetable fermentations typically begin with heterofermentative LAB, such as *Leuconostoc mesenteroides* and end with the most acid-resistant homofermentative LAB, usually *Lactobacillus plantarum* [1,52,53]. *Lactobacillus plantarum* is able to tolerate a lower internal pH than other LAB, and this feature may allow it to predominate in the terminal stages of most vegetable fermentations [54]. During the fermentation of cucumbers and cabbage, hexose sugars, including glucose and fructose, are typically converted to lactic acid by homofermentative LAB via the Embden–Myerhof–Parnas pathway, while the heterofermentative LAB will produce a combination of lactic acid and acetic acid or ethanol, along with CO_2 via the phospho-ketolase pathway [55]. When fructose is present, LAB can use this sugar as an electron acceptor, producing mannitol, which subsequently can be converted anaerobically to lactic acid with an appropriate electron acceptor [56]. In cucumber fermentation where malate is present, *L. plantarum* and other LAB have been found to carry out a decarboxylation of malate to produce lactic acid and CO_2 [57]. This one-step reaction occurs via malolactic enzyme, and is analogous to the amino acid decarboxylation reactions described below [119]. During the reaction, a proton is taken up from the surrounding medium, which helps to buffer cellular pH and causes the pH in the surrounding medium to rise.

14.3.1 FERMENTATION CHEMISTRY

In the U.S., commercial cucumber fermentations are typically carried out with 5 to 6% NaCl, while cabbage fermentations are carried out with 2 to 3% NaCl [51]. During the growth of LAB in vegetable fermentations, a variety of antimicrobial metabolic end products are produced, including organic acids, peroxides, amines, thiols, bacteriocins, and other enzymes and compounds [1,4,5,58–61]. These inhibitory compounds begin to accumulate in the initial stages of fermentation. A combination of several factors, including organic acids from the fermentation (up to 2 to 3% organic acids may be produced), complete fermentation of available sugar, terminal pH values around 3 to 3.5, and salt, can serve to destroy most vegetative bacterial cells, including human pathogens. Desirable textural and nutritional properties of the fermented vegetables may be maintained during storage in the fermentation brine for extended periods of time (a year or more) without refrigeration.

14.4 ACIDIFIED VEGETABLES

For nonfermented, acidified vegetable products, acetic acid is commonly used as an acidulant. At a concentration of 3.6% or greater, acetic acid-acidified foods can be preserved without the addition of other antimicrobial agents or use of heat treatments [62,63]. For pickled pepper products, acidification with 2% acetic acid to pH values around 3.2 was found to prevent microbial growth for 6 months or more [64]. In general, preservation by organic acids alone results in products that can only be consumed in small amounts, as condiments, or as ingredients in other foods. Many acidified vegetable products contain between 0.5 and 2% acetic acid and are pasteurized to prevent spoilage, as well as to ensure safety. For nonfermented pickled vegetables, the combination of heat treatments, acid, and sugar concentration (for sweet pickles) serves to prevent microbial growth. Fresh-pack cucumber pickle products typically contain between 0.5 and 1% acetic acid. A recommended pasteurization procedure consists of heating to an internal temperature to 74°C for 15 minutes [65].

Both acidified and fermented vegetable products have enjoyed an excellent safety record with few or no reported cases of foodborne disease resulting from consumption of these products. Recently, however, there have been reports of disease outbreaks in juice products with pH values below 4.0, in the same range as many fermented and acidified vegetable products. *Escherichia coli* O157:H7 and salmonella serotypes have caused serious illness and death from the consumption of apple cider and orange juice [66,67]. These disease outbreaks have raised questions about the safety of acidified and fermented vegetable products. While pathogenic microorganisms have not been found to grow in these products due to the low pH (typically below 4.0), these microorganisms may adapt to acid conditions and survive for extended periods [68]. Acid types and concentrations vary considerably for acidified foods. Factors affecting acid inhibition of microbial pathogens include the pH of the product, as well as specific effects of the acid or acid anion on cellular enzymes or membranes, and the ability of bacteria to transport protons and organic acids out of the cell interior [69–72].

14.4.1 DEFINITIONS AND REGULATIONS FOR ACID AND ACIDIFIED FOODS

Acid foods are defined in the U.S. Code of Federal Regulations (21 CFR part 114) as foods that have a natural pH value at or below 4.6. These foods include fermented vegetables; vegetable fermentation is considered a "field process" and typically results in a product with a final pH below 4.6. A pH value of 4.6 is used in the definition of acid foods because this is a limiting pH at or below which *C. botulinum* spore outgrowth and neurotoxin toxin production is prevented [73]. Foods with pH values above 4.6 are defined as low-acid foods, and, when packaged in hermetically sealed containers, must be made commercially sterile as defined in 21 CFR part 113. Acidified foods are defined in 21 CFR part 114 as foods to which acid or acid food ingredients have been added that have a water activity (a_w) greater than 0.85 and have a finished equilibrium pH value at or below 4.6. The regulation requires producers of acidified foods to verify that the final equilibrium pH is maintained at or below 4.6 to ensure safety. This regulation governing acidified foods in the U.S. was promulgated by the U.S. Food and Drug Administration (FDA) in 1979. At that time, vegetative pathogenic microorganisms were not considered to be a significant risk for acidified or fermented food products. Included in the regulation, however, is the requirement for a heat process "to the extent that is sufficient" to destroy vegetative cells of microorganisms of public health significance or those of nonhealth significance capable of reproducing in the product. The regulations governing acidified foods are, therefore, based primarily on the pH needed to prevent botulism, and do not include any specification about the type or concentration of acid needed to meet the pH requirement.

In a study of beef carcass wash water, a treatment with 0.2% (33.3 mM) acetic acid and a pH of approximately 3.7 showed that an *E. coli* O157:H7 strain survived for up to 14 days at 15°C, while cell numbers dropped about 4 log cycles [74]. In that study, competitive microflora were also present and could have influenced the survival of the *E. coli* strains. A statistical analysis of several published studies showed that, under typical storage conditions for apple cider (which typically has a pH value less than 4.0 and contains malic acid), the acid conditions alone were not sufficient to ensure a 5 log reduction in the cell numbers of *E. coli* [75]. From these and other studies [68,75–79], it is clear that the potential for *E. coli* to survive for extended periods in acidified vegetable products with a pH below 4 clearly exists, and pasteurization for some acidified food products may be needed to ensure safety.

14.4.2 Pathogenic Bacteria

After recent outbreaks of *E. coli* O157:H7 in apple cider and salmonella in orange juice [66,67], the FDA in 2001 proposed that all new process filings (which are required for the production of acidified foods) should include a heating or pasteurization step. Of primary concern was *E. coli* O157:H7 because of its low infectious dose and lethal sequelae which can result from infection [80,81]. *Escherichia coli* and other food pathogens have been shown to have inducible acid resistance mechanisms [76,82–85]. If only pH is considered, acid-resistant pathogens might, therefore, pose a potential threat to acidified foods. It is likely that the organic acids present in these products have contributed to their excellent safety record because some acidified products have been produced safely for many years without heat treatments [84], although quantitative measurements of the independent effects of organic acids and pH on the killing of pathogens in these products are lacking. In response to the pathogen outbreaks in juice products, 21 CFR part 120 was promulgated in 2001. This regulation mandated a HACCP (hazard analysis critical control point) system with a processing step designed to deliver the equivalent of a 5 log reduction in target pathogen populations in juices. Typically, a heat pasteurization process is used, based on thermal destruction time data for inactivation of *E. coli* O157, which was found to be the most heat- and acid-resistant pathogen in fruit juices [86]. In recent experiments (Breidt, unpublished data), the thermal resistance of *E. coli* O157:H7 and *L. monocytogenes* was found to be identical under the conditions typical of acidified pickle products, and salmonella strains were significantly less heat-resistant. Similarly, salmonella was found to be less heat resistant than *L. monocytogenes* or *E. coli* O157:H7 in fruit juices [86]. For the variety of acidified vegetable products currently available, the time and temperature needed to ensure a 5 log or greater reduction (although a 5 log reduction is not currently mandated by existing federal regulations) in numbers of microbial pathogens will depend on the type and concentration of organic acid present, the composition of the brine or suspending medium during heating, heat resistance of the microorganisms, and other factors.

Some pickled pepper products with high concentrations of acetic acid (greater than 2% acetic acid) and pH values around 3.1 to 3.3 may not need a heat treatment to ensure the destruction of acid-resistant pathogens because sufficient acid is present. In a study of firmness retention with unpasteurized pickled peppers, which typically have pH values around 3.1 to 3.3, and cucumbers, using 2 to 5% acetic acid, microbiological stability was achieved for a 6-month period [64] for all products tested. A heat process is typically not used for these pickled peppers because sliced peppers are susceptible to softening during pasteurization. Historically, pasteurization treatments were designed to prevent spoilage by LAB in brined vegetables and inactivate softening enzymes. Currently, most commercial acidified vegetable products with pH values between 3.3 and 4.1 are produced using a pasteurization process to prevent spoilage. In addition, low water activity and preservatives can reduce the amount of acetic acid needed for preservation. A preservation prediction chart showing the effects of acid and sugar in preventing the growth of spoilage yeasts in sweet pickles was developed in the 1950s [62]. The acid concentrations that will ensure the death of microbial pathogens for many acidified foods remain to be determined.

14.5 ORGANIC ACIDS AND DESTRUCTION OF PATHOGENS

Organic acid preservatives have widespread application for preventing food spoilage and contribute to the manufacture of safe food products [87–89]. The survival or death of pathogenic bacteria in acid and acidified foods has been investigated in a variety of products, including apple cider [68,90–93], mayonnaise, dressings and condiments [76,84,94,95], and fermented meats [96–98]. The mechanism of action of organic acids is commonly attributed to acidification of the cytoplasm of target cells, but also to intracellular accumulation of anions [99]. The protonated form of weak acid preservatives may diffuse across microbial cell membranes and then dissociate in the cell cytoplasm, releasing protons and anions because the intracellular pH must be maintained at a higher value than the external environment. Internal acid anion concentrations may correlate with the cessation of growth. Goncalves *et al.* [100] found that the specific growth rate of *L. rhamnosus* approached zero at approximately 4 molar lactate (anion), with pH values between 5.0 and 6.8. In vegetable fermentations, *L. plantarum* was found to tolerate a lower internal pH than other LAB and, therefore, would have lower acid anion concentrations.

Data on the relative effects of various organic acids and preservatives on the inhibition of microbial pathogens are often conflicting in the scientific literature. For example, Young and Foegeding [101] showed that with equal initial pH values in brain–heart infusion broth ranging from 4.7 to 6.0 and on an equimolar basis, the order of effectiveness in inhibiting the growth of *L. monocytogenes* for three weak organic acids was acetic > lactic > citric. However, when based on initial undissociated acid concentrations, the order was reversed. Ostling and Lindgren [102] determined MIC values for the inhibition of *L. monocytogenes* by lactic, acetic, and formic acids. They found lactic acid was the most inhibitory over a range of pH values from 4.2 to 5.4, with an MIC value of less than 4 mM (protonated acid) for aerobic growth and less than 1 mM for anaerobic growth. They used cells grown in glucose-containing nutrient broth and reported MIC values for the protonated acid as no growth for 5 days. Similar MIC values for the inhibition of growth of *Listeria innocua* were reported as 217 mM sodium lactate at pH 5.5, corresponding to about 5 mM protonated lactic acid [103], and 4.7 mM protonated lactic acid in another study [7]. Buchanan and Edelson [104] looked at the effects of a variety of organic acids on *E. coli* O157:H7 at a fixed concentration of 0.5% and pH 3.0. They examined the effects of citric, malic, lactic, and acetic acids on the viability of this organism; variables included growth phase and the presence or absence of glucose in the growth medium. The ability of the cells to survive when held in an acid solution varied in a strain-dependent manner. For nine strains, lactic acid was the most effective at reducing the viable cell population, and HCl was the least effective [104]. This study clearly demonstrated that strain-to-strain variability, as well as growth conditions (induction of acid resistance by growth in the presence of glucose), must be considered in studies of the effects of weak acids and low pH on *E. coli*.

The effect of acetate on *E. coli* O157:H7 was investigated by Diez-Gonzalez and Russell [70,105]. They investigated intracellular pH, acetate anion accumulation, glucose consumption rates, and intracellular potassium concentrations. They showed that *E. coli* O157:H7 cells could divide in the presence of about twice as much intracellular acetate anion (80 vs. 160 mM) as *E. coli* K12. In cells grown at a constant pH of 5.9, *E. coli* O157:H7 lowered its internal pH to close to 6.0 and accumulated significantly less anion when compared to *E. coli* K12, which kept a constant internal pH of 7. To test the theory that acetate acted as an uncoupler (i.e., ferrying protons across the *E. coli* cell membrane), Diez-Gonzales and Russell [105] compared the effects of acetate and the uncoupler carbonylcyanide-*m*-chlorophenylhydrazone (CCCP). They found that the effects of acetate and CCCP differed, specifically in reference to intracellular ATP concentrations of *E. coli* O157:H7. Acetate had very little or no effect on intracellular ATP, even at concentrations greater than 200 mM, while about 10 mM CCCP reduced intracellular ATP concentrations by about 50%. These and similar experiments showed that acetate was having effects other than simply acting as an uncoupler on *E. coli* O157:H7. It was also apparent from these studies that *E. coli* O175:H7 and *E. coli* K12 regulate internal pH differently.

14.5.1 SPECIFIC EFFECTS OF ACIDS

A complicating factor in the study of acid inhibition of microorganisms is that protonated acids and pH (which are interdependent variables linked by the Henderson–Hasselbalch equation for common conditions) may both independently inhibit growth [106], or they may interact. Tienungoon *et al.* [107] modeled the probability of growth of *L. monocytogenes* using a logistic regression procedure with a function relating specific growth rate to temperature, water activity, pH, lactic acid, and lactate ion concentrations. They found that their equation accurately predicted conditions allowing growth using their own laboratory data, as well as examples from the literature [107]. They presented no data, however, on the growth/no growth interface for *L. monocytogenes*, based on protonated acid and pH; they cited a lack of independent data sets available in the literature.

To address the safety concerns of the FDA and the acidified foods industry, Breidt *et al.* [79] investigated the specific effects of organic acids independent of pH. This study was made possible by using gluconic acid as a noninhibitory low pH buffer. While gluconic acid has been investigated for use as an antimicrobial agent in meats [108,109], it has not proven to be as effective as acetic or lactic acid. The antimicrobial effects of gluconic acid solutions were found to be primarily due to pH rather than to specific effects of the acid itself [79]. No change in the log reduction time (D value) was observed over a 100-fold range of gluconic acid concentrations (Figure 14.2).

By using gluconic acid as a noninhibitory buffer, the inhibitory effects of pH alone were compared with the combined effects of pH and acetic acid, while holding ionic strength, temperature, and other variables constant [79]. As expected, survival of *E. coli* O157:H7 was reduced with the addition of acetic acid at concentrations typically found in acidified foods, and with

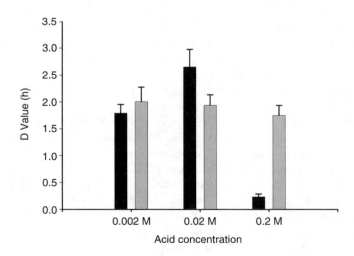

FIGURE 14.2 Effects of acetic and gluconic acid on the destruction of *Escherichia coli* O157:H7 (cocktail of strains). The log reduction times (D values) for acetic (black bars) and gluconic (gray bars) acids at 0.002, 0.02, and 0.2 M concentrations in water at 25°C and pH 3.1. The error bars indicate the upper 95% confidence intervals. No statistically significant difference was detected for the gluconic acid D values ($p > 0.05$). (From Breidt, F., Hayes, J.S., and McFeeters, R.F., *J. Food Prot.*, 67, 12–18, 2004.)

increasing temperature for a given pH and ionic strength. Gluconic acid may have wider application as a noninhibitory buffer for similar experiments with other organic acids.

In addition to the proposed mechanisms for the effects of weak acids on microorganisms mentioned above (acidification of the cytoplasm and intracellular accumulation of anion), the effectiveness of these compounds may be modulated by additional factors. Examples of these include: specific effects of the acid or acid anion on cellular enzymes or membranes, the internal buffering capacity of cells, proton pumping at the expense of cellular ATP, and facilitated transport of acid molecules, among others. To investigate the relative importance of these effects for the inhibition of yeasts with sorbic acid, Stratford and Anslow [71] compared the effects of acids with similar pK values (acetic acid, $pK = 4.76$; sorbic acid, $pK = 4.74$) and used structural analogs of sorbic acid that have similar lipophilic properties. Interestingly, a variety of structural analogs, including aldehydes and alcohols, had similar MIC values to sorbic acid (which was 3 mM) for the inhibition of a *Saccharomyces cerevisiae* strain, and a survey of yeast strains showed sorbate resistance correlated with ethanol tolerance [71]; they proposed that sorbic acid acted specifically on yeast membranes. Krebs *et al.* [69] examined glycolysis intermediates in yeast cells treated with benzoate and showed an increase in the intracellular concentrations of glucose 6-phosphate and fructose 6-phosphate, while frucotose 1,6-bisphosphate and triose phosphate concentrations were reduced. The specific inhibition of phosphofructokinase, however, could be attributed to a lack of ATP required for the function of this enzyme [69]. Alakomi *et al.* [72] showed that lactic acid had a specific membrane effect on Gram-negative bacteria. They found that lactic acid could sensitize *E. coli* O157:H7, pseudomonas, and salmonella to lytic agents such as detergents and lysozyme, presumably by disrupting the outer membrane. Lactic acid (5 mM, pH 3.5) was found to have a greater ability to liberate lipopolysaccharides from the outer membrane of *Salmonella* serovar Typhimurium than a 1 mM EDTA solution under similar conditions [72]. The effect of sorbate on the germination of *C. botulinum* spores was investigated [110]. This study indicated that sorbate inhibited spore outgrowth by disrupting the cell membrane after the start of germination. In addition to membrane effects, organic acids may have a variety of other possibly minor effects on the inhibition of microorganisms. A review by Shelef [88] cites additional effects of lactate salts on the inhibition of microorganisms. These effects include lowering water activity, chelating iron, and the inhibition of lactate dehydrogenase.

14.5.2 Genetic Regulation of Acid Resistance

Induction of acid resistance genes in *E. coli* can be accomplished by growing cells statically to stationary phase in media containing an excess of glucose, resulting in a pH of about 5.5 [104]. The acid resistance systems in *E. coli* and other pathogenic bacteria are also subject to crosstalk, or regulation by

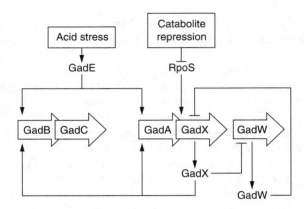

FIGURE 14.3 Regulatory network governing *gadA/BC* expression and glutamate-dependent acid resistance. (Adapted from Ma, Z., Gong, S., Richard, H., Tucker, D.L., Conway, T., and Foster, J.W., *Mol. Microbiol.*, 49, 1309–1320, 2003. With permission.)

stresses other than acid [111–113]. A clear example of this crosstalk is exhibited by *E. coli* O157:H7 in which acid resistance is induced in response to heat stress [114], and heat tolerance is induced in response to acid stress [115]. Crosstalk can be mediated by two-component (sensor–effector) regulatory systems used by bacteria, where sensor kinases phosphorylate noncognate regulatory proteins [116,117]. The precise nature of the signal(s) recognized by the cells for controlling acid resistance remains unclear, although considerable research has been carried out investigating genes induced by exposure to acid and other stresses.

Escherichia coli has several known inducible acid resistance systems that allow the organism to respond to the presence of organic acids and low pH in the environment [118,119]. The most well studied system uses decarboxylation of glutamic acid as a means for modulating internal pH [120]. The system consists of two inducible proteins, glutamate decarboxylase (GadA and an isozyme GadB), and an antiport transporter (GadC) for glutamate and the decarboxylated product of glutamate, gamma-aminobutyric acid. The genetic regulation of this system has been found to be quite complex (Figure 14.3). RpoS, a sigma factor produced in response to stress, mediates expression of two regulatory proteins, GadW and GadX, that control expression of the decarboxylase and transport proteins [121]. In addition, there is a two-component regulatory system that responds to (unidentified) external acid signals and can cause expression of the proteins of the glutamate decarboxylase system through the action of another regulatory protein, GadE [118,122]. The other acid resistance systems include arginine and lysine decarboxylase systems [119] similar to the glutamic acid system and a glucose-repressed, acid-induced system also controlled by RpoS which does not require external amino acids [83]. Inducible acid resistance mechanisms have been observed in a variety of other food pathogens, including *Salmonella* spp., *L. monocytogenes, Shigella flexneri, B. cereus*, and others [111,123–126]. As the details of gene regulation of acid resistance of microbial food pathogens become clearer, strategies may be devised to help prevent the survival of these pathogens in acidified foods.

14.6 CONCLUSIONS

Preservation of vegetables by fermentation is one of the earliest and most widespread technologies developed by humans. Fermented and acidified vegetable products are produced and consumed in every culture and society around the world, usually based on traditional processing methods. This is because the products produced are safe even in the absence of refrigerated storage, due to the inhibitory metabolites, primarily organic acids produced by lactic acid bacteria. The lactic acid bacteria may also be used to control spoilage of fresh vegetable products. The factors influencing microbial competition during fermentation or spoilage of fresh vegetable products have proved to be difficult to understand, but biocontrol strategies have the potential to ensure the safety and control the microbial ecology of food spoilage for many types of nonfermented foods. Significant challenges remain, however, in understanding the mode of action of organic acids in killing bacterial pathogens, and how those pathogens respond and adapt to acid challenge.

ACKNOWLEDGMENTS

This work was supported in part by a research grant from Pickle Packers International, Inc., St. Charles, IL. The author thanks Mr. Jim Cook of the M.A. Gedney Company, Mr. Mike Woller of Dalton's Best Maid Products, Inc., Dr. Roger McFeeters of USDA/ARS and NC State University, and Dr. Henry Fleming, formerly of USDA/ARS and Professor Emeritus NC State University, for helpful advice and contributions to this chapter, and Ms. Dora Toler for excellent secretarial assistance.

REFERENCES

1. Caplice, E. and Fitzgerald, G.F., Food fermentations: role of microorganisms in food production and preservation, *Int. J. Food Microbiol.*, 50, 131–149, 1999.
2. Lee, C.-H., *Fermentation Technology in Korea*, Korea University Press, Seoul, Korea, 2001, pp. 23–71.
3. Ross, R.P., Morgan, S., and Hill, C., Preservation and fermentation: past, present and future, *Int. J. Food Microbiol.*, 79, 3–16, 2002.
4. DeVuyst, L. and Vandamme, E.J., *Antimicrobial Potential of Lactic Acid Bacteria. Bacteriocins of Lactic acid Bacteria*, Blackie Academic and Professional, London, 1994, pp. 91–142.

5. Holzapfel, W.H., Geisen, R., and Schillinger, U., Biological preservation of foods with reference to protective cultures, bacteriocins, and food-grade enzymes, *Int. J. Food Microbiol.*, 24, 343–362, 1995.

6. Vescovo, M., Torriani, S.,Orsi, C., Macchiarlol, F., and Scolari, G., Application of antimicrobial-producing lactic acid bacteria to control pathogens in ready-to-use vegetables, *J. Appl. Bacteriol.*, 81, 113–119, 1996.

7. Breidt, F. and Fleming, H.P., Modeling the competitive growth of *Listeria monocytogenes* and *Lactococcus lactis* in vegetable broth, *Appl. Environ. Microbiol.*, 64, 3159–3165, 1998.

8. Harp, E. and Gilliland, S.E., Evaluation of a selected strain of *Lactobacillus delbrueckii* subsp. *lactis* as a biological control agent for pathogens on fresh-cut vegetables stored at 7C, *J. Food Prot.*, 66, 1013–1018, 2003.

9. Lund, B.M., Ecosystems in vegetable foods. *J. Appl. Bacteriol. Symp. Suppl.*, 73, 115S–126S, 1992.

10. Nguyen-the, C. and Carlin, F., The microbiology of minimally processed fresh fruits and vegetables, *Crit. Rev. Food Sci. Nutri.*, 34, 371–401, 1994.

11. Fleming, H.P., Etchells, J.L., and Costilow, R.N., Microbial inhibition by an isolate of *Pediococcus* from cucumber brines, *Appl. Microbiol.*, 30, 1040–1042, 1975.

12. Beuchat, L.R., Pathogenic microorganisms associated with fresh produce, *J. Food Prot.*, 59, 204–216, 1996.

13. Beuchat, L.R., Ecological factors influencing survival and growth of human pathogens on raw fruits and vegetables, *Microb. Infec.*, 4, 413–423, 2002.

14. Brackett, R.E., Incidence, contributing factors, and control of bacterial pathogens in produce, *Postharvest Biol. Technol.*, 15, 305–311, 1999.

15. Taormina, P.J. and Beuchat, L.R., Behavior of enterohemorrhagic *Escherichia coli* O157:H7 on alfalfa sprouts during the sprouting process as influenced by treatments with various chemicals, *J. Food Prot.*, 62, 850–856, 1999.

16. Kurdziel, A.S., Wilkinson, N., Langton, S., and Cook, N., Survival of poliovirus on soft fruit and salad vegetables, *J. Food Prot.*, 64, 706–709, 2001.

17. Mukherjee, A., Speh, D., Dyck, E., and Diez-Gonzalez, F., Pre-harvest evaluation of coliforms, *Escherichia coli, Salmonella,* and *Escherichia coli* O157:H7 in organic and conventional produce grown by Minnesota farmers, *J. Food Prot.*, 67, 894–900, 2004.

18. McMahon, M.A.S. and Wilson, I.G., The occurrence of enteric pathogens and *Aeromonas* species in organic vegetables, *Int. J. Food Microbiol.*, 70, 155–162, 2001.

19. Johannessen, G.S., Froseth, R.B., Solemdal, L., Jarp, J., Wasteson, Y., and Rorvik, L.M., Influence of bovine manure as fertilizer on the bacteriological quality of organic iceberg lettuce, *J. Appl. Microbiol.*, 96, 787–794, 2004.

20. Reina, L.D., Fleming, H.P., and Breidt, F., Bacterial contamination of cucumber fruit through adhesion, *J. Food Prot.*, 65, 1881–1887, 2002.

21. Ukuku, D.O. and Fett, W.F., Relationship of cell surface charge and hydrophobicity to strength of attachment of bacteria to cantaloupe rind, *J. Food Prot,.* 65, 1093–1099, 2002.

22. Kumar, C.G. and Anand, S.K., Significance of microbial biofilms in food industry: a review, *Int. J. Food Microbiol.*, 42, 9–27, 1998.

23. Bower, C.K. and Daeschel, M.A., Resistance responses of microorganisms in food environments, *Int. J. Food Microbiol.*, 50, 33–34, 1999.

24. Sapers, G.M., Efficacy of washing and sanitizing methods for disinfection of fresh fruit and vegetable products, *Food Technol. Biotechnol.*, 39, 305–311, 2001.

25. Blackman, I.C. and Frank, J.F., Growth of *Listeria monocytogenes* as a biofilm on various food-processing surfaces, *J. Food Prot.*, 59, 827–831, 1996.

26. Hood, S.K. and Zottola, E.A., Isolation and identification of adherent Gram-negative microorganisms from four meat-processing facilities, *J. Food Prot.*, 60, 1135–1138, 1997.

27. Reina, L.D., Fleming, H.P., and Humphries, E.G., Microbiological control of cucumber hydrocooling water with chlorine dioxide, *J. Food Prot.*, 58, 541–546, 1995.

28. Buchanan, R.L., Edelson, G., Miller, R.L., and Sapers, G.M., Contamination of intact apples after immersion in an aqueous environment containing *Escherichia coli* O157:H7, *J. Food Prot.*, 62, 444–450, 1999.

29. Riordan, D.C.R., Sapers, G.M., Hankinson, T.R., Magee, M., Matttrazzo, A.M., and Annous, B.A., A study of U.S. orchards to identify potential sources of *Escherichia coli* O157:H7, *J. Food Prot.*, 64, 1320–1327, 2001.

30. Daeschel, M.A. and Fleming, H.P., Entrance and growth of lactic acid bacteria in gas-exchanged, brined cucumbers, *Appl. Environ. Microbiol.*, 42, 1111–1118, 1981.

31. Seo, K.H. and Frank, J.F., Attachment of *Escherichia coli* O157:H7 to lettuce leaf surface and bacterial viability in response to chlorine treatment as demonstrated by confocal scanning laser microscopy, *J. Food Prot.*, 62, 3–9, 1999.

32. Takeuchi, K. and Frank, J.F., Penetration of *Escherichia coli* O157:H7 into lettuce tissues as affected by inoculum size and temperature and the effect of chlorine treatment on cell viability, *J. Food Prot.*, 63, 434–440, 2000.

33. Hintlian, C.B. and Hotchkiss, J.H., Comparative growth of spoilage and pathogenic organisms on modified atmosphere-packaged cooked beef, *J. Food Prot.*, 50, 218–223, 1987.

34. Hao, Y.-Y. and Brackett, R.E., Influence of modified atmosphere on growth of vegetable spoilage bacteria in media, *J. Food Prot.*, 56, 223–228, 1993.

35. Bennik, M.H., Smid, E.J., Rombouts, F.M., and Gorris, L.G.M., Growth of psychrotrophic foodborne pathogens in a solid surface model system under the influence of carbon dioxide and oxygen, *Food Microbiol.*, 12, 509–519, 1995.

36. Francis, G.A., Thomas, C., and O'Beirne, D., The microbiological safety of minimally processed vegetables, *Int. J. Food Sci. Technol.*, 34, 1–22, 1999.

37. Gilliland, S.E. and Speck, M.L., Inhibition of psychrotrophic bacteria by lactobacilli and pediococci in nonfermented refrigerated foods, *J. Food Sci.*, 40, 903–905, 1975.

38. Raccach, M., Baker, R.C., Regenstein, J.M., and Mulnix, E.J., Potential application of microbial antagonism to extended storage stability of a flesh type food, *J. Food Sci.*, 44, 43–46, 1979.

39. Breidt, F. and Fleming, H.P., Using lactic acid bacteria to improve the safety of minimally processed fruits and vegetables, *Food Technol.*, 51, 44–51, 1997.

40. Vescovo, M., Orsi, C., Scolari, G., and Torriani, S., Inhibitory effect of selected lactic acid bacteria on microflora associated with ready-to-use vegetables, *Lett. Appl. Microbiol.*, 21, 121–125, 1995.

41. Marshall, D.L. and Schmidt, R.H., Growth of *Listeria monocytogenes* at 10°C in milk pre-incubated with selected pseudomonads, *J. Food Prot.*, 51, 277–282, 1989.

42. Freedman, D.J., Kondo, J.K., and Willrett, D.L., Antagonism of food-borne bacteria by *Pseudomonas* spp.: a possible role for iron, *J. Food Prot.*, 52, 484–489, 1989.

43. Janisiewicz, W.J., Conway, W.S., and Leverentz, B., Biological control of postharvest decays of apple can prevent growth of *Escherichia coli* O157:H7 in apples, *J. Food Prot.*, 62, 1372–1375, 1999.

44. Liao, C.-H. and Fett, W.F., Analysis of native microflora and selection of strains antagonistic to human pathogens on fresh produce, *J. Food Prot.*, 64, 1110–1115, 2001.

45. Del Campo, J., Carlin, F., and Nguyen-The, C., Effects of epiphytic *Enterobacteriaceae* and pseudomonads on the growth of *Listeria monocytogenes* in a model media, *J. Food Prot.*, 64, 721–724, 2001.

46. Schuenzel, K.M. and Harrison, M.A., Microbial antagonists of foodborne pathogens on fresh minimally processed vegetables, *J. Food Prot.*, 65, 1909–1915, 2002.

47. Cheng, C.-M., Doyle, M.P., and Luchansky, J.B., Identification of *Pseudomonas fluorescens* strains isolated from raw pork and chicken that produce siderophores antagonistic towards foodborne pathogens, *J. Food Prot.*, 58, 1340–1344, 1995.

48. Simon, N., Coulanges, V., Andre, P., and Vidon, D.J.M., Utilization of exogenous siderophores and natural catechols by *Listeria monocytogenes*, *Appl. Environ. Microbiol.*, 61, 1643–1645, 1995.

49. Buchanan, R.L. and Bagi, L.K., Microbial competition: effect of *Pseudomonas fluorescens* on the growth of *Listeria monocytogenes*, *Food Microbiol.*, 16, 523–529, 1999.

50. Leistner, L. and Gorris, L.G.M., Food preservation by hurdle technology, *Trends Food Sci. Technol.*, 6, 41–46, 1995.

51. Fleming, H.P., Kyung, K.H., and Breidt, F., Vegetable fermentations, in *Biotechnology*, Rehm, H.J. and Reed, G., Eds., VCH Publishers, New York, 1995, pp. 629–661.

52. Pederson, C.S. and Albury, M.N., The Sauerkraut Fermentation, New York State Agr. Expt. Sta. Technical Bulletin 824, Geneva, New York, 1969.

53. Buckenhuskes, H.J., *Fermented Vegetables. Food Microbiology, Fundamentals and Frontiers,* Doyle, M.P., Beuchat, L.R., and Montville, T.J., Eds., American Society for Microbiology, Washington D.C., 1997, pp. 595–609.

54. McDonald, L.C., Fleming, H.P., and Hassan, H.M., Acid tolerance of *Leuconostoc mesenteroides* and *Lactobacillus plantarum*, *Appl. Environ. Microbiol.*, 56, 2120–2124, 1990.

55. Gottschalk, G., *Bacterial Metabolism,* 2nd ed., Springer-Verlag, New York, 1986, pp. 208–282.

56. McFeeters, R.F. and Chen, K.-H., Utilization of electron acceptors for anaerobic mannitol metabolism by *Lactobacillus plantarum.* Compounds which serve as electron acceptors, *Food Microbiol.*, 3, 73–81, 1986.

57. McFeeters, R.F., Fleming, H.P., and Daeschel, M.A., Malic acid degradation and brined cucumber bloating, *J. Food Sci.*, 49, 999–1002, 1984.

58. Lindgren, S.E. and Dobrogosz, W.J., Antagonistic activities of lactic acid bacteria in food and feed fermentations, *FEMS Microbiol. Rev.*, 87, 146–164, 1990.
59. Ray, B., Cells of lactic acid bacteria as food biopreservatives, in *Food Biopreservatives of Microbial Origin*, Ray, B. and Daeschel, M., Eds., CRC Press, Boca Raton, FL, 1992, pp. 81–101.
60. Vandenberg, P.A., Lactic acid bacteria, their metabolic products and interference with microbial growth, *FEMS Microbiol. Rev.*, 12, 221–238, 1993.
61. Adams, M.R. and Nicolaides, L., Review of the sensitivity of different foodborne pathogens to fermentation, *Food Control*, 8, 227–239, 1997.
62. Bell, T.A. and Etchells, J.L., Sugar and acid tolerance of spoilage yeasts from sweet-cucumber pickles, *Food Technol.*, 6, 468–472, 1952.
63. Campbell-Platt, G. and Anderson, K.G., *Pickles, Sauces and Salad Products*, Food Industries Manual, Van Nostrand-Reinhold, Ranken, MD, 1988.
64. Fleming, H.P., Thompson, R.L., and McFeeters, R.F., Firmness retention in pickled peppers as affected by calcium chloride, acetic acid, and pasteurization, *J. Food Sci.*, 58, 325–330, 356, 1993.
65. Monroe, R.J., Etchells, J.L., Pacilio, J.C., Borg, A.F., Wallace, D.H., Rogers, M.P., Turney, L.J., and Schoene, E.S., Influence of various acidities and pasteurizing temperatures on the keeping quality of fresh-pack dill pickles, *Food Technol.*, 23, 71–77, 1969.
66. CDC, Outbreak of *Escherichia coli* O157:H7 infections associated with drinking unpasteurized commercial apple juice: British Columbia, California, Colorado, and Washington, *Morb. Mortal. Weekly Rep.*, 45, 975, 1996.
67. CDC, Outbreak of *Salmonella* serotype muenchen infections associated with unpasteurized orange juice: United States and Canada, *Morbid. Mortal. Weekly Rep.*, 48, 582–585, 1999.
68. Hsin-Yi, C. and Chou, C.-C., Acid adaptation and temperature effect on the survival of *E. coli* O157:H7 in acidic fruit juice and lactic fermented milk product, *Int. J. Food Microbiol.*, 70, 189–195, 2001.
69. Krebs, H.A., Wiggins, D., and Stubbs, M., Studies on the mechanism of the antifungal action of benzoate, *Biochem. J.*, 214, 657–663, 1983.
70. Diez-Gonzalez, F. and Russell, J.B., The ability of *Escherichia coli* O157:H7 to decrease its intracellular pH and resist the toxicity of acetic acid, *Microbiology*, 143, 1175–1180, 1997.
71. Stratford, M. and Anslow, P.A., Evidence that sorbic acid does not inhibit yeast as a classic "weak acid preservative", *Lett. Appl. Microbiol.*, 27, 203–206, 1998.
72. Alakomi, H. L., Dkytta, E., Saarela, M., Mattila-Sandholm, T., Latva-Kala, K., and Helander, I.M., Lactic acid permeabilizes Gram-negative bacteria by disrupting the outer membrane, *Appl. Environ. Microbiol.*, 66, 2001–2005, 2000.
73. Ito, K.A., Chen, J.K., Lerke, P.A., Seeger, M.L., Unverferth, J.A., Effect of acid and salt concentration on the growth of *Clostridium botulinum* spores, *Appl. Environ. Microbiol.*, 32, 121–124, 1976.
74. Stopforth, J.D., Samelis, J., Sofos, J.N., Kendall, P.A., and Smith, G.C., Influence of organic acid concentration on survival of *Listeria monocytogenes*

and *Escherichia coli* O157:H7 in beef carcass wash water and on model equipment surfaces, *Food Microbiol.*, 20, 651–660, 2003.

75. Duffy, S. and Schaffner, D.W., Modeling the survival of *Escherichia coli* O157:H7 in apple cider using probability distribution functions for quantitative risk assessment, *J. Food Prot.*, 64, 599–605, 2001.

76. Tsai, Y.-W. and Ingham, S.C., Survival of *Escherichia coli* O157:H7 and *Salmonella* spp. in acidic condiments, *J. Food Prot.*, 60, 751–755, 1997.

77. Brudzinski, L. and Harrison, M.A., Influence of incubation conditions on survival and acid tolerance response of *Escherichia coli* O157:H7 and non-O157:H7 isolates exposed to acetic acid, *J. Food Prot.*, 61, 542–546, 1998.

78. McKellar, R.C. and Lu, X., A probability of growth model for *Escherichia coli* O157:H7 as a function of temperature, pH, acetic acid, and salt, *J. Food Prot.*, 64, 1922–1928, 2001.

79. Breidt, F., Hayes, J.S., and McFeeters, R.F., The independent effects of acetic acid and pH on the survival of *Escherichia coli* O157:H7 in simulated acidified pickle products, *J. Food Prot.*, 67, 12–18, 2004.

80. Doyle, M.P., *Escherichia coli* O157:H7 and its significance in foods, *Int. J. Food Microbiol.*, 12, 289–302, 1991.

81. Griffin, P.M. and Tauxe, R.V., The epidemiology of infections caused by *Escherichia coli* O157:H7, other enterohemorrhagic *E. coli* and the associated hemolytic uremic syndrome, *Epidemiol. Rev.* 13, 60–98, 1991.

82. O'Driscoll, B., Gahan, C.G.M., and Hill, C., Adaptive acid tolerance response in *Listeria monocytogenes*: isolation of an acid-tolerant mutant which demonstrates increased virulence, *Appl. Environ. Microbiol.*, 62, 1693–1698, 1996.

83. Castanie-Cornet, M.-P., Penfound, T.A., Smith, D., Elliptt, J.F., and Foster, J.W., Control of acid resistance in *Escherichia coli*, *J. Bacteriol.*, 181, 3525–3535, 1999.

84. Smittle, R.B., Microbiological safety of mayonnaise, salad dressings, and sauces produced in the United States: a review, *J. Food Prot.*, 63, 1144–1153, 2000.

85. Tetteh, G.L. and Beuchat, L.R., Sensitivity of acid-adapted and acid-shocked *Shigella flexneri* to reduced pH achieved with acetic, lactic, and propionic acids, *J. Food Prot.*, 64, 975–981, 2001.

86. Mazzotta, A.S., Thermal inactivation of stationary-phase and acid-adapted *Escherichia coli* O157:H7, *Salmonella*, and *Listeria monocytogenes* in fruit juices, *J. Food Prot.*, 64, 315–320, 2001.

87. Sofos, J.N., Current microbiological considerations in food preservation, *Int. J. Food Microbiol.*, 19, 87–108, 1993.

88. Shelef, L.A., Antimicrobial effects of lactates: a review, *J. Food Prot.*, 57, 445–450, 1994.

89. Brul, S. and Coote, P., Preservative agents in foods. Mode of action and microbial resistance mechanisms, *Int. J. Food Microbiol.*, 50, 1–17, 1999.

90. Miller, L.G. and Kaspar, C.W., *Escherichia coli* O157:H7 acid tolerance and survival in apple cider, *J. Food Prot.*, 57, 460–464, 1994.

91. Silk, T.M. and Donnelly, C.W., Increased detection of acid-injured *Escherichia coli* O157:H7 in autoclaved apple cider by using nonselective repair on trypticase soy agar, *J. Food Prot.*, 60, 1483–1486, 1997.

92. Uljas, H.E. and Ingham, S.C., Survival of *Escherichia coli* O157:H7 in synthetic gastric fluid after cold and acid habituation in apple juice or trypticase soy broth

acidified with hydrochloric acid or organic acids, *J. Food Prot.*, 61, 939–947, 1998.

93. Roering, A.M., Luchansky, J.B., Ihnot, A.M., Ansay, S.E., Kaspar, C.W., and Ingham, S.C., Comparative survival of *Salmonella typhimurium* DT 104, *Listeria monocytogenes*, and *Escherichia coli* O157:H7 in preservative-free apple cider and simulated gastric fluid, *Kint. J. Food Microbiol.*, 46, 263–269, 1999.

94. Raghubeer, E.V., Ke, J.S., Campbell, M.L., and Meyer, R.S., Fate of *Escherichia coli* O157:H7 and other coliforms in commercial mayonnaise and refrigerated salad dressings, *J. Food Prot.*, 58, 13–18, 1995.

95. Mayerhauser, C.M., Survival of enterohemorrhagic *Escherichia coli* O157:H7 in retail mustard, *J. Food Prot.*, 64, 783–787, 2001.

96. Glass, K.A., Loeffelholz, J.M., Ford, J.P., and Doyle, M.P., Fate of *Escherichia coli* O157:H7 as affected by pH or sodium chloride and in fermented, dry sausage, *Appl. Environ. Microbiol.*, 58, 2513–2516, 1992.

97. Faith, N.G., Le Countour, N.S., Alvarenga, M.B., Calicioglu, M., Buege, D.R., and Luchansky, J.B., Viability of *Escherichia coli* O157:H7 in ground and formed beef jerky prepared at levels of 5 and 20% fat and dried at 52, 57, 63, and 68 degrees C in a home-style dehydrator, *Int. J. Food Microbiol.*, 41, 213–221, 1998.

98. Pond, T.J., Wood, D.S., Mumin, I.M., Barbut, S., and Griffiths, M.W., Modeling the survival of *Escherichia coli* O157:H7 in uncooked, semidry, fermented sausage, *J. Food Prot.*, 64, 759–766, 2001.

99. Russell, J.B., Another explanation of the toxicity of fermentation acids at low pH: anion accumulation versus uncoupling, *J. Appl. Bacteriol.*, 73, 363–370, 1992.

100. Goncalves, L.M.D., Ramos, A., Almeida, J.S., Xavier, A.M.R.B., and Carrondo, M.J.T., Elucidation of the mechanism of lactic acid growth inhibition and production in batch cultures of *Lactobacillus rhamnosus, Appl. Microbiol. Biochem.*, 48, 346–350, 1997.

101. Young, K.M. and Foegeding, P.M., Acetic, lactic and citric acids and pH inhibition of *Listeria monocytogenes* Scott A and the effect on intracellular pH, *J. Appl. Bacteriol.*, 74, 515–520, 1993.

102. Ostling, C.E. and Lindgren, S.E., Inhibition of enterobacteria and *Listeria* growth by lactic, acetic, and formic acids, *J. Appl. Bacteriol.*, 75, 18–24, 1993.

103. Houtsma, P.C., Kusters, B.J.M., de Wit, J.C., Rombouts, F.M., and Zwietering, M.H., Modelling growth rates of *Listeria innocua* as a function of lactate concentration, *Int. J. Food Microbiol.*, 24, 113–123, 1994.

104. Buchanan, R.L. and Edelson, S.G., pH-dependent stationary-phase acid resistance response of enterohemorrhagic *Escherichia coli* in the presence of various acidulants, *J. Food Prot.*, 62, 211–218, 1999.

105. Diez-Gonzalez, F. and Russell, J.B., Effects of carbonylcyanide-*m*-chlorophenylhydrazone (CCCP) and acetate on *Escherichia coli* O157:H7 and K-12: uncoupling versus anion accumulation, *FEMS Microbiol. Lett.*, 151, 71–76, 1997.

106. Passos, F.V., Fleming, H.P., Ollis, D.F., Hassan, H.M., and Felder, R.M., Modeling the specific growth rate of *Lactobacillus plantarum* in cucumber extract, *Appl. Microbiol. Biotechnol.*, 40, 143–150, 1993.

107. Tienungoon, S., Ratkowsky, D.A., McMeekin, T.A., and Ross, T., Growth limits of *Listeria monocytogenes* as a function of temperature, pH, NaCl, and lactic acid, *Appl. Environ. Microbiol.*, 66, 4979–4987, 2000.

108. Garcia Zepeda, C.M., Kastner, C.L., Willard, B.L., Phebus, R.K., Schwenke, J.R., Fijal, B.A., and Prasai, R.K., Gluconic acid as a fresh beef decontaminant, *J. Food Prot.*, 57, 956–962, 1994.

109. Stivarius, M.R., Pohlman, F.W., McElyea, K.S., and Apple, J.K., The effects of acetic acid, gluconic acid and trisodium citrate treatment of beef trimmings on microbial, color and odor characteristics of ground beef through simulated retail display, *Meat Sci.*, 60, 245–252, 2002.

110. Blocher, J.C. and Busta, F.F., Multiple modes of inhibition of spore germination and outgrowth by reduced pH and sorbate, *J. Appl. Bacteriol.*, 59, 469–478, 1985.

111. Abee, T. and Wouters, J.A., Microbial stress response in minimal processing, *Int. J. Food Microbiol.*, 50, 65–91, 1999.

112. Rowbury, R.J., Cross-talk involving extracellular sensors and extracellular alarmones gives early warning to unstressed *Escherichia coli* of impending lethal chemical stress and leads to induction of tolerance responses, *J. Appl. Microbiol.*, 90, 677–696, 2001.

113. Zook, C.D., Busta, F.F., and Brady, L.J., Sublethal sanitizer stress and adaptive response of *Escherichia coli* O157:H7, *J. Food Prot.*, 64, 767–769, 2001.

114. Wang, G. and Doyle, M.P., Heat shock response enhances acid tolerance of *Escherichia coli* O157:H7, *Lett. Appl. Microbiol.*, 26, 31–34, 1998.

115. Ryu, J.-H. and Beuchat, L.R., Changes in heat tolerance of *Escherichia coli* O157:H7 after exposure to acidic environments, *Food Microbiol.*, 16, 447–458, 1999.

116. Verhamme, D.T., Arents, J.C., Postma, P.W., Crielaard, W., and Hellingwerf, K.J., Investigation of *in vivo* cross-talk between key two-component systems of *Escherichia coli, Microbiology*, 148, 69–78, 2002.

117. Alves, R. and Savageau, M.A., Comparative analysis of prototype two-component systems with either bifunctional or monofunctional sensors: differences in molecular structure and physiological function, *Mol. Microbiol.*, 48, 25–51, 2003.

118. Ma, Z., Gong, S., Richard, H., Tucker, D.L., Conway, T., and Foster, J.W., GadE (YhiE) activates glutamate decarboxylase-dependent acid resistance in *Escherichia coli* K-12, *Mol. Microbiol.*, 49, 1309–1320, 2003.

119. Diaz-Gonzalez, R. and Karaibrahimoglu, Y. Comparison of the glutamate-, arginine- and lysine-dependent acid resistance systems in *Escherichia coli* O157:H7, *J. Appl. Microbiol.*, 96, 1237–1244, 2004.

120. Smith, D.K., Kassam, T., Singh, B., and Elliott, J.F., *Escherichia coli* has two homologus glutamate decarboxylase genes that map to distinct loci, *J. Bacteriol.*, 174, 5820–5826, 1992.

121. Ma, Z.R.H., Tucker, D.L., Conway, T., and Foster, J.W., Collaborative regulation of *Escherichia coli* glutamate-dependent acid resistance by two AraC-like regulators, GadX and GadW (YhiW), *J. Bacteriol.*, 184, 7001–7012, 2002.

122. Masuda, N. and Church, G.M., Regulatory network of acid resistance genes in *Escherichia coli, Mol. Microbiol.*, 48, 699–712, 2003.

123. Foster, J.W. and Spector, M.P., How *Salmonella* survive against the odds, *Ann. Rev. Microbiol.*, 49, 145–174, 1995.

124. Browne, N. and Dowds, B.C.A., Acid stress in the food pathogen *Bacillus cereus, J. Appl. Microbiol.*, 92, 404–414, 2002.

125. Koutsoumanis, K.P., Kendall, P.A., and Sofos, J.N., Effect of food processing-related stresses on acid tolerance of *Listeria monocytogenes*, *Appl. Environ. Microbiol.*, 69, 7514–7516, 2003.
126. Tetteh, G.L. and Beuchat, L.R., Survival, growth and inactivation of acid-stressed *Shigella flexneri* as affected by pH and temperature, *Int. J. Food Microbiol.*, 87, 131–138, 2003.

Section IV

Interventions to Reduce Spoilage and Risk of Foodborne Illness

15 HACCP: A Process Control Approach for Fruit and Vegetable Safety

William C. Hurst

CONTENTS

15.1   INTRODUCTION

Hazard Analysis Critical Control Point (HACCP) has been called a logical, cost-effective, systematic approach toward food safety management. It involves the identification, evaluation, and control of potential hazards that

339

might contaminate fruits and vegetables during handling between the farm and consumer. The basic objective of HACCP is to ensure consistent and safe food production. By identifying in advance potential problems in a fruit or vegetable operation and establishing control measures at those stages critical for food safety, potential microbiological, chemical, or physical hazards can be reduced, prevented, or eliminated. HACCP is based on two important concepts in safe food production: prevention and documentation [1]. HACCP is a tool that can determine how and where safety hazards may exist in a food operation and how to prevent their occurrence. Once located, these hazards may then be confirmed and controlled through documentation (record-keeping) procedures.

During HACCP implementation and maintenance, an understanding of the production system in relation to process control is of paramount importance. There are many sources of variation that must be recognized and reduced to achieve consistent control over product quality and safety. Properly defining and documenting a process is an important first step toward understanding and gaining control over the process. Summers has identified a process as simply taking inputs and performing value-added activities on those inputs to create an output [2]. Take, for example, the fresh-cut (minimally processed) produce process for making a salad product. Inputs such as raw materials (lettuce, red cabbage, carrots), people (plant workforce), machines (processing equipment), and methods (machine settings, quality control, etc.) perform value-added activities (grading, cutting, washing, drying, and packaging) to transform whole produce into a 3/8 inch chopped salad product contained in a 14-ounce bag. Complementary to this production process is the need to ensure consistent product safety. Thus HACCP serves as that portion of the establishment's overall process control system that focuses on safety [3].

The need and value of process control technology and its integration into the HACCP process is addressed in this chapter. Although HACCP is often referred to as a preventive system, from a statistical standpoint HACCP would be more appropriately described as a means of minimizing the variability of safety parameters in a processing system [3]. However, in order to achieve consistency of operation, process control techniques must be coupled with HACCP principles to keep those parameters being monitored under control and within safety limits. This chapter also focuses on how a statistical approach to safety, such as statistical process control (SPC), can be effectively applied to the HACCP process.

15.2 WHAT IS HACCP?

HACCP is a science-based, objective, and proactive method of assuring food safety by focusing on hazard identification and control at its source. Originally developed in the early 1960s by the Pillsbury Company, the U.S. Army Natick Laboratories, and the National Aeronautics and Space Administration (NASA) to develop safe food for astronauts, HACCP is derived from failure mode and effect analysis (FMEA). This is an engineering system that looks at a product, all of its components and manufacturing stages and asks what can go wrong within the total system [4].

HACCP-like controls for thermally canned foods were first mandated by the U.S. Food and Drug Administration (FDA) in 1973, followed by regulations concerning acidified foods (marinated vegetables) that likewise mandated HACCP tools [5,6]. The National Advisory Committee on Microbiological Criteria (NACMCF) prepared the first official HACCP document in 1989, which was subsequently issued by the FDA as a proposed regulation for recommending HACCP as the food safety program of choice for the entire U.S. food industry [7,8]. In 2001 the FDA issued its mandatory juice HACCP regulations that apply to the production of all fresh fruit and vegetable juices [9].

Although not yet mandated by the FDA, HACCP for the fresh-cut produce industry presents some unique challenges because there is no definitive kill step for pathogens (e.g., retorting, pasteurization, acidification) in the processing operation. Instead, HACCP must incorporate a series of intervention steps or hurdles, such as using antimicrobial agents in flume wash water, applying modified atmosphere packaging techniques, and following consistent good sanitation and low-temperature management practices to retard pathogen growth [10–12]. Today, HACCP implementation is recommended or required by FDA/U.S. Department of Agriculture (USDA) throughout the food industry in the U.S. [13]. HACCP has achieved international acceptance with the recognition by the World Health Organization (WHO) as the most effective means of controlling foodborne disease [14].

Traditionally, the fruit and vegetable industry has used the concept of end-line product testing to attempt to ensure a high level of safety and quality of products. Today the defect detection approach has given way to defect prevention whereby a process is monitored as a product is being manufactured to determine when adjustments are required to maintain stability and where a change is needed to reduce inherent variability [15]. This prevention approach is in harmony with HACCP, which also focuses on prevention, whereby potential hazards are identified and controlled within the food processing environment to prevent unsafe products from being made [16].

While prevention is obviously superior to detection, it is not always easy to accomplish. This is because all processes are inherently unpredictable in nature and gravitate toward the natural force of entropy or disorder [17]. The only way to overcome the effects of entropy is to utilize statistical tools to find and eliminate the causes of disorder in the production system so the process can be brought back under control.

The National Academy of Sciences (NAS) drafted a report in 2003 that included science-based tools for food safety regulatory use in ensuring safe food production for the American consumer. It recommended the establishment of performance standards that would guide fruit and vegetable producers and processors to an appropriate means of ensuring food safety when operating under a HACCP system [18]. The committee also recommended the use of SPC as the most appropriate scientific tool for monitoring performance standards in fruit and vegetable operations and for developing the science-based food safety criteria that processors will need to verify and validate their HACCP programs.

15.3 APPLYING THE HACCP CONCEPT

The HACCP concept is implemented through a logical sequence of activities, known as the HACCP study, that utilizes the seven principles of HACCP (Table 15.1) to produce a HACCP plan.

By definition, a HACCP plan is a document prepared in accordance with the principles of HACCP to ensure control of hazards which are significant for food safety in the segment of the food chain under consideration [19]. A HACCP system is a system that identifies, evaluates, and controls hazards that are significant for food safety. Thus, the HACCP study should yield a HACCP plan which is then implemented as the HACCP system [20].

Design, implementation, and maintenance of a HACCP plan are not an easy job. When Pillsbury first decided to implement HACCP, the CEO publicly stated that all raises, promotions, and evaluations would be based on developing and implementing HACCP to ensure safe food production [21]. Now, that was a strong statement of support! Employees ultimately determine the success or failure of HACCP. Therefore, training programs are essential to develop a positive attitude about food safety and to help empower personnel to maintain the HACCP program. Implementing HACCP takes time;

TABLE 15.1
Seven Principles of HACCP

Principle 1	Conduct a *hazard analysis*. Construct a flow diagram of the steps of a process to determine where significant hazards exist and what control measures should be instituted
Principle 2	Determine *critical control points* (CCPs) required to control the identified hazards. CCPs are any steps where hazards can be prevented, eliminated or reduced to acceptable levels
Principle 3	Establish *critical limits* (CLs). These are specifications (target values and tolerances) that must be met to ensure that CCPs are under control
Principle 4	Establish procedures to *monitor* CCPs. These are used to assess when a process must be adjusted to maintain CCP control
Principle 5	Establish *corrective actions* to be taken when monitoring indicates that a particular CCP is not under control
Principle 6	Establish *verification* procedures for determining whether the HACCP program is working correctly
Principle 7	Establish *documentation* procedures concerning all activities with records appropriate to these principles and their application

From USDA-FSIS, Hazard Analysis and Critical Control Point Principles and Application Guidelines, 1997 (http://www.fsis.usda.gov/OPHS/nacmcf/past/JFP0998.pdf).

experience has shown that installation and implementation can take from six months to two years.

Today, there are numerous manufacturers promoting computer software programs, manuals, and services that can aid in the development of generic HACCP systems. While these models are useful tools to demonstrate how to create a HACCP plan, it must be emphasized that they should not be used "out of the box," as the generic HACCP plan may not be applicable for every facility, processing line, or product. Since each HACCP plan is process-specific, the plan must be tailored to address the unique aspects of the production, including processing and preparation operations, equipment being used, the foods being prepared, and the training of personnel [22].

15.4 PREREQUISITES FOR HACCP

It has been pointed out by Sperber *et al.* that HACCP cannot exist as a stand-alone food safety program [23]. Rather, it must be supported by a strong foundation of prerequisite programs. While not a formal part of HACCP, the prerequisite programs are written, implemented procedures that address operational conditions and provide the documentation to help an operation run more smoothly to maintain comprehensive safety assurance. Standard operating procedures (SOPs), good agricultural practices (GAPs), good manufacturing practices (GMPs), and sanitation standard operating procedures (SSOPs) are the basis for establishing a comprehensive sanitation program in a fruit and vegetable processing plant or packinghouse facility.

SOPs are step-by-step instructions that outline how an operation is to be carried out in such a way that ensures that all processing steps critical for product safety are accomplished in an orderly fashion. SOPs provide the detailed framework and safety continuum between agricultural production and commercial processing of fruit and vegetable products. GAPs are a collection of HACCP-like principles that have been extended to the on-farm production and postharvest handling activities in order to minimize microbial contamination of fresh fruits and vegetables [24]. They ensure that the fresh produce industry, for which the traditional HACCP methods do not necessarily fit, will have a systematic and proactive method to reduce potential product contamination and thereby ensure safety [25]. GMPs, sanctioned by the FDA, are the minimal sanitary requirements that must be met by a food packer or processor to ensure safe and wholesome food for interstate commerce [26]. GMPs are broadly written, based on federal regulations, general in nature, are not intended to be plant-specific, do not establish deviation limits, and do not describe corrective action requirements [27]. SSOPs focus more narrowly on specific procedures that allow a fruit or vegetable processing plant to achieve sanitary process control in its daily operation. SSOPs can be categorized into two types: preoperational and operational SSOPs. Preoperational SSOPs are the sanitary procedures carried out prior to the start of production each day. Operational SSOPs refer to sanitary actions taken during production to prevent product contamination or adulteration [19]. More extensive information on these programs as applied to fresh produce have been reviewed by Gorny and Hurst [27,28].

15.5 PLANNING AND CONDUCTING AN HACCP STUDY

Before the actual HACCP plan design begins, several preliminary tasks must be accomplished: (1) identifying the scope of the study, (2) assembling the HACCP team and providing adequate training, (3) describing the product for which the HACCP plan is being developed and its intended use, and (4) developing and verifying a flow chart of the operation required to produce the product. It is important to remember that a separate HACCP plan must be developed for each product or processing line in the plant operation. This means that each HACCP plan will be plant-specific and must be uniquely tailored for each packer or processor.

Task 1: setting limits for the HACCP study. First, the terms of reference for the HACCP study should be established at the outset to define where the HACCP plan begins and ends. The model for discussion in this chapter considers only the preparation of shredded iceberg lettuce at a fresh-cut processing plant. Where complex production operations of a product are concerned, it can be simpler (and safer) to break down the handling chain into smaller segments (e.g., lettuce production, distribution to the consumer) and link the operations together later to form the overall HACCP system [20].

Task 2: the HACCP team. A HACCP coordinator with good communication skills and knowledge of HACCP techniques should be appointed to lead, coordinate, and build the HACCP team. Selection of team members should be based on their working knowledge of the entire process and their ability to contribute unique aspects of the operation toward ensuring the safety of the product. Team members should be multidisciplinary and multifunctional. The HACCP team may include a microbiologist, quality assurance and sanitation personnel, production operations, engineering and maintenance, purchasing or procurement, marketing and sales, and on-line personnel [4].

Task 3: product description. The HACCP team must first develop a complete description of the product under study. Information on key parameters that determine safety must be included (e.g., specific refrigeration temperature for fresh-cut produce, pasteurization temperature for processed juices, pH value for acidified vegetables). A complete description detailing its form, size, packaging and storage requirements, shelf life, instructions for use, and intended consumer must be defined, as demonstrated for fresh-cut lettuce in Table 15.2.

Task 4: flow chart of process. A flow chart can be a picture worth a thousand words when used to describe in clear and simple terms the steps

TABLE 15.2
Product Description, Use, and Distribution

Common name	Shredded lettuce; prepared from refrigerated iceberg lettuce; trimmed, cored, and cut; washed in a solution of potable water and chlorine
Type of package	Packed in food-grade plastic bags, 8 oz to 10 lb units
Length of shelf life, at what temperature?	Optimum shelf life of 14 days if refrigerated at 34 to 38°F (1.1 to 3.3°C)
Labeling instructions	Bag and/or box contains "processed on" or "use by" date
Where will it be sold?	Foodservice operations and retail markets
Intended use and consumer	For use in salads and sandwiches by foodservice customers; prepackaged units for in-house use by consumer
Special distribution control	Product distributed under refrigeration, stored in refrigeration at 34 to 38°F (1.1 to 3.3°C)

From IFPA Technical Committee, *HACCP for the Fresh-Cut Produce Industry*, IFPA, Alexandria, VA, 2000. With permission.

involved in a process. A block-type flow diagram is typically used for simplicity of understanding by the HACCP team. Its purpose is to facilitate hazard analysis and assist in the identification of critical control points (CCPs). Also, the flow diagram serves as a record of the operation and a future guide for employees, regulators, and customers who must understand the process [1].

Once the flow diagram has been completed, it needs to be verified for accuracy and completeness. Ideally, confirmation should be made by having HACCP team members "walk the process" whereby the flow diagram is compared with what actually happens, as it happens [20]. This also should include confirmation of activities during the night shift or weekend running of the operation.

15.6 CONDUCTING A HAZARD ANALYSIS/RISK ASSESSMENT STUDY (HACCP PRINCIPLE 1)

Using the flow diagram and description of the product as a guide, the HACCP team must conduct a hazard analysis of the process as the first step in the formal HACCP plan design. By definition, a hazard is a biological, chemical, or physical agent in, or condition of, food with the potential to cause an adverse health effect [19]. The purpose of hazard analysis is to develop a list of hazards at each step of the process that are of such significance that they are reasonably likely to cause illness or injury if not effectively controlled. For fruit and vegetable component operations to which HACCP may be applied, these hazards may be introduced as inputs during the crop production process, during the plant process, or as outputs from the process and the final product. Consideration of the myriad of conceivable hazards that might contaminate fresh and processed fruit and vegetable products is not within the scope of this chapter. However, extensive listings of hazards have been reviewed [19,20,29]. Based on epidemiological data and industry experience, the primary safety issue in fresh-cut processing is microbiological contamination.

Hazard analysis consists of a two-part process: hazard identification and hazard evaluation. During hazard identification, the HACCP team generates a list of all potential hazards associated with each step in the process, by following the flow diagram. Brainstorming and Pareto diagrams are two problem-solving tools that can aid in this process. Brainstorming stimulates creative, exhaustive thinking that can help team members by identifying every conceivable hazard so that none are missed. Pareto diagrams allow the team to prioritize identified hazards based on their relative importance to safety [30].

Step two of hazard analysis is hazard evaluation. Each identified hazard must be carefully considered based on its severity in the extent of exposure and likely occurrence (risk) in a product. For example, *Listeria monocytogenes* and *Clostridium botulinum* are both potential microbiological hazards in the fresh-cut produce industry. However, *L. monocytogenes*, due to its ubiquity, is a greater health risk to consumers of fresh-cut produce, even though *C. botulinum* would be considered a more severe hazard based on mortality rates [27]. Potential listeria contamination within the processing and/or storage environment of fruit and vegetable facilities is a constant threat to products since there is a zero tolerance regulation for this hazard in ready-to-eat (RTE) foods.

Another problem-solving tool, the cause-and-effect diagram, can be utilized to find ways to minimize environmental hazards such as listeria. Constructing a cause-and-effect diagram of a process requires the HACCP team to itemize all possible locations where a problem or hazard could occur, to look for all causes of a problem, and to collect data to evaluate the possible risk. A cause-and-effect diagram highlighting the processing environment as the possible source of listeria contamination to fruit and vegetable products would show the effect or problem (e.g., listeria contamination in a fresh-cut plant) at the end of a horizontal arrow or spine. Primary causes (e.g., poor sanitation in the processing room, produce cooler) are represented by oblique arrows entering the spine. Secondary causes (e.g., condensate on overhead pipes, electrical conduits, floors, in drains, and on refrigeration units) are represented by perpendicular arrows off the primary cause arrow [27].

An important component of hazard evaluation is risk assessment, which is defined as a scientifically based process of four activities: (1) hazard identification, (2) exposure assessment, (3) dose–response assessment, and (4) risk characterization. A comprehensive discussion of risk assessment with relationship to HACCP cannot be covered in this chapter, but excellent resources to consult include Tapia *et al.* [31] and Forsythe [32].

Once all potential hazards have been identified and their risks to the final consumer of the product evaluated, the HACCP team must consider what preventive measures are to be used to control the hazard. Preventive measures are those actions or activities that are required to control hazards, eliminate hazards, or reduce their effect to an acceptable level. More than one control may be required for a specific hazard that occurs at different parts of the production process. For example, if the hazard is *Listeria monocytogenes* on raw fruit or vegetables, which are consistently heat-treated in the in the manufacture of a processed juice, then pasteurization could be the appropriate control measure for this hazard. If the same hazard arises from environmental contamination during ingredient assembly of a refrigerated product which is given no heat treatment, such as fresh-cut produce, other control measures (environmental sanitation, personnel sanitation, refrigeration management) would be required.

It cannot be overemphasized that a thorough hazard analysis is essential to the design of an effective HACCP plan. If this is not done correctly, and the significant hazards requiring control within the plan are not properly identified and evaluated, the HACCP program will not be effective, regardless of how well the plan is followed [33].

15.7 USING SPC TO ENSURE HACCP CONTROL

SPC emphasizes building product safety and quality into the manufacturing process by focusing on the process rather than the product. SPC tools are used to measure and interpret different kinds of variation that affect the behavior of a process. An important part of dealing with variation in a process control system is to have knowledge of the extent of variability. The critical point in process control is not to eliminate all variation, but to have any variation that is present in the process to be stable and predictable.

The key to controlling safety and quality in manufacturing is to understand how to recognize the different types of variation present in a process. Common cause variation is the inherent, random chance events expected in every process which fluctuate in a normal, predictable manner. Variation is small in magnitude and therefore difficult to locate and eliminate from a process. Special cause variation is sporadically induced variation that impacts a process, causing large fluctuations that are easily discernable and therefore can be effectively eliminated from a process [34] (see Figure 15.1). The basic objective of SPC is to use valid statistical methods to identify the existence of special causes of variation and to eliminate them from a process. This will produce a stable, constant-cause system which can be measured and controlled [35].

Although SPC is the most effective tool to achieve process control in an operation, most processes do not naturally operate in a state of control. They tend to deteriorate over time. Process control is defined as the functioning of an operation within predetermined statistical limits, such that only common cause (controlled) variation is occurring among its manufactured products [36]. While a process may be in control initially, it will not remain there. The only way to determine whether a process is in or out of control is

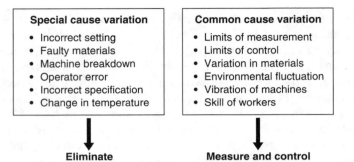

Special cause variation	Common cause variation
• Incorrect setting	• Limits of measurement
• Faulty materials	• Limits of control
• Machine breakdown	• Variation in materials
• Operator error	• Environmental fluctuation
• Incorrect specification	• Vibration of machines
• Change in temperature	• Skill of workers
Eliminate	Measure and control

FIGURE 15.1 Distinguishing between special and common cause variation. (Adapted from Cullen, J. and Hollingum, J., *Implementing Total Quality*, IFS Publications, Bedford, U.K., 1987.)

to interpret the data gathered from the process with the appropriate statistical tool. That tool is the *statistical control chart.*

Control chart methods provide an objective and statistically valid means to assess the nature of ongoing processes, and, as such, are particularly applicable to HACCP monitoring [3]. Control chart theory is based upon the notion that the parameter being measured, when in statistical control, will vary normally (e.g., only common cause variation) about a central value [37]. Control chart methodology is the only SPC tool that can distinguish between common cause (inherent) and special cause (unnatural) variation in a process. The control chart allows the highlighting of special cause variation, if present, when monitoring a process. If the special cause variation source can be found and eliminated in the process, then the process will exhibit only common cause variation. Only when common cause variation is the only source of variation present is the process in a state of statistical control. What makes statistical control so important? The essence of statistical control is predictability. A process is predictable when it is in a state of statistical control, and it is unpredictable when it is not in a state of statistical control [38].

So how can HACCP be made a more effective prediction tool for safety hazards in the production process? The key lies in integrating SPC methodology into the HACCP plan. Achieving process control in a HACCP monitoring system will assist the processor in systematically and predictably demonstrating control of identified safety parameters [39]. It will also provide warning signs signifying out-of-control status so that corrective action can be taken to return the system to the established safety limits [27].

15.8 IDENTIFYING AND STABILIZING VARIABILITY AT CCPs (HACCP PRINCIPLE 2)

Once all significant hazards have been identified through hazard analysis, they must be controlled at some point in the process. Determining the identity and location of CCPs is the second principle in HACCP design and is really the heart of the HACCP plan. A CCP is a step at which control can be applied and is essential to prevent or eliminate a food safety hazard, or to reduce it to an acceptable level [7]. For every significant hazard identified, one or more CCPs must be designated to control or eliminate the hazard. For example, the thermal process given to canned vegetables at the retort step would be designated as a CCP in the low-acid foods industry, while proper acidification or pH adjustment in the brine kettle step would be the CCP in the acidified foods industry. Sometimes it is not possible to eliminate a potential microbiological hazard, only to minimize it to an acceptable level. In the fresh-cut produce industry, proper water chlorination at levels ≥ 1 ppm free available chlorine in flumes and dip tanks becomes the CCP [27].

Unfortunately, there is no simple, clear-cut answer to the question of how many CCPs a HACCP plan may need and where should they be located. It depends on plant layout and design, the product being produced, the ingredients used, equipment age and condition, processing methods employed, and, especially, the effectiveness of the prerequisite programs implemented. Often an SSOP or SOP can be incorporated to control a hazard rather than a CCP. To keep HACCP programs plant-friendly and sustainable, Bernard recommends that the number of CCPs be kept to a minimum, and none should be redundant [33]. Redundancy will also add to the cost of record keeping. Experience has shown that HACCP plans that are unnecessarily cumbersome will likely be the ones that fail.

However, many points in a flow diagram not identified as CCPs may be considered control points. A control point (CP) is any step at which biological, physical, or chemical factors can be controlled [7]. Many types of control points can exist in fruit and vegetable operations, including those that address quality control (color, flavor, texture), sanitary control (SSOPs, GMPs), maintenance (calibration of equipment), and process control (fill weights, seal closures).

Pinpointing the right CCPs is the most crucial and problematic aspect of an effective HACCP program [40]. Therefore, a common strategy to facilitate the proper identification of CCPs is to use the CCP decision tree (Figure 15.2) [7,41]. The decision tree consists of four questions that are asked for each process step for which hazards have been identified during hazard analysis. The answer to each question will direct the process of elimination and ultimately lead to a decision as to whether a CCP or CP is required at that step. A benefit of the CCP decision tree is that it forces and facilitates HACCP team discussion and teamwork and ensures a consistent approach to every hazard at each step [42]. However, as pointed out by Wedding, this is not a perfect tool and is not a substitute for common sense and process knowledge, because complete reliance on the decision tree may lead to false conclusions [41].

To determine the kind of variability that might exist at a specific CCP, data of the parameter used to maintain control at this location must be collected and statistically analyzed. For example, the conditions that influence variability at a CCP, like pasteurization temperature for fresh juices, must be

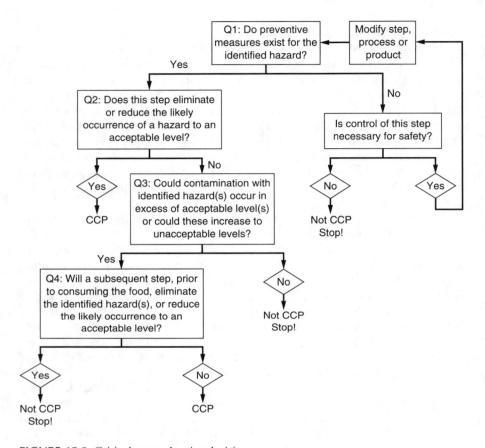

FIGURE 15.2 Critical control point decision tree.

understood and controlled within an acceptable range to ensure that a safe product is manufactured. Processes with a high degree of variability, especially when that variability is not recognized or understood, are more likely to produce unacceptable and possibly hazardous food [3].

Before data collection begins, the appropriate control chart must be determined for the parameter to be evaluated at the CCP. Two principal categories of control charts are employed in SPC work: variable and attribute. Variable control charts use actual measurements (e.g., temperatures, chlorine concentration, oxidation–reduction potential (ORP), pH values) for charting. Attribute control charts use pass–fail information (metal inclusion, cull fruit presence, foreign objects) for charting. Smith presents a good description of the different types of charts in each category [43].

To create the chart, individual data, normally arranged into subgroups, are sampled from the process. The average value of the data is then calculated and becomes the centerline of the chart. Using statistical formulae specific for each chart type, upper and lower trial control limits are calculated.

They describe the spread of the process. Finally, the individual (or averaged) measured values are plotted on the control chart. Once the chart is constructed, it presents a picture of the types of variation occurring in the process over the time at which the samples were taken. If one or more plotted points exceed either trial control limit, special cause variation has become a part of the process, forcing it out of statistical control. If a cause can be assigned to each value exceeding the control limit, then it can be discarded from the data and new control limits can be computed from the remaining data. However, if no cause can be found and corrected, then the points cannot be removed from the chart [2]. Once the assignable causes have been eliminated, the revised control charts should be in control. If a process shows only common cause variation present, it is stable, and the process of improvement can begin.

When first implemented, SPC will do a good job of finding areas of high variability (special cause variation) in a process. This results in readily demonstrated points exceeding the control limits. However, as more problems are solved, those remaining will be more subtle in their variation [44].

When an unusual number of nonrandom points produces a pattern on a control chart, none being beyond the control limits, this signifies that the process is unstable and on the verge of going out of control. While a dozen or more of these patterns may occur in a process, Evans has characterized the five most common ones, as shown in Figure 15.3 [45]: (a) shift — seven or more consecutive points on one side of the center line of a control chart; (b) run — a pattern of seven points consecutively climbing or falling in a control chart; (c) cycling — short repeated patterns of points having alternate high peaks and low valleys on a control chart; (d) instability — unnatural and erratic swings on both sides of the chart over time with points often lying near or on the control limits; and (e) stratification — 14 or more consecutive points hugging the center line on the control chart. When these patterns occur, it is a warning signal that something has gone wrong in the process and immediate action is needed to avoid loss of control.

15.9 CONDUCTING PROCESS CAPABILITY ANALYSES TO VERIFY CRITICAL LIMITS (HACCP PRINCIPLE 3)

The third step in HACCP plan development is to set safety boundaries, or critical limits, for each CCP identified in the hazard analysis. A critical limit (CL) is defined as a maximum or minimum value to which a biological, chemical, or physical parameter must be controlled at a CCP to prevent, eliminate, or reduce to an acceptable level the occurrence of a food safety hazard [7]. CLs are individual values that signify whether the variation of the control measure implemented at the CCP is capable of remaining within its safety boundaries. It is important to note that CLs cannot be arbitrarily set based on the variation in a process. They are not control limits; they must

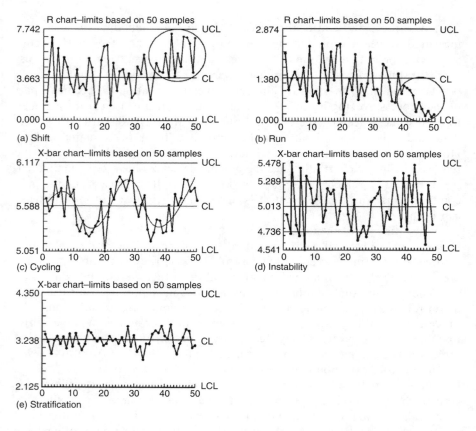

(a) Shift
(b) Run
(c) Cycling
(d) Instability
(e) Stratification

FIGURE 15.3 Interpreting control chart patterns. (Adapted from Evans, J.R., *Statistical Process Control for Quality Improvement: A Training Guide to Learning SPC*, 1st ed. Copyright 1991. Reprinted by permission of Pearson Education, Inc., Upper Saddle River, NJ.)

be scientifically determined. In many cases, the appropriate CLs may not be readily apparent or available to HACCP team members. Wedding has listed some sources to consult for this information, including scientific research articles, government documents, trade association guidelines, in-plant studies, university extension publications, and industry experts [46]. If outside sources are used to establish CLs, they should be documented and become part of the HACCP plan.

Once CLs based on scientific data have been determined for each CCP, a capability study must be conducted on the HACCP process to ensure it can be realistically and consistently maintained within these defined limits. As noted by Evans, the process must first be in a state of statistical control before performing a process capability study [45]. The major function of a capability analysis is to determine by measurement how well the control measure used at that CCP is functioning when compared to the specifications set at the CL [43]. Establishing CLs that may be outside the capability of the process will ultimately jeopardize the integrity of the entire HACCP plan [47].

Several texts reference how to conduct a process capability study [2,43]. Assessment of process capability is required to determine the relationship between the natural process variation and specified tolerances. Thus, individual temperature readings for producing safe juices should always operate within the CLs for safe pasteurization temperature. Evans has expressed process capability as the ratio of the tolerance width to the natural process variation. In the context of HACCP, this would be defined as shown in Figure 15.4 [45]. As noted by the CL definition, only one-sided limits are necessary for most HACCP capability studies. When this is the case, the formulae in Figure 15.5 apply. Examples of when this might be appropriate include the maximum pH allowed for an acidified vegetable product where no heat treatment is applied during the process and the product is stored in ambient temperature, or the minimum scheduled process temperature allowed for a low-acid canned vegetable to ensure safety [3]. When a CL is violated, it signals that an unsafe product may have been manufactured at this CCP. Immediate action must be taken to bring the CCP back into its CL range. Also, any product manufactured at the time the CL was violated must be held for evaluation and/or reprocessing.

When CLs have been set for all CCPs, the task is to keep the parameter being measured in control within the established tolerances. This may or may not be an easy job depending on the kind of variation in the process.

$$C_P = \frac{\text{tolerance width of CL}}{\text{natural variation}} = \frac{CL_U - CL_L}{6\sigma}$$

Where:
C_P = quotient of tolerated variation
CL_U = upper critical limit of CCP
CL_L = lower critical limit of CCP
6σ = actual process variation, assuming a normal distribution

FIGURE 15.4 Calculating upper and lower critical limits at a CCP.

$$C_{P_U} = \frac{CL_U - \bar{x}}{3\sigma} \qquad \text{or} \qquad C_{P_L} = \frac{CL_L - \bar{x}}{3\sigma}$$

Where:

C_{P_U} = process capability in relationship to the upper Critical Limit

C_{P_L} = process capability in relationship to the lower Critical Limit

CL_U = upper Critical Limit
$\bar{x}$ = process average
3σ = only right side of normal distribution

CL_L = lower Critical Limit
$\bar{x}$ = process average
3σ = only left side of normal distribution

FIGURE 15.5 Calculation formulae for a process requiring only one critical limit.

Establishing operating limits is a practical means to help prevent routine violation of the CLs [46]. Operating limits are criteria that are more stringent than CLs and are established at a level that would be reached before the critical limit is violated [48]. Process adjustment should be taken when the operating limit is exceeded to avoid loss of control and the need to take corrective action at the CL.

15.10 ESTABLISHING SPC MONITORING PROCEDURES (HACCP PRINCIPLE 4)

Selection of the correct monitoring system is an essential part of any HACCP study because it is what the HACCP team relies upon to maintain control at the CCPs. By definition, monitoring is a planned sequence of observations or measurements to assess whether a CCP is under control and produces an accurate record for future use in verification [7]. Monitoring serves to (1) track the operation of a process and enable the identification of trends toward a CL that may trigger process adjustment; (2) identify when and where there was a loss of control (a deviation occurred at the CCP) such that corrective action is needed; and (3) provide written documentation of the process control system [48].

The HACCP team will be responsible for designing the monitoring activities at each CCP, as well as training the individual(s) who will carry out the monitoring. These individuals should have the authority to stop the operation and to take corrective action if the CL is violated. All records and documents associated with monitoring CCPs should be recorded, signed, and dated by the person doing the monitoring, and, where necessary, assessed by a designated manager with overall responsibility for the food product [4]. Procedures must identify what control measures will be monitored, how frequently monitoring should be performed, what procedures will be used, and who will perform the monitoring [27].

Physical (e.g., temperature, ORP millivolts, metal detection) and chemical (e.g., chlorine, pH, acidity) monitoring systems are always the preferred methods of monitoring because of their ease and real-time data feedback. Monitoring systems may be continuous (e.g., recording on a continuous circular chart the pasteurization temperatures of juice filled into bottles) or discontinuous (e.g., measuring residual free chlorine in fresh-cut vegetable flume water at specific intervals). Monitoring equipment systems may be online (e.g., temperature and ORP probes, metal detectors) or offline (e.g., chlorine test kits, determination of titratable acidity, water activity measurements). The equipment chosen for CCP monitoring must have the degree of sensitivity to control hazards accurately. Daily calibration or standardization is necessary, and records should be maintained on these procedures, to become a part of the support documentation for the HACCP plan [27]. While microbiological testing is not suitable for controlling CCPs, the above measurements (excluding metal detection) can serve as an indirect measure of microbiological control at the CCPs.

Statistical control charting is ideally suited for HACCP monitoring of designated CCPs, because it provides an early warning signal of the need for corrective action before a CCP is violated. In terms of process control, however, all statistical control charts are not created equal. Variable control charts are much better than attribute charts in detecting an impending change in a process. This is because variable charts use quantitative data measurements while attribute charts work with qualitative data. Variable charts can pinpoint the relative position of plotted points within a CL such that if there is a move toward the boundary, or if an unusual pattern of points signals there is trouble in the process, corrective action can be taken immediately before the CL is compromised. In contrast, attribute charts utilize a pass/fail system of data gathering, and cannot signal a change until the problem has already occurred. Therefore attribute charts are poor tools for anticipating process change [49].

It is important to note, however, that process control may not be HACCP control. If the common cause variation of the parameter monitoring a CCP is too great, the process may exceed the CL. Thus, a process in statistical control may not be capable of producing a safe product. Likewise, the parameter monitoring a CCP (ORP in millivolts) may be within the CL, but not in statistical control. In fact, any one of four scenarios may exist, as demonstrated in Figure 15.6 [27]. Thus, to ensure product safety, the important point is that SPC limits must have less variability than HACCP limits [50].

It must be remembered that any statistical chart that relies on plotted data averages may obscure extreme values that could pose a health hazard [37]. While plotted averages for a CCP may be within critical limits, individual values may be above or below the CL for safety. For this reason, it is recommended to first monitor CCPs using individual values plotted on individual/moving range charts to be certain they can remain within their predetermined CLs [49]. Once process stability has been achieved, then one can proceed to construct average/range charts. These are better indicators of any process shift that may occur for a CCP within the CL.

The availability of numerous SPC software packages has increased significantly in the last 15 years, because of listings and reviews in publications such as *Quality Progress, Journal of Quality Technology* from the American Society for Quality (ASQ), as well as from vendors on the Internet. Much of the fruit and vegetable industry has moved from charts on paper and clipboards to computer-controlled processing. It is now possible to load raw data into software that creates control charts and performs capability studies with great speed and accuracy. However, even though this technology is time and cost saving, there is the danger that operators do not understand the theoretical background on which these programs are based and therefore can draw inappropriate conclusions based on computer results. As Cullen and Hollingum point out, a computer will carry out complicated calculations

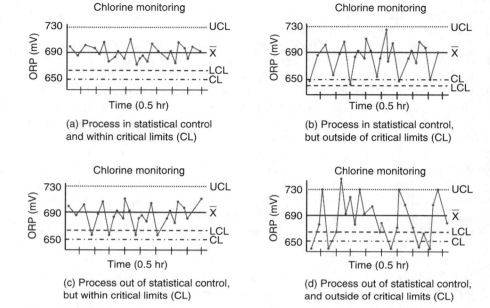

(a) Process in statistical control and within critical limits (CL)

(b) Process in statistical control, but outside of critical limits (CL)

(c) Process out of statistical control, but within critical limits (CL)

(d) Process out of statistical control, and outside of critical limits (CL)

FIGURE 15.6 Monitoring chlorine concentration in ORP (mV) using SPC and HACCP methodology. (Adapted from Hurst, W.C., Safety aspects of fresh-cut fruits and vegetables, in *Fresh-Cut Fruits and Vegetables: Science, Technology and Market*, Lamikanra, O., Ed., CRC Press, Boca Raton, FL, 2002. With permission.)

very quickly, but unless the user fully understands the significance of the figures and graphs generated, the user can easily move to some extremely misleading conclusions [35]. Computers are no substitute for a thorough training in the fundamentals of SPC.

15.11 DETERMINING CORRECTIVE ACTION PROCEDURES (HACCP PRINCIPLE 5)

Since the natural forces of entropy cause all HACCP processes to deteriorate toward a state of disorder, deviations from CLs will occur, and corrective actions will be needed [38]. By definition, a corrective action must be taken when a CL deviation is identified by monitoring a CCP [7]. Tompkin pointed out that corrective action involves four activities: (1) bringing the process back into its CL through process adjustment; (2) determining and correcting the cause of the deviation; (3) determining the disposition of the noncompliant product; and (4) recording the corrective action taken and the disposition of the noncompliant product [51].

When CLs are violated at a CCP, predetermined (developed in advance for each CCP and included in the HACCP plan) corrective action procedures must be initiated. Slade noted that a corrective action should take care

TABLE 15.3
Steps for Determining Product Disposition

1. Determine if the product presents a safety hazard, based on:
   (a) Expert evaluation
   (b) Biological, chemical, or physical testing
2. If no hazard exists, the product may be released.
3. If a potential hazard exists, determine if the product can be:
   (a) Reworked or reprocessed
   (b) Diverted for an alternative use
4. If potentially hazardous product cannot be handled as described in step 3, the product must be destroyed

of the immediate (short-term) problem as well as provide a long-term solution [9]. It may be necessary to determine the root cause of the deviation to prevent future recurrence. A CL failure that was not anticipated or one that reoccurs should result in the adjustment of the process or product and a re-evaluation of the HACCP plan. Because of the great diversity in fruit and vegetable products, and variation in the equipment used, type of processing, raw materials, etc., specific corrective actions must be developed for each CCP, according to the parameters of processing. When a deviation occurs and nonconforming product is produced, there are four steps for determining product disposition, as outlined in Table 15.3 [9].

Individuals who have a thorough understanding of the product, process, and HACCP plan should be assigned responsibility for writing the corrective action procedures and overseeing that the corrective actions are implemented [33]. Likewise, a log entry for each corrective action procedure should identify the person responsible for taking action to control product safety. HACCP plan records should contain a separate file in which all deviations and corresponding actions are maintained in an organized fashion.

15.12 CONFIRMING HACCP IMPLEMENTATION THROUGH VERIFICATION ACTIVITIES (HACCP PRINCIPLE 6)

Verification and validation comprise principle 6 of the HACCP study and must be carried out for each identified CCP in the HACCP plan. Verification is defined as the application of tests, procedures, and evaluations, in addition to monitoring, that determine the validity of the HACCP plan and that the system is operating as written [7]. Verification deals with implementation of the HACCP plan. It must be a routine part of the daily production process. The purpose of verification is to confirm through documentation that food safety has been achieved at the CCPs according to the implemented HACCP plan [52].

Major verification activities include plant audits, calibration of instruments and equipment, CCP records review, targeted sampling, and microbiological testing [48]. The objective of audit verification is to compare actual practices with what is contained in the HACCP plan. Audits can be performed by a member of the HACCP team, plant management, outside experts or consultants, regulatory agencies, and customers. If the calibration of instruments and equipment is not done on a scheduled and frequent basis, significant deviations at a CCP might go unnoticed, thus creating a potential health hazard. If this happens, the CCP would be considered out of control since the last documented acceptable calibration. The CCP record review involves examining two types of records generated at each CCP: monitoring and corrective action. These records are valuable management tools, providing documentation that CCPs are operating within established safety parameters and that deviations are being handled in a safe and appropriate manner. Verification also includes targeted sampling and microbiological testing. Vendor compliance can be checked by targeted sampling when receipt of material is a CCP, and purchase specifications are relied on as control limits. Microbiological testing can be used as a verification tool to determine if the overall operation is under control.

In 1986 the International Commission on Microbiological Specifications for Food (ICMSF) published a statistically based acceptance sampling plan for testing microbiological hazards in foods [53]. Although acceptance sampling utilizes random sample collection and has good statistical validity, the nonrandom distribution of pathogens and low probability for detection gives no guarantee that their number is below a safe level or that they are absent from fruit and vegetable products [54]. While acknowledging this inherent flaw, acceptance sampling is based on sound science and therefore can serve as an important verification tool in the design and validation of process control through monitoring [3]. Microbiological results obtained for a food product can tell about the process: whether the process at the time of sampling was good (under control) or bad (out-of-control) in its abatement of microorganisms and/or human pathogens. Also, to improve verification, the stringency of acceptance sampling plans can be increased by modifying the operating characteristic (OC) curves associated with these plans to increase sample size draw or lower acceptance numbers. These more rigorous criteria for acceptance will ensure a higher probability for making the correct decision to pass or fail inspection by these plans [37]. As noted by Leaper, some end product testing, particularly for verification purposes, will always be required by customers to document product safety to consumers [4].

Validation of the HACCP plan should be performed whenever there are indications that a process is unstable or out of control, or whenever there is a change in product, formulation, or processing equipment. HACCP revalidation should be performed on a periodic basis, even if no changes have occurred in the process, so the plan will retain its support base [33].

15.13 ESTABLISHING DOCUMENTATION AND RECORD KEEPING (HACCP PRINCIPLE 7)

Establishing accurate documentation and efficient record keeping is essential to the successful application of HACCP. Documentation demonstrates that the principles of HACCP have been correctly applied. Records provide the written evidence that all HACCP activities were carried out as specified. Record keeping, admittedly, can be a boring, tedious task; yet, in the words of a USDA inspector, if it was not written down, as far as he was concerned, it was not done! Experience has demonstrated that inadequate or inefficient record keeping is a major reason for HACCP audit failures.

All HACCP records should be kept in a separate master file so that only product/process safety is reviewed during a HACCP audit. Software systems are available to assist in the documentation of HACCP plans and keeping of records. The FDA requires that HACCP records be kept on file for at least one year from the date of production for refrigerated foods (e.g., fresh-cut produce). Although record keeping may appear to be a burden, there are sound reasons for this activity which will benefit the processor, including the following [27]:

1. Documents that all CCPs are within CLs to ensure product safety.
2. Provides the only reference available for produce traceability once the product leaves the plant.
3. Documents corrective actions taken when CLs were exceeded.
4. Provides a monitoring tool so process adjustments can be made to prevent loss of control.
5. Provides data for review during regulatory, customer, and third-party auditing.
6. Provides demonstrable evidence that procedures and processes were followed in strict accordance with the written HACCP plan.

Record keeping includes records that go beyond those that are tabulated on a day-to-day basis. NACMCF endorses the maintenance of four types of records [7]:

1. Summary of the hazard analysis — includes records on the HACCP team's deliberations on the rationale for determining hazards and control measures.
2. The HACCP plan for each product, including records of the product description, distribution and end use, verified flow diagram, and all HACCP plan summaries addressing the seven required components.
3. Support documentation — CCP records, CL records monitoring and corrective action records, verification and validation records.

TABLE 15.4
Excerpt of HACCP Plan Summary Page for Fresh-Cut Lettuce

Process step; CCP no.	Biological/chemical/physical hazard description	Critical limits	Monitoring procedures/frequency/person responsible	Corrective actions/person responsible	Verification procedures/person responsible	HACCP records
Water flume wash; CCP 1B	Biological: *L. monocytogenes*; *E. coli* O157:H7; *Salmonella* spp., *Shigella* spp., other microbial pathogens	Potable water containing ≥1 ppm free residual chlorine for 30 seconds; at pH ≤7.0	Prior to start of processing and each 30 minutes thereafter, QC personnel will monitor free chlorine using standardized test kit, and a calibrated pH meter will be used to monitor pH three times per shift	QC personnel will adjust water chemistry to maintain pH and chlorine added to maintain CL; held product will be rewashed and CL deviations noted in log	QC personnel will maintain chlorine, pH, temperature monitoring logs; CCP deviations/corrective action logs, calibration logs for thermometer, pH test and chlorine test kit used; microbiological tests will be run on finished product at least once per year to validate pathogen absence. HACCP plan will be revalidated at least one per year	HACCP coordinator reviews all HACCP records weekly, HACCP coordinator will conduct calibration tests; plant manager reviews records daily, state food inspector/FDA audits, customer audits/internal/consultant audit once per year, all records kept at least one year; random sampling/testing product to ensure process verification

From IFPA Technical Committee, *HACCP for the Fresh-Cut Produce Industry*, IFPA, Alexandria, VA, 2000. With permission.

4. Daily operational records — includes records generated daily detailing control of the HACCP process for each CCP (specifically, monitoring, corrective action, and verification).

HACCP records can be quite diverse and may include procedures for monitoring, calibrating, corrective action, and verifying and validating a CCP. An example of the records necessary to maintain and document control at the water flume wash step CCP of a fresh-cut lettuce operation is given in Table 15.4.

15.14 SUMMARY

To meet the challenges of today's food safety issues, the FDA has increased its emphasis toward programs that are proactive and prevention oriented. The most comprehensive, science-based program to date for reducing pathogen contamination in fruit and vegetable products is HACCP. It has become the industry standard for ensuring food safety. Unlike past traditional approaches which relied on end-product testing, HACCP focuses on continuous control and monitoring of CCPs to ensure safety all along the production and processing continuum. Because all product and processing operations tend to vary over time, however, it becomes important to be able to identify and quantify the type of variation present in them. Unfortunately, an inherent weakness of HACCP is that it can neither identify variation within a process nor provide any advanced warning as to when a CCP has a high probability of exceeding its CL, causing loss of control within the safety zone. If HACCP is to be a truly effective prevention tool, it must be linked to appropriate procedures that both monitor and verify that a process can remain in control and safe.

The reliability and effectiveness of HACCP as a safety tool can be greatly strengthened by the incorporation of statistical quality control (SQC) methods, namely SPC and acceptance sampling, into its structure. SPC is an objective, quantitative, and statistically valid means of predicting CCP control during monitoring. In SPC, the data generated can be used on a continuous basis to assess whether any unacceptable trends are developing over a period of time at a CCP and whether the process is in statistical control. In similar fashion, acceptance sampling brings the scientific method to HACCP verification activities. It can validate and verify that a process is not only operating in a safe zone of control but is producing a safe product.

Integration of SPC into a HACCP program will provide several benefits. First, it will bring about a culmination to any processor's/packer's HACCP plan in that statistically valid control charts will demonstrate to customers evidence of product safety. Second, it will provide an on-going and continuous improvement of all processes which will have a positive impact on the company's "bottom line." Third, it will satisfy future government regulations that are moving toward the requirement that a fresh produce processor/packer or fresh-cut processor be able to document compliance with product performance standards.

REFERENCES

1. Marriott, N.G., The role of HACCP in sanitation, in *Principles of Food Sanitation*, Marriott, N.G., Ed., Aspen Publishers, Gaithersburg, MD, 1999, p. 75.
2. Summers, D.C.S., *Quality*, 2nd ed., Prentice-Hall, Upper Saddle River, NJ, 2000.
3. International Commission on Microbiological Specifications for Food (ICMSF), Microorganisms, in *Foods 7: Microbiological Testing in Food Safety Management*, Kluwer Academic/Plenum, New York, 2002.
4. Leaper, S., Ed., *HACCP: A Practical Guide*, 2nd ed., Campden and Chorleywood Food Research Association, Chipping Campden, U.K., 1997.
5. Lopez, A., *A Complete Course in Canning, Vol. 1: Basic Information on Canning*, 12th ed., Canning Trade, Baltimore, MD, 1987.
6. U.S. Food and Drug Administration (FDA), Acidified Foods and Low Acid Canned Foods in Hermetically Sealed Containers, Federal Register, March 16, 1979.
7. National Advisory Committee on Microbiological Criteria for Foods (NACMCF), Hazard analysis and critical control point principles and application guidelines, *J. Food Prot.* 6, 762, 1998.
8. U.S. Food and Drug Administration (FDA), Food and Safety Assurance Program: Development of Hazard Analysis Critical Control Points (HACCP): Proposed Rule, Federal Register, August 4, 1994.
9. Slade, P., Ed., *Juice HACCP Training Curriculum*, Juice HACCP Alliance, Food Processors Institute, Washington D.C., 2002.
10. Beuchat, L., *Surface Decontamination of Fruits and Vegetables Eaten Raw: A Review*, Food Safety Unit, World Health Organization, Brussels, Belgium, 1998.
11. Gorris, L.G.M. and Tauscher, B., Quality and safety aspects of novel minimal processing technologies, in *Processing Foods: Quality Optimization and Process Assessment*, Oliveira, F.A.R., and Oliveira, J.C., Eds., CRC Press, Boca Raton, FL, 1999, p. 325.
12. Hurst, W.C., and Schuler, G.A., Fresh produce processing: an industry perspective, *J. Food Prot.*, 55, 824, 1992.
13. Hui, Y.H., Nip, W.K., and Gorham, J.R., The FDA's GMPs, HACCP and the food code, in *Food Plant Sanitation*, Hui, Y.H. *et al.*, Eds., Marcel Dekker, New York, 2003, p. 31.
14. Mortimore, S. and Wallace, C., *HACCP: A Food Industry Briefing Series*, Blackwell Science, Malden, MA, 2001.
15. Hoffer, P., Moving to prevention: an industry transition, *Quality Prog.*, 18, 24, 1985.
16. U.S. Food and Drug Administration (FDA), Food and Safety Assurance Program: Development of Hazard Analysis Critical Control Points (HACCP): Proposed Rule, Federal Register, August 4, 1994.
17. Skrabec, Q.R., The transition from 100% inspection to process control, *Quality Prog.*, 22, 35, 1989.
18. National Academy of Sciences (NAS), *Scientific Criteria to Ensure Safe Food*, National Academies Press, Washington D.C., 2003.
19. Stevenson, K.E., and Bernard, D.T., *HACCP: A Systematic Approach to Food Safety*, Food Processors Institute, Washington D.C., 1999.

20. Early, R., Use of HACCP in fruit and vegetable production and post-harvest pretreatment, in *Fruit and Vegetable Processing: Improving Quality*, Jongen, W., Ed., CRC Press, Boca Raton, FL, 2002.

21. Stier, R.F. and Blumenthal, M.M., Will HACCP be carrot or stick?, *Dairy Food Environ. Sanit.*, 15, 616, 1995.

22. Bryan, F.L., HACCP approach to food safety past, present and future, *Food Testing Anal.*, 5, 13, 1999.

23. Sperber, W.H. *et al.*, The role of prerequisite programs in managing a HACCP system, *Dairy Food Environ. Sanit.*, 18, 418, 1998.

24. Center for Food Safety and Applied Nutrition (CFSAN), *Guidance for Industry Guide to Minimize Microbial Food Safety Hazards for Fresh Fruits and Vegetables*, U.S. Food and Drug Administration and U.S. Department of Agriculture, Washington D.C., 1998 (available online at http://vm.cfsan.fda. gov/~dms/prodguid.html in English, French, Portuguese, Spanish, and Arabic).

25. Stier, R.F. and Nagel, N.E., Ensuring safety in juices and juice products: good agricultural practices, in *Beverage Quality and Safety*, Foster, T. and Vasavada, P.C., Eds., CRC Press, Boca Raton, FL, 2003, p. 1.

26. Troller, J.A., Sanitation, in *Food Processing*, 2nd ed., Academic Press, New York, 1993.

27. Hurst, W.C., Safety aspects of fresh-cut fruits and vegetables, in *Fresh-Cut Fruits and Vegetables: Science, Technology and Market*, Lamikanra, O., Ed., CRC Press, Boca Raton, FL, 2002, p. 45.

28. Gorny, J.R., Ed., *Food Safety Guidelines for the Fresh-Cut Produce Industry*, 4th ed., International Fresh-Cut Produce Association, Alexandria, VA, 2001.

29. Rhodehamel, J.E., Overview of biological, chemical and physical hazards, in *HACCP Principles and Applications*, Pierson, M.D. and Corlett, D.A., Jr., Eds., AVI/Van Nostrand Reinhold, New York, 1992, p. 8.

30. Rao, A. *et al.*, *Total Quality Management: A Cross Functional Perspective*, John Wiley, New York, 1996.

31. Tapia, M.S., Martinez, A., and Diaz, R.V., Tools for safety control: HACCP, risk assessment, predictive microbiology and challenge tests, in *Minimally Processed Fruits and Vegetables*, Tapua, M.S. and Lopez-Malo, A., Eds., Aspen Publishers, Gaithersburg, MD, 2000, p. 79.

32. Forsythe, S.J., *The Microbiological Risk Assessment of Food*, Blackwell Science, Malden, MA, 2002.

33. Bernard, D.T., Hazard analysis and critical control point system: use in controlling microbiological hazards, in *Food Microbiology: Fundamentals and Frontiers*, Doyle, M.P., Beuchat, L.R., and Montville, T.J., Eds., American Society of Microbiology Press, Washington D.C., 1997, p. 740.

34. Joiner, B.L. and Gaudard, M.A., Variation, management and W. Edwards Deming, *Quality Prog.*, 23, 29, 1990.

35. Cullen, J. and Hollingum, J., *Implementing Total Quality*, IFS Publications, Bedford, U.K., 1987.

36. Hubbard, M.R., *Statistical Quality Control for the Food Industry*, 3rd ed., Kluwer Academic/Plenum, New York, 2003.

37. International Commission on Microbiological Specifications for Food (ICMSF), Microorganisms, in *Foods 5: Applications of the Hazard Analysis Critical Control Point (HACCP) System to Ensure Microbiological Safety and Quality*, Blackwell Scientific Applications, Oxford, U.K., 1988.

38. Wheeler, D.J. and Chambers, D.S., *Understanding Statistical Process Control,* 2nd ed., SPC Press, Knoxville, TN, 1992.
39. Grigg, N.P., Statistical process control in U.K. food production: an overview, *Br. Food J.,* 100, 371, 1998.
40. Demetrakakes, P., Pinpointing critical control points for HACCP success, *Food Processing,* 58, 24, 1997.
41. Wedding, L.M., Critical control points, in *HACCP: A Systematic Approach to Food Safety,* Stevenson, K.E. and Bernard, D.T., Eds., Food Processors Institute, Washington D.C., 1999, p. 81.
42. Mortimore, S. and Wallace, C., *HACCP: A Practical Approach,* Chapman & Hall, New York, 1994.
43. Smith, G.M., *Statistical Process Control and Quality Improvement,* 3rd ed., Prentice-Hall, Upper Saddle River, NJ, 1998.
44. Clements, R.R., *Statistical Process Control and Beyond,* Robert E. Krieger, Malabar, FL, 1988.
45. Evans, J.R., *Statistical Process Control for Quality Improvement,* Prentice-Hall, Englewood Cliffs, NJ, 1991.
46. Wedding, L.M., Critical limits, in *HACCP: A Systematic Approach to Food Safety,* Stevenson, K.E. and Bernard, D.T., Eds., Food Processors Institute, Washington D.C., 1999, p. 85.
47. Keener, L., HACCP: a view to the bottom line, *Food Saf. Mag.,* 8, 20, 2002.
48. Lockwood, D.W., Beattie, S., and Morris, W.C., *Southeastern Regional Apple Cider Safety Workshop Manual,* Cooperative Extension Service, University of Tennessee, Knoxville, TN, 1998.
49. Surak, J.C., Cawley, J.L, and Hussain, H., Integrating HACCP and SPC, *Food Quality Mag.,* 5, 41, 1998.
50. VanSchothorst, M. and Jongeneel, S., Line monitoring, HACCP and food safety, *Food Control,* 5, 107, 1994.
51. Tompkin, R.B., Corrective action procedures for deviations from the critical control point critical limit, in *HACCP Principles and Applications,* Pierson, M.D. and Corlett, D.A., Eds., AVI/VanNostrand Reinhold, New York, 1992, p. 72.
52. Prince, G., Verification of the HACCP program, in *HACCP: Principles and Applications,* Pierson, M.D. and Corlett, D.A., Jr., Eds., AVI/Van Nostrand Reinhold, New York, 1992.
53. International Commission on Microbiological Specifications for Food (ICMSF), Microorganisms, in *Foods 2: Sampling for Microbiological Analysis: Principles and Specific Applications,* 2nd ed., University of Toronto Press, Toronto, Canada, 1986.
54. Doores, S., Ed., *Food Safety: Current Status and Future Needs,* American Society of Microbiology, Washington D.C., 1999.

16 Effect of Quality Sorting and Culling on the Microbiological Quality of Fresh Produce

Susanne E. Keller

CONTENTS

16.1 INTRODUCTION

Quality sorting and culling of fresh produce is performed to separate damaged or decayed produce from undamaged sound produce. The principal motivation for such sorting is financial. Consumers are not likely to purchase fresh fruit or vegetables that are noticeably damaged or decayed. Produce destined for further processing, rather than fresh consumption, will also demand a better price if quality is higher. Poor-quality produce will result in higher processing cost, greater losses, and shorter shelf life of the final product.

Although the principal motivation for quality sorting may be financial, another more important motivation should be the desire to provide a safe and nutritious product. Damaged and decayed produce can have substantially higher levels of microorganisms than undamaged sound produce. Storing or processing such damaged and decayed produce with sound produce may result in the spread of spoilage organisms, resulting in further losses and lower quality of finished products. In addition, foodborne pathogens can find greater ingress in damaged and decayed produce, resulting in a significant increase in the risk of foodborne illnesses. Consequently, the prompt removal of unsound produce will impact not only on costs, but the safety of the product as well, be that fresh or processed produce.

16.2　GRADE STANDARDS

In general, produce is sorted according to established standards. Standards for many types of produce are established by the U.S. Department of Agriculture (USDA) and can be obtained at http://www.ams.usda.gov/standards. Standards can vary for produce from different areas. Oranges produced in Florida have somewhat different standards than oranges produced in California. Differences in grade standards are related to different growing conditions and climate which influences characteristics such as sugar and acid levels. The highest grades are generally represented by produce that is free of blemishes, cuts and bruises, and any decay.

Numerous environmental and mechanical factors can influence product grade. Growing conditions and weather can dramatically affect produce quality. The consequences of environmental and mechanical factors for apples were examined by Baugher et al. [1]. Economic losses due to various defects resulting from both environmental and mechanical factors were measured for apples at nine packinghouses. Severe drought and high temperatures caused a significant increase in losses due to undersized fruit, cork spot, and spray injury. Some types of defects, particularly bacterial soft rot of fresh fruits and vegetables, have been associated with an increased incidence of pathogen contamination. Wells and Butterfield found higher salmonella contamination (59%) in wash water from fruit and vegetables that were affected by bacterial soft rot [2]. Wash water used for healthy produce had a lower incidence of salmonella contamination (33%). Other defects, such as undersized fruit, would appear to be primarily cosmetic and therefore of less importance in relation to spoilage or food safety. However, research suggests that lower quality produce, regardless of defect type, is more prone to spoilage. In a study on the effect of tomato grade on subsequent spoilage, it was determined that lower grade tomatoes inoculated with *Erwinia carotovora* subsp. carotova had higher levels of spoilage and infection after 14 days in storage than higher grade tomatoes [3]. Although the tomatoes were inoculated with *E. carotovora*, 82.4% of the infection was due to *Alternaria alternate*, and only 17.6% was due to the bacterium. The bacterium did, however, increase decay in lower grade tomatoes to a greater extent than higher grade. This study suggested surface blemishes of any type promoted postharvest decay.

16.3　EFFECTIVENESS OF GOOD AGRICULTURAL PRACTICES (GAPs)

Good agricultural practices (GAPs) fundamentally influence the level of microflora on produce and products made from produce. In 1998 the U.S. Food and Drug Administration (FDA) published a document in conjunction with the USDA entitled Guide to Minimize Microbial Food Safety Hazards for Fresh Fruits and Vegetables. This document was published as a draft on April 13 and as a final version on October 26, 1998 [4,5]. The document outlines hazards associated with common agricultural practices such as irrigation water, application of manure, field and facility sanitation, packing, and transportation. Although this document does not specifically address sorting and culling, inclusion of damaged and decayed produce amplifies the risks described in it.

The risk of subsequent spoilage and invasion of foodborne pathogens can be reduced through careful handling and appropriate phytosanitary measures including appropriate sorting and culling. The majority of spoilage organisms found on fruits and vegetables at harvest are essentially opportunistic organisms that require physical injury or excessive softening to gain entry to host tissue [6]. Many plant pathogens will infect adjacent areas of healthy produce once they have become established in damaged produce. In addition, such damaged and rotted areas particularly in ordinarily acidic tissue allow growth and survival of foodborne pathogens that might otherwise not survive. *Escherichia coli* O157:H7 survival was enhanced in bruised apples [7]. The bruised tissue was found to have significantly higher pH values that allowed growth of the pathogen. Survival of *E. coli* O157:H7 was also enhanced when apples were wounded and inoculated with the plant pathogen *Glomerella cingulata* [8]. This enhanced survival and growth was again attributed to increases in pH that accompanied infection by *G. cingulata*.

In addition to the increased risk of infection with foodborne pathogenic organisms, decayed produce may be contaminated with toxins produced by the invading microorganisms. Patulin is one mycotoxin that is produced primarily by *Penicillium expansium*, responsible for blue mold rot on apples, pears, and other fruit. Levels of patulin in apple juice have been correlated to the level of decay found on the apples [9]. Clearly, improper sanitation and handling of damaged produce where pathogens have greater ingress to internal tissue would represent a greater risk of foodborne illness than the use of sound produce. Consequently, proper sorting and removal of any damaged and decayed produce is an essential prerequisite in the prevention of foodborne illnesses.

16.4 EFFECTIVENESS OF SORTING AND SORTING METHODS

That removal of damaged and decayed fruit and vegetables prior to storage or subsequent processing should reduce microbial loads and thus reduce any risk of foodborne illness would seem intuitive. However, data to document the actual level of reductions achieved by sorting and culling are not always available. In data provided by the Florida Department of Citrus, juice microflora was $4.5 \pm 0.7 \log \text{CFU/ml}$ with no grading, $3.7 \pm 1.0 \log \text{CFU/ml}$ with light grading, and $2.2 \pm 0.6 \log \text{CFU/ml}$ with moderate grading [10]. Light grading was defined as splits removed, 0.5 to 1% of fruit stream; and moderate grading was splits, peel plugs, and significant blemishes removed, 2 to 3% of incoming fruit stream.

Sorting and removing poor-quality apples during cider production results in a reduction in aerobic counts in the final cider. In one recent study, cider was produced from seven different apple varieties. Production variables included method of harvest, quality sorting, and storage. Cider from fresh ground-harvested fruit, considered to be lowest quality and greatest risk, had significantly greater numbers of aerobic microflora (4.89 log CFU/ml) than any cider from tree-harvested fruit (Figure 16.1) [11]. Yeast and mold populations were also elevated in ground-harvested fruit (Figure 16.2). Cider from unsorted and unculled fruit had an average of 3.45 log CFU/ml, whereas cider from sorted and culled apples had an average of 2.88 log CFU/ml ($p < 0.05$) [11]. As with ground-harvested fruit, yeast and mold populations were elevated in cider from apples that were not sorted and culled prior to cider production.

Along with the examination of microflora levels in sorted or unsorted apples, the cider in the same study was also tested for changes in the level of patulin [12]. Patulin levels varied depending on storage and variety. No patulin was detected in cider produced from any fresh tree-harvested fruit. However,

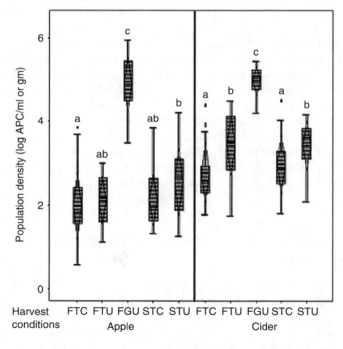

FIGURE 16.1 Differences in Tukey box plots of aerobic plate counts (APC) of pooled apple varieties due to harvest conditions, storage, and culling. Means sharing a letter were not statistically different at the $p < 0.05$ level. FTC = fresh, tree-harvested, culled; FTU = fresh, tree-harvested, unculled; FGU = fresh, ground-harvested, unculled; STC = stored, tree-harvested, culled; STU = stored, tree-harvested, unculled. (Modified from Keller, S.E. *et al.*, *J. Food Prot.*, 67, 2240, 2004. With permission.)

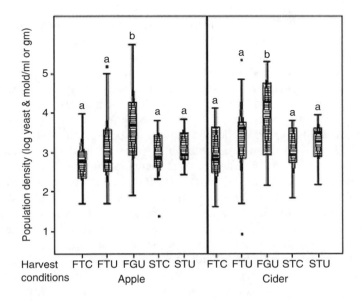

FIGURE 16.2 Differences in Tukey box plots of yeast and mold populations in pooled apple varieties due to harvest conditions, storage, and culling. Means sharing a letter were not statistically different at the $p < 0.05$ level. FTC = fresh, tree-harvested, culled; FTU = fresh, tree-harvested, unculled; FGU = fresh, ground-harvested, unculled; STC = stored, tree-harvested, culled; STU = stored, tree-harvested, unculled. (Modified from Keller, S.E. *et al.*, *J. Food Prot.*, 67, 2240, 2004. With permission.)

cider pressed from controlled atmosphere (CA) stored apples did contain significant levels of patulin. These patulin levels were significantly reduced in most varieties when the apples were culled prior to cider production. Chapter 13 contains a more in-depth review of patulin in apple cider.

Sorting and culling are included in a 1999 survey of industry practices published by the USDA [13]. This report covers 30 fruit and vegetable commodities in 14 states. The report focuses on fruit and vegetable commodities that are predominately consumed raw, as these represent a substantially greater risk than produce that undergoes further processing. Of all the produce in the survey, the vast majority was harvested by hand. For fruit acres, 94% was harvested by hand only, whereas for vegetable acres, 87% was harvested by hand. The majority of packers in this survey manually sorted both fruit and vegetables. It should be noted, however, that produce in this survey was intended for the fresh (consumed raw) market, and for many specific types of produce in the survey, data were unavailable. Nonetheless, the survey reported that 100% of apples, table grapes, broccoli, celery, cucumbers, cantaloupes, and honeydew melons were manually sorted.

That manual sorting and culling are used for the majority of fresh produce is not surprising. Mechanical handling and sorting can result in substantial damage, increased losses, and may incur considerable expense to implement.

In a study on grade-lowering defects in grapefruit, mechanical injury levels were found to increase from pregrading areas (5.6%) to final grading (8.6%) and to final packaging (12.3%) [14]. Increased injury was also related to increased levels of bacterial soft rot in Bell peppers [15].

Despite such reported problems, mechanical handling and sorting have an advantage in speed. For manual sorting, fruit or vegetables are sorted either in the field or at the packinghouse. Most frequently, a sorting table and/or conveyor are used. Care must be taken that the speed of the conveyor is slow enough to allow individual examination by personnel of each piece. Most mechanical systems consist of two basic elements, a conveyor and some means of separation, based on specific fruit or vegetable characteristics such as size, weight, color, and/or the presence of defects. The simplest of such instruments sorts using uniformity of size and shape by means of appropriate holes or sizers. Such simple sorting machines still require considerable worker attention, but can significantly increase sorting speed, particularly for less fragile produce. Such instruments would not, however, detect other defects.

More elaborate and modern instruments that sort by color and defects using camera-based systems and computer technology are available commercially. Produce is conveyed in traditional fashion to a scanning area where it is scanned and analyzed using computer programs to determine appropriate grade. Use of such a method was described by Leemans et al. for apples [16]. Leemans et al. reported correct classification up to 78% for Golden Delicious apples and 72% for Jonagold. Classification was based on external quality parameters such as color and texture.

Both manual sorting and machine sorting based on camera methods detect defects and characteristics external to the fruit or vegetable. However, some defects may be small and difficult to detect. Other defects may be internal and not detectable using any surface examination methods. More recently, research has been directed at the development of methods that may allow the detection of such internal defects in a nondestructive manner. These new methods are based primarily on X-ray imaging, magnetic resonance imaging (MRI), or near-infrared spectroscopy (NIR) spectroscopy [17–24].

Some produce types are not particularly suitable for machine sorting, frequently because of their fragile nature. For such produce, researchers look for novel means of assessing quality that may require less manual manipulation of the produce. One such method that has potential is the use of an electronic system that measures aromatic volatile gas emissions. Such a method was employed by Simon et al. to assess the quality of blueberries [25]. The method successfully detected differences in maturation levels and detected damaged fruit in closed containers of blueberries.

16.5 IMPACT ON FOOD SAFETY

The *Guide to Minimize Microbial Food Safety Hazards for Fresh Fruits and Vegetables*, jointly published by the FDA and the USDA, clearly describes where the greatest risks of foodborne pathogen contamination are likely to occur. It makes the assumption that damaged and decayed produce will not be included during any further processing or packaging. The consequence of such inclusion would be significantly higher microbial loads, shorter shelf life, and a substantially increased risk of infiltration of pathogens internally in the produce.

There is no intervention currently available that, if applied solely to the surface of the fruit or vegetable, will destroy any internal foodborne pathogen. Consequently, produce with internal foodborne pathogens that is minimally processed or consumed raw will present a substantial risk of foodborne illness. It is therefore critical to prevent the initial contamination of fresh produce so that internal contamination does not occur and that internally contaminated produce is not used for the fresh market. Appropriate quality sorting and culling of damaged and decayed produce are essential to reducing the risk that such potentially contaminated fruit will enter the fresh market.

The critical nature of appropriate sorting and culling was recognized by the FDA in the final HACCP (hazard analysis critical control points) juice regulation [26]. Because of the risk of pathogen infiltration into fresh produce, all juice must receive a treatment that will give a 5-log reduction in the pertinent pathogen. The treatment must be applied post-extraction, so all juice is treated, with the exception of citrus products. For citrus products, the 5-log treatment may be applied to the surface of the fruit. The risk of infiltration by pathogens in sound and intact citrus fruit was believed to be too small to merit a mandated treatment to the extracted juice. However, it is important to note that the legislation clearly states that such surface treatment must be applied only to sound, intact fruit and after the fruit has been sorted and culled.

Although the incidence of foodborne illness caused by the consumption of fruits and vegetables is low, there is some evidence that this risk is increasing [27]. Part of the increase may be due to the increased consumption of fresh produce, perceived by many consumers to be a healthy food choice. With an increase in consumption of fresh produce comes increased risk particularly if that produce is not handled in a safe and sanitary manner. Appropriate quality sorting and culling to ensure damaged and potentially contaminated produce does not enter the fresh market is not just an economically sound approach, it is an important component in the prevention of foodborne illness.

REFERENCES

1. Baugher, T.A., Hogmire, H.W.J., and Lightner, G.W., Determining apple packout losses and impact on profitability, *Appl. Agric. Res.*, 5, 343, 1990.
2. Wells, J.M. and Butterfield, J.E., *Salmonella* contamination associated with bacterial soft rot of fresh fruits and vegetables in the marketplace, *Plant Dis.*, 81, 867, 1997.
3. Blender, R.J., Sargent, S.A., Brecht, J.K., and Bartz, J.A., Effect of tomato grade on incidence of decay during simulated shipping, *Proc. Fla. State Hortic. Soc.*, 105, 119, 1992.

4. U.S. Food and Drug Administration, Draft guidance for industry: guide to minimize microbial food safety hazards for fresh fruits and vegetables, *Fed. Regist.*, 63, 18029, 1998.

5. FDA, Guide to Minimize Microbial Food Safety Hazards for Fresh Fruits and Vegetables, U.S. Food and Drug Administration, 1998.

6. Chesson, A., Maceration in relation to post-harvest handling and processing of plant material, *J. Appl. Bacteriol.*, 48, 1, 1980.

7. Dingman, D.W., Growth of Escherichia coli O157:H7 in bruised apple (Malus domestica) tissue as influences by cultivar, date of harvest, and source, *Appl. Environ. Microbiol.*, 66, 1077, 2000.

8. Riordan, D.C.R., Sapers, G.M., and Annous, B.A., The survival of *Echerichia coli* O157:H7 in the presence of *Penicillium expansium* and *Glomerella cingulata* in wounds on apple surfaces, *J. Food Prot.*, 63, 1637, 2000.

9. Kadakal, C. and Nas, S., Effect of apple decay proportion on the patulin, fumaric acid, HMF and other apple juice properties, *J. Food Saf.*, 22, 17, 2002.

10. Kelsey, F. and Pao, S., Factsheet 2, Fresh Citrus Juice. Key Factors in Determining Fresh Citrus Juice Quality and Safety, Vol. 2003, Florida Department of Citrus, 1999.

11. Keller, S.E., Chirtel, S.J., Merker, R.I., Taylor, K.T., Tan, H.L., and Miller, A.J., Influence of fruit variety, harvest technique, quality sorting, and storage on the native microflora of unpasteurized apple cider, *J. Food Prot.*, 67, 2240, 2004.

12. Jackson, L.S., Beacham-Bowden, T., Keller, S.E., Adhikari, C., Taylor, K.T., Chirtel, S.J., and Merker, R.I., Apple quality, storage, and washing treatments affect patulin levels in apple cider, *J. Food Prot.*, 66, 618, 2003.

13. U.S. Department of Agriculture, Fruit and Vegetable Practices: 1999, Agricultural Statistics Board NASS, USDA, 2001, p. 1.

14. Miller, W.M. and Burns, J.K., Grade lowering defects and grading practices for Indian River grapefruit, *Proc. Fla. State Hortic. Soc.*, 105, 129, 1992.

15. Carballo, S.J., Blankenship, S.M., Ritchie, D.F., and Boyette, M.D., Comparison of packing systems for injury and bacterial soft rot on Bell pepper fruit, *HortScience*, 4, 269, 1994.

16. Leemans, V., Magein, H., and Destain, M.F., On-line fruit grading according to their external quality using machine vision, *Biosyst. Eng.*, 83, 397, 2002.

17. Bull, C.R. and McFarlane, N.J.B., X-ray detection of subsurface damage in fruits and vegetables, *Postharvest News and Information*, 9, 29N, 1998.

18. Clark, C.J. and Burmeister, D.M., Magnetic resonance imaging of browning development in "Braeburn" apple during controlled-atmosphere storage under high CO_2, *HortScience*, 34, 915, 1999.

19. Clark, C.J., McGlone, V.A., and Jordan, R.B., Detection of brownheart in "Braeburn" apple by transmission NIR spectroscopy, *Postharvest Biol. Technol.*, 28, 87, 2003.

20. Fraser, D.G., McGlone, V.A., Kunnemeyer, R. and Jordan, R.B., NIR (Near Infra-Red) light penetration into an apple, *Postharvest Biol. Technol.*, 22, 191, 2001.

21. Gonzalez, J.J., Valle, R.C., Bobroff, S., Biasi, W.V., Mitcham, E.J., and McCarthy, M.J., Detection and monitoring of internal browning developments in "Fuji" apples using MRI, *Postharvest Biol. Technol.*, 22, 179, 2001.

22. Keener, K.M., Stroshine, R.L., and Nyenhuis, J.A., Evaluation of low field (5.40 MHz) proton magnetic resonance measurements of Dw and T2 as methods of non-destructive quality evaluation of apples, *J. Am. Soc. Horticult. Sci.*, 124, 289, 1999.
23. Schatzki, T.F., Haff, R.P., Young, R., Can, I., Le, L.C., and Toyofuku, N., Defect detection in apples by means of X-ray imaging, *Trans. Am. Soc. Agric. Eng.*, 40, 1407, 1997.
24. Upchurch, B.L., Throop, J.A., and Aneshansley, D.J., Detecting internal breakdown in apples using interactance measurements, *Postharvest Biol. Technol.*, 10, 15, 1997.
25. Simon, J.E., Hetzroni, A., Bordelon, B., Miles, G.E., and Charles, D.J., Electronic sensing of aromatic volatiles for quality sorting of blueberries, *J. Food Sci.*, 61, 967, 1996.
26. U.S. Food and Drug Administration, 21 CFR Part 120, Hazard analysis and critical control point (HACCP); procedures for the safe and sanitary processing and importing of juice, Final rule, *Fed. Regist.*, 66, 6137, 2001.
27. Anon., Analysis and Evaluation of Preventive Control Measures for the Control and Reduction/Elimination of Microbial Hazards on Fresh and Fresh-Cut Produce, Vol. 2003, U.S. Food and Drug Administration, 2001.

17 Washing and Sanitizing Treatments for Fruits and Vegetables

Gerald M. Sapers

CONTENTS

Mention of trade names or commercial products is solely for the purpose of providing specific information and does not imply recommendation or endorsement by the U.S. Department of Agriculture.

17.1   INTRODUCTION

The detection of human pathogens in fresh produce and occurrence of outbreaks of foodborne illness associated with contaminated produce, as documented in previous chapters, represent serious public health problems. Contamination of fruits and vegetables with human pathogens or organisms causing spoilage also has important economic consequences. Consequently, it is in the interests of the produce industry to develop interventions to reduce the risk of microbial contamination. If contamination is likely during crop production or harvest, it is usually better to reduce this risk by avoidance of contamination sources through implementation of good agricultural practices (GAPs). However, this is not always possible, and in such situations the grower/shipper or processor must depend on washing and sanitizing treatments as a second line of defense. If produce contamination occurs postharvest and contamination sources cannot be eliminated through improvements in plant layout, implementation of good manufacturing practices (GMPs), and improvements in plant sanitation, then washing and sanitizing of produce and equipment become the first line of defense. The subject of washing and sanitizing technology for fresh produce has been reviewed previously [1–3].

In this chapter we review the efficacy, advantages, and disadvantages of conventional washing and sanitizing agents for fresh fruits and vegetables. We also examine the regulatory status of interventions for decontamination of produce and equipment. We examine the types of equipment available for treatment application and their performance. We briefly consider some of the factors that limit the efficacy of cleaning and sanitizing agents and methods of treatment. We examine the potential of new treatments for produce decontamination. We also consider the problem of decontamination of fresh fruits and vegetables in foodservice situations or in the home. This chapter does not examine vapor-phase treatments, surface pasteurization, nonthermal physical treatments, or biological control methods, all of which are covered elsewhere in the book.

17.2   CONVENTIONAL WASHING TECHNOLOGY

17.2.1   WASHING AGENTS

17.2.1.1   Chlorine

Most freshly harvested fruits and vegetables are washed by the grower, packer, or processor to remove soil, plant debris, pesticide residues, and microorganisms from the commodity surface. This may be accomplished by spraying or immersion in water or solutions containing one of a number of cleaning or sanitizing agents, using equipment designed for each particular commodity type, e.g., leafy vegetables, root vegetables, fruit vegetables, tree fruits, or melons. Chlorine is the most widely used sanitizing agent for fresh produce. It may be added to wash water as Cl_2 gas or, more commonly, as sodium or calcium hypochlorite. In water, at pH levels and concentrations used on produce, these chlorine sources are converted to hypochlorous acid and hypochlorite ion in a ratio determined by the solution pH [4,5]. At pH 6.0, roughly 97% of the unreacted chlorine is hypochlorous acid, whereas, at pH 9.0, 97% is hypochlorite ion. The antimicrobial activity of these solutions is due largely to hypochlorous acid rather than to hypochlorite.

The concentration of chlorine in a wash solution is sometimes expressed as total available chlorine (or total residual chlorine = combined residual chlorine + free residual chlorine), based on the calculated amount present in the added hypochlorite or chlorine, or determined by oxidation of KI to I_2, which may not be indicative of the actual potency as a sanitizer because of the inclusion of reaction products such as monochloramine which are not very effective as sanitizers. Preferably, the chlorine concentration can be expressed as free available (or residual) chlorine, the sum of hypochlorous acid and hypochlorite ion concentrations [5]. The total or free chlorine concentration can be monitored by means of test kits, based on colorimetry (www.chemetrics.com; www.emscience.com, www.hach.com), or by measurement of the oxidation–reduction potential (ORP). Chlorine is highly reactive with certain types of compounds in organic materials and soils that are leached or washed from fruits and vegetables. If this chlorine sink is excessive, the free chlorine concentration will be depleted rapidly. Computerized ORP systems that monitor the pH and chlorine concentration can be used to control the level of chlorine in a wash tank in such situations (www.pulsein struments.net; numerous other suppliers listed on www.globalspec.com).

Use levels of chlorine will depend on allowable levels, the commodity, and the anticipated microbial load. The U.S. Food and Drug Administration (FDA) specifies a use level for washing fruits and vegetables not to exceed 0.2% when followed by a potable water rinse [6]. The U.S. Environmental Protection Agency (EPA) exempts calcium hypochlorite "from the requirement of a tolerance when used preharvest or postharvest in solution on all raw agricultural commodities" [7]. The concentration range of 50 to 200 ppm is commonly used for most commodities. However, as much as 20,000 ppm calcium hypochlorite may be used to sanitize alfalfa seeds intended for sprout production because of the failure of other treatments to disinfect adequately seeds and sprouts, and the high risk that sprouts grown from contaminated seeds may be a source of salmonella or *Escherichia coli* O157:H7 outbreaks [8–10].

Chlorine is highly effective for inactivating planktonic cells of bacteria, yeasts, molds, and viruses, although bacterial and fungal spores are considerably more resistant [5]. However, chlorine is less effective for inactivating bacteria attached to produce surfaces or embedded within the product [11–18]. Typically, population reductions of native microflora on produce surfaces or of human pathogens on inoculated produce are no greater than 2 logs (99%). While such reductions can greatly reduce spoilage, they are insufficient to ensure safety in the event of contamination with human pathogens.

The activity of chlorinated water may be increased by the addition of an acidulant or buffer so that the pH is shifted from an alkaline value (about pH 9) to the neutral to slightly acidic range (pH 6 to 7), thereby increasing the proportion of hypochlorous acid in the equilibrium mixture. Organic acids such as citric acid or mineral acids such as phosphoric or hydrochloric acid can be used for this purpose. If the solution pH is too low (e.g., below pH 4), hypochlorous acid may be converted to free chlorine which is subject to off-gassing. This will result in a loss of activity and may be potentially hazardous. Additionally, equipment corrosion is enhanced as pH levels drop below as well as rise above neutrality. Unpublished data obtained at the Eastern Regional Research Center indicated that hypochlorite solutions acidified with a mineral acid were more stable than solutions acidified with citric acid [19]. Buffers for hypochlorite solutions are available commercially (www.cerexagri.com).

The effectiveness of chlorine in inactivating microorganisms on produce may be enhanced by adding a surfactant to the solution so that it can penetrate into the irregular crevices and pores on produce surfaces where microorganisms may lodge and escape contact with a sanitizer. Several commercial surfactant formulations have been developed for this purpose (www.cerexagri.com/usa/Markets/Cleaners). Addition of a nonionic surfactant improved the efficacy of chlorine against decay fungi in pears [20,21]. Washing formulations containing sodium hypochlorite, buffers, and surfactants have been described by Park et al. [22] and marketed by Bonagra Technologies under the name Safe-T-Washed^{TM} (www.bonagra.com). The efficacy of chlorine in reducing the microbial flora of shredded iceberg lettuce was increased by elevating the solution temperature to 47°C [23]. However, no greater reduction of non-pathogenic E. coli (ATCC 25922) populations on inoculated apples was obtained when apples were washed at 50 or 60°C compared to 20°C using 200 ppm Cl_2 (added as sodium hypochlorite), adjusted to pH 6.5 with citric acid [19].

Chlorine's major advantages are its broad spectrum of antimicrobial activity, ease of application, and low cost. However, chlorine is highly corrosive and may damage stainless steel equipment after prolonged exposure. Its other major disadvantages are rapid depletion in the presence of a high organic load [24], and the potential carcinogenicity and mutagenicity of its reaction products with organic constituents of foods [25–27]. This is a matter of concern to processors, regulators, and consumers [28]. For these reasons, and the desirability of obtaining greater population reductions, the development of alternative sanitizing agents has been an active area of research, and a limited number of agents suitable for use on fresh produce have been commercialized.

Electrolyzed water, a technology developed largely in Japan [29,30], is really a special case of chlorination. This technology is discussed in detail in Chapter 22.

17.2.1.2 Alternatives to Chlorine

A number of commercial detergent formulations have been developed for washing fruits and vegetables. In addition, three approved sanitizing agents

TABLE 17.1
Advantages and Disadvantages of Commercially Available Sanitizing Agents for Washing Fresh Fruits and Vegetables

Sanitizing agent	Use level (ppm)	Advantages	Disadvantages
Chlorine	50–200	Easy to apply, inexpensive, effective against all microbial forms, not affected by hard water, easy to monitor, FDA approved	Decomposed by organic matter, reaction products may be hazardous, corrosive to metals, irritating to skin, activity pH-dependent, population reductions limited to <1–$2\,logs$
Ozone	0.1–2.5	More potent antimicrobial than chlorine, no chlorinated reaction products formed, economical to operate, self-affirmed GRAS, but FDA review possible, activity not pH-dependent	Requires on-site generation, requires good ventilation, phytotoxic at high concentrations, corrosive to metals, difficult to monitor, higher capital cost than chlorine, no residual effect, population reductions limited to <1–$2\,logs$
Chlorine dioxide	1–5	More potent than chlorine, activity not pH-dependent, fewer chlorinated reaction products formed than with Cl_2, effective against biofilms, FDA approved, residual antimicrobial action, less corrosive than Cl_2 or O_3	Must be generated on-site, explosive at high concentrations, not permitted for cut fruits and vegetables, population reductions limited to <1–$2\,logs$, generating systems expensive
Peroxyacetic acid	≤ 80	Broad spectrum antimicrobial action, no pH control required, low reactivity with soil, effective against biofilms, FDA approved, no hazardous breakdown products, no on-site generation required, monitoring not difficult, available at safe concentration	Population reduction limited to <1–$2\,logs$, strong oxidant, concentrated solutions may be hazardous

are available as alternatives to chlorine: chlorine dioxide (or acidified sodium chlorite), ozone, and peroxyacetic acid. The advantages and disadvantages of the agents described in the following sections are compared in Table 17.1.

17.2.1.2.1 Detergent Formulations

Among the detergents approved by the FDA for washing produce are sodium *n*-alkylbenzenesulfonate, sodium dodecylbenzenesulfonate, sodium mono- and dimethyl naphthalenesulfonates, sodium 2-ethylhexyl sulfate, and others [6]. These formulations may be neutral in pH, acidic due to the presence of citric or phosphoric acid, or alkaline because of the addition of sodium or potassium hydroxide. Major suppliers of detergent formulations for produce cleaning include Cerexagri (formerly Elf Atochem N.A., Inc., source of Decco products) (800-221-0925; www.cerexagri.com), Microcide, Inc. (www.microcideinc.com), and Alex C. Fergusson, Inc. (800-345-1329; www.afcocare.com).

These products are designed to remove soil and pesticide residues from produce and do not contain antimicrobial agents *per se*. Relatively little information is available concerning the ability of these products to remove or inactivate microorganisms attached to produce surfaces. However, their use can result in significant population reductions. Sapers *et al.* reported that some commercial washing formulations could achieve population reductions as great as 1 to 2 logs in decontaminating apples inoculated with a non-pathogenic *E. coli*, comparable to reductions obtained with hypochlorite [16]. When these products were applied at 50°C instead of at ambient temperature, a 2.5 log reduction was obtained. Wright *et al.* [31] reported similar efficacy with a commercial phosphoric acid fruit wash and with a 200 ppm hypochlorite wash, each applied to apples inoculated with *E. coli* O157:H7. Kenney and Beuchat [32] compared the efficacy of representative commercial cleaning agents in removing or inactivating *E. coli* O157:H7 and *S. muenchen* on spot-inoculated apples. They obtained reductions as great as 3.1 logs with an alkaline product and as great as 2.7 logs with an acidic product, reductions generally being greater with salmonella. Raiden *et al.* [33] compared the efficacy of water, sodium lauryl sulfate, and Tween 80 in removing *Salmonella* spp. and *Shigella* spp. from the surface of inoculated strawberries, tomatoes, and leaf lettuce. They obtained high removal rates but concluded that the detergents were no more effective than water. However, this result may have been a reflection of the brief time interval (1 hour) between inoculation and treatment, which may have been insufficient for strong bacterial attachment. In nature, the interval between preharvest contamination and postharvest application of a wash may be days or weeks, sufficient time for strong attachment and even biofilm formation.

In a study of cantaloupe rind decontamination, Sapers *et al.* [34] reported reductions in the total aerobic plate count of about 1.3 logs when the rind was washed with a 1% solution of a commercial produce wash containing dodecylbenzene sulfonic acid and phosphoric acid (pH 2) at 50°C. Sequential washing with this product followed by treatment with 1% hydrogen peroxide, both at 50°C, resulted in a 3.1 log reduction. Both washes extended the shelf life of fresh-cut cantaloupe prepared from the treated melons. No significant population reductions were obtained when the cantaloupe rind was washed with aqueous solutions of sodium dioctyl sulfosuccinate or sodium 2-ethylhexyl sulfate.

17.2.1.2.2   Chlorine Dioxide

Solutions of chlorine dioxide and acidified sodium chlorite have been used commercially as alternatives to chlorine for sanitizing fresh produce.

Chlorine dioxide is considered to be efficacious against many classes of microorganisms [5]. Chlorine dioxide and acidified sodium chlorite are approved by the FDA for use on fresh produce [35,36], but chlorine dioxide is not permitted for use on fresh-cut products. Chlorine dioxide must be generated on-site, usually by reaction of sodium chlorite with an acid or chlorine gas. Information concerning various proprietary generating and stabilizing systems are available from suppliers such as Vulcan Chemical (800-873-4898), Alcide Corp. (Sanova®; www.alcide.com/sanova), CH2O Inc. (Fresh-Pak™; www.ch2o.com), Rio Linda Chemical Co., Inc. (916-443-4939), Bio-Cide International, Inc. (Oxine®; www.biocide.com), International Dioxcide (www.idiclo2.com), Alex C. Fergusson (800-345-1329; www.afco care.com), CDG Technology, Inc. (www.cdgtechnology.com), and others. Unlike chlorine, chlorine dioxide is claimed to be effective over a broad range of pH levels, more resistant to neutralization by the organic load, and unlikely to produce trihalomethanes (see Oxine Technical Data Sheet; www.bio-cide.com). Chlorine dioxide also is claimed to be less corrosive than chlorine and to be effective against bacteria in biofilms. However, generation of chlorine dioxide by reaction of sodium chlorite with acid or Cl_2 must be carefully controlled to avoid production of high concentrations of ClO_2 gas which can be toxic and explosive (MSDS for IVR-San 15 sodium chlorite; www.ch2o.com). Additionally, unlike chlorine, chlorine dioxide dissolves in water as a gas and is subject to off-gassing if the water is moving or used in washers. In that situation, special venting would be required to prevent worker discomfort.

The efficacy of chlorine dioxide in disinfecting produce is comparable to that of chlorine. Published reports indicate that chlorine dioxide and related products were potentially effective in preventing potato spoilage by *Erwinia carotovora* [37], reducing populations of *E. coli* O157:H7, *S.* Montevideo, and poliovirus on inoculated strawberries [38], reducing the population of *E. coli* O157:H7 on inoculated apples (but at a treatment level 16 times the recommended concentration) [39], and suppressing decay in pears [40]. Treatments were less effective in suppressing microbial growth on the surface of cucumbers [41]. Fett obtained only a 1 log reduction in alfalfa sprouts irrigated with acidified sodium chlorite [42]. Population reductions of *L. monocytogenes* on uninjured surfaces of inoculated green bell peppers, washed with ClO_2 solution (3 mg/l), were about 2 logs greater than could be achieved with a water wash, but reductions were negligible on injured surfaces [43]. In contrast, these investigators obtained population reductions of 7.4 and 3.6 logs on uninjured and injured surfaces of peppers, respectively, using a ClO_2 gas treatment (see Chapter 18).

17.2.1.2.3 Ozone

The efficacy of ozone in killing human pathogens and other microorganisms in water is well established [44], and it is widely used as an alternative to chlorine in municipal water treatment systems and for production of bottled water

[45]. Ozone is effective in killing food-related microorganisms [46] and has been approved for use on foods by the FDA [47]. Potential applications of ozone in disinfecting foods have been reviewed [48,49]. Ozone is effective in reducing bacterial populations in flume and wash water and may have some applications as a chlorine replacement in reducing microbial populations on produce [50,51]. Ozone treatment was effective in suppressing decay of table grapes by *Rhizopus stolonifer* [52]. Use levels of 0.5 to 4.0 µg/ml are recommended for wash water and 0.1 µg/ml for flume water [53,54].

However, not all ozone treatments show high efficacy. Ozone treatment of fresh-cut lettuce, inoculated with a mixture of natural microflora, yielded reductions of only 1.1 logs [18]. Treatment of lettuce, inoculated with *Pseudomonas fluorescens*, with 10 µg/ml of ozone for 1 minute achieved less than a 1 log population reduction [50]. While ozone treatment of apples inoculated with *E. coli* O157:H7 was effective in reducing populations on the surface (3.7 log reduction), reductions were < 1 log in the stem and calyx regions [55]. Ozone treatment of pears (5.5 µg/ml water for 5 minutes) was ineffective in reducing postharvest fungal decay [56]. Population reductions obtained by ozone treatment of alfalfa seeds inoculated with *E. coli* O157:H7 were only marginally better than those for water-treated controls [57]. In another study, ozone treatment of alfalfa seeds, inoculated with *L. monocytogenes*, was ineffective in reducing the population of this pathogen, while treatment of inoculated alfalfa sprouts reduced the *L. monocytogenes* population by < 1 log and was phytotoxic to the sprouts [58]. These results are probably a reflection of the difficulty in contacting and inactivating bacteria attached to produce surfaces in inaccessible sites (see Chapters 2 and 3).

One of the major advantages claimed for ozone is the absence of potentially toxic reaction products. However, ozone must be adequately vented to avoid worker exposure [48]. Ozone has to be generated on-site by passing air or oxygen through a corona discharge or UV light [48]. A number of commercial systems for generating ozonated water for produce washing are available. Information about commercial ozone generators is available on-line from Air Liquide (www.airliquide.com), Praxair, Inc. (www.praxair.com), Novazone (www.novazone.net), Pure Ox (www.pureox.com), Osmonics, Inc. (www.osmonics.com/food), Ozonia North America, Inc. (www.ozonia.com), Lynntech, Inc. (www.lynntech.com), Clean Air & Water Systems, Inc. (360-394-1525), Electric Power Research Institute (EPRI; www.epri.com), and others. For information about ozone gas disinfection treatments, see Chapter 18.

17.2.1.2.4 Peroxyacetic Acid

Peroxyacetic acid (peracetic acid) is an equilibrium mixture of the peroxy compound, hydrogen peroxide, and acetic acid [59–61]. The superior antimicrobial properties of peroxyacetic acid are well known [59]. Peroxyacetic acid is approved by the FDA for addition to wash water at concentrations not to exceed 80 ppm [6]. Under EPA regulations, an exemption from the requirements of a tolerance was established for peroxyacetic acid as an antimicrobial treatment for fruits and vegetables at concentrations up to 100 ppm [62]. Much higher concentrations are permitted for sanitizing food contact surfaces [63]. Peroxyacetic acid decomposes into acetic acid, water, and oxygen, all harmless residuals.

Peroxyacetic acid is recommended for use in treating process water, but Ecolab, one of the major suppliers, is also claiming substantial reductions in microbial populations on fruit and vegetable surfaces [64]. However, company literature provides insufficient information on methodology to assess treatment efficacy (www.ecolab.com/initiatives/foodsafety). Population reductions for aerobic bacteria, coliform bacteria, and yeasts and molds on fresh-cut celery, cabbage, and potatoes treated with 80 ppm peroxyacetic acid were less than 1.5 logs [65]. Addition of 40 ppm Tsunami 100 (the Ecolab peroxyacetic acid product) to the irrigation water used during sprout propagation did not suppress the outgrowth of the native microflora [42]. Treatment with 100 ppm Tsunami reduced the population of *E. coli* O157:H7 and *S.* Montevideo on inoculated strawberries by about 97% [38]. Several published studies have looked at the efficacy of peroxyacetic acid against *E. coli* O157:H7 on inoculated apples. Attempts to disinfect apples, inoculated with *E. coli* O157:H7, by washing with 80 ppm peroxyacetic acid 30 minutes after inoculation resulted in a 2 log reduction compared to a water wash [31]. However, in another study where inoculated apples were held for 24 hours before washing (allowing more time for attachment), an 80 ppm peroxyacetic acid treatment reduced the *E. coli* O157:H7 population by less than 1 log; at 16 times the recommended concentration, a 3 log reduction was obtained [39]. Sapers *et al.* [16] reported similar results with apples inoculated with a nonpathogenic *E. coli*. Like ozone and chlorine dioxide, low concentrations of peroxyacetic acid are effective in killing pathogenic bacteria in aqueous suspension [59]. Addition of octanoic acid to peroxyacetic acid solutions increased efficacy in killing yeasts and molds in fresh-cut vegetable process waters but had little effect on population reductions on fresh-cut vegetables [65].

Peroxyacetic acid is a strong oxidizing agent and can be hazardous to handle at high concentrations, but not at strengths marketed to the produce industry. Peroxyacetic acid is available at various strengths from Ecolab, Inc. (www.ecolab.com), FMC Corp. (www.fmcchemicals.com), and Solvay Interox (www.solvayinterox.com).

17.2.2 WASHING EQUIPMENT

17.2.2.1 Types of Washers

Washing equipment for produce is designed primarily for removal of soil, debris, and any pesticide residues from the harvested commodity. The design of most commercial equipment has not taken into account requirements for the reduction of microbial populations on produce surfaces although this is a desirable goal of washing.

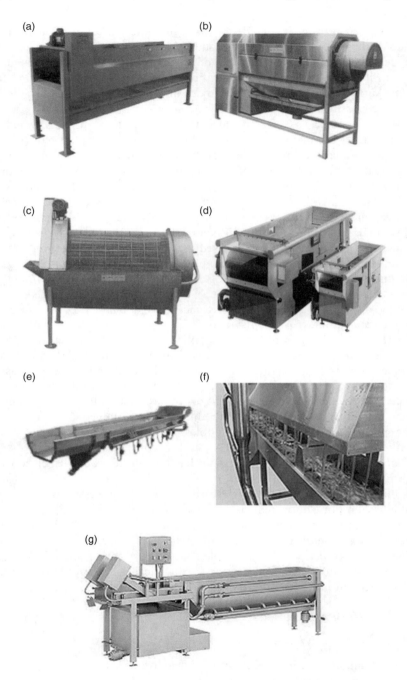

FIGURE 17.1 Commercial washing equipment for fruits and vegetables: (a) flat-bed brush washer; (b) U-bed brush washer; (c) rotary washer; (d) pressure washer; (e, f) flume washers; (g) helical washer.

Numerous types of washers have been developed for cleaning fresh fruits and vegetables, varying in complexity from a garden hose used for cleaning apples prior to farm-scale cider production (an unsatisfactory procedure due to lack of control) to sophisticated systems employing rotating brushes and applying heated water under pressure with agitation. The more common types of commercial washers for produce include dump tanks, brush washers, reel washers, pressure washers, hydro air agitation wash tanks, and immersion pipeline washers (Figure 17.1). Major suppliers of such equipment are listed on the Postharvest Resources website of the University of Florida (http://postharvest.ifas.ufl.edu). The choice of washer for a particular commodity will depend on such characteristics of the commodity as shape, size, and fragility. It is obvious that equipment requirements are quite different for cut lettuce than for tomatoes or potatoes.

17.2.2.2 Efficacy of Washers

The efficacy of commercial flat-bed and U-bed brush washers in removing or inactivating a nonpathogenic *E. coli* on artificially contaminated apples was investigated by Annous *et al.* [66] and Sapers [3]. These studies demonstrated that the *E. coli* population could be reduced by about 1 log (90%) by passage of the apples through a dump tank with minimal agitation (Table 17.2). However, further cleaning of the apples in a flat-bed brush washer had little further effect on the *E. coli* population, irrespective of the cleaning or sanitizing agent used (water, 200 ppm Cl_2, 1% acidic detergent, 8% trisodium phosphate, 5% H_2O_2). Similar results were obtained with a U-bed brush washer. Subsequent studies by the investigators showed that the bacteria that had attached in the

TABLE 17.2

Decontamination of Apples Inoculated with *E. coli* (Strain K12) with Sanitizing Washes Applied in a Flat-Bed Brush Washer

		E. coli ($\log_{10}$ CFU/g)[a]		
Wash treatment	Temp. (°C)	Before dump tank	After dump tank	After brush washer
Water	20	5.49 ± 0.09	4.92 ± 0.37	4.81 ± 0.26
	50	5.49 ± 0.09	5.03 ± 0.15	4.59 ± 0.08
200 ppm Cl_2	20	5.87 ± 0.07	5.45 ± 0.05	5.64 ± 0.23
8% Na_3PO_4	20	5.49 ± 0.09	5.02 ± 0.43	4.98 ± 0.02
	50	5.49 ± 0.09	5.02 ± 0.08	4.75 ± 0.45
1% acidic detergent	50	5.87 ± 0.07	5.49 ± 0.03	5.42 ± 0.50
5% H_2O_2	20	5.87 ± 0.07	5.46 ± 0.40	5.27 ± 0.09
	50	5.87 ± 0.07	5.54 ± 0.31	5.49 ± 0.10

[a] Mean of four determinations $\pm$ standard deviation.

From Annous, B.A. *et al.*, *J. Food Prot.*, 64, 159, 2001. Reprinted with permission. Copyright International Association for Food Protection, Des Moines, IA.

TABLE 17.3
Distribution of _E. coli_ (ATCC 25922) on Surfaces of Inoculated Apples Before and After Washing with 5% H_2O_2 at 50°C

| | Log_{10} (CFU/cm^2) | | | |
| | 24 h after inoculation | | 72 h after inoculation | |
Location	Inoculated	Washed	Inoculated	Washed
Skin except at calyx and stem ends	4.77	2.05	4.37	1.63
Skin at calyx end of core	7.26	5.20	6.79	4.46
Skin on stem end of core	6.63	5.06	5.61	4.89

From Sapers, G.M. _et al._, _J. Food Sci._, 65, 529, 2000. Reprinted with permission.

relatively inaccessible stem and blossom ends of the apples, or were internalized within the latter region, survived washing while _E. coli_ attached elsewhere on the apple surface were readily inactivated (Table 17.3). Greater efficacy was obtained when the apples were washed by full immersion in a sanitizing solution with vigorous agitation [67].

Gagliardi _et al._ [68] examined commercial practices for washing melons produced in the Rio Grande River Valley of Texas. They reported little or no reduction in the population of coliforms, fecal coliforms, enterococci, and fecal enterococci in cantaloupes and honeydew melons that were washed with water in a tank and then spray rinsed on a conveyor line. Use of chlorinated water in the secondary rinse appeared to reduce the populations of fecal coliforms and fecal enterococci but not total coliforms and enterococci. Laboratory-scale washing studies with cantaloupes that had been dip-inoculated with _Salmonella_ Stanley or a nonpathogenic _E. coli_ (ATCC 25922) demonstrated that the population reductions obtained by immersion of the melons in 200 ppm Cl_2 or 5% H_2O_2 decreased as the time interval between inoculation and washing increased from 24 hours to 5 days [69,70]. However, the efficacy of these treatments in inactivating _L. monocytogenes_ on inoculated cantaloupes was not dependent on the length of storage between inoculation and treatment [71]. Sapers _et al._ obtained minimal inactivation of _E. coli_ B-766 (a surrogate for _S._ Poona) when dip-inoculated cantaloupes were immersed in 300 ppm Cl_2 for 3 minutes [72]. Apparently, cantaloupes are especially difficult to disinfect, even if fully immersed in the sanitizing solution. This may be due to the movement, attachment, and possible biofilm formation by the targeted bacteria within inaccessible pores in the netting so that contact between the sanitizing solution and the attached bacteria is minimal. This is borne out by the success of treatments with 5% H_2O_2 at 70°C or near boiling water where heat penetration contributes to the efficacy of the antimicrobial treatment [73] (see Chapter 10). Such treatments can greatly reduce the risk of transfer of human pathogens from the rind surface to the flesh during fresh-cut processing.

17.2.3 FACTORS LIMITING THE EFFICACY OF WASHING

The action of commercial washing agents and equipment in removing or inactivating microorganisms on fresh produce is not well understood. In general, microbial populations on produce surfaces are not easily detached or inactivated for a number of reasons discussed in Chapters 2 and 3. Briefly, the microbial contaminants may become strongly attached to the produce surface by physical forces within a short time of contamination or incorporated within a biofilm over a longer time period. Microbial contaminants may be located in a protected attachment site, e.g., a cut surface, puncture, or pore, where a wash solution cannot reach. Microorganisms also may become internalized within the commodity either during crop production or when submerged in water in a packing plant dump tank or flume as a consequence of infiltration driven by a negative temperature differential or by hydrostatic pressure. Consequently, the inaccessible population will escape direct contact with a cleaning or sanitizing agent in a commercial washer. These conditions are discussed in greater detail in an earlier review article [3] and in Chapter 3.

17.3 NOVEL WASHING TECHNOLOGY

Because the commercially available alternatives to chlorine discussed above generally cannot achieve population reductions of human pathogens on contaminated produce much in excess of 2 logs, which is insufficient to ensure safety, a number of experimental treatments have been examined to obtain greater efficacy. The efficacy and regulatory status of some of these experimental treatments are described in the following.

17.3.1 HYDROGEN PEROXIDE

Hydrogen peroxide is a highly effective antimicrobial agent against bacteria but is less active against yeasts, fungi, and viruses [59]. Characteristics and potential food applications of hydrogen peroxide as a sanitizer for produce were recently reviewed by the author [74]. Hydrogen peroxide may be considered as a potential alternative to chlorine. Numerous studies have demonstrated the efficacy of dilute hydrogen peroxide in sanitizing fresh produce including mushrooms [75–77], apples [16,67,78], melons [34,69,70,73], eggplant, and sweet red pepper [80]. In side-by-side comparisons, dilute (1 to 5%) hydrogen peroxide washes were at least as effective as 200 ppm chlorine [16,79]. When applied to apples with vigorous agitation at an elevated temperature (50 to 60°C), population reductions approaching 3 logs were obtained [67]. However, temperatures exceeding 60°C could not be used without inducing browning of the apple skin. Hydrogen peroxide treatments were ineffective in decontaminating sprouts [42] or the seeds used to produce sprouts [81].

While treatment with hydrogen peroxide vapor can reduce microbial populations on grapes [82], melons [83], and prunes [84], required treatment times are long compared to the application of a dilute hydrogen peroxide dip [85]. The vapor treatments proved to be ineffective with apples [86] and produced discolorations with mechanically damaged berries [85].

The regulatory status of hydrogen peroxide as a washing agent for produce is unclear. The FDA has jurisdiction if the washing treatment is applied as part of a processing operation, while the EPA has jurisdiction if the treatment is applied to a raw commodity. While fresh produce clearly falls within the EPA regulations, fresh-cut produce is under FDA regulations. However, if the wash treatment is applied to the raw produce before cutting, and if this operation is carried out in a receiving area, separate from the processing room, it would appear that EPA regulations apply. Under FDA regulations, hydrogen peroxide is GRAS (generally recognized as safe) for some specified food applications, provided that residual H_2O_2 is removed "by appropriate physical and chemical means during processing," but the regulation does not cover hydrogen peroxide as a washing or sanitizing agent for produce [87]. According to an Agency Response Letter (GRAS notice no. GRN 000014, May 26, 1999) a petition to the FDA to amend the regulation would be required to seek approval for a new application (in this case, reduction of the microbial load on onions prior to dehydration; http://vm.cfsan.fda.gov/~rdb). Peroxyacetic acid formulations, which contain low levels of hydrogen peroxide (59 ppm), are approved by the FDA for use in washing fruits and vegetables [6]. A higher concentration is permitted if the formulation is used to sanitize food contact surfaces [63] Under EPA regulations, postharvest hydrogen peroxide applications to produce as an antimicrobial treatment are exempt from the requirements of a tolerance if the concentration is $\leq 1\%$ per application [88].

The presence of residual hydrogen peroxide should not represent an obstacle to use of this agent as a produce sanitizer. Most fruits and vegetables contain sufficient catalase to permit rapid breakdown of residual peroxide to water and oxygen. Peroxide residues could not be detected in mushrooms, apples, or cantaloupes following hydrogen peroxide wash treatments [16,34,77].

Information on hydrogen peroxide applications can be obtained from FMC Corp. (www.fmcchemicals.com), Solvay Interox (www.solvayinterox. com), US Peroxide (h2o2.com), and Degussa Corp. (www.degussa.com). BiosSafe Systems (www.biosafesystems.com) is marketing a formulation containing hydrogen peroxide and peroxyacetic acids (Storox®) for sanitizing fruits and vegetables; the recommended maximum use level is 0.27%.

17.3.2 TRISODIUM PHOSPHATE AND OTHER ALKALINE WASHING AGENTS

Trisodium phosphate (TSP) has been marketed by Rhodia Specialty Phosphates (www.rhodia-phosphates.com) as an antimicrobial rinse (AvGard®,

Assur-Rinse®) to reduce human pathogen populations on processed beef and poultry. TSP is classified as GRAS by the FDA [89].

The antimicrobial activity of TSP probably is due to its high pH (pH 12) which disrupts the cytoplasmic membrane [90,91]. Highly alkaline washes based on sodium and potassium hydroxide (pH 11 to 12) resulted in 3 log reductions in the population of a nonpathogenic *E. coli* on surface-inoculated oranges [92]. A 30-minute dip in 0.25% calcinated calcium suspension, another highly alkaline product derived from oyster shells (pH 10), reduced the native bacterial population on cucumbers by about 2 logs [93]. In a more recent study, Bari *et al.* [94] reported population reductions exceeding 5 logs on tomatoes that had been surface inoculated with *E. coli* O157:H7, salmonella strains, or *L. monocytogenes* and treated with 0.5% calcinated calcium. These exceptionally high population reductions (for a wash) may be a reflection of the brief interval (30 minutes) between inoculation and treatment used by these investigators. Sapers *et al.* [67] obtained population reductions approaching 3 logs when apples that had been dip-inoculated with *E. coli* (ATCC 25922) were washed with 5% hydrogen peroxide, followed by brushing the calyx and stem areas with a paste of calcinated calcium; the population reduction was < 2 logs with only the peroxide wash. TSP solutions (12 to 15%) were highly effective in reducing *S.* Montevideo populations on inoculated tomato surface but failed to inactivate completely this organism in the tomato core tissue [95]. Survival in the latter tissue probably resulted from bacterial infiltration. Sapers *et al.* [78] reported a 2 log reduction in a nonpathogenic *E. coli* strain on inoculated apples washed with 4% TSP at 50°C. A 1% TSP wash reduced the population of *E. coli* O157:H7 and *S.* Montevideo on strawberries by 93 and 96%, respectively [38]. Treatment of lettuce with 2% TSP was ineffective in killing *L. monocytogenes* [14]. Addition of 0.3% TSP to the irrigation water was ineffective in reducing the native microflora on alfalfa sprouts [42]. TSP was reported to be highly effective in inactivating *E. coli* O157:H7 in biofilms but less effective against *S.* Typhimurium and *L. monocytogenes* in biofilms [96].

17.3.3 ORGANIC ACIDS

Organic acids such as lactic and acetic acids are effective antibacterial agents [97] and are classified by the FDA as GRAS [98,99] (21CFR184.1005; 21CFR184.1061). Lactic acid dips and sprays are used commercially to decontaminate animal carcasses containing *E. coli* O157:H7, *L. monocytogenes*, and salmonella [100] (see additional information from Purac America, Inc., www.purac.com). Lactic acid rinses might have applications for the decontamination of fruits and vegetables. A 5% acetic acid wash was reported to reduce the population of *E. coli* O157:H7 on inoculated apples by about 3 logs [31]. In another study, apples that had been inoculated with *E. coli* O157:H7 were treated with 5% acetic acid at 55°C for as long as 25 minutes. While the *E. coli* population was greatly reduced in the apple skin and stem areas, as many as 3 to 4 logs survived in the calyx tissue [101]. In a more recent study, application of 2.4% acetic acid to apple disks that had been inoculated with *S. mbandaka* or *S.* Typhimurium resulted in population reductions of 1.1 and 1.4, respectively [102]. However, the combination of 5% acetic acid with 5% hydrogen peroxide yielded a population reduction approaching 4 logs. It is not clear whether organic acid treatments would produce off-flavors or discoloration in treated produce.

17.3.4 OTHER EXPERIMENTAL ANTIMICROBIAL WASHING AGENTS

Cetylpyridinium chloride (CPC) is being marketed as Cecure® for use in oral hygiene products and may have application as an antimicrobial rinse for fresh produce and other foods. Yang *et al.* [103] reported population reductions in the range 1 to 2 logs for *S.* Typhimurium and *E. coli* O157:H7 on inoculated fresh-cut lettuce, treated by spraying with 0.3% CPC. Similar reductions were obtained with strawberries inoculated with *E. coli* O157:H7 or *S.* Montevideo and immersed in 0.1% CPC at 43°C [38]. However, regulatory approval for this agent has not yet been obtained (www.safefoods.net/cecure.htm). Activated lactoferrin, which prevents attachment of bacteria to meat, is approved by the FDA and USDA for application to beef as a carcass rinse [104] (also see www.activinlf.com). However, there are no reports of its applicability to fruits and vegetables. Silver and copper ions are known to exert antimicrobial activity against bacteria in water [105], and ion generators have been marketed for disinfection of water in swimming pools, irrigation systems, and various other commercial applications (Tew Manufacturing Corp., 800-380-5839; T.P. Technology plc, www.tarn-pure.com). Application of this technology to produce packing lines and dump tanks at recommended levels of 0.50 ppm copper and 0.035 to 0.05 ppm silver has been proposed (Tew Manufacturing Corp.), but published efficacy data are lacking, and the regulatory status of such applications is unclear.

17.3.5 SYNERGISTIC TREATMENT COMBINATIONS

Certain combinations or sequences of treatments may show synergism in inactivating or detaching microbial contaminants on produce. Such behavior might be anticipated if the individual treatments have different modes of action, e.g., cell membrane disruption and oxidation. Several examples of promising combination treatments have been reported: the sequential washing of cantaloupes with detergents and hydrogen peroxide [34] and the application of an acetic acid–hydrogen peroxide combination to inoculated apple disks [102]. Lin *et al.* [106] investigated the inactivation of *E. coli* O157:H7, *S. enterica* serotype Enteritidis, and *L. monocytogenes* by combinations of hydrogen peroxide and lactic acid and hydrogen peroxide with mild heat. Further research in this area may yield treatment combinations that show greater efficacy towards bacteria located in punctures or pores or incorporated in biofilms on produce surfaces.

17.4 FOODSERVICE AND HOME APPLICATIONS

While conventional sanitizing agents, applied to produce with commercial-scale washing equipment, have the capability of achieving 1 to 2 log population reductions in contaminated produce, this option is not generally available for foodservice and consumer applications. Consumers and operators of delicatessens, restaurants, and other foodservice establishments do not have the technical skills or knowledge to prepare the more potent sanitizer solutions used commercially nor do they have access to commercial washing equipment. Duff *et al.* [107] developed an economic model to evaluate the potential cost-effectiveness of a disinfection program that targets high-risk food preparation activities in household kitchens. They concluded that such a program would be cost-effective. What options are available to consumers and foodservice managers so that they can provide some meaningful level of protection to their families or customers?

17.4.1 FDA RECOMMENDATIONS

The FDA advises consumers to: "Wash all fresh fruits and vegetables with cool tap water immediately before eating. Don't use soap or detergents. Scrub firm produce, such as melons and cucumbers, with a clean produce brush. Cut away any bruised or damaged areas before eating." Consumers are also advised to:

> Wash surfaces often. Cutting boards, dishes, utensils, and counter tops should be washed with hot soapy water and sanitized after coming in contact with fresh produce, or raw meat, poultry, or seafood. Sanitize after use with a solution of 1 teaspoon of chlorine bleach in 1 quart of water. Don't cross contaminate. Use clean cutting boards and utensils when handling fresh produce. If possible, use one clean cutting board for fresh produce and a separate one for raw meat, poultry, and seafood. During food preparation, wash cutting boards, utensils, or dishes that have come into contact with fresh produce, raw meat, poultry, or seafood. Do not consume ice that has come in contact with fresh produce or other raw products (www.fda.gov/bbs/topics/ANSWERS/2002/ANS01167.html/).

In the situation where a particular fruit or vegetable is suspect, more specific advice is provided. For example, in response to an outbreak of hepatitis A in green onions, the FDA recommended: "Cook green onions thoroughly. This minimizes the risk of illness by reducing or eliminating the virus. Cook in a casserole or sauté in a skillet" and "Cook sprouts. This significantly reduces the risk of illness" [108]. While a kill step is undoubtedly effective, it would not be applicable to many fruits and vegetables that would no longer be considered "fresh" if subjected to a cook or full blanch and would lose their appeal to consumers. Washing produce without a sanitizer is not likely to achieve the population reductions that can be obtained with commercial sanitizing agents and equipment.

17.4.2 OTHER OPTIONS

Alternative methods of surface sanitizing cantaloupes were examined by Barak et al. [109]. They reported reductions in the bacterial load of 70, 80, and 90% by scrubbing the melons with a vegetable brush in tap water, washing with soap, and dipping in 150 ppm sodium hypochlorite, respectively. However, a three-compartment sanitation method comprising washing with an antimicrobial soap, scrubbing with a brush in tap water, and immersion in a hypochlorite solution resulted in a 99.8% reduction. Population reductions exceeding 5 logs were obtained on cut iceberg lettuce, inoculated with E. coli CDC1932, by washing with diluted vinegar (1.9% acetic acid); in contrast, washing with diluted bleach solution (180 ppm available chlorine) and lemon juice (0.6% citric acid) yielded 1.6 and 2.1 log reductions, respectively [110]. However, the vinegar treatment resulted in some product damage. Application of a solution containing 1.5% lactic acid and 1.5% hydrogen peroxide as a 15-minute soak at 40°C was reported to yield greater than 5 log reductions in the population of E. coli O157:H7, Salmonella enteritidis, and Listeria monocytogenes on spot-inoculated apples, oranges, and tomatoes [111]. However, in both studies, the surviving bacteria were recovered by a rinsing procedure such that only unattached, exposed cells were being recovered and not bacteria that were embedded in fruit tissues or biofilms or attached to fruit surfaces. This may have yielded unrealistically high population reductions. Smith et al. [112] evaluated a commercial peroxyacetic acid formulation intended for food-service applications (Victory produce wash; Ecolab, St. Paul, MN; www.ecolab.com) for reducing the bacterial load on lettuce; small reductions (~1 log) were obtained. Lukasik et al. [38] compared various washing treatments, including consumer-oriented products (detergents, Fit® and Healthy Harvest) on inoculated strawberries; population reductions for E. coli O157:H7, S. montevideo, and several viruses were between 1 and 2 logs. Parnell and Harris [113] compared water, sodium hypochlorite, and vinegar as consumer washes for reducing salmonella on spot-inoculated apples. Population reductions obtained with vinegar and chlorine washes were 2 to 3 logs greater than reductions obtained with water. Treatment with sodium hypochlorite and vinegar yielded comparable reductions in the population of natural microbiota of lettuce [114]. A study of consumer acceptance of a home use antibacterial solution for sanitizing apples indicated that consumers would be unwilling to use a procedure requiring the 15-minute heat and soak step [115]. Venkitanarayanan et al. [116] reported that an electrolyzed water treatment was effective in inactivating foodborne pathogens on smooth plastic kitchen cutting boards. They did not investigate scarred cutting boards which might be expected in a kitchen or foodservice situation.

17.4.3 COMMERCIAL EQUIPMENT AND WASH FORMULATIONS FOR HOME OR FOODSERVICE USE

Some manufacturers of commercial equipment for sanitizing produce have developed small-scale units suitable for consumer and foodservice use. Systems based on use of electrolyzed water are being marketed by Sterilox Technologies, Inc. (www.steriloxtechnologies.com) and Hoshizaki America, Inc. (www.hoshizakiamerica.com). Small-scale systems based on ozone are being marketed by Sterilion Ltd (www.performancesystems.com/medical.htm) and UltrOzone (UC Davis Postharvest Technology Center; 1-866-21-OZONE).

A number of commercial fruit and vegetable wash formulations intended for consumer use are being marketed, but little information is available about their performance in reducing microbial populations. Fit$^®$, a produce wash produced by Procter & Gamble Co. and marketed for a number of years, did show some antimicrobial activity in addition to removing dirt, wax, and other residues [117,118], although no claims were made by the company that the consumer product had antimicrobial activity. They did make such a claim for a "Pro Line Fit" intended for commercial rather than consumer use. Fit is now marketed by HealthPro Brands, Inc. (www.healthprobrands.com). JohnsonDiversey markets a Hard Surface Sanitizer/Fruit & Vegetable Wash (Product 4444) claimed to have antimicrobial activity (www.jwp.com/jwp/ProdInfo.nsf/; click on foodservice, then sanitizers). Another product with documented antimicrobial activity is Pro-San$^®$, previously marketed as Vegi-Clean$^®$ (www.microcideinc.com/prosan.htm). A product derived from oranges and other GRAS ingredients and claimed to have antibacterial properties is marketed under the name CitroBio for postharvest processing, use in retail misting systems, or as a produce wash for consumers (www.citrobio.com). Grapefruit seed extract (Citricidal$^®$) is reputed to have antimicrobial properties (www.biochemresearch.com) and is being marketed as a consumer-use cleaner and disinfectant for fruits and vegetables (www.pureliquidgold.com). Other produce washes include: Veggie Wash$^®$ marketed by Beaumont Products (www.citrusmagic.com), Nature Clean Fruit & Veggie Wash (claimed to remove bacteria) (www.smallplanetinc.com, www.healthyhomeservices.ca, www.frankross.com), CleanGreens! (www.cleangreensinc.com), and Organiclean (www.organiclean.com).

In addition to these commercial products, recipes for fruit and vegetable washes can be found on the internet. Typical examples include diluted 3% hydrogen peroxide (www.wellnesstoday.com), and vinegar and 3% hydrogen peroxide sprays applied individually to produce (http://myexecpc.com/~mjstouff/articles/vinegar.html). One source suggests use of 35% hydrogen peroxide around the house, a potentially dangerous recommendation; specific uses for produce treatment call for use of 3 or 5% solutions (http://h2o2hydrogenperoxide.com/additrion.html).

17.5 CONCLUSIONS

The efficacy of conventional washing technology in reducing populations of human pathogens and other microorganisms on fresh produce surfaces is limited to 1 to 2 logs, a significant improvement compared to the unwashed produce but insufficient to ensure food safety. Incremental improvements in washing efficacy can be obtained through buffering, addition of surfactants, temperature elevation, full immersion, and washing with vigorous agitation. However, greater population reductions cannot be obtained because of the strength of microbial attachment to produce and location of attached microorganisms in inaccessible sites. Approved alternatives to chlorine may provide certain technical advantages and avoid disadvantages such as formation of toxic reaction products, but differences in antimicrobial efficacy are small. Washing agents developed for foodservice or home use may exhibit antimicrobial activity, but safe and uniform application may be problematic without the controls available for large-scale produce packing and processing applications. Microbial reduction benefits claimed by many purveyors of home-use formulations, especially those marketed via the internet, are unsubstantiated. Experimental washing agents, if found to be technically and economically feasible, or synergistic sequences or combinations of treatments may provide addition gains in efficacy over current technology, but attainment of high levels of safety such as afforded by a 5 log reduction in pathogen populations is unrealistic. Use of other technologies such as surface pasteurization or irradiation may be required to reach this level of safety.

ACKNOWLEDGMENTS

The author thanks Prof. Jerry A. Bartz at the University of Florida in Gainesville and Prof. William C. Hurst at the University of Georgia for their thorough and constructive review of this chapter.

REFERENCES

1. Beuchat, L.R., Surface Decontamination of Fruits and Vegetables Eaten Raw: A Review, Food Safety Unit, World Health Organization, WHO/FSF/FOS/98.2, 1998.
2. FDA, Methods to reduce/eliminate pathogens from fresh and fresh-cut produce, in Analysis and Evaluation of Preventative Control Measures for the Control and Reduction/Elimination of Microbial Hazards on Fresh and Fresh-cut Produce, U.S. Food and Drug Administration, Center for Food Safety and Applied Nutrition, 2001, chap. V (www.cfsan.fda.gov/~comm/ift3-5.html).
3. Sapers, G.M., Washing and sanitizing raw materials for minimally processed fruit and vegetable products, in *Microbial Safety of Minimally Processed Foods*, Novak, J.S., Sapers, G.M., Juneja, V.K., Eds., CRC Press, New York, 2003, chap. 11.

4. White, G.C., *Handbook of Chlorination and Alternative Disinfectants*, 4th ed., John Wiley, New York, 1998.
5. Dychdala, G.R., Chlorine and chlorine compounds, in *Disinfection, Sterilization, and Preservation*, 4th Ed., Block, S.S., Ed., Lea & Febiger, Philadelphia, 1991, chap. 7.
6. 21CFR173.315, Chemicals Used in Washing or to Assist in the Peeling of Fruits and Vegetables, Code of Federal Regulations Title 21, Part 173, Section 173.315.
7. 40CFR180.1054, Calcium Hypochlorite; Exemption from the Requirement of a Tolerance, Code of Federal Regulations Title 40, Part 180, Section 180.11054.
8. FDA, Guidance for industry: reducing microbial food safety hazards for sprouted seeds and guidance for industry: sampling and microbial testing of spent irrigation water during sprout production, *Fed. Register*, 64, 57893, 1999.
9. Suslow, T.V. *et al.*, Detection and elimination of *Salmonella* Mbandaka from naturally contaminated alfalfa seed by treatment with heat or calcium hypochlorite, *J. Food Prot.*, 65, 452, 2002.
10. Fett, W.F., Reduction of *Escherichia coli* O157:H7 and *Salmonella* spp. on laboratory-inoculated mung bean seed by chlorine treatment, *J. Food Prot.*, 65, 848, 2002.
11. Brackett, R.E., Antimicrobial effect of chlorine on *Listeria monocytogenes*, *J. Food Prot.*, 50, 999, 1987.
12. Zhuang, R.-Y., Beuchat, L.R., and Angulo, F.J., Fate of *Salmonella montevideo* on and in raw tomatoes as affected by temperature and treatment with chlorine, *Appl. Environ. Microbiol.*, 61, 2127, 1995.
13. Wei, C.I. *et al.*, Growth and survival of *Salmonella montevideo* on tomatoes and disinfection with chlorinated water, *J. Food Prot.*, 58, 829, 1995.
14. Zhang, S. and Farber, J.M., The effects of various disinfectants against *Listeria monocytogenes* on fresh-cut vegetables, *Food Microbiol.*, 13, 311, 1996.
15. Beuchat, L.R. *et al.*, Efficacy of spray application of chlorinated water in killing pathogenic bacteria on raw apples, tomatoes, and lettuce, *J. Food Prot.*, 61, 1305, 1998.
16. Sapers, G.M., Miller, R.L., and Mattrazzo, A.M., Effectiveness of sanitizing agents in inactivating *Escherichia coli* in Golden Delicious apples, *J. Food Sci.*, 64, 734, 1999.
17. Pirovani, M.E. *et al.*, Survival of *Salmonella hadar* after washing disinfection of minimally processed spinach, *Lett. Appl. Microbiol.*, 31, 143, 2000.
18. Garcia, A., Mount, J.R., and Davidson, P.M., Ozone and chlorine treatment of minimally processed lettuce, *J. Food Sci.*, 68, 2747, 2003.
19. Annous, B.A. and Sapers, G.M., unpublished data, 2001.
20. Spotts, R.A., Use of surfactants with chlorine to improve pear decay control, *Plant Dis.*, 66, 725, 1982.
21. Spotts, R.A. and Cervantes, L.A., Effects of the nonionic surfactant Ag-98 on three decay fungi of Anjou pear, *Plant Dis.*, 71, 240, 1987.
22. Park, D.L., Rua, S.M., Jr., and Acker, R.F., Direct application of a new hypochlorite sanitizer for reducing bacterial contamination on foods, *J. Food Prot.*, 54, 960, 1991.
23. Delaquis, P.J. *et al.*, Effect of warm, chlorinated water on the microbial flora of shredded iceberg lettuce, *Food Res. Int.*, 32, 7, 1999.

24. Pirovani, M.E., Guemes, D.R., and Piagnetini, A.M., Predictive models for available chlorine depletion and total microbial count reduction during washing of fresh-cut spinach, *J. Food Sci.*, 66, 860, 2001.

25. Chang, T.-L., Streicher, R., and Zimmer, H., The interaction of aqueous solutions of chlorine with malic acid, tartaric acid, and various fruit juices, A source of mutagens, *Anal. Lett.*, 21, 2049, 1988.

26. Hidaka, T. *et al.*, Disappearance of residual chlorine and formation of chloroform in vegetables treated with sodium hypochlorite, *Shokuhin Eiseigaku Zasshi*, 33, 267, 1992.

27. Richardson, S.D., Scoping the chemicals in your drinking water, *Today's Chemist at Work*, 3, 29, 1994.

28. Tsai, L.-S., Randall, V.G., and Schade, J.E., Chlorine uptake by chicken frankfurters immersed in chlorinated water, *J. Food Sci.*, 58, 987, 1993.

29. Izumi, H., Electrolyzed water as a disinfectant for fresh-cut vegetables, *J. Food Sci.*, 64, 536, 1999.

30. Bari, M.L. *et al.*, Effectiveness of electrolyzed acidic water in killing *Escherichia coli* O157:H7, *Salmonella enteritidis*, and *Listeria monocytogenes* on the surface of tomatoes, *J. Food Prot.*, 66, 542, 2003.

31. Wright, J.R. *et al.*, Reduction of *Escherichia coli* O157:H7 on apples using wash and chemical sanitizer treatments, *Dairy Food Environ. Sanit.*, 20, 120, 2000.

32. Kenney, S.J. and Beuchat, L.R., Comparison of aqueous commercial cleaners for effectiveness in removing *Escherichia coli* O157:H7 and *Salmonella muenchen* from the surface of apples, *Int. J. Food Microbiol.*, 74, 47, 2002.

33. Raiden, R.M. *et al.*, Efficacy of detergents in removing *Salmonella* and *Shigella* spp. from the surface of fresh produce, *J. Food Prot.*, 66, 2210, 2003.

34. Sapers, G.M. *et al.*, Anti-microbial treatments for minimally processed cantaloupe melon, *J. Food Sci.*, 66, 345, 2001.

35. 21CFR173.300, Chlorine Dioxide, Code of Federal Regulations Title 21, Part 173, Section 173.300.

36. 21CFR173.325, Acidified Sodium Chlorite Solutions, Code of Federal Regulations 21, Part 173, Section 173.325.

37. Tsai, L.-S., Huxsoll, C.C., and Robertson, G., Prevention of potato spoilage during storage by chlorine dioxide, *J. Food Sci.*, 66, 472, 2001.

38. Lukasik, J. *et al.*, Reduction of poliovirus 1, bacteriophages, *Salmonella* Montevideo, and *Escherichia coli* O157:H7 on strawberries by physical and disinfectant washes, *J. Food Prot.*, 66, 188, 2003.

39. Wisniewsky, M.A. *et al.*, Reduction of *Escherichia coli* O157:H7 counts on whole fresh apples by treatment with sanitizers, *J. Food Prot.*, 63, 703, 2000.

40. Spotts, R.A. and Peters, B.B., Chlorine and chlorine dioxide for control of d'Anjou pear decay, *Plant Dis.*, 64, 1095, 1980.

41. Costilow, R., Uebersax, M.A., and Ward, P.J., Use of chlorine dioxide for controlling microorganisms during the handling and storage of fresh cucumbers, *J. Food Sci.*, 49, 396, 1984.

42. Fett, W.F., Reduction of native microflora on alfalfa sprouts during propagation by addition of antimicrobial compounds to the irrigation water, *Int. J. Food Microbiol.*, 72, 13, 2002.

43. Han, Y. *et al.*, Reduction of *Listeria monocytogenes* on green peppers (*Capsicum annuum* L.) by gaseous and aqueous chlorine dioxide and water washing and its growth at 7°C, *J. Food Prot.*, 64, 1730, 2001.

44. Wickramanayake, G.B., Disinfection and sterilization by ozone, in *Disinfection, Sterilization, and Preservation*, 4th ed., Block, S.S., Ed., Lea & Febiger, Philadelphia, 1991, chap. 10.

45. Graham, D.M., Use of ozone for food processing, *Food Technol.*, 51, 72, 1997.

46. Restaino, L. *et al.*, Efficacy of ozonated water against various food-related microorganisms, *Appl. Environ. Microbiol*, 61, 3471, 1995.

47. 21CFR173.368, Ozone, Code of Federal Regulations Title 21, Part 173, Section 173.368.

48. Xu, L., Use of ozone to improve the safety of fresh fruits and vegetables, *Food Technol.*, 53, 58, 1999.

49. Khadre, M.A., Yousef, A.E., and Kim, J.-G., Microbiological aspects of ozone applications in food: a review, *J. Food Sci.*, 66, 1242, 2001.

50. Kim, J.-G., Yousef, A.E. and Chism, G.W., Use of ozone to inactivate microorganisms on lettuce, *J. Food Saf.*, 19, 17, 1999.

51. Smilanick, J. *et al.*, Control of spores of postharvest fungal pathogens of produce with ozonated water, Abstracts of Papers Presented at 2000 IFT Annual Meeting, Dallas, TX, June 11–14, 2000, abstr. 47–6.

52. Sarig, P. *et al.*, Ozone for control of post-harvest decay of table grapes caused by *Rhizopus stolonifer*, *Physiol. Mol. Plant Pathol.*, 48, 403, 1996.

53. Zagory, D. and Hurst, W.C., Eds., *Food Safety Guidelines for the Fresh-cut Produce Industry*, International Fresh-cut Produce Association, Alexandria, VA, 1996.

54. Strasser, J., Ozone Applications in Apple Processing, Tech Application, Electric Power Research Institute, Inc., Palo Alto, CA, 1998.

55. Achen, M. and Yousef, A.E., Efficacy of ozone against *Escherichia coli* O157:H7 on apples, *J Food Sci.*, 66, 1380, 2001.

56. Spotts, R.A. and Cervantes, L.A., Effect of ozonated water on postharvest pathogens of pear in laboratory and packinghouse tests, *Plant Dis.*, 76, 256, 1992.

57. Sharma, R.R. *et al.*, Inactivation of *Escherichia coli* O157:H7 on inoculated alfalfa seeds with ozonated water and heat treatment, *J. Food Prot.*, 65, 447, 2002.

58. Wade, W.N. *et al.*, Efficacy of ozone in killing *Listeria monocytogenes* on alfalfa seeds and sprouts and effects on sensory quality of sprouts, *J. Food Prot.*, 66, 44, 2003.

59. Block, S.S, Peroxygen compounds, in *Disinfection, Sterilization, and Preservation*, 4th ed., Block, S.S., Ed., Lea & Febiger, Philadelphia, 1991, chap. 9.

60. Solvay Interox, Inc., MSDS for Proxitane® EQ Liquid Sanitizer, 2002 (www.solvayinterox.com).

61. FMC Corp. Active Oxidants Division, VigorOx® Liquid Sanitizer and Disinfectant Technical Brochure, 2003 (www.fmcchemicals.com).

62. 40CFR180.1196, Peroxyacetic Acid; Exemption from the Requirement of a Tolerance, Code of Federal Regulations Title 40, Part 180, Section 180.1196.

63. 21CFR178.1010, Sanitizing Solutions, Code of Federal Regulations 21, Part 178, Section 178.1010, paragraph (b)(38).

64. Ecolab, Inc., Catching the wave, *Food Qual.*, 4, 51, 1997.

65. Hilgren, J.D. and Salverda, J.A., Antimicrobial efficacy of a peroxyacetic/octanoic acid mixture in fresh-cut vegetable process waters, *J. Food Sci.*, 65, 1376, 2000.

66. Annous, B.A. *et al.*, Efficacy of washing with a commercial flat-bed brush washer, using conventional and experimental washing agents, in reducing populations of *Escherichia coli* on artificially inoculated apples, *J. Food Prot.*, 64, 159, 2001.

67. Sapers, G. M. *et al.*, Improved anti-microbial wash treatments for decontamination of apples, *J. Food Sci.*, 67, 1886, 2002.

68. Gagliardi, J.V. *et al.*, On-farm and postharvest processing sources of bacterial contamination to melon rinds, *J. Food Prot.*, 66, 82, 2003.

69. Ukuku, D.O., Pilizota, V., and Sapers, G.M., Influence of washing treatment on native microflora and *Escherichia coli* population of inoculated cantaloupes, *J. Food Saf.*, 21, 31, 2001.

70. Ukuku, D.O. and Sapers, G.M., Effect of sanitizer treatments on *Salmonella* Stanley attached to the surface of cantaloupe and cell transfer to fresh-cut tissues during cutting practices, *J. Food Prot.*, 64, 1286, 2001.

71. Ukuku, D.O. and Fett, W., Behavior of *Listeria monocytogenes* inoculated on cantaloupe surfaces and efficacy of washing treatments to reduce transfer from rind to fresh-cut pieces, *J. Food Prot.*, 65, 924, 2002.

72. Sapers, G.M. and Jones, D.M., unpublished data, 2004.

73. Ukuku, D.O., Pilizota, V., and Sapers, G.M., Effect of hot water and hydrogen peroxide treatments on survival of *Salmonella* and microbial quality of whole and fresh-cut cantaloupe, *J. Food Prot.*, 67, 432, 2004.

74. Sapers, G.M., Hydrogen peroxide as an alternative to chlorine for sanitizing fruits and vegetables, FoodInfo Online Features, IFIS Publishing, July 23, 2003 (http://foodsciencecentral.com/library.html#ifis/12433).

75. Sapers, G.M. *et al.*, Enzymatic browning control in minimally processed mushrooms, *J. Food Sci.*, 59, 1042, 1994.

76. Sapers, G.M. *et al.*, Structure and composition of mushrooms as affected by hydrogen peroxide wash, *J. Food Sci.*, 64, 889, 1999.

77. Sapers, G.M. *et al.*, Shelf-life extension of fresh mushrooms (*Agaricus bisporus*) by application of hydrogen peroxide and browning inhibitors, *J. Food Sci.*, 66, 362, 2001.

78. Sapers, G.M. *et al.*, Factors limiting the efficacy of hydrogen peroxide washes for decontamination of apples containing *Escherichia coli*, *J. Food Sci.*, 65, 529, 2000.

79. Sapers, G.M. and Sites, J.E., Efficacy of 1% hydrogen peroxide wash in decontaminating apples and cantaloupe melons, *J. Food Sci.*, 68, 1793, 2003.

80. Fallik, E. *et al.*, Postharvest hydrogen peroxide treatment inhibits decay in eggplant and sweet red pepper, *Crop Prot.*, 13, 451, 1994.

81. Weissinger, W.R. and Beuchat, L.R., Comparison of aqueous chemical treatments to eliminate *Salmonella* on alfalfa seeds, *J. Food Prot.*, 63, 1475, 2000.

82. Forney, C.F. *et al.*, Vapor phase hydrogen peroxide inhibits postharvest decay of table grapes, *HortSci.*, 26, 1512, 1991.

83. Aharoni, Y., Copel, A., and Fallik, E., The use of hydrogen peroxide to control postharvest decay on "Galia" melons, *Ann. Applied Biol.*, 125, 189, 1994.

84. Simmons, G.F. *et al.*, Reduction of microbial populations on prunes by vapor-phase hydrogen peroxide, *J. Food Prot.*, 60, 188, 1997.

85. Sapers, G.M. and Simmons, G.F., Hydrogen peroxide disinfection of minimally processed fruits and vegetables, *Food Technol.*, 52, 48, 1998.

86. Sapers, G.M. *et al.*, Vapor-phase decontamination of apples inoculated with *Escherichia coli*, *J. Food Sci.*, 68, 1003, 2003.

87. 21CFR184.1366, Hydrogen Peroxide, Code of Federal Regulations Title 21, Part 184, Section 184.1366.

88. 40CFR180.1197, Hydrogen Peroxide; Exemption from the Requirement of a Tolerance, Code of Federal Regulations Title 40, Part 180, Section 180.1197.

89. 21CFR182.1778, Sodium Phosphate, Code of Federal Regulations Title 21, Part 182, Section 182.1778.

90. Mendonca, A.F., Amoroso, T.L., and Knabel, S.J., Destruction of Gram-negative food-borne pathogens by high pH involves disruption of the cytoplasmic membrane, *Appl. Environ. Microbiol.*, 60, 4009, 1994.

91. Sampathkumar, B., Khachatourians, G.G., and Korber, D.R, High pH during trisodium phosphate treatment causes membrane damage and destruction of *Salmonella enterica* Serovar Enteritidis, *Appl. Environ. Microbiol.*, 69, 122, 2003.

92. Pao, S., Davis, C.L., and Kelsey, D.F., Efficacy of alkaline washing for the decontamination of orange fruit surfaces inoculated with *Escherichia coli*, *J. Food Prot.*, 63, 961, 2000.

93. Isshiki, K. and Azuma, K., Microbial growth suppression in food using calcinated calcium, *JARQ*, 29, 269, 1995.

94. Bari, M.L. *et al.*, Calcinated calcium killing of *Escherichia coli* O157:H7, *Salmonella*, and *Listeria monocytogenes* on the surface of tomatoes, *J. Food Prot.*, 65, 1706, 2002.

95. Zhuang, R.-Y. and Beuchat, L.R., Effectiveness of trisodium phosphate for killing *Salmonella montevideo* on tomatoes, *Lett. Appl. Microbiol.*, 232, 97, 1996.

96. Somers, E.B., Schoeni, J.L., and Wong, A.C.L., Effect of trisodium phosphate on biofilm and planktonic cells of *Campylobacter jejuni*, *Escherichia coli* O157:H7, *Listeria monocytogenes* and *Salmonella typhimurium*, *Int. J. Food Microbiol.*, 22, 269, 1994.

97. Foegeding, P.M. and Busta, F.F., Chemical food preservatives, in *Disinfection, Sterilization, and Preservation*, 4th ed., Block, S.S., Ed., Lea & Febiger, Philadelphia, 1991, chap. 47.

98. 21CFR184.1005, Direct Food Substances Affirmed as Generally Recognized as Safe, Listing Of Specific Substances Affirmed as GRAS, Acetic Acid, Code of Federal Regulations, Title 21, Part 184, Subpart B, Section 184.1005.

99. 21CFR.184.1061, Direct Food Substances Affirmed as Generally Recognized as Safe, Listing of Specific Substances Affirmed as GRAS, Lactic Acid, Code of Federal Regulations, Title 21, Part 184, Subpart B, Section 184.1061.

100. Castillo, A. *et al.*, Lactic acid sprays reduce bacterial pathogens on cold beef carcass surfaces and in subsequently produced ground beef, *J. Food Prot.*, 64, 58, 2001.

101. Delaquis, P.J., Ward, S.M., and Stanich, K., Evaluation of pre-pressing sanitary treatments for the destruction of *Escherichia coli* O157:H7 on apples destined for production of unpasteurized apple juice, Technical Report No. 9901, Agriculture and Agri-Food Canada, Pacific Agri-Food Research Centre, Summerland, BC V0H 1Z0, 2000.

102. Liao, C.-H., Shollenberger, L.M., and Phillips, J.G., Lethal and sublethal action of acetic acid on *Salmonella* in vitro and on cut surfaces of apple slices, *J. Food Sci.*, 68, 2793, 2003.

103. Yang, H. *et al.*, Efficacy of cetylpyridinium chloride on *Salmonella* Typhimurium and *Escherichia coli* O157:H7 in immersion spray treatment of fresh-cut lettuce, *J. Food Sci.*, 68, 1008, 2003.

104. Naidu, A.S., Activated lactoferrin: a new approach to meat safety, *Food Technol.*, 56, 40, 2002.

105. Yahya, M.T. *et al.*, Disinfection of bacteria in water systems by using electrolytically generated copper:silver and reduced levels of free chlorine, *Can. J. Microbiol.*, 36, 109, 1990.

106. Lin, C.-M. *et al.*, Inactivation of *Escherichia coli* O157:H7, *Salmonella enterica* Serotype Enteritidis, and *Listeria monocytogenes* by hydrogen peroxide and lactic acid and by hydrogen peroxide with mild heat, *J. Food Prot.*, 65, 1215, 2002.

107. Duff, S.B. *et al.*, Cost-effectiveness of a targeted disinfection program in household kitchens to prevent foodborne illnesses in the United States, Canada, and the United Kingdom, *J. Food Prot.*, 66, 2103, 2003.

108. FDA Talk Paper, Consumers Advised That Recent Hepatitis A Outbreaks Have Been Associated With Green Onions, U.S. Food and Drug Administration, Nov. 15, 2003 (www.fda.gov/bbs/topics/ANSWERS/2003/ASNS01262.html; accessed Aug. 14, 2004).

109. Barak, J.D., Chue, B., and Mills, D.C., Recovery of surface bacteria from and surface sanitization of cantaloupes, *J. Food Prot.*, 66, 1805, 2003.

110. Vijayakumar, C. and Wolf-Hall, C.E., Evaluation of household sanitizers for reducing levels of *Escherichia coli* on iceberg lettuce, *J. Food Prot.*, 65, 1646, 2002.

111. Venkitanarayanan, K.S. *et al.*, Inactivation of *E. coli* O157:H7, *Salmonella* Enteritidis and *Listeria monocytogenes* on apples, oranges, and tomatoes by lactic acid with hydrogen peroxide, *J. Food Prot.*, 65, 100, 2002.

112. Smith, S. *et al.*, Efficacy of a commercial produce wash on bacterial contamination of lettuce in a food service setting, *J. Food Prot.*, 66, 2359, 2003.

113. Parnell, T.L. and Harris, L.J., Reducing *Salmonella* on apples with wash practices commonly used by consumers, *J. Food Prot.*, 66, 741, 2003.

114. Nascimento, M.S. *et al.*, Effects of different disinfection treatments on the natural microbiota of lettuce, *J. Food Prot.*, 66, 1697, 2003.

115. McWatters, K.H. *et al.*, Consumer acceptance of raw apples treated with an antibacterial solution designed for home use, *J. Food Prot.*, 65, 106, 2002.

116. Venkitanarayanan, K.S. *et al.*, Inactivation of *Escherichia coli* O157:H7 and *Listeria monocytogenes* on plastic kitchen cutting boards by electrolyzed oxidizing water, *J. Food Prot.*, 62, 857, 1999.

117. Beuchat, L.R. *et al.*, Development of a proposed standard method for assessing the efficacy of fresh produce sanitizers, *J. Food Prot.*, 64, 1103, 2001.

118. Harris, L.J., Efficacy and reproducibility of a produce wash in killing *Salmonella* on the surface of tomatoes assessed with a proposed standard method for produce sanitizers, *J. Food Prot.*, 64, 1477, 2001.

18 Gas-/Vapor-Phase Sanitation (Decontamination) Treatments

Richard H. Linton, Yingchang Han, Travis L. Selby, and Philip E. Nelson

CONTENTS

18.1 INTRODUCTION

Since the 1950s chemical gases, such as ethylene oxide, propylene oxide, formaldehyde, and β-propiolactone, have been used to sterilize medical products and biological preparations that are not compatible with heat or radiation sterilization. In the 1980s sterilization of these products using chemical gases, such as chlorine dioxide (ClO_2), ozone, and hydrogen peroxide vapor or plasma, emerged as a new technology [1–4]. For the purpose of this chapter, the gaseous form is considered as the direct application of the gas phase. The vapor form is considered as the application of a vaporized chemical from a liquid starting material.

In more recent years applications of such gaseous chemical disinfectants are gaining interest in the food industry for reducing microorganisms. Researchers have successfully used ClO_2 gas to sterilize bulk orange juice storage tanks [5] and have also found that ozone gas can eliminate insects in grain storage facilities without harming food quality or the environment [6,7]. Other studies have focused on potential applications of gaseous or vapor-phase antimicrobials for decontamination of fruits and vegetables. Such antimicrobials being studied include ClO_2 gas, ozone, allyl isothiocyanate vapor, hydrogen peroxide vapor, acetic acid vapor, and natural volatile compounds (methyl jasmonate, *trans*-anethole, carvacrol, cinnamic aldehyde, eugenol, linalool, and thymol). ClO_2 and ozone gases are examples of promising technologies which have been shown to lead to high microbial reductions. This chapter reviews the efficacy of these antimicrobials in inactivation of microorganisms on fruits and vegetables.

18.2 CHLORINE DIOXIDE GAS

18.2.1 PHYSICAL, CHEMICAL, AND SAFETY PROPERTIES OF ClO_2 GAS

ClO_2 is a neutral compound of chlorine in the +IV oxidation state. It disinfects by oxidation; however, it does not chlorinate. The major oxidation/reduction is:

$$ClO_{2(aq)}+e^- = ClO_2^- \quad (E_o = -0.954\,V) \tag{1}$$

ClO_2 gas is usually mixed with nitrogen or air and has a yellowish green color. It has an odor similar to chlorine and sodium hypochlorite and can be easily detected at levels as low as about 0.1 ppm in air. ClO_2 gas has a density 2.4 times that of air. It is unstable as a gas and can decompose to chlorine and oxygen with noise, heat, flame, and a minor pressure wave at low concentrations. Selected physical, chemical, and safety properties of ClO_2 gas are summarized in Table 18.1. ClO_2 is also highly water soluble and does not

TABLE 18.1
Physical, Chemical, and Safety Properties of Chlorine Dioxide Gas and Ozone

Property	Chlorine dioxide gas	Ozone
Molecular weight	67.45 g/mol	48.00 g/mol
Oxidation potential	−0.954 V	−2.07 V
Melting point	−59°C	−192.7°C
Boiling point	11°C	−111.9°C
Density, 25°C, 760 mmHg	2.757 g/l	1.962 g/l
Vapor pressure	10 kPa at −34.3°C; 100 kPa at 10.5°C	10 kPa at −139.7°C; 100 kPa at −111.5°C
Solubility limit	~3 g/l, aqueous, 25°C, 34.5 mmHg; ~20 g/l, aqueous, 0–5°C, 70–100 mmHg	~0.57 g/l, 100% ozone, 20°C, 760 mmHg
Heat of formation at 298.15 K	89.1 kJ/mol	142.7 kJ/mol
Entropy at 298.15 K	263.7 J/mol/K	238.9 J/mol/K
Heat capacity at 298.15 K	46.0 J/mol/K	39.2 J/mol/K
Heat of vaporization at boiling point	30 kJ/mol	Unknown
Heat of solution	6.6 kcal/mol	Unknown
Explosion velocity, pure gas	1250 m/s	Unknown
Explosion velocity, in air	50 m/s	Unknown
Explosion concentration, in air or >76 mmHg	>10% v/v	Unknown
Permissible exposure limit (PEL) specified by the Occupational Safety and Health Administration (OSHA), in air	0.1 ppm	0.1 ppm
Short-term exposure limit specified by OSHA, in air, for 15 min	0.3 ppm	Unknown

Modified from Lide, D.R., *CRC Handbook of Chemistry and Physics*, 82nd ed., CRC Press, Boca Raton, FL, 2001; Occupational Safety and Health Administration, Air Contaminant Exposure Standards, 29 CFR 1910.1000, chap. XVII, 6, July 1, 1991; Gates, D., *The Chlorine Dioxide Handbook*, American Water Works Association, Denver, CO, 1998; EPA, Ozone, in *Guidance Manual Alternative Disinfectants and Oxidants*, Environmental Protection Agency, 1999, chap. 3; Kim, J., Yousef, A.E., and Dave, S., *J. Food Prot.*, 62, 1071, 1999.

hydrolyze readily but remains in solution as a dissolved gas [13]. ClO_2 cannot be compressed or stored commercially as a gas because it decomposes with time and is highly explosive at high concentration ($>10\%$ in air) or under pressure. Therefore, ClO_2 is generated on-site. The permissible exposure limit (PEL) or time-weighted average (TWA) of ClO_2 gas in air is 0.1 ppm, specified by the Occupational Safety and Health Administration [9].

18.2.2 ANTIMICROBIAL PROPERTIES OF AQUEOUS AND GASEOUS ClO_2

ClO_2 is strong oxidizing agent that has broad and high biocidal effectiveness. Aqueous ClO_2 effectively inactivates bacteria [14–18] including human pathogens [19–21] and bacterial spores [14,24], viruses [22,23], and algae [25]. ClO_2 has approximately 2.5 times the oxidation capacity of chlorine [15] and has been shown to produce a bactericidal effect equivalent to seven times that of chlorine at the same concentration in poultry processing water [26]. Advantages of ClO_2 over chlorine also include effectiveness at low concentration, nonconversion to chlorophenols that result in residual odors and off-flavors, ability to remove chlorophenols already present from other sources, effectiveness at high and low pH values, and inability to react with ammonia or humic acid to produce harmful chloramines and trihalomethanes [27]. Therefore, the use of ClO_2 as an alternative disinfectant to chlorine is attractive not only in the drinking water industry but also in the food industry.

Gaseous ClO_2 has been used successfully for sterilization of medical implements in the medical science area for years [2,3,28]. More recently gaseous ClO_2 was used to decontaminate *B. anthracis* contaminated areas of the Hart senate office building and the Brentwood postal sorting facility in Washington D.C. [29] (Table 18.2). In recent years additional applications of gaseous ClO_2 in the food industry have been studied. Research has demonstrated that ClO_2 gas is highly effective in reducing foodborne pathogens on fruit and vegetable surfaces [31–34], spoiled orange juice isolates from epoxy-coated storage tank surfaces [5], and bacillus spores on paper, plastic, epoxy-coated stainless steel, and wood surfaces [30] (Table 18.2). These results have demonstrated that ClO_2 gas treatments are a promising surface decontamination technology which could be applicable to the food industry. The efficacy and potential applications of ClO_2 gas treatment for decontamination of fruit and vegetables are reviewed later in this chapter.

18.2.3 ClO_2 GAS GENERATION

ClO_2 gas is most often generated based on the reaction between chlorine gas and sodium chlorite. The principal reaction can be described as:

$$2NaClO_2 + Cl_2 \rightarrow 2ClO_2 + 2NaCl \tag{2}$$

Figure 18.1 shows an example of a ClO_2 gas generation system from CDG Technology, Inc. (Bethlehem, PA; http://www.cdgtechnology.com). Chlorine

TABLE 18.2
Efficacy of ClO$_2$ Gas Treatments in Inactivation of Microorganisms on Different Surfaces

Surface	Pathogenic microorganism	ClO$_2$ gas treatment conditions	Total population reduction/surface	Ref.
Paper strips and aluminum foil cups	*Bacillus subtilis* spores	Continuous, >40 mg/l, 1 h, 27°C, 60% RH	>6 log CFU	2
Analytical paper disk	*Bacillus subtilis niger* spores	Continuous, 6–7 mg/l, 1 h, 23°C, 70–75% RH	>6 log CFU	3
Paper, plastic, wood	*Bacillus thuringiensis* spores	Batch, 15 mg/l, 30 min, 22°C, >90% RH	>5 log CFU	30
Paper in Tyvek/film pouch	*Bacillus subtilis* spores	Continuous, 5 mg/l, 30 min, room temperature, 70% RH	>6 log CFU	28
Epoxy-coated stainless steel	*Lactobacillus* spp., *Penicillium* spp., *S. cerevisiae*	Batch, 10 mg/l, 30 min, 22°C, >90% RH	>6 log CFU	5

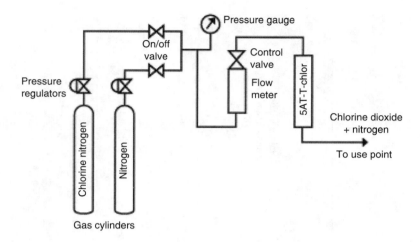

FIGURE 18.1 Schematic of a CDG Gas:Solid^{TM} ClO$_2$ gas generation system. (From CDG Technology, Inc. With permission.)

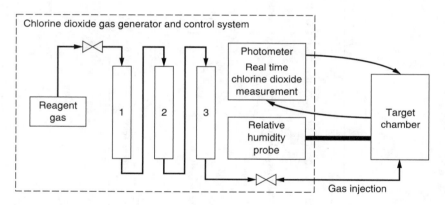

FIGURE 18.2 Schematic of a ClorDiSys ClO$_2$ gas generation and control system. (From ClorDiSys Solutions, Inc. With permission.)

gas (4%) in nitrogen flows into a Saf-T-Chlor^{TM} reactor cartridge containing thermally stable sodium chlorite; approximately 140 g and 1300 g ClO$_2$ gas are produced by a bench- and a pilot-scale generator, respectively. Another example of a ClO$_2$ gas generator is manufactured by ClorDiSys Solutions, Inc. (Lebanon, NJ, http://www.clordisys.com) which uses a similar approach, but produces approximately 900 g ClO$_2$ gas by flowing 2% chlorine gas in nitrogen through three sodium chlorite cartridges (Figure 18.2).

ICA TriNova, LLC (Forest Park, GA) has developed a Z-Series ClO$_2$ technology that involves generating ClO$_2$ gas by mixing two dry solids: a ClO$_2$ precursor and an activator (Figure 18.3). Sodium chlorite or sodium chlorate is used as the ClO$_2$ precursor in either crystalline or impregnated forms. The

FIGURE 18.3 Z-series products for ClO_2 generation. (From ICA TriNova, LLC. With permission.)

activator is a granular porous solid impregnated with an acid or with an acid precursor. ClO_2 gas is produced by a disproportionation reaction as the two dry solids are mixed:

$$4H^+ + 5NaClO_2 \rightarrow 4ClO_2 + NaCl + 4Na^+ + 2H_2O \quad (3)$$

The Z-series products can produce approximately 0.1 mg to 50 g of ClO_2. This product comes packaged in sachets, small tubs, and buckets based on different applications.

18.2.4 GENERAL GAS/VAPOR TREATMENT SYSTEMS

Batch and continuous ClO_2 gas treatment systems have been developed and used for decontamination of fruits and vegetables [31–34,36–38,40–42]. Typically, the batch system (Figure 18.4) includes a ClO_2 gas generator (such as a bench-scale CDG Technology or ClorDiSys Solutions generator), a treatment chamber, a diaphragm vacuum pump, and a thermo-hydro recorder. Produce samples are placed on expanded stainless steel shelves inside the chamber. ClO_2 gas is generated from the generator and is stored in a Teflon storage bag before being injected into the chamber using a gas-sampling syringe. The injected volume of ClO_2 gas (known concentration) is determined based on required ClO_2 gas concentration for the treatment chamber volume. The injected ClO_2 gas is distributed and circulated in the chamber by a diaphragm vacuum pump. Relative humidity (RH) and temperature inside the chamber are monitored using a thermo-hydro recorder. To measure ClO_2 gas concentration, a certain amount (5 to 60 ml) of ClO_2 gas–air mixture is taken out of the chamber using a gas-sampling syringe and dissolved in

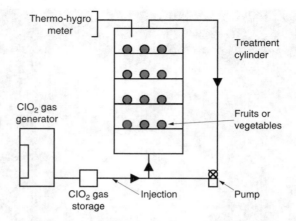

FIGURE 18.4 Schematic of a batch laboratory ClO_2 gas treatment system for fruits and vegetables. (Patent pending, Purdue University.)

deionized water, followed by analysis using a DPD (N,N-diethyl-p-phenylene-diamine) colorimetric analysis kit and a VVR photometer (CHEMetrics, Inc., Calverton, VA).

A continuous laboratory-scale ClO_2 gas treatment system mainly consists of a bench-scale generator, a treatment chamber, a ClO_2 gas dilution panel, a diaphragm vacuum pump, an ultrasonic humidifier, a wireless thermo-hygrometer, and a continuous ClO_2 gas monitor. ClO_2 gas from the generator is diluted with filtered air and passes through the treatment chamber. Relative humidity in the chamber is controlled using an ultrasonic humidifier. During the treatment, ClO_2 gas concentration is continuously monitored using the continuous ClO_2 monitor (Interscan Corp., Chatsworth, CA; http://www.gasdetection.com). An automated continuous pilot-scale ClO_2 gas treatment system has also been developed (Figure 18.5). ClO_2 gas is generated using a Mindox-M generator from ClorDiSys Solutions, Inc. Fruit and vegetables are placed on two movable shelves in a 400 l stainless steel chamber. After preconditioning (humidifying) the chamber using a humidifier (50 to 95%), the ClO_2 gas is fed in and circulated by a gas blower. The gas concentration, RH, and pressure relief are continuously monitored and controlled by the Mindox-M generator. Each treatment cycle is programmed and run automatically. After the treatment, the products may be washed by spraying filtrated water inside the chamber for 5 to 10 minutes to reduce residual ClO_2, if any residuals remain on the product and/or in the chamber.

Concentrations of ClO_2 are reported either as ppm in volume or mg/l. When using ClO_2 as a solution, 1 ppm is equivalent to approximately 1 mg/l. However, when ClO_2 is used as a gas, 1 mg/l ClO_2 is equivalent to approximately 332 ppm in volume under standard conditions (0°C, 1 atm) and 362 ppm under normal conditions (25°C, 1 atm). These conversions are based on the ideal gas law. In this chapter, the conversion factor of $1\,mg/l = 362\,ppm$ is used.

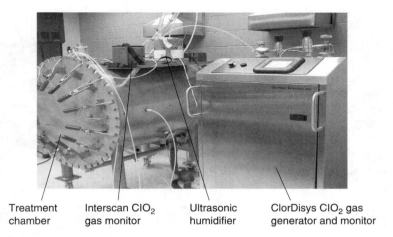

| Treatment chamber | Interscan ClO_2 gas monitor | Ultrasonic humidifier | ClorDisys ClO_2 gas generator and monitor |

FIGURE 18.5 Continuous pilot-scale ClO_2 gas treatment system for fruits and vegetables. (Patent pending Purdue University.)

18.2.5 MECHANISMS FOR MICROBIAL INACTIVATION

The mechanism of microbial inactivation by aqueous ClO_2 has been explored, but it is not fully understood. ClO_2 has been shown to react readily with amino acids (cysteine, tryptophan, and tyrosine), but not with viral ribonucleic acid (RNA) [44,45]. Bernarde et al. [16] suggested that the primary inactivation mechanism was the disruption of protein synthesis. However, Roller et al. [46] indicated that the inhibition of protein might not be the primary inactivation mechanism. The increase of the permeability of the outer membrane was considered as another mechanism due to reactions of the outer membrane protein and lipids with ClO_2 [13,45,47]. Berg et al. [48] reported that gross cellular damage involving significant leakage of intracellular macromolecules did not occur for ClO_2-treated E. coli, but the cells lost control of potassium efflux, which may be the primary lethal physiological event. Because ClO_2 gas is highly water soluble, it may inactivate microorganisms in a similar way as the aqueous form of ClO_2. However, ClO_2 gas and/or its radicals may directly diffuse or penetrate into microbial cells to cause damage. Therefore, to understand completely the inactivation mechanism of aqueous and/or gaseous ClO_2, extensive research is needed.

18.2.6 FACTORS INFLUENCING ClO_2 GAS TREATMENT

The efficacy of ClO_2 gas treatment for decontamination of produce is affected by gas concentration, exposure time, RH, temperature, cut or intact surfaces, and microbial inoculation sites. The effects of ClO_2 gas concentration (0.1 to 0.5 mg/l), RH (55 to 95%), treatment time (7 to 135 minutes) and temperature (5 to 25°C) on inactivation of E. coli O157:H7 on green peppers have been

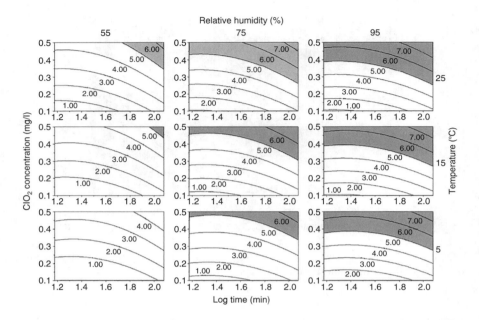

FIGURE 18.6 Effects of ClO_2 gas concentration (0.1 to 0.5 mg/l), RH (55 to 95%), time (15 to 135 minutes), and temperature (10 to 20°C) on log reductions of *E. coli* O157:H7 on inoculated green peppers. Shaded areas represent reductions greater than 5 logs. Log time at 1.2, 1.4, 1.6, 1.8, and 2.0 correspond to 15.8, 25.1, 39.8, 63.1, and 100 minutes, respectively. (From Han, Y. *et al.*, *J. Food Prot.*, 64, 1128, 2001. With permission.)

studied using response surface methodology [34]. A predictive model developed in this study suggests that increasing ClO_2 gas concentration, treatment time, RH, and temperature all significantly ($P < 0.01$) increased the inactivation of *E. coli* O157:H7. Contour plots (Figure 18.6) for *E. coli* O157:H7 log reduction, generated using a predictive model, showed a good comprehensive picture of the model, in which the shaded areas indicate treatment conditions that give greater than a 5 log reduction. ClO_2 gas concentration was the most important factor in the predictive model followed by time, RH, and then temperature. The interaction between ClO_2 gas concentration and RH indicated a synergistic effect. Other research also indicates that a higher RH results in a higher bacterial inactivation rate by ClO_2 gas [1–3,5]. High RH can moisturize treatment surfaces possibly leading to a thin layer of water droplets. These water droplets can further absorb and dissolve large amounts of ClO_2 gas due to high water solubility. Thus, the localized and concentrated ClO_2 contributes to microbial inactivation on the surfaces of fruits and vegetables. This may explain why ClO_2 gas is unique and more effective compared to other gaseous disinfectants.

Similar to aqueous sanitation treatments, ClO_2 gas treatment may also be less effective in reducing microorganisms on cut produce surface compared

TABLE 18.3
Log Reduction of *E. coli* O157:H7 Inoculated on Uninjured and Injured Green Pepper Surfaces After ClO_2 Gas Treatments for 30 min at 20°C Under 90 to 95% Relative Humidity

Sample[a]	Log reduction after different ClO_2 gas treatments[b]			
	0.15 mg/l	0.30 mg/l	0.60 mg/l	1.2 mg/l
Uninjured surface	$2.90 \pm 0.09_{Ad}$	$3.99 \pm 0.07_{Ac}$	$7.27 \pm 0.68_{Ab}$	$8.04 \pm 0_{Aa}{}^{c}$
Injured surface	$1.67 \pm 0.08_{Bd}$	$1.87 \pm 0.03_{Bc}$	$3.03 \pm 0.02_{Bb}$	$6.45 \pm 0.02_{Ba}$

[a] The initial populations of *E. coli* O157:H7 on surface-uninjured and surface-injured green peppers were $7.9 \pm 0.29 \log CFU/5g$.
[b] Values in the same column with different uppercase subscript letters are significantly different ($P < 0.05$). Values in the same row with different lower subscript letters are significantly different ($P < 0.05$).
[c] No viable *E. coli* O157:H7 was detected using the end-point method after 1.2 mg/l ClO_2 gas treatments.
From Han, Y. *et al., Food Microbiol.,* 17, 643, 2000. With permission.

to an uncut or intact surface. Studies [31] have shown that injuries to the wax layer, the cuticle, and underlying tissue layers of green pepper surfaces increased bacterial adhesion, growth, and resistance to washing and ClO_2 gas treatments. Han *et al.* [32] reported that ClO_2 gas treatments (0.15 to 1.2 mg/l ClO_2) had significantly higher reductions for inoculated *E. coli* O157:H7 on uninjured green pepper surfaces than on injured surfaces ($P < 0.05$) (Table 18.3). Microphotographs obtained using confocal laser scanning microscopy (CLSM) illustrated that bacteria preferentially attached to injured surfaces, where the bacteria could be protected from the treatment. Similar results have been reported for inactivation of *Listeria monocytogenes* on uninjured and injured green pepper surfaces by aqueous and gaseous ClO_2 treatments [33].

Microbial inoculation sites also influenced the efficacy of ClO_2 gas treatments. Du *et al.* [40,41] inoculated *L. monocytogenes* and *E. coli* O157:H7 on three sites of an apple: stem cavity, calyx cavity, and pulp skin. After ClO_2 gas treatments, bacteria on the pulp skin were less resistant to the treatment compared to the other two sites. At each inoculation site, however, bacterial inoculation levels (at 6, 7, and $8 \log CFU/site$) did not affect log reductions after treatment [40].

To determine accurately the efficacy of ClO_2 gas treatments, appropriate bacterial recovery and enumeration methods should be used. In evaluating the effectiveness of a sanitation treatment on these pathogens in fruits and vegetables, one of the common problems is overestimation of the treatment effectiveness, because sublethally injured cells have not been taken into account since direct selective plating enumeration methods are normally used. Han *et al.* [35] reported that a membrane-transferring surface-plating method

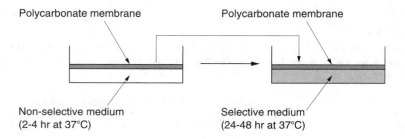

Polycarbonate membrane

Polycarbonate membrane

Non-selective medium
(2-4 hr at 37°C)

Selective medium
(24-48 hr at 37°C)

FIGURE 18.7 Schematic of a membrane-transferring surface-plating method for enumeration of ClO_2-treated bacteria.

was better for recovering uninjured and ClO_2-injured *E. coli* O157:H7 and *L. monocytogenes* on green peppers after ClO_2 gas treatments compared to traditional methods using direct surface-plating and overlay surface-plating. In this method (Figure 18.7), a 100 µl bacterial sample is first spread over a polycarbonate filter membrane (0.4 µm pore size, 90 mm diameter) (Osmonics Co., Westboro, MA) previously placed on the surface of a tryptic soy agar plate. Plates are incubated at 37°C for 2–4 hours to repair injured cells. Then the membranes are gently transferred onto appropriate selective media using sterile tweezers, followed by further incubation for 20 to 40 hours at 37°C. The membrane-transferring surface-plating method is also able to quantify low levels ($<2\log CFU/ml$) of surviving bacteria by using a filtration procedure to concentrate bacterial populations in test samples [41]. An end-point determination method has been successfully used to detect a complete inactivation of inoculated bacteria on apple surfaces after ClO_2 gas treatments [40,41]. This method is useful to validate the efficacy of a ClO_2 gas treatment when known levels of bacteria are applied to produce surfaces.

18.2.7 EFFICACY IN REDUCING MICROORGANISMS ON DIFFERENT PRODUCE SURFACES

The efficacy of batch and continuous ClO_2 gas treatments in reducing pathogenic and spoilage microorganisms on several fruits and vegetables, including green peppers, apples, potatoes, strawberries, cantaloupes, and lettuce, has been evaluated (Table 18.4). Results show that more than a 5 log reduction of selected pathogens such as *E. coli* O157:H7, *L. monocytogenes*, and *Salmonella* spp. can be achieved on these produce surfaces while maintaining acceptable quality, except for lettuce leaves where leaf discoloration was noted. Gaseous ClO_2 has been shown to be a more effective sanitizer for fruits and vegetables than aqueous ClO_2 and chlorinated water wash. More than $6\log CFU$ of *L. monocytogenes* on uninjured green pepper surfaces and $3.5\log CFU$ on injured surfaces were inactivated using a batch ClO_2 treatment system with $3\,mg/l$ ClO_2 gas treatment for 10 minutes at 20°C and 90 to 95% RH. However, a $3\,mg/l$ aqueous ClO_2 treatment for 10 minutes at 20°C achieved

TABLE 18.4
Efficacy of ClO_2 Gas Treatments in Inactivation of Microorganisms on Different Produce Surfaces

Surface	Pathogenic microorganism	ClO_2 gas treatment conditions	Log reduction	Ref.
Green pepper	*E. coli* O157:H7	Batch, 0.6 mg/l, 30 min, 22°C, >90% RH	7.3 log CFU/site	31, 32, 34
	L. monocytogenes		6.3 log CFU/site	33
	Salmonella spp.	Continuous, 0.6 mg/l, 10 min, 22°C, >90% RH	5.5 log CFU/site	37
Apple	*E. coli* O157:H7	Batch, 4.0 mg/l, 10 min, 22°C, >90% RH	5.5 log CFU/site	41
	E. coli	Batch, 0.3 mg/l, 3 h, 4°C	4.4 log CFU/g	43
	L. monocytogenes	Batch, 4.8 mg/l, 30 min, 22°C, >90% RH	4.8 log CFU/site	40
Strawberry	*E. coli* O157:H7	Batch, 4.0 mg/l, 30 min, 22°C, >90% RH	5.1 log CFU/site	36
	L. monocytogenes		5.3 log CFU/site	
	E. coli O157:H7	Continuous, 3.0 mg/l, 10 min, 22°C,	>6 log CFU/site	38
	L. monocytogenes	>90% RH	>6 log CFU/site	
Cantaloupe	*Salmonella* spp.	Continuous, 6.0 mg/l, 10 min, 22°C, >90% RH	4.5 log CFU/site	42
Lettuce	*L. monocytogenes*	Batch, 0.2 mg/l, 30 min, 22°C, >90% RH	2.0 log CFU/25g	49
Potato	*Erwinia carotvora*, natural flora (aerobic bacteria, yeast, and mold)	Vaporized ClO_2 by purging air through 500 ppm acidified Oxine at 6 l/h for 1 h	3–4 log CFU/g	39

only 0.4 and 3.7 log reductions on injured and uninjured green pepper surfaces, respectively, while water washing for 10 minutes showed 0.4 and 1.4 log reductions, respectively [33]. Costilow *et al.* [50] also found that up to 105 mg/l aqueous ClO_2 failed to reduce the population of microorganisms present on fresh cucumbers. Zhang and Farber [51] reported that ClO_2 solution treatment (5 mg/l, 10 minutes) at 22°C resulted in only a 0.8 log reduction of *L. monocytogenes* on both cut lettuce and cut cabbage. The mechanism of why aqueous ClO_2 is not as effective as the gaseous form needs further study.

A continuous ClO_2 gas treatment showed a higher efficacy compared to that of a batch treatment. Using a batch system (Figure 18.4), a 4.0 mg/l ClO_2 gas treatment for 30 minutes at 22°C and under 90% RH achieved an approximately 5 log reduction of *E. coli* O157:H7 and *L. monocytogenes* on strawberry surfaces [36]. With a continuous treatment system, more than a 6 log reduction of *E. coli* O157:H7 and *L. monocytogenes* on strawberries was achieved using a 3.0 mg/l ClO_2 gas treatment for 10 minutes [38].

18.2.8 Effects of ClO_2 Gas Treatment on Quality of Produce

Effects of ClO_2 gas treatments on quality of produce have not been extensively studied. However, some researchers have shown minimal quality effects on several types of produce, including green peppers [37], apples [40,41,43], cantaloupes [42], strawberries [38], and potatoes [39]. Han *et al.* reported [37,38] that no aerobic microorganisms (aerobic plate count, APC) were detected after treatment of strawberries with 3.0 mg/l ClO_2 gas for 10 minutes followed by a 1-week storage period at 4°C and after treatment of green peppers with 0.6 mg/l ClO_2 gas for 10 minutes followed by a 4-week storage period. Additionally, the color of both strawberry and green pepper surfaces did not change significantly ($p > 0.05$) during the storage period after ClO_2 gas treatments.

Residues of ClO_2 and chlorite on strawberries treated with 3 mg/l ClO_2 gas for 10 minutes were 0.19 ± 0.33 mg ClO_2/kg and 1.17 ± 2.02 mg Cl_2/kg; while after 1 week of storage no ClO_2 residues were detected, and residual chlorite levels were reduced to 0.07 ± 0.12 mg Cl_2/kg. Residues of ClO_2 and chlorite on peppers were 0.13 ± 0.05 mg ClO_2/kg and 0.39 ± 0.49 mg Cl_2/kg after treatment with 0.6 mg/l ClO_2 gas for 10 minutes and 0.02 ± 0.04 mg ClO_2/kg and 0 ± 0 mg Cl_2/kg after a 4-week further storage of treated products. No significant color changes ($p > 0.05$) were observed after 5.5 mg/l ClO_2 gas treatment for both strawberries and green peppers. Tsai *et al.* [39] found that, after treating potatoes with vaporized ClO_2 by purging air through 100 ppm ClO_2 solution for 1 hour, the residuals of chlorite and chlorate were less than 0.07 ppm, measured using ion chromatography. Additionally, no significant color changes ($p > 0.05$) were observed on cantaloupes after 5.5 mg/l ClO_2 gas treatment [42] and on apples after 7.8 mg/l ClO_2 gas treatment for 30 minutes [41]. However, Sapers *et al.* [43] observed darkening of lenticels on apples when treated with 0.3 mg/l ClO_2 gas for 20 hours. Discoloration (bleaching effect) of lettuce [49] and green cap on strawberries [38] was observed. It appears that ClO_2 gas treatment may have a negative effect on the color of leafy vegetables. However, more studies evaluating the effects of ClO_2 treatments on leafy vegetables are needed.

18.3 OZONE GAS

18.3.1 Properties of Ozone

Ozone is a gas at ambient (22°C) and refrigerated (4 to 7°C) temperatures. This gas is colorless with a pungent odor readily detectable at concentrations as low as 0.02 to 0.05 ppm (v/v). Important properties of ozone are summarized in Table 18.1. Ozone is partially soluble in water (12.9 mg/l ozone can be dissolved in the water when 3% ozone gas flows through water at 20°C), and solubility can be affected by partial pressure, flow rate of ozone, temperature, purity of water, and contact time [52]. Ozone has the unique property of autodecomposition, producing numerous free radical species, the most prominent being the hydroxyl free radical. Ozone reacts with organic and inorganic compounds in aqueous solutions either directly with whole molecular ozone or by its radicals [53]. Ozone is a powerful oxidant with an oxidation potential of -2.07 V. The half-life of molecular ozone in air is relatively long ($\sim$12 hours), but in aqueous solutions the half-life can be as short as seconds when organic compounds exist. Decomposition of ozone is so rapid in water that its antimicrobial properties take place mainly at the microbial surface [54]. The PEL of ozone in the workplace is specified by the Occupational Safety and Health Administration (OSHA) at 0.1 ppm in air on an 8 hours/day basis, for a 40-hour work week.

18.3.2 POTENTIAL APPLICATIONS OF OZONE IN THE FOOD INDUSTRY

Ozone is a strong antimicrobial agent in both gaseous and aqueous phases. It is well known that ozone is an effective disinfectant in water and wastewater treatments [55]. Ozone is 1.5 times more effective as an antimicrobial agent than chlorine. Additionally, ozone is much more effective for a wider spectrum of microorganisms than chlorine and other disinfectants [56]. It reacts up to 3000 times faster than chlorine with organic materials and produces no harmful decomposition products [57]. Numerous studies have focused on the inactivation of pathogens and spores by aqueous ozone, including *Cryptosporidium parvum* [58,59], *Giardia* spp. [60], bacillus and clostridium spores [61, 62], *Salmonella* Typhimurium, *E. coli*, *Yersinia enterocolitica*, *Pseudomonas aeruginosa*, *Staphylococcus aureus*, and *L. monocytogenes* [63].

Applications of ozone for enhancing the microbiological safety and quality of foods have been reviewed by Kim *et al.* [12] and Khadre *et al.* [52]. Most applications for the food industry focus on the use of ozonated water for sanitation of food-contact surfaces and foods, including fruits and vegetables, meat and poultry, fish, cheese, and eggs. Although ozonated water has been shown to be an alternative to chlorinated water for decontamination of produce, its effectiveness in reducing microorganisms is less than 3 log CFU per gram or surface [12,52,65].

For more than half a century gaseous ozone as a disinfectant and/or preservative has also been applied in many areas in the food industry. Such applications include preservation of perishable foods, including fruits, vegetables, meat, poultry products, fish, cheese, spices, and eggs [12,52,56,66], grains [6,7,67], spices [12,68], decontamination of packaging materials [64,69], and decontamination of environmental air in food plants [12].

18.3.3 GENERATION OF OZONE

Ozone is commonly produced from oxygen or air by utilizing ultraviolet (UV) light or corona discharge generators. UV light systems use radiation at 185 nm wavelength emitted from high-transmission UV lamps. These systems are relatively low cost and do not require dry air for ozone production. The corona discharge generators can produce larger concentrations of ozone, up to 4%, compared to UV light systems. They mainly consist of two electrodes separated by a dielectric or nonconducting material, providing a narrow discharge gap. When a high-voltage alternating current is applied across this gap, the air or oxygen passing through the gap is partially ionized, and the oxygen molecules are dissociated. The split oxygen atoms combine with other oxygen molecules to form ozone. Dried air is required for this system to prevent corrosion of metal surfaces. Ozone can also be produced by cold plasma and electro-chemical methods [12]. Some commercial ozone generator manufacturers include IN USA, Inc. (Needham, MA), Lenntech, Inc. (College Station, TX), Ozone Solution, Inc. (Sioux Center, IA), Ozomax Ltd (Quebec, Canada), Longevity Resources, Inc. (British Columbia, Canada), and Ozone Services & Yanco Industries Ltd (Burton, B.C., Canada).

18.3.4 TREATMENT SYSTEMS

A continuous laboratory-scale O_3 gas treatment system (Figure 18.8) has been used for decontamination of fruits and vegetables [66]. This system includes a corona discharge ozone generator (Clear Water Tech, Inc., San Luis Obispo, CA), a 10 l Irvine Plexiglass treatment chamber with a stainless steel shelf, a humidifier, a diaphragm vacuum pump, a thermo-hydro recorder, an ozone neutralization unit, and a continuous ozone monitor (model 450H, Advanced Pollution Instrumentation, Inc., San Diego, CA). Ozone gas from the genera-tor is first humidified by flowing it through water in a 125 ml gas-washing bottle. The gas then continuously passes through the treatment chamber.

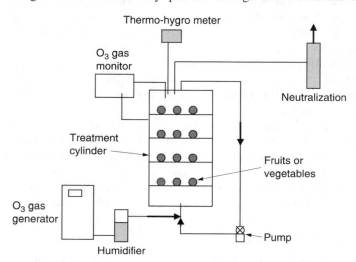

FIGURE 18.8 Schematic of a continuous laboratory ozone gas treatment system for decontamination of fruits and vegetables.

Meanwhile, the gas in the chamber is circulated by the pump. The ozone gas is finally neutralized by passing through a solution of reducing agents, such as sodium sulfite. The concentration of ozone gas is continuously monitored using a high-concentration ozone monitor. The concentration of ozone in the air can be recorded as mg/l or ppm. One mg/l ozone gas is approximately equivalent to 467 ppm in volume under standard conditions ($0°C$, 1 atm) and 509 ppm under normal conditions ($25°C$, 1 atm).

Ozone concentration can be measured using the indigo colorimetric method, which has been approved by the Committee on Standard Methods for the Examination of Water and Wastewater in 1988 [70]. For in-line monitoring gaseous ozone, a UV spectrophotometric method can be used [12], such as continuous monitors manufactured by Advanced Pollution Instrumentation, Inc. (San Diego, CA) and IN USA (Needham, MA).

18.3.5 MECHANISMS OF INACTIVATION OF MICROBES

The oxidizing mechanisms of ozone may involve direct reactions of molecular ozone and free radical-mediated destruction [12,52]. Inactivation of microorganisms by ozone may be due to the oxidation of a number of cellular components. The oxidation and disruption of cell membranes is considered to be one of the most important inactivation mechanisms. Ozone can oxidize polyunsaturated fatty acids, membrane-bound enzymes, glycoproteins, and glycolipids, and cause a decrease in cell permeability and disruption of normal cellular activity [52,71,72]. It has also been reported that bacterial inactivation may be due to inactivation of cellular enzymes, such as dehydrogenating enzymes in *E. coli* cells [73], β-galactosidase in the cytoplasm and alkaline phosphatases in the periplasm of *E. coli* [74], and destruction of genetic materials, such as DNA of *E.coli* [75,76], circular plasmid DNA [77], phage DNA and RNA [78,79], and viral DNA and RNA [52,80]. However, there is very limited information about microbial inactivation mechanisms by gaseous ozone, and the primary mechanism by both aqueous and gaseous ozone still needs to be clearly identified and investigated.

18.3.6 FACTORS INFLUENCING SANITATION TREATMENT BY OZONE GAS

The efficacy of gaseous ozone in reducing microbes on produce can be affected by many factors, such as ozone concentration, treatment time, temperature, RH, and surface properties of produce [66,81–83].

The effects of ozone gas concentration (2 to 8 mg/l), RH (60 to 90%), and treatment time (10 to 40 minutes) on inactivation of *E. coli* O157:H7 on green peppers were studied [66]. Ozone gas concentration, RH, and treatment time were all significant ($P < 0.01$) factors for the inactivation of *E. coli* O157:H7 (Figure 18.9). Among the three factors, the effect of ozone gas concentration was the greatest. The interaction between ozone gas concentration and RH exhibited a significant and synergistic effect ($P < 0.05$).

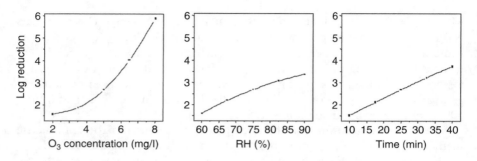

FIGURE 18.9 Effects of ozone gas concentration, % RH, and treatment time on log reductions of *E. coli* O157:H7 on green peppers at 22°C. (From Han, Y. *et al.*, *J. Food Sci.*, 67, 1188, 2002. With permission.)

Ishizaki *et al.* [82] reported that the efficacy of ozone gas for inactivation of bacillus spores on filter paper increased as ozone concentration (0 to 3.0 g/l), time (0 to 6 hours), and RH increased (54 to 90%). The spores were more resistant on a glass fiber filter than on filter paper. At a RH of 50% or below, there was no appreciable decrease in the number of survivors within 6 hours exposure to 3.0 mg/l ozone. Other researchers also reported that RH is an important factor for microbial inactivation by ozone gas, and ozone is less effective to inactivating dehydrated microorganisms [64].

Moreover, Liew and Prange [84] reported that temperature played an important role in the storage of carrots treated by ozone gas. The ozone concentration at 2°C was higher than that at 16°C, hence providing a greater reduction in fungal growth rate on carrots. A different linear effect of ozone concentration was found for each temperature. Significant ($P < 0.05$) effects of temperature and linear and quadratic effects of ozone on the growth rate of *Botrytis cinerea* on carrots were also observed.

18.3.7 EFFICACY IN REDUCING FOODBORNE MICROORGANISMS ON PRODUCE SURFACES

The efficacy of ozone gas in reducing foodborne pathogens and spoilage microorganisms has been studied on produce, including green peppers, carrots, black peppers, grapes, strawberries, and lettuce (Table 18.5). High concentrations (>1000 ppm or 2 mg/l) of gaseous ozone treatments have been shown to be extremely effective in reducing *E. coli* O157:H7 [85] and bacillus spores [82] on filter paper. Han *et al.* [66] found that a greater than 5 log reduction (CFU/site) of *E. coli* O157:H7 on green pepper surface could be achieved by a continuous ozone gas treatment at 5 mg/l for 25 minutes under >70% RH and at 22°C. Sarig *et al.* [87] reported that microbial counts on grapes were significantly reduced after an 8 mg/l ozone gas treatment for 20 minutes. Zhao and Cranston [68] treated ground black peppers with 6.7 mg/l ozonized air (6 l/min). They found a greater than 3 log reduction of *E. coli* and *Salmonella* spp. after

TABLE 18.5
Efficacy of Ozone Gas Treatments in Reducing Microorganisms on Different Surfaces

Surface	Microorganism	Ozone gas treatment conditions	Efficacy	Ref.
Filter paper	*Bacillus subtilis* spores	Continuous gas flow, 3 mg/l (1527 ppm) for 1 h at 95% RH and 22°C	>5 log CFU/paper reduction	82
Filter paper	*E. coli* O157:H7	Continuous gas flow at 3 l/min for 5 min, 1000 ppm at 4°C	>5 log CFU/paper reduction	85
Green pepper	*E. coli* O157:H7	Continuous gas treatment, 5 mg/l (2545 ppm) for 25 min at >70% RH and 22°C	>5 log CFU/sample reduction	66
Black pepper	*E. coli, Salmonella* spp., *Penicillium* spp., *Aspergillus* spp.	Continuous gas sparge at 6 l/min for 10–60 min, 6.7 mg/l (3410 ppm) at room temperature	3–6 log CFU/g reduction	68
Carrot	*Botrytis cinerea*	Continuous gas flow at 0.5 l/min for 8 h daily for 28 days, 60 ppm at 2, 8, or 16°C	50% reduction of daily growth rate	84
Blackberry	Fungi	Gas in storage room, 0.3 ppm for 12 days at 2°C and 90% RH	No visible fungal decay	86
Lettuce	Total aerobic plate count	Bubbling gaseous ozone (4.9%, v/v; 0.5 l/min) in a lettuce–water mixture (1:20, w/w) for 5 min	1.9 log CFU/g reduction	64
Black peppercorn	*E. coli, Salmonella* spp., *Penicillium* spp., *Aspergillus* spp.	Bubbling gaseous ozone (6.7 mg/l; 6 l/min) in a peppercorn–water mixture (2:5, w/w) for 10 min	3–4 log CFU/g reduction	68

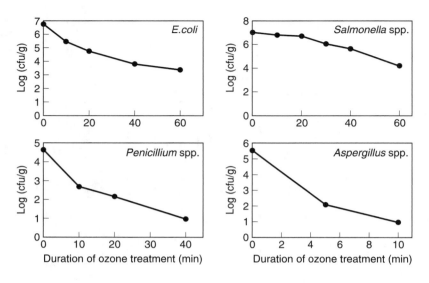

FIGURE 18.10 Kinetics of microbial reduction in ground black pepper after 6.7 mg/l ozone gas treatment at a flow rate of 6 l/min. (From Zhao, J. and Cranston, P.M., *J. Sci. Food Agric.*, 68, 11, 1995. With permission.)

60 minute exposure and more than a 3 to 4 log reduction of *Penicillium* spp. and *Aspergillus* spp. after 10 minutes. Figure 18.10 shows the kinetics of microbial reduction in ground black pepper after the ozone gas treatment. Zhao and Cranston [68] also treated black peppercorns by bubbling gaseous ozone (6.7 mg/l; 6 l/min) in a peppercorn–water mixture (2:5, w/w) for 10 minutes. This treatment reduced the microbial population by 3 to 4 log numbers. Kim *et al.* [64] reported that bubbling gaseous ozone (4.9%, v/v; 0.5 l/min) in a lettuce–water mixture under sonication and high-speed stirring was the most effective ozonation method that inactivated up to a 1.9 log of the natural microbial load in 5 minutes.

The effectiveness of ozone gas treatments is summarized in Table 18.5. Low concentrations (<100 ppm) of ozone gas treatments with long exposure times (days) have been used for growth inhibition and inactivation of spoilage microorganisms on fresh fruits and vegetables. Barth *et al.* [86] found that fungal development was suppressed when blackberries were stored at 2°C for 12 days in the air with 0.3 ppm ozone, with 20% of control fruits showing decay. Growth rates of yeast surviving ozone treatment were markedly decreased under longer exposure times and higher concentrations, RH, and temperatures [83]. Constant exposure to ozone throughout storage has also been reported to be effective in inhibiting storage pathogens on lemons and oranges at 1 ppm (14°C, 85% RH) [88] and peaches at 0.25 ppm (4 to 15°C) [89]. However, ozone was reported to be ineffective in preventing fungal decay in strawberries after 4-day treatment with 0.35 ppm ozone at 20°C [90]. Other early reports also indicated that ozonated apples, cantaloupes, and cranberries demonstrated more decay or damage than those not ozonated [91–94].

18.3.8 EFFECTS OF OZONE GAS TREATMENTS ON QUALITY OF FRUITS AND VEGETABLES

Ozone gas can be used to prevent fungal decay and rot of fruits and vegetables during cold storage [55]. Those products include bananas, citrus fruits, apples, berries, peaches, and potatoes. Ozone can also retard the ripening process of fruit and vegetables by oxidation of ethylene released during storage. Ewell [95] indicated that the shelf life of strawberries, raspberries, and grapes could be doubled when 2 to 3 ppm ozone is applied continuously for a few hours per day. Barth et al. [86] reported that 0.3 ppm treatment suppressed fungal development for 12 days at 2°C, and did not cause observable injury or defects on thornless blackberries. By the 12th day, anthocyanin content and surface color were maintained; however, peroxidase activity was reduced. Norton et al. [92] found that 0.6 ppm ozone at 15°C was effective in controlling fungus rot on Early Black and Howe varieties of cranberries, but caused weight loss and quality damage by the second and third week. Perez et al. [90] also reported that ozone was ineffective in preventing fungal decay in strawberries and also was detrimental to strawberry aroma after 4-day treatment with 0.35 ppm ozone at 20°C. Another study by Kute et al. [96] evaluating ozone-treated strawberries suggested that 0.3 or 0.7 ppm ozone did not affect the ascorbic acid levels, but significantly increased total soluble solid levels after 1 week of treatment and storage. The shelf life of apples treated by 2 to 3 ppm ozone for a few hours per day could be increased by several weeks, but damage on apples was observed with 10 ppm ozone [97]. The shelf life of potatoes exposed to 3 ppm ozone could be extended to as long as 6 month at 6 to 14°C and 93 to 97% RH [98]. Liew and Prang [84] observed some physiological and quality changes in ozone-treated carrots, such as higher respiration rate, electrolyte leakage, and lower color, compared to the control samples. Skog and Chu [99] studied the effect of ozone on quality of fruit and vegetables in cold storage. They found that 0.04 ppm ozone treatment under 95 to 98% RH appears to have the potential for extending the storage life of broccoli and seedless cucumber at 3°C. Response to ozone was minimal for mushrooms stored at 4°C and cucumbers stored at 10°C. The ethylene level in vegetable storage rooms was reduced from 1.5–2 ppm to a nondetectable level after the 0.04 ppm ozone treatment. The treatment did not affect the quality of apples and pears. Artes-Hernandez et al. [100] studied the effects of ozone enriched air treatment for improving quality of seedless table grapes during cold storage at 0°C for 60 days followed by 7 days of shelf life at 15°C in air. Compared to control grapes, gas-treated samples had superior texture and visual appearance after the shelf life study.

18.4 ALLYL ISOTHIOCYANATE GAS

18.4.1 PROPERTIES

Allyl isothiocyanate (AITC) is the major component in essential oils of cruciferous plants, such as black mustard seeds (*Brassica nigra*), brown mustard seeds (*Brassica juncea*), cabbage (*Brassica oleracia*), and horseradish (*Armoracia rusticana*) [101,102]. AITC is a specific compound from the isothiocyanate (ITC) group that has been shown to have bactericidal, bacteristatic, and antifungal activities. AITCs are released upon disruption or injury of plant tissue due to hydrolysis of glucosinolates by cell wall-bound myrosinase [103,104]. Other important properties of ITC compounds are their high volatility, extreme pungency, and low water solubility.

18.4.2 MECHANISMS AND FACTORS INFLUENCING SANITATION TREATMENT

The proposed mechanisms for inactivation and inhibition of microbes have focused on the nonspecific inactivation of enzymes through cleavage of disulfide bonds of proteins; interference with specific enzymes (carriers) in the electron transport chain, such as cytochrome C oxidase; uncoupling oxidative phosphorylation; nonspecific reactions with key enzymes or proteins; and reactions with free amino groups [104–106]. Even though there are several proposed antimicrobial mechanisms, the true antimicrobial mechanism(s) of ITCs are not fully understood.

18.4.3 EFFICACY IN REDUCING FOODBORNE PATHOGENS ON PRODUCE SURFACES

Gaseous and vaporized AITC has been shown to be an effective antimicrobial agent for produce and grain surfaces (Table 18.6). Table 18.6 shows different treatment parameters (antimicrobial concentration, treatment time and temperature, container size) to achieve a log reduction of $3 \log CFU/g$ or higher. Additionally, data from this table suggest that microorganisms are more difficult to inactivate on a rough surface than on a smooth surface.

18.4.4 EFFECTS OF AITC VAPOR TREATMENT ON QUALITY OF FRUITS AND VEGETABLES

Detrimental effects on product quality (lettuce texture and crispness) have been shown when lettuce is treated with high doses ($>300 \mu l$) of AITC vapor [102]. Additionally, Lin *et al.* [102] showed that tomato and apple texture did not change when exposed to low levels of AITC. However, when the apples and tomatoes were exposed to high levels of AITC vapors ($>300 \mu l$), texture started to soften within 2 days following treatment. Residual pungent odor and off-flavors on the produce existed for 12 hours when product was treated with

TABLE 18.6
Efficacy of Allyl Isothiocyanate in Reducing Foodborne Pathogens on Produce Surfaces

Surface	Microorganism	Treatment conditions	Efficacy	Ref.
Lettuce	*Listeria monocytogenes*	400 µl in a 1 gal bag with 100 g lettuce; 96 h	~4 log CFU/g	102
	Salmonella Montevideo	400 µl in a 1 gal bag with 100 g lettuce; 48 h	8 log CFU/g	
	E. coli 0157:H7	400 µl in a 1 gal bag with 100 g lettuce; 48 h	7 log CFU/g	
Tomato skin	*Salmonella* Montevideo	400 µl in a 4 l container with 3 tomatoes; 24 and 48 h	5 log CFU/g	102
Tomato scar	*Salmonella* Montevideo	500 µl in a 4 l container with 3 tomatoes; 48 h	5 log CFU/g	102
Apple stem scar	*E. coli* O157:H7	600 µl in a 4 l container with 3 apples; 48 h	3 log CFU/g	102
Wet alfalfa seeds	*E. coli* O157:H7	50 µl /950 ml jar; 24 h; 25°C	~4 log CFU/g	107
Alfalfa sprouts	Salmonella mixture	200 and 500 mg/l; 48 h at 10°C	7 log CFU/g	108
Alfalfa seeds	Salmonella mixture	1000 mg/l; 7 h at 60°C	~2 log CFU/g	108

AITC vapors $< 300\,\mu l$; however, when high levels of AITC vapors ($500\,\mu l$) were used, off-flavors and odors lasted up to 24 hours after product treatment [102]. An additional side effect that has been seen after treatment of alfalfa seeds with AITC (up to 500 mg/l) is reduced seed germination [108]. Treatment of alfalfa seeds with 200 and 500 mg AITC/l of air adversely affected sensory quality attributes (appearance, color, aroma, and overall acceptance) [108].

18.5 OTHER GASES/VAPORS

18.5.1 HYDROGEN PEROXIDE VAPOR

The use of hydrogen peroxide (H_2O_2) has been approved by the U.S. Food and Drug Administration (FDA) as an antimicrobial for several food processing applications, such as disinfection of aseptic packaging materials and equipment, processing of cheese and modified whey, and production of thermophile-free starch [109]. Washes with aqueous H_2O_2, or commercial sanitizers containing H_2O_2, have been studied in laboratories for disinfection and/or decontamination of fresh or minimally processed fruits and vegetables [109–115]. Additionally, vapor-phase H_2O_2 has been evaluated for the decontamination of fruits and vegetables, due to its sporicidal properties against bacillus spores and food spoilage microbes [116–118]. Forney *et al.* [119] treated table grapes inoculated with *Botrytis cinerea* spores with H_2O_2 vapor generated from 30 to 35% H_2O_2 solution at 40°C for 10 minutes. After the treatment, and a 24-hour storage period at 10°C, the number of germinable spores was reduced by 60% or more, and the incidence of decay was also reduced after 12 day storage at $10°C$. In another study, Simmons et al. [120] applied $3.1 \text{ mg/l } H_2O_2$ vapor to dried prunes for 0 to 60 minutes at ambient room temperature (20 to $26°C$). More than $2 \log CFU/g$ reductions in total plate count (TPC), yeast, and mold counts were observed after the treatment for 10 minutes. However, some oxidation damage was observed in prunes treated for greater than 20 minutes. Other studies have shown that H_2O_2 vapor is effective in reducing spoilage microbial counts on whole cantaloupes, raisins, and walnut nut meat [109]. Sapers et al. [121] also evaluated the use of H_2O_2 vapor to extend the shelf life of various other fresh and fresh-cut commodities. Produce samples were exposed to H_2O_2 vapor for 2 to 15 minutes at injection rates of 2.5 or 5 g of H_2O_2/min. The treatment appeared to delay or diminish the severity of bacterial soft rot in fresh cut cucumber, green bell pepper, and zucchini. No effect on spoilage of fresh-cut broccoli, carrot, cauliflower, or celery, or fresh raspberries and strawberries was observed. More recently, Sapers et al. [110] studied decontamination of apples using H_2O_2 vapor generated from 35% H_2O_2 solution at $150°C$ in a pressurized vessel. The treatment led to a $1.7 \log CFU/g$ reduction of E. coli on apples when three treatment cycles were applied over 5 minutes, which was less than those that would be obtained with a 5% H_2O_2 wash [112].

18.5.2 ACETIC ACID VAPOR

Acetic acid is known for its preservative properties [122] and has been used extensively in foods such as pickles, salad dressing, tomato products, and mustards. Vaporized acetic acid has also shown biocidal effects for decontamination of fruits and vegetables. Researchers have demonstrated that fumigation with acetic acid vapor or vinegar vapor could control postharvest decay of fruits and vegetables such as apples, grapes, stonefruit (peaches, nectarines, and apricots), strawberries, oranges, kiwifruit, tomatoes, and coleslaw made from cabbage [123–127]. They also demonstrated that fumigation treatments with 242 ppm (v/v) gaseous acetic acid in air for 24 hours at $22°C$ or for 12 hours at $45°C$ could reduce 3 to $5 \log CFU/g$ Salmonella Typhimurium, E. coli O157:H7, and Listeria monocytogenes on mung bean seed without significant reduction of seed germination rates [128]. Recently, Sapers et al. [43] evaluated the use of pressurized acetic acid vapor to decontaminate apples inoculated with E. coli. In their vacuum–pressure cycle system, a vacuum (508 to 686 mmHg) was applied to a 6 l stainless steel treatment, where inoculated apples were placed, and then acetic acid vapor generated from glacial acetic acid at $60°C$ was applied to the vessel to achieve a pressure of 34.5, 68.9, or 103.4 kPa. The treatment time was 5 to 30 minutes, and temperatures ranged from 40 to $60°C$. After the treatment, the apples were immediately rinsed with tap water for 30 seconds, observed for treatment-induced injury, and prepared for microbiological analysis. They found that the vapor treatment with three vacuum–pressure cycles at $60°C$ provided population reductions exceeding $3.5 \log CFU/g$. However, these conditions induced discoloration. When a total treatment time of 5 minutes was used, log reductions increased with increasing treatment temperature and the number of vacuum–pressure cycles applied. Treatment pressure did not appear to affect the bacterial log reduction. During storage for several hours, the treated Golden Delicious apples developed dark lesions surrounding the lenticels which penetrated several millimeters into the flesh beneath the skin, indicating that acetic acid vapor treatment may not be useful for apples under these conditions.

18.5.3 OTHER NATURAL PLANT VOLATILES

Some natural plant volatiles, such as methyl jasmonate, *trans*-anethole, carvacrol, cinnamic aldehyde, eugenol, linalool, and thymol, have been used as antimicrobials in reducing microbial contamination and extending shelf life of fruits and vegetables [129]. Methyl jasmonate (MJ) is known for its properties to enhance resistance to chilling temperature of fruits and vegetables [130]. MJ vapor, in combination with modified atmosphere packaging, can reduce loss of firmness, fungal decay, and development of chilling injury and increase retention of organic acids in papayas [131]. MJ vapor from a 10^{-4} or 10^{-5} mol source in a 1 l container also retarded deterioration of celery sticks for 2 weeks at 10°C and reduced the bacterial load by approximately 3 logs after 1 week storage [130]. Wang [132] reported that postharvest quality of raspberries was enhanced with treatments of MJ, AITC, tea tree oil, or absolute ethyl alcohol during storage at 10°C. Wang and Buta [133] also found that 2.24 to 22.4 µl/l MJ vapor maintained good quality of fresh-cut kiwifruit for up to 3 weeks, as did absolute ethanol and isopropyl alcohol.

Weissinger *et al.* [108] evaluated nine natural volatile compounds for their ability to destroy salmonella on alfalfa seeds and sprouts. In this study, vapor-phase acetic acid, AITC, *trans*-anethole, carvacrol, cinnamic aldehyde, eugenol, linalool, MJ, or thymol were applied to inoculated alfalfa seeds at 1000 mg/l of air concentration for 1, 3, and 7 hours at 60°C. Only acetic acid, cinnamic aldehyde, and thymol caused significant reductions in salmonella populations ($>3 \log CFU/g$) compared to the untreated control ($1.9 \log CFU/g$) after treatment for 7 hours. Treatment of seeds at 50°C for 12 hours with acetic acid (100 and 300 mg/l) and thymol or cinnamic aldehyde (600 mg/l) led to a $1.7 \log CFU/g$ reduction of salmonella on seeds without affecting germination percentage. Treatment of seeds at 50°C with AITC (100 and 300 mg/l) and cinnamic aldehyde or thymol (200 mg/l) did not significantly reduce populations compared with the untreated control. Seed germination percentage was largely unaffected by treatment with gaseous acetic acid, AITC, cinnamic aldehyde, or thymol for up to 12 hours at 50°C. Acetic acid at 200 and 500 mg/l reduced an initial population of $7.50 \log CFU/g$ of alfalfa sprouts by 2.33 and $5.72 \log CFU/g$, respectively, within 4 days at 10°C, whereas AITC at 200 and 500 mg/l reduced populations to undetectable levels. However, both treatments caused deterioration in sensory quality. Treatment of sprouts with 1 or 2 mg/l AITC also adversely affected sensory quality and did not reduce salmonella populations after 11 days of exposure at 10°C.

18.6 PRESENT AND FUTURE APPLICATIONS OF GASEOUS/VAPOR-PHASE ANTIMICROBIALS FOR DECONTAMINATION OF FRESH PRODUCE

As an alternative to traditional sanitizers and fumigation agents, gaseous/ vapor-phase antimicrobials certainly have potential for use to improve microbial safety and to extend shelf life of fresh and minimally processed fruits and vegetables. Some of the possible applications are presented below.

Technologies could be designed and optimized to treat produce in larger scale operations within a treatment chamber, a storage room, or a continuous belt tunnel using ClO_2 or ozone gas that is continuously supplied by generators and monitored throughout the process. The treated produce may be most appropriate in production of fresh-cut products or as ingredients used in juice production. One such application is an automated continuous ClO_2 gas treatment system that is being studied and evaluated for decontamination of produce at Purdue University (Figure 18.5). Natural Sterilization & Fumigation, Inc. (Sparks, NV) has also developed a PureOx® ozone gas treatment system for decontamination of fruits and vegetables. Products are treated in a vacuum chamber with controlled ozone generation. McCabe [134] has also patented a continuous process apparatus and method to treat delicate vegetables, such as potato chips and dehydrated onions, by ozone gas. In this system, products are continuously passed through a treatment zone containing ozone. The products, air, and byproducts (oxygen) of the ozone treatment are separated from the ozone by gravity in a separation zone located above the treatment zone.

Treatments could also be designed and optimized to treat produce in a chamber or a storage room using vapor-phase antimicrobials, such as ClO_2. Tsai *et al.* [39] evaluated a system for the prevention of potato spoilage during storage using vaporized ClO_2. In this system, ClO_2 is vaporized by purging air through different concentrations of acidified Oxine (ACD, Bio-Cide International Inc., Norman, OK) solutions, in which ClO_2 is the active ingredient. ClO_2 vapor is delivered to a treatment chamber or storage facility where the ClO_2 treatment is applied to the potatoes based on a calculation dealing with the amount of ClO_2 loss in the ACD solution. Using this system with 500 ppm ACD Oxine solution, a 3 to 4 log reduction of *Erwinia carotvora* and natural flora (aerobic bacteria, yeast, and mold) were found after purging air through the Oxine solution at a flow rate of 6 l/hour for 1 hour.

A closed chamber or package in which ClO_2 gas is introduced could also be developed. Using this approach, the gas could be slowly released by sachets, such as sachets that are currently manufactured by ICA TriNova LLC (Figure 18.3). One advantage of this technology is that there is no need for costly and space-consuming ClO_2 gas generating systems. Sapers *et al.* [43] studied decontamination of apples in a 24.6 l high-density polyethylene pail using sachets that generate 0.03 to 0.3 mg/l ClO_2 gas in the pail. They found a 3.24 and 4.42 log CFU/g reduction of *E. coli* after treatment with 0.3 mg/l ClO_2

for 3 hours at 4 and 20°C, respectively, while product quality was minimally affected. After 20 hours' exposure, the microbial reduction approached nearly 5 logs, but darkening of lenticels was also observed. Lee et $al.$ [135] evaluated the same ClO_2 sachets, which slowly released 11, 18, and 26 mg ClO_2 gas within 0.5, 1, and 3 hours, respectively, for the decontamination of lettuce in a 20 l chamber. They reported a 4 to 6 log reduction of $E.$ $coli$, $L.$ $monocytogenes$, and $Salmonella$ Typhimurium on lettuce leaves.

There may also be an opportunity to treat produce within a plastic film package where the packaging material is designed to allow for a slow and continuous release of the antimicrobial gas or vapor in the package. Bernard Technologies Inc. (BTI, Chicago, IL) has developed two proprietary, controlled-release technologies called Microsphère® and Microlite® that focus on preventing and eliminating biological contamination. Both the Microsphère and Microlite patented sustained release systems [136] enable the creation of an active Microatmosphère® environment that inhibits the growth of microorganisms and provides an extended shelf life for fruits and/or vegetables. BTI's Microsphère-containing films have been affirmed as a GRAS product by the FDA and U.S. Department of Agriculture (USDA) for use in preserving fresh fruits, vegetables, meat, and poultry products.

The effects of combining antimicrobial vapors or gases with a modified atmosphere packaging strategy to extend the shelf life of fruits and vegetables could also be explored further. Ozen et $al.$ [69] studied the effects of ozone gas on mechanical, thermal, and barrier properties of several plastic films used in food packaging, including linear low-density polyethylene (LLDPE), oriented polypropylene (OPP), and biaxially oriented nylon (BON) films. Exposure to 2.1 to 4.3 mg/l ozone gas for 2 to 24 hours caused a decrease in tensile strength of OPP, decreased melting temperature of OPP and BON, and reduced oxygen permeability of LLDPE and BON.

18.7 REGULATORY CONSIDERATIONS

Although the FDA has approved the use of aqueous ClO_2 (3 ppm residual) to wash fruits and vegetables, it has not granted permission to use the gaseous form of ClO_2 as a sanitizer for decontaminating produce. It is assumed that gaseous ClO_2 treatment would be very similar to aqueous ClO_2, producing similar oxychloro byproducts (chlorine, ClO_2, chlorate, and chlorite) as an aqueous ClO_2 treatment. Based on existing scientific literature and data related to exposure to drinking water containing 1 mg/kg of the oxidant species (chlorine, ClO_2, chlorate, and chlorite), the use of aqueous chlorine dioxide is not likely to lead to chemical byproducts that are harmful to human health [10]. The research group at Purdue University has shown that residuals of ClO_2 and chlorite on green peppers were not detectable after a 4-week storage period and after a 1-week storage period for strawberries [37,38]. However, in order to pursue regulatory approval for the use of ClO_2 gas for decontamination of fresh and cut produce, more information and research are needed on production of byproducts on fruit and vegetable surfaces after ClO_2 treatment.

In the food industry, ozone was first permitted by the FDA to disinfect bottled water [137]. In 1997 ozone was affirmed GRAS status as a disinfectant and/or sanitizer for broad-based food usage by an expert panel sponsored by the Electric Power Research Institute (EPRI) in the U.S. [57]. In 2001 the FDA approved the use of ozone as a direct food additive for the treatment, storage, and processing of foods in gaseous and aqueous phases. Acetic acid and hydrogen peroxide have been considered as GRAS food additives by the FDA; however, currently there are no regulations in the U.S. on the usage of their vaporous forms or of other natural plant volatiles (AITC, MJ, *trans*-anethole, carvacrol, cinnamic aldehyde, eugenol, linalool, and thymol) as antimicrobial agents for preservation or sanitation of fruits and vegetables. Regulatory approval of the usage of these antimicrobials for produce will depend upon many different factors. Certainly, treatment effectiveness and efficacy data, worker safety and human health exposure data, and industrial and commercial needs will all play very important roles as this technology moves forward.

REFERENCES

1. Rosenblatt, D.H., Rosenblatt, A.A., and Knapp, J.E., Use of Chlorine Dioxide Gas as a Chemosterilizing Agent, U.S. Patent 4,504,442, 1985.
2. Rosenblatt, D.H., Rosenblatt, A.A., and Knapp, J.E., Use of Chlorine Dioxide Gas as a Chemosterilizing Agent, U.S. Patent 4,681,739, 1987.
3. Jeng, D.K. and Woodworth, A.G., Chlorine dioxide gas sterilization under square-wave conditions, *Appl. Environ. Microbiol.,* 56, 514, 1990.
4. Parisi, A.N. and Young, W.E., Sterilization with ethylene oxide and other gases, in *Disinfection, Sterilization, and Preservation,* 4th ed., Seymour, S.B., Ed., Lea and Febiger, Malvern, PA, 1991, chap. 22.
5. Han, Y. *et al.,* Experimental model for inactivation of spoilage micro-organisms on storage tank surfaces by chlorine dioxide gas, *Food Microbiol.,* 16, 53, 1999.
6. Kells, S.A. *et al.,* Efficacy and fumigation characteristics of ozone in stored maize, *J. Stored Products Res.,* 37, 371, 2001.
7. Mendez, F. *et al.,* Penetration of ozone into columns of stored grains and effects on chemical composition and processing performance, *J. Stored Products Res.,* 39, 33, 2003.
8. Lide, D.R., *CRC Handbook of Chemistry and Physics,* 82nd ed., CRC Press, Boca Raton, FL, 2001.
9. Occupational Safety and Health Administration, Air Contaminant Exposure Standards, 29 CFR 1910.1000, chap. XVII, 6, July 1, 1991.
10. Gates, D., *The Chlorine Dioxide Handbook,* American Water Works Association, Denver, CO, 1998.
11. EPA, Ozone, in *Guidance Manual Alternative Disinfectants and Oxidants,* Environmental Protection Agency, 1999, chap. 3.

12. Kim, J., Yousef, A.E., and Dave, S., Application of ozone for enhancing the microbiological safety and quality of foods: a review, *J. Food Prot.*, 62, 1071, 1999.

13. Aieta, E.M. and Berg, J.D., A review of chlorine dioxide in drinking water treatment, *J. Am. Water Works Assoc.*, 78, 62, 1986.

14. Ridenour, G.M., Ingols, R.S., and Armbruster, E.H., Sporicidal properties of chlorine dioxide, *Water Sewage Works*, 96, 279, 1949.

15. Benarde, M.A. *et al.*, Efficiency of chlorine dioxide as a bactericide, *J. Appl. Microbiol.*, 13, 776, 1967.

16. Benarde, M.A., *et al.*, Kinetics and mechanism of bacterial disinfection by chlorine dioxide, *J. Appl. Microbiol.*, 15, 257 1967.

17. Harakeh, M.S., Illescas, A., and Matin, A., Inactivation of bacteria by Purogene, *J. Appl. Bacteriol.*, 64, 459, 1988.

18. Foschino, R. *et al.*, Bactericidal activity of chlorine dioxide against *Escherichia coli* in water and on hard surfaces, *J. Food Prot.*, 61, 668, 1998.

19. Harakeh, M.S., Berg, J.C., and Matin, A., Susceptibility of chemostat-grown *Yersinia enterocolitica* and *Klebsiella pneumoniae* to chlorine dioxide, *Appl. Environ. Microbiol.*, 49, 69, 1985.

20. Korich, D.G. *et al.*, Effects of ozone, chlorine dioxide, chlorine, and monochloramine on *Cryptosporidium parvum* oocyst viability, *Appl. Environ. Microbiol.*, 56, 1423, 1990.

21. Roberts, R.G. and Reymond, S.T., Chlorine dioxide for reduction of postharvest pathogen inoculum during handling of tree fruits, *Appl. Environ. Microbiol.*, 60, 2864, 1994.

22. Noss, C.I. and Olivier, V.P., Disinfecting capabilities of oxychlorine compounds, *Appl. Environ. Microbiol.*, 50, 1162, 1985.

23. Chen, Y. and Vaughn, J.M., Inactivation of human and simian rotaviruses by chlorine dioxide, *Appl. Environ. Microbiol.*, 56, 1363, 1990.

24. Foegeding, P.M., Hemstapat, V., and Giesbrecht, F.G., Chlorine dioxide inactivation of *Bacillus* and *Clostridium* spores, *J. Food Sci.*, 51, 197, 1986.

25. White, G.C., *Handbook of Chlorination*, Van Nostrand Reinhold, New York, 1972, p. 596.

26. Lillard, H.S., Levels of chlorine dioxide of equivalent bactericidal effect in poultry processing water, *J. Food Sci.*, 44, 1594, 1979.

27. Elphick, A., The growing use of chlorine dioxide, *Processing*, March 24, 1998.

28. Eylath, A. *et al.*, Successful sterilization using chlorine dioxide gas, *BioProcess International*, August, 2, 2003.

29. Heilprin J., Fumigation to Kill Final Anthrax Spores, Dec. 2001, http://www.the dailycamera.com/news/terror/dec01/02aanth.html.

30. Han, Y. *et al.*, Decontamination of *Bacillus thuringiensis* spores on selected surfaces by chlorine dioxide gas, *J. Environ. Health*, 66, 16, 2003.

31. Han, Y. *et al.*, The effects of washing and chlorine dioxide gas on survival and attachment of *Escherichia coli* O157:H7 to green pepper surfaces, *Food Microbiol.*, 17, 521, 2000.

32. Han, Y. *et al.*, Inactivation of *Escherichia coli* O157:H7 on surface-uninjured and -injured green pepper (*Capsicum annuum* L.) by chlorine dioxide gas as demonstrated by confocal laser scanning microscopy, *Food Microbiol.*, 17, 643, 2000.

33. Han, Y. *et al.*, Reduction of *Listeria monocytogenes* on green peppers (*Capsicum annuum*) by gaseous and aqueous chlorine dioxide and water washing, and its growth at 7°C, *J. Food Prot.*, 64, 1730, 2001.

34. Han, Y. *et al.*, Response surface modeling for the inactivation of *Escherichia coli* O157:H7 on green peppers (*Capsicum annuum* L.) by chlorine dioxide gas treatments, *J. Food Prot.*, 64, 1128, 2001.

35. Han, Y. *et al.*, A comparison of methods for recovery of chlorine dioxide-injured *Escherichia coli* O157:H7 and *Listeria monocytogenes*, *Food Microbiol.*, 19, 201, 2002.

36. Han, Y., Linton, R.H., and Nelson, P.E., Inactivation of *Escherichia coli* O157:H7 and *Listeria monocytogenes* on strawberry by chlorine dioxide gas, Annual Meeting of Institute of Food Technologists, Anaheim, CA, 2002.

37. Han, Y. *et al.*, Effects of chlorine dioxide gas treatment on microbial safety and quality of green peppers, Annual International Food Safety and Quality Conference, Orlando, FL, Nov. 5–7, 2003.

38. Han, Y. *et al.*, Decontamination of strawberries using batch and continuous chlorine dioxide gas treatments. *J. Food Prot.*, 67, 2450, 2004.

39. Tsai, L.S, Huxsoll, C., and Robertson, G., Prevention of potato spoilage during storage by chlorine dioxide, *J. Food Sci.*, 66, 472, 2001.

40. Du, J., Han, Y., and Linton, R.H., Inactivation of *Listeria monocytogenes* spotted onto different apple surfaces using chlorine dioxide gas, *Food Microbiol.*, 19, 481, 2002.

41. Du, J., Han Y., and Linton, R.H., Efficacy of chlorine dioxide gas in reducing *Escherichia coli* O157:H7 on apple surfaces, *Food Microbiol.*, 20, 583, 2003.

42. Rosentrader, R. *et al.*, Inactivation of *Salmonella* spp. on the outer surfaces of whole cantaloupes using chlorine dioxide gas, Annual Meeting of Institute of Food Technologists (IFT), Chicago, IL, 2003.

43. Sapers, G.M. *et al.*, Vapor-phase decontamination of apples inoculated with Escherichia coli, *J. Food Sci.*, 68, 1003, 2003.

44. Noss, C.I., Dennis, W.H., and Olivieri, V.P., Reactivity of chlorine dioxide with nucleic acids and proteins, in *Water Chlorination: Environmental Impact and Health Effects,* Jolley, R.L. *et al.*, Eds., Lewis Publishers, Chelsea, MI, 1983.

45. Olivieri, V.P. *et al.*, Mode of action of chlorine dioxide on selected viruses, in *Water Chlorination: Environmental Impact and Health Effects,* Jolley, R.L. *et al.*, Eds., Lewis Publishers, Chelsea, MI, 1985.

46. Roller, S.D. *et al.*, Mode of bacterial inactivation by chlorine dioxide, *Water Res.*, 14, 635, 1980.

47. Ghandbari, E.H. *et al.*, Reaction of chlorine and chlorine dioxide with free fatty acids, fatty acid esters, and triglycerides, *Water Chlorination: Environmental Impact and Health Effects,* Jolley, R.L. *et al.*, Eds., Lewis Publishers, Chelsea, MI, 1983.

48. Berg, J.D., Roberts, P.V., and Matin, A., Effect of chlorine dioxide on selected membrane functions of *Escherichia coli*, *J. Appl. Bacteriol.*, 60, 213, 1986.

49. D'lima, C.B. and Linton, R.H., Inactivation of *Listeria monocytogenes* on lettuce by gaseous and aqueous chlorine dioxide and chlorinated water, Annual Meeting of Institute of Food Technologists, Anaheim, CA, 2002, abstr. 15D-4.

50. Costilow, R.N., Uebersax, M.A., and Ward, P.J., Use of chlorine dioxide for controlling microorganisms during handling and storage of fresh cucumbers, *J. Food Sci.*, 49, 396, 1984.

51. Zhang, S. and Farber, J.M., The effects of various disinfectants against *Listeria monocytogenes* on fresh-cut vegetables, *Food Microbiol.*, 13, 311, 1996.

52. Khadre, M.A., Yousef, A.E., and Kim, J.G., Microbiological aspects of ozone application in foods: a review, *J. Food Sci.*, 66, 1242, 2001.

53. Staehelin, J. and Hoigné, J., Decomposition of ozone in water in the presence of organic solutes acting as promoters and inhibitors of radicals chain reactions, *Environ. Sci. Technol.*, 19, 1206, 1985.

54. Hoigne, J., and Bader, H., Ozonation of water: role of hydroxyl radicals as oxidizing intermediates, *Science*, 190, 782, 1975.

55. Rice, R.G. *et al.*, Use of ozone in drinking water treatment, *J. Am. Water Works Assoc.*, 73, 44, 1981.

56. Xu, L., Use of ozone to improve the safety of fresh fruits and vegetables, *Food Technol.*, 53, 58, 1972.

57. Graham, D.M., Use of ozone for food processing, *Food Technol.*, 51, 72, 1997.

58. Liyanage, L.R.J., Finch, G.R., and Belosevic, M., Sequential disinfection of *Cryptosporidium parvum* by ozone and chlorine dioxide, *Ozone Sci. Eng.*, 19, 409, 1997.

59. Finch, G.R. *et al.*, Ozone inactivation of *Cryptosporidium parvum* in demand free phosphate buffer determined by in vitro excystation and animal infectivity, *Appl. Environ. Microbiol.*, 59, 4203, 1993.

60. Finch, G.R. *et al.*, Comparison of *Giardia lamblia* and *Giardia muris* cyst inactivation by ozone, *Appl. Environ. Microbiol.*, 59, 3674, 1993.

61. Rickloff, J.R., An evaluation of the sporicidal activity of ozone, *Appl. Environ. Microbiol.*, 53, 683, 1987.

62. Foegeding, P.M., Ozone inactivation of *Bacillus* and *Clostridium* spore populations and the importance of the spore coat to resistance, *Food Microbiol.*, 2, 123, 1985.

63. Restaino, L. *et al.*, Efficacy of ozonated water against various food-related microorganisms, *Appl. Environ. Microbiol.*, 61, 3471, 1995.

64. Kim, J., Yousef, A.E., and Chism, G.W., Use of ozone to inactivate microorganisms on lettuce, *J. Food Saf.*, 19, 17, 1999.

65. Cherry, J.P., Improving the safety of fresh produce with antimicrobials, *Food Technol.*, 53, 54, 1999.

66. Han, Y. *et al.*, Response surface modeling for the inactivation of *Escherichia coli* O157:H7 on green peppers (*Capsicum annuum*) by ozone gas treatments, *J. Food Sci.*, 67,1188, 2002.

67. Allen, B., Wu, J., and Doan, H., Inactivation of fungi associated with barley grain by gaseous ozone, *J. Environ. Sci. Health*, B38, 617, 2003.

68. Zhao, J. and Cranston, P.M., Microbial decontamination of black pepper by ozone and the effect of the treatment on volatile oil constitutes of the spice, *J. Sci. Food Agric.*, 68, 11, 1995.

69. Ozen, B.F., Floros, J.D., and Nelson, P.E., Effects of ozone gas on mechanical, thermal, and barrier properties of plastic films used in food packaging, Annual Meeting of Institute of Food Technologists, New Orleans, LA, abstr. 2001.

70. Greenberg, A.E., Clesceri, L.S., and Eaton, A.D., 4500-ClO$_2$ C. Amperometric method I, in *Standard Methods for the Examination of Water and Wastewater*,

18th ed., American Public Health Association, Washington D.C., 1992, pp. 4-55, 4-56.

71. Scott, D.B.M. and Lesher, E.C., Effect of ozone on survival and permeability of *Escherichia coli, J. Bacteriol.,* 85, 567, 1963.
72. Komanapalli, I.R. and Lau, B.H.S., Ozone-induced damage of *Escherichia coli* K-12, *Appl. Microbiol. Biotechnol.,* 46, 610, 1996.
73. Ingram, M. and Haines, R.B., Inhibition of bacterial growth by pure ozone in the presence of nutrients, *J. Hyg.,* 47, 146, 1949.
74. Takamoto, Y., Maeba, H., and Kamimura, M., Changes in survival rate of enzyme activities and in *Escherichia coli* with ozone, *Appl. Microbiol. Biotechnol.,* 37, 393, 1992.
75. Prat, R., Nofre, C., and Cier, A., Effects de l'hypochlorite de sodium, de l'ozone et des radiations ionisontes dur les constituants pyrimidiques d' *Escherichia coli, Ann Inst. Pasteur Paris,* 114, 595, 1968.
76. Scott, D.B.M., The effect of ozone on nucleic acids and their derivatives, in *Aquatic Application of Ozone,* Blogoslawski, W. J. and Rice, R.G., Eds., International Ozone Institute, Syracuse, NY, 1975, p. 226.
77. Hamelin, C., Production of single- and double-stranded breaks in plasmid DNA by ozone, *Oncol. Biol. Phys,* 11, 253, 1985.
78. Mura, C. and Chung, Y.S., In vitro transcription assay of ozonated T7 phage DNA, *Environ. Mol. Mutagen.,* 16, 44, 1990.
79. Kim, C.K., Gentile, D.M., and Sproul, O.J., Mechanism of ozone inactivation of bacteriophage f2, *Appl. Environ. Microbiol.,* 39, 210, 1980.
80. Roy, D., Chian, E.S.K., and Engelbrecht, R.S., Kinetics of enteroviral inactivation by ozone, *J. Environ. Eng. ASCE,* 107, 887, 1981.
81. Sakurai, M. *et al.,* Several factors affecting ozone gas sterilization, *Biocontrol Sci.,* 8, 69, 2003.
82. Ishizaki, K, Shinriki, N., and Matsuyama, H., Inactivation of *Bacillus* spores by gaseous ozone, *J. Appl. Bacteriol.,* 60, 67, 1986.
83. Naitoh, S., Studies on the application of ozone in food preservation: microbial properties of ozone in the gas phase to yeast, *J. Antibac. Antifung. Agents Japan,* 21, 341, 1992.
84. Liew, C.L. and Prange, R.K., Effect of ozone and storage temperature on postharvest diseases and physiology of carrots, *J. Am. Soc. Hort. Sci.,* 119, 563, 1994.
85. Yuan, T.C., Boisrobert, C., and Steiner, E., Ozone bactericidal efficacy: effect of organic compounds, Annual Meeting of Institute of Food Technologists, IFT, Chicago, IL, Book of Abstracts, 9-18, p. 15.
86. Barth, M.M. *et al.,* Ozone storage effects on anthocyanin content and fungal growth in blackberries, *J. Food Sci.,* 60, 1286, 1995.
87. Sarig, P. *et al.,* Ozone for control of post-harvest decay of table grapes caused by *Rhizopus stolonifer, Physiol. Mol. Plant Pathol.,* 48, 403, 1996.
88. Harding, R.P., Effect of ozone on *Pencillium* mold decay and sporulation, *Plant Dis. Rep.,* 52, 245, 1968.
89. Ridley, J.D. and Sims, E.T., The response of peaches to ozone during storage, *S.C. Agr. Exp. Station Tech. Bull.,* 1027, 24, 1967.
90. Perez A.G. *et al.,* Effects of ozone treatment on postharvest strawberry quality, *J. Agric. Food Chem.,* 47, 1652, 1999.
91. Barger, W.R. *et al.,* A comparison of fungicidal treatments for the control of decay in California cantaloupes, *Phytopathology,* 38, 1019, 1948.

92. Norton, J.S., Charig, A.J., and Demoranville, I.E., The effect of ozone on storage of cranberries, *Proc. Am. Soc. Hort. Sci.,* 93, 792, 1968.

93. Schomer, H.A. and McCulloch, L.P., Ozone in relation to storage of apples, *U.S. Dept. Agric. Circ.,* 765, 1948.

94. Spalding, D.H., Effects of ozone atmospheres on spoilage of fruits and vegetables after harvest, *U.S. Dept. Agric. Mark. Res. Rep.,* 801, 1968.

95. Ewell, A.W., Ozone and its applications in food preservation, in *Refrigeration Application Data Book,* 2nd ed., Sect. II, American Society of Refrigeration Engineers, Menasha, WI, 1940, p. 1990203.

96. Kute, K.M., Zhou, C., and Barth, M.M., The Effect of Ozone Exposure on Total Ascorbic Acid Activity and Soluble Solids Contents in Strawberry Tissue, Annual Meeting of Institute of Food Technologists, 1995, p. 82.

97. Kuprianoff, J., The use of ozone in cold storage of fruits, *Z. Kältetechnik.,* 10, 1, 1953.

98. Baranovskaya, V.A. *et al.,* Use of ozone gas sterilization during storage of potatoes and vegetables, *Ovoshchesus. Promst.,* 4, 10, 1979.

99. Skog, L.J. and Chu, C.L., Effect of ozone on qualities of fruits and vegetables in cold storage, *Can. J. Plant Sci.,* 81, 773, 2001.

100. Artes-Hernandez, F., Aguayo, E., and Artes, F., Alternative atmosphere treatments for keeping quality of Autumn seedless table grapes during long-term cold storage, *Postharvest Biol. Technol.,* 31, 59, 2004

101. Isshiki, K. *et al.,* Preliminary examination of ally isothiocyanate vapor for food preservation, *Biosci. Biotech. Biochem.,* 56, 1476, 1992.

102. Lin, C. *et al.,* Bactericidal activity of isothiocyanate against pathogens on fresh produce, *J. Food Prot,* 63, 25, 2000.

103. Fenwick, G.R., Heany, R.R.K., and Mullin, W.J., Glucosinolates and their breakdown products in foods and food plants, *Crit. Rev. Food Sci. Nutr.,* 18, 123, 1982.

104. Delaquis, P.J. and Sholberg, P.L., Antimicrobial activity of gaseous allyl isothiocyanate, *J. Food Prot.,* 60, 943, 1997.

105. Kojima, M. and Ogawa, K., Studies on the effects of isothiocyanates and their analogues on microorganisms. I. Effect of isothiocyanates on the oxygen uptake of yeasts, *J. Ferment. Technol.,* 49, 740, 1952.

106. Block, S.S., *Disinfection, Sterilization and Preservation,* 3rd ed., Lea and Febiger, Philadelphia, 1983.

107. Park, C.M., Taormina, P.J., and Beuchat, L.R., Efficacy of allyl isothiocyanate in killing enterohemorrhagic *Escherichia coli* O157:H7 on alfalfa seeds, *Int. J. Food Microbiol.,* 56, 13, 2000.

108. Weissinger, W.R., Mcwatters, K.H., and Beuchat, L.R., Evaluation of volatile chemical treatments for lethality to *Salmonella* on alfalfa seeds and sprouts, *J. Food Prot.,* 64, 442, 2001.

109. Sapers, G.M. and Simmons, G. F., Hydrogen peroxide disinfection of minimally processed fruits and vegetables, *Food Technol.,* 52, 48, 1998.

110. Sapers, G.M. and Sites, J.E., Efficacy of 1% hydrogen peroxide wash in decontaminating apples and cantaloupe melons, *J. Food Sci.,* 68, 1793, 2003.

111. Smith, S. *et al.,* Efficacy of a commercial produce wash on bacterial contamination of lettuce in a food service setting, *J. Food Prot.,* 66, 2359, 2003.

112. Sapers, G.M. *et al.,* Improved antimicrobial wash treatment for decontamination of apples, *J. Food Sci.,* 67, 1886, 2002.

113. Sapers, G.M., Efficacy of washing and sanitizing methods for disinfection of fresh fruit and vegetable products, *Food Technol. Biotech.*, 39, 305, 2001.

114. Sapers, G.M. *et al.*, Shelf-life extension of fresh mushrooms (Agaricus bisporus) by application of hydrogen peroxide and browning inhibitors, *J. Food Sci.*, 66, 362, 2001.

115. Sapers, G.M. *et al.*, Factors limiting the efficacy of hydrogen peroxide washes for decontamination of apples containing *Escherichia coli*, *J. Food Sci.*, 65, 529, 2000.

116. Klapes, N.A. and Vesley, D., Vapor-phase hydrogen peroxide as a surface decontaminant and sterilant, *Appl. Environ. Microbiol.*, 56, 503, 1990.

117. Wang, J. and Toledo, R.T., Sporicidal properties of mixtures of hydrogen peroxide vapor and hot air, *Food Technol.*, 40, 60, 1986.

118. Toledo, R.T., Escher, F.E., and Ayres, J.C., Sporicidal properties of hydrogen peroxide against food spoilage organisms, *Appl. Microbiol.*, 26, 592, 1973.

119. Forney, C.F. *et al.*, Vapor phase hydrogen peroxide inhibits postharvest decay of table grapes, *HortSci.*, 26, 1512, 1991.

120. Simmons, G.F. *et al.*, Reduction of microbial populations on prunes by vapor-phase hydrogen peroxide, *J. Food Prot.*, 60, 188, 1997.

121. Sapers, G.M., Miller, R.L., and Simmons, G.F., Effects of Hydrogen Peroxide Treatments on Fresh-Cut Fruits and Vegetables, Annual Meeting of Institute of Food Technologists, Anaheim, CA, June 3–7, 1995.

122. Foegeding, P.M. and Busta, F.F., Chemical food preservatives, in *Disinfection, Sterilization, and Preservation*, 4th ed., Block, S.S., Ed., Lea and Febiger, Malvern, PA, 1991, chap. 47, p. 807.

123. Sholberg, P.L. and Gaunce, A.P., Fumigation of fruit with acetic acid to prevent postharvest decay, *HortSci.*, 30, 1271, 1995.

124. Sholberg, P.L. and Gaunce, A.P., Fumigation of stonefruit with acetic acid to control postharvest decay, *Crop Prot.*, 15, 681, 1996.

125. Sholberg, P.L., Reynolds, A.G., and Gaunce, A.P., Fumigation of table grapes with acetic acid to prevent postharvest decay, *Plant Dis.*, 80, 1425, 1996.

126. Delaquis, P.J., Graham, H.S., and Hocking, R., Shelf-life of coleslaw made from cabbage treated with gaseous acetic acid, *J. Food Proc. Preserv.*, 21, 129, 1997.

127. Sholberg, P.L. *et al.*, The use of vinegar vapor to reduce postharvest decay of harvested fruit, *HortSci.*, 35, 898, 2000.

128. Delaquis, P.J., Sholberg, P.L., and Stanich, K., Disinfection of mung bean seed with gaseous acetic acid, *J. Food Prot.*, 62, 953, 1999.

129. Ippolito, A. and Nigro, F., Natural antimicrobials in postharvest storage of fresh fruits and vegetables, in *Natural Antimicrobials for the Minimal Processing of Foods*, Roller, S., Ed., Woodhead, Cambridge, U.K. and CRC Press, Boca Raton, FL, 2003, chap. 10.

130. Buta, J.G. and Moline, H.E., Methyl jasmonate extends shelf life and reduces microbial contamination of fresh-cut celery and peppers, *J. Agric. Food Chem.*, 46, 1253, 1998.

131. Gonzalez-Aguilar, G.A., Buta, J.G., and Wang, C.Y., Methyl jasmonate and modified atmosphere packaging (MAP) reduce decay and maintain postharvest quality of papaya "Sunrise", *Postharvest Biol. Technol.*, 28, 361, 2003.

132. Wang, C.Y., Maintaining postharvest quality of raspberries with natural volatile compounds, *Int. J. Food Sci. Technol.*, 38, 869, 2003.

133. Wang, C.Y. and Buta, J.G., Maintaining quality of fresh-cut kiwifruit with volatile compounds, *Postharvest Biol. Technol.*, 28, 181, 2003.

134. McCabe, B.C.J., Ozone Treatment System for Food, U.S. Patent 4,549, 477, 1985.
135. Lee, S.Y., Costello, M.J., and Kang, D.H., Gaseous Chlorine Dioxide as a Sanitizer of Lettuce Leaves, IFT Annual Meeting, 2003, abstr. 104-7.
136. Wellinghoff, S.T., Chlorine Dioxide Generating Polymer Packaging Films, U.S. Patent 5,360,609, 1994.
137. Food and Drug Administration, Beverages: bottled water; final rule, *Fed. Reg.* 60, 57075, 1995.

19 Modified Atmosphere Packaging

B.G. Werner and J.H. Hotchkiss

CONTENTS

19.1 INTRODUCTION

Modified atmosphere packaging (MAP) has been successfully and widely used commercially, for both whole and fresh-cut or minimally processed (MP) fruits and vegetables, as a packaging strategy for maintenance of product safety and

437

extension of shelf life. MAP technology and systems development has striven to parallel the increasing demand for longer shelf life, improved food safety, and expanded variety of convenient MP ready-to-eat (RTE) or ready-to-use (RTU) fruits and vegetables. MAP systems have been shown to have the potential to increase the shelf life of specific produce commodities by 50 to 400% [1], particularly when used in concert with hurdle technologies such as active packaging, postharvest handling strategies and other newly evolving technologies in an integrated approach.

19.1.1 DEFINITIONS

MAP generally utilizes an internal package atmosphere of something other than air (air can be approximated as $<0.1\%$ CO_2, 21% O_2, 78% N_2) in a hermetically sealed package of suitable permeability in order to extend product shelf life and maintain food safety. While other gases have been explored for use in MAP systems, O_2, CO_2, and N_2 are most commonly employed; O_2 levels are commonly reduced below and CO_2 increased above atmospheric levels (with a balance of N_2) in order to reduce the commodity respiration rate, retard ripening and senescence, and reduce microbial activity. Microorganisms are affected indirectly by reductions in ripening and senescence and directly by restriction of O_2 and antimicrobial activity of CO_2; superatmospheric O_2 has also been shown to be antimicrobial but is not currently commonly employed.

MAP is a dynamic process where environmental and packaging characteristics and the contained product interact to create an equilibrated internal atmosphere (EMA). The EMA is achieved when the rate of O_2 consumption and CO_2 generation as a result of respiration by a particular commodity equals the rate of gas transmission through the packaged material. Generally, an EMA of 3 to 6% O_2 and 2 to 10% CO_2 achieves microbial control and extension of shelf life for a wide variety of whole and MP produce, although other atmospheres are also used with commodities that are not physiologically sensitive to high O_2 or CO_2. Package EMA can be created actively, where a target internal atmosphere is established initially upon packaging by actively flushing with the desired atmosphere, or, more commonly, passively, where the package atmosphere is allowed to reach the desired gas mixture around the commodity during the course of storage at a particular temperature; a longer time period is required to achieve a target EMA for passive than for active MAP.

The EMA is dependent upon both extrinsic and intrinsic factors, including product respiration rate, packaging film permeability to gases and water vapor, package dimensions, and fill weight. The intrinsic respiratory activity of produce is in turn influenced by the particular commodity, the cultivar, stage of maturity at harvest, type of tissue, mass, condition, and whether the product is whole or minimally processed. The mass or produce size indirectly influences respiratory activity by affecting O_2 diffusion rates into tissues, which subsequently directly influence respiration rates; alternatively, the stage of maturity or age has a direct impact on metabolic activities and rates. The extent of film permeability to gases per unit thickness, the effect of relative humidity on this permeability, the package surface area, seal integrity, free volume inside the package, and relative humidity around the package will also affect the EMA achieved. Temperature is the most influential extrinsic factor to consider as it affects both commodity respiration rate and film gas and water vapor permeability. MAP systems will generally be exposed to a dynamic environment during distribution, storage, display, and consumer purchase. Thus, while a particular MAP system should be optimized for a particular storage temperature, the effect of significant temperature changes on the system should be considered [2].

19.2 ANTIMICROBIAL ACTIVITY OF MAP GASES

19.2.1 CO_2

Dissolved CO_2 has been found to inhibit microbial growth [3–5], affecting the lag phase (λ), maximum growth rate (μ_{max})m and/or maximum population (N_{max}) densities reached; levels in excess of 5% in MAP systems have been found to be bacteriostatic [6]. The mode of action, although not yet fully understood, is thought to be due to a number of effects, including changes in intracellular pH, alteration of microbial protein and enzyme structure and function, and alteration of cell membrane function and fluidity. The partial pressure and concentration of CO_2, package headspace gas volume, temperature, pH, water activity, specific microorganism and growth phase, and growth medium (produce commodity) all influence the inhibitory effect of CO_2. The antimicrobial effect of CO_2 is enhanced as temperatures decrease and CO_2 becomes increasingly soluble.

Low to moderate levels of CO_2 have been shown to inhibit growth of many common aerobic produce spoilage bacteria. Moderate CO_2 levels of 20 to 60% have been found to reduce the μ_{max} and N_{max} of *Pseudomonas* spp. and *Moraxella* spp., two predominant spoilage bacteria found on produce [7]. Low CO_2 levels below 20% were found to primarily increase λ, with slight reductions in μ_{max} and no changes in N_{max}. CO_2 is not antimicrobial towards all microorganism strains or species, and may in some cases actually promote growth. *Lactobacillus* spp. are generally unaffected by CO_2; however, some levels can enhance growth, and 100% CO_2 environments have inhibited growth of some strains. In absence of O_2, it has been generally shown that the growth and toxin production of *Clostridium botulinum* is only minimally affected by CO_2 concentrations less than 50%; 100% CO_2 has been reported to delay toxin production compared to a 100% N_2 atmosphere [1] and decreases growth at 5 and 10°C [8]. Levels of 10% CO_2 have been found to be inhibitory to growth of *Yersinia enterocolitica* while 40% CO_2 increased λ, and 100% CO_2 both increased λ and decreased μ_{max} [1]. There is no agreement on the effect of CO_2 on *Listeria monocytogenes*; however, generally it has been found that CO_2 does not affect or in some cases promotes growth. *L. monocytogenes* has been found to grow well under atmospheres of both 100% N_2 and 3% O_2/97% N_2; growth was enhanced by increasing levels of CO_2 in either atmosphere [9].

19.2.2 SUPERATMOSPHERIC O_2

Superatmospheric O_2 as an MAP atmosphere is a new concept not yet commercially in use due to an incomplete understanding of its effects on MAP systems and mode of action towards microbial populations. Additionally, O_2 in excess of 25% is considered explosive, and practical safety issues that need to be employed in its use may not be feasible. Conventional MAP systems that commonly target an EMA of 3 to 6% O_2 may be exposed to fluctuating temperatures or temperature abuse conditions during handling, resulting in complete or near depletion of O_2. Under these conditions, growth of some pathogens such as *C. botulinum* may be enhanced due to anaerobiosis, or unrestricted growth of psychrotrophic facultative anaerobic pathogens such as *L. monocytogenes* may occur due to removal of competitive aerobic microorganisms. Under certain atmospheric conditions, *Staphylococcus aureus*, *Vibrio* spp., *E. coli*, *Bacillus cereus*, and *Enterococcus faecalis* have also been shown to grow with restricted or zero O_2 [10]. As an alternative MAP atmosphere strategy, high oxygen atmospheres (typically above 70%) that surpass optimal levels for growth of aerobes (21%) and anaerobes (0 to 2%) could generally result in growth inhibition of both anaerobic and aerobic microorganisms, resolving some of the food safety issues possible with lower O_2 EMA.

Few reports exist of the effects on specific microorganisms of superatmospheric O_2 in MAP systems, and some data are conflicting. Jacxsens and others [11], in a study on RTE mushrooms, grated celeriac, and shredded chicory endive, found that growth of *Pseudomonas fluorescens*, *Candida lambica*, *Botrytis cinerea*, *Aspergillus flavus*, and *Aeromonas caviae* was retarded by high O_2 MAP atmospheres (70, 80, or 95% O_2, bal. N_2), an effect that increased with increasing levels of O_2; increasing O_2 levels extended λ of *L. monocytogenes*. Using an agar-surface model system, Amanatidou and others [12] found 90% O_2 (bal. N_2) extended λ of *L. monocytogenes* and *Salmonella typhimurium*, reduced μ_{max} of *E. coli* and *S. enteritidis*, and significantly increased μ_{max} of *P. fluorescens*, *E. agglomerans*, *Candida guilliermondii*, and *C. sake*. Combined applications of high O_2 and 10 to 20% CO_2 generally both increased λ and reduced N_{max} for all strains tested. On mixed vegetable salad, Allende and others [13] found that yeast growth was stimulated by MAP atmospheres of 95 kPa O_2 while growth of psychrotrophic bacteria and *L. monocytogenes* was unaffected. Generally, greater levels of lactic acid bacteria were found on the mixed salad during storage under conventional MAP gas mixtures than with superatmospheric O_2 MAP. Salads treated with superatmospheric O_2 also exhibited a longer shelf life, retaining acceptable visual characteristics longer than conventional MAP treatments; the authors did not report whether any other significant organoleptic changes occurred. While superatmospheric O_2 has the potential to extend shelf life and maintain produce marketable qualities, these effects may vary depending upon the commodity. Wszelaki and Mitcham [14] found that superatmospheric storage of Camarosa strawberries resulted in acceptable product firmness, titratable acidity, external color, ethylene production, respiration, and soluble solids, but unacceptable odors and flavors developed as a result of increased production of volatile fermentative metabolites (ethanol, acetaldehyde, and ethyl acetate).

The mode of action of high O_2 is thought by some [12] to be due to oxidative stress and reduction of cell viability due to the generation of intracellular reactive oxygen species such as peroxides or superoxides. Some microorganisms may adapt by producing radical scavengers or inducing O_2 decomposing enzymes; repair proteins have been identified for *S. typhimurium*, *E.coli*, and *L. lactis* [11]. It is clear that significantly more work is needed to examine and clarify the effects on the growth parameters of individual spoilage microorganisms and food pathogens of superatmospheric O_2, alone or in combination with CO_2, applied to different produce commodities and MAP systems. Additionally, a more complete understanding of the underlying basic biological mechanisms of superatmospheric O_2 is necessary prior to successful commercialization of this technology [15].

19.3. PACKAGING AND FILMS FOR MAP PRODUCE SYSTEMS

19.3.1 FILM PERMEABILITY AND CO_2/O_2 PERMSELECTIVITY

The specific gases employed, the gas permeability coefficients of the film, and the EMA desired will in part dictate the packaging films utilized. Concentrations of the two most commonly metabolically active gases employed in MAP, O_2 and CO_2, will impact produce quality and levels of both should be optimized within the package. CO_2/O_2 permselectivity, the ratio of CO_2 to O_2 permeation coefficients, will vary for different films and can be selected or altered to concurrently optimize levels of both CO_2 and O_2 in MAP systems [16]. Most commercial packaging films available have CO_2/O_2 permselectivities between 4 and 8, allowing greater diffusion of CO_2 than of O_2 [17]; anaerobiosis and less than optimal CO_2 levels can result, depending upon the commodity and MAP system employed. Nitrogen is metabolically inert but can also be important as a filler gas to prevent package collapse. MAP produce subjected to temperature abuse or temperature changes along the normal distribution chain may result in increased respiration and depletion of in-package O_2; subsequently, the effective target EMA will not be maintained, and premature spoilage or increased food safety risks may occur [18]. High-barrier films that further restrict diffusion of CO_2 can result in excessive buildup of CO_2 as well as anaerobiosis, altered EMA, and lowered product safety and quality. Thus, there is a demand for films with $CO_2:O_2$ permselectivities closer to 1, or engineered packaging systems that use novel technologies or films to increase effectively or finely manipulate the O_2 flux. Particularly for higher respiring produce, permselectivites >2 increase the likelihood that the atmosphere will rapidly become anaerobic [16]. This is exacerbated as storage temperature increases. Polymer technology has been developed (Landec Corporation, Inc., CA) that allows the O_2 transmission rate of films to increase more rapidly than the CO_2 transmission rate in response to temperature, resulting in an adjustable CO_2/O_2 permeability ratio [19].

Use of a composite film comprising ethylene vinyl acetate (EVA), low-density polyethylene (LDPE), and oriented polypropylene (OPP) can enhance or improve gas permeability characteristics. Shredded cabbage and grated carrot stored in this composite material achieved an extension of shelf life of 2 to 3 days over that achieved in OPP alone [20]. Films have been developed with pores (micro-perforations) or holes (macro-perforations) to increase the O_2 transmission rate, resulting in more equal rates of movement of O_2 and CO_2 between the internal package atmosphere and external atmosphere, achieving permeability ratios of CO_2 to O_2 near 1. Micro-perforated films may be appropriate for products with a high respiration rate, such as strawberries, where finer control of package atmosphere is desired, where internal package O_2 depletion is a concern, or where temperature fluctuations may be anticipated.

The number, perimeter, and total effective area of perforations affects the rate of gas exchange [21,22]; the application and level of atmosphere control will determine these factors. Lee and others [21] developed a model to describe and predict changes in atmosphere and humidity in micro-perforated packages, verifying the model on refrigerated MAP peeled garlic. The number, cross-sectional area, and placement of perforations as well as the thickness of the film affect the EMA attained and alter gas and moisture exchange rates across package film. Long, narrow channels are more difficult for the gases to move through than wide, short-path perforations. Perforated films can be applied as an overwrap to a nonpermeable formed container or attached as a patch or label to a selectively permeable or nonpermeable bag. Used as a patch or label, only a small percentage of the package area is utilized to achieve the desired effects, creating a wide range of gas atmospheres, particularly when used in combination with a selectively permeable film.

19.3.2 ACTIVE PACKAGING: ANTIMICROBIAL FILMS

19.3.2.1 Synthetic Polymer Films

Antimicrobial packaging is an active packaging strategy that can serve a variety of distinct barrier functions in MAP systems, such as package self-sterilization, sterilization of produce, or reduction of growth of spoilage organisms and/or pathogens on packaged produce. Synthetic polymer films are most commonly researched for this functionality. Antimicrobial packaging films can be grouped into two general categories, nonmigratory and migratory. Nonmigratory packaging incorporates antimicrobials into the polymer or immobilizes them on the surface of the film in such a way that the compounds are not released; food must be in direct and intimate contact with such films in order for antimicrobial activity to occur. Migratory antimicrobial packaging incorporates the antimicrobial into or on the surface of the film in such a way that migration can occur to the food product where activity then occurs; unlike nonmigratory antimicrobial films, the antimicrobial becomes part of the foodstuff that eventually is ingested. Migratory antimicrobials can be released in aqueous solution (e.g., nisin, organic acids), or as a vapor (e.g., allylisothiocyanate, chlorine dioxide). The latter method of release is perhaps most suited to MAP systems where headspace between package and product is maintained and intimate contact between package and product does not typically occur. Release of migratory preservatives must be finely controlled for useful and effective activity that persists over the course of a defined or desired package shelf life.

The development of antimicrobial films has been significantly restricted by the legal status of antimicrobial compounds available for food contact use or as food preservatives or additives; currently only a limited number of such approved compounds exist, and approval varies among countries. Silver-substituted zeolites, a broad-spectrum high-activity antimicrobial with low human toxicity, has been extensively used commercially in Japan as a thin laminant on packaging film surfaces; its use on food contact surfaces in the European Union and the U.S., however, is unclear [23]. Some U.S. Food and Drug Administration (FDA) generally regarded as safe (GRAS) materials that have been considered for use as antimicrobials in synthetic polymer films include organic acids (benzoic, lactic, propionic, malic, succinic, tartaric, sorbic), enzymes (lactoperoxidase, lactoferrin, lysozyme, chitinase, glucose oxidase, ethanol oxidase), isothiocyanates (allylisothiocyanate), bacteriocinsc (nisin, pediocin, sakacin, subtilin, carnocin), and essential oils (thymol, cinnamic acid, eugenol) [6,24]. Natural plant extracts such as grapefruit seed extract have been shown to be effective against *Staphylococcus aureus* and *E. coli* on MAP lettuce and bell pepper, and when used in combination with imazalil could also provide protection against growth of molds, yeasts, and lactic acid bacteria [15]. Antimicrobials that are volatile, such as chlorine dioxide and allylisothiocanates, have an advantage in that they can be distributed within the closed package.

Any antimicrobial must not only be approved for food use, it must also be compatible with the packaging material and the package/film manufacturing process as well as maintain activity in the particular food matrix and MAP system [23]. Thus, different strategies may be employed in creating antimicrobial films and designing antimicrobial packaging systems from synthetic polymers versus more natural materials as are used in edible and biodegradable films.

19.3.2.2 Edible and Biodegradable Films

Biodegradable and edible films mainly comprise one or more proteins, lipids, or polysaccharides; each of these base materials has unique strengths and weaknesses as packaging materials and vehicles for antimicrobial compounds, and is thus usually employed in combinations. Polysaccharide (cellulose, gums, starch) or protein (gelatin, corn zein, soy protein, whey, etc.) films are highly sensitive to moisture and are poor barriers to water vapor; however, they exhibit suitable mechanical and optical properties. Films composed of lipids (waxes, lipids) have good water vapor barrier characteristics, but do not exhibit suitable mechanical and optical properties (are opaque and may be brittle). Wheat gluten and soy protein isolate films are effective O_2 barriers at low relative humidity, but have limited vapor barrier ability. Addition of lipid components to protein-based films improves the characteristics of both materials by optimizing both permeability to moisture and structural strength. Polarity of these natural films will determine compatibility with a particular antimicrobial and application or incorporation method.

Use of antimicrobials in edible films and coatings concentrates active compounds at the produce surface where protection is needed; thus small levels of additive are needed. This type of treatment is attractive to the increasing population of consumers who desire minimally preserved fruits and vegetables. Antimicrobials as edible film or coating components must be approved for food use. Waxes, with incorporated antimicrobials such as imazalil and benomyl, have been successfully used to minimize water loss and to inhibit microbial growth through gas exclusion and wound protection on fruit; however, benomyl is not an FDA GRAS substance and cannot be directly applied to food, and imazalil has limited FDA approval.

Some biodegradable film components such as chitosan naturally exhibit antimicrobial properties. Chitosan, a polysaccharide derived from shellfish and some fungi, has been found to exhibit broad antimicrobial activity towards a range of yeasts, molds, and bacteria and thus shows potential for application in MAP systems. Lee and others [25] found that chitosan, applied as a paper packaging coating, inhibited growth of *E. coli* O157:H7 in orange juice. Srinivasa and others [26] found that a chitosan film employed as a lid on a cardboard container for MAP storage of whole mango fruit inhibited fungal growth and extended shelf life of the product from 9 to 18 days at 27°C as compared to use of LDPE as an overwrap.

As with synthetic films, O_2 permeabilities of edible films and coatings generally can be very low and CO_2/O_2 permselectivities can be quite high. In a selection of edible films [27] including pectin, wheat gluten, chitosan and bilayer gluten, and beeswax films, at 25°C the CO_2/O_2 permselectivity ranged between 6 and 28.4, and O_2 permeability ranged between 2 and 258.8 pO_2 ml mm/(m^2 day atm). Thus the same risk for development of anoxic conditions and reduced food safety conditions exists for edible natural films as for synthetic films. Appropriate combinations of coatings and antimicrobial compounds may compensate for these effects. For example, a wax-based coating may incorporate nisin to reduce the risk of growth of *Clostridium botulinum* and/or *L. monocytogenes* [15].

19.4. AN INTEGRATED APPROACH: MULTIPLE BARRIERS AND MAP

19.4.1 BACKGROUND

Active packaging strategies and other technologies may be used in MAP systems, where multiple barriers combining two or more technologies at inhibitory levels provide integrated and enhanced control of microbial growth. Barrier technologies may be selected to serve different roles, such as maintenance of activity even under temperature abuse conditions or failure of MAP atmospheres. Barrier technologies may reduce initial microbial populations on produce prior to packaging and MAP storage, or may be selected specifically to reduce the incidence of a target pathogen of concern.

The path of produce from field to the point of packaging involves many stages where handling and environments can be controlled and optimized to avoid contamination with pathogens or spoilage organisms and to reduce initial microbial load. Good hygiene practices during harvest and storage, optimal postharvest storage and transportation/distribution temperatures, and HACCP (hazard analysis critical control point) implementation during processing are basic steps that have been historically incorporated into a multiple barrier approach to microbial control and should include the use of MAP. Use of one or more active packaging technologies, discussed in the previous section, can be excellent additions to a multiple barrier approach; addition of biopreservatives, antagonistic or protective microbial cultures, inclusion of gas absorbers or generators, ultraviolet C (UVC) treatments, and combination atmosphere technologies such as CA and MAP or dual MAP packaging systems can also serve as effective barrier technologies.

19.4.2 BIOPRESERVATION AND PROTECTIVE CULTURES

A growing demand for minimally preserved or preservative-free fresh produce has led to a search for alternatives to more traditionally used food antimicrobial compounds. Biopreservation, the use of antagonist or protective cultures, has shown potential for extension of produce storage life; natural microflora such as lactic acid bacteria (LAB) or bacterial metabolic byproducts such as organic acids or bacteriocins can serve as natural inhibitors of spoilage organisms. Application of antagonist organisms along with use of MAP technology and additional microbial control strategies can be synergistic in effect. For example, a prepackaging application to sweet cherry of the antagonist yeast *Cryptococcus infirmo-miniatus* (CIM) Pfaff and Fell followed by modified atmosphere storage at 2.8°C for 20 days or −0.5°C for 42 days resulted in significant reduction of the causal agent of brown rot, *Monilinia fructicola* G. Wint., an effect that was enhanced when a preharvest application of propiconazole was incorporated [28]. Prepackage application of organic acids to vegetables or fruits such as melon, papaya, or avocado, which are typically low-acid, results in a pH decline that is inhibitory to groups of spoilage organisms that grow best under neutral or near neutral pH environments.

Protective cultures such as LAB can be found naturally on produce, and thus may not significantly alter the typical or expected product taste or cause significant spoilage if applied directly. The most studied LAB bacteriocinogenic strains include *Lactococcus lactis*, *Pediococcus acidilactici*, and *Lactobacillus sakei*, which produce the antimicrobials nisin, pediocin, and sakacin, respectively [29]. Bacteriocins produced by LAB are typically antimicrobial towards Gram-positive spoilage organisms and pathogens, including *L. monocytogenes* and *C. botulinum*. Bacteriocins do not affect Gram-negative bacteria, which are a major spoilage group of concern; however LAB and associated bacteriocins may be used to target specific Gram-positive pathogens of concern or be used in combination with other technologies that reduce Gram-negative bacterial growth for a broader overall range of microbial control.

Some LAB strains may not grow well enough on produce at refrigeration temperatures in order to produce levels of bacteriocin necessary for antimicrobial activity; additionally, bacteriocins can be inactivated by bacterial proteolytic enzymes or by binding to food components, and target bacteria may become resistant. Bennik and others [30] isolated bacteriocinogenic strains of *Pediococcus parvulus* and *Enterococcus mundtii* from MAP endive and evaluated the ability of these strains to produce bacteriocin on mung bean sprouts at refrigeration temperatures of 4 to 8°C. *E. mundtii* was able to produce the bacteriocin mundticin on inoculated mung beans stored under MAP (1.5% O_2, 20% CO_2, balance N_2) at 8°C, while *P. parvulus* did not survive under these conditions. When mundticin was extracted and used as a dip (200 BU ml^{-1}) or incorporated into an alginate film (200 BU ml^{-1}) on mung bean, the bacteriocin exhibited antimicrobial activity under refrigerated MAP storage [30]. Cai and others [31] isolated a strain of *Lactococcus lactis* subsp. lactis from mung bean sprouts that contained a gene for nisin-Z, an antilisterial compound. This isolate could survive on fresh-cut RTE Caesar salad at levels of 8 $\log_{10}$ CFU/g at 3 to 4.5°C for up to 20 days and could grow at 4°C and produce nisin-Z at 5°C. When co-incubated with 2 $\log_{10}$ CFU/g cells of *Listeria monocytogenes* on salad, *L. monocytogenes* populations were reduced by 1 to 1.4 $\log_{10}$ CFU/g after 10 days' storage at 7 and 10°C. Thus bacteriocinogenic strains should be assessed for their ability to grow and produce bacteriocin on a target commodity under specific MAP conditions and storage temperatures. Additionally, bacteriocins directly applied should be assessed for their persistence and activity on a specific commodity during MAP shelf life.

Other microorganisms have been found to exhibit antimicrobial activity due to competition for nutrients, rapid growth rates, or production of inhibitory metabolites. Enterobacteriacea have been found to limit growth of *L. monocytogenes* on endive, most likely due to competition for nutrients. Mixed populations of nonbacteriocinogenic strains of *Lactobacillus brevis* and *Leuconostoc citrium* have been shown to inhibit competitively the growth of

L. monocytogenes on MAP MP vegetables [32]; *Enterobacter cloacae* and *E. agglomerans* were also found to be competitively inhibitive. This research group also found in challenge studies with MAP lettuce that increasing CO_2 atmospheres decreased this inhibitory effect; when CO_2 levels increased from 5 to 10 to 20%, a delayed inhibitory effect was increasingly observed [33]. The use of nonpathogenic strains naturally found on produce that competitively inhibit spoilage organisms and/or pathogens on produce under MAP storage conditions warrants further study as a promising biopreservative hurdle strategy.

19.4.3 O_2/CO_2 ABSORBERS AND GENERATORS

Passively achieved MAP systems may be slow to reach a target EMA, creating a sufficient lag time for significant growth of psychrotrophic aerobic spoilage organisms such as *Pseudomonas* spp. O_2 scavengers incorporated into packaging materials as sheets, labels, trays, or films can be used as an active strategy to more rapidly reach EMA. Commercially available O_2 scavengers such as Ageless® (Mitsubishi Gas Chemical Co., Japan) and Freshpax® (Multisorb Technologies, Inc., USA) [23] are based on iron oxidation.

O_2 scavenging technology has been used successfully in MAP stored bakery and dairy products, and applications in MAP stored produce are being explored. Charles and others [34] created a mathematical model based on the respiration rate of produce, film permeability, and oxygen absorption kinetics of the scavenger. Validation using LDPE pouch packaged tomato and a commercial iron-based O_2 scavenger system at 20°C showed that target EMA was actively established within 50 hours; without the absorber, the EMA was passively reached within 100 hours. When using O_2 absorbers, the possibility exists that anaerobiosis may occur. In order to optimize MAP produce safety, more information is needed about how O_2 scavengers function or respond in different MAP environments with different commodities [10].

Alternatively, CO_2 generators can be used to achieve high levels of CO_2 (60 to 80%), which can inhibit microbial growth on produce surfaces. CO_2 generators may pose a safety risk; moderate to high levels of CO_2 will inhibit growth of aerobic spoilage organisms that usually warn consumers of spoilage, and growth of pathogens may be enhanced due to lack of competition and the altered environment [1,10].

19.4.4 PRETREATMENTS AND MISCELLANEOUS STRATEGIES

Treatment of produce with methyl jasmonate prior to MAP has been found to be successful in suppressing fungal decay in a number of commodities, including fresh-cut celery and peppers, grapefruit, papaya, strawberries, zucchini squash, mango, and avocado [35]; the effects and mode of action of jasmonates in reducing disease development differ among various crops and pathogens. Synergistic activity between methyl jasmonate treatments and MAP has been found for several commodities, including papaya. Gonzalez-Aguilar and others [35] found that exposure of papaya to methyl jasmonate vapor (10^{-5} or 10^{-4} M) for 16 hours at 20°C inhibited growth of *Collectotrichum gloeosporioides* and fruit decay in papaya, an effect that enhanced a MAP treatment of 14 to 32 days at 10°C followed by 4 days at 20°C in a modified atmosphere of 3 to 5 kPa O_2 and 6 to 9 kPa CO_2.

As a treatment prior to MAP packaging, nonionizing, artificial UVC radiation has the potential to be effective in reducing the initial microbial load on produce, providing shelf life extension. UVC has been shown to damage microbial DNA, an effect that weakens or kills microbial cells. Some bacteria have been found to utilize repair mechanisms to overcome DNA damage, and some cells may mutate. Thus typically UVC treatment results in a reduction of microbial load but not complete sterilization. Allende and Artes [13] found that treatments of 254 nm UVC doses up to 8.14 kJ/m^2 on Red Oak Leaf lettuce, subsequently stored at 5°C for 9 to 10 days, significantly decreased the growth of psychrotrophic bacteria, yeast, and coliforms. UVC has been shown to reduce postharvest diseases and decay in a variety of whole produce including strawberries, apples, carrots, sweet potatoes, zucchini squash, tomatoes, and onions [13,36]. These different produce typically are smooth surfaced and simple in shape; UVC would not be fully effective on produce with naturally convoluted, rough, or inaccessible surfaces, as radiation would not penetrate into shadowed regions of these types of surfaces.

Some antimicrobial compounds have been found naturally in fruits and vegetables and can be used as additional hurdles in MAP systems. Raw carrots produce compounds antimicrobial towards *L. monocytogenes*, an antimicrobial effect that is more pronounced in shredded than in whole carrots, and is absent in cooked carrots. Mixing shredded raw carrots with other MP vegetables that did not produce the antilisterial compounds resulted in overall reductions in populations of *L. monocytogenes* during storage, and a coleslaw mix of shredded carrot and cabbage stored under MAP conditions had less spoilage than either product stored individually [37]. Red chicory has been found to inhibit growth of *Pseudomonas* spp. and *Aeromonas hydrophylla* [38], and capsaicinoids found in green bell pepper were hypothesized to be antimicrobial towards *Shigella* spp. [39]. More knowledge is needed about antimicrobial compounds naturally found in fresh fruits and vegetables in order to utilize their benefits.

19.5 MICROBIOLOGY OF MAP FRUITS AND VEGETABLES

19.5.1 MINIMALLY PROCESSED FRUITS AND VEGETABLES

MP produce includes fresh fruits and vegetables that may be washed, chopped, trimmed, peeled, sliced, or shredded prior to packaging and storage at refrigeration temperatures. There is increasing consumer demand for MP produce, due to the level of convenience offered by pre-use processing and availability as a fresh RTE or RTU food. MP produce typically is not washed or cooked prior to ingestion, increasing the risk of food poisoning. Thus the level of quality and safety of MP produce must be quite high for the shelf life achieved. MAP has great potential as a strategy to achieve this goal, and much research and development effort has been initiated to develop useful MAP systems for a variety of MP produce.

Commercial MAP systems have been developed for a wide range of whole fruits and vegetables. However, these same systems cannot be used in parallel commodities that have been processed; processed produce deteriorates and metabolizes much differently from whole produce. Processing steps such as chopping, slicing, and dicing rupture tissues and cells, releasing nutrients and degradative enzymes such as oxidases. Plant cells are less physically resistant to microbial invasion, nutrients are made more available for microbial growth, respiration rates increase, and surface area increases, allowing for greater incidence of spoilage. Processing thus significantly reduces shelf life, producing a highly perishable product compared to whole fruits and vegetables; whole produce that may be stored for several weeks under refrigerated MAP storage when processed may only have a 1- to 2-day shelf life.

It is a challenge to create a MP produce commodity that exhibits high quality and safety over a reasonable amount of time for feasible distribution and sale. MAP technology has the potential to provide adequate shelf life for MP produce, particularly when used in combination with additional hurdle or control strategies. MAP strategies must be created for each specific commodity and preparation method, as commodity characteristics and indirect effects of preparation steps can influence package EMA, microbial growth, and shelf life. Allende and others [40] looked at microbial levels on commercial fresh processed red Lollo Rosso lettuce after reception and processing steps of shredding, washing, draining, rinsing, centrifugation, and packaging and found that shredding, rinsing, and centrifugation significantly increased bacterial counts. Improvements made to reduce microbial levels during each of these three steps resulted in further extensions to shelf life when the product was stored under MAP conditions. Others have found that some processing methods can increase the respiration activity of some produce commodities by 1.2- to 7-fold or more. Hand peeled carrots exhibited a 15% increase in respiration rate while machine peeled carrots exhibited a 100% increase in respiration rate; the respiration rate during storage also differed between the two processing methods [20]. Pretel and others [41] found significantly different respiration rates and mesophilic bacterial growth on MAP-stored enzymatically peeled versus manually separated oranges; manually separated oranges generally exhibited a higher level of bacterial growth and production of CO_2 than enzymatically peeled oranges. These differences could be ameliorated to some extent by manipulating packaging film permeabilities and storage temperature, achieving similar shelf lives. MP fruits will pose different storage challenges from vegetables, due to differences in inherent composition, physiology, biochemistry, and microbiology as well as differences in processing procedures and equipment. Thus the MAP strategy must be matched to not only a particular commodity, and whether whole or processed, but also to the specific processing method.

19.5.2 SPOILAGE ORGANISMS AND COMMODITY SHELF LIFE

The spoilage microorganisms present on produce in MAP storage systems will be influenced by the particular commodity and by the atmospheres and temperatures employed. Initially, Gram-negative bacteria predominate in the microflora of typically low-acid vegetables while LAB, molds, and yeasts predominate on high-acid fruits. Indigenous microflora on vegetables that cause spoilage include a majority of Gram-negative bacteria, predominantly *Pseudomonas* spp., *Enterobacter* spp., and *Erwinia* spp. as well as *Flavobacterium* spp. and *Xanthomonas* spp. and Gram-positive LAB such as *Leuconostoc mesenteroides* and *Lactobacillus* spp. Indigenous yeasts and molds that cause spoilage include *Cryptococcus* spp., *Candida* spp., *Rhodotorula* spp., *Fusarium* spp., *Rhizopus* spp., *Cryptococcus* spp., *Botrytis* spp., *Mucor* spp., and *Penicillium* spp., among others.

Zagory [38] reported that for a majority of fresh vegetables, *Pseudomonas* spp. comprised 50% or more of the total initial spoilage microflora in MAP stored product. Jacxsens and others [42] reported that MAP spoilage of leafy greens and cucumber was primarily due to growth of members of the Enterobacteriacea family while spoilage of celeriac and green bell peppers was due to LAB and yeasts. The diversity of spoilage organisms initially found on MAP produce upon packaging may dynamically change during the course of shelf life and establishment of EMA. Bennik and others [43] found lowest counts of pseudomonads under 0% O_2 compared to 21% O_2 atmospheres, irrespective of CO_2 levels. Pseudomonads were predominant at 21% O_2, while enterics were more predominant under 0% O_2. Differences in sensitivities to modified atmospheres among strains, availability of nutrients, nutrient requirements, and/or physiological state of the produce can result in shifts in microbial populations during storage. Bennik and others [44] examined the microbial composition of MP mung bean sprouts and chicory endive stored under MAP (atmospheres of 1.5 or 21% O_2 with 0, 5, 20 or 50% CO_2) at 8°C. On mung bean sprouts, the predominant species before and after storage were *Enterobacter cloacae*, *Pantoea agglomerans*, *Pseudomonas fluorescens*, *Ps. viridilivida*, and *Ps. corrugata*. Predominant species on chicory endive before storage were *Rahnella aquaatilis* and several *Pseudomonas* spp.; after storage, *E. vulneris* and *Ps. fluorescens* predominated.

Generally, MAP utilizing mixed atmospheres of $O_2/CO_2/N_2$ is most inhibitory towards aerobic bacteria and molds and may not inhibit or only minimally inhibit many spoilage yeasts and LAB. Exceptions have been discovered using specific MAP atmospheres and produce commodities. Martinez-Ferrer and others [45] found that MAP atmospheres of 4% O_2,

10% CO_2, balance N_2 significantly reduced total yeast populations on prepared mango and pineapple stored for up to 30 days at 5°C, as compared with storage under vacuum or air. Piga and others [46] found that in cactus pear fruit stored for 9 days at 4°C under atmospheres of 17%CO_2, <1% O_2, balance N_2, fungal mycelia were not visible on produce surfaces, but these conditions did not inhibit mold growth; molds increased in number from 20 to 5×10^2 CFU/g.

19.5.3 PATHOGENIC ORGANISMS AND SHELF LIFE

19.5.3.1 Food Safety Risk of MAP Produce

When competitive microflora are eliminated by MAP atmospheres, some pathogens may grow unimpeded. Certain MAP systems can produce anoxic conditions which, while inhibiting growth of spoilage organisms such as aerobic bacteria and molds, can allow growth of obligate anaerobic pathogens such as nonproteolytic *C. botulinum* even at refrigeration temperatures; temperature abuse conditions that increase product respiration can also result in anaerobiosis. Where high levels of CO_2 alone restrict growth of susceptible microorganisms, selection of pathogens that can survive under these conditions may also occur [47]. At 13 or 22°C, CO_2 was reported to not inhibit growth of *E. coli* on shredded lettuce; in fact, atmospheres of 5% O_2 and 30% CO_2 (balance N_2) actually enhanced growth over storage in air. Atmospheres containing 40 to 50% CO_2 were only slightly inhibitory towards *Yersinia enterocolitica* at 4°C, although inhibition increased as storage temperature decreased [9]. Bennik and others [44] observed extended λ for this psychrotrophic pathogen under conditions of 50% CO_2 and 21% O_2 (balance N_2), but no effect under decreasing CO_2 concentrations of 5 or 20%. Their results suggest that typically employed MAP conditions of 1 to 5% O_2 and 5 to 10% CO_2 at 8°C may not inhibit growth of the pathogens *Aeromonas hydrophila*, *L. monocytogenes*, or cold-tolerant strains of *Bacillus cereus*.

The behavior of a particular pathogen in a MAP system is influenced by the fruit or vegetable type as well as by the nature of the particular microbial strain. Francis and O'Beirne [48] found that acid-adapted *L. monocytogenes* grew on mung bean sprouts at 8°C and atmospheres of 2 to 5% O_2 and 0 to 15% CO_2, while nonacid-adapted strains did not grow. In further work [49] these authors assessed the effects of vegetable type and strain on survival and growth of different pathogens under different modified atmospheres at 4 and 8°C. Different growth responses were observed between strains of *E. coli* O157:H7 on different RTU vegetables (lettuce, swedes, dry coleslaw, soybean sprouts), while no difference was observed on these same vegetables among multiple strains of *L. monocytogenes*.

Different methods can be utilized to determine the safety of foods stored under specific MAP systems. Challenge studies, where survival and growth of inoculated pathogens are followed over time, can be performed in isolation or in combination with natural produce microflora. Challenge studies using

C. botulinum can be used to examine toxin production as well as occurrence of spoilage. Predictive models can be generated to determine microbial growth or toxin development in produce. To consider the interactions of fluctuating populations and ratios of spoilage organisms and pathogens, a safety index ratio may be used to indicate relative spoilage and pathogenicity. The ratio of a specific pathogenic organism to a food spoilage organism over MAP storage time can be used as a practical safety index, created for any pathogen of concern where levels required to produce illness may significantly differ. Such an index would not represent an absolute measurement of the safety of a food product: it would quantitatively depict the relationship between spoilage and pathogenicity [47].

19.5.3.2 Psychrotrophic Pathogens

A potential consumer safety risk may occur due to MAP inhibition of the aerobic microorganisms that usually warn consumers of spoilage, resulting in reduction in growth competition, and creation of an altered environment, allowing enhanced or unrestricted growth of anaerobic or facultative anaerobic pathogens capable of growing under MAP conditions at refrigeration temperatures. These include *L. monocytogenes*, *C. botulinum*, *Yersinia enterocolitica*, and *Aeromonas hydrophila*.

L. monocytogenes, ubiquitous in the environment, is naturally found on many fruits and vegetables. A facultative anaerobe capable of growing under temperatures as low as $-1.5°C$ and under CO_2-enriched environments, this pathogen can feasibly grow on MAP refrigerated produce. Beuchat [50] reported little inhibitory effect of MAP at 4 to 15°C on growth of *L. monocytogenes* on broccoli, cauliflower, and asparagus. Bennik and others [37] found that the extent of growth of *L. monocytogenes* on chicory endive was not influenced by MAP atmospheres; the initial inoculum level, cultivar of chicory endive, and population of competitive spoilage microorganisms were primary growth influences. *L. monocytogenes* grew better on chicory disinfected with chlorine than on chicory left untreated prior to MAP storage, most likely due to reduction of competitive indigenous microflora after treatment. Francis and O'Beirne [33] found that survival and growth of *L. innocua* (as a model of *L. monocytogenes*) was affected by the indigenous microflora; *Enterobacter cloacae* and LAB reduced growth of *L. innocua* while pseudomonads had little effect. Thus MAP treatments have the potential to change the dynamics of microbial populations and alter product safety. Berrang and others [51] increased shelf life and reduced spoilage microorganisms of cut asparagus, broccoli, and cauliflower by MAP; however, in later studies they found that the growth of *L. monocytogenes* and *A. hydrophila* was unaffected. Thus shelf life extension feasibly allowed a longer time period for the pathogens to grow by removing competitive microorganisms [52,53].

Aeromonas spp. generally grow at temperatures between 1 and 45°C, and under low O_2 atmospheres. *Aeromonas* spp. can grow at low temperatures under vacuum but are inhibited by high concentrations of CO_2. Researchers have found *A. hydrophila* to be present on 100% of 12 different produce items surveyed, recovering the pathogen from green salad, coleslaw, salad samples, and mixed salad greens [9]. *A. hydrophila* was found to survive but not grow on vegetable salads stored under MAP at 4°C, but rapidly grew at 15°C. Others [54] have found that *A. hydrophila* would grow on cucumber slices but not on mixed lettuce under MAP conditions at 2°C. Bennik and others [44] found that growth of *A. hydrophila* was the same under MAP conditions of 1.5 or 21% O_2; μ_{max} decreased with increasing CO_2 concentrations; however, N_{max} was not affected until CO_2 levels were above 50%.

At 4°C, *Y. enterocolitica* has been found to grow in air, under vacuum, and under MAP and atmospheres containing 40 to 50% CO_2 [9]. Yersinia can grow at temperatures as low as 1°C with a 40-hour doubling time. Farber [1] reported that 10% CO_2 stimulated growth of *Y. enterocolitica*, but 40% CO_2 increased λ, and 100% CO_2 increased λ and decreased μ_{max}.

C. botulinum poses a significant food safety risk in MAP produce, as previously discussed in this chapter. MAP conditions that extend product shelf life may create an organoleptically acceptable consumer product, but may pose a food safety hazard not immediately visible to the consumer. Macura and others [55] found that anaerobic conditions developed under a range of MAP atmospheres and temperatures employed for storage of ginseng roots; at 10°C, *C. botulinum* toxin was detected in roots while overall product quality was still acceptable. Nonproteolytic strains of *C. botulinum* have a growth potential between 3.3 and 45°C and are minimally affected by CO_2 concentrations <50% [15]. Toxigenesis has been detectable at 8 and 5°C [9] and at O_2 concentrations up to 10%; however, it has been reported that toxin production is dependent as well on the produce commodity [56]. Of all vegetables tested (butternut, onion, mixed greens, lettuce, rutabaga), at 5°C nonproteolytic strains could only produce toxin on butternut squash; proteolytic strains could produce toxin on all vegetables tested at temperatures ≥15°C. While acidic environments such as those produced by high CO_2 packaging atmospheres, acid treatments, and/or low pH produce can inhibit growth of *C. botulinum*, microbial diversity and dynamics may increase product pH. Growth of mold on tomato may increase typical product pH from about 4 to 5–9 [57], creating microenvironments suitable for growth of *C. botulinum*. Thus the effects of atmospheres, temperatures employed, temperature abuse, influences of other organisms, as well as commodity type should be assessed when designing MAP systems to reduce risk of food poisoning due to *C. botulinum*.

19.5.3.3 Other Pathogens of Concern

Salmonella, *E. coli*, and shigella can survive but are unable to grow at temperatures lower than 4°C; however, growth can resume if temperature abuse occurs. At higher refrigeration temperatures, growth may be possible, and atmospheres used in MAP will differently influence growth parameters of these pathogens. Amanatidou and others [12] found that 8°C and 10 to 20% CO_2 in N_2 reduced the growth rate of *Salmonella enteritidis* but had no effect on the growth rate of *S. typhimurium* or *E. coli*. Combined atmospheres of high O_2 and CO_2 increased λ and decreased μ_{max} and N_{max} for *S. enteritidis* and *E. coli*, but had no effect on *S. typhimurium*. High O_2 alone reduced μ_{max} or N_{max} for *S. typhimurium* and *S. enteritidis*, but had little effect on *E. coli*. Another study found that *Shigella sonnei* and *S. flexneri* growth was not affected by atmospheres of 3% O_2 and 5 to 10% CO_2 at 7 and 12°C. Growth and survival was predominantly affected by the type of vegetable tested (grated carrot, chopped bell pepper, mixed lettuce) [47] as well as by temperature. Francis and O'Beirne [49] determined that survival and growth of *E. coli* O157:H7 under MAP was dependent upon type of vegetable (swedes, lettuce, soybean sprouts, dry coleslaw mix), temperature, atmosphere, and strain; the pathogen grew better under atmospheres of 30% CO_2 and 5% O_2 compared to air, and 9 to 12% CO_2 and 2 to 4% O_2 were not inhibitory at 8°C. Reduction of storage temperatures from 8 to 4°C prevented growth and reduced survival of *E. coli*. Others [58] have shown that *E. coli* O157:H7 can survive on fresh-cut apples under $>15\%$ CO_2 at abusive temperatures (15 and 20°C). These studies emphasize the important effect of temperature in maintaining MAP produce safety.

Campylobacter jejuni requires 5% O_2, 10% CO_2, and 85% N_2 for optimal growth, atmospheres that may commonly occur in MAP systems. Even under refrigeration temperatures, MAP atmospheres may create conditions more hospitable to the survival of this pathogen than under air; populations of *C. jejuni* on cilantro, green pepper, and romaine lettuce packaged under MAP for 15 days at 4°C were reported to be reduced by 2 $\log_{10}$ CFU/g by day 9 while a much greater reduction of 3 to 4 $\log_{10}$ CFU/g occurred under air and vacuum storage [15]. *Campylobacter* spp. have a low infective dose (100 to 500 CFU); while they do not typically grow below 30°C, they can survive. Thus, if the pathogen was initially present on the commodity upon packaging, only short intervals of temperature abuse in a typical MAP atmosphere might be needed to allow enough growth of the pathogen to cause food poisoning. Phillips [59] reported that 22.2% of mixed salad vegetable MAP products tested were contaminated with between 80 and 170 CFU/g *Campylobacter* spp., levels that could potentially produce illness, particularly if several grams of salad were ingested.

19.5.4 MICROBIAL ECOLOGY OF MAP SYSTEMS

Interactions between indigenous microflora and pathogens on produce generally have not been well studied. Indigenous LAB can be antagonistic due to organic acid production, generation of H_2O_2, bacteriocin production, or competition for nutrients. Naturally present in low numbers on vegetables, they can reach high numbers in MAP where high levels of CO_2 are employed. MAP gas combinations may be manipulated to encourage growth of these antagonists, which may indirectly control growth of spoilage organisms or pathogens. Research has shown that growth of some organisms can result in enhanced growth of others. *Salmonella* spp. co-inoculated with a soft-rot bacterium or *Pseudomonas* spp. on potato, carrot, and pepper grew significantly better than when inoculated alone [57]. Pathogens may grow on biofilms naturally formed on produce, where the environment may be altered such that it is more favorable for microbial growth compared to the direct produce surface. Work has shown that *L. monocytogenes* can grow on multi-species biofilms on meat; comparable studies have not yet been performed on produce. Biofilms have been found to constitute between 10 and 40% of bacterial populations on endive and parsley, and more work is needed to determine the extent of biofilm development and microbial interactions at the biofilm surface on other whole produce [57].

19.6. MATHEMATICAL PREDICTIVE MODELING

The evaluation of a MAP system for any given commodity should be accomplished by a systematic and comprehensive approach by first establishing an initial predictive theoretical model to represent and manipulate underlying principles, followed by validation through empirical study. Empirical study unsupported by theoretical models and performed through trial and error is a lengthy and expensive process that does not take into consideration microbial ecology, product safety, or interaction of underlying MAP system variables; shelf life and associated product quality are the primary factors examined. Mathematical models of interactions among variables that affect MAP package atmospheres have been proposed and used to design MAP systems. However, improvements are needed to create more comprehensive models, and little work has been done to create models for different MP fruits and vegetables.

Additionally, models are needed for more sophisticated MAP systems such as multiple pack nested or multiple commodity packaging, multiple barrier systems, or where the behavior of specific MAP system variables may be expected to be different, as with perforated films [60], films incorporating gas scavengers or generators, or systems utilizing superatmospheric O_2 or novel gas mixtures. Models should generally consider the packaging internal and external environments, the product under storage, and the storage gases and packaging materials employed. Temperature, product respiration rate (both consumption of O_2 and production of CO_2), product weight, package headspace, film permeability to gases and water vapor, film surface area and thickness, product diffusion resistance, and product tolerance to low O_2 and high CO_2 are all important variables to consider. Typically, a film with specific gas and water vapor characteristics is selected to achieve the particular target EMA for a stored product having a specific respiration rate, at a specific temperature. Predictive models should be validated by empirical studies incorporating the particular produce commodity and testing one or more variables to optimize.

19.7 FUTURE DIRECTIONS

The future direction of MAP system design must rely on significant advances in understanding the controlling variables and underlying factors influencing product safety and shelf life, particularly for MP fruits and vegetables. A broader knowledge base is needed to more fully understand the microbial ecology of MAP stored MP produce and the effects of processing and different atmospheres employed. More information is needed about the microbiological safety of MAP whole and MP produce, the effects of MAP on growth of psychrotrophic pathogens, interactive effects of microorganisms in MAP produce systems, effects of MAP system failures, and effects of varying storage conditions and temperature abuse. The interactions between MAP and other preservation methods should be defined to enable development of effective multiple barrier preservation systems. As we come to a more comprehensive understanding, more effective and applicable predictive models will be constructed for designing MAP systems for high-quality, safe produce.

REFERENCES

1. Farber, J.M., Microbiological aspects of modified-atmosphere packaging technology: a review, *J. Food Prot.*, 54, 58, 1991.
2. Jayas, D.S. and Jeyamkondan, S., Modified atmosphere storage of grains, meats, fruits and vegetables, *Biosyst. Eng.*, 82, 235, 2002.
3. Daniels, J.A., Krishnamurthi, R., and Rizvi, S.S.H., A review of effects of carbon dioxide on microbial growth and food quality, *J. Food Prot.*, 48, 532, 1985.
4. Devlieghere, F., Debevere, J., and Van Impe, J., Concentration of carbon dioxide in the water-phase as a parameter to model the effect of a modified atmosphere on microorganisms, *Int. J. Food Microbiol.*, 43, 105, 1998.
5. Devlieghere, F. and Debevere, J., Influence of dissolved carbon dioxide on the growth of spoilage bacteria, *Lebens.-Wissen. Technol.*, 33, 531, 2000.
6. Hotchkiss, J.H. and Banco, M.J., Influence of new packaging technologies on the growth of microorganisms in produce, *J. Food Prot.*, 55, 815, 1992.
7. Cutter, C.N., Microbial control by packaging: a review, *Crit. Rev. Food Sci. Nutr.*, 42, 151, 2002.
8. Gibson, A.M. *et al.*, The effect of 100% CO_2 on the growth of nonproteolytic *Clostridium botulinum* at chill temperatures, *Int. J. Food Microbiol.*, 54, 39, 2000.
9. Francis, G.A., Thomas, C., and O'Beirne, D., Review article: the microbiological safety of minimally processed vegetables, *Int. J. Food Sci. Technol.*, 34, 1, 1999.
10. Suppakul, P. *et al.*, Active packaging technologies with an emphasis on antimicrobial packaging and its applications, *J. Food Sci.*, 68, 408, 2003.
11. Jacxsens, L. *et al.*, Effect of high oxygen modified atmosphere packaging on 7 microbial growth and sensorial qualities of fresh-cut produce, *Int. J. Food Microbiol.*, 71, 197, 2001.
12. Amanatidou, A., Smid, E.J., and Gorris, L.G.M., Effect of elevated oxygen and carbon dioxide on the surface growth of vegetable-associated micro-organisms, *J. Appl. Microbiol.*, 86, 429, 1999.

13. Allende, A. and Artes, F., Combined ultraviolet-C and modified atmosphere packaging treatments for reducing microbial growth of fresh processed lettuce, *Lebensmittel-Wissenschaft und Technologie*, 36, 779, 2003.

14. Wszelaki, A.L. and Mitcham, E.J., Effects of superatmospheric oxygen on strawberry quality and decay, *Postharvest Biol. Technol.*, 20, 125, 2000.

15. U.S. Food and Drug Administration, Center for Food Safety and Applied Nutrition, Subject: Microbiological Safety of Controlled and Modified Atmosphere Packaging of Fresh and Fresh-Cut Produce, in Analysis and Evaluation of Preventative Control Measures for the Control and Reduction/ Elimination of Microbial Hazards on Fresh and Fresh-Cut Produce, http:// www.cfsan.fds.gov/~comm/ift3-6.html, 2001.

16. Al-Ati, T. and Hotchkiss, J.H., The role of packaging film permselectivity in modified atmosphere packaging, *J. Agric. Food Chem.*, 51, 4133, 2003.

17. Alique, R., Martinez, M.A., and Alonso, J., Influence of the modified atmosphere packaging on shelf life and quality of Navalinda sweet cherry, *Eur. Food Res. Technol.*, 217, 416, 2003.

18. Rai, D.R., Oberoi, H.S., and Baboo, B., Modified atmosphere packaging and its effect on quality and shelf life of fruits and vegetables: an overview, *J. Food Sci. Technol.*, 39, 199, 2002.

19. Brecht, J.K. *et al.*, Maintaining optimal atmosphere conditions for fruits and vegetables throughout the postharvest handling chain, *Postharvest Biol. Technol.*, 27, 87, 2003.

20. Ahvenainen, R., New approaches in improving the shelf life of minimally processed fruit and vegetables, *Trends Food Sci. Technol.*, 7, 179, 1996.

21. Lee, D.S., Kang, J.S., and Renault, P., Dynamics of internal atmosphere and humidity in perforated packages of peeled garlic cloves, *Int. J. Food Sci. Technol.*, 35, 455, 2000.

22. Cliffe-Byrnes, V., McLaughlin, C.P., and O'Beirne, D., The effects of packaging film and storage temperature on the quality of a dry coleslaw mix packaged in a modified atmosphere, *Int. J. Food Sci. Technol.*, 38, 187, 2003.

23. Vermeiren, L. *et al.*, Developments in the active packaging of foods, *Trends Food Sci. Technol.*, 10, 77, 1999.

24. Vermeiren, L., Devlieghere, F., and Debevere, J., Effectiveness of some recent antimicrobial packaging concepts, *Food Add. Contam.*, 19 (Suppl.), 163, 2002.

25. Lee, C.H. *et al.*, Wide-spectrum antimicrobial packaging materials incorporating nisin and chitosan in the coating, *Packag. Technol. Sci.*, 16, 99, 2003.

26. Srinivasa, P.C. *et al.*, Storage studies of mango packed using biodegradable chitosan film, *Eur. Food Res. Technol.*, 215, 504, 2002.

27. Guilbert, S., Gontard, N., and Gorris, L.G.M., Prolongation of the shelf-life of perishable food products using biodegradable films and coatings, *Lebens.-Wissen.Technol.*, 29, 10, 1996.

28. Spotts, R.A., Cervantes, L.A., and Facteau, T.J., Integrated control of brown rot of sweet cherry fruit with a preharvest fungicide, a postharvest yeast, modified atmosphere packaging, and cold storage temperature, *Postharvest Biol. Technol.*, 24, 251, 2002.

29. Devlieghere, F., Vermeiren, L., and Debevere, J., New preservation technologies: possibilities and limitations, *Int. Dairy J.*, 14, 273, 2004.

30. Bennik, M.H.J. *et al.*, Biopreservation in modified atmosphere stored mungbean sprouts: the use of vegetable-associated bacteriocinogenic lactic acid bacteria to control the growth of *Listeria monocytogenes*, *Lett. Appl. Microbiol.*, 28, 226, 1999.

31. Cai, Y., Ng, L.-K., and Farber, J.M., Isolation and characterization of nisin-producing *Lactococcus lactis* subsp. *lactis* from bean-sprouts, *J. Appl. Microbiol.*, 83, 499, 1997.

32. Francis, G.A. and O'Beirne, D., Effects of the indigenous microflora of minimally processed lettuce on the survival and growth of *Listeria innocua*, *Int. J. Food Sci. Technol.*, 33, 477, 1998.

33. Francis, G.A. and O'Beirne, D., Effects of storage atmosphere on *Listeria monocytogenes* and competing microflora using a surface model system, *Int. J. Food Sci. Technol.*, 33, 465, 1998.

34. Charles, F., Sanchez, J., and Gontard, N., Active modified atmosphere packaging of fresh fruits and vegetables: modeling with tomatoes and oxygen absorber, *J. Food Sci.*, 68, 1736, 2003.

35. Gonzalez-Aquilar, G.A., Buta, J.G., and Wang, C.Y., Methyl jasmonate and modified atmosphere packaging (MAP) reduce decay and maintain postharvest quality of papaya "Sunrise", *Postharvest Biol. Technol.*, 28, 361, 2003.

36. Allende, A. and Artes, F., UV-C radiation as a novel technique for keeping quality of fresh processed "Lollo Rosso" lettuce, *Food Res. Int.*, 36, 739, 2003.

37. Beuchat, L.R., *Listeria monocytogenes*: incidence on vegetables, *Food Control*, 7, 223, 1996.

38. Zagory, D., Effects of post-processing handling and packaging on microbial populations. *Postharvest Biol. Technol.*, 15, 313, 1999.

39. Bagamboula, C.F., Uyttendaele, M., and Debevere, J., Growth and survival of *Shigella sonnei* and *S. flexneri* in minimal processed vegetables packed under equilibrium modified atmosphere and stored at 7°C and 12°C, *Food Microbiol.*, 19, 529, 2002.

40. Allende, A., Aguayo, E., and Artes, F., Microbial and sensory quality of commercial fresh processed red lettuce throughout the production chain and shelf life, *Int. J. Food Microbiol.*, 91, 109, 2004.

41. Pretel, M.T. *et al.*, The effect of modified atmosphere packaging on "ready-to-eat" oranges, *Lebens-Wissen. Technol.*, 31, 322, 1998.

42. Jacxsens, L. *et al.*, Relation between microbiological quality, metabolite production and sensory quality of equilibrium modified atmosphere packaged fresh-cut produce, *Int. J. Food Microbiol.*, 83, 263, 2003.

43. Bennik, M.H.J. *et al.*, The influence of oxygen and carbon dioxide on the growth of prevalent Enterobacteriaceae and *Pseudomonas* species isolated from fresh and controlled-atmosphere-stored vegetables, *Food Microbiol.*, 15, 459, 1998.

44. Bennik, M.H. *et al.*, Growth of psychrotrophic foodborne pathogens in a solid surface model system under the influence of carbon dioxide and oxygen, *Food Microbiol.*, 12, 509, 1996.

45. Martinez-Ferrer, M. *et al.*, Modified atmosphere packaging of minimally processed mango and pineapple fruits, *J. Food Sci.*, 67, 3365, 2002.

46. Piga, A. *et al.*, Influence of storage temperature on shelf-life of minimally processed cactus pear fruits, *Lebens-Wissen. Technol.*, 33, 15, 2000.

47. Hintlian, C.B. and Hotchkiss, J.H., The safety of modified atmosphere packaging: a review, *J. Food Technol.*, 40, 70, 1986.

48. Francis, G.A. and O'Beirne, D., Effects of vegetable type, package atmosphere and storage temperature on growth and survival of *Escherichia coli* O157:H7 and *Listeria monocytogenes*, *J. Indust. Microbiol. Biotechnol.*, 27, 111, 2001.

49. Francis, G.A. and O'Beirne, D., Effects of acid adaptation on the survival of Listeria monocytogenes on modified atmosphere packaged vegetables, *Int. J. Food Sci. Technol.*, 36, 477, 2001.

50. Bennik, M.H.J. *et al.*, Microbiology of minimally processed, modified-atmosphere packaged chicory endive, *Postharvest Biol. Technol.*, 9, 209, 1996.

51. Berrang, M.E., Brackett, R.E., and Beuchat, L.R., Microbial, color and textural qualities of fresh asparagus, broccoli and cauliflower stored under controlled atmosphere, *J. Food Prot.*, 53, 391, 1990.

52. Berrang, M.E., Brackett, R.E., and Beuchat, L.R., Growth of *Aeromonas hydrophila* on fresh vegetables stored under a controlled atmosphere, *Appl. Environ. Microbiol.*, 55, 2167, 1989.

53. Berrang, M.E., Brackett, R.E., and Beuchat, L.R., Growth of *Listeria monocytogenes* on fresh vegetables stored under a controlled atmosphere, *J. Food Prot.*, 52, 702, 1989.

54. Jacxsens, L., Devlieghere, F., and Debevere, J., Temperature dependence of shelf-life as affected by microbial proliferation and sensory quality of equilibrium modified atmosphere packaged fresh produce, *Postharvest Biol. Technol.*, 26, 59, 2002.

55. Macura, D., McCannel, A.M., and Li, M.Z.C., Survival of *Clostridium botulinum* in modified atmosphere packaged fresh whole North American ginseng roots, *Food Res. Int.*, 34, 123, 2001.

56. Austin, J.W., Dodds, K.L., Blanchfield, B., and Farber, J.M., Growth and toxin production by *Clostridium botulinum* on inoculated fresh-cut packaged vegetables, *J. Food Prot.*, 61, 324, 1998.

57. Beuchat, L.R., Ecological factors influencing survival and growth of human pathogens on raw fruits and vegetables, *Microbes Infect.*, 4, 413, 2002.

58. Gunes, G. and Hotchkiss, J.H., Growth and survival of *Escherichia coli* O157:H7 on fresh-cut applies in modified atmospheres at abusive temperatures, *J. Food Prot.*, 65, 1641, 2002.

59. Phillips, C.A., The isolation of *Campylobacter* spp. from modified atmosphere packaged foods, *Int. J. Environ. Health Res.*, 8, 215, 1998.

60. Paul, D.R. and Clarke, R., Modeling of modified atmosphere packaging based on designs with a membrane and perforations, *J. Membr. Sci.*, 208, 269, 2002.

20 Hot Water Treatments for Control of Fungal Decay on Fresh Produce

Elazar Fallik

CONTENTS

20.1 INTRODUCTION

Fresh fruits and vegetables have been a part of the human diet since the dawn of history, while farmers and food sellers have been concerned about losses since agriculture began. While fruits and vegetables have always provided variety in the diet through differences in color, shape, taste, aroma, and texture [1], their full nutritional importance has only been recognized in recent times.

Contamination of fresh produce with pathogenic agents may occur at any point during production, harvesting, packing, processing, distribution, or marketing. Therefore, all fresh harvested commodities need to be free of disease agents, insects, synthetic chemicals, and cleaned of dirt or dust before being sent to the markets. The problem of how much food is lost after harvest

---

This chapter is dedicated to the late Mr. Erwin Fisher: an expert on scanning electron microscopy analysis, who contributed significantly to the understanding of the mode of action of hot water rinsing and brushing.

by inefficient processing, spoilage, insects and rodents, or other factors takes on greater importance as world food demand grows. Marketing of produce has also benefited from an international trend towards fresh natural foods, which are perceived to be superior to processed foods and to contain fewer chemical additives.

Although fresh produce is generally not considered a common source of foodborne illness, the incidence of this problem is increasing [2]. In recent years the number of cases of illness linked with eating fruits and vegetables has risen from 2% to about 8% of reported cases. The increased incidence may be related to changing patterns of food consumption, recognition of new means for transmission of disease organisms, emergence of pathogens that can cause infections at very low doses, an expectation that most foods distributed in any country are safe, and/or a perception that foodborne illness does not occur at home.

For many years chemical treatments have been the basis for ensuring postharvest quality [3]. Although government authorities in each country regulate fungicide use to ensure that chemicals are not toxic at the concentrations used [4], there is still growing concern and apprehension by the public about the use of synthetic pesticides. Pressure is building for the use of alternative "nonchemical" means of disease control by the horticultural and agricultural industries.

Several chemical-free technologies to extend the storage and shelf life of fresh produce are being investigated. Among these technologies are modified atmosphere packaging [5], irradiation [6,7], use of materials that are generally regarded as safe (GRAS), such as bicarbonate salts [8] or hydrogen peroxide [9,10], hypobaric treatment [11], biological control [12], or prestorage heat treatments [13]. Heat treatment appears to be one of the most promising means for postharvest control of decay [13,14]. Prestorage heat treatments to control decay development during storage and marketing period are often applied for a relatively short time (minutes), because the targeted decay-causing agents are found on the surface or in the first few cell layers under the skin of the fruit or vegetable [14]. Heat treatments against pathogens may be applied to the fresh harvested produce in several ways: by hot water dips, by vapor heat, by hot dry air [13], and by a short hot water rinse and brush [15,16]. Hot water is an effective heat transfer medium and, when properly circulated through the load of fruit, establishes a uniform temperature profile more quickly than either vapor or dry heat [17]. Hot water treatments were originally used to control fungal diseases such as brown rot (*Phytophthora* spp.) on citrus fruits [18,19], but their use has been extended to achieve disinfestations from insects and alleviation of physiological deterioration [13].

Several reviews have been published on the effect of both dry and wet heat on maintenance of quality in fresh harvested crops [13,14,17,20,21]. This chapter summarizes recent research on the technologies used in hot water treatments and their effects on decay development in fruits, vegetables, and minimally processed products.

20.2 TECHNOLOGIES

There are three basic designs for administering hot water treatments: the *batch system*, the *continuous system* (hot water treatment, HWT) [22], and the *hot water rinsing and brushing system* (HWRB) [15].

Most of the hot water immersion treatment facilities that are in commercial use currently are of the batch system. In this system, baskets of produce are loaded onto a platform, which is then lowered into the hot water immersion tank, where the fruits or vegetables remain at the prescribed temperature for a certain time before being taken out, usually by means of an overhead hoist. In the continuous system the produce is submerged (either loosely or in a wire or plastic mesh basket) on a conveyor belt, which moves slowly from one end of the hot water tank to the other. The belt speed is set to ensure that the produce is submerged for the required length of time. This system requires an instrument to monitor the speed of the conveyor belt.

The main components in these two systems are an insulated treatment tank of several hundred liters, a heat exchange unit operated by gas, diesel, or electricity, a pump and water circulation system to provide uniform water temperatures throughout the treatment process and to avoid the formation of cool pockets during treatment, and temperature sensors to control and monitor water temperature during treatment [22,23]. An inexpensive hot water immersion system can be assembled easily; the machinery can even be made mobile with little difficulty [23,24].

In contrast to hot water immersion, a new technology based on a brief HWRB for simultaneous cleaning and disinfestation of fresh produce was first introduced commercially in 1996 [20]. The fourth generation of the HWRB machine (Figure 20.1) contains 18 to 22 parallel brushes, all of which are controlled by a single motor. All components in the machine, including the hot water tank (300 to 500 l), are made from stainless steel materials. The produce is prewashed by nonrecycled tap water (ambient temperature) for about 5 to 10 seconds while revolving on cylindrical brushes. A speed-adjustable conveyor belt is connected to the simultaneous cleaning and disinfecting stage, and controls the duration of exposure to hot water, which is heated with a thermostatically controlled gas or electric heating element. Fruit or vegetables are rinsed with the pressurized hot water, from nozzles that point down either vertically or at predetermined angles onto the produce, which rolls on brushes made from medium-soft synthetic bristles. The produce is exposed to water at temperatures between 48 and 63°C for 10 to 25 seconds, depending upon produce type and cultivar [25–34]. The water is filtered and then recycled, while being supplemented from time to time with new water to compensate for loss due to evaporation, adsorption, spilling, etc. At the end of the hot washing treatment, forced-air fans, or hot forced air is used to dry the produce inside a 4 to 6 m long tunnel for 1 to 2 minutes. The estimated cost of the HWRB system, including the drying tunnel, is $15,000 to $30,000 for units with washing capacities of 1 to 25 tons/h (further information can be obtained from the author).

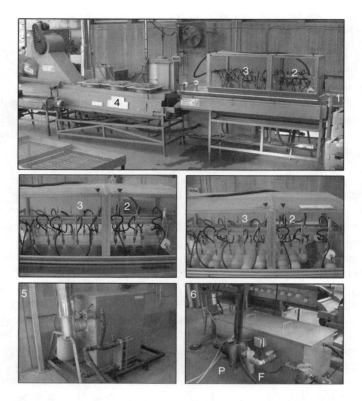

FIGURE 20.1 Hot water rinsing and brushing machine, fourth generation, made of stainless steel: (1) conveyor; (2) tap water rinsing and brushing unit; (3) hot water rinsing and brushing unit with adjustable nozzles, with and without fruits. Water is recycled. (4) Drying tunnel equipped with forced air fans; (5) heat unit (120,000 kcal); (6) hot water container equipped with water pump (P) to pressurize and recycle the hot water and filter (F).

20.3 HEAT TREATMENTS

20.3.1 *In Vitro* Studies

There is considerable variation in sensitivity to high temperature among various fungi [14]. Vegetative cells and conidia of most fungi are inactivated when exposed to a temperature of 60°C for 5 to 10 minutes *in vitro* [35]. Spore germination and germ tube elongation *in vitro* was inversely related to the duration of exposure or the range of temperature used [36]. Hot water treatments were found to be ineffective in killing dormant spores [37,38]. Ranganna *et al.* [39] reported that bacterial infection (*Erwinia caratovora*) was more sensitive to hot water treatments than was the fungal pathogen *Fusarium solani*. *Colletotrichum gloeosporiodes* was more heat sensitive than *Dothiorella dominicana* in mango fruits [40]. *Botrytis cinerea* was found to be more susceptible to hot water treatment than *Alternaria alternata* in sweet bell pepper [36]. The effective time to kill 50% of the spores (ET_{50} germination) for *B. cinerea* was 3.2, 1.5, and 0.8 minutes at 45, 50, and 55°C, and for *A. alternata* was 8.8, 4.2, and 1.4 minutes, respectively, at those temperatures. The ET_{50} for germ tube elongation for *B. cinerea* was 2.6, 0.9, and 0.5 minutes at 45, 50, and 55°C, and for *A. alternata* was 7.2, 2.5, and 1.6 minutes at the same temperatures [36]. Percentage spore germination of *A. alternata* and *F. solani* was inversely proportional to the length of exposure to 55 and 60°C. Exposing fungal spores of *A. alternata* and *F. solani* to 60°C for about 15 seconds, *in vitro*, resulted in 48 and 42% reduction in spore germination, respectively [27]. The ET_{50} for *A. alternata* was 25 and 16 seconds at 55 and 65°C, respectively, whereas for *F. solani* the ET_{50} was 18 seconds at 60°C. None of the temperature/time regimes tested completely inhibited spore germination, although *A. alternata* was slightly more susceptible to heat treatment than *F. solani* [27]. A minimum exposure period of 20 seconds at 56°C was required to inhibit *Penicillium digitatum* spore germination *in vitro* [30]. No surviving spores of *B. cinerea* were observed after 15 minutes at 45°C [7]. *Monilinia fructigena* was more sensitive and a thermal treatment of 3 minutes at 45°C resulted in complete spore inactivation [7]. *In vitro* studies showed *Monilinia fructicola* to be more sensitive than *Penicillium expansum* to high temperature (60°C for 20 seconds) [41].

The viability of five pathogens was decreased by treatment with hot water when tested *in vitro*. *Polyscytalum pustulans* was most sensitive and *Rhizoctonia solani* least sensitive. The temperatures that killed 50% (LT_{50}) and 95% (LT_{95}) of spores of *Penicillium digitatum* in 15 seconds were about 5.2°C higher than those for arthrospores of *Geotrichum citri-aurantii*, and those for spores of *P. digitatum* in 30 seconds were about 3.4°C higher [34].

20.3.2 *IN VIVO* STUDIES

Fruit responses to heat treatments depend on the condition of the fruit prior to treatment, the commodity concerned, the temperature and duration of treatment, as well as the mode of heat application. The physiological responses of different fruits, vegetables, or flower species to hot water treatments can vary by season and growing location, and can be due to differences in climate, soil type, season, production practices, fruit maturity at harvest, and fruit size [24,42–45]. Hot water treatments prevent rot development in numerous temperate, subtropical, and tropical fruits and vegetables [13,14,16,17,46,47].

A dip time of 1 to 2 minutes in 55°C water for Galia melons (*Cucumis melo* L.) was an optimal antifungal treatment, while a higher temperature or longer exposure time resulted in heat injury to the fruits [48]. However, the optimal HWRB treatment to reduce decay while maintaining fruit quality after prolonged storage and marketing was 59 ± 1°C for 15 seconds [15].

Hot water treatment at 44 or 46°C for 15 minutes delayed *Botrytis cinerea* proliferation on artificially inoculated or naturally infected strawberries

(*Fragaria* x *ananassa* Duch. Tudla) [49,50]. Dipping strawberries inoculated with botrytis at 63°C for 12 seconds, followed by controlled atmosphere (CA) storage (15 kPa CO_2) was found to reduce rot development during a short or long storage regime [51]. Peaches and nectarine infected with *Monilinia fructicola* were immersed in hot water at 46 or 50°C for 2.5 minutes to control decay. These treatments reduced the incidence of decayed fruit from 82.8 to 59.3 and 38.8%, respectively [52]. Mature green plums (*Prunus salicina* Lindl. cv. Friar) were treated in water at 40, 45, 50, and 55°C for 40, 35, 30, and 25 minutes, respectively, and stored at 0°C for 35 days plus 9 days of ripening at 20 to 25°C. Decay symptoms were retarded in fruits treated at 45 and 50°C, while decay symptoms were severe in fruits of the control and those treated at 55 and 40°C [53].

HWRB at 55°C for 15 seconds significantly reduced decay development in *Penicillium expansum*-inoculated apple fruit after 4 weeks at 20°C, or in naturally infected (*P. expansum*) apple fruit after prolonged storage of 4 months at 1°C plus 10 days at 20°C [27]. Recently, Lunardi *et al.* [54] reported that prestorage hot water immersion at 47°C for 3 minutes significantly reduced rot development caused by white rot (*Botryosphaeria dothidea*) on Fuji apples after prolonged CA storage. *In vivo* studies of inoculation of peach and nectarine fruit with *Monilinia fructicola* followed by HWRB at 55 or 60°C for 20 seconds gave 70 and 80% decay inhibition, respectively, compared with the control [41]. The inhibition percentages of *M. fructicola* with HWRB were similar, if HWRB was applied shortly after inoculation or 24 hours later. In contrast, the sensitivity of *P. expansum* spores inoculated into wounds increased when the fruit were treated with HWRB 24 hours after the inoculation, compared with treatment just after inoculation [41]. Treating fruit with HWRB at 60°C for 20 seconds and then dipping them into a cell suspension (10^8 cells/ ml) of *Candida* spp. 24 hours after inoculation with *P. expansum* reduced decay development by 60% compared with the controls, but did not reduce rot development caused by *M. fructicola* [41].

Many types of fresh produce from the Solanaceae benefit from hot water treatments. Potato tubers were dipped at 55°C for 5 minutes in a commercial continuous hot water treatment plant [55]. The frequency of eyes colonized by *Pseudomonas pustulans, Helminthosporium solani*, and *Rhizoctonia solani* was reduced to virtually zero and the effect persisted on tubers subsequently stored at 4 and at 15°C for up to 16 weeks. Results with *Colletotrichum coccodes* were inconclusive. Treatment suppressed *Penicillium* spp. which, however, rapidly colonized the eyes during storage, leading to higher contamination levels in the treated than in the untreated tubers. With tubers inoculated with *Phoma foveata*, good control was achieved when the incubation period before treatment was 10 days but not when the fungus was more established 42 days after inoculation [55]. Potato tubers inoculated with *Erwinia carotovora* and *Fusarium solani* were safely stored for 12 weeks at either 8 or 18°C without spoilage if dipped in a 57.5°C hot water bath for 20 to 30 minutes [39].

The effectiveness of hot water dipping on the control of grey mould, caused by *Botrytis cinerea*, and black mould, caused by *Alternaria alternata*, and on sweet red pepper quality was investigated. Dipping naturally infected or artificially inoculated fruit at 50°C for 3 minutes completely inhibited or significantly reduced decay development caused by *B. cinerea* and *A. alternata*, respectively [36]. Heat damage was observed on fruit dipped for 5 minutes at 50°C, or at 55°C for 1 minute or longer [36]. Significant differences in the incidence of decay were found between temperatures and in the time–temperature interaction, but not between times of dipping [36]. Treatment of peppers with hot water at 53°C for 4 minutes was found to be effective in reducing decay after 14 and 28 days of storage at 8°C. Treatment at 45°C for 15 minutes was less effective in maintaining pepper quality during storage [56,57]. HWRB of red and yellow sweet bell pepper cultivars at 55°C for about 12 seconds significantly reduced decay incidence while maintaining quality compared both to untreated control and to most other commercial treatments [25].

A hot water dip at 50°C for 2 minutes was more effective than 1 kGy of gamma radiation in reducing *Botrytis cinerea* and *Rhizopus stolonifer* decay in inoculated light-red tomatoes (*Lycopersicon esculentum* Mill. cv F-121) [58]. Mature green tomatoes (*L. esculentum* Mill. cv Sunbeam) were treated in water for 1 hour at 27, 39, 42, 45, or 48°C, and then ripened for 14 days at 20 or 2°C. Treatment at 42°C reduced decay by 60%, whereas other water temperatures were less effective [59]. HWRB of freshly harvested tomatoes at 52°C for 15 seconds, or dipping the fruit at 52°C for 1 minute, significantly reduced decay development and chilling injury after 3 weeks' storage at 2 or 12°C and additional 5 days at 20°C [33]. These two prestorage heat treatments did not affect other quality parameters such as fruit firmness, total soluble solids, or acidity [33].

Fresh broccoli floret (*Brassica oleracea* L. Italica group) is a highly perishable fresh vegetable when held at ambient temperatures; it becomes unmarketable within 1 to 3 days. Immersion in hot water at 50 or 52°C for 2 minutes was most effective in controlling decay development and reducing yellowing [75].

Several recent studies have shown that hot water immersion for 2 to 3 minutes at 53°C significantly reduced decay development in a wide variety of citrus cultivars [43,61–64]. Hot water dipping at 53°C for 3 minutes gave beneficial effects on decay control of Tarocco blood oranges (*Citrus sinensis* L. Obsek) fruit harvested in February and March but was detrimental for fruit harvested in April [43]. No decay symptoms were detected after 90 days' storage at 4°C in March grapefruit treated with 45°C water for 3 hours [65]. A hot water dip for 2 minutes at 52 to 53°C inhibited the development of decay in lemons inoculated with *P. digitatum* [63]. When the fruit were heat treated 1 day after inoculation, no decay occurred within 6 days after the treatment, whereas fruit dipped in water at 25°C reached 100% decay after 4 days. When the hot water dip was applied 2 days after inoculation, 90% of the treated fruit remained healthy [63]. The optimal HWRB treatment to reduce decay development while maintaining fruit quality of kumquat was 55°C for 20 seconds [28]. When organic citrus fruits were treated with HWRB at 56, 59, and 62°C for 20 seocnds after artificial inoculation with *P. digitatum*, decay development in infected wounds was reduced to 20, 5, and less than 1%, respectively, of that in untreated control fruits or fruits treated with tap water [30]. A 20-second HWRB treatment at 59 or 62°C reduced decay in Star Ruby grapefruit that was artificially inoculated with *P. digitatum*, by 52 and 70%, respectively, compared with control unwashed fruit. Tap water wash (~20°C) or HWRB at 53 or 56°C were ineffective [31]. Green mold incidence caused by *P. digitatum* was reduced from 97.9 and 98% on untreated lemons and oranges, respectively, to 14.5 and 9.4% by a brief 30-second HWRB treatment at 62.8°C [34].

Hot water immersion also inhibits decay development on tropical and subtropical fruits. Dipping Kensington Pride mango fruits in hot water at 52°C for 5 minutes together with the fungicide benomy gave good control of stem end rot caused by *Dothiorella dominicana* and *Lasiodiplodia theobromae* [40]. Treatments consisting of hot water only or hot water followed by the fungicide prochloraz gave only partial control of stem-end rot. All treatments gave good control of anthracnose caused by *Colletotrichum gloeosporioides* [40]. The effectiveness of different postharvest treatments to control different levels of quiescent infections of *Alternaria alternata* causing alternaria rot in mango fruits during storage was compared. A combined HWRB treatment at 48 to 62°C (depending on the cultivar) for 15 to 20 seconds with 225 mg/ml prochloraz was the most effective treatment for control of alternaria rot in fruit with a high relative quiescent infected surface [66]. However, the effectiveness of postharvest HWRB and prochloraz applications are dependent on the quiescent infected area of the fruit by *A. alternata* at harvest [66]. Hot water treatments reduced body rot caused by *Colletotrichum* spp. in ripe avocado fruit with 40 and 41°C for 30 minutes [67]. However, treatment at 42°C for 30 minutes increased body rot compared to the other HWTs in one season, but there was no benefit of HWT times longer than 30 minutes [67]. Stem rot caused mainly by *Dothiorella* spp. was also reduced by HWT at 40 and 41°C [67]. HWRB at 55°C for 20 seconds significantly reduced chemical use (prochloraz) to control decay development caused mainly by *Penicillium* spp. in litchi fruit [29].

Food safety has become a very important issue for fresh and minimally processed products [68]. Minimal processing of vegetables provides convenience to the food industry and retail consumers, but may result in limited shelf life and marketing because of undesirable physiological and pathological changes [69,70]. Very little research has been done to evaluate the efficacy of hot water treatment on minimally processed products. A 4-minute water wash of green onion at 52.5°C reduced the aerobic plate count by 1 to 2 logs compared with water wash at 20°C [70]. A similar reduction in microbial population of soybean sprouts and watercress after a 30-second water dip at 60°C was reported by Park *et al.* [71].

Immersing spot-inoculated apple fruits at 80 and 95°C for 15 seconds produced a reduction of more than 5 log in *Escherichia coli* O157:H7 [72]. Several pasteurization procedures for alfalfa (*Medicago sativa*) seeds were investigated to disinfect completely inoculated *Escherichia coli* (Migula) Castellani and Chalmers ATCC 25922 [73]. Hot water treatments (85°C for 9 seconds) were equally or more effective than 20,000 ppm calcium hypochlorite treatments, yielding a reduction of 2 log CFU/g [73]. Li *et al.* [74] reported that the population of *Listeria monocytogenes* on cut iceberg lettuce treated at 50°C for 90 seconds steadily increased throughout storage at 5°C for up to 18 days.

20.3.3 HEAT DAMAGE

Hot water immersion or rinsing while brushing may result in commodity damage, which typically is manifested as browning on fruit surfaces, uneven ripening, breakdown of the fruit flesh, and even enhanced rot development if the technique is not properly applied [13]. Incidence of HWT-associated damage varied between regions, harvest dates, and orchards [76]. Immersion of guavas (*Psidium guajava* L.) for 35 minutes in water at $46.1 \pm 0.2°C$ delayed ripening by 2 days, but increased susceptibility to decay [77]. Immersion of Marsh grapefruit at 48°C for 2 or 3 hours significantly increased decay ($>30\%$). However, if fruit were treated at 45°C for 3 hours, no decay symptoms were detected after 90 days' storage at 4°C [56].

Heat treatment has been associated with increased susceptibility to decay in a number of crops, such as nectarine [78] and papaya [79]. The increase in the incidence of rot is likely to be due to pathogens invading areas on the fruit injured by the heat treatment. Hot water dips at 45°C for 2.5 minutes did not control mold development caused by *P. digitatum* in clementines [78]. The significant water loss and softness of fruit dipped at 55°C for 5 minutes was due to heat damage causing cracks and pitting on the surface of the treated fruit and the expansion and collapse of the hypoderm cells [36]. HWRB at 60°C for 25 seconds caused heat damage as irregular reddish pits of 0.5 to 1.5 mm in diameter [16]. Heat damage on HWRB-treated apples (60°C for 15 seconds) appeared as round brown sunken pits [32].

Hot water dips can change some biochemical properties in minimally processed (peeled and trimmed) onions. Dipping prepeeled onions in 80°C for 1 minute resulted in irreversible membrane damage [80]. Dipping heads of fresh broccoli (*Brassica oleracea* L. Italica group cv. Paragon) at 52°C for 3 minutes enhanced off-odor development and caused visual damage to newer buds [81].

Hot water treatment of mandarins at 56 and 58°C for 3 minutes induced heat damage in the form of rind browning [62]. Hot water treatments for 10 minutes at 48°C and below were noninjurious to both yellow and green lemon fruit, with injury in the form of lesions on the rind beginning to occur at 50°C (yellow) and 52°C (green), and increasing in severity up to 58°C, the highest temperature tested. Emanation of d-limonene increased correspondingly with increasing injury. Green lemons were injured more severely than yellow and tended to release more d-limonene, especially at higher temperatures [82]. Heat damage was evident in avocado fruit as hardening of the skin when fruit ripened [83]. However, Obenland and Aung [84] reported that sodium chloride at a concentration of 200 mM reduced hot water damage in nectarine cultivars by effectively reducing the amount of water entering the fruit during hot water treatment.

20.4　MODE OF ACTION

Heat treatments can interact directly or indirectly with pathogens and/or fresh harvested produce via several responses. The efficacy of heat on pathogens is usually measured by reduced viability of the heated propagules [14,40]. Heat effects may be lethal or sublethal, and pathogen kill is not always proportional to the temperature–time product of the treatment [85,86]. Heat treatments may cause changes in nuclei and cell walls, denature proteins, destroy mitochondria and outer membranes, disrupt vacuolar membranes, and form gaps in the spore cytoplasm, which lead to reducing inoculum level [14].

Applying HWRB to melon and citrus fruits resulted in a 3 to 4 log reduction of the total microbial colony forming units (CFU) of the epiphytic microorganism population, compared to untreated control fruit [27,30,34]. Scanning electron microscopy (SEM) showed that HWRB removed fungal spores from the fruit surface, and partially or entirely sealed natural openings in the epidermis (Figure 20.2) [25,27,30,33]. As a result of heat treatments that reduce fungal viability, the effective inoculum concentration that causes decay development is reduced, thus reducing rot development [86]. In addition, sealing epidermal cracks with heat treatment could reduce sites of fungal penetration into the fruit, thus reducing decay incidence [13]. Schirra and D'Hallewin [62] and Ben-Yehoshua [64] reported that hot water dips of grapefruit and mandarins redistributed the epicuticular wax layers, which sealed or partially sealed cracks, thus improving physical barriers to pathogen invasion.

Hot water immersion or rinse was found to inhibit ripening processes as measured by relatively low respiration rate and ethylene evolution, and slow color development, compared with nonheated control fruit. In addition, heat treatment prevented postharvest geotropic curvature of vegetables [13,87,88]. Fruits that ripen slowly have less susceptibility to fungal attack during storage [47]. Ben-Yehoshua [64] reported that heat treatment induced resistance of grapefruits to decay caused by *P. digitatum* by delaying the breakdown and disappearance of preformed antifungal compounds. The heated citrus fruit had higher concentrations of the phytoalexin scoparone, which in turn was correlated with antifungal activity in the fruit extract [64]. However, a 10-minute hot water dip caused a faster decrease of antifungal compounds and the earlier appearance of rot symptoms in treated avocado, compared to nonheated fruit [89].

Hot water treatments caused a delay in spore germination and fungal growth in citrus fruit [37]. This was explained by a building up or improvement

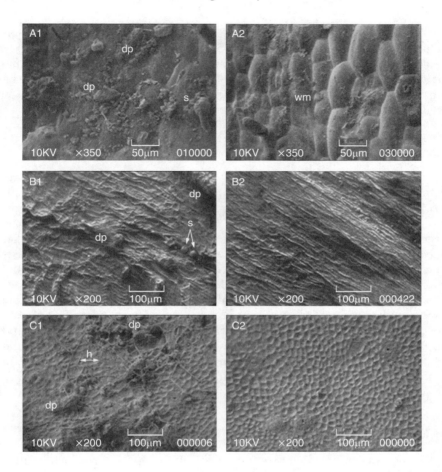

FIGURE 20.2 Scanning electron micrographs of untreated tomato (A1), pepper calyx (B1), and melon (C1) compared to HWRB-treated tomato (A2), pepper calyx (B2), and melon (C2) (s, fungal spores; h, hyphae; dp, dirt particles).

in the defense systems against pathogens that slowed fungal development as a result of heat-induced changes in the fruit tissue. Heat treatments induced the biosynthesis of lignin-like polymers that were bound to walls of cells adjacent to wound sites in citrus peel [64]. HWRB treatment of 52°C for 15 seconds induced tomato resistance when fruit was artificially inoculated with *B. cinerea* 24 hours after treatment [90]. HWRB treatment at 62°C for 20 seconds was most effective in inducing disease resistance against green mold (*P. digitatum*) when grapefruit was inoculated after 1 and 3 days after treatment [91]. The HWRB treatment induced the accumulation of proteins that cross-reacted with heat shock proteins and of proteins that cross-reacted with chitinase and 1,3-glucanase antibodies [61]. The increases in the accumulation of chitinase and glucanase proteins, which were detected 1 and 3 days after

HWRB treatment, may be part of the complex fruit resistance mechanisms induced by this technology [31,91]. The mode of action of the hot water dip in reducing the decay of lemon fruit is partly related to the temporary thermal inhibition of pathogen growth that allowed the infected fruit to build up production of lignin-like material at the inoculation site, followed later by accumulation of the phytoalexins scoparone and scopolitin [63].

20.5 CONCLUSIONS

Interest in alternative methods for postharvest decay control of horticultural crops in order to minimize pre- or postharvest treatments with agrochemicals has been growing continuously. Prestorage heat treatment is one of the most promising and simple technologies to reduce rot development on fresh harvested fruits, vegetables, and minimally processed products. However, further research is needed to determine the most heat-sensitive life-stage of disease-causing agents of economic importance and to obtain the intrinsic kinetics information of this life stage for developing hot water treatment protocols. There is also a special need to obtain information regarding time–temperature effects on the quality of fresh harvest produce. A better understanding of both direct and indirect modes of action of heat treatments on pathogens and on fresh produce tissue will enable development of optimal, successful, and relatively cheap hot water dip or rinsing treatments and equipment that will control decay-causing agents without affecting the overall quality of the fruit or vegetable.

ACKNOWLEDGMENTS

This paper was prepared under grant no. 406-0813-03 from the Chief Scientist of the Ministry of Agriculture, Israel, with a contribution from the Agricultural Research Organization, The Volcani Center, Bet Dagan, Israel, no. 402/04. The author thanks Dr. Joshua D. Klein for his critical review and Michael Fallik for the graphic work.

REFERENCES

1. Kays, S.J., *Postharvest Physiology of Perishable Plant Products*, 2nd ed., AVI Book, Van Nostrand Reinhold, New York, 1997, p. 532.
2. Li-Cohen, A.E. and Bruhn, C.M., Safety of consumer handling of fresh produce from the time of purchase to the plate: a comprehensive consumer survey, *J. Food Prot.*, 65, 1287, 2002.
3. Eckert, J.W. and Ogawa, J.M., The chemical control of postharvest diseases: deciduous fruits, berries, vegetables and root/tuber crops, *Ann. Rev. Phytopathol.*, 26, 433, 1988.
4. How, B.R., *Marketing Fresh Fruits and Vegetables*, AVI Book, Van Nostrand Reinhold, New York, 1991, p. 336.

5. Fonseca, S.C., Oliveira, A.R., and Brecht, J.K., Modelling respiration rate of fresh fruits and vegetables for modified atmosphere packages: a review, *J. Food Eng.*, 52, 99, 2002.

6. Follett, P.A. and Sanxter, S.S., Longan quality after hot-water immersion and X-ray irradiation quarantine treatments, *HortSci.*, 37, 571, 2002.

7. Marquenie, D., Lammertyn, J., Geeraerd, A.H., Soontjens, C., Van Impe, J.F., Nicolai, B.M., and Michiels, C.W., Inactivation of conidia of *Botrytis cinerea* and *Monilinia fructigena* using UV-C and heat treatment, *Int. J. Food Microbiol.*, 74, 27, 2002.

8. Larrigaudiere, C., Pons, J., Torres, R., and Usall, J., Storage performance of clementines treated with hot water, sodium carbonate and sodium bicarbonate dips, *J. Hort. Sci. Biotechnol*, 77, 314, 2002.

9. Fallik, E., Aharoni, Y., Grinberg, S., Copel, A., and Klein, J.D., A postharvest hydrogen peroxide treatment inhibits decay in eggplant and sweet red pepper, *Crop Prot.*, 13, 451, 1994.

10. Sapers, G.M. and Simmons, G.F., Hydrogen peroxide disinfection of minimally processed fruits and vegetables, *Food Technol.*, 52, 48, 1998.

11. Romanazzi, G., Nigro, F., Ippolito, A., and Salerno, M., Effect of short hypobaric treatments on postharvest rots of sweet cherries, strawberries and table grapes, *Postharvest Biol. Technol.*, 22, 1, 2001.

12. Conway, W.S., Janisiewicz, W.J., Klein, J.D., and Sams, C.E., Strategy for combining heat treatments, calcium infiltration, and biological control to reduce postharvest decay of "Gala" apples, *HortSci.*, 34, 700, 1999.

13. Lurie, S., Postharvest heat treatments of horticultural crops, *Hortic. Rev.*, 22, 91, 1998.

14. Barkai-Golan, R. and Phillips, D.J., Postharvest heat treatment of fresh fruits and vegetables for decay control, *Plant Dis.*, 75, 1085, 1991.

15. Fallik, E., Aharoni, Y., Yekutieli, O., Wiseblum, A., Regev, R., Beres, H., and Bar-Lev, E., A Method for Simultaneously Cleaning and Disinfecting Agricultural Produce, Israel Patent Application No. 116965, 1996.

16. Fallik, E., Tuvia-Alkalai, S., Copel, A., Wiseblum, A., and Regev, R., A short water rinse with brushing reduces postharvest losses: 4 years of research on a new technology, *Acta Hortic.*, 553, 413, 2001.

17. Couey, H.M., Heat treatment for control of postharvest diseases and insect pests of fruits, *HortSci.*, 24, 198, 1989.

18. Fawcett, H.S., Packing house control of brown rot, *Citrograpth*, 7, 232, 1922.

19. Brooks, C. and McColloch, C., Some storage diseases of grapefruit, *J. Agric. Res.*, 52, 319, 1936.

20. Klein, J.D. and Lurie, S., Postharvest heat treatment and fruit quality, *Postharvest News Info.*, 2, 15, 1991.

21. Klein, J.D. and Lurie, S., Heat treatments for improved postharvest quality of horticultural crops, *HortTechnol.*, 2, 316, 1992.

22. Animal and Plant Health Inspection Service (APHIS), T103: high temperature forced air, *Plant Protection and Quarantine Treatment Manual*, United States Department of Agriculture, Hyattsville, MD, 1996, p. 5.51.

23. Tsang, M.M.C., Hara, A.H., Hata, T.Y., Hu, B.K.S., Kaneko, R.T., and Tenbrink, V., Hot-water immersion unit for disinfestation of tropical floral commodities, *Appl. Eng. Agric.*, 11, 397, 1995.

24. Sharp, J.L., Hot-water immersion appliance for quarantine research, *J. Econ. Entomol.*, 82, 189, 1989.

25. Fallik, E., Grinberg, S., Alkalai, S., Yekutieli, O., Wiseblum, A., Regev, R., Beres, H., and Bar-Lev, E., A unique rapid hot water treatment to improve storage quality of sweet pepper, Postharvest Biol. Technol., 15, 25, 1999.

26. Prusky, D., Fuchs, Y., Kobiler, I., Roth, I., Weksler, A., Shalom, Y., Fallik, E., Zaurberman, G., Pesis, E., Akerman, M., Yekutieli, O., Wiseblum, A., Regev, R., and Artes, L., Effect of hot water brushing, prochloraz treatment and waxing on the incidence of black spot decay caused by *Alternaria alternata* in mango fruit, *Postharvest Biol. Technol.*, 15, 165, 1999.

27. Fallik, E., Aharoni, Y., Copel, A., Rodov, R., Tuvia-Alkalai, S., Horev, B., Yekutieli, O., Wiseblum, A., and Regev, R., A short hot water rinse reduces postharvest losses of "Galia" melon, *Plant Pathol.*, 49, 333, 2000.

28. Ben-Yehoshua, S., Peretz, J., Rodov, V., Nafussi, B., Yekutieli, O., Wiseblum, A., and Regev, R., Postharvest application of hot water treatment in citrus fruits: the road from the laboratory to the packing-house, *Acta Hortic.*, 518, 19, 2000.

29. Lichter, A., Dvir, O., Rot, I., Akerman, M., Regev, R., Wiseblum, A., Fallik, E., Zauberman, G., and Fuchs, Y., Hot water brushing: an alternative method to SO_2 fumigation for color retention of litchi fruits, *Postharvest Biol. Technol.*, 18, 235, 2000.

30. Porat, R., Daus, A., Weiss, B., Cohen, L., Fallik, E., and Droby, S., Reduction of postharvest decay in organic citrus fruit by a short hot water brushing treatment, *Postharvest Biol. Technol.*, 18, 151, 2000.

31. Porat, R., Pavoncello, D., Peretz, Y., Weiss, B., Cohen, L., Ben-Yehoshua, S., Fallik, E., Droby, S., and Lurie, S., Induction of resistance against *Penicillium digitatum* and chilling injury in Star Ruby grapefruit by a short hot water brushing treatment, *J. Hort. Sci. Biotechnol.*, 75, 428, 2000.

32. Fallik, E., Tuvia-Alkalai, S., Feng, X., and Lurie, S.,. Ripening characterization and decay development of stored apples after a short prestorage hot water rinsing and brushing, *Innov. Food Sci. Emerg. Technol.*, 2, 127, 2001.

33. Ilic, Z., Polevaya, Y., Tuvia-Alkalai, S., Copel, A., and Fallik, E., A short prestorage hot water rinse and brushing reduces decay development in tomato, while maintaining its quality, Trop. Agric. Res. Ext., 4, 1, 2001.

34. Smilanick, J.L., Sorenson, D., Mansour, M., Aieyabei, J., and Plaza, P., Impact of a brief postharvest hot water drench treatment on decay, fruit appearance, and microbe populations of California lemons and oranges, *HortTechnol.*, 13, 333, 2003.

35. Civello, P.M., Martinez, G.A., Chavas, A.R., and Anon, M.C., Heat treatments delay ripening and postharvest decay of strawberry fruit, *J. Agric. Food Chem.*, 45, 4589, 1997.

36. Fallik, E., Grinberg, S., Alkalai, S. ,and Lurie, S., The effectiveness of postharvest hot water dips on the control of gray and black moulds in sweet red pepper (*Capsicum annuum*), *Plant Pathol.*, 45, 644, 1996.

37. Dettori, A., D'Hallewin, G., Agabbio, M., Marceddu, S., and Schirra, M., SEM Studies on *Penicillium italicum* – "Star Ruby" Grapefruit Interactions as Affected by Fruit Hot Water Dipping, Proceedings of the VIII International Citrus Congress, Vol. 2, Sun City Resort, South Africa, May 12–17, 1996, p. 1158.

38. D'Hallewin, G., Dettori, A., Marceddu, S., and Schirra, M., Evoluzione dei processi infettivi di *Penicillium digitatum* Sacc. In vivo e in vitro dopo immersione in acqua calda, *Italus Hortus.*, 4, 23, 1997.

39. Ranganna, B., Raghaven, G.S.V., and Kushalappa, A.C., Hot water dipping to enhance storability of potatoes, *Postharvest Biol. Technol.*, 13, 215, 1998.

40. Rappel, L.M., Cooke, A.W., Jacobi, K.K., and Wells, I.A., Heat treatments for postharvest disease control in mangoes, *Acta Hortic.*, 291, 362, 1991.

41. Karabulut, O.A., Cohen, L., Wiess, B., Daus, A., Lurie, S., and Droby, S., Control of brown rot and blue mold of peach and nectarine by short hot water brushing and yeast antagonists, *Postharvest Biol. Technol.*, 24, 103, 2002.

42. Shellie, K.C. and Mangan, R., Disinfestation: effect of non-chemical treatments on market quality of fruit, in *Postharvest Handling of Tropical Fruit*, ACIAR Proceedings No. 50, Champ, B.C., Highly, E., and Johnson, G.I. Eds., ACIAR, Canberra, Australia, 1994, p. 304.

43. Schirra, M., Agabbio, M., D'Hallewin, G., Pala, M., and Ruggiu, R., Response of Tarocco oranges to picking date, postharvest hot water dips, and chilling storage temperature, *J. Agric. Food Chem.*, 45, 3216, 1997.

44. Jacobi, K.K., MacRae, E.A., and Hetherington, S.E., Effect of fruit maturity on the response of "Kensington" mango fruit to heat treatment, *Aust. J. Exp. Agric.*, 41, 793, 2001.

45. Jacobi, K.K., MacRae, E.A., and Hetherington, S.E., Postharvest heat disinfection treatments of mango fruit, *Sci. Hortic.*, 89, 171, 2001.

46. Ferguson, I.B., Ben-Yehoshua, S., Mitcham, E.J., McDonald, R.E., and Lurie, S., Postharvest heat treatments: introduction and workshop summary, *Postharvest Biol. Technol.*, 21, 1, 2000.

47. Schirra, M., D'hallewin, G., Ben-Yehoshua, S., and Fallik, E., Host–pathogen interaction modulated by heat treatment, *Postharvest Biol. Technol.*, 21, 71, 2000.

48. Teitel, D.C., Barkai-Golan, R., Aharoni, Y., Copel, A., and Davidson, H., Toward a practical postharvest heat treatment for "Galia" melons, *Sci. Hortic.*, 45, 339, 1991.

49. Garcia, J.M., Aguilera, C., and Albi, M.A., Postharvest heat-treatment on Spanish strawberry (Fragaria x ananassa cv. Tudla), *J. Agric. Food Chem.*, 43, 1489, 1995.

50. Garcia, J.M., Aguilera, C., and Jimenez, A.M., Gray mold in and quality of strawberry fruit following postharvest heat treatment, *HortSci.*, 31, 255, 1996.

51. Wszelaki, A.L. and Mitcham, E.J., Effect of combinations of hot water dips, biological control and controlled atmospheres for control of gray mold on harvested strawberries, *Postharvest Biol. Technol.*, 27, 255, 2003.

52. Margosan, D.A., Smilanick, J.L., Simmons, G.F., and Henson, D.J., Combination of hot water and ethanol to control postharvest decay of peaches and nectarines, *Plant Dis.*, 81, 1405, 1997.

53. Abu-Kpawoh, J.C., Xi, Y.F., Zhang, Y.Z., and Jin, Y.F., Polyamine accumulation following hot-water dips influences chilling injury and decay in "Friar" plum fruit, *J. Food Sci.*, 67, 2649, 2002.

54. Lunardi, R., Sanhueza, R.M.V., and Bender, R.J., Imersao em agua quente on controle pos-colheita da podridao branca em macas cv. Fuji. (Postharvest control of white rot on cv. Fuji apples by hot water heat treatment), *Fitopathol. Bras.*, 28, 431, 2003.

55. Dashwood, E.P., Burnett, E.M., and Perombelon, M.C.M., Effect of a continuous hot water treatment of potato tubers on seedborne fungal pathogens, *Potato Res.*, 34, 71, 1991.

56. Gonzalez-Aguilar, G.A., Cruz, R., Baez, R., and Wang, C.Y., Storage quality of bell peppers pretreated with hot water and polyethylene packaging, *J. Food Qual.*, 22, 287, 1999.

57. Gonzalez-Aguilar, G.A., Gayosso, L., Cruz, R., Fortiz, J., Baez, R., and Wang, C.Y., Polyamines induced by hot water treatments reduce chilling injury and decay in pepper fruit, *Postharvest Biol. Technol.*, 18, 19, 2000.

58. Barkai-Golan, R., Padova, R., Ross, I., Lapidot, M, Davidson, H., and Copel, A., Combined hot water and radiation treatments to control decay of tomato fruits, *Sci. Hortic.*, 56, 101, 1993.

59. McDonald, R.E., McCollum, T.G. and Baldwin, E.A., Temperature of hot water treatments influences tomato fruit quality following low-temperature storage, *Postharvest Biol. Technol.*, 16, 147, 1999.

60. Rodov, V., Ben-Yehoshua, S., Albagli, R., and Fang, D.Q., Reducing chilling injury and decay of stored citrus fruit by hot water dips, *Postharvest Biol. Technol.*, 5, 11, 1995.

61. Gonzalez-Aguilar, G.A., Zacarias, L., Mulas, M., and Lafuente, M.T., Temperature and duration of water dips influence chilling injury, decay and polyamine content in "Fortune" mandarins, *Postharvest Biol. Technol.*, 12, 61, 1997.

62. Schirra, M. and D'Hallewin, G., Storage performance of Fortune mandarins following hot water dips, *Postharvest Biol. Technol.*, 10, 229, 1997.

63. Nafussi, B., Ben-Yehoshua, B., Rodov, V., Peretz, J., Ozer, B.K., and D'Hallewin, G., Mode of action of hot-water dip in reducing decay of lemon fruit, *J. Agric. Food Chem.*, 49, 107, 2001.

64. Ben-Yehoshua, S., Effects of postharvest heat and UV applications on decay, chilling injury and resistance against pathogens of citrus and other fruits and vegetables, *Acta Hortic.*, 599, 159, 2003.

65. El-Shiekh, A.F., Effect of different postharvest hot water treatments on quality and storability of "marsh" grapefruit, *Gartenbauwissenschaft*, 61, 91, 1996.

66. Prusky, D., Shalom, Y., Kobiler, I., Akerman, M., and Fuchs, Y., The level of quiescent infection of *Alternaria alternata* in mango fruits at harvest determines the postharvest treatment applied for the control of rots during storage, *Postharvest Biol. Technol.*, 25, 339, 2002.

67. Hofman, P.J., Stubbings, B.A., Adkins, M.F., Meiburg, G.F., and Woolf, A.B., Hot water treatments improve "Hass" avocado fruit quality after cold disinfestation, *Postharvest Biol. Technol.*, 24, 183, 2002.

68. Sapers, G.M., Efficacy of washing and sanitizing methods, *Food Technol. Biotechnol.*, 39, 305, 2001.

69. Saltveit, M., Physical and physiological changes in minimally processed fruits and vegetables, in *Phytochemistry of Fruit and Vegetables*, Tomas'-Barberan, F.A. and Robins, R.J., Eds., Clarendon Press, Oxford, 1997, p. 205.

70. Cantwell, M.I., Hong, G., and Suslow, T.V., Heat treatments control extension growth and enhance microbial disinfection of minimally processed green onions, *HortSci.*, 36, 732, 2001.

71. Park, W.P., Cho, S.H., and Lee, D.S., Effect of minimal processing operations on the quality of garlic, green onion, soybean sprouts and watercress, *J. Sci. Food Agric.*, 77, 282, 1998.

72. Fleischman, G.J., Bator, C., Merker, R., and Keller, S.E., Hot water immersion to eliminate *Escherichia coli* O157:H7 on the surface of whole apples: thermal effects and efficacy, *J. Food Prot.*, 64, 451, 2001.

73. Enomoto, K., Takizawa, T., Ishikawa, N., and Suzuki, T., Hot-water treatments for disinfecting alfalfa seeds inoculated with *Escherichia coli* ATCC 25922, *Food Sci. Technol. Res.*, 8, 247, 2002.

74. Li, Y., Brackett, R.E., Chen, J., and Beuchat, L.R., Mild heat treatment of lettuce enhances growth of *Listeria monocytogenes* during subsequent storage at 5°C or 15°C, *J. Appl. Microbiol.*, 92, 269, 2002.

75. Forney, C.F., Hot-water dips extend the shelf life of fresh broccoli, *HortSci.*, 30, 1054, 1995.

76. Smith, K.J. and Lay-Yee, M., Response of "Royal Gala" apples to hot water treatment for insect control, *Postharvest. Biol. Technol.*, 19, 111, 2000.

77. McGuire, R.G., Market quality of guavas after hot water quarantine treatments and application of carnauba wax coating, *HortSci.*, 32, 271, 1997.

78. Lay-Yee, M. and Rose, K.L., Quality of "Fantasia" nectarines following forced-air heat treatments for insect disinfestations, *HortSci.*, 29, 663, 1994.

79. Lay-Yee, M., Clare, G.K., Petry, R.J., Fullerton, R.A., and Gunson, A., Quality and disease incidence of "Waimanalo Solo" papaya following forced-air heat treatments, *HortSci.*, 33, 878, 1998.

80. Lee, H.H., Hong, S.I., Han, Y.S., and Kim, D., Effect of hot water treatment on biochemical changes in minimally processed onion, *Food Sci. Biotechnol.*, 12, 445, 2003.

81. Forney, C.F. and Jordan, M.A., Induction of volatile compounds in broccoli by postharvest hot-water dips, *J. Agric. Food Chem.*, 46, 5295, 1998.

82. Obenland, D.M., Fouse, D.C., Aung, L.H., and Houck, L.G., Release of d-limonene from non-injured and injured lemons treated with hot water and low temperature, *J. Hort. Sci.*, 71, 389, 1996.

83. Woolf, A.B. and Laing, W.A., Avocado fruit skin fluorescence following hot water treatments and pretreatments, *J. Am. Soc. Hort. Sci.*, 121, 147, 1996.

84. Obenland, D.M. and Aung, L.H., Sodium chloride reduces damage to nectarines caused by hot water treatments, *Postharvest. Biol. Technol.*, 12, 15, 1997.

85. Roebroeck, E.J.A, Jansen, M.J.W., and Mes, J.J., A mathematical model describing the combined effect of exposure time and temperature of hot-water treatments on survival of gladiolus corms, *Ann. Appl. Biol.*, 119, 89, 1991.

86. Trapero-Casas, A. and Kaiser, J.W., Influence of temperature, wetness period, plant age, and inoculum concentration on infection and development of ascochyta blight of chickpea. *Phytopathology*, 82, 589, 1992.

87. Paull, R.E. and Chen, N.J., Heat treatment prevents postharvest geotropic curvature of asparagus spears (*Asparagus officinalis* L.), *Postharvest Biol. Technol.*, 16, 37, 1999.

88. Paull, R.E. and Chen, N.J., Heat treatment and fruit ripening, *Postharvest Biol. Technol.*, 21, 21, 2000.

89. Plumbley, R.A., Prusky, D., and Kobiler, I., The effect of hot water treatment on the levels of antifungal diene and quiescence of *Colletotrichum gloeosporioides* in avocado fruits, *Plant Pathol.*, 42, 116, 1993.

90. Fallik, E., Ilic, Z., Tuvia-Alkalai, S., Copel, A., and Polevaya, Y., A short hot water rinsing and brushing reduces chilling injury and enhance resistance against *Botrytis cinerea* in fresh harvested tomato, *Adv. Hortic. Sci.*, 16, 3, 2002.

91. Pavolcello, D., Lurie, S., Droby, S., and Porat, R., A hot water treatment induces resistance to *Penicillium digitatum* and promotes the accumulation of heat shock and pathogenesis-related proteins in grapefruit flavedo, Physiol. Plant., 111, 17, 2001.

21 Surface Pasteurization with Hot Water and Steam

Bassam A. Annous and Michael F. Kozempel

CONTENTS

21.1 INTRODUCTION

The demand by consumers for fresh and fresh-cut fruits and vegetables has steadily increased due to nutritious qualities associated with fresh produce and the convenience of ready-to-eat fresh foods. This increased demand has resulted in increased per capita consumption of fresh produce. Although

---

Mention of trade names or commercial products in this chapter is solely for the purpose of providing specific information and does not imply recommendation or endorsement by the U.S. Department of Agriculture.

fresh produce is generally considered safe, it has been implicated in numerous foodborne outbreaks in recent years. The Centers for Disease Control and Prevention reported that foodborne outbreaks associated with fresh produce doubled between the period 1973 to 1987 and 1988 to 1992 [1]. Contamination of fresh produce, often grown on the ground and/or in areas adjacent to animal production, with human pathogens may occur during growth, harvesting, handling, and processing. Conventional washing and sanitizing treatments have limited efficacy in inactivating and/or removing pathogens on the surface of produce. Survival of human pathogens and other bacteria during washing and sanitizing treatments is attributed to their attachment to inaccessible sites on produce surfaces such as within the netting of a cantaloupe [2], infiltration within the stem scar of tomatoes and the calyx region of apples [3,4], and incorporation into biofilms, as seen with apples [3], cantaloupes [2], and leaf surfaces [5,6]. Inadequate decontamination of fresh produce can result in the survival of human pathogens on the surface with the possibility of subsequent transfer of the pathogen from the surface, such as the rind of a cantaloupe or the peel of an orange, to the flesh during fresh-cut processing or juice extraction, respectively. Thus, the safety of fresh and fresh-cut produce in supermarkets and salad bars, as well as the safety of freshly squeezed unpasteurized juices, especially those served in fresh juice bars, is of concern. Although experimental approaches to washing produce, such as vacuum infiltration of sanitizers and application of abrasives during washing, have resulted in greater microbial reductions compared to conventional treatments [7], these new treatments are not capable of adequately inactivating the pathogenic bacteria in their protective attachment states on produce surfaces. Furthermore, inactivation of sanitizing agents by organic material such as soil and debris in the washing solution, prior to contact with microorganisms, may limit their sanitizing effectiveness [8].

An alternative approach to chemical sanitizers is surface pasteurization with steam or hot water. Of all the agents used to sanitize the surface of foods, water is probably the most readily acceptable to the public.

21.2 SURFACE PASTEURIZATION WITH HOT WATER

Unlike chemical sanitizers that only affect the surface of produce, hot water (heated potable city water) washing can inactivate bacteria below the produce surface [8], and thus is potentially more effective than chemical washes [2,8,9]. Hot water immersion provides excellent heat transfer between the produce and the heating medium [10] and can quickly establish a uniform temperature profile on the surface of produce [2,10]. Hot water surface pasteurization has been used to control insects and is the most effective method for destroying microorganisms, including postharvest plant pathogens that cause spoilage (Chapter 20). While surface pasteurization, using hot water or steam, has been shown to be effective in reducing levels of human pathogens on the surface of meat and poultry [11,12] and intact eggs [13], it has only limited use in the fresh and fresh-cut produce industries. Fresh fruits and vegetables investigated

TABLE 21.1
**Effect of Washing Treatment (2 minutes) on Log Reduction[a] in *Escherichia coli*
O157:H7 Cell Concentration Applied to the Skin Region of Apples**

Washing solution	Inoculated control[b] ($\log_{10}$ CFU/g)	Washing temperature	
		25°C ($\log_{10}$ CFU/g)	60°C ($\log_{10}$ CFU/g)
Tap water	6.37	3.71 ± 0.25 AB	4.23 ± 1.24 AB
5% hydrogen peroxide	5.24	3.97 ± 1.20 AB	3.74 ± 0.68 AB
1200 ppm Sanova[c]	5.49	4.38 ± 0.45 AB	4.83 ± 0.75 A
400 ppm chlorine (pH[d] = 6.5)	5.39	3.00 ± 1.23 ABC	4.84 ± 0.15 A
Acidified electrolyzed water	4.65	1.64 ± 0.19 C	4.07 ± 0.37 AB

[a] Log reduction = mean cell population of untreated inoculated control (duplicate samples) minus mean cell population following washing treatment (duplicate samples). Means with no letter in common are significantly different at $p < 0.05$.
[b] Mean populations of untreated inoculated control samples.
[c] Sanova (acidified sodium chlorite) solution was prepared according to the manufacturer's specifications.
[d] The pH of the chlorine solution was adjusted to 6.5 using concentrated hydrochloric acid.

for surface pasteurization include apples, melons, mangoes, lemons, oranges, cucumbers, pears, tomatoes, and alfalfa seeds.

Immersion of apples in hot water or sanitizing solutions (60°C for 2 minutes) resulted in $\geq 4 \log$ CFU/g reductions in *Escherichia coli* O157:H7 populations inoculated on the skin surface (Table 21.1). However, these treatments were not effective in inactivating cells inoculated in inaccessible sites (stem and calyx) of apples (Table 21.2 and Table 21.3) [3,4]. Fleischman *et al.* [14] reported similar results for surface pasteurization of apples using water at 95°C for up to 60 seconds. Hot water immersion of apples can result in heat damage resulting in browning of the skin at temperatures above 60°C and softening of the subsurface flesh above 70 to 80°C [7,15].

Reductions in *Salmonella* Poona populations on cantaloupe surfaces were $\geq 5 \log$ CFU/cm^2 following commercial-scale hot water immersion at 76°C for 3 minutes (Table 21.4) [2]. Also, this hot water commercial-scale treatment maintained the fresh quality and increased the shelf life of this commodity. The use of laboratory-scale hot water or heated hydrogen peroxide treatments (70 or 97°C for 1 minute) to inactivate salmonella cells on cantaloupe rind surface resulted in fresh-cut product with enhanced microbiological qualities [16] and extended shelf life [17]. Hot water immersion (70°C for 2 minutes or 80°C for 1 minute) was shown to be effective in reducing populations of *E. coli* O157:H7 on orange surfaces [18]. Also, hot water (≥ 57°C for 5 minutes) immersion was effective in reducing populations of *S.* Stanley on alfalfa seeds [19].

Hot water treatment of a variety of fruits and vegetables greatly improves their microbiological quality and shelf life, while maintaining their sensory qualities. Over-processing of produce, however, can significantly reduce seed germination and cause thermal injury to apples and to juice extracted from

TABLE 21.2
Effect of Washing Treatments (2 minutes) on Log Reduction[a] in *Escherichia coli* O157:H7 Cell Concentration Applied to the Calyx Region of Apples

Washing solution	Inoculated control[b] ($\log_{10}$ CFU/g)	Washing temperature	
		25°C ($\log_{10}$ CFU/g)	60°C ($\log_{10}$ CFU/g)
Tap water	6.71	0.19 ± 0.18 AB	0.43 ± 0.15 AB
5% hydrogen peroxide	5.64	0.39 ± 0.08 AB	0.80 ± 0.44 AB
1200 ppm Sanova[c]	5.80	0.48 ± 0.09 AB	1.06 ± 0.14 A
400 ppm chlorine (pH[d] = 6.5)	6.11	0.66 ± 0.37 AB	0.95 ± 0.28 A
Acidic electrolyzed water	5.18	$-0.04^{e} \pm 0.20$ B	$-0.09^{e} \pm 0.28$ B

[a] Log reduction = mean cell population of untreated inoculated control (duplicate samples) minus mean cell population following washing treatment (duplicate samples). Means with no letter in common are significantly different at $p < 0.05$.
[b] Mean populations of untreated inoculated control samples.
[c] Sanova (acidified sodium chlorite) solution was prepared according to the manufacturer's specifications.
[d] The pH of the chlorine solution was adjusted to 6.5 using concentrated hydrochloric acid.
[e] Negative numbers indicate no reduction in cell populations was detected following washing treatment.

TABLE 21.3
Effect of Washing Treatment (2 minutes) on Log Reduction[a] in *Escherichia coli* O157:H7 Cell Concentration Applied to the Stem Region of Apples

Washing solution	Inoculated control[b] ($\log_{10}$ CFU/g)	Washing temperature	
		25°C ($\log_{10}$ CFU/g)	60°C ($\log_{10}$ CFU/g)
Tap water	6.37	$-0.10^{c} \pm 0.12$ D	$0.11 \pm 0.0.12$ D
5% hydrogen peroxide	5.50	1.83 ± 0.17 AB	0.96 ± 0.72 BC
1200 ppm Sanova[d]	5.66	2.24 ± 0.68 A	2.04 ± 0.62 AB
400 ppm chlorine (pH[e] = 6.5)	6.53	0.49 ± 0.51 CD	1.56 ± 0.26 ABC
Acidic electrolyzed water	5.19	$-0.20^{c} \pm 0.27$ D	$-0.30^{c} \pm 0.30$ D

[a] Log reduction = mean cell population of untreated inoculated control (duplicate samples) minus mean cell population following washing treatment (duplicate samples). Means with no letter in common are significantly different at $p < 0.05$.
[b] Mean populations of untreated inoculated control samples.
[c] Negative numbers indicate no reduction in cell populations was detected following washing treatment.
[d] Sanova (acidified sodium chlorite) solution was prepared according to the manufacturer's specifications.
[e] The pH of the chlorine solution was adjusted to 6.5 using concentrated hydrochloric acid.

TABLE 21.4
Efficacy of Surface Pasteurization Process Using Hot Water Immersion on *Salmonella* Poona Populations[a] on Inoculated Cantaloupes[b]

	Storage temperature	
Treatment[c]	4°C	20°C
2 h control	3.66 ± 0.43	3.66 ± 0.43
24 h control	3.31 ± 0.16	5.54 ± 0.09
76°C for 3 min	0.10 ± 0.00^d	0.16 ± 0.08^d
Room temperature wash for 3 min	4.23 ± 0.32	5.08 ± 0.20

[a] *S.* Poona populations were selectively isolated on XLT4 agar medium, and reported as log CFU/cm^2 rind.
[b] Data are reported as the mean $\pm$ standard deviation for three separate cantaloupes.
[c] Cantaloupes were dip inoculated with *S.* Poona for 5 min, allowed to air dry under biosafety cabinet for 2 h, and were stored at either room temperature or 4°C for 24 h prior to washing treatments.
[d] Although two of three cantaloupes tested showed no survivors, $0.1 \log CFU/cm^2$ (minimum detection level) was used in place of no survivors for determining the mean and standard deviation.

treated oranges. These adverse effects can be controlled by limiting treatment temperatures and times. Since individual commodities have different thermal tolerances, the hot water immersion treatment should be tailored to each commodity. While the rind of a cantaloupe [2] and the peel of an orange [18] effectively insulate the flesh from thermal damage at temperatures above 70°C, the peel of an apple does not protect the flesh from thermal damage at temperatures above 60°C [7]. Accordingly, the tolerance to hot water immersion over a range of temperatures must be determined for individual commodities at different maturity stages [15].

Following hot water immersion, produce should be rapidly cooled to reduce the risk of heat damage to the commodity [2]. This cooling process must be carefully controlled, for it is known to induce infiltration of the cooling solution, including any possible contaminating microorganisms, into the commodity [20–23]. Therefore, the cooling water to be used for this purpose should be free of human pathogens and spoilage microorganisms. A forced cold air tunnel could be used for rapid cooling of the commodity. The use of sanitizing agents during washing treatments, which includes a hot water wash, is recommended to reduce the microbial load in the washing solution. This prevents possible cross contamination in the washing tank, which could result in internalization during the subsequent cooling treatment.

21.3 SURFACE PASTEURIZATION WITH STEAM

Steam is a gas — gaseous water. Because water vapor molecules are many orders of magnitude smaller (about $2 \times 10^{-4}\,\mu m$) than bacterial cells such as salmonella (4 µm long and 0.7 µm thick), and the mean free path length of water vapor molecules ($0.4\,\mu m$) is smaller than bacterial cells, steam should be able to enter any crevices or pores that bacteria can enter [24]. Steam is a unique fluid for pasteurizing food surfaces. It is sufficiently hot to kill virtually all bacterial vegetative cells on contact. However, much like hot water, treatment with steam may damage heat-sensitive foods like fruits and vegetables.

Much of the research on the use of steam for surface pasteurization has been on meat rather than fruits and vegetables. Of course, the meat-related research can be relevant to fruits and vegetables, but meats, except poultry, are generally more thermally resistant and forgiving than fruits and vegetables. Unfortunately, most of the information found on steam treatment of fruits and vegetables is not in the peer-reviewed literature but on web sites and company brochures.

In 1970 Klose and Bayne [25] experimented with steam to kill bacteria on the surface of chicken. Chicken samples were hung inside a three-necked flask, and steam was introduced under vacuum at 70 to 75°C. They obtained a 3 log reduction of naturally present bacteria with a 2-minute exposure, and a 5 log reduction after 16 minutes. Unfortunately, treatment above 60°C resulted in partial cooking of the outer layers of the samples.

In a follow-up study, Klose *et al.* [26] developed a cylindrical metal vacuum chamber to treat whole chicken carcasses with steam. Reductions of 3 logs of inoculated *S.* Typhimurium were achieved by application of subatmospheric pressure steam at 75°C for 4 minutes. However, "the cooked breast meat was almost twice as tough for steam treated as for controls (5.4 versus 3.0 kg shear) and was similarly judged by a trained taste panel," presumably because the surface was cooked.

Davidson *et al.* [27] used a double-walled steel plate steam chamber to treat whole chicken carcasses and chicken parts with 180 to 200°C steam for 20 seconds. They realized a 1 to 2 log reduction of the aerobic plate count (APC) on whole carcasses and breasts. The kill on legs and wings was 2 logs. They reported "evidence of fat separation in the skin and a lightly cooked appearance of skin and exposed muscles."

Steam has been used commercially as a surface treatment for meats [28]. Nutsch et al. [29] reported that the bacterial reduction in a commercial beef processing plant using atmospheric pressure steam for 6 or 8 seconds was 1.35 logs.

When steam is brought into contact with food surfaces, it displaces the air while compressing a very thin film of air against the food surface. This film of air insulates the food surface against direct contact by the steam. The steam is hot enough to kill bacteria instantly, but to do so it must transfer its thermal energy to the bacterial cell. With a film of air present, the steam cannot contact the bacteria directly and must transfer the energy across the compressed air film to the bacteria. This is a relatively slow process compared to condensation of steam directly onto the bacteria cell walls. The process is so slow, in fact, that the steam will cook the surface before killing the bacteria, which is detrimental to the quality of thin-skinned and heat-sensitive commodities. However, for some thick-skinned fruits and vegetables which are destined for subsequent processing, such as for production of juice or fresh-cuts, this might not be a problem since the thermal injury would not extend into the edible portion of the commodity.

In the following sections, new steam surface pasteurization technologies applicable to fresh produce are described.

21.3.1 THERMOSAFE PROCESS

Thermosafe is a patented [30,31] process of Biosteam Technologies, Inc. that uses condensing steam to kill bacteria on the surface of fruits and vegetables. Steam raises the surface temperature of fruits and vegetables to a preset value for a preset hold time. Chilled water follows the steam treatment to quench cooking. Bacterial reductions of 5 logs or greater can be realized with this process. The resultant product is acceptable for produce destined for further processing. It is not acceptable for the fresh food market because the steam cosmetically degrades the surface.

21.3.1.1 Process Operation

The equipment is relatively inexpensive and mechanically simple. The process consists of a chamber or steam tunnel which is designed to be integrated into a fruit or vegetable process line. The fruit or vegetable enters the chamber, usually on a conveyer, or the produce can be inserted batch mode. Pressurized saturated steam is injected through vents into the chamber to bring the surface temperature up to $74°C$. Surface temperature can be monitored by contacting the surface with a thermocouple or by inserting a thermocouple into the produce 6 mm below the surface. Since this might not be reliable or practical in a continuous operation, surface temperature can also be monitored with a remote infrared sensor. After reaching $74°C$, steam injection continues for a 60-second hold time. The actual time and temperature can be adjusted, but a surface temperature of $84°C$ degrades the organoleptic properties of fruits and vegetables. Following the hold period, chilled water at 2 to $5°C$ quenches the surface for another 60 seconds. The unit comes with its own steam supply and self-contained water system, including chilled water, making it easy to install and operate.

21.3.1.2 Process Effectiveness

Food Safety Net Services Ltd conducted a large-scale validation study of the process [32]. The study concluded that the "process can be effective in reducing the microbial load of *Listeria monocytogenes*, *Salmonella* spp., *E. coli* O157:H7, and more thermoduric *Lactobacillus* spp. on the surface of fruits and vegetables by at least 5 logs and in some cases up to 9 logs." The data on cantaloupes show a 5 log reduction for salmonella and *E. coli*, a 7 log reduction for listeria, and a 4 log reduction for *Lactobacillus*. For oranges the data show almost total kill, 9 logs, for salmonella, listeria, and *E. coli* and an 8 log reduction for *Lactobacillus*. For apples, there was a 5 log reduction for salmonella, listeria, and *E. coli* and a 7 log reduction for *Lactobacillus*. Using a combination of steam and hot air, bacterial reductions in excess of 7 logs were realized on peppers.

21.3.1.3 Product Quality

The quality of the interior portions of treated products was successfully maintained. Sensory evaluations were made on citrus juice that was extracted from fruit heated in the range of 65 to 88°C. The results of triangle tests indicated no significant differences ($p < 0.05$) in flavor between treated and untreated product. Therefore, the Thermosafe process effectively pasteurizes the surface of the tested fruits and vegetables with no significant sensory damage to the interior. These fruits and vegetables are suitable for further processing into processed products such as juice but typically would not be suitable for the fresh food market [33].

21.3.2 University of Bristol Process

The University of Bristol investigated the use of pressurized steam, atmospheric pressure steam, and vacuum steam for reducing the bacterial contamination of meat and fruits and vegetables [34]. The process consists of three stages: (1) noncondensable gases (air) are removed with vacuum, (2) steam is applied to the surface of the produce to reach a pasteurization temperature, and (3) the surface is evaporatively cooled under vacuum to quench cooking. Steam, with its high latent heat of condensation, gives a rapid rise in the surface temperature which minimizes thermal exposure time.

21.3.2.1 Process Operation

The pressure and subatmospheric pressure process systems consist of a steam boiler, a processing chamber, and a vacuum pump. The system operates in batch mode. The pressure process chamber is 1 m by 0.6 m in diameter. The flushing action tends to remove noncondensable gases. The subatmospheric steam chamber is 0.45 m high by 0.3 m in diameter. Steam is injected in the top, and air and condensate are removed from the bottom.

With a chamber pressure of 2.3 bar, product exposure time was a nominal 90 seconds. Initial vacuum time was of the order of 10 minutes, and the evaporative cooling was about 5 minutes. Exposure times in the atmospheric pressure process were 2 to 6 seconds. Times for the subatmospheric steam process were not reported.

21.3.2.2 Process Effectiveness

Although the pressure pilot plant process was applied to peppers and soft fruits, no detailed results were reported. The manufacturer reported 1.2 to 3.4 log reductions in APC of beef treated at temperatures of 100, 120, or 135°C, depending on whether it was lean or fat. In a comparison of various decontamination methods, pressurized steam gave a 1.5 to 5 log reduction in APC on peppers. Reductions in APC for chilled raspberries and blackberries were 3 to 5 logs. The subatmospheric steam unit was used for peppers, apples, and lettuce. Bacterial reductions up to 2 logs were achieved after exposure for 10 seconds at 65°C, and up to 4 logs were achieved following exposure to 80 to 85°C for 40 seconds.

21.3.2.3 Product Quality

There is no published assessment of the quality for the processed produce. Because of the temperature and time conditions of treatment, the authors suspect that the products should be suitable for further processing.

21.3.3 VENTILEX CONTINUOUS STEAM STERILIZING SYSTEM

The Ventilex process uses saturated steam to decontaminate or sterilize herbs, spices, and seeds [35]. The product enters a horizontal steam chamber through a patented rotary valve, designed to prevent buildup of product within the valve. Once in the chamber, the herbs, spices, or seeds are contacted by saturated steam for a given time appropriate to reduce or eliminate pathogenic bacteria. The treated product exits through a second rotary valve. Following steam treatment, the herbs, spices, or seeds are dried and cooled.

21.3.3.1 Process Operation

A description of the process is available at the Ventilex web site [35]. The process is a high-temperature/short-time treatment for herbs, spices, and seeds using saturated steam. Small particle products such as these tend to clump in valves and clog the process. This is prevented by using a continuous scraping action within the valve to dislodge any adhering product.

The treatment chamber is horizontal with a vibrating belt. The belt moves the material through the chamber in plug flow at a set speed to achieve the desired residence time. The frequency of the vibrating belt is variable and governs the flow rate that determines the dwell time for each product. Products are treated with saturated steam over the range 107 to 123°C. Typical treatment times are 25 to 50 seconds depending on the commodity, contamination level, and final use of the product. The treated product drops off the vibrating belt into a second rotary valve using the same scraping action. The herbs, spices, or seeds then go to a fluidized bed dryer/cooler. The condensed steam flashes off, and the herbs, spices, or seeds are dried with indirectly heated sterile air.

21.3.3.2 Process Effectiveness

According to the manufacturer, products treated with this process often have APC counts below 1 log. Salmonella is eliminated. Mold and yeast populations

TABLE 21.5
Effectiveness of the Ventilex Process on Natural Microbial Flora (log CFU/g) of Paprika and Rosemary

Commodity		Aerobic plate count	Aerobic spore formers	Enterobacteriaceae	*Bacillus cereus*	Yeast	Molds
Paprika	Before treatment	6	6	3	4	2	2
	After treatment	3	3	—	<2	—	—
Rosemary	Before treatment	5	5	4	—	3	4
	After treatment	<2	—	—	—	—	—

Note: —, not determined.
From van Gelder, A., Personal communication, 2003.

are below 2 logs. Spores of *Bacillus cereus*, *Clostridium perfringens*, and *Staphylococcus aureus* have counts below 2 log. Specific results for paprika and rosemary are shown in Table 21.5 [36]. There is a 3 log or better reduction in total APC and aerobic spore formers. Yeasts and molds are essentially eliminated.

21.3.3.3   Product Quality

There is little or no product degradation. For example, the color value for paprika differed from the control by only 2 ASTA units after treatment (reduced from 95 to 93). Volatile oils for rosemary did not change (0.7 ml/100 g). In addition, the process inactivates enzymes.

21.3.4   VACUUM–STEAM–VACUUM (VSV) PROCESS

Thermal damage is a major problem for steam surface-pasteurized fruits and vegetables destined for the fresh market. Conventional wisdom seems to dictate that if the steam exposure time is sufficient to kill the bacteria, the produce is thermally damaged. The treated produce may be suitable for the processed fruit or vegetable market but not for the fresh market. If fresh quality is to be maintained by using a shorter exposure time, the bacterial population will not be sufficiently reduced. One solution to this problem is the U.S. Department of Agriculture's (USDA) novel VSV process [37].

To circumvent the problem of thermal damage, the film of air and moisture on the commodity surface is removed so that steam can rapidly contact the bacteria directly. It is a simple concept but difficult to achieve in practice. One approach was the concept proposed by Morgan *et al.* [24,38,39]. In this method, the food is exposed to vacuum to remove air and moisture. Next, saturated steam is applied to the surface. When the saturated steam contacts the product, it condenses to form a water film on the fruit or vegetable surface which impedes further bacteria reduction. Therefore, the food is exposed to a vacuum again to remove the condensate and to evaporatively cool the surface. Kozempel *et al.* [40] showed that cycling between vacuum and steam to remove the condensate enhanced the population reduction of *Listeria innocua* on hot dogs. This concept of alternating vacuum and steam is the basis of the VSV process.

Initial research used a stainless steel device consisting of a rotor and stator. The 150 mm long and 150 mm in diameter [24,38] rotor was turned rapidly around its horizontal axis, stopping at precisely determined angular positions, exposing the sample alternately to vacuum or steam. A 25 mm × 75 mm × 75 mm deep treatment chamber was milled into the surface of the rotor.

The treatment consisted of four steps: (1) air was removed by exposure to vacuum; (2) the sample was flushed with low-temperature saturated steam (this flush was later abandoned); (3) the sample was exposed to pressurized saturated steam; and (4) the sample was evaporatively cooled with vacuum. Bacterial reductions on chicken meat inoculated with nonpathogenic *L. innocua* were about 2 to 2.5 logs. Steam exposure time was 0.1 to 0.2 seconds [24,38].

This prototype proved the concept, but was not practical with actual fruits and vegetables such as cantaloupes. For mechanical reasons it was preferable to move the machinery and not the food sample. Therefore, a new prototype pilot plant unit was designed and fabricated. The surface intervention processor was designed to process chicken carcasses, specifically broilers. However, the design is also suitable for many fruits and vegetables, especially cantaloupes. The performance requirements of a surface intervention processor are to accept the individual food sample and enclose it in a chamber within a rotor; to evacuate that chamber; to pressurize the chamber with steam; to vacuum cool it; and, finally, to eject the sample into a clean environment. The simplest execution of this prototype, one chamber in one rotor, was designed and constructed [41]. Figure 21.1 shows the processor, and Figure 21.2 shows details of the product treatment section. The chamber is cylindrical, about 200 mm in diameter and 240 mm deep, and is provided with an 8-inch ball valve.

To admit vacuum or steam into the closed chamber, two opposing 200 mm holes were bored through the stator at right angles to both the axis of rotation of the ball and to the centerline of the open chamber. Two platter valves, consisting of a flat disk rotating against an inlet header that holds polyetheretherketone (PEEK) seals, were close-coupled to the 200 mm ports. Each disk contained two holes, which when stopped at one of the ports in the inlet header permitted steam flow into or vacuum evacuation from the treatment chamber. Multiple holes reduced the rotor angular movement necessary for valve action and increased the cross-sectional area for gas flow. Each disk was programmed independently and moved by its own servomotor. To expose all

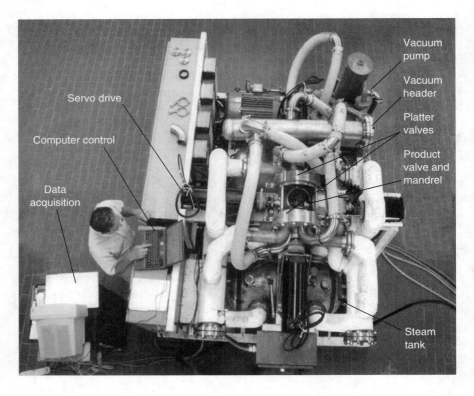

FIGURE 21.1 Vacuum–steam–vacuum processor.

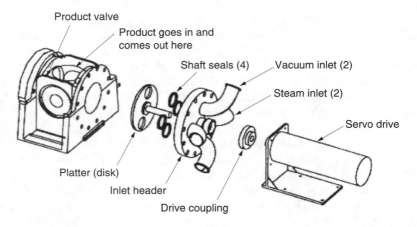

FIGURE 21.2 Schematic of the product treatment section of the Vacuum–steam–vacuum processor.

exterior surfaces of the test sample to treatment, a screen was installed at the midpoint of the treatment chamber to hold the sample.

21.3.4.1 Process Operation

Each sample of fruit or vegetable is manually inserted into the treatment chamber of the VSV processor. A computer-controlled servomotor is used to rotate the ball valve 90° to seal the chamber from the atmosphere. The platter valves rotate to expose alternately the sample to vacuum, then steam, and then vacuum again. With multiple cycles, the sequence of vacuum, then steam, is repeated multiple times. After treatment, the ball valve rotates back 90° to expose the sample to the atmosphere. After treatment, the fruit or vegetable sample is removed manually with sterile gloves.

21.3.4.2 Process Effectiveness

Three different kinds of produce (uninoculated) were processed to assess thermal damage and bacterial reduction. The commodities were chosen to represent aerial fruits (grapefruits), fruits growing on the ground (cantaloupes), and vegetables that grow in the ground (carrots). Table 21.6 summarizes the results. There was no visual thermal damage, and the bacterial population reductions were 3.4 to >5 logs, but these process conditions were not optimized. The optimum conditions give maximum bacteria reduction with minimal or no thermal damage to the product. Therefore, a series of optimization experiments were conducted to determine the best processing conditions [41]. Beets were substituted for the in-ground crop because the shape of the treatment chamber was more amenable to spherical foods and tended to chop off the ends of carrots. (A VSV processor for carrots or other cylindrical crops would require a differently shaped treatment chamber.) Papayas were added to the list of products tested. Table 21.7 lists the optimum process condition and bacterial reduction for cantaloupes, grapefruits, papayas, and beets. Because of the high natural bacteria count on beets, they were not inoculated; cantaloupes, grapefruits, and papayas were inoculated with *Listeria innocua*.

TABLE 21.6
Initial Results for the VSV Intervention Process for Uninoculated Produce

Commodity	Steam temperature (°C)	Control	Population reduction
Carrots	130	5.6	>5.0
Grapefruits	130	3.6	3.6
Cantaloupes	138	5.6	3.4

Note: Vacuum time $= 0.25$ sec, and steam time $= 0.25$ sec.

TABLE 21.7
Optimization Results for the VSV Surface Intervention Process on Inoculated and Uninoculated Produce

Commodity	Steam temp. ($°C$)	Steam time (sec)	Vacuum time (sec)	Number of cycles	Bacterial reduction (log CFU/ml)
Cantaloupes	143	0.1	0.1	2	3.4 L. innocua
Grapefruits	138	0.1	0.1	2	3.6 L. innocua
Papayas	138	0.2	0.1	2	3.6 L. innocua
Beets	143	0.2	0.1	3	2.5 aerobic plate count

TABLE 21.8
Application of the VSV Surface Intervention Process to Other Fruits and Vegetables

Commodity	Bacteria	Steam temp ($°C$)	Steam time per cycle (sec)	No. of cycles	Control (log CFU/ml)	Treated (log CFU/ml)	Bacterial reduction (log CFU/ml)
Mangoes	L. innocua	138	0.1	2	5.4	1.4	4.0
Avocados	L. innocua	138	0.1	2	4.1	1.0	3.1
Kiwis	L. innocua	138	0.1	3	6.4	1.6	4.8
Bananas	—[a]	104	0.1	1			Mutilated
Carrots	Aerobic plate count	138	0.1	3	5.7	1.6	4.1
Cucumbers	Aerobic plate count	138	0.1	3	5.4	1.6	3.8
Peaches	L. innocua	138	0.1	2	5.0	1.4	3.6
Cauliflower	—[a]	127	0.1	1			Color change
Broccoli	—[a]	116	0.1	1			Color change
Peppers	—[a]	116	0.1	1			Mutilated

Note: Vacuum times $= 0.1$ sec.
[a] No microbiology analyses were performed on products that were thermally damaged.

Several tropical fruits were tested at the general optimum conditions of $138°C$ steam for 0.1 seconds using one, two, or three cycles and a vacuum time of 0.1 seconds. The results for kiwis, mangoes, and avocados are listed in Table 21.8. All samples were inoculated with *L. innocua* for 10 minutes and dried under ambient conditions for 1 hour. The log reduction for kiwis was 4.8; for mangoes, 4.0; and for avocados, 3.1.

Another tropical fruit, banana, was tested. However, the process caused the peel to split and the fruits to darken immediately. Milder conditions ($104°C$ for 0.1 seconds and one cycle) were tried with green bananas, but the samples still were destroyed. No microbiological analyses were performed on products that were thermally damaged.

Other products processed without success were peppers, broccoli, and cauliflower. When subjected to vacuum, the peppers exploded. Upon exposure to steam the delicate florets on broccoli turned a bright green indicative of blanching or heat treatment. Although the flower part of cauliflower was essentially unscathed, the stalk and the remnants of the leaves turned bright green as in blanching.

Other fruits and vegetables were tested at the conditions stated above. The results are listed in Table 21.8. The bacterial reduction (APC) on uninoculated carrots was $4.1 \log CFU/ml$. Treatment of cucumbers, inoculated for 10 minutes with *L. innocua* and dried at ambient conditions for 1 hour, resulted in $3.8 \log CFU/ml$ reduction with three cycles. Peaches were inoculated for 10 minutes with *L. innocua* and were allowed to dry for 1 hour under ambient conditions. Using two cycles, the reduction for *L. innocua* was $3.6 \log CFU/ml$ with no thermal damage.

In addition to bacteria, some insects such as red scale infest the surface of fruits. Red scale is a major problem on citrus fruits. Currently, methyl bromide is used to eliminate insects such as red scale, but the impending loss of methyl bromide in 2005 requires alternative methods of quarantine treatment for disinfestations of produce imported or exported each year.

The VSV process was used to process lemons infested with red scale [42]. No scale insects survived the process. The process resulted in 100% kill of insects at all stages of development. As a bonus, up to 96% of first molt scales were physically removed, but the process was much less effective in removing other stages from the fruit, especially those that had advanced beyond the second instar. However, the process was completely effective in killing the scales.

21.3.4.3 Product Quality

To date, evaluation of thermal damage has been only qualitative. Except for bananas, broccoli, and peppers that were not amenable to this process, there was no thermal damage observed. Most of the treated produce samples (uninoculated) were consumed and found to be indistinguishable from the untreated controls.

21.4 CONCLUSIONS

Steam and hot water surface pasteurization are both promising technologies that are capable of achieving more than 5 log reductions in target pathogens as well as greatly reducing populations of spoilage microorganisms on the surface of fruits and vegetables. However, hot water immersion treatment of fresh produce appears to be a gentler process and has better control over the surface temperature of produce during treatment as compared to steam treatment. Steam processes are acceptable treatments for produce intended for further processing due to thermal damage of the produce surface. The VSV process produces good results with a number of commodities with bacterial reductions up to $4.8 \log CFU/ml$ (Table 21.8), depending on the fruit or the vegetable. The VSV is a rapid process requiring less than 2 seconds for treatment and with little or no thermal damage.

Even though highly promising, surface pasteurization technology is in need of further research to determine thermal penetration profiles and heat sensitivity at different temperatures for individual commodities. There is also a need to obtain thermal inactivation data for human pathogens of concern, attached to surfaces of commodities that have subsurface sites (e.g., pores) and other sites providing protection as well as exposed sites. Furthermore, research is needed to determine the temperature–time effect of surface pasteurization on sensory qualities, storability, and processability of fresh produce at different maturity stages. Results from such research would enable the development of cheap, safe, and environmentally sound disinfection treatments for controlling pathogens and/or spoilage microorganisms on fresh produce.

REFERENCES

1. Buck, J.W., Walcott, R., and Beuchat, L.R., Recent trends in microbiological safety of fruits and vegetables, http://www.apsnet.org/online/feature/safety/, 2003.
2. Annous, B.A, Burke, A., and Sites, J.E., Surface pasteurization of whole fresh cantaloupes inoculated with *Salmonella* Poona or *Escherichia coli*, *J. Food Prot.*, 67, 1876, 2004.
3. Annous, B.A. *et al.*, Efficacy of washing with a commercial flatbed brush washer, using conventional and experimental washing agents, in reducing populations of *Escherichia coli* on artificially inoculated apples, *J. Food Prot.*, 64, 159, 2001.
4. Annous, B.A. and Burke, A., Unpublished data, 2001.
5. Carmichael, I. *et al.*, Bacterial colonization and biofilm development on minimally processed vegetables, *J. Appl. Microbiol. Symp. Suppl.*, 85, 45S, 1999.
6. Fett, W.F., Naturally occurring biofilms on alfalfa and other types of sprouts. *J. Food Prot.* 63, 625, 2000.
7. Sapers, G.M. *et al.*, Improved antimicrobial wash treatments for decontamination of apples, *J. Food Sci.*, 67, 1886, 2002.
8. Breidt, E., Hayes, J.S., and Fleming, H.P., Reduction of microflora of whole pickling cucumbers by blanching, *J. Food Sci.*, 65, 1354, 2000.
9. Lichter, A. *et al.*, Hot water brushing: an alternative method to SO_2 fumigation for color retention of litchi fruits, *Postharvest Biol. Technol.*, 18, 235, 2000.
10. Couey, H.M., Heat treatment for control of postharvest diseases and insect pests of fruits, *HortScience*, 24, 198, 1989.
11. Purnell, G., Mattick, K., and Humphrey, T., The use of "hot wash" treatments to reduce the number of pathogenic and spoilage bacteria on raw retail poultry, *J. Food Eng.*, 62, 29, 2004.
12. Whyte, P., McGill, K., and Collins, J.D., An assessment of steam pasteurization and hot water immersion treatments for the microbiological decontamination of broiler carcasses, *Food Microbiol.*, 20, 111, 2003.

13. Schuman, J.D. *et al.*, Immersion heat treatments for inactivation of *Salmonella enteritidis* with intact eggs, *J. Appl. Microbiol.*, 83, 438, 1997.
14. Fleischman, G.J. *et al.*, Hot water immersion to eliminate *Escherichia coli* O157:H7 on the surface of whole apples: thermal effects and efficacy, *J. Food Prot.*, 64, 451, 2001.
15. Lurie, S. *et al.*, Postharvest heat treatment of apples to control San Jose scale (*Quadraspidiotus perniciosus* Comstock) and blue mold (*Penicillium expansum* Link) and maintain fruit firmness, *J. Am. Soc. Hort. Sci.*, 123, 110, 1998.
16. Ukuku, D., Pilizota, V., and Sapers, G.M., Effect of hot water and hydrogen peroxide treatments on survival of *Salmonella* and microbial quality of whole and fresh-cut cantaloupe, *J. Food Prot.*, 67, 432, 2004.
17. Sapers, G.M. *et al.*, Antimicrobial treatments for minimally processed cantaloupe melon, *J. Food Sci.*, 66, 345, 2001.
18. Pao, S. and Davis, C.L., Enhancing microbiological safety of fresh orange juice by fruit immersion in hot water and chemical sanitizers, *J. Food Prot.*, 62, 756, 1999.
19. Jaquette, C.B., Beuchat, L.R., and Mahon, B.E., Efficacy of chlorine and heat treatment inn killing *Salmonella stanley* inoculated onto alfalfa seeds and growth and survival of the pathogen during sprouting and storage, *Appl. Environ. Microbiol.*, 62, 2212, 1996.
20. Bartz, J.A., Infiltration of tomatoes immersed at different temperatures to different depths in suspensions of *Erwinia cartovora* subsp. *cartovora*, *Plant Dis.*, 66, 302, 1982.
21. Buchanan, R.L. *et al.*, Contamination of intact apples after immersion in an aqueous environment containing *Esherichia coli* O157:H7, *J. Food Prot.*, 62, 444, 1999.
22. Eblen, B.S. *et al.*, Potential for internalization, growth, and survival of *Salmonella* and *Escherichia coli* O157:H7 in oranges, *J. Food Prot.*, 67, 1578, 2004.
23. Zhuang, R.-Y., Beuchat, L.R., and Abgulo, F.J., Fate of *Salmonella Montevideo* on and in raw tomatoes as affected by temperature and treatment with chlorine, *Appl. Environ. Microbiol.*, 61, 2127, 1999.
24. Morgan, A.I. *et al.*, Surface pasteurization of raw poultry meat by steam, *Lebensmittel Wissenschaft und Technologie*, 29, 447, 1996.
25. Klose, A.A. and Bayne, H.G., Experimental approaches to poultry meat surface pasteurization by condensing vapors, *Poultry Sci.*, 49, 504, 1970.
26. Klose, A.A. *et al.*, Pasteurization of poultry meat by steam under reduced pressure, *Poultry Sci.*, 50, 1156, 1971.
27. Davidson, C.M., D'Aoust, J.Y., and Allewell, W., Steam decontamination of whole and cut-up raw chicken, *Poultry Sci.*, 64, 765, 1985.
28. Wilson, R.C. *et al.*, Apparatus for Steam Pasteurization of Food, U.S. Patent 6,019,033, Feb. 1, 2000.
29. Nutsch, A.L. *et al.*, Evaluation of a steam pasteurizing process in a commercial beef processing facility, *J. Food Prot.*, 60, 485, 1997.
30. Tottenham, D.E. and Purser, D.E., Apparatus and Method for Food Surface Microbial Intervention and Pasteurization, U.S. Patent 6,153,240, Nov. 28, 2000.
31. Tottenham, D.E. and Purser, D.E., Apparatus and Method for Food Microbial Intervention and Pasteurization, U.S. Patent 6,350,482, Feb. 26, 2002.
32. Biosteam Technologies, Inc., www.biosteamtech.com.

33. Pao, S. and Kelsey, F., Hot Water and Steam Application as a Means for Surface Microbial Reduction of Citrus Fruit, Factsheet 6, Fresh Citrus Juice, www.fdocitrus.com/factsheet_6.htm.

34. Evans, J., Ed., Newsletter, Decontamination Issue, Food Refrigeration and Process Engineering Research Centre, University of Bristol, Langford, Bristol, U.K., April 1999.

35. Anon., Decontamination of Spices, Herbs, Seeds, and Other Organic Materials, Ventilex USA, http://www.ventilex.net/Steam%20Sterilization.htm.

36. van Gelder, A., Personal communication, 2003.

37. Kozempel, M.F., Goldberg, N., and Craig, J.C., Jr., Development of a new process to reduce bacteria on solid foods without thermal damage: the VSV (Vacuum/Steam/Vacuum) process, *Food Technol.*, 57, 30, 2003.

38. Morgan, A.I., Radewonuk, E.R., and Scullen O.J., Ultra high temperature, ultra short time surface intervention of meat, *J. Food Sci.*, 61, 1216, 1996.

39. Morgan, A.I., Method and Apparatus for Treating and Packaging Raw Meat, U.S. Patent 5,281,428, 1994.

40. Kozempel, M. *et al.*, Rapid hot dog surface pasteurization using cycles of vacuum and steam to kill *Listeria innocua*, *J. Food Prot.*, 63, 17, 2000.

41. Kozempel, M. *et al.*, Optimization and application of the vacuum/steam/ vacuum surface intervention process to fruits and vegetables, *Innov. Food Sci. Emerg. Technol.*, 3, 63, 2002.

42. Fuester, R.W. *et al.*, A novel non-chemical method for quarantine treatment of fruits: California red scale on citrus, *J. Econ. Entomol.*, 97, 1861, 2004.

22 Novel Nonthermal Treatments

Dongsheng Guan and Dallas G. Hoover

CONTENTS

22.1 INTRODUCTION

22.1.1 NONTHERMAL PROCESSING METHODS

In recent years there has been a growing and sustained interest in the group of food processing methods commonly referred to as the nonthermal processing technologies. These processes usually employ a somewhat novel application of energy to reduce or eliminate problematic microorganisms without the generation of high levels of heat normally found in conventional thermal processing. Ideally with limited exposure to high temperatures, the food possesses sensory qualities and nutrient content more closely resembling the raw, fresh, or minimally processed counterpart while retaining the desired shelf life and safety.

22.1.2 ADVANTAGES AND DISADVANTAGES OF APPLICATION

In the thermal processing of foods, heat inactivates both microorganisms and enzymes to extend the shelf life of the treated foods. With application and incorporation of nonthermal processing methods the same microorganisms and enzymes in the food are usually targeted with the same expectation of inactivation or control; however, the actual mechanisms of microbial inactivation and protein denaturation are usually different when comparing thermal processing to nonthermal processing methods. Different mechanisms of inactivation mean the rates of inactivation are different and the degree of effectiveness is also usually different. Consequently, the usual primary goal in adapting the nonthermal methods for commercial use is to maintain the same high degree of safety enjoyed with thermal processing while minimizing changes to the desirable sensory qualities and nutrition in the product. That can be an immense challenge depending on the food product. Each nonthermal process is somewhat distinct with its own set of limiting factors that can include an inability to denature browning enzymes, limited or nonexistent inactivation of viruses or bacterial endospores, regulatory hurdles, high costs for equipment and maintenance, and a lack of background information usually provided by past industrial experience. Examples of areas of insufficient information for these new technologies include standardization of industrial process procedures, and methods and surrogate identification for process validations. These factors make the commercialization of new food products utilizing nonthermal processes complex and daunting.

With the exploding demand for fresh fruits and vegetables by consumers in North America comes the necessary importation of produce from areas of the world where crops can be grown and harvested all year round. For example, the Produce Marketing Association stated that imports of fresh produce increased from 13.8 billion pounds in 1993 to 20.2 billion pounds in 2000. The quality of fruit in winter is now nearly equivalent to fruit sold in the warmer months due to significant improvements in storage, transportation, and distribution. Unfortunately, the incidence of foodborne illness also coincides with this increased bounty of fresh fruits and vegetables (both domestic and imported). For example, in the marketplace the most dangerous food related to acute illness is not derived from meat, milk, eggs, or seafood, but is a plant product: sprouts [1]. In fact, while meat is the most regulated and monitored food commodity, in comparison fruits and vegetables receive only a fraction of government attention. Federal health surveillance of foodborne diseases from 1993 to 1997 documented 2,751 outbreaks that involved 12,537 individual cases of foodborne illness related to contaminated fruits and vegetables, compared with 6,709 cases involving meat products. Outbreaks linked to fruits and vegetables are often the result of fecal contamination caused by inadequate hygiene during production and harvest. Imported fruits and vegetables now appear to harbor exotic and emerging parasites once unknown in North America, and traditional pathogens, such as salmonellae, hemorrhagic *Escherichia coli*, and *Listeria monocytogenes*, can be anticipated in both domestic and imported produce as a consequence of poor agricultural practices involving the composting and distribution of manure. Landmark outbreaks involving cyclospora in Guatemalan raspberries, salmonella-contaminated sprouts, *E. coli* in fresh cider, and Mexican scallions tainted with hepatitis A have become well-known reference points to American consumers who worry about the safety of the foods they eat.

The aim of this chapter is to furnish an overview of the use of nonthermal processing technologies applied for the preservation of fruits, vegetables, and their byproducts, most notably juices. The nonthermal processing methods that are covered in this chapter include high hydrostatic pressure processing, and applications of ionizing irradiation, high-intensity pulsed electric fields, ultrasonic waves, and electrolyzed oxidizing water.

22.2 HIGH HYDROSTATIC PRESSURE PROCESSING (HPP)

22.2.1 INTRODUCTION

22.2.1.1 Definition and Historical Perspective

Historically, applications of HPP for the preservation of foods were first conducted by Bert Hite who adapted high-pressure processing to a variety of foods and beverages in the late 1890s and early 20th century [2,3]. From Hite's work until the 1980s, only scattered attempts were made to investigate the potential commercial application of HPP to foods. With the sustained demand for high-quality foods that are minimally processed and additive-free in developed countries, HPP has attracted interest from research institutions, food companies, and regulatory agencies over the last two decades in the pursuit of producing better quality foods more economically.

HPP subjects foods to pressures between 100 and 800 MPa with exposure times ranging from a millisecond pulse to over 20 minutes, although most commercial treatments times are 7 minutes or preferably less [4,5]. The temperatures of products and media during pressure processing can be below $0°C$ or above $100°C$, depending on the product requirements; however, current commercial HPP uses ambient temperatures.

The first commercialized food product employing pressure preservation was fruit preserves marketed in Japan in 1991. These pressure-treated jams and jellies continue to be sold in Japan in addition to salad dressings and a wide range of fruit juices. Pressure-treated guacamole successfully entered the U.S. marketplace in 2001, followed by HPP salsa. These products have been available nationwide, but pressurized guacamole and salsa remain most popular in the southwestern region of the U.S. In 2004 it is anticipated that pressure-processed chopped onions will be sold as an ingredient in premium salad dressings. This onion product will be available in fresh chopped form months after the normal season ends. They will be available in 8 oz stand-up, resealable bags and as 2.5 lb pouches for club and superstores. The intended refrigerated shelf life is 45 days, but 90 days of storage has been demonstrated without detrimental quality changes. Pressure-processed chopped onions are said to have a "sweeter taste that isn't bitter, with a fresher, crunchier texture." Also anticipated in 2004 from a Canadian venture is applesauce and applesauce/fruit blends packaged as eat-on-the-go single-serve flexible tubes: "Our Way of Preserving Nature," and from Mexico, fruit "smoothie" products for North American distribution. This company produces juices, nectars, and bottled drinks.

22.2.1.2 Equipment

Typical HPP equipment usually consists of a pressurized vessel, two end closures at each end of the vessel, a low-pressure pump, an intensifier to generate higher pressures, and system controls. HPP systems can be designed to treat either unpackaged liquid foods semicontinuously or packaged foods in a batch manner. A schematic of a batch HPP system is presented in Figure 22.1.

22.2.1.3 Critical Processing Factors

Critical process factors in HPP include, but are not limited to, treatment pressure, holding time at pressure, come-up time to achieve treatment

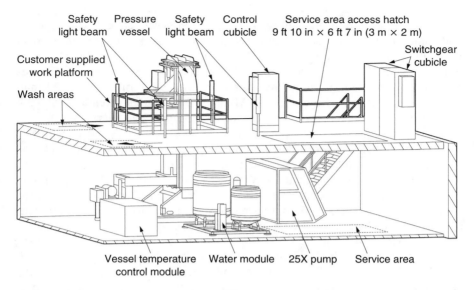

FIGURE 22.1 Fresher Under Pressure™ high-pressure processing systems (215 L) from Avure Technologies Inc., a wholly owned subsidiary of Flow International Corporation.

pressure, decompression time, initial temperature of food products, treatment temperature, the temperature distribution in the vessel at pressure as a result of adiabatic heating, product properties (e.g., pH, composition and water activity, food compressibility), packaging material, and the microbiota of the food [6]. Package size and shape are not critical factors in process determination because pressure acts instantaneously and uniformly throughout the chamber and the food mass. For pulsed pressure processing, additional process factors are pulse shape (i.e., the waveform), frequency, and pulse pressure magnitudes.

22.2.2 INACTIVATION OF PROBLEMATIC MICROORGANISMS

22.2.2.1 Spores and Vegetative Bacteria

Most fungal conidiospores and ascospores can usually be inactivated at pressures between 300 and 450 MPa at ambient temperature, but exceptions exist. For example, in a study on dormant *Talaromyces macrosporus* ascospores, mild treatment (200 to 500 MPa, 20°C) activated dormant ascospores but caused little or no inactivation. Higher pressures (500 to 700 MPa, 20°C) were required to inactivate the ascospores; however, a treatment of 700 MPa after 60 minutes only reduced the spore population by less than $2 \log_{10}$ units, indicating the resistance of the ascospores to high pressure [7].

Hayashi [8] found that pressures of 200 MPa effectively killed yeasts and molds in freshly squeezed orange juice at ambient temperature. Ogawa *et al.* [9] stated that after HPP (400 MPa and 23°C) both freshly squeezed orange juice and orange juice that had been inoculated with yeasts and molds showed no increase in total counts after 17 months of storage at 4°C. Aleman *et al.* [10] reported treatment at 340 MPa for 15 minutes extended the shelf life of fresh-cut pineapple. Populations of Gram-negative bacteria, yeasts, and molds could be reduced by at least $1 \log_{10}$ at pressures of 300 and 350 MPa for inoculated lettuce and tomatoes, even though the tomato skins loosened and peeled away and lettuce browned in this range of pressures [11]. Raso *et al.* [12] used pressure to inactivate ascospores and vegetative cells of *Zygosaccharomyces bailii* suspended in apple, orange, pineapple, cranberry, and grape juices. HPP at 300 MPa for 5 minutes reduced the population of vegetative cells and ascospores by almost $5 \log_{10}$ units and 0.5 to $1 \log_{10}$ units, respectively.

Parish [13] applied pressures between 500 and 350 MPa to nonpasteurized Hamlin orange juice (pH 3.7) inoculated with *Saccharomyces cerevisiae* and found that the D values for ascospores and vegetative cells inoculated into the pasteurized orange juice were from 4 to 76 sec and from 1 to 38 sec, respectively. The native microbiota in the orange juice had D values from 3 to 74 sec in the pressure range of 500 and 350 MPa. The corresponding z values were 123, 106 and 103 MPa for ascospores, vegetative cells, and native microbiota, respectively.

Zook *et al.* [14] used fruit juices and a model juice buffer (pH 3.5 to 5.0) as suspension media to determine pressure inactivation kinetics of *S. cerevisiae* ascospores. Approximately 0.5×10^6 to 1.0×10^6 ascospores/ml were pressurized at 300 to 500 MPa in juice or buffer. D values were 8 sec to 10.8 min at 500 and 300 MPa, respectively; the corresponding z values were 115 and 121 MPa. No differences ($P \geq 0.05$) in D values (at constant pressure) or z values among buffers or juices at any pH were determined, suggesting little influence of pH in this range.

Bacterial endospores are the most difficult life forms to eliminate with hydrostatic pressure; spores of bacillus have been exposed to $> 1,724$ MPa (250,000 psi) and remained viable [15]. Applications of pressure alone will not inactivate bacterial endospores. Hurdle technology that utilizes pressure in combination with other process technologies (including pressure pulsing) are proposed to improve spore inactivation rates. Mild elevated heat (e.g., 40 to 55°C) with pressure treatment is required for substantial reduction of spore loads [16,17]. Sterilization requires higher temperatures resulting in a definite cooked appearance of the food.

Bacterial spores have been shown to demonstrate variable pressure resistances with respect to sporulation conditions. The anhydrous structure and dimensions of the spore are believed to contribute to the pressure resistance of bacterial spores, causing a major challenge to produce shelf-stable low-acid food products [18]. Spores can germinate at different combinations of temperature and pressure [19]. The initiation of spore germination results in loss of resistance. Two-exposure treatments (i.e., twin pressure pulse) have been proposed to enhance the inactivation of spores by HPP [20]. The concept is that the first exposure at low pressure results in spore germination, and the second exposure at a higher pressure inactivates the germinated spores and vegetative cells. Unfortunately, it appears that not all spores are germinated by pressure and not all germinated spores appear to be inactivated by pressure [21].

Oh and Moon [22] investigated the effect of pH on the initiation of spore germination and inactivation of *Bacillus cereus* KCTC 1012 spores using pressures to 600 MPa. The pH of the sporulation medium affected inactivation of *B. cereus* spores under pressure more than the suspension medium pH. *B. cereus* spores obtained through sporulation at pH 6.0 showed greater resistance to pressure than those sporulated at pH 7.0 and 8.0 at 20, 40, and 60°C.

To date, successful commercial preservation of foods utilizing HPP depends upon the use of post-treatment refrigeration or a product pH below 4.5 to block the germination of spores of *C. botulinum* and other sporeforming bacteria. Production of commercially sterile low-acid foods such as meat, milk, and vegetables must overcome the extreme pressure resistance of spores.

Similar as found in fungi, vegetative forms of bacteria are normally more easily inactivated by pressure than spores. Linton *et al.* [23] investigated the inactivation of a pressure-resistant strain of *Escherichia coli* O157:H7 (NCTC 12079) in orange juice over the pH range 3.4 to 5.0. The sterile orange juices were adjusted to various pH levels (3.4, 3.6, 3.9, 4.5, and 5.0) and inoculated with *E. coli* O157:H7 at 10^8 CFU/ml. A $6\log_{10}$ inactivation was obtained after 5 minutes at 550 MPa and 20°C at every pH evaluated except pH 5.0 ($\sim$5.5 $\log_{10}$); this pressure combined with mild heat (30°C) resulted in a $6\log_{10}$ inactivation at pH 5.0.

There were considerable variations in bacterial pressure resistance in different types of fruit juices. Teo *et al.* [24] reported HPP treatment at low temperature (15°C) had the potential to inactivate *E. coli* O157:H7 strains. A three-strain cocktail of *E. coli* O157:H7 (SEA13B88, ATCC 43895, and 932) was found to be most sensitive to pressure in grapefruit juice (8.3 $\log_{10}$ reduction) and least sensitive in apple juice (0.4 $\log_{10}$ reduction) when pressurized at 615 MPa for 2 minutes at 15°C. The resistance difference might come from the various pH values and the presence of natural antimicrobials in different fruit juices.

Wuytack *et al.* [25] applied pressures of 250, 300, 350, and 400 MPa to reduce the microbial loads of garden cress, sesame, radish, and mustard seeds that were immersed in water and treated at 20°C for 15 minutes. The percentages of seeds germinating on water agar were recorded to 11 days after pressure treatment. Radish and garden cress seeds were the most pressure-sensitive and pressure-resistant types, respectively. For example, after a 250 MPa treatment, radish seeds displayed 100% germination nine days later than untreated controls, while garden cress seeds attained 100% germination one day after the controls. Garden cress seeds were inoculated with suspensions of seven different kinds of bacteria (starting inocula 10^7 CFU/g). Treatment at 300 MPa for 15 minutes and 20°C resulted in $6\log_{10}$ reductions of *Salmonella* Typhimurium, *E. coli* MG1655, and *Listeria innocua*, $>4\log_{10}$ reductions of

Shigella flexneri and the pressure-resistant stain *E. coli* LMM1010, and a $2\log_{10}$ reduction of *Staphylococcus aureus*; however, *Enterococcus faecalis* was not inactivated.

Ramaswamy *et al.* [26] applied 150 to 400 MPa to apple juices inoculated with *E. coli* 29055 at 25°C for 0 to 80 minutes. The surviving cells with and without injury were differentiated through the use of brain–heart infusion agar (BHIA) and violet–red bile agar (VRBA). It was found that D values of *E. coli* decreased with an increase in pressure, and pressure D values from BHIA (survivors including injured cells) were higher than from VRBA (survivors excluding injured cells), indicating that a greater number of cells were initially injured than killed with HPP treatment. The associated z values (pressure range to result in a decimal change in D values) were 126 and 140 MPa on BHIA and VRBA, respectively.

22.2.2.2 Viruses

The first examination of the pressure sensitivity of viruses was by Giddings *et al.* [27] who found that a 920 MPa exposure was required to inactivate tobacco mosaic virus (TMV). Since that early work, it now appears that most human viruses are substantially more pressure-sensitive than TMV. Human immuno-deficiency viruses (HIV) can be reduced by 10^4 to 10^5 viable particles after exposure to 400 to 600 MPa for 10 minutes [28], but some viruses can be inactivated at even lower levels of pressures. For example, Brauch *et al.* [29] reported that pressures of 300 to 400 MPa significantly killed bacteriophage (ϕx, λ and T4), and Shigehisa *et al.* [30] found that an $8\log_{10}$ plaque-forming unit (PFU) population of herpes simplex virus type 1 could be eliminated by a 10-minute exposure to 400 MPa, and a $5\log_{10}$ PFU population of human cytomegalovirus was inactivated by a 10-minute exposure to 300 MPa. Shigehisa *et al.* [31] later reported that a $5.5\log_{10}$ tissue culture infectious dose of HIV type 1 was eliminated after a 10-minute exposure to 400 MPa at 25°C.

According to Kingsley *et al.* [32], a $7\log_{10}$ PFU/ml hepatitis A virus (HAV) stock in tissue culture medium was reduced to nondetectable levels after exposure to >450 MPa for 5 minutes. Titers of HAV were reduced in a time- and pressure-dependent manner between 300 and 450 MPa, but poliovirus titer was unaffected by a 5-minute treatment at 600 MPa. Salts had a protective effect on viruses because dilution with seawater increased the pressure resistance of HAV. Experiments involving RNase protection indicated that viral capsids might remain intact during pressure treatment, suggesting that inactivation was due to subtle alterations of viral capsid proteins. A $7\log_{10}$ tissue culture infectious dose of feline calicivirus, a Norwalk virus surrogate, was completely inactivated by exposure to 275 MPa or above after 5 minutes, indicating that HAV and feline calicivirus could be inactivated by pressure.

Currently, there are few publications available addressing pressure inactivation of viruses in fruit and vegetables products. It can be anticipated that investigations will more closely evaluate inactivation of viruses by HPP given the recent food safety issues concerning fecally contaminated fresh produce.

22.2.2.3 Parasites

Human feces are not just a source of human viruses, but also a source of human parasites. Raw fruits and vegetables can become fecally contaminated with parasites that include the protozoans *Giardia intestinalis, Cryptosporidium parvum, Cyclospora cayetanensis*, and the helminth parasites *Fasciola hepatica, Ascaris lumbricoides*, and *Ascaris suum* [33]; however, few articles are available regarding pressure inactivation of parasites in or on fresh fruits and vegetables. Slifko *et al.* [34] applied 550 MPa to apple and orange juices in which *Cryptosporidium parvum* oocysts were suspended. After a 30-second exposure, *C. parvum* oocysts were inactivated by at least $3.4 \log_{10}$, and an exposure to 550 MPa for more than 60 seconds efficiently rendered the oocysts nonviable and noninfectious.

Recently, HPP was used to inactivate parasites from muscle tissues and fish. A pressure of 200 MPa for 10 minutes inactivated all anisakis larvae isolated from fish tissues either in distilled water or in a physiological isotonic solution between 0 and 15°C; when exposed to 140 MPa for 1 hour, all larvae were killed [35]. Dong *et al.* [36] pressure-inactivated *Anisakis simplex* larvae inoculated in king salmon and arrowtooth flounder. Complete kill of the larvae (ranging from 13 to 118) contained in fish fillets was obtained by treatments of 414 MPa for 0.5 to 1 minute, 276 MPa for 1.5 to 3 minutes, and 207 MPa for 3 minutes; however, it was stated that the application of HPP to raw fish was limited because of the significant whitening of the flesh of HPP-treated fish fillets ($P < 0.05$).

22.2.3 SUMMARY

Because of its capacity to inactivate pathogenic microorganisms with minimal application of heat and quality loss, HPP is continuing to gain attention as one of the viable alternative nonthermal methods to thermal processing. HPP-treated products maintain the nutritional and sensory quality with extended shelf life. Standardization and commercialization of HPP seem very promising for a range of food and beverage products.

22.3 IRRADIATION

22.3.1 INTRODUCTION

Studies on the effect of ionizing radiation upon living organisms started after the discoveries of X-rays in 1895 and radioactivity in 1896. The first patent for the use of irradiation as a food processing technology was filed in 1905, but sustained effort to use radiation to preserve foods did not begin until the end of World War II. The first commercial use of food irradiation occurred

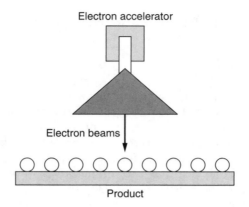

FIGURE 22.2 Schematic of an electron accelerator.

in 1957 on a spice in Germany [37]. Irradiation is now widely used to inhibit tuber sprouting, delay fruit and vegetable ripening, control insects in fruits and grains, and reduce parasites in products of animal origin [38]. Through the interaction of chemically active species (i.e., free radicals) induced from irradiation and direct damage to microbial DNA caused by high-energy particles, irradiation is also used to reduce or eliminate foodborne microorganisms [39].

Typical ionizing radiation facilities use either gamma rays from the radioactive isotopes ^{60}Co or ^{137}Cs or electron beams as well as X-rays generated in electron accelerators [40]. Strict safety measures are required for gamma ray facilities due to continuous emission from ^{60}Co and ^{137}Cs, such as use of thick concrete walls to construct the irradiation chamber. In contrast, electron accelerators, as shown in Figure 22.2, have few leakage problems because they produce no high-energy electrons when not in use. As recommended by a joint FAO/IAEA/WHO expert committee on food irradiation (JECFI) in 1980, the absorbed dose or amount of energy absorbed by a food product has a limit of 10 kGy (1 Gy is a dose equal to 1 J/kg of absorbing material). After reviewing toxicological, nutritional, and microbiological data on foods irradiated at doses over 10 kGy, the committee concluded that foods are both safe and nutritious to consumers when irradiated to any dose adequate to obtain the intended technological objective; however, most foods exposed to dosages above 10 kGy will lose sensory quality to some extent [41].

22.3.2 APPLICATION TO FRUITS, VEGETABLES, AND JUICES

22.3.2.1 Spores and Vegetative Bacteria

Al-Bachir [42] investigated the effect of irradiation (0 to 2.5 kGy) on the quality of two cultivars of Syrian grapes (*Vitis vinifera*) stored at 1 to 2°C for two weeks. The irradiation treatment decreased spoilage caused by *Botrytis cinerea* and improved the quality of both varieties. The optimum doses were 0.5 to 1.0 kGy for Helwani grapes and 1.5 to 2.0 kGy for Baladi grapes. The storage periods were extended by 50% after irradiation at optimal doses for both varieties.

Aziz and Moussa [43] studied the effect of irradiation on the viable population of fungi and production of mycotoxins in randomly collected fruits that included strawberries, apricots, plums, peaches, grapes, dates, figs, apples, pears, and mulberries. Analysis of these fruits detected the mycotoxins penicillic acid, patulin, cyclopiazonic acid, citrinin, ochratoxin A, and aflatoxin B_1. Irradiation of fruits at doses of 1.5 and 3.5 kGy significantly decreased the total fungal counts compared with unirradiated controls. The corresponding occurrence of mycotoxins in fruits decreased with increasing irradiation dose and was not detected after treatments at 5.0 kGy.

Niemira *et al.* [44] irradiated frozen broccoli, corn, lima beans, and peas at subfreezing temperatures ranging from -20 to $-5°C$ and determined the influence of irradiation temperature on quality factors of frozen vegetables as well as irradiation sensitivity of inoculated *L. monocytogenes*. The irradiation resistance of *L. monocytogenes* changed significantly with the type of vegetable and the treatment temperature. The levels of irradiation necessary to reduce the bacterial population by 90% (D values) for *L. monocytogenes* increased with decreasing temperature for all the vegetables that were evaluated. D values ranged from 0.505 kGy for broccoli to 0.613 kGy for corn at $-5°C$ and from 0.767 kGy for lima beans to 0.916 kGy for peas at $-20°C$.

Lettuce inoculated with 1×10^7 CFU/g of acid-adapted *E. coli* 0157:H7 was chlorinated at 200 μg/ml and irradiated at 0.15, 0.38, or 0.55 kGy by Foley *et al.* [45]. The viability of *E. coli* 0157:H7, aerobic mesophiles, yeast, and molds was measured over 10 days. Chlorination alone reduced counts of *E. coli* 0157:H7 by 1 to $2 \log_{10}$ CFU/g. Chlorination combined with irradiation at 0.55 kGy produced $5.4 \log_{10}$ reductions in *E. coli* 0157:H7 levels. When stored at 1 and 4°C after irradiation at 0.55 kGy, standard plate counts and yeast and mold counts were reduced by $2.5 \log_{10}$ CFU/g for samples storage on day 17 without obvious softening of the lettuce or any other adverse effect on sensory quality.

Niemira *et al.* [46] irradiated leaf pieces and leaf homogenate of endive (*Cichorium endiva*) inoculated with *L. monocytogenes* (pathogen) or *Listeria innocua* (nonpathogenic surrogate). Similar radiation sensitivity was obtained for the two strains, but *L. innocua* was more sensitive to irradiation in leaf homogenate than on the leaf surface. A dose of 0.42 kGy reduced *L. monocytogenes* on inoculated endive by 99%; however, the pathogen grew after 5 days of refrigerated storage until it exceeded the bacterial levels of the control after 19 days of storage, but a dose of 0.84 kGy, equivalent to a 99.99% reduction, suppressed *L. monocytogenes* throughout refrigerated storage. When increasing the doses up to 1.0 kGy, no significant change of color was observed for endive leaves taken either from the leaf edge or the leaf midrib. Dose tolerances for acceptable texture of leaf edge and midrib material were a maximum of 1.0 and 0.8 kGy, respectively.

Niemira et al. [47] also irradiated orange juices with varying levels of turbidity and inoculations with *Salmonella* Anatum, *Salmonella* Infantis, *Salmonella* Newport, or *Salmonella* Stanley at 2°C. Neither the resistance of each isolate (D value) nor the pattern of relative resistance among isolates was altered in orange juice. *S.* Anatum ($D = 0.71$ kGy) was significantly more resistant than the other species in orange juice, followed by *S.* Newport ($D = 0.48$ kGy), *S.* Stanley ($D = 0.38$ kGy), and *S.* Infantis ($D = 0.35$ kGy).

Van Gerwen [48] analyzed the irradiation resistance of both spores and vegetative bacteria based on the data available from the literature. As expected, spores were found to have significantly higher D values with an average of 2.48 kGy compared to the average D value for most vegetative bacteria of 0.76 kGy. The notoriously radiation-resistant nonpathogenic vegetative bacterium *Deinococcus radiodurans*, had the highest D value of 10.4 kGy. The average irradiation resistances for spores and vegetative bacteria were further estimated to be 2.11 and 0.42 kGy after excluding specific conditions showing extreme D values.

22.3.2.2 Parasites

According to Dubey et al. [49], outbreaks of cyclospora-associated gastroenteritis in humans have been epidemiologically linked to the ingestion of fecally contaminated fruits (raspberries), vegetables (lettuce), or herbs (basil). Also, cryptosporidium oocysts have been demonstrated on vegetables. Unsporulated and sporulated *T. gondii* oocysts were used as a model system to determine the effect of irradiation on fruits contaminated with other coccidia such as cyclospora or cryptosporidium. Unsporulated oocysts of *T. gondii* irradiated at 0.4 to 0.8 kGy sporulated, but were not infective to mice; however, sporulated oocysts irradiated at doses greater than 0.4 kGy were able to encyst. Sporozoites were infective but not capable of inducing a viable infection in mice. *T. gondii* was detected in histological sections of mice up to 5 days, but not 7 days after feeding oocysts irradiated at 0.5 kGy. Raspberries inoculated with sporulated *T. gondii* oocysts were rendered nonviable after irradiation at 0.4 kGy. An irradiation of 0.5 kGy was recommended in this study to inactivate coccidian parasites on fruits and vegetables.

22.3.2.3 Viruses

Against viruses the effectiveness of irradiation is dependent on the size of the virus, the suspension medium and/or type of food product, and the temperature of exposure [50,51]. Because of their smaller size and genetic makeup (often single-stranded RNA), most viruses are more resistant to irradiation than bacteria, parasites, or fungi [51]. According to Bidawid et al. [52], only a few studies have examined the efficiency of irradiation on viruses in or on food products, including work on poliovirus in fish fillets [53], coxsackievirus B in ground beef [54], and rotavirus and HAV in clams and oysters [55]. Lettuce and strawberries were inoculated with HAV and irradiated with doses ranging between 1 and 10 kGy at ambient temperature [52]. Plaque assays of HAV after irradiation showed a linear pattern of inactivation, i.e., a linear decrease in virus titer occurred when the irradiation dose was increased. Data analysis by a linear model indicated that D values were 2.72 ± 0.05 and 2.97 ± 0.18 kGy for HAV in lettuce and strawberries, respectively. These data were similar to those reported of 2.0 kGy for HAV in both clams and oysters [53]. No noticeable deterioration was observed in the texture and appearance of lettuces and strawberries, even at the highest dose of 10 kGy.

22.3.3 SUMMARY

Ionizing (gamma) radiation can be used to control microbiological spoilage agents and pathogens and parasites and viruses in fruits, vegetables, and juices, while increasing the shelf life without major damage to the physical or chemical properties (texture, appearance, and sensory palatability); however, the widespread use of irradiation is still limited mostly by concerns from consumers, costs, and effect on the product quality. Irradiation, possibly in combination with other processes such as mild heat or modified atmosphere packaging, can provide a suitable means to improve produce safety.

22.4 PULSED ELECTRIC FIELDS IN JUICE PROCESSING

22.4.1 INTRODUCTION

Pulsed electric field (PEF) processing applies high voltage pulses to foods located between a series of electrode pairs. The electrical fields (generally at 20 to 80 kV/cm) are achieved through capacitors that store electrical energy from DC power supplies. During PEF treatment, the applied electric field increases membrane permeability of microbial cells by either forming transmembrane pores (electrical breakdown) or temporarily destabilizing the lipid bilayer and proteins of cell membranes (electroporation), causing inactivation of microorganisms [56–58].

PEF units usually consist of three major parts: the PEF generation unit, treatment chamber, and process control system (Figure 22.3). A PEF can be generated in the form of exponentially decaying, square-wave, bipolar, instant-charge-reversal, or oscillatory pulses, depending on the circuit design of the generating device. The treatment chamber in a PEF unit holds two electrodes in position with insulating materials that form a chamber. Electrochemical reactions can occur at the electrode surfaces, causing partial electrolysis of medium solution, electrode corrosion, and introduction of small particles of electrode material into the liquid medium [59]. Use of very short pulses or bipolar pulses is recommended to avoid the cumulative buildup of charges and thus minimize electrode corrosion.

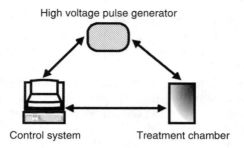

FIGURE 22.3 Schematic of a PEF unit.

Another major issue for the design of PEF treatment chambers is to provide a relatively uniform electric field. For example, uniform electric fields can be achieved with parallel plate electrodes if the distance between the electrodes is sufficiently smaller than the electrode surface dimension. For operation safety, pressure relief devices are necessary to avoid the destruction, or the explosion, of PEF chambers due to the possibility of the buildup of pressure, which can arise either from the expansion of dissolved air or partial vaporization promoted from local heating within the PEF chamber after a spark [60].

Critical processing factors for PEF include electric field intensity, pulse width, treatment time, temperature, and pulse wave shape. The induced potential difference across the cell membrane of a microorganism is proportional to the applied electric field (electroporation theory). A lethal effect to living cells is observed when the induced potential or transmembrane electric potential exceeds by a large margin a critical value of approximately 1 V. Qin *et al.* [61] stated that microbial inactivation increases with an increase in the electric field intensity above the critical transmembrane potential. Pulse width influences the critical electric field and the efficiency of microbial inactivation [62]. Treatment time, defined as the product of pulse numbers and pulse duration, affects microbial inactivation when either of the two variables changes [63]. The efficiency for microbial inactivation varies with different pulse wave shapes. Oscillatory pulses are the least efficient for microbial inactivation; square wave pulses are more efficient than exponential decaying pulses; and bipolar pulses are more lethal than monopolar pulses [64,65]. Treatment temperatures can change cell membrane fluidity and permeability, thus affecting the susceptibility of cells to mechanical disruption [66]. Targeted microorganisms (type, growth stage, and initial concentration) and properties of the PEF treatment medium (pH, conductivity, and medium ionic strength) also influence the microbial inactivation efficiency. Critical processing factors of PEF treatment need to be monitored and recorded to ensure the microbiological safety of the processed food products while maintaining food quality with acceptable energy efficiency.

22.4.2 Application to Juices

PEF research had been mostly focused on the inactivation of microorganisms suspended in foods, including semisolid and liquid foods such as pea soup, milk, liquid eggs, and juices, particularly orange and apple juices. Sitzmann [67] obtained a $3 \log_{10}$ reduction of native microbiota for freshly squeezed orange juice using a continuous PEF process with an electric field of $15\,kV/cm$. There was no significant change in quality. Zhang et al. [68] found that total aerobic counts of reconstituted orange juice were reduced 3- to 4-$\log_{10}$ cycles when treated with an integrated PEF pilot plant system operating at less than $32\,kV/cm$. Raso et al. [69] investigated PEF inactivation of ascospores and vegetative cells of Zygosaccharomyces bailii suspended in apple, orange, pineapple, cranberry, and grape juices. Two pulses of 32 to $36.5\,kV/cm$ decreased the population of vegetative cells or ascospores 3.5 to $5 \log_{10}$ cycles for each fruit juice studied. Evrendilek et al. [70] treated fresh apple juice inoculated with E. coli O157:H7 and E. coli 8739 using bipolar PEF. A $5 \log_{10}$ reduction was obtained for each culture when the treatment temperature was below $35°C$. The lethality for fresh apple cider inoculated with E. coli O157:H7 was also reported for PEF treatment with instant charge reversal pulses [71]. In orange juice, McDonald et al. [72] inactivated Leuconostoc mesenteroides, E. coli, and L. innocua by as much as $5 \log_{10}$ cycles at $30\,kV/cm$ and $50°C$. A maximum of $2.5 \log_{10}$ cycle reduction was achieved for Saccharomyces cerevisiae ascospores at $50\,kV/cm$ and $50°C$.

The synergy of PEF, pH, water activity, ionic strength, temperature, antimicrobial agents (e.g., nisin, lysozyme), and other combinations of hurdle technology (ozone treatment or high hydrostatic pressure) can increase microorganism inactivation [73–75]. Using the hurdle approach, Hodgins et al. [76] studied the effect of temperature, acidity, and number of pulses on microbial inactivation in orange juice. A $6 \log_{10}$ reduction in the natural microbiota was obtained under optimal conditions consisting of 20 pulses of an electric field of $80\,kV/cm$, at pH 3.5, and a temperature of $44°C$ with a dose of 100 U nisin/ml. The process was most influenced by a change in temperature ($p < 0.0001$). There was a 97.5% retention of vitamin C along with a 92.7% reduction in pectinmethylesterase activity after PEF treatment. The shelf life of the orange juice was at least 28 days when stored at $4°C$ without aseptic packaging. Gas chromatography revealed no significant differences in aroma compounds before and after pulsing.

Liang et al. [77] applied PEF to pasteurized and freshly squeezed orange juices (with and without pulp) and determined the reduction of Salmonella Typhimurium at moderately high temperatures ($< 60°C$). The effect of antimicrobial compounds (nisin and lysozyme) was examined. PEF treatment ($90\,kV/cm$ and 20 pulses) did not have a notable effect on cell viability or injury until the temperature reached $46°C$ or above. Presence of nisin, lysozyme, or a mixture of nisin and lysozyme increased cell viability loss by an additional 0.04 to $2.75 \log_{10}$ cycles for PEF treatment. The combination of nisin and lysozyme had a more pronounced bactericidal effect than did either nisin or lysozyme alone.

The inactivation of enzymes also accompanies pasteurization of juices using PEF technology. In general, higher electric field strengths and longer total treatment times are more effective for the purpose of enzyme inactivation [78]. Decrease of polyphenoloxidase (PPO) activity in peach juices was reported to follow an exponential decay kinetic model [79]. For those orange juices that had a similar shelf life (196 days at 4°C) after thermally processing (90°C and 90 seconds) and PEF processing (40 kV/cm and 97 milliseconds), ascorbic acid, flavor, and color of PEF-treated juice were found to be superior to that of thermally processed juice ($P < 0.05$) [78]. This is also true for tomato juices processed either by PEF at 40 kV/cm (57 milliseconds) or thermally processed (92°C and 90 seconds) and stored at 4°C for 112 days [80]. In both cases, sensory evaluations indicated that the flavor of PEF-processed juices was preferred to that of thermally processed juices ($P < 0.01$) [78,80].

22.4.3 SUMMARY

PEF technology is effective in the inactivation of microorganisms in liquid foods, particularly in orange and apple juices, without significant changes to sensory quality. Advantageous preservative effects of PEF with other process treatments, including ozone treatment, pressure, pH, water activity, ionic strength, temperature, antimicrobial agents (e.g., nisin, lysozyme), has been demonstrated.

22.5 ULTRASONIC WAVES FOR PRESERVATION OF FRUIT AND VEGETABLE PRODUCTS

22.5.1 INTRODUCTION AND DESCRIPTION OF PROCESS

The oscillation of a vibrating body can cause a periodic disturbance that travels through an elastic medium (air, ground, or water) and radiates outward in straight lines in the form of a pressure wave perceived as sound. Based on whether or not it can be heard by the human ear, sound can be divided into communication waves (audible) and ultrasonic waves (or ultrasound, inaudible). Ultrasound, having little or no effect on the ear even at high intensities, vibrates at frequencies greater than 20 kHz and is produced by a transducer, which contains a piezoelectric substance such as a quartz crystal oscillator and converts high-frequency electric current (an input of energy) into vibrating ultrasonic waves (an output of energy) with a fixed relationship.

Ultrasound was first developed in World War II to locate submerged objects. Additional uses have been developed for industrial applications in the field of nondestructive testing, cleaning, welding, and sonochemistry [81].

22.5.2 MICROBIAL INACTIVATION

The fluctuating pressures induced by an ultrasonication process produce and break microscopic bubbles, creating micromechanical shocks to disrupt cellular structural and functional components up to the point of cell lysis [82]. Intracellular cavitations make ultrasound capable of inactivating microorganisms [83]. The inactivation effect depends on the control of critical factors including the amplitude of ultrasonic waves, the exposure or contact time with the microorganisms, the type of microorganism, the volume of food to be processed, the composition of the food, and the temperature of treatment [82]. The mechanism of inactivation of vegetative bacteria appears to be intracellular cavitations that lead to cellular lysis, but ultrasound alone has no effect on spores. Cavitations may play an auxiliary role and allow ultrasound to assist other methods in spore inactivation. This limits the singular use of ultrasound as a preservation method, requiring the use of a combination of ultrasound with other preservation processes (e.g., heat and mild pressure) for industrial applications.

Palacios et al. [84] examined the effect of ultrasound on the heat resistance of spores. After ultrasound treatment (20 kHz, 120 W, 12°C, 30 minutes), several substances were detected to be released from *B. stearothermophilus* spores to the surrounding aqueous medium, including calcium, dipicolinic acid, a glycopeptide of 7 kDa, fatty acids, acyl glycerols, and glycolipids (but no phospholipids). The release of low-molecular-weight substances from the spore protoplast and the consequent modification of its hydration state led to the heat resistance reduction.

The presence of spoilage bacteria, yeasts, molds, and the occasional pathogen on fresh produce is not uncommon. Seymour et al. [85] examined the potential of ultrasound in cleaning minimally processed fruits and vegetables. Cut iceberg lettuce (100 g) inoculated with *S.* Typhimurium (10^6 CFU/g) was washed for 10 minutes with tap water (control), a 25 ppm free chlorine dip only, ultrasound (32 to 40 kHz, 10 to 15 W/l) only, and ultrasound combined with the 25 ppm free chlorine dip. The control reduction was $0.7 \log_{10}$, while reductions of $1.6 \log_{10}$ and $1.7 \log_{10}$ were obtained from washing treatment by ultrasound and chlorine individually. Reductions obtained from the combined washing treatment were 2.6 to $2.7 \log_{10}$, corresponding to a 99.8% reduction in total bacteria. The cleaning action of cavitations appeared to remove cells attached to the surface of fresh produce, rendering the pathogens more susceptible to the sanitizer. The frequency of ultrasound (25, 32 to 40, 62 to 70 kHz) showed no significant effect on decontamination efficiency ($P > 0.69$).

Yeasts such as *S. cerevisiae* and *Zygosaccharomyces* spp., including *Z. bailii* and *Z. rouxii*, and pathogenic bacteria like *L. monocytogenes* can cause significant spoilage and affect the safety of nonpasteurized fruit juice products. Traditional thermal pasteurization methods can detrimentally affect the organoleptic properties when they are used to extend the shelf life for fruit juices. Mincz et al. [86] explored the potential use of ultrasound combined with refrigeration to extend the shelf life of fresh juice products. Freshly squeezed lemon and pineapple juices inoculated with *S. cerevisiae, Z. bailii, Z. rouxii,* and *L. monocytogenes* were immediately sonicated at 20 kHz at 45°C and an amplitude of 95 μm. The treated samples were stored in sterilized glass containers (10 ml) at 7°C for 15 days. The microbial population and color of the inoculated samples was monitored at preset intervals during storage. The combined ultrasound and refrigeration treatment significantly suppressed microbial growth in fresh lemon and pineapple juices with improved color retention. No significant change in pH and a_w was observed.

22.5.3 SUMMARY

Ultrasonic waves have the potential to inactivate microorganisms in fruits, vegetables, and juices. In most cases, a combination of ultrasound with other preservation processes is probably more realistic to achieve an effective degree of microbial inactivation. Further efforts are required for its application as a component in a commercially feasible preservation process.

22.6 ELECTROLYZED WATER

22.6.1 INTRODUCTION

Electrolyzed oxidizing (EO) water, also called strongly oxidizing water, strongly acidic electrolyzed water, or acidic oxidative potential water, has recently attracted interest in medicine, agriculture, and food processing for purposes of sanitation.

An EO water generator usually contains a power supply, and a pair of electrodes (i.e., anode and the cathode) installed in two individual cells that hold sodium chloride solutions separated by a specialized membrane (Figure 22.4). When electrolyzing saline solutions, hydrogen is generated in the cathode side and chlorine is generated in the anode side. The chlorine further reacts with water to form HOCl and HCl. EO water is then produced in the cell installed with anodes and electrolyzed reducing (ER) water is produced in the cell installed with cathodes.

EO water contains free chlorine and has a high oxidation–reduction potential (ORP, above 1000 mV) and low pH (around 2.3). ER water exhibits a high pH (above 11.0), low redox potential (RP, below 800 mV), low levels of dissolved oxygen, and high levels of dissolved molecular hydrogen. The chlorine gas, HOCl, and OCl^- ions contained in EO water contribute to the availability of uncombined chlorine radicals or free available chlorine, the primary component responsible for the disinfection ability of EO water [87–89]. A generator without a separating membrane produces water at pH 6.8 because HCl formed on the anode side neutralizes NaOH on the cathode side [90]. EO water can be preserved for one year under shaded and sealed conditions [91], but EO water becomes inert after three days when exposed to light.

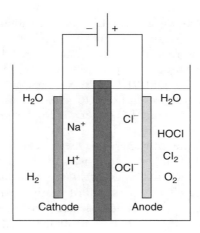

FIGURE 22.4 Schematic of an EO water generator.

22.6.2 APPLICATION AS A NOVEL DISINFECTANT FOR FRUITS AND VEGETABLES

EO water has been extensively applied to fresh or fresh-cut vegetables and fruits because of its strong bactericidal effects [88,92]. Izumi [93] utilized EO water to reduce the total microbial counts of fresh-cut carrots, bell peppers, spinach, and potatoes and found microbial counts on all cuts reduced by 0.6 to $2.6 \log_{10}$ CFU/g. The bactericidal effect of EO water increased with available chlorine in the range of 15 to 50 ppm for fresh-cut carrots, spinach, or cucumbers. Tissue pH, surface color, and general appearance of fresh-cut vegetables were not affected after treatment.

Park *et al.* [94] examined the efficacy of EO water and acidified chlorinated water (45 ppm residual chlorine) against *E. coli* O157:H7 and *L. monocytogenes* on lettuce. Each leaf was surface-inoculated and immersed in 1.5 l of EO or acidified chlorinated water for up to 3 minutes at 22°C. Compared to water washes, a 3-minute EO water rinsing significantly decreased mean populations of *E. coli* O157:H7 and *L. monocytogenes* by 2.4 and $2.7 \log_{10}$ CFU per lettuce leaf, respectively ($p \leq 0.05$). There was no significant difference between the bactericidal activity of EO water and acidified chlorinated water ($p \geq 0.05$), and no obvious quality change was observed during two weeks of storage after washing.

Fresh tomatoes were surface-inoculated with *E. coli* O157:H7, *S.* Enteritidis, *L. monocytogenes*, and nonpathogenic *E. coli* and rinsed in neutral EO water for up to 60 seconds [95]. EO water rinsing reduced the surface population from $5 \log_{10}$ to $< 1 \log_{10}$ CFU/cm^2 independent of the type of microorganism and treatment time. No cells were detected in the washing solution. There was no significant difference in organoleptic qualities compared to untreated tomatoes.

Ratna and Demirci [96] applied EO water to alfalfa seeds and sprouts inoculated with a five-strain cocktail of nalidixic acid-resistant *E. coli* O157:H7. Reductions were in the range 0.2 to 1.6 and 1.1 to $2.7\log_{10}$ CFU/g for treated seeds and sprouts, respectively, corresponding to a percentage reduction of 38.2 to 97.1 and 91.1 to 99.8%, respectively. Germination of the treated seeds was reduced from 92 to 49% when soaking time and the electric current used to generate EO water were increased. No visible damage occurred to the sprouts.

Besides the inactivation of bacteria, EO water can also serve as an effective fungicide on fruits and the foliage and flowers of bedding plants. Al-Haq *et al.* [97] immersed peach inoculated with *Monilina fructicola* in EO water up to 5 minutes to examine its effectiveness against postharvest brown rot. EO water did not control brown rot in wound-inoculated peaches, but did reduce the incidence and severity in nonwounded inoculated fruit. No chlorine-induced phytotoxicity was observed on treated products. In this case, EO water was an effective surface sanitizer that delayed disease development. Al-Haq *et al.* [98] also evaluated the effects of EO water on suppressing fruit rot of pear caused by *Botryosphaeria berengeriana*. Pears with wounds necessary to cause "bot rot" were inoculated with spore suspensions of *B. berengeriana* and immersed in EO for 10 minutes. No chlorine-induced phytotoxicity was observed on the treated fruits, and EO water suppressed the incidence and severity of disease, suggesting that EO water can be used as surface sanitizer to possibly reduce postharvest fungal rot development.

22.6.3 SUMMARY

EO water has demonstrated disinfecting ability against bacteria and fungi. EO water is easy to use and environmental friendly; however, further study regarding quality changes after EO water rinsing is necessary for commercial use on fresh-cut fruits and vegetables.

22.7 FINAL REMARKS AND FUTURE PERSPECTIVES

The success of any fruit and vegetable preservation technology depends on a complete and correct understanding of the reasonable causes of food spoilage and associated foodborne illness. To ensure safety and prolong shelf life, conventional thermal processing is a mainstay of the food industry; however, driving forces from the market, and development and introduction of nonthermal processing treatments by the food industry and research institutions have provided a well-accepted platform to deliver safe products processed at lower temperatures with minimum quality losses. The potential use of each of the described nonthermal treatments in combination with other procedures, including established manipulations employing temperature, water activity adjustment, oxidation–reduction potential and pH controls, and modified atmosphere packaging, is worth considering in food product development. A good indication of this potential is the commercial success now being realized with pressurization methods to preserve fruits and vegetables. With pressure-treated jams, jellies, juices, salsa, guacamole, and chopped onions available on the market, one can assume that more products will follow.

ACKNOWLEDGMENT

The authors are with the Department of Animal & Food Sciences, University of Delaware, Newark, DE. The authors wish to acknowledge the support provided by the USDA (grant no. 2001-35201-09947).

REFERENCES

1. Bruhn, C., Personal communication, 2004.
2. Hite, B.H., The effect of pressure in the preservation of milk, *W. Va. Univ. Agric. Exp. Stn. Morgantown Bull.*, 58, 15, 1899.
3. Hite, B.H., Giddings, N.J., and Weakly, C.E., The effects of pressure on certain microorganisms encountered in the preservation of fruits and vegetables, *W. Va. Agric. Exp. Stn. Morgantown Bull.*, 146, 1, 1914.
4. Hoover, D.G., Pressure effects on biological systems, *Food Technol.*, 47, 150, 1993.
5. Hoover, D.G. *et al.*, Biological effects of high hydrostatic pressure on food microorganisms, *Food Technol.*, 43, 99, 1989.
6. Farkas, D.F. and Hoover, D.G., High pressure processing, in Kinetics of Microbial Inactivation for Alternative Food Processing Technologies, *J. Food Sci. Suppl.*, 47, 2000.
7. Reyns, K.M.F.A., Veraverbeke, E.A., and Michiels, C.W., Activation and inactivation of *Talaromyces macrosporus* ascospores by high hydrostatic pressure, *J. Food Prot.*, 66, 1035, 2003.
8. Hayashi, R., Application of high pressure to food processing and preservation: philosophy and development, in *Engineering and Food*, Spiess, W.E.L and Schubert, H., Eds., Elsevier, London, 1989, p. 815.
9. Ogawa, H. *et al.*, Effect of hydrostatic pressure on sterilization and preservation of various kinds of citrus juice, in *Pressure Processed Food: Research and Development*, Hayashih, R., Ed., San-sai Publishing, Kyoto, Japan, 1990.
10. Aleman, G.D. *et al.*, Ultra-high pressure pasteurization of fresh cut pineapple, *J. Food Prot.*, 57, 931, 1994.
11. Arroyo, G., Sanz, P.D., and Prestamo, G., Effect of high pressure on the reduction of microbial populations in vegetables, *J. Appl. Microbiol.*, 82, 735, 1997.
12. Raso, J. *et al.*, Inactivation of *Zygosaccharomyces bailii* in fruit juices by heat, high hydrostatic pressure and pulsed electric fields, *J. Food Sci.*, 63, 1042, 1998.
13. Parish, M.E., High pressure inactivation of *Saccharomyces cerevisiae*, endogenous microflora and pectinmethylesterase in orange juice, *J. Food Prot.*, 18, 57, 1998.
14. Zook, C.D. *et al.*, High pressure inactivation kinetics of *Saccharomyces cerevisiae* ascospores in orange and apple juices, *J. Food Sci.*, 64, 533, 1999.

15. Timson, W.J. and Short, A.J., Resistance of microorganisms to hydrostatic pressure, *Biotechnol. Bioeng.* 12, 139, 1965.
16. Seyderhelm, I. and Knorr, D., Reduction of *Bacillus stearothermophilus* spores by combined high pressure and temperature treatments, *ZFL Eur. Food Sci.*, 43, 17, 1992.
17. Kakugawa, K. *et al.*, Thermal inactivating behavior of *Bacillus stearothermophilus* under high pressure, in *High Pressure Bioscience and Biotechnology*, Hayashi, R. and Balny, C., Eds., Elsevier Science, Amsterdam, 1996, p. 171.
18. Palou, E. *et al.*, High pressure treatment in food preservation, in *Handbook of Food Preservation*, Shafiur Rahman, M., Ed., Marcel Dekker, New York, 1999, chap. 19.
19. Paidhungat, M., *et al.*, Mechanisms of induction of germination of *Bacillus subtilis* spores by high pressure, *Appl. Envrion. Microbiol.*, 68, 3172, 2002.
20. Heinz, V. and Knorr, D., High pressure germination and inactivation kinetics of bacterial spores, in *High Pressure Food Science, Bioscience and Chemistry*, Isaacs, N.S., Ed., Royal Society of Chemistry, Cambridge, U.K., 1998.
21. Ludwig, H., Van Almsick, G., and Sojka, B., High pressure inactivation of microorganisms, in *High Pressure Bioscience and Biotechnology*, Hayashi, R. and Balny, C., Eds., Elsevier Science, Amsterdam, 1996, p. 237.
22. Oh, S. and Moon, M.J., Inactivation of *Bacillus cereus* spores by high hydrostatic pressure at different temperatures, *J. Food Prot.*, 66, 599, 2003.
23. Linton, M., McClements, J.M.J., and Patterson, M.F., Inactivation of *Escherichia coli* O157:H7 in orange juice using a combination of high pressure and mild heat, *J. Food Prot.*, 62, 277, 1999.
24. Teo, A.Y.L., Ravishankar, S., and Sizer, C.E., Effect of low-temperature, high-pressure treatment on the survival of *Escherichia coli* O157:H7 and *Salmonella* in unpasteurized fruit juices, *J. Food Prot.*, 64, 1122, 2001.
25. Wuytack, E.Y. *et al.*, Decontamination of seeds for seed sprout production by high hydrostatic pressure, *J. Food Prot.*, 66, 918, 2003.
26. Ramaswamy, H.S., Riahi, E., and Idziak, E., High-pressure destruction kinetics of *E. coli* (29055) in apple juice, *J. Food Sci.*, 68, 1750, 2003.
27. Giddings, N.J., Allard, H.A., and Hite, B.H., Inactivation of the tobacco mosaic virus by high pressure, *Phytopathology*, 19, 749, 1929.
28. Otake, T. *et al.*, Effects of high hydrostatic pressure treatment of HIV infectivity, in *High Pressure Research in Bioscience and Biotechnology*, Heremans, K., Ed., Leuven University Press, Leuven, 1997, p. 223.
29. Brauch, G., Haensler, U., and Ludwig, H., The effect of pressure on bacteriophages, *High Pressure Res.*, 5, 767, 1990.
30. Shigehisa, T. *et al.*, Effects of high hydrostatic pressure on characteristics of pork slurries and inactivation of microorganisms associated with meat and meat products, *Int. J. Microbiol.*, 12, 207, 1991.
31. Shigehisa, T. *et al.*, Inactivation of HIV in blood plasma by high hydrostatic pressure, in *High Pressure Bioscience and Biotechnology*, Hayashi, R. and Balny, C., Eds., Elsevier Science, Amsterdam, 1996, p. 273.
32. Kingsley, D.H. *et al.*, Inactivation of hepatitis A virus and a calicivirus by high hydrostatic pressure, *J. Food Prot.*, 65, 1605, 2002.
33. Robertson, L.J. and Gjerde, B., Occurrence of parasites on fruits and vegetables in Norway, *J. Food Prot.*, 64, 1793, 2001.
34. Slifko, T.R. *et al.*, Effect of high hydrostatic pressure on *Cryptosporidium parvum* infectivity, *J. Food Prot.*, 63, 1262, 2000.

35. Molina-Garcia, A.D. and Sanz, P.D., Anisakis simplex larva killed by high hydrostatic-pressure processing, *J. Food Prot.*, 65, 383, 2002.
36. Dong, F.M., Cook, A.R., and Herwig, R.P., Research Note: high hydrostatic pressure treatment of finfish to inactivate *Anisakis simplex*, *J. Food Prot.*, 66, 1924, 2003.
37. Diehl, J.F., Food irradiation: past, present and future, *Radiat. Phys. Chem.*, 63, 211, 2002.
38. Lacroix, M. and Ouattara, B., Combined industrial processes with irradiation to assure innocuity and preservation of food products: a review, *Food Res Int.*, 33, 719, 2000.
39. Urbain, W.M., *Food Irradiation*, Academic Press, Orlando, FL, 1986.
40. Patterson, M.F. and Loaharanu, P., Irradiation, in *The Microbiological Safety and Quality of Food*, Lund, B.M., Baird-Parker, T.C., and Gould, G.W., Eds., Aspen, Gaithersburg, MD, 2000, p. 65.
41. WHO, High Dose Irradiation, Wholesomeness of Food Irradiated With Doses Above 10 kGy, report of a Joint FAO/IAEA/WHO Study Group, World Health Organization Technical Report Series No. 890, Geneva, 1999.
42. Al-Bachir, M., Effect of gamma irradiation on storability of two cultivars of Syrian grapes (*Vitis vinifera*), *Radiat. Phys. Chem.*, 55, 81, 1999.
43. Aziz, N.H. and Moussa, L.A.A., Influence of gamma-radiation on mycotoxin producing moulds and mycotoxins in fruits, *Food Control*, 13, 281, 2002.
44. Niemira, B.A., Fan, X., and Sommers, C.H., Irradiation temperature influences product quality factors of frozen vegetables and radiation sensitivity of inoculated *Listeria monocytogenes*, *J. Food Prot.*, 65, 1406, 2002.
45. Foley, D.M. *et al.*, Reduction of *Escherichia coli* O157:H7 in shredded iceberg lettuce by chlorination and gamma irradiation, *Radiat. Phys. Chem.*, 63, 391, 2002.
46. Niemira, B.A. *et al.*, Ionizing radiation sensitivity of *Listeria monocytogenes* ATCC 49594 and *Listeria innocua* ATCC 51742 inoculated on endive (*Cichorium endiva*), *J. Food Prot.*, 66, 993, 2003.
47. Niemira, B.A. *et al.*, Irradiation inactivation of four *Salmonella* serotypes in orange juices with various turbidities, *J. Food Prot.*, 64, 614, 2001.
48. van Gerwen, S.J.C., A data analysis of the irradiation parameter D_{10} for bacteria and spores under various conditions, *J. Food Prot.*, 62, 1024, 1999.
49. Dubey, J.P. *et al.*, Effect of gamma irradiation on unsporulated and sporulated *Toxoplasma gondii* oocysts, *Int. J. Parasitol.*, 28, 369, 1998.
50. Patterson, M.F., Food irradiation and food safety, *Rev. Med. Microbiol.*, 4, 151, 1993.
51. Farkas, J., Irradiation as a method for decontaminating food: a review, *Int. J. Food Microbiol.*, 44, 189, 1998.
52. Bidawid, S., Farber, J.M., and Sattar, S.A., Inactivation of hepatitis A virus (HAV) in fruits and vegetables by gamma irradiation, *Int. J. Food Microbiol.*, 57, 91, 2000.
53. Heidelbaugh, N.D. and Giron, D.J., Effect of processing on recovery of poliovirus from inoculated foods, *Food Sci.*, 34, 239, 1969.
54. Sullivan, R. *et al.*, Gamma radiation inactivation of coxsackievirus B-2, *Appl. Microbiol.*, 26, 14, 1973.
55. Mallett, J.C., Potential of irradiation technology for improving shellfish sanitation, *J. Food Saf.*, 11, 231, 1991.

56. Zimmermann, U. and Benz, R., Dependence of the electrical breakdown voltage on the charging time in valonia utricularis, *J. Membr. Biol.*, 53, 33, 1980.

57. Zimmermann, U., Electrical breakdown, electropermeabilization and electrofusion, *Rev. Physiol. Biochem. Pharmacol.*, 105, 175, 1986.

58. Castro, A.J., Barbosa-Cánovas, G.V., and Swanson, B.G., Microbial inactivation of foods by pulsed electric fields, *J. Food Process. Pres.*, 17, 47, 1993.

59. Morren, J., Roodenburg, B., and de Haan, S.W.H., Electrochemical reactions and electrode corrosion in pulsed electric field (PEF) treatment chambers, *Innov. Food Sci. Emerg. Technol.*, 4, 285, 2003.

60. Zhang, Q.H., Barbosa-Cánovas, G.V., and Swanson, B.G., Engineering aspects of pulsed electric field pasteurization, *J. Food Eng.*, 25, 261, 1995.

61. Qin, B.L. *et al.*, Inactivating microorganism using a pulsed electric field continuous treatment system, *IEEE Trans. Ind. Appl.*, 34, 43, 1998.

62. Schoenbach, K.H. *et al.*, The effect of pulsed electric fields on biological cells: experiments and applications, *IEEE Trans. Plasma Sci.*, 25, 284, 1997.

63. Sale, A.J.H. and Hamilton, W.A., Effects of high electric fields on microorganisms: I. Killing of bacteria and yeast, *Biochim. Biophys. Acta*, 148, 781, 1967.

64. Ho, S.Y. *et al.*, Inactivation of *Pseudomonas fluorescens* by high voltage electric pulses, *J. Food Sci.*, 60, 1337, 1995.

65. Qin, B.L. *et al.*, Inactivation of microorganisms by pulsed electric fields with different voltage waveforms, *IEEE Trans. Dielec. Insul.*, 1, 1047, 1994.

66. Hülsheger, H., Pottel, J., and Niemann, E.G., Killing of bacteria with electric pulses of high field strength, *Radiat. Environ. Biophys.*, 20, 53, 1981.

67. Sitzmann, V., High voltage pulse techniques for food preservation, in *New Methods for Food Preservation*, Gould, G.W., Ed., Blackie Academic and Professional, London, 1995, p. 236.

68. Zhang, Q.H., Qiu, X., and Sharma, S.K., Recent development in pulsed electric field processing, in *New Technologies Yearbook*, National Food Processors Association, Washington D.C., 1997, p. 31.

69. Raso, J. *et al.*, Inactivation of *Zygosaccharomyces bailii* in fruit juices by heat, high hydrostatic pressure and pulsed electric fields, *J. Food Sci.*, 63, 1042, 1998.

70. Evrendilek, G.A., Zhang, Q.H., and Richiter, E.R., Inactivation of *Escherichia coli* O157:H7 and *Escherichia coli* 8739 in apple juice by pulsed electric fields, *J. Food Prot.*, 62, 793, 1999.

71. Iu, J., Mittal, G.S., and Griffiths, M.W., Reduction in levels of *Escherichia coli* O157:H7 in apple cider by pulsed electric fields, *J. Food Prot.*, 64, 964, 2001.

72. McDonald, C.J. *et al.*, Effects of pulsed electric fields on microorganisms in orange juice using electric field strengths of 30 and 50 kV/cm, *J. Food Sci.*, 65, 984, 2000.

73. Terebiznik, M.R. *et al.*, Combined effect of nisin and pulsed electric fields on the inactivation of *Escherichia coli.*, *J. Food Prot.*, 63, 741, 2000.

74. Pol, I.E. *et al.*, Influence of food matrix on inactivation of *Bacillus cereus* by combinations of nisin, pulsed electric field treatment, and carvacrol, *J. Food Prot.*, 64, 1012, 2001.

75. Unal, R., Kim, J.G., and Yousef, A.E., Inactivation of *Escherichia coli* O157:H7, *Listeria monocytogenes*, and *Lactobacillus leichmannii* by combinations of ozone and pulsed electric field, *J. Food Prot.*, 64, 777, 2001.

76. Hodgins, A.M., Mittal, G.S., and Griffiths, M.W., Pasteurization of fresh orange juice using low-energy pulsed electrical field, *J. Food Sci.*, 67, 2294, 2002.

77. Liang, Z., Mittal, G.S., and Griffiths, M.W., Inactivation of *Salmonella* Typhimurium in orange juice containing antimicrobial agents by pulsed electric field, *J. Food Prot.*, 65, 1081, 2002.

78. Min, S. *et al.*, Commercial-scale pulsed electric field processing of orange juice, *J. Food Sci.*, 68, 1265, 2003.

79. Giner, J. *et al.*, Inactivation of peach polyphenoloxidase by exposure to pulsed electric fields, *J. Food Sci.*, 67, 1467, 2002.

80. Min, S., Min, S.K., and Zhang, Q.H., Inactivation kinetics of tomato juice lipoxygenase by pulsed electric fields, *J. Food Sci.*, 68, 1995, 2003.

81. Mason, T.J. and Lorimer, J.P., *Applied Sonochemistry: Uses of Power Ultrasound in Chemistry and Processing*, 1st ed., Wiley-VCH, Berlin, 2002, p. 314.

82. Hoover, D.G., Ultrasound, in Kinetics of Microbial Inactivation for Alternative Food Processing Technologies, *J. Food Sci. Suppl.*, 93, 2000.

83. Hughes, D.E. and Nyborg, W.L., Cell disruption by ultrasound, *Science*, 38, 108, 1962.

84. Palacios, P. *et al.*, Study of substances released by ultrasonic treatment from *Bacillus stearothermophilus* spores, *J. Appl. Bacteriol.*, 71, 445, 1991.

85. Seymour, I.J. *et al.*, Ultrasound decontamination of minimally processed fruits and vegetables, *Int. J. Food Sci. Technol.*, 37, 547, 2002.

86. Mincz, M., Guerrero, S., and Alzamora, S.M., Effectiveness of ultrasound combined with refrigeration on extending shelf-life of fresh fruit juices, 30G-20, presented at IFT Annual Meeting, Anaheim, CA, June 15–19, 2002.

87. McPherson, L.L., Understanding ORP's role in the disinfection process, *Water Eng. Manage.*, 140, 29, 1993.

88. Hayashibara, T., Kadowaki A., and Yuda N., A study of the disinfection/microbiocidal effects of electrolyzed oxidizing water, *Japn. J. Med. Technol.*, 43, 555, 1994.

89. Guan, D. and Li, L., Studies on the production of strongly oxidized water by electrolysis, *J. China Agric. Univ.*, 2, 109, 1997.

90. Hirano, H. and Ueda, O., Characteristics of electrolyzed neutral water possibility of the practical use for food hygiene, *Shokuhin Kogyo*, 40, 25, 1997.

91. Koseki, S. and Itoh, K, Fundamental properties of electrolyzed water, *Nippon Shokuhin Kagaku Kokago Kaishi*, 47, 390, 2000.

92. Venkitanarayanan, K.S. *et al.*, Efficacy of electrolyzed oxidizing water for inactivating *Escherichia coli* O157:H7, *Salmonella enteritidis*, and *Listeria monocytogenes*, *Appl. Environ. Microbiol.*, 65, 4276, 1999.

93. Izumi, H., Electrolyzed water as a disinfectant for fresh-cut vegetables, *J. Food Sci.*, 64, 536, 1999.

94. Park, C.M. *et al.*, Pathogen reduction and quality of lettuce treated with electrolyzed oxidizing and acidified chlorinated water, *J. Food Sci.*, 66, 1368, 2001.

95. Deza, M.A., Araujo, M., and Garrido, M.J., Inactivation of *Escherichia coli* O157:H7, *Salmonella enteritidis* and *Listeria monocytogenes* on the surface of tomatoes by neutral electrolyzed water, *Lett. Appl. Microbiol.*, 37, 482, 2003.

96. Ratna, R.S. and Demirci, A., Treatment of *Escherichia coli* O157:H7 inoculated alfalfa seeds and sprouts with electrolyzed oxidizing water, *Int. J. Food Microbiol.*, 86, 231, 2003.

97. Al-Haq, M.I. *et al.*, Fungicidal effectiveness of electrolyzed oxidizing water on postharvest brown rot of peach, *Hort. Sci.*, 36, 1310, 2001.

98. Al-Haq, M.I. *et al.*, Disinfection effects of electrolyzed oxidizing water on suppressing fruit rot of pear caused by *Botryosphaeria berengeriana*, *Food Res. Int.*, 35, 657, 2002.

23 Biological Control of Microbial Spoilage of Fresh Produce

Julien Mercier and Pamela G. Marrone

CONTENTS

23.1 INTRODUCTION

With the registration and commercialization of biocontrol products such as Aspire (*Candida oleophila*) and BioSave (*Pseudomonas syringae*) in the U.S.

and YieldPlus (*Cryptococcus albidus*) in South Africa, biological control has become a new tool for managing storage diseases, which until recently were only controlled by chemical and cultural means. With the large volume of recent publications on this subject from several on-going research programs, it is possible that more products will be submitted to regulatory agencies and commercialized in the future. The needs for alternatives to chemical control have in a large part been responsible for this effort. Not only is there demand for more organic or "chemical-free" produce, but also many countries have lowered the residue tolerance for many chemical pesticides, putting pressure on exporters to have low residues on the fruit they ship abroad. Furthermore, risk assessments for fungicides make postharvest applications less attractive for pesticide companies, as this might cause them to limit the amounts that can be sold in larger, more profitable markets, such as field crops. Finally, with the prevalence of fungicide resistance in certain pathogen populations [1], biofungicides can be used to manage such resistance and help extend the commercial life of some chemical products. This chapter provides an overview of the different uses of biocontrol agents for the management of postharvest decay, discussing their possibilities and limitations, as well as possible modes of action.

23.2 APPROACHES TO BIOCONTROL IN POSTHARVEST SITUATIONS

23.2.1 USE OF NATURALLY OCCURRING ANTAGONISTS FOR COLONIZATION OF INFECTION SITES

23.2.1.1 Postharvest Applications

So far, the biological control of postharvest diseases with naturally occurring microorganisms has relied essentially on the inundative approach, that is, the mass introduction of a microbe to establish an antagonistic population on wounds and other possible infections sites on fruits or tubers. Little work has been done on other approaches such as the enhancement of the existing surface microflora. While many antagonistic microbes were initially selected from the fruit microflora [2], their natural populations are likely to be too low to have a significant impact on plant pathogens [3]. Mass introduction of yeasts or bacteria permits achieving an instant antagonistic population that would never be attained under normal conditions. This approach has several advantages since a drench or spray treatment of several commodities can easily be applied as the harvested commodities are brought in from the field before packing or storage. If the process is performed in a timely fashion, most wounds sustained during harvesting and handling should become protected before significant pathogen development can occur. Treatment after harvest can also be more economical than treating a whole orchard or field and allows use of more concentrated cell suspensions.

The purpose of antagonists is to colonize rapidly possible infection sites and protect them from infections. Usually, populations of effective antagonists increase rapidly initially and stabilize thereafter. Such a colonization pattern can be seen in fruit wounds treated with yeasts such as *Candida oleophila* [3,4], *Cryptococcus albidus* [5], *C. laurentii* [6], and *Pichia membranefaciens* [7] and bacteria such as *Pseudomonas syringae* [2]. However, wounds can have a narrow window for optimal colonization as they dry out [8]. This requires that application takes place as soon as possible after harvest and handling, as antagonists usually have little curative activity [2]. Also, wounds on oil glands of citrus fruit were found to be more difficult to colonize by *C. oleophila*, resulting in poor decay control [4].

Research has shown that there may be limitations to the postharvest use of antagonists [2], although results comparable to those obtained with synthetic fungicides can sometimes be achieved [9]. Often, a lack of curative activity is the main problem, as efficacy becomes much reduced or nil when antagonists arrive on wounds after pathogens [2,5,10]. Also, antagonists may have different efficacy depending on the fruit species [11] or the type of decay [12]. These limitations likely result from the mode of action of a given antagonist or differing ability to colonize and establish on various commodities. For these reasons, antagonistic yeasts or bacteria cannot be considered as "silver bullets," and as biological systems are more sensitive to environmental conditions than chemical agents. Depending on the disease system, improvements in formulation or combination with other treatments may help increase decay control and meet the requirements of the horticultural industry. Most probably, further improvements in efficacy are likely to come from a better understanding of the mode of action and ecology of these antagonists.

23.2.1.2 Preharvest Applications

While most of the research on naturally occurring antagonists has focused on postharvest treatment, there have been positive reports recently on the use of preharvest applications of antagonists to control postharvest diseases [13–15]. Such application can be done periodically during the growth of the fruit, up to the day of harvest. Field application can help achieve an early colonization of possible infection sites and reduce incipient infections from the field. Also, it can make biological control possible in crops that are too fragile or incompatible with postharvest drenching or spraying, such as grapes and soft fruits. However, field application of antagonists will expose them to possibly adverse environmental conditions such as desiccation and solar radiation, which they will have to withstand in order to be effective. In this situation, the selection process must be different than for postharvest application and take into account the ability of antagonists to survive on the intact fruit surface. Benbow and Sugar showed that certain yeasts are naturally adapted to field conditions and could maintain their populations on pear fruit for three weeks [13]. Using another strategy, Teixidó *et al.* used culture media with low water activity to help *Candida sake* adapt to water stress [15]. Such physiologically modified yeasts were better adapted to colonize apples in the orchard. Our own field research on preharvest applications with *Bacillus subtilis* (Serenade® AS) has shown promise on stone fruit for control of monilinia. On apricots artificially inoculated after harvest with Monilinia, there was only 8% infection when sprayed preharvest with Serenade (applied as 4 Qt in 100 gal of water per acre) compared to 52% infection in untreated fruit. A rate of 4 oz per 100 gal of Elite 45WP (tebuconazole) had 0% infection. In a second test, there was 14% infection with Elite, 28% with Serenade, and 71% in the untreated fruit.

Inoculated trials are the most severe test, and in an actual postharvest situation the results are likely to be better. A trial was conducted with the University of California at Davis on Bing cherries inoculated postharvest with brown rot (monilinia) and gray mold (*Botrytis cinerea*) after preharvest treatment (air blast sprayer, 100 gal water per acre) with Serenade (6 lb) and an adjuvant (Sylgard) and chemical pesticides. The incidence of brown rot decay with the Serenade treatment was 5.6%, compared to 4.5% for triflumizole (Procure 4SC) (rate of 12 fl oz), 0.4% for iprodione (Rovral) 4F (1 Qt), and 32% in the untreated fruit. Against gray mold, there was a 1% incidence of decay (percent of fruit with decay) with Serenade treatment compared to 0% incidence for both Procure and Rovral, and 2.9% incidence for untreated fruit.

23.2.1.3 Possible Mechanisms for Biocontrol

There has been no systematic study of the mode of action of any given postharvest biocontrol agent, and most possible inhibition mechanisms remain unproven at this time. Antagonists could act in passive ways, simply using space or nutrients needed by pathogens, or directly interact with the pathogen to cause inhibition through parasitism or the synthesis of inhibitory molecules, such as antibiotics or hydrolytic enzymes. Finally, the triggering of defense responses in the host, resulting in enhanced resistance, could also be part of the biocontrol mechanism. It is quite possible that in many cases, a number of active and passive mechanisms are involved and act together which make it even more difficult to decipher the basis of the biocontrol phenomenon.

Many antagonistic yeasts that are good wound colonizers are not associated with any obvious inhibitory mechanism. By simply colonizing and forming a cell layer on the surface of wounds, these antagonists may act by competitive or pre-emptive exclusion, blocking access to the infection site. This mechanism is difficult to prove but its occurrence is plausible, especially when a colonization period is required for the antagonist to be effective. Along with competitive exclusion, competition for nutrients is often claimed in the absence of other more obvious or active mechanisms and is supported by the fact that yeasts or yeast-like organisms were able to remove amino acids or sugars in nutrient wells or in wounds [16,17]. Also, the addition of nutrients was shown to cancel antagonistic activity [18]. However, in many cases, the question remains as to whether nutrients in wounds are really limiting for pathogens. Furthermore, yeasts that effectively remove nutrients in wounds are not necessarily good antagonists [16]. As in leaf surface bacteria, where nutrient-regulated reporter genes were used to study nutrient consumption on leaves [19,20], such molecular tools in antagonists or pathogens could be useful for elucidating the question of nutrient competition in wounds.

The involvement of active mechanisms such as antibiotic production by *Pseudomonas syringae* [21], production of cell wall degrading enzymes by *Aureobasidium pullulans* [18] or *Pichia anomala* [22], or attachment to pathogens by various bacteria and yeasts [23,24] have been associated with biocontrol activity on fruits. Again, the definite role of these mechanisms in biocontrol is difficult to demonstrate, and the use of molecular tools might be the best approach to elucidate the role of those antifungal factors. Such a molecular approach was used by Grevesse *et al.* to elucidate the role of β-1,3-glucanase produced by the antagonistic yeast *P. anomala* [25]. The biocontrol activity of the yeast remained unaffected by the shut down of β-1,3-glucanase production from the disruption of a gene involved in the production of the enzyme, thus dismissing its role in antagonism.

So far, the induction of disease resistance in stored fruits and vegetables by antagonists has been little studied. In most cases, it is not known whether biocontrol agents can induce such defense responses. Defense enzymes such as β-1,3-glucanase, chitinase, and peroxidase were induced in apple wounds by *A. pullulans* [26]. In oranges, Arras reported the accumulation of the phytoalexins scoparone and scopoletin in response to *Candida famata* [27]. While these defenses could contribute at least in part to the biocontrol activity, the importance of induced resistance in postharvest biocontrol remains unknown and its possible role is yet to be demonstrated. It is possible that many more antagonists can trigger defense responses and enhance host resistance. It is likely that biocontrol action relying on induced defenses would be rather host-specific, as harvested fruits and vegetables vary in their ability to respond to elicitors and produce defense responses. More advances in the development of biofungicides are likely to come when we better understand mechanisms of biological control on stored commodities.

23.2.2 USE OF MUTANT PATHOGEN STRAINS

Little work has been done on the use of pathogens for biological control. Mutant or attenuated plant pathogens could be used to induce disease resistance or to compete against wild pathogen populations on the host. One promising avenue for the control of aflatoxin production in grain, nuts, and dry fruit involves the use of *Aspergillus flavus* strains that cannot produce toxins [28,29]. These strains are applied in the field, where they compete and help suppress the wild populations of *Aspergillus* spp. While we know that localized inoculation with *Sclerotinia sclerotiorum* and *Botrytis cinerea* can induce systemic disease resistance in cold-stored carrot [30,31], to our knowledge, there has been no study on the use of attenuated pathogens for decay control in fresh fruits and vegetables.

23.2.3 BIOLOGICAL FUMIGATION

The production of volatile antibiotics is rare among microorganisms and has been reported only in a few soilborne organisms such as *Trichoderma* spp. and *Bacillus* spp. [32,33]. A recently discovered fungus, *Muscodor albus*, produces about 28 volatile compounds, mainly alcohol, ester, ketone, and acid derivatives, which together can inhibit or kill fungi, bacteria, and oomycetes [34]. The fungus, which was isolated from a cinnamon tree in Honduras, was described as a new genus and is related to endophytes of the family Xylariaceae (Ascomycetes) [35]. Recently, the possibility of controlling postharvest decay by biological fumigation with *M. albus* was demonstrated by Mercier and Jiménez [36]. Biofumigation was performed passively by placing a grain culture of the fungus in the presence of inoculated fruits. Diseases controlled by such biofumigation treatment were gray mold of apples and grapes, caused by *Botrytis cinerea*, blue mold of apples, caused by *Penicillium expansum*, brown rot of peaches, caused by *Monilinia fructicola*, and green mold and sour rot of lemons, caused by *P. digitatum* and *Geotrichum citri-aurantii*, [36,37]. Several storage pathogens belonging to species of botrytis, colletotrichum, geotrichum, monilinia, penicillium, and rhizopus were also killed *in vitro* by exposure to volatile compounds produced by a potato dextrose agar colony of *M. albus* [36]. This suggests that pathogens are not merely inhibited but are killed in fruit wounds. In some cases, there was effective decay control when biofumigation was performed 24 hours after inoculation. Fumigation at low storage temperature was also effective in grapes [37] and apples (J. Mercier, unpublished data). Besides controlling postharvest decay, biofumigation with *M. albus* also reduced populations of pathogenic bacteria such as *Escherichia coli* O157:H7, salmonella serotypes, *Shigella* spp., and *Listeria monocytogenes* on the surface of cucurbit and tomato fruits [38]. Biofumigation treatment with *M. albus* would be applicable to different stages of storage and shipping in most commodities. It could also be used with commodities that are too fragile to be handled or to receive a liquid fungicide treatment, such as strawberries or grapes. Exposure to the volatiles does not cause any off-flavor of the treated produce in informal taste tests. There were no detectable volatile compounds in the skin of any fruit tested (apples, peaches).

23.3 ADVANTAGES AND LIMITATIONS OF POSTHARVEST BIOCONTROL

23.3.1 ADVANTAGES OF POSTHARVEST BIOCONTROL

Biocontrol agents have the following advantages:

1. *Complex mode of action: low chance for resistance development.* Biocontrol agents work in complex ways, rather than having a single site of action as for some chemical pesticides. For example, a

biocontrol agent may act via antagonism and antibiosis (production of multiple compounds by the organism). If used in a program with chemical pesticides as tank mixes, biocontrol agents could delay development of resistance to chemical products, which has been documented and can be a problem in commercial production [39,40].

2. *Reduction or elimination of chemical residues.* The production of fruit and vegetables is now a global enterprise, and products are shipped around the world for export markets. As such, chemical residues are a consumer concern as well as nontariff trade barriers. Products can be rejected due to the presence of residues of specific chemical products not allowed in importing countries or because residue levels exceed limits allowed by importing countries and companies [41]. In addition, the Food Quality Protection Act of 1996 requires that the registration and use of a product take into account the amount of pesticide residue that occurs in foods eaten frequently by children. If a product is also used preharvest, the "risk cup" (total amount of allowable residues on all crops) from postharvest use may result in restrictions of quantity and frequency of applications for preharvest use. This may not be economically feasible to companies who make more money on preharvest markets. Biocontrol agents are exempt from residue tolerances and thus not subject to international rules regulating chemical residues.

3. *Safety to workers and the environment.* Worker safety is a significant worldwide concern. Postharvest handling of fruits and vegetables may result in worker exposure to chemical residues of products that are carcinogens or acutely toxic. For example, certain governments in Central America have strongly encouraged banana producers to reduce the use of chemicals due to worker exposure [42]. Use of biocontrol agents will reduce the worker exposure to these toxic chemicals. Waste chemicals from fruit dipping operations are an environmental issue. Biocontrol agents are biodegradable, leave no chemical residues, and do not pollute the environment and ground water.

23.3.2 DISADVANTAGES OF BIOCONTROL AGENTS

Biocontrol agents have not met with significant commercial success for post-harvest applications and have remained relegated to small niches despite the market need for methods to reduce the development of resistance to chemical products. Biocontrol agents have the following disadvantages:

1. *Special handling and timing and use restrictions.* Some biocontrol agents cannot be tank mixed with other fungicide products because the fungicides will kill (and thus deactivate) the biocontrol agent. This limits their use and negates the most positive reason to use

them: resistance management. Also, when a biocontrol agent is an antagonist to the pathogen, timing the application is crucial. The microorganism must be applied shortly before or at the same time as exposure to the pathogen for efficacy. This reduces the agent's practicality in packinghouses.

2. *Efficacy*. Efficacy reports are mixed largely because users expect to use a biocontrol product like a chemical without special regard to timing and handling. Efficacy can equal or approach chemicals if the biocontrol agent is used exactly as required and directed, with an understanding and appreciation of the mechanism of action.

3. *Shelf stability*. Some biocontrol agents do not produce spores and consist only of bacterial cells or fungal mycelia, and as a result the shelf life of a biocontrol product may not match that of a chemical product. Maintaining a high degree of viability and efficacy in formulated biocontrol agents can be a challenging task. As with most pesticides, biofungicides are commonly dehydrated into powder or concentrated into liquid formulations. Such processes can have a negative impact on disease control efficacy. As most information on the formulation of biopesticides is proprietary, there is limited literature on the subject. For example, freeze-drying of *Candida sake* caused cell mortality and greatly affected control of blue mold of apple, compared to that obtained with fresh cells [43]. Protective agents and additives, such as skimmed milk, peptone, or lactose, at the time of freeze-drying and rehydration can somewhat help reduce cell mortality of *C. sake* [43]. As mentioned above, growing yeast cells at low osmotic potential can be used for better desiccation adaptation [15].

23.4 ENHANCING BIOCONTROL ACTIVITY

23.4.1 IN COMBINATION WITH OTHER TREATMENTS

Under the experimental conditions used by some researchers, there have been occurrences of disappointing results when the currently registered biocontrol products were used as stand-alone treatments [44–46]. Research on biocontrol organisms, whether for pre- or postharvest uses, is typically reductionist — testing one product versus another in stand-alone trials. When the biocontrol agent does not perform equally well as the chemical in these stand-alone trials, the biocontrol agent may be pronounced inferior, and that is usually the end of the story. Research shows that mixtures of biocontrol agents with other biorational products can provide excellent efficacy not seen in stand-alone trials. For example, Conway *et al.* showed that they could eliminate postharvest decay caused by *Colletotrichum acutatum* on apples with a combination of either of two antagonists and heat [47]. Either antagonist alone eliminated decay caused by *Penicillium expansum*, but the two were more effective together. Combinations with chemical agents should be tested as well to prevent resistance, increase efficacy, and reduce the amount of chemicals that could be used. Continued funding for research of this type will increase the probability that biocontrol agents will be more widely adopted for postharvest uses. For this reason, combination with other treatments might be envisaged to achieve more consistent and robust disease control. Biocontrol agents are not necessarily incompatible with chemical fungicides, which can often be used at reduced rate in such situations. The use of several antagonistic bacteria and yeasts, including the agents in BioSave and Aspire, has been shown to be compatible with postharvest chemical fungicides at full or reduced rates, and such combinations have resulted in higher efficacy than biocontrol agents alone [5,7,12,44–46,48,49]. Lowering the use rates of chemical fungicides, while helping reduce chemical residues on fruit, can also compensate for possible shortcomings of biofungicides against early or incipient infections and sensitivity to environmental conditions.

The combination of biocontrol agents with other storage technologies is not only desirable, but necessary. Biocontrol agents should be compatible with low temperature and controlled or modified atmosphere storage, as these methods are commonly used to extend the shelf life of fruits and vegetables. So far, reports of combining antagonistic yeasts with controlled or modified atmospheres have been positive, improving the performance of yeasts in high CO_2 and low O_2 compared to storage in air [11,12]. Several antagonistic yeasts are capable of growth and survival on fruit held in cold storage [3,5,49,50], which make them compatible for use on commodities that require rapid chilling and storage at low temperatures. Combination of biocontrol agents with other cultural disease control methods like fruit curing [51] or heat treatment [52,53] could also be advantageous depending on the commodity. Finally, the combination of two or more antagonists with different characteristics, which could be considered as mixtures of microorganisms, can be more effective in controlling decay than single agents [3,10]. While it has been suggested that the postharvest environment can be manipulated somewhat to accommodate biological control, it is likely that the most successful biofungicides will be effective under the normal handling and storage conditions used for each commodity.

23.4.2 IMPROVEMENT IN FORMULATION

The addition of nontoxic substances in biofungicide formulations or to spray preparations could also be used to improve disease control. Although their mode of action is often poorly understood, these "inert" ingredients can act in several ways, such as reducing disease susceptibility, inhibiting pathogens, or improving wound colonization or the inhibitory action of antagonists. Combining yeasts with calcium salts [5,7,54], sodium bicarbonate [55,56], or ammonium molybdate [56,57] resulted in better control of fruit diseases. Since salts are inexpensive additives, more research should be done on their mode of action and their compatibility with biofungicides.

Other chemicals such as sugar analogs like 2-deoxy-D-glucose [58] and chitosan [59] also improved decay control by yeasts. Some amino acids can enhance biocontrol activity and improve colonization of wounds by antagonists [50,60]. Thus, further enhancement of the activity of biocontrol agents can be expected with benign chemicals likely to be suitable for use on food crops.

23.4.3 SCREENING AND SELECTION OF THE MICROORGANISM

Relatively few microorganisms have been screened for postharvest use. AgraQuest (Davis, CA) screened over 20,000 microorganisms of diverse taxonomies for preharvest uses and found many microorganisms that equal or approach the efficacy of chemical pesticides. The same approach could be taken with screening for postharvest use.

There are many microorganisms that have a long shelf life due to hardy spores (bacillus, some actinomycetes, and some fungi). Efforts to date have focused largely on microorganisms wherein the living cells act as antagonists by competing with the plant pathogen, as opposed to microorganisms that work through other mechanisms such as antibiosis. Microorganisms are known to produce compounds that can disrupt the membranes of plant pathogens [61]. By releasing these compounds, such microbes can be effective in tank mixes and perform more like chemical pesticides.

23.4.4 COLLABORATIVE RESEARCH AMONG INDUSTRY, UNIVERSITY RESEARCHERS, GOVERNMENT, AND GROWERS/PACKERS

Canada's Biocontrol Network (www.biocontrol.ca) is a model of how to increase adoption of biocontrol agents through a coordinated research and testing effort by all stakeholders.

> We work on replacing pesticides with effective and economically viable biocontrol treatments based on the natural enemies of insect pests and disease pathogens, used in coordinated and synergistic ways. We establish effective and mutually beneficial research partnerships with the private sector (pest management products and services companies and grower associations) as well as other stakeholders in the field of plant protection.

The Biocontrol Network efforts are organized around a seven-stage product-oriented process:

1. Identification of users' needs
2. Ecological studies
3. Screening

4. Product development: production
5. Product development: efficacy
6. Product development: environmental impact and toxicology
7. Registration (if needed), marketing, training, and education

If a similar approach were taken in the U.S. or other countries, adoption of new biological control agents would increase, to the benefit of the environment, workers, and consumers.

23.5 REGULATORY PROCESS FOR BIOCONTROL AGENTS

23.5.1 U.S. ENVIRONMENTAL PROTECTION AGENCY (EPA)

Under FIFRA (Federal Insecticide, Fungicide, and Rodenticide Act), biological pesticides (also known as biopesticides) are regulated by the EPA's Biopesticide Pollution and Prevention Division. Biopesticides fall into the categories of "microbials" and "biochemicals."

Microbial pesticides contain a microorganism (e.g., a bacterium, fungus, virus, or protozoan) as the active ingredient. The most widely used microbial pesticides are various types of the bacterium *Bacillus thuringiensis*, or Bt. Biochemical pesticides are naturally occurring substances that control pests by nontoxic mechanisms. Biochemicals include products, such as pheromones, that interfere with pests' growth or mating patterns, and certain plant growth regulators that increase the productivity of many crops.

Products can be registered as biochemicals provided the active ingredients are natural or derived from a natural source, show no direct toxic effects, and have a specific, nontoxic mode of action. Biopesticides (microbials and biochemicals) can take considerably less time and money to bring to market than synthetic chemicals (3 years and \$3 million to \$6 million versus >10 years and \$185 million for synthetic chemicals).

Each new biopesticide must go through Tier I toxicology, ecotoxicology, and end product (final formulation) tests. If there are no direct toxic effects in these tests, no further testing is required.

The following data are required for registering a biopesticide active ingredient:

- Product chemistry, batch analysis
- Microbiology/human pathogens
- Acute toxicity/pathogenicity
- Ecological effects (nontarget birds, fish, invertebrates, insects, plants)
- Primary dermal and eye irritation

The following data are required for registering a biopesticide "end use" (formulated):

- Product chemistry/storage stability
- "Acute 6-pack":
    Acute oral LD50
    Acute dermal LD50
    Primary eye irritation
    Primary dermal irritation
    Hypersensitivity
    Acute inhalation

The FFDCA (Federal Food, Drug, and Cosmetic Act) requires a tolerance for all chemical pesticides (a limit on the amount of chemical allowed in a fresh or processed food). All biopesticides registered so far have been exempt from tolerance because of their safety.

23.5.2 CALIFORNIA AND INTERNATIONAL REGULATIONS

California EPA's Department of Pesticide Regulation (DPR) regulates biopesticides. Currently, biopesticides are given favorable "fast track" treatment. A biopesticide application submitted concurrently to the U.S. EPA and DPR may be approved by both agencies at the same time, or DPR approval may come before the EPA approval. Sales cannot occur until there is federal approval, however. California currently requires efficacy data for approval. The EPA may ask for efficacy data, but typically does not.

Canada has coordinated its biopesticide regulations with the U.S. so that biopesticides can be registered concurrently in the U.S. and Canada if a registrant chooses to submit the applications at the same time. Otherwise, a Canadian registration will take one year or more for approval after submitting a separate application. In Mexico registration can take six months to years after U.S. approval, although there is currently a NAFTA (North American Free Trade Association) harmonization initiative. In Japan there is an EPA-like tiered system, which takes approximately 12 months after registration submission for approval. In Europe, the European Union is harmonizing biopesticide regulations, and the regulations for biopesticides are becoming more favorable, but registration still takes several years. Most countries outside the U.S. require two years of official field trials conducted by government entities, making the timeline longer than the U.S. process.

23.6 CONCLUDING REMARKS

Progress on postharvest biocontrol has been accomplished in the last 20 years, but biocontrol agents have much more unrealized potential. A better knowledge of antagonist modes of action, screening of additional microorganisms, improvements in biofungicide formulation, and the ecology of antagonists on fruits and vegetables will help in the design of better application strategies and ensure more reliable disease control. Land grant and U.S. Department of Agriculture research should increase their focus on practical implementation of biocontrol agents in integrated programs to reduce the chemical load on postharvest commodities. Setting up a postharvest biocontrol network like the Canadian network can provide a dedicated focus on discovery, development, and adoption.

REFERENCES

1. Holmes, G.J. and Eckert, J.W., Sensitivity of *Penicillium digitatum* and *P. italicum* to postharvest citrus fungicides in California, *Phytopathology*, 89, 716–721, 1999.
2. Janisiewicz, W.J. and Korsten, L., Biological control of postharvest diseases of fruits, *Annu. Rev. Phytopathol.*, 40, 411–441, 2002.
3. Mercier, J. and Wilson, C.L., Colonization of apple wounds by naturally occurring microflora and introduced *Candida oleophila* and their effect on infection by *Botrytis cinerea* during storage, *Biol. Cont.*, 4, 138–144,1994.
4. Brown, G.E., Davis, C., and Chambers, M., Control of citrus green mold with Aspire is impacted by the type of injury, *Postharvest Biol. Technol.*, 18, 57–65, 2000.
5. Fan, Q. and Tian, S., Postharvest biological control of grey mold and blue mold on apple by *Cryptococcus albidus* (Saito) Skinner, *Postharvest Biol. Technol.*, 21, 341–350, 2001.
6. Roberts, R.G., Postharvest biological control of gray mold of apple by *Cryptococcus laurentii. Phytopathology*, 80, 526–530, 1990.
7. Qing, F. and Shiping, T., Postharvest biological control of Rhizopus rot of nectarine fruits by *Pichia membranefaciens, Plant Dis.*, 84, 1212–1216, 2000.
8. Mercier, J. and Wilson, C.L., Effect of wound moisture on the biocontrol by *Candida oleophila* of gray mold rot (*Botrytis cinerea*) of apple, *Postharvest Biol. Technol.*, 6, 9–15, 1995.
9. Janisiewicz, W.J. and Jeffers, S.N., Efficacy of commercial formulation of two biofungicides for control of blue mold and gray mold of apples in cold storage, *Crop Prot.*, 16, 629–633, 1997.
10. Janisiewicz, W.J., Biocontrol of postharvest diseases of apples with antagonistic mixtures, *Phytopathology*, 78, 194–198, 1988.
11. Tian, S.P., Fan, Q., Xu, Y., and Liu, H.B., Biocontrol efficacy of antagonist yeasts to gray mold and blue mold on apples and pears in controlled atmospheres, *Plant Dis.*, 86, 848–853, 2002.
12. Spotts, R.A., Cervantes, L.A., Facteau, T.J., and Chand-Goyal, T., Control of brown rot and blue mold of sweet cherry with preharvest iprodione, postharvest *Cryptococcus infirmo-miniatus*, and modified atmosphere packaging, *Plant Dis.*, 82, 1158–1160, 1998.
13. Benbow, J.M. and Sugar, D., Fruit surface colonization and biological control of postharvest diseases of pear by pre-harvest yeast applications, *Plant Dis.*, 83, 839–844, 1999.

14. Ippolito, A. and Nigro, F., Impact of pre-harvest applications of biological control agents on postharvest diseases of fresh fruits and vegetables, *Crop Prot.*, 19, 715–723, 2000.

15. Teixidó, N., Viñas, I., Usall, J., and Magan, N., Control of blue mold of apples by preharvest application of *Candida sake* grown in media with different water activity, *Phytopathology*, 88, 960–964, 1998.

16. Filonow, A.B., Role of competition for sugars by yeasts in the biocontrol of gray mold of apple, *Biocontrol Sci. Technol.*, 8, 243–256, 1998.

17. Janisiewicz, W.J, Tworkoski, T.J., and Sharer, C., Characterizing the mechanism of biological control of postharvest diseases on fruits with a simple method to study competition for nutrients, *Phytopathology*, 90, 1196–1200, 2000.

18. Castoria, R., De Curtis, F., Lima, G., Caputo, L., Pacifico, S., and De Cicco, V., *Aureobasidium pullulans* (LS-30) an antagonist of postharvest pathogens of fruits: study of its modes of action, *Postharvest Biol. Technol,.* 22, 7–17, 2001.

19. Leveau, J.H.J. and Lindow, S.E., Appetite of an epiphyte: quantitative monitoring of bacterial sugar consumption in the phyllosphere, *PNAS*, 98, 3446–3453, 2001.

20. Miller, W.G., Brandl, M.T., Quiñones, B., and Lindow, S.E., Biological sensor for sucrose availability: relative sensitivities of various reporter genes, *Appl. Environ. Microbiol.*, 67, 1308–1317, 2001.

21. Bull, C.T., Wadsworth, M.L., Sorensen, K.N., Tekemoto, J.Y., Austin, R.K., and Smilanick, J.L., Syringomycin E produced by biological control agents controls green mold of lemons, *Biol. Cont.*, 12, 89–95, 1998.

22. Jijakli, M.H. and Lepoivre, P., Characterization of an exo-β-glucanase produced by *Pichia anomala* strain K, antagonist of *Botrytis cinerea* on apples, *Phytopathology*, 88, 335–343, 1998.

23. Cook, D.W.M., Long, P.G., Ganesh, S., and Cheah, L.-H., Attachment microbes antagonistic against *Botrytis cinerea*: biological control and scanning electron microscope studies in vivo, *Ann. Appl. Biol.*, 131, 503–518, 1997.

24. Wisniewski, M., Biles, C., Droby, S., McLauglin, R., Wilson, C.L., and Chalutz, E., Mode of action of the postharvest biocontrol yeast *Pichia guilliermondii*. Characterization of attachment to *Botrytis cinerea*, *Physiol. Mol. Plant Pathol.*, 39, 259–267, 1991.

25. Grevesse, C., Lepoivre, P., and Jijakli, M.H., Characterization of the exoglucanase encoding gene *PaEXG2* and study of its role in the biocontrol activity of *Pichia anomala* Strain K, *Phytopathology*, 93, 1145–1152, 2003.

26. Ippolito, A., El-Ghaouth, A., Wilson, C.L., and Wisniewski, M., Control of postharvest decay of apple fruit by *Aureobasidium pullulans* and induction of defense responses, *Postharvest Biol. Technol.*, 19, 265–272, 2000.

27. Arras, G., Mode of action of an isolate of *Candida famata* in biological control of *Penicillium digitatum* in orange fruits, *Postharvest Biol. Technol.*, 8, 191–198,1996.

28. Doster, M., Michailides, T., Cotty, P., Doyle, J., Morgan, D., Boeckler, L., Felts, D., and Reyes, H., Aflatoxin Control in Figs: Biocontrol and New Resistant Cultivars, in Proceedings 3rd Fungal Genomics, 4th Fumonisin, and 16th Aflatoxin Elimination Workshops, Savannah, GA, Oct. 13–15, 2003, p. 86.

29. Michailides, T., Doster, M., Cotty, P., Morgan, D., Boeckler, L., Felts, D., and Reyes, H., Aflatoxin Control in Pistachios: Biocontrol Using Atoxigenic Strains, in Proceedings 3rd Fungal Genomics, 4th Fumonisin, and 16th Aflatoxin Elimination Workshops, Savannah, GA, Oct. 13–15, 2003, p. 87.

30. Mercier, J. and Arul, J., Induction of systemic disease resistance in carrot roots by pre-inoculation with storage pathogens, *Can. J. Plant Pathol.*, 15, 281–283, 1993.

31. Mercier, J., Roussel, D., Charles, M.-T., and Arul, J., Systemic and local responses associated with UV- and pathogen-induced resistance to *Botrytis cinerea* in stored carrots, *Phytopathology*, 90, 981–986, 2000.

32. Dennis, C. and Webster, J., Antagonistc properties of specie-groups of *Trichoderma*. Part II. Production of volatile antibiotics, *Trans. Br. Mycol. Soc.*, 57, 41–48, 1971.

33. Fiddaman, P.J. and Rossall, S., The production of antifungal volatiles by *Bacillus subtilis*, *J. Appl. Bacteriol.*, 74, 119–126, 1993.

34. Strobel, G.A., Dirkse, E., Sears, J., and Markworth, C., Volatile antimicrobials from *Muscodor albus*, a novel endophytic fungus, *Microbiology*, 147, 2943–2950, 2001.

35. Worapong, J., Strobel, G., Ford, E.J., Li, J.Y. Baird, G., and Hess, W.M., *Muscodor albus* anam. sp. Nov., an endophyte from *Cinnamomum zeylanicum*, *Mycotaxon*, 79, 67–79, 2001.

36. Mercier, J. and Jiménez, J., Control of fungal decay of apples and peaches by the biofumigant fungus *Muscodor albus*, *Postharvest Biol. Technol.*, 31, 1–8, 2004.

37. Mercier, J. and Smilanick, J.L., Control of green mold and sour rot of lemons and gray mold rot of grapes by biofumigation with *Muscodor albus*, *Phytopathology*, 93, S61, 2003 (abstr.).

38. Suslow, T.V., deFreita, P.M., and Mercier, J., Efficacy of the mycofumigant Arabesque™ (*Muscodor albus*) in postharvest pathogen control on fruit–vegetables, ISHS Meeting, Postharvest Symposium, Verona, Italy, June 6–10, 2004.

39. Prusky, D., Bazak, M., and BenArie, R., Development, persistence, survival, and strategies for control of thiabendazole-resistant strains of *Penicillium expansum* on pome fruit, *Phytopathology*, 75, 877–882, 1985.

40. Kuferman, E., Postharvest applied chemicals to pears: a survey of pear packers in Washington, Oregon, and California, *Tree Fruit Postharvest J.*, 9, 2–24, 1998.

41. Gullino, M.L. and Kuijers, L.A.M., Social and political implications of managing plant diseases with restricted fungicides in Europe, *Annu. Rev. Phytopathol.*, 32, 559–579, 1994.

42. Mlot, C., Greening the world's most popular fruit, *National Wildlife*, Feb./Mar., 18–20, 2004.

43. Abadias, M., Teixidó, N., Usall, J., Benabarre, A., and Viñas, I., Viability, efficacy, and storage stability of freeze-dried biocontrol agent *Candida sake* using different protective and rehydration media, *J. Food Prot.*, 64, 856–861, 2001.

44. Brown, G.E. and Chambers, M., Evaluation of biological products for the control of postharvest diseases of Florida citrus, *Proc. Fla. State Hortic. Soc.*, 109, 278–282, 1996.

45. Droby, S., Cohen, L., Daus, A., Weiss, B. Horev, B., Chalutz, E., Katz, H., Keren-Tzur, M., and Shachnai, A., Commercial testing of Aspire: a yeast

preparation for the biological control of postharvest decay of citrus, *Biol. Control*, 12, 97–101, 1998.

46. Sugar, D. and Spotts, R.A., Control of postharvest decay in pear by four laboratory-grown yeasts and two registered biocontrol products, *Plant Dis.*, 83, 155–158, 1999.

47. Conway, W.S., Leverentz, B., Janisiewicz, W., Saftner, R.A., and Camp, M.J., Improving biocontrol using antagonist mixtures with heat and/or sodium bicarbonate to control post harvest decay of apple fruit, *Phytopathology*, 94, S20 (abstr.), 2004.

48. Pusey, P.L., Wilson, C.L., Hotchkiss, M.W., and Franklin, J.D., Compatibility of *Bacillus subtilis* for postharvest control of peach brown rot with commercial fruit waxes, dicloran, and cold-storage conditions, *Plant Dis.*, 70, 587–590, 1986.

49. Usall, J., Teixidó, N., Torres, R., Ochoa de Eribe, X., and Viñas, I., Pilot tests of *Candida sake* (CPA-1) applications to control postharvest blue mold on apple fruit, *Postharvest Biol. Technol.*, 21, 147–156, 2001.

50. Vero, S., Mondino, P., Burgueño, J., Soubes, M., and Wisniewski, M., Characterization of biocontrol activity of two yeast strains from Uruguay against blue mold of apple, *Postharvest Biol. Technol.*, 26, 91–98, 2002.

51. Cook, D.W.M., Long, P.G., and Ganesh, S., The combined effect of delayed application of yeast biocontrol agents and fruit curing for the inhibition of the postharvest pathogen *Botrytis cinerea* in kiwifruit, *Postharvest Biol. Technol.*, 16, 233–243, 1999.

52. Janisiewicz, W.J, Leverentz, B., Conway, W.S., Saftner, R.A., Reed, A.N., and Camp, M.J., Control of bitter rot and blue mold of apples by integrating heat and antagonist treatments on 1-MCP treated fruit stored under controlled atmosphere conditions, *Postharvest Biol. Technol.*, 29, 129–143, 2003.

53. Leverentz, B., Janisiewicz, W.J., Conway, W.S., Saftner, R.A, Fuchs, Y., Sam, C.E.. and Camp, M.J., Combining yeasts or a bacterial biocontrol agent and heat treatment to reduce postharvest decay of "Gala" apples, *Postharvest Biol. Technol.*, 21, 87–94, 2000.

54. McLaughlin, R.J., Wisniewski, M.E., Wilson, C.L., and Chalutz, E., Effect of inoculum concentration and salt solutions on biological control of postharvest diseases of apple with *Candida* sp., *Phytopathology*, 80, 456–461, 1990.

55. Gamagae, S.U., Sivakumar, D., Wilson Wijeratnam, R.S., and Wijesundera, R.L.C., Use of sodium bicarbonate and *Candida oleophila* to control anthracnose in papaya during storage, *Crop. Prot.*, 22, 775–779, 2003.

56. Wan, Y.K., Tian, S.P., and Qin, G.Z., Enhancement of biocontrol activity of yeasts by adding sodium bicarbonate or ammonium molybdate to control postharvest disease of jujube fruits, *Lett. Appl. Microbiol.*, 37, 249–253, 2003.

57. Nunes, C., Usall, J., Teixidó, N., Abadias, M., and Viñas, I., Improveed control of postharvest decay of pear by the combination of *Candida sake* (CPA-1) and ammonium molybdate, *Phytopathology*, 92, 281–287, 2002.

58. El-Ghaouth, A., Smilanick, J.L., Wisniewski, M., and Wilson, C.L., Improved control of apple and citrus fruit decay with a combination of *Candida saitoana* and 2-deoxy-D-glucose, *Plant Dis.*, 84, 249–253, 2000.

59. El-Ghaouth, A., Smilanick, J.L., and Wilson, C.L., Enhancement of the performance of *Candida saitoana* by the addition of glucochitosan for the control of postharvest decay of apple and citrus fruit, *Biocontrol Sci. Technol.*, 19, 103–110, 2000.
60. Janisiewicz, W.J, Usall, J., and Bors, B., Nutritional enhancement of biocontrol of blue mold on apples, *Phytopathology*, 82, 1364–1370, 1992.
61. Marrone, P.G., An effective biofungicide with novel modes of action, *Pesticide Outlook*, 213, 193–194, 2002.

Section V

Microbiological Evaluation of Fruits and Vegetables

24 Sampling, Detection, and Enumeration of Pathogenic and Spoilage Microorganisms

Larry R. Beuchat

CONTENTS

24.1 INTRODUCTION

Fruits and vegetables can become contaminated with spoilage and pathogenic microorganisms at several points from the field through to the time they are consumed. Given sufficient time at an appropriate temperature, some pathogens can grow on produce to populations exceeding 10^7 CFU/g, resulting in increased risks of human infections. Outbreaks of human illnesses associated with the consumption of raw fruits and vegetables and unpasteurized fruit juices have been documented with increased frequency in recent years. [1–3]. Conditions affecting survival and growth of pathogens and spoilage microorganisms on raw produce have been studied extensively. A wide range of chemical and physical treatments have been evaluated for their effectiveness in killing microorganisms on raw fruits and vegetables [4]. Substantial variations in conditions used to prepare inoculum, methods for inoculation, storage of samples after inoculation, and application of treatments have been used. Procedures used to sample, detect, and enumerate pathogens and spoilage in microorganisms on raw produce have also varied across laboratories, making it difficult to compare the results.

There is a need to develop and validate standard methods to determine accurately the presence and numbers of pathogenic and spoilage bacteria, yeasts, molds, parasites, and viruses on raw fruits and vegetables. These

TABLE 24.1

Considerations When Developing Standard Method(s) for Determining the Efficacy of Sanitizers in Killing Pathogenic Microorganisms, and Survival and Growth of Pathogens on Raw Fruits and Vegetables

Type of produce
    Whole or cut
    Washed, brushed, waxed, or oiled
    Botanical part (fruit, leaf, stem, flower, root, tuber)
Pathogen of interest
    Gram-negative or Gram-positive bacteria, parasite, or virus; mixture of strains or a single strain
    Marker or no marker
    Conditions for preparing inoculum
    Number of cells in inoculum
Procedure for inoculation
    Composition of carrier
    Temperature of produce and inoculum
    Dip, spray, or spot inoculum
    Temperature and relative humidity between time of inoculation, testing, and analysis
Procedure for evaluating test condition
    Define treatment, condition, or sanitizer
    Method for measurement of concentration and activity
    Temperature of produce and treatment condition or sanitizer
    Dipping, spraying, fogging, or atmospheric
    Agitated, rubbed, or static condition during exposure
    Time of exposure of inoculated produce to sanitizer or condition
    Ratio of sanitizer to produce sample
    Blending, homogenizing, macerating, or washing
    Time of treatment
    Composition of neutralizer (for sanitizer studies)
    Detection and enumeration media
    Conditions for incubating plates and broth
    Confirmation procedures
Reporting results
    Number of replicates and samples/replicate
    CFU/g, CFU/cm^2, CFU/piece, fraction negative
    Appropriate statistical analysis and interpretation

methods can then be used in studies focused on determining survival and growth characteristics in challenge studies and efficacy of antimicrobial treatments in killing specific pathogenic and spoilage microorganisms that may be present on raw produce. The objective would be to develop, validate, and recommend, through an appropriate authoritative body, a basic experimental protocol or protocols that could be modified according to specific applications to various groups of fruits and vegetables.

Some of the factors that should be considered when developing a standard method(s) for determining the effectiveness of sanitizers in killing microorganisms or, in the case of challenge studies, to determine the survival and growth characteristics of microorganisms on raw fruits and vegetables are listed in Table 24.1. These include the type of produce to be examined, anticipated population of pathogenic or spoilage microorganism or group of microorganisms to be used in the inoculum or naturally present on produce, composition of the carrier for the inoculum, and conditions for storing produce between the time of inoculation and treatment or sampling. The time produce is exposed to chemical or physical treatment, the temperature of the produce and treatment solution, procedures for washing produce after treatment, and procedures for removing and enumerating viable cells of pathogenic and spoilage microorganisms after treatment should be standardized.

Modifications of a basic analytical method for groups of fruits and vegetables may be necessary to enable the most accurate detection or enumeration of microorganisms of interest and to determine accurately the efficacy of sanitization treatments. These modifications will be necessary for a yet to be determined number of groups of fruits and vegetables to be defined according to similarities and differences in surface morphology and hydrophobicity, internal tissue composition, and conditions of processing, e.g., washing, brushing, or waxing, to which they had been previously subjected. Observations on current methods and those under development, with options and suggestions concerning directions that might be taken to establish standard methods to detect accurately or enumerate pathogenic and spoilage microorganisms on raw fruits and vegetables, are presented here.

24.2 PATHOGEN OR SPOILAGE MICROORGANISM UNDER STUDY

An evaluation of the efficacy of treatments to sanitize fruits and vegetables or the appropriateness of experimental protocols for challenge studies to determine the survival or growth characteristics of pathogenic or spoilage microorganisms must be preceded by the development and validation of standard methods for detection and enumeration. Broad considerations that need to be assessed in evaluating or developing standard protocols for pathogenic and spoilage microorganisms must be addressed.

24.2.1 MEDIA FOR ROUTINE MICROBIOLOGICAL ANALYSES

Methods have been developed and validated to analyze raw and processed foods of animal origin and processed foods of plant origin for the presence and populations of pathogenic and spoilage microorganisms. Methods for analysis of raw fruits and vegetables, in contrast, are not well defined. Procedures for enrichment and direct plating of raw produce samples in the U.S. are generally modifications of those outlined for processed fruits and vegetables in the *Bacteriological Analytical Manual* (BAM) of the U.S. Food and Drug Administration [5] or the *Compendium of Methods for the Microbiological Examination of Foods* published by the American Public Health Association (APHA) [6]. A 25 g analytical unit diluted at a 1:9 ratio (weight:volume) of sample:diluent or sample:broth is prescribed for most foods. The food and diluent or broth are then mixed, swirled, soaked, or blended before withdrawing samples for direct plating or incubated for a specified time at a given temperature for preenrichment or enrichment. In part because of the lack of a standard method(s) to analyze raw fruits and vegetables for pathogens and spoilage microorganisms, researchers have modified BAM and APHA methods to suit their needs to analyze specific types of produce, often without validation of the efficiency of detection or recovery of the test microorganism or group of microorganisms under investigation. Substantial variations in methods to select and process samples for preenrichment, enrichment, and direct plating have been used to analyze raw produce for the presence and populations of pathogens. To illustrate these variations, Table 24.2 gives some examples of procedures used by researchers to analyze produce for salmonella.

The inability of injured or stressed cells of pathogenic bacteria to resuscitate and grow on selective media is too often not recognized. Selective media recommended for foods other than raw fruits and vegetables have been formulated and evaluated for that purpose. The same media may not perform well for detection and enumeration of pathogens or other microorganisms on produce. The number of stressed salmonella recovered from tomatoes [24,31,32], lettuce [33], alfalfa sprouts and seeds [32,35], and parsley [33], for example, is significantly less on selective versus nonselective media. Recovery of *Escherichia coli* O157:H7 and *Listeria monocytogenes* from tomatoes [31], lettuce, and parsley [33] treated with sanitizers has also been shown to be less on selective media.

While progress has been made in developing media for enumerating yeasts and molds in a wide range of foods and beverages [36], less attention has been given to evaluating the performance of diluents in removing and dispersing fungal propagules on fruits and vegetables. Standard methods for determining yeast and mold counts in foods recommend 0.1% peptone as a diluent [5,37,38]. A study done by Beuchat *et al.* [39] was aimed at determining the retention of viability of mycoflora recovered in seven diluents used to wash seven types of raw fruits as affected by composition of diluents.

TABLE 24.2
Examples of Variations in Weight, Diluent Composition, and Volume, and Processing Methods Used to Prepare Raw Fruits and Vegetables to Analyze for Populations of Salmonella

Produce type (no.)	Weight (g)	Diluent or wash solution Type[a]	Vol. (ml)	Process type	Time	Ref.
Alfalfa seeds	10	NB	90	Stomach	1 min	7
	5 ± 1	BPB	45	Stomach	90 sec	8
Alfalfa sprouts	50	PW	450	Stomach	2 min	9
	50	PW	100	Hand massage	1 min	10
	25	PBS	25	Stomach	5 min	11
Apples						
Whole (1)	180	PW	20	Hand rub	40 sec	12
Skins	5 apples	PBS	250	Homogenize	Not stated	13
Broccoli	25	BPW	225	Stomach	2 min	14
Cantaloupe	1 (whole)	BPBT	200	Shake	10.5 min	15
	25	PW	75	Blend	1 min	16
Carrot	20–30	PWT	99	Shake	1 min	17
Lettuce	50	PW	50	Shake	20 sec	12
	20	BPB	180	Blend	30 sec low speed plus 30 sec high speed	18
	25	BPW	225	Stomach	Not stated	19
	10	PWS	90	Stomach	2 min	20
	15–25	PW	200	Stomach	2 min	21
Melons	25	BPB	18	Stomach	1 min	22
Radish	50–75	PWT	90	Shake	1 min	17
Spinach	50	PW	50	Stomach	1 min	23
Strawberry	25–30	Several	30	Stomach or shake	2 min; 25 min	24
Tomatoes						
Whole (1)	75	PW	20	Hand massage	2 min	25
	180–200	PW	20	Hand rub	40 sec	26
	110–140	PW	20	Hand rub	1 min	27
Chopped	50	PW	50	Stomach	1 min	27
Small pieces	20	Saline	180	Stomach	30 sec	28
Various						
Vegetables (9)	25	BPW	225	Stomach	2 min	29
Fruits/vegetables (401)	10–600	Water	500	Agitate	30 min	30

[a] BPB, Butterfield's phosphate buffer (pH 7.2); BPBT, Butterfield's phosphate broth + 1% Tween 80; BPW, buffered peptone water; NB, neutralizing broth; PBS, phosphate-buffered saline (pH 7.2); PW, peptone water (0.1%); PWS, peptone water (0.1%) + 0.85% sodium chloride; PWT, peptone water (0.1%) + 1% Tween 80.

The performance of recovery media for supporting colony development was also evaluated. The composition of diluents had little effect on the number of yeasts and molds recovered from fruits. Dichloran rose bengal chloramphenicol agar and plate count agar supplemented with chloramphenicol were equivalent in supporting colony formation. A study to determine the effect of diluent composition on recovery of yeasts from grape juice and passion fruit pulp showed that dilution in 0.1% peptone gave higher counts compared to dilution in distilled water, saline, or phosphate buffers [40].

Media selective for yeasts and molds found on raw fruits and vegetables, as well as other types of foods, have been reviewed [36,41]. Formulations have been developed to select for acid-resistant yeasts and detection of proteolytic and lipolytic enzyme activity. Media for detecting mycotoxigenic molds and selecting for xerophilic and xerotolerant fungi have been developed. Release of tissue juices from produce, particularly those with high sugar content, followed by drying on the produce surface can result in reduced water activity that is selective for these yeasts and molds. Some considerations when analyzing produce and other foods for the presence of xerophilic fungi are recommended by Beuchat and Hocking [42].

24.2.2 SELECTION OF TEST STRAINS FOR SANITIZER EFFICACY AND CHALLENGE STUDIES

The strain or strains of a particular microorganism selected for studies designed to determine the efficacy of a decontamination treatment or survival and growth in challenge studies are extremely important. The use of well-characterized reference strains enhances the comparative assessment of a given method among laboratories. Five or more strains, preferably recently isolated from produce or other plant materials, and from patients suffering from illness associated with consumption of a raw fruit or vegetable, are preferred. Approximately equal populations of each strain in a mixed inoculum should be used. If there are differences in the ability of one or more of these strains to survive or grow on produce subjected to various environmental conditions during storage, or if there are differences in susceptibility to decontamination treatments, the most robust strain(s) will prevail. If only one strain is used in the inoculum, it should be first evaluated against several other strains for its ability to survive or grow under the proposed test conditions. The use of a single strain that may be less tolerant to test conditions could result in an inaccurate assessment of the behavior of the test microorganism. Strains used to prepare mixed-strain suspensions should be examined for potential reactions against each other that may be caused by bactoriocins, killer proteins, and other inhibitors they may produce.

Test microorganisms should be cultured in a standard broth or on a defined agar medium at a specific temperature for a specific time. The temperature at which microorganisms are grown for preparing inocula should be representative of the temperature at which they had grown before contaminating produce or the temperature at which inoculated produce will be stored after inoculation, in the case of a challenge study. Several transfers of cultures should be made preceding the day of inoculum preparation. The time elapsed between the last transfer and collecting cells to prepare the inoculum will depend on the test microorganism. Although this practice may result in strains with reduced environmental stress tolerance as a result of adaptation to a nutrient-rich medium, it is desirable to prepare inoculum of uniform cell type. The type of study being conducted should be considered in terms of potential impact of genetic selection of test strains on the predictive value of the results. This is particularly important for pathogenic bacteria that may have originated from diverse sources such as clinical specimens, foods, or the environment. Stationary phase bacterial and yeast cells are generally more tolerant than are logarithmic growth phase cells to environmental stresses [43]. For this reason, cells in stationary growth phase should be used in studies to develop optimum procedures to assess their behavior on or in inoculated produce.

The use of markers such as antibiotic resistance may be desirable to facilitate the recovery of cells in enrichment broth or counting colonies on selective or nonselective direct plating media. Otherwise, these media may support the growth of large numbers of background microflora which interfere with growth of the test microorganism. Adaptation of Gram-negative pathogens to nalidixic acid (50 μg/ml) has been used to achieve this objective. In a study to determine survival of five strains of nalidixic acid-resistant and refampicin-resistant *Salmonella* Poona on cantaloupes it was observed that average reductions in the number of control and antibiotic-resistant cells were not significantly different ($P > 0.05$) [44]. Resistance of test cells to rifampicin (80 μg/ml) can also be successfully used as a marker, particularly for isolating pathogens from inoculated fruits and vegetables that have significant adhering soil. Plasmid-borne or chromosomally stabilized markers such as fluorescent proteins with various chromophoric properties have also been used. It is important to assess the impact of markers on the growth rate, stress tolerance, and recovery efficiency of cells on enumeration media before subjecting them to sanitizer efficacy or challenge studies. Characterization of the stability of the marker over at least ten generations, without selection, is needed for challenge studies in which growth may occur and for recovery methods that include preenrichment or enrichment procedures.

Different strains of the same bacterial species may release byproducts that inhibit or kill other strains. Colicins produced by *Escherichia coli* and killer toxins produced by some species of yeasts are examples. When an inoculum containing several strains of the same microorganism is used, each strain should be tested for its potential to inhibit all other strains in the inoculum. This can be done by cross streaking cultures of individual strains on an appropriate agar medium and examining incubated plates for inhibition of growth at intersections of the streaked cultures.

Determination of the survival characteristics of viruses and parasites on produce poses unique problems. Survival of enteric viruses has been studied but obstacles still remain in standardizing methodology to determine the efficacy of sanitizers in killing or removing viruses that may occasionally contaminate produce. Freshly isolated viruses such as hepatitis A and noroviruses are not culturable, and thus cannot be propagated in sufficient quantities to prepare inocula, nor can they be quantitated by plaque assay. A few viruses, e.g., poliovirus, have been adapted to grow in tissue culture in the laboratory and can be quantitated by plaque assays. Although some of these viruses belong to the same family, they can vary greatly in their level of resistance to chemical and physical stresses. Adapted strains may be representative of their parental wild type but not other members in the same family. Propagation of foodborne viruses has been limited to only a few adapted strains. The behavior of these strains may or may not be similar in behavior to other isolates of the same virus. Rapid molecular methods to detect viruses in foods are sensitive but cannot be used to quantitate viruses or to differentiate between infectious and noninfectious strains [45]. These attributes pose unique challenges in developing and standardizing methodology for detecting and quantitating viruses on produce.

Survival of parasites on raw produce as affected by treatment with sanitizers or exposure to various environmental stress factors during storage has not been well defined. A major constraint to investigating survival characteristics is the limited supply of oocysts [46]. Infected humans are the only source of *Cyclospora cayetanensis* oocysts in quantities needed in studies to determine susceptibility to stress or lethal conditions that may be imposed by decontamination treatment or storage conditions. The lack of sensitive laboratory methods for quantitating and assessing the viability of oocysts hampers progress in developing methods to determine the efficacy of sanitization treatments and the influence of processing, packaging, and storage conditions on their survival.

Vehicles of pathogens and spoilage microorganisms for contaminating fruits and vegetables include dust, rain water, irrigation water, sewage, soil, feces, decayed plant material, contact surfaces, workers at any point from harvesting through preparation in foodservice, and home settings [47] Vegetative cells, spores, cysts, and other propagules of microorganisms are likely to be entrapped in organic material. To simulate practical conditions of surface contamination of produce, the carrier for the inoculum should contain organic material. Horse serum (5%) and aqueous peptone solution (0.1%) have been used as carriers with fairly defined composition in studies to determine the efficacy of sanitizers. Buffer solutions and other carriers containing salts or other chemicals that could be detrimental to cells after the inoculum has dried on the surface of test produce are not recommended for use as carriers. Cells in broth cultures of bacteria or yeasts should be washed in peptone water and resuspended in the organic carrier shortly before using as an inoculum. For challenge studies designed to determine the survival or growth of pathogenic and spoilage microorganisms on or in produce, the carrier may provide a source of nutrients, thus complicating interpretation of results. The use of two carriers, one with and one without organic material (deionized or distilled water) for test cells may be useful in generating information to enable the effects of carrier nutrients on survival and growth of test microorganisms to be discerned.

The desired population of test cells in the inoculum depends on the objective of the study. Two or three levels of inocula, ranging from 10^0 to $10^7 \, CFU/g$ or CFU/cm^2, may be applied to facilitate the determination of efficiency of retrieval, efficacy of sanitizers, or survival and growth during subsequent storage. High numbers of cells in the inoculum are needed in decontamination studies to enable measurement of several $\log_{10}$ reductions in population. Challenge studies require inocula containing low numbers of cells to enable measurement of growth during storage under conditions simulating practices to which produce is subjected in commercial distribution, retail, foodservice, and home settings.

24.3 TYPES OF PRODUCE AND METHODS FOR PREPARING SAMPLES

A single method to remove efficiently microbial cells or spores from all types of fruits and vegetables for the purpose of detection or enumeration would be ideal, but this may not be an achievable goal. Differences in size, shape, and surface morphology of fruits and vegetables complicate the protocol. The ratio of surface area to weight of individual produce items varies substantially, raising the need to establish a basis (CFU/g or CFU/cm^2) to be used to record and report data.

The procedure for preparing the sample for analysis may affect the efficiency of retrieval of microbial cells from produce, as well as dispersal before preenrichment, enrichment, or direct plating. Homogenization of a standard weight of a fruit or vegetable using a standard volume of diluent would be a simple procedure for selecting sample size and method of preparation of samples. Problems, however, may be associated with homogenized, blended, or macerated plant tissues. These include the potential lethal effect of naturally occurring antimicrobial compounds against pathogens or other microflora targeted for detection or enumeration. When microbial cells on the surface of produce tissues come in contact with organic acids or other antimicrobials naturally present in tissue fluid, or produced in the form of phytoalexins as a result of rupture of cells or invasion with insects or molds, death may occur [48].

Acids and phenolic compounds are naturally present in plant stems, leaves, flowers, and fruits. These compounds may interfere with detection and enumeration of pathogenic and spoilage microorganisms. The low pH of produce tissues, particularly those in many fruits, is attributable to a wide range of organic acids they may contain. Garlic, onion, and leek are probably the most widely consumed vegetables that have antimicrobial activity. Allicin, a diallyl thiosulfate, is not present in intact tissues but is produced when the tissues are disrupted. Plant tissues used largely as seasoning agents may also be inhibitory to pathogenic and spoilage microorganisms (Table 24.3). Spices

TABLE 24.3
Plants Used Largely as Seasoning Agents That Also Contain Antimicrobials

Achiote	Cinnamon	Licorice	Rosemary
Allspice (pimenta)	Citronella	Mace	Sage
Angelica	Clove	Marjoram	Sassafras
Anise	Coriander	Musky bugle	Savory
Basil (sweet)	Dill	Mustard	Spearmint
Bay (laurel)	Elecampane	Nutmeg	Star anise
Bergamot	Fennel	Onion	Tarragon (estragon)
Calmus	Fenugreek	Oregano	Thyme
Cananga	Garlic	Paprika	Turmeric
Caraway	Ginger	Parsley	Vanillin
Cardamom	Horseradish	Pennyroyal	Verbena
Celery	Leek	Peppermint	Wintergreen
Chenopodium	Lemongrass	Pimento	

and herbs prepared from plant parts owe some of their desired sensory attributes to these antimicrobials [49]. Like organic acids, these compounds are released from cut tissues and may be lethal to microorganisms naturally occurring or intentionally inoculated onto fruits and vegetables.

Compounds involved in plant defense mechanisms have been classified as prohibitins, inhibitins, postinhibitins, or phytoalexins, depending on preinfection or postinfection factors [50]. These compounds may also kill or inhibit test microorganisms or microflora naturally present on produce. Table 24.4 lists some antimicrobials other than major flavor and aroma compounds that are known to be naturally present in raw fruits and vegetables or produced as a result of breakage of tissues or infections with bacteria and molds. These antimicrobials may interfere with detection and enumeration of bacteria capable of causing human illnesses.

Several studies have investigated inhibitory or lethal activities of naturally occurring antimicrobials against foodborne pathogenic and spoilage microorganisms. A study to compare washing in 0.1% peptone, stomaching, and homogenizing for their influence on recovery of salmonella inoculated onto 26 types of fruits, vegetables, and herbs, revealed that, overall, no significant differences in recovery of the pathogen could be attributed to a particular sample processing method [52]. In an attempt to determine if exposure of salmonella to low pH of tissue fluids as a result of stomaching or homogenizing samples was lethal to the pathogen, the 26 types of produce were arbitrarily separated into groups based on pH. Significantly higher percent recoveries were obtained in produce in the pH 5.53 to 5.99 range compared to produce in lower pH ranges. Reduced percent recoveries from herbs (pH 5.94 to 6.34) were attributed in part to antimicrobials released from plant cells during sample preparation. The mean pH of herbs was 6.08. Lethality caused by antimicrobials other than acids in herbs may have been masked by the minimal inhibitory affect of this slightly acidic pH, causing a reduction in the number

TABLE 24.4
Antimicrobials Other Than Major Flavor and Aroma Compounds That Are Naturally Present in Edible Plant Tissues or Produced as a Result of Infection or Rupture of Tissues

Common name	Botanical name	Antimicrobial produced
Alfalfa	*Medicago sativa*	Medicarpin
Apple	*Malus* spp.	Phloretin, hydroxybenzoic acid, anthocyanidins
Avocado	*Persea* spp.	Borbonol
Beet (red)	*Beta vulgaris*	Beta vulgarin
Broad bean	*Vica faba*	Wyerone acid
Cabbage	*Brassica oleracea*	Rapine, sinigrin
Carrot	*Daucus carota*	Falcarindiol, 6-methoxymellein
Chick pea	*Cicer aretietum*	Medicarpin
Eggplant	*Solanum melongena*	Aubergenone
French bean	*Phaseollus vulgaris*	Phaseollin
Garlic	*Allium sativum*	Allyl sulfoxides
Grape	*Vitis* spp.	Yiniferin
Mulberry	*Morus alba*	Mulberrofuran, albafuran, moracin
Olive	*Olea europaea*	Oleuropein
Onion	*Allium cepa*	Protocatechoic acid
Parsley	*Petroselinum* spp.	Begapten, graveolone, isopimpinellin, psoralen, xanthotoxin
Passion fruit	*Passiflora mollissima*	Passicol
Pea (shoot)	*Pisum sativa*	Pisatin
Peanut	*Arachis hypogaea*	Resveratrol
Pear	*Pyrus* spp.	Arbutin
Pepper (sweet)	*Capsicum annuum*	Capsidiol
Pigeon pea	*Cajanus cajan*	Stilbene-2-carboxylic acid, glyceollin
Potato	*Solanum tuberosum*	Rishitin, hydroxylubimin, caffeic acid, scopoletin, α-tomatine
Radish	*Raphanus sativus*	Raphanin
Sweet potato	*Ipomoea batatas*	Impomeamarone
Soybean	*Glycine max*	Pterocarpan, glyceollin, daidzein, coumestrol
Tomato	*Lycopersicon esculentum*	Alkaloids
Yam	*Discorea rotundata*	Hicicol, isobatasin

Adapted from Walker, J.R.L., Antimicrobial compounds in food plants, in *Natural Antimicrobial Systems and Food Preservation*, Dillon, V.M. and Board, R.G., Eds., CAB International, Wallingford, U.K., 1994, p. 181; Whitehead, I.M. and Threlfall, D.R., *J. Biotechnol.*, 26, 63, 1992.

of salmonella recovered. While blending, homogenizing, or macerating may be acceptable in preparing samples of some types of fruits and vegetables, a simple surface washing without rupturing of plant cells may be required for other types.

The presence of inhibitory or protective residues from crop management practices and variations in surface morphology unique to specific fruits or vegetables should also be a consideration when selecting a procedure for preparing samples for analysis. Microorganisms may be most effectively retrieved by washing the surface of fruits and vegetables such as tomatoes, mangoes, avocados, watermelons, oranges, and other produce with a relatively smooth, rigid surface. Even produce with hard, apparently blemish-free surfaces, however, can harbor microorganisms in areas that are not easily accessible by washing or homogenization [53]. Infiltration of microbial cells into stomata, lenticels, broken trichomes, and cracks in the skin surface can occur. The porous stem scar tissue of tomatoes, for example, offers a relatively easy port of entry for microorganisms compared to intact skin [27,54]. Microorganisms are also known to partition into the cut tissues of produce. Infiltration of *E. coli* O157:H7 into cut tissue of lettuce is affected by temperature [55]. Microorganisms harbored in subsurface and other protected areas should be considered when selecting a method to retrieve them from produce tissues for the purpose of detection and enumeration.

Sonication of samples may be an alternative method for removal of microorganisms from the surface of produce with minimal tissue disruption, although this approach has not been thoroughly researched. Seymour *et al.* [20] evaluated the use of ultrasound to promote decontamination of raw vegetables. Cavitation caused by treatment appeared to enhance the release of *Salmonella* Typhimurium. Release of salmonellae and *E. coli* O157:H7 from inoculated alfalfa seeds is enhanced by ultrasound treatment [56]. These observations suggest that ultrasound treatment, perhaps in combination with other methods for removing microorganisms from produce tissues, may result in a more accurate assessment of populations. Tissues of leafy and floret vegetables, strawberries, raspberries, blackberries, and other produce with complex surface tissues are easily ruptured by rubbing, thus exposing surface microflora to stress conditions imposed by reduced pH or other factors associated with tissue juice. Agitation using a mechanical shaker or by manually shaking in a wash fluid with standard composition and volume for a set period of time may be the most suitable method for removing microbial cells from these produce items.

To avoid too many modifications of a standard sample preparation protocol, the sample weight or number of pieces and the volume of wash solution or diluent should be standardized for each type or group of produce. Results of analysis can be reported as CFU/g of sample or be converted to CFU/cm^2 using a conversion table listing estimated values for specific fruits and vegetables categorized as spheres, cylinders, two-sided planes, or perhaps other geometric shapes. An alternative would be to calculate microbial populations on the basis of CFU/piece of fruit or vegetable, although this method has little meaning if the weight of each produce piece is not reported.

Removal and disposal of microbial cells from surface tissues of fruits and vegetables that have been mechanically cleaned by brushing or that have been waxed or oiled may be more difficult, compared with retrieval from untreated produce. Microorganisms entrapped in bruised tissue, waxes, and oils may be more difficult to remove and disperse in homogenates or wash fluids, resulting in an underestimation of populations. Dip inoculation of bruised, unwaxed apples in a suspension of *E. coli* O157:H7 is known to result in lodging and infiltration of cells in broken tissues, the waxy cutin layer, and lenticels (Figure 24.1). Cells can be harbored in lenticels at depths up to 24 μm, making their retrieval difficult [57]. It may be necessary to modify the sample preparation protocol to maximize release of cells from tissues, as well as from

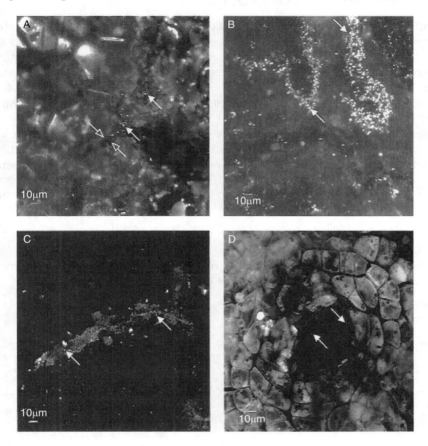

FIGURE 24.1 (Color insert follows page 594) Confocal laser scanning microscopy (CLSM) images showing attachment of *E. coli* O157:H7 to various sites on the surface of apples. (A) Bruised tissue of unwashed, unrubbed, bruised apple at a junction (4.8 μm depth) between wax platelets: edge of wax platelets (open arrow); most cells attached to the edge of the wax platelets (filled arrow). (B) Bruised tissue of unwashed, unrubbed, bruised apple at a junction (6.6 μm depth) between wax platelets: heavy colonization of junctions between wax platelets (arrow). (C) Bruised tissue of unwashed, rubbed, bruised apple showing a cuticular crack on surface of apple (16.6 μm depth): cells are trapped within the cuticular crack (arrow). (D) Bruised tissue of unwashed, unrubbed, bruised apple showing a lenticel (9.2 μm depth): cells are within lenticel (arrow) at a depth of 20.6 μm below the surface of the apple. (From Kenney, S.J., Burnett, S.L., and Beuchat, L.R., *J. Food Prot.*, 64, 132, 2001. With permission. Copyright International Association for Food Protection, Des Moines, IA.)

the naturally occurring waxy cutin layer and waxes or oils that may be applied to enhance appearance or extend the shelf life of some fruits and vegetables.

24.4 PROCEDURES FOR INOCULATION

Surface inoculation of fruits and vegetables with pathogens or spoilage microorganisms can be done by dipping or spraying with a suspension of cells or by applying a known volume of suspension containing a known population (spot inoculation). Some of the advantages and disadvantages of these methods are listed in Table 24.5. The inoculation method should ideally simulate various contamination events and postcontamination conditions prior to washing and sanitization. If contamination of produce is suspected to occur by an immersion process, dipping the produce in the test cell suspension may be an appropriate method for inoculation. A problem associated with inoculation by dipping or spraying is that the number of cells actually applied or adhering to the produce is not known. The population remaining on the surface and in subsurface tissues after dip or spray inoculation is not consistent among

TABLE 24.5
Some Advantages and Disadvantages of Dip, Spot, and Spray Inoculation Methods for Determining the Efficacy of Sanitizer Washes on Produce

Inoculation method	Advantages	Disadvantages
Dip	Mimics contamination from highly contaminated irrigation, run-off, or flume water	Volume of inoculum and number of cells delivered to each produce item are unknown. Some inoculum may be internalized, complicating interpretation of data. Large volumes of high inoculum containing populations of pathogens are difficult to manage safely, even in highly experienced laboratories
Spot	Delivered volume of inoculum known. Population in inoculum can be accurately calculated. Efficacy of sanitizer can be compared on different tissues within the same produce item. Most consistent inoculum applied of three inoculation methods	May not reflect contamination that would occur from contaminated irrigation, run-off, or flume water
Spray	Mimics contamination from aerosols	Accurate delivery of inoculum is difficult, especially with smaller produce items. Aerosols generated are difficult to manage safely, even in highly experienced laboratories

various types of produce and between different types of tissues on the same produce. Infiltration of the inoculum into cavities, wounded tissues, or porous areas on the produce surface, e.g., cut tissue, stem scar tissue, stomata, or lenticils [58,59], can result in conditions that may inhibit or enhance growth. Cells lodged in these areas may be protected against contact with sanitizers, particularly those with little or no surfactant activity. Analysis of produce inoculated by dip or spray methods requires a large number of units for each treatment, as random error values can be unpredictably large. Thus, efficiency of recovery or $\log_{10}$ changes in viable populations during subsequent storage or as a result of treatment with a sanitizer cannot be accurately calculated. Alternatively, fruits and vegetables can be inoculated by applying a known volume of cell suspension, e.g., up to 100 µl, of known population to the surface. Spot inoculation represents contamination from a point source such as contact with soil, workers' hands, or surfaces of equipment and has been recommended for testing the efficacy of sanitizers in killing foodborne pathogens on tomatoes [31], lettuce, and parsley [33].

Temperature differentials between produce and inocula can affect the number of cells that infiltrate tissues. A negative differential, i.e., when the temperature of the produce is higher than the temperature of the inoculum, can result in enhanced infiltration of microbial cells [27,54,60]. A standard temperature at which both the produce and the inoculum are adjusted before inoculation should be selected for sanitizer efficacy studies. Otherwise, exposure of test cells to sanitizers and the efficiency of retrieval of cells may be affected.

In studies to determine the efficiency of retrieval of cells or efficacy of sanitizers, the inoculum applied to produce should be dried for a set period of time at a controlled temperature and relative humidity before treatment is applied and samples are analyzed. Fluctuations in temperature and relative humidity should be minimized between the time of drying and treatment or analysis. Three or more replicate experiments, each including four or more samples for each set of test parameters in each replicate, should be done. More samples may be needed, depending on specific objectives. Negative controls should always be included.

24.5 EFFICIENCY OF RETRIEVAL

Development of a standard protocol for detecting or enumerating a specific microorganism or group of microorganisms on or in produce should include experiments to validate the efficiency of recovery based on a known number of cells applied. This can be done using a known volume of inoculum containing a known number of test cells. Although some cells may die during the drying period following application of inoculum, efficiency of retrieval can be more accurately measured using spot inoculation than dip or spray inoculation, which do not enable measurement of the number of cells adhering to the produce.

The efficiency of retrieval of microbial cells naturally occurring on produce is not easily determined, simply because the actual number of retrievable cells is not known [61]. The presence of a surfactant in peptone water used to remove pathogens from produce, e.g., cantaloupes, may enhance the number detected [44]. A comparison of various combinations of sample weights, wash fluids, diluents, homogenization or washing treatments, and neutralizers (in the case of chemical sanitizer tests) should be made before choosing test parameters that give the highest percentage of viable microorganisms recovered. Some protocols have been demonstrated to be more efficient than others, and a single basic protocol should be selected for analysis of specific fruits, vegetables, or groups of produce in all laboratories.

24.6 EFFICACY OF DECONTAMINATION TREATMENT

A protocol for efficient recovery of pathogens or groups of microorganisms from fruits and vegetables must be established before proceeding with experiments designed to determine the efficacy of treatment with sanitizers or changes in populations as affected by storage conditions. Procedures for chemical decontamination should use standard weight-to-volume ratios (produce:treatment solution or atmosphere), whether applied as a dip, spray, or fog. A standard concentration of treatment solution applied for a standard time at a standard temperature, followed by neutralization of the active component using a standard volume and concentration of neutralizer should be defined. Whether the produce should be static, agitated, or hand rubbed during chemical treatment should be stated. Agitation, e.g., by placing the produce and treatment solution on a mechanical shaker or manually shaking, should be standardized. Conditions for separating the produce from the chemical treatment solution, washing with a specific neutralizer, and subsequent homogenization or washing in a specific volume of a given diluent should also be standardized. Controls that will reveal the effect of rinsing after treatment should also be included.

For physical decontamination treatments, standardization of conditions, e.g., temperature, irradiation, or pressure, would facilitate comparison of observations across laboratories. A neutralization step is not necessary in a standard protocol to measure the efficacy of physical treatments but, like protocols for determining the efficacy of chemical sanitizer treatments, standardization of diluent composition, ratio of produce weight:diluent volume, homogenization or washing procedure, preenrichment, enrichment, and direct plating media, and incubation conditions is necessary.

24.7 PROCEDURES FOR DETECTION AND ENUMERATION

The selection of preenrichment, enrichment, and/or direct plating media, as well as conditions for incubation and procedures for confirmation of isolates will differ, depending on the microorganism or group of microorganisms targeted for detection or enumeration. Media, incubation conditions, and confirmation techniques selected for each microorganism or group of microorganisms that may be inoculated onto or naturally present in produce should be the same across laboratories. Optimum protocols for retrieving pathogens and nonpathogens from fruits and vegetables may differ, depending upon whether analysis of the surface, tissue, or a composite of both is desirable. Washing, rubbing, blending, homogenizing, stomaching, macerating, and grinding, or a combination of one or more of these procedures, are among the choices to process samples for preenrichment, enrichment, or direct plating. One piece of fruit or vegetable, several pieces, or only a portion of the whole or cut produce may be selected for analysis, but the procedure needs to be standardized in terms of sample weight and/or excision technique. The composition and pH of the diluent and ratio of diluent to sample need to be consistent, at least within each type of fruit and vegetable. The time and temperature for processing samples for preenrichment, enrichment, or direct plating should be standardized. The likelihood of stressed or injured microbial cells being present on or in fruits and vegetables should be recognized, and appropriate resuscitation conditions should be considered and applied. Repair of cells on the surface of produce that, for example, may be debilitated by desiccation or as a result of exposure to a harsh chemical environment, is important if these cells are to be detected or enumerated. Adjustment of the pH of homogenates of highly acidic fruits and vegetables may be necessary to protect microorganisms against exposure to potentially lethal conditions during preparation of samples for inoculation of recovery media.

24.8 NUMBER OF SAMPLES ANALYZED AND REPORTING THE RESULTS

Conditions intrinsic to fruits and vegetables, as well as variation in types and numbers of microorganisms and amount of soil and organic matter present on produce surfaces, are variable, necessitating a standard procedure for selecting samples for sanitizer efficacy or challenge studies. A sufficient number of replicates with a sufficient number of whole fruits or vegetables, or cut produce samples will be necessary to enable appropriate types of statistical analysis to be applied to the data generated. The experimental design should enable statistical analysis to be done at a level rigorous enough to deal with the complexities associated with microbiological testing. Traditional methods of bacteriological or mycological analysis of foods and beverages report results on the basis of CFU/g, CFU/ml, or CFU/cm^2. Treatment with sanitizers or application of processing technologies may be designed to achieve a certain $\log_{10}$ reduction in the number of a specific pathogen, several pathogens, or a spoilage microorganism, based on weight or volume of the product. Substantial variation in the weight-to-surface area ratio (g:cm^2)

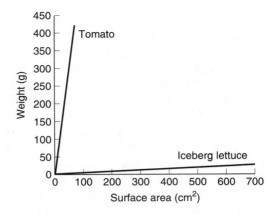

FIGURE 24.2 Relationship between weight and surface area of tomato fruit and lettuce leaf. (From Beuchat, L.R., Farber, J.M., Garrett, E.H., Harris, L.J., Parish, M.E., Suslow, T.V., and Busta, F.F., *J. Food Prot.*, 64, 1079, 2001. With permission. Copyright International Association for Food Protection, Des Moines, IA.)

can exist among various types of produce, making this approach unreasonable in terms of assessing $\log_{10}$ reductions against risk of illness that may result from consumption of a given weight of produce. Relationships between the weight (g) and surface area (cm^2) of iceberg lettuce and tomato (Figure 24.2) illustrate this point. Recognizing that the weight:surface area will vary, depending on the thickness of the lettuce leaf and variations in shape of both vegetables, this figure simply shows that large differences in ratios in weight:surface area can exist among fruits and vegetables. Ratios for other fruits and vegetables with geometric configurations other than a two-sided plane (lettuce) or a sphere (tomato) would fall somewhere between these extremes. A decontamination process designed to achieve, for example, a $3 \log_{10}$ reduction in CFU/g of lettuce or tomato would theoretically result, respectively, in approximately 0.11 and $18 \log_{10}$ reductions in CFU/cm^2; a $3 \log_{10}$ reduction in CFU/cm^2 of lettuce or tomato would result, respectively, in approximately 79 and $0.5 \log_{10}$ reductions in CFU/g [62].

A standard procedure for calculating and reporting populations of microorganisms on fruits and vegetables needs to be established and, if guidelines or limits for maximum populations of pathogens are to be considered, calculation should be done on a standard basis (CFU/g or CFU/cm^2). The number of $\log_{10}$ reductions in CFU resulting from a processing or decontamination treatment should likewise be based on a standard procedure for calculation. Regardless of the procedure used, if guidelines or limits and $\log_{10}$ reductions for specific pathogens are established, differences in weight and geometric configuration of fruits and vegetables should be considered. Data need to be subjected to appropriate statistical analysis to determine significant differences in populations of pathogens or spoilage microorganisms recovered from produce that has been subjected to various treatment or storage conditions.

REFERENCES

1. Institute of Food Technologists, Analysis and Evaluation of Preventive Control Measures for the Control and Reduction/Elimination of Microbial Hazards on Fresh and Fresh-Cut Produce, Report of the IFT for the Food and Drug Administration of the U.S. Department of Health and Human Services, 2001 (www.cfsan.fda.gov/~comm/ift3-toc.html).
2. National Advisory Committee on Microbiological Criteria for Foods, Microbiological safety evaluations and recommendation on fresh produce, *Food Control*, 10, 117, 1999.
3. Nguyen-the, C. and Carlin, F., Fresh and processed vegetables, in *The Microbiological Safety and Quality of Food*, Lund, B., Baird-Parker, T.C., and Gould, G.W., Eds., Aspen, Gaithersburg, MD, 2000, p. 620.
4. Beuchat, L.R., *Surface Decontamination of Fruits and Vegetables Eaten Raw: A Review*, WHO/FSF/FOS/98.2, 1998.
5. U.S. Food and Drug Administration, *United States Food and Drug Administration Bacteriological Analytical Manual*, 8th ed., AOAC International, Gaithersburg, MD, 1998.
6. Pouch Downes, F. and Ito, K., Eds., *Compendium of Methods for the Microbiological Examination of Foods*, 4th ed., American Public Health Association, Washington D.C., 2001.
7. Stan, S.D. and Daeschel, M.A., Reduction of *Salmonella* enterica on alfalfa seeds with acidic electrolyzed oxidizing water and enhanced uptake of acidic electrolyzed oxidizing water into seeds by gas exchange, *J. Food Prot.*, 66, 2017, 2003.
8. Thayer, D.W., Boyd, G., and Fett, W.F., γ-Radiation decontamination of alfalfa seeds naturally contaminated with *Salmonella* Mbandaka, *J. Food Sci.*, 68, 177, 2003.
9. Castro-Rosas, J. and Escartin, E.F., Incidence of germicide sensitivity of *Salmonella typhi* and *Vibrio cholerae* O1 in alfalfa sprouts, *J. Food Saf.*, 19, 137, 1999.
10. Jaquette, C.B., Beuchat, L.R., and Mahon, B.E., Efficacy of chlorine and heat treatment in killing *Salmonella stanley* inoculated onto alfalfa seeds and growth and survival of the pathogen during sprouting and storage, *Appl. Environ. Microbiol.*, 62, 2212, 1996.
11. Gandhi, M. and Matthews, K.R., Efficacy of chlorine and calcinated calcium treatment of alfalfa seeds and sprouts to eliminate *Salmonella*, *Int. J. Food Microbiol.*, 87, 301, 2003.
12. Beuchat, L.R., Nail, B.V., Adler, B.B., and Clavero, M.R.S., Efficacy of spray application of chlorinated water in killing pathogenic bacteria on raw apples, tomatoes, and lettuce, *J. Food Prot.*, 61, 1305, 1998.
13. Liao, C.-H. and Sapers, G.M., Attachment and growth of *Salmonella* Chester on apple fruits in vivo response of attached bacteria to sanitizer treatments, *J. Food Prot.*, 63, 876, 2000.
14. Wang, H., Li, Y., and Slavik, M.F., Efficacy of cetylpyridinium chloride in immersion treatment for reducing populations of pathogenic bacteria on fresh-cut vegetables, *J. Food Prot.*, 64, 2071, 2001.
15. Barak, J.D., Chue, B., and Mills, D.C., Recovery of surface bacteria from and surface sanitization of cantaloupes, *J. Food Prot.*, 66, 1805, 2003.

16. Ukuku, D.O. and Sapers, G.M., Effect of sanitizer treatments on *Salmonella* Stanley attached to the surface of cantaloupe and cell transfer to fresh-cut tissues during cutting practices, *J. Food Prot.*, 64, 1286, 2001.

17. Natvig, E.E., Ingham, S.C., Ingham, B.H., Cooperbrand, L.R., and Roper, T.R., *Salmonella enterica* serovar typhimurium and *Escherichia coli* contamination of root and leaf vegetables grown in soils with incorporated bovine manure, *Appl. Environ. Microbiol.*, 68, 2737, 2002.

18. Lin, C.-M., Kim, J., Du, W.-X., and Wei, C.-I., Bactericidal activity of isothiocyanate against pathogens on fresh produce, *J. Food Prot.*, 63, 25, 2000.

19. Soriano, J.M., Rico, H., Maltó, J.C., and Manes, J., Assessment of the microbiological quality and wash treatments of lettuce in university restaurants, *Int. J. Food Microbiol.*, 58, 123, 2000.

20. Seymour, I.J., Burfoot, D., Smith, R.L., Cox, L.A., and Lockwood, A., Ultrasound decontamination of minimally processed fruits and vegetables, *Int. J. Food Sci. Technol.*, 37, 547, 2002.

21. Koseki, S., Yoshida, K., Kamitani, Y., and Itoh, K., Influence of inoculation method, spot inoculation site, and inoculation size on the efficacy of acidic electrolyzed water against pathogens on lettuce, *J. Food Prot.*, 66, 2010, 2003.

22. Golden, D.A., Rhodehamel, E.J., and Kautter, D.A., Growth of *Salmonella* spp. in cantaloupe, watermelon, and honeydew melons, *J. Food Prot.*, 56, 194, 1993.

23. Pirovani, M.E., Güemes, D.R., DiPentima, J.H., and Tessi, M.A., Survival of *Salmonella hadar* after washing disinfection of minimally processed spinach, *Lett. Appl. Microbiol.*, 31, 143, 2000.

24. Lukasik, J., Bradley, M.L., Scott, T.M., Hsu, W.-Y., Farrah, S.R., and Tamplin, M.L., Elution, detection, and quantification of polio I, bacteriophages, *Salmonella* Montevideo, and *Escherichia coli* O157:H7 from seeded strawberries and tomatoes, *J. Food Prot.*, 64, 292, 2001.

25. Guo, X., Chen, J., Beuchat, L.R., and Brackett, R.E., PCR detection of *Salmonella enterica* serotype Montevideo in and on raw tomatoes using primers derived from *hilA*, *Appl. Environ. Microbiol.*, 66, 5248, 2000.

26. Beuchat, L.R., Harris, L.J., Ward, T.E., and Kajs, T.M., Development of a proposed standard method for assessing the efficacy of fresh produce sanitizers, *J. Food Prot.*, 64, 1103, 2001.

27. Zhuang, R.Y., Beuchat, L.R., and Angulo, F.J., Fate of *Salmonella montevideo* on and in raw tomatoes as affected by temperature and treatment with chlorine, *Appl. Environ. Microbiol.*, 61, 2127, 1995.

28. Asplund, K. and Nurmi, E., The growth of salmonellae in tomatoes, *Int. J. Food Microbiol.*, 13, 177, 1991.

29. Odumoru, J.A., Mitchell, S.J., Alves, D.M., Lynch, J.A., Yee, A.J., Wang, S.L., Styliadis, S., and Farber, J.M., Assessment of the microbiological quality of ready-to-use vegetables for health-care food services, *J. Food Prot.*, 60, 954, 1997.

30. Wells, J.M. and Butterfield, J.E., *Salmonella* contamination associated with bacterial soft rot of fresh fruits and vegetables in the marketplace, *Plant Dis.*, 81, 867, 1997.

31. Lang, M.M., Harris, L.J., and Beuchat, L.R., Evaluation of inoculation method and inoculum drying time for their effects on survival and efficiency of recovery of *Escherichia coli* O157:H7, *Salmonella*, and *Listeria monocytogenes* inoculated on the surface of tomatoes, *J. Food Prot.*, 67, 732, 2004.

32. Wade, W.N. and Beuchat, L.R., Metabiosis of proteolytic moulds and *Salmonella* in raw, ripe tomatoes, *J. Appl. Microbiol.*, 95, 437, 2003.
33. Lang, M.M., Harris, L.J., and Beuchat, L.R., Survival and recovery of *Escherichia coli* O157:H7, *Salmonella*, and *Listeria monocytogenes* on lettuce and parsley as affected by method of inoculation, time between inoculation and analysis, and treatment with chlorinated water, *J. Food Prot.*, 67, 1092, 2004.
34. Fett, W.F. and Cooke, P.H., Reduction of *Escherichia coli* O157:H7 and *Salmonella* on laboratory inoculated alfalfa seed with commercial citrus-related products, *J. Food Prot.*, 66, 1158, 2003.
35. Weissinger, W.H., McWatters, K.H., and Beuchat, L.R., Evolution of volatile chemical treatments for lethality to *Salmonella* on alfalfa seeds and sprouts, *J. Food Prot.*, 64, 442, 2001.
36. Beuchat, L.R., Media for detecting and enumerating yeasts and moulds, in *Culture Media for Food Microbiology*, 2nd ed., Corry, J.E.L., Curtis, G.D.W., and Baird, R.M., Eds., Elsevier, Amsterdam, 2003, p. 369.
37. Beuchat, L.R. and Cousin, M.A., Yeasts and molds, in *Compendium of Methods for the Microbiological Examination of Foods*, Pouch Downes, F. and Ito, K., Eds., American Public Health Association, Washington D.C., 2001, p. 209.
38. Pitt, J.I., Hocking, A.D., Swanson, R.A., and King, A.D., Recommended methods for mycological examination of foods, in *Modern Methods in Food Mycology*, Samson, R.A., Hocking, A.D., Pitt, J.I., and King, A.D., Eds., Elsevier, Amsterdam, 1992, p. 365.
39. Beuchat, L.R., Scouten, A.J., and Jablonska, J., Influence of composition of diluent on populations of yeasts and moulds recovered from raw fruits, *Lett. Appl. Microbiol.*, 35, 399, 2002.
40. Mian, M.A., Fleet, G.H., and Hocking, A.D., Effect of diluent type on viability of yeasts enumerated from foods or pure culture, *Int. J. Food Microbiol.*, 35, 103, 1997.
41. Beuchat, L.R., Progress in conventional methods for detection and enumeration of foodborne yeasts, *Food Technol. Biotechnol.*, 36, 267, 1998.
42. Beuchat, L.R. and Hocking, A.D., Some considerations when analyzing foods for the presence of xerophilic fungi, *J. Food Prot.*, 53, 984, 1990.
43. Jay, J.M., *Modern Food Microbiology*, 6th ed., Aspen, Gaithersburg, MD, 2000, p. 679.
44. Beuchat, L.R. and Scouten, A.J., Factors affecting survival, growth, and retrieval of *Salmonella* Poona on intact and wounded cantaloupe rind and in stem scar tissue, *Food Microbiol.*, 21, 683, 2004.
45. Jaykus, L.A., Detection of human enteric viruses in foods, in *Foodborne Disease Handbook: Viruses, Parasites, Pathogens, and HACCP*, Vol. 2, Sattar, S., Ed., Marcel Dekker, New York, 2000, p. 137.
46. Herwaldt, B.L., *Cyclospora cayetanensis*: a review, focusing on the outbreaks of cyclosporiasis in the 1990s, *Clin. Infect. Dis.*, 31, 1040, 2000.
47. Beuchat, L.R., Pathogenic microorganisms associated with fresh produce, *J. Food Prot.*, 59, 204, 1996.
48. Sofos, J.N., Beuchat, L.R., Davidson, P.M., and Johnson, E.A., Naturally Occurring Antimicrobials in Food, Council for Agricultural Science and Technology, Task Force report no. 132, 1998.
49. Beuchat, L.R., Antimicrobial properties of spices and their essential oils, in *Natural Antimicrobial Systems in Food Preservation*, Board, R.G. and Dillon, V.M., Eds., CAB International, Wallingford, U.K., 1994, p. 167.

50. Walker, J.R.L., Antimicrobial compounds in food plants, in *Natural Antimicrobial Systems and Food Preservation*, Dillon, V.M. and Board, R.G., Eds., CAB International, Wallingford, U.K., 1994, p. 181.

51. Whitehead, I.M. and Threlfall, D.R., Production of phytoalexins by plant tissue cultures, *J. Biotechnol.*, 26, 63, 1992.

52. Burnett, A.B. and Beuchat, L.R., Comparison of sample preparation methods for recovering *Salmonella* from raw fruits, vegetables, and herbs, *J. Food Prot.*, 64, 1459, 2001.

53. Burnett, S.L. and Beuchat, L.R., Human pathogens associated with raw produce and unpasteurized juices, and difficulties in decontamination, *J. Ind. Microbiol. Biotechnol.*, 25, 281, 2000.

54. Bartz, J.A., Washing fresh fruits and vegetables: Lessons from treatment of tomatoes and potatoes with water, *Dairy Food Environ. Sanit.*, 19, 853, 1999.

55. Takeuchi, K. and Frank, J.F., Penetration of *Escherichia coli* O157:H7 into lettuce tissues as affected by inoculum size and temperature and the effect of chlorine treatment on cell viability, *J. Food Prot.*, 63, 434, 2000.

56. Scouten, A.J. and Beuchat, L.R., Combined effects of chemical, heat and ultrasound treatments to kill *Salmonella* and *Escherichia coli* O157:H7 on alfalfa seeds, *J. Appl. Microbiol.*, 92, 668, 2002.

57. Kenney, S.J., Burnett, S.L., and Beuchat, L.R., Location of *Escherichia coli* O157:H7 on and in apples as affected by bruising, washing, and rubbing, *J. Food Prot.*, 64, 132, 2001.

58. Buchanan, R.L., Edelson, S.G., Miller, R.L., and Sapers, G.M., Contamination of intact apples after immersion in an aqueous environment containing *Escherichia coli* O157:H7, *J. Food Sci.*, 62, 444, 1999.

59. Seo, K.H., and Frank, J.F., Attachment of *Escherichia coli* O157:H7 to lettuce leaf surface and bacterial viability in response to chlorine treatment as demonstrated by using confocal scanning laser microscopy, *J. Food Prot.*, 62, 3, 1999.

60. Burnett, S.L., Chen, J., and Beuchat, L.R., Attachment of *Escherichia coli* O157:H7 to the surface and internal structures of apples as detected by confocal scanning laser microscopy, *Appl. Environ. Microbiol.*, 66, 4679, 2000.

61. Beuchat, L.R., Ecological factors influencing survival and growth of human pathogens on raw fruits and vegetables, *Microbes Infect.*, 4, 413, 2002.

62. Beuchat, L.R., Farber, J.M., Garrett, E.H., Harris, L.J., Parish, M.E., Suslow, T.V., and Busta, F.F., Standardization of a method to determine the efficacy of sanitizers in inactivating human pathogenic microorganisms on raw fruits and vegetables, *J. Food Prot.*, 64, 1079, 2001.

25 Rapid Detection of Microbial Contaminants

Daniel Y.C. Fung

CONTENTS

25.1 INTRODUCTION

Rapid methods and automation in microbiology is a dynamic area in applied microbiology dealing with the study of improved methods for the isolation, early detection, characterization, and enumeration of microorganisms and their products in clinical, food, industrial, and environmental samples. In the past 20 years this field has emerged into an important subdivision of the general field of applied microbiology and is gaining momentum nationally and internationally as an area of research and application to monitor the numbers, kinds, and metabolites of microorganisms related to food spoilage, food preservation, food fermentation, food safety, and foodborne pathogens. Medical microbiologists began involvement with rapid methods around the mid-1960s. In the 1970s developments started to accelerate and continued to do so into the 1980s, 1990s, and up to the present day. Food microbiologists were lagging about 10 years behind the medical microbiologists but in the past decade they have greatly increased their activities in this field [1].

565

This chapter presents rapid microbiological methods for food in general with emphasis in fruit and vegetable microbiology.

25.2　SAMPLE PREPARATION AND TREATMENTS

One of the most important steps for successful microbiological analysis of any material is sample preparation. With the advancement of microbiological techniques and miniaturization of kits and test systems to ever smaller sizes, proper sample preparation becomes critical. Chapter 24 discusses in detail various sample preparation, detection, and enumeration methods for fruits and vegetables. Some novel methods are discussed in this chapter.

Fruits and vegetables are considered solid food. The most efficient method to prepare the samples for enumeration and detection is to use the Stomacher instrument where a known weight of solid sample is placed in the stomacher bag, and a volume of sterile diluent is added to make a 1:10 dilution of the sample. Then the sample is "stomached" for one to two minutes before an aliquot is taken out for viable cell count, differential count, or pathogen count and detection. Dr. Anthony Sharpe invented the Stomacher about 25 years ago, and now more than 40,000 units are in use worldwide. Recently he introduced a new instrument called the Pulsifier (Microgen BioProducts Ltd, Surrey, U.K.) for dislodging microorganisms from foods without excessively breaking the food structure. The Pulsifier has an oval ring that can house a plastic bag with sample and diluent. When the instrument is activated the ring will vibrate vigorously for a predetermined time (30 to 60 seconds). During this time microorganisms on the food surface or in the food will be dislodged into the diluent with minimum destruction of the food. Fung *et al.* [2] evaluated the Pulsifier against the Stomacher with 96 food items (including beef, pork, veal, fish, shrimp, cheese, peas, a variety of vegetables, cereal, and fruits) and found that both systems gave essentially the same viable cell count in the food, but the "pulsified" samples were much clearer than the "stomached" samples. Kang *et al.* [3] found that the Pulsifier and Stomacher had a correlation coefficient of 0.971 and 0.959 for total aerobic count and coliform count, respectively, with 50 samples of lean meat tissues. More recently, Wu *et al.* [4] made a comprehensive study of the Pulsifier versus the Stomacher on 30 vegetables and reported no difference in total count and coliform between the two methods (Table 25.1 for total count). However, there were distinct differences in the liquids between the methods with pulsified samples having less turbidity, less total solids, and higher pH than the stomached samples (Table 25.2). The superior quality of microbial suspensions with minimum food particles and inhibitors from the Pulsifier has positive implications for general microbial analysis such as ease of pipetting samples and ease of filtration through bacteriological membrane filters, as well as for techniques such as adenosine triphosphate (ATP) bioluminescence tests, DNA/RNA hybridization, polymerase chain reaction (PCR) amplifications, enzymatic assays, etc.

TABLE 25.1
Comparison of Total Viable Cell Counts Obtained from Stomached and Pulsified Samples of All Vegetables

	Log CFU/g Pulsifier (P)	Stomacher (S)	(P – S)	(P/S) ratio
1. Head lettuce	4.78	5.09	−0.31	0.94
2. Onions, green	6.20	6.18	0.02	1.00
3. Fresh carrot	5.83	5.85	−0.02	1.00
4. Radish-radicchio	6.07	6.04	0.03	1.00
5. Parsley	5.96	6.16	−0.20	0.97
6. Green leaf lettuce	5.85	6.13	−0.28	0.95
7. Romaine	5.82	4.03	1.79	1.44
8. Red leaf lettuce	6.12	6.12	0.00	1.00
9. Boston lettuce	5.99	5.79	0.20	1.03
10. Spinach	5.76	6.11	−0.35	0.94
11. Endive	6.11	4.82	1.29	1.27
12. Orange pepper	4.14	4.36	−0.22	0.95
13. Cucumber	6.03	4.25	1.78	1.42
14. Celery hearts	5.04	4.39	0.65	1.15
15. Broccoli	5.61	4.90	0.71	1.14
16. Cauliflower	3.29	5.31	−2.02	0.62
17. Snow peas	4.91	5.70	−0.79	0.86
18. Turnips	5.41	5.58	−0.17	0.97
19. Cilantro	6.07	6.10	−0.03	1.00
20. Zucchini squash	6.20	6.01	0.19	1.03
21. Rhubarb	5.65	4.38	1.27	1.29
22. Parsnips	6.12	6.12	0.00	1.00
23. Asparagus	6.10	5.72	0.38	1.07
24. Cabbage	3.54	2.31	1.23	1.53
25. Chinese cabbage	5.74	5.84	−0.10	0.98
26. Green beans	5.34	4.21	1.13	1.27
27. Tomato (hot house)	2.01	2.31	−0.30	0.87
28. Potato (baking)	5.75	5.48	0.27	1.05
29. Walla Walla jumbo yellow onion	0	0	0	0.00
30. Eggplant	5.64	5.91	−0.27	0.95
Average			0.20	1.02

From Wu, V.C.H., Jitareerat, P., and Fung, D.Y.C., *J. Rapid Methods Automat. Microbiol.*, 11, 145, 2003. With permission from Food Nutrition Press, Trumbull, CT.

25.3 TOTAL VIABLE CELL COUNT METHODOLOGIES

One of the most important factors concerning food quality, food spoilage, food safety, and potential implication of foodborne pathogens is the total viable cell count of food, water, food contact surfaces, air, and environments in food plants. The conventional standard plate count method has been in use for the past 100 years in applied microbiology. The method involves sample preparation, dilution, and plating with a nonselective or selective agar,

TABLE 25.2
Comparison of Chemical Attributes of the Liquid Obtained from Stomached (S) and Pulsified (P) Samples of All Vegetables

	Turbidity[a]		pH		TSS[b]	
	P	S	P	S	P	S
1. Head lettuce	0.01	0.073	6.66	6.48	0.2	0.2
2. Onions, green	0.031	0.148	6.46	6.29	0.4	0.6
3. Fresh carrot	0.016	0.365	6.76	6.61	0.25	0.9
4. Radish-radicchio	0.021	0.037	6.42	6.61	0.1	0.4
5. Parsley	0.033	0.322	6.94	6.46	0.15	0.4
6. Green leaf lettuce	0.063	0.192	6.88	6.6	0.05	0.4
7. Romaine	0.063	0.222	6.75	6.39	0.15	0.4
8. Red leaf lettuce	0.138	0.32	6.55	6.55	0.4	0.3
9. Boston lettuce	0.047	0.177	6.73	6.53	0.4	0.3
10. Spinach	0.282	0.56	6.66	6.45	0.4	0.4
11. Endive	0.102	0.283	6.63	6.45	0.4	0.5
12. Orange pepper	0.001	0.125	6.57	5.64	0.22	0.7
13. Cucumber	0.003	0.095	6.67	6.21	0.4	0.5
14. Celery hearts	0.003	0.123	6.78	6.45	0.2	0.5
15. Broccoli	0.024	0.271	6.96	6.86	0.3	0.5
16. Cauliflower	0.004	0.078	6.94	6.96	0.3	0.5
17. Snow peas	0.012	0.227	6.78	6.44	0.5	1.0
18. Turnips	0.001	0.019	6.81	6.64	0.3	0.6
19. Cilantro	0.14	0.765	6.67	6.4	0.1	0.7
20. Zucchini squash	0.003	0.094	6.85	6.73	0.3	0.5
21. Rhubarb	0.079	0.224	4.66	3.75	0.3	0.5
22. Parsnips	0.284	1.158	5.51	6.49	0.5	0.8
23. Asparagus	0.041	0.154	6.49	6.47	0.4	0.7
24. Cabbage	0.001	0.034	6.74	6.6	0.2	0.4
25. Chinese cabbage	0.011	0.01	6.68	6.57	0.22	0.2
26. Green beans	0	0.232	6.83	6.94	0.22	0.7
27. Tomato (hot house)	0.023	0.05	5.07	4.89	0.5	0.4
28. Potato (baking)	0.018	0.247	6.84	6.59	0.3	0.6
29. Walla Walla jumbo yellow onion	0.013	0.042	6.74	6.15	0.4	0.8
30. Eggplant	0.082	0.47	6.72	6.3	0.22	0.8
Average	0.052	0.237	6.56	6.35	0.30	0.55

[a] Optical density measurement.
[b] Total soluble solids.
From Wu, V.C.H., Jitareerat, P., and Fung, D.Y.C., *J. Rapid Methods Automat. Microbiol.*, 11, 145, 2003. With permission from Food Nutrition Press, Trumbull, CT.

incubating the plates at 35°C, and counting the colonies after 48 hours. There is a great variety of factors to be considered. These include plating media, incubation time and temperature and incubation environment, and volumes to be plated. The operation of the conventional standard plate count method, although simple, is time-consuming both in terms of execution and data collection. Also, this method utilizes a large number of test tubes, pipettes, dilution bottles, dilution buffer, sterile plates, incubator space, and related disposable materials and requires resterilizing and clean up of reusable materials for further use.

Several methods have been developed, tested, and used effectively in the past 20 years as alternatives to the standard plate count method. Most of these methods were first designed to perform viable cell counts and relate the counts to standard plate counts. Later, coliform count, fecal coliform count, and yeast and mold counts were introduced. Further developments in these systems include differential counts, pathogen counts, and even pathogen detection after further manipulations. Many of these methods have been extensively tested in many laboratories throughout the world and went through AOAC (Association of Official Analytical Chemists) International collaborative study approvals. The aim of these methods is to provide reliable viable cell counts of food and water in more convenient, rapid, simple, and cost effective alternative formats, compared to the cumbersome standard plate count method.

The spiral plating method is an automated system to obtain viable cell count (Spiral Biotech, Bethesda, MD). By use of a stylus, this instrument can spread a liquid sample on the surface of a prepoured agar plate (selective or nonselective) in a spiral shape (the Archimedes spiral) with a concentration gradient starting from the center and decreasing as the spiral progresses outward on the rotating plate. The volume of the liquid deposited at any segment of the agar plate is known. After the liquid containing microorganisms is spread, the agar plate is incubated overnight at an appropriate temperature for the colonies to develop; the colonies appearing along the spiral pathway can be counted either manually or electronically. The time for plating a sample is only several seconds compared to minutes used in the conventional method. Also, using a laser counter an analyst can obtain an accurate count in a few second as compared with a few minutes in the tiring procedure of counting colonies by the naked eye. The system has been used extensively in the past 20 years with satisfactory microbiological results from many products including meat, poultry, seafood, vegetables, fruits, dairy products, and spices. Manninen et al. [5] evaluated the spiral plating system against the conventional pour plate method using both manual count and laser count and found that the counts were essentially the same for bacteria and yeast. Newer versions of the spiral plater were introduced as Autoplater (Spiral Biotech, Bethesda, MD) and Whitley Automatic Spiral Plater (Microbiology International, Rockville, MD). With these automatic instruments an analyst needs only to present the liquid sample, and the instrument completely and automatically processes the sample, including resterilizing the unit for the next sample.

The Isogrid system (Neogen, Lansing, MI) consists of a square filter with hydrophobic grids printed on the filter to form 1600 squares for each filter. A food sample is first weighed, homogenized, diluted, and enzymatically treated, then passed through the filter assisted by vacuum. Microbes are trapped in the squares on the filter. The filter is then placed on prepoured nonselective or selective agar and then incubated for a specific time and temperature. Since a growing microbial colony cannot migrate over the hydrophobic material, all colonies are confined to a square shape. The analyst can then count the squares as individual colonies. Since there is a chance that more than one bacterium is trapped in one square, the system has a most probable number (MPN) conversion table to provide statistically accurate viable cell counts. Automatic instruments are also available to count these square colonies in seconds. This method also has been used to test a great variety of foods in the past 20 years.

Petrifilm (3M Co., St. Paul, MN) is an ingenious system involves appropriate rehydratable nutrients embedded in a series of films in the unit. The unit is little larger than the size of a credit card. To obtain viable cell count, the protective top layer is lifted, and 1 ml of liquid sample is introduced to the center of the unit, and then the cover is replaced. A plastic template is placed on the cover to make a round mold. The rehydrated medium will support the growth of microorganisms after suitable incubation time and temperature. The colonies are counted directly in the unit. This system has a shelf life of over one year in cold storage. The attractiveness of this system is that it is simple to use, small in size, has a long shelf-life, does not require agar preparation, and provides easy-to-read results. Recently the company also introduced a Petrifilm counter so that an analyst only needs to place the Petrifilm with colonies into the unit, and the unit will automatically count and record the viable cell count in the computer. The manual form of the Petrifilm has been used for many food systems and is gaining international acceptance as an alternative to the standard plate count method.

Redigel system (3M Co., St. Paul, MN) consists of tubes of sterile nutrient with a pectin gel in the tube but no conventional agar. This liquid system is ready for use, and no heat is needed to "melt" the medium since there is no agar in the liquid. After an analyst mixes 1 ml of liquid sample with the liquid in the tube, the resultant contents are poured into a special Petri dish coated with calcium. The pectin and calcium will react and form a gel which will solidify in about 20 minutes. The plate is then incubated at the proper time and temperature, and the colonies can be counted the same way as the conventional standard plate count method.

The four methods described above have been in use for approximately 20 years. Chain and Fung [6] made a comprehensive evaluation of all four methods against the conventional standard plate count method on 7 different foods, 20 samples each, and found that the alternative systems and the conventional method were highly comparable at an agreement of $r = 0.95$. In the same study these researchers also found that the alternative systems cost less than the conventional standard plate count method.

A newer alternative method, the SimPlate system (BioControl, Bellevue, WA), has 84 wells imprinted in a round plastic plate. After the lid is removed, a diluted food sample (1 ml) is dispensed onto the center landing pad, and 10 ml of rehydrated nutrient liquid, provided by the manufacturer, is poured onto the landing pad. The mixture (food and nutrient liquid) is distributed evenly into the wells by swirling the SimPlate in a gentle, circular motion.

Excessive liquid is absorbed by a pad housed in the unit. After 24 hours of incubation at 35°C, the plate is placed under ultraviolet (UV) light. Positive fluorescent wells are counted and the number is converted in the MPN table to determine the number of bacteria present in the SimPlate. The method is simple to use with minimum amount of preparation. A 198-well unit is also available for samples with high counts. Using different media, the unit can also make counts of total coliforms and *E. coli* counts, as well as yeast and mold counts.

The above methods are designed to count aerobic microorganisms. To count anaerobic microorganisms, one has to introduce the sample into the melted agar, and after solidification the plates need to be incubated in an enclosed anaerobic jar. In the anaerobic jar, oxygen is removed by the hydrogen generated by the gas pack in the jar to create an anaerobic environment. After incubation, the colonies can be counted and reported as anaerobic count of the food. The method is simple but requires expensive anaerobic jars and disposable gas packs. It is of concern that the interior of the jar needs almost an hour to become anaerobic. Some strict anaerobic microorganisms may die during this one-hour period of reduction of oxygen. Fung and Lee [7] developed a simple anaerobic double-tube system which is easy to use and provides instant anaerobic condition for the cultivation of anaerobes from foods. In this system, the desired agar ($\sim$23 ml) is first autoclaved in a large test tube (OD 25×150 mm). When needed, the agar is melted and tempered at 48°C. A liquid food sample (1 ml) is added into the melted agar. A smaller sterile test tube (OD 16×150 mm) is inserted into the large tube with the food sample and the melted agar. By so doing, a thin film is formed between the two test tubes. The unit is tightly closed by a screw cap. The entire unit is placed into an incubator for the colonies to develop. No anaerobic jar is needed for this simple anaerobic system. After incubation, the colonies developing in the agar film can be counted and provide an anaerobic count of the food being tested. The Fung double-tube system has been used extensively for applied anaerobic microbiology in the author's laboratory for more than 20 years [8,9]. Recently, the author tested the double-tube method for *Clostridium perfringens* in recreational waters, and he was able to obtain anaerobic *C. perfringens* counts in about 6 to 8 hours from the time of sampling to the time of reading the results. By combining the Isogrid system with the double-tube method, the author can test volumes of waters ranging from 1 to 100 ml or more.

The above-mentioned methods are designed to grow colonies to visible sizes for enumeration and report the data as CFU per gram, milliliter, or square centimeter of the food being tested.

A few "real time" viable cell count methods have been developed and tested in recent years. These methods rely on using "vital" stains to stain "live" cells or ATP detection of live cells. All these methods need careful sample preparation, filtration, selection of dyes and reagents and instrumentation. Usually the entire systems are quite costly. However, they can provide results in one shift (8 hours) and can handle a large number of samples.

The direct epifluorescent filter techniques (DFET) method has been tested for many years and is in use in the U.K. for raw milk quality assurance programs. In this method, the microorganisms are first trapped on a filter and then the filter is stained with acridine orange dye. The slide is then observed by UV microscopy. "Live" cells usually fluoresce orange-red, orange-yellow, or orange-brown whereas "dead" cells fluoresce green. The slide can be read manually or by a semiautomated counting system marketed by Bio-Foss, which can provide a viable cell count in less than an hour.

The Chemunex Scan RDI system (Monmouth Junction, NJ) involves filtering cells on a membrane and staining cells with vital dyes (Fluorassure). After approximately 90 minutes of incubation (for bacteria), the membrane with stained cells is read in a scanning chamber that can scan and count fluorescing viable cells. This system has been used to test disinfecting solutions against such organisms as *Pseudomonas aeruginosa*, *Serratia marcescens*, *Escherichia coli*, and *Staphylococcus aureus* with satisfactory results.

The MicroStar system (Millipore Corporation, Billerica, MA) utilizes ATP bioluminescence technology by trapping bacteria in a specialized membrane (Milliflex). Individual live cells are trapped in the matrix of the filter and grow into microcolonies. The filter is then sprayed with permeabilizing reagent in a reaction chamber to release ATP. The bioluminescence reagent is sprayed onto the filter. Live cells will give off light due to the presence of ATP, the light is measured using a CCD camera, and thus the fluorescent particles (live cells) are counted.

These are new developments in staining technology, ATP technology, and instrumentation for viable cell counts. The application of these methods for the food industry is still in the evaluation stage. The future looks promising.

25.4 ADVANCES IN MINIATURIZATION AND DIAGNOSTIC KITS

Identification of microorganisms constituting normal flora, spoilage organisms, foodborne pathogens, starter cultures, etc., in food microbiology is an important part of microbiological manipulations. Conventional methods, dating back more than 100 years, utilize large volumes of medium (10 ml or more) to test for a particular characteristic of a bacterium (e.g., lactose broth for lactose fermentation by *Escherichia coli*). Inoculating a test culture into these individual tubes one at a time is also very cumbersome. According to Hartman [10], over the years many microbiologists have devised vessels and smaller tubes to reduce the volumes used for these tests. This author has systematically developed many miniaturized methods to reduce the volume of reagents and media (from 5 to 10 ml down to about 0.2 ml) for microbiological testing in a convenient microtiter plate which has 96 wells arranged in an 8×12 format. The basic components of the miniaturized system are the commercially sterilized microtiter plates for housing the test cultures, a multiple inoculation device, and containers to house solid media (large Petri dishes) and liquid media (in another series of microtiter plates with 0.2 ml of liquid per well). The procedure involves placing liquid cultures (pure cultures) to be studied into sterile wells of a microtiter plate ($\sim$0.2 ml for each well) to form a master plate. Each microtiter plate can hold up to 96 different cultures, 48 duplicate cultures, or various combinations as desired. The cultures are then transferred using a sterile multipoint inoculator (96 pins protruding from a template) to solid or liquid media. Sterilization of the inoculator is accomplished by alcohol flaming. Each transfer represents 96 separate inoculations in the conventional method. After incubation at an appropriate temperature, the growth of cultures on solid media or liquid media can be observed and recorded, and the data can be analyzed. These methods are ideal for studying large numbers of isolates or for research involving challenging large numbers of microbes against a host of test compounds. Using this miniaturized system the author has characterized thousands of bacterial cultures isolated from meat and other foods, studied the effects of organic dyes against bacteria and yeasts, and performed challenge studies of various compounds against microbes with excellent results.

Other scientists also have miniaturized many systems and developed them into diagnostic kits in the late 1960s and early 1970s. Diagnostics systems such as API, Enterotube, Minitek, Crystal ID, MicroID, RapID, Biolog, and VITEK systems are currently available. Most of these systems were first developed for identification of enterics (salmonella, shigella, proteus, enterobacter, etc.). Later, many of these companies expanded the capacity of their diagnostic systems to identify nonfermentors, anaerobes, Gram-positive organisms, and even yeast and molds. Originally, an analyst needed to read the color reaction of each well in the diagnostic kit and then use a manual identification code to "key" out the organisms. Recently, diagnostic companies have developed automatic readers interfaced with a computer to provide rapid and accurate identification of the unknown cultures.

The most successful and sophisticated miniaturized automated identification system is the VITEK system (bioMerieux, Hazelwood, MO) which utilizes a plastic card containing 30 tiny wells in each of which there is a different reagent. The unknown pure culture in a liquid form is "pressurized" into the wells in a vacuum chamber, and then the cards are placed in an incubator for a period of time ranging from 4 to 12 hours. The instrument periodically scans each card and compared the color changes or gas production of each tiny well with the database of known cultures. VITEK can identify a typical *Escherichia coli* culture in 2 to 4 hours. Each VITEK unit can automatically scan 120 cards or more simultaneously. There are a few thousand VITEK units currently in use in the world, and the database is especially good for clinical isolates.

Biolog system (Hayward, CA) is also a miniaturized system using the microtiter format for growth and reaction information. Pure cultures are first isolated on agar and then suspended in a liquid to the appropriate density ($\sim$6 log cell/ml). The culture is then dispensed into a microtiter plate containing different carbon sources in 95 wells and one nutrient control well.

The plate with the pure cultures is then incubated overnight, after which the microtiter plate is removed, and the color pattern of the wells with carbon utilization is observed and compared with profiles of typical patterns of microbes using computer software to obtain identification. This system is very ambitious and tries to identify more than 1400 genera and species of environmental, food, and medical isolates from major groups of Gram-positive, Gram-negative, and other organisms. There is no question that miniaturization of microbiological methods has saved much material and operational time and has provided needed efficiency and convenience in diagnostic microbiology. The systems developed by the author and others can be used in many research and developmental laboratories for studying large numbers of cultures. These miniaturized systems and diagnostic kits can be used efficiently in identifying isolates from fruits and vegetables.

The conventional viable cell count method and the MPN (3- or 5-tube MPN) procedure have been used extensively for water and food testing for almost 100 years. The conventional methods are too cumbersome, time consuming, and utilize too many tubes, plates, and media. More than 30 years ago, Fung and Kraft [11] miniaturized the viable cell count procedure by diluting the samples in the microtiter plate using 0.025 ml size calibrated loops in 1:10 dilution series. One can simultaneously dilute 12 samples to 8 series of 1:10 dilutions in a matter of minutes. After dilution, the samples can be transported by a calibrated pipette and spot plating 0.025 ml on agar; one conventional agar plate can accommodate 4 to 8 spots. After incubation, colonies in the spots can be counted, and the number of viable cells in the original sample can be calculated since all the dilution factors are known. The accepted range of colonies to be counted in one spot is 10 to 100. The conventional agar plate standard is from 25 to 250 colonies per plate. This procedure actually went through an AOAC International collaborative study with satisfactory results [12]. However, the method has not received much attention and is waiting to be "rediscovered" in the future.

In a similar vein, Fung and Kraft [13] also miniaturized the MPN method in the microtiter plate by diluting a sample in a 3-tube miniaturized series. In one microtiter plate one can dilute 4 samples, each in triplicate (3-tube MPN), to 8 series of 1:10 dilution. After incubation, the turbidity of the wells is recorded, and a modified 3-tube MPN table can be used to calculate the MPN of the original sample. This procedure has recently received renewed interests in the scientific community.

Walser [14] in Switzerland reported the use of an automated system for microtiter plate assay to perform classic MPN of drinking water. He used a pipetting robot equipped with sterile pipetting tips for automatic dilution of the samples. After incubation, the robot placed the plate in a microtiter plate reader and obtained MPN results with the use of a computer. The system can cope with low or high bacterial load from 0 to 20,000 colonies per milliliter. This system takes out the tediousness and personnel influences on routine microbiological work and can be applied to determine MPN of fecal organisms in water as well as other microorganisms of interest in food microbiology.

Irwin *et al.* [15] in the U.S. also worked on a similar system using a modified Gauss–Newton algorithm and a 96-well micro-technique for calculating MPN using Microsoft EXCEL spreadsheets. These improvements are possible today compared with the original work of the author in 1969 because: (1) automated instruments are now available in many laboratories to dispense liquid into the microtiter plate and automated dilution instruments are also available to facilitate rapid and aseptic dilutions of samples; (2) automated readers of microtiter wells are now commonplace to read efficiently turbidity, color, and fluorescence of the liquid in the wells for calculation of MPN; and (3) elegant mathematic models, computer interpretations and analysis, and printout of data are now available which the author could not have envisioned back in 1969.

25.5 IMMUNOLOGICAL TESTING

The antigen and antibody reaction has been used for decades for detecting and characterizing microorganisms and their components in medical, food, and diagnostic microbiology. This reaction is the basis for serotyping bacteria such as salmonella, *Escherichia coli* O157:H7, and *Listeria monocytogenes*. These antibodies can be polyclonal (a mixture of several antibodies in the antisera which can react with different sites of the antigens) or monoclonal (only one pure antibody in the antiserum which will react with only one epitope of the antigens). Both polyclonal and monoclonal antibodies have been used extensively in applied food microbiology. There are many ways to perform antigen–antibody reactions, but the most popular format in recent years has been the "sandwich" enzyme-linked immunosorbant assay, popularly known as the ELISA test.

Briefly, antibodies (e.g., anti-salmonella antibody) are fixed on a solid support (e.g., wells of a microtiter plate). A solution containing a suspect target antigen (e.g., salmonella) is introduced to the microtiter well. If the solution has salmonella antigens, it will be captured by the immobilized antibodies.

After washing away food debris and excess materials, another anti-salmonella antibody complex is added into the solution. The second anti-salmonella antibody will react with another part of the trapped salmonella. This second antibody is linked with an enzyme such as horseradish peroxidase. After another washing to remove debris, a chromagen complex such as tetramethylbenzidine and hydrogen peroxide is added. The enzyme will react with the chromagen and will produce a colored compound that will indicate that the first antibody has captured salmonella. If all the reaction procedures are done properly and the liquid in a microtiter well exhibits a color reaction, then the sample is considered positive for salmonella.

This procedure is simple to operate and has been used for decades with excellent results. It should be emphasized that these ELISA tests need about a million cells to be reactive, and therefore, before performing the ELISA tests, the food sample has to go through an overnight incubation so that the target organism reaches a detectable level. The total time to detect pathogens by these systems includes the enrichment time of the target pathogens (ca. 24 hrs).

Many diagnostic companies (such as BioControl, Organon Teknika, and Tecra) have marketed ELISA test kits for foodborne pathogens and toxins (e.g., salmonella, *Escherichia coli*) and toxins (e.g., staphylococcal enterotoxins). However, the time involved in sample addition, incubating, washing and discarding of liquids, adding of another antibody complex, washing, and, finally, adding of reagents for color reaction all contribute to the inconvenience of the manual operation of the ELISA test. Recently several companies have completely automated the entire ELISA procedure.

VIDAS (bioMerieux, Hazelwood, MO) is an automated system which can perform the entire ELISA procedure automatically and can complete an assay in 45 minutes to 2 hours, depending on the test kit. Since VIDAS utilizes a more sensitive fluorescent immunoassay for reporting the results, the system is named enzyme-linked fluorescent assay (ELFA). All the analyst needs to do is to present to the reagent strip a liquid sample of an overnight enriched sample. The reagent strip contains all the necessary reagents in a ready-to-use format. The instrument will automatically transfer the sample into a plastic tube called the solid phase receptacle (SPR) which contains antibodies to capture the target pathogen or toxin. The SPR will be automatically transferred to a series of wells in succession to perform the ELFA test. After the final reaction, the result can be read, and interpretation of a positive or negative test will be automatically determined by the instrument. Presently, VIDAS can detect listeria, *Listeria monocytogenes*, salmonella, *E. coli* O157, staphylococcal enterotoxin, and campylobacter. Its manufacturers also market an immuno-concentration kit for salmonella and *E. coli* O157. Currently more than 13,000 VIDAS units are in use internationally.

BioControl (Bellevue, WA) markets an enzyme immunoassay (EIA) system called Assurance EIA which can be adapted to automation for high-volume testing. Assurance EIA is available for salmonella, listeria, *E. coli* O157:H7, and campylobacter. Diffchamb (Hisings Backa, Sweden) has a high-precision liquid delivery system that can be used to perform a variety of ELISA tests depending on the pathogens to be tested. Tecra OPUS (International BioProducts, Redmond, WA) and Bio-Tek (Highland Park, VT) instruments can also perform ELISA tests automatically as long as the proper reagents are applied to the system. Many ELISA test kits are now highly standardized and the test can be performed automatically to increase efficiency and reduce human errors.

Another exciting development in immunology is the use of lateral flow technology to perform antigen–antibody tests. In this system, the unit has three reaction regions. The first well contains antibodies to react with target antigens. These antibodies have color particles attached to them. A liquid sample (after overnight enrichment) is added to this well, and if the target organism (e.g. *E. coli* O157:H7) is present, it will react with the antibodies. The complex will migrate laterally by capillary action to the second region which contains a second antibody designed to capture the target organism. If the target organism is present, the complex will be captured, and a blue line will form due to the color particles attached to the first antibody. Excess antibodies will continue to migrate to the third region which contains another antibody that reacts with the first antibody (which has now become an antigen) and will form a blue color band. This is a "control" band indicating that the system is functioning properly. The entire procedure takes only about 10 minutes. This is truly a rapid test!

Neogen (Lansing, MI; Reveal system) and BioControl (Bellevue, WA; VIP system) are the two main companies marketing this type of system for *E. coli* O157, salmonella, and listeria. Merck KGaA (Darmstadt, Germany) developed similar systems using gold particles in the reagent to increase the sensitivity of the test.

A number of interesting methods utilizing growth of the target pathogen are also available to detect antigen–antibody reactions. The BioControl 1-2 test (BioControl, Bellevue, WA) is designed to detect motile salmonella from foods. In this system, the food sample is first preenriched for 24 hours in a broth, and then 0.1 ml is inoculated into one of the chambers in an L-shaped system. The chamber contains selective enrichment liquid medium for salmonella. There is a small hole connecting the liquid chamber with a soft agar chamber through which salmonella can migrate. An opening on the top of the soft agar chamber allows the analyst to deposit a drop of polyvalent anti-H antibodies against flagella of salmonella. The antibodies move downward in the soft agar due to gravity and diffusion. If salmonella is present, it will migrate throughout the soft agar. As the salmonella and the anti-H antibodies meet, they will react and form a visible V-shaped "immuno-band." The presence of the immunoband indicates the presumptive positive for salmonella in the food sample. This reaction occurs after overnight incubation of the unit. This system is easy to use and interpret, and it has gained popularity because of its simplicity.

Tecra (Roseville, Australia) developed a detection system (Unique Salmonella) that combines immuno-capturing, growth of the target pathogen, and an ELISA test in a simple-to-use self-contained unit. The food is first preenriched in a liquid medium overnight and an aliquot is added into the first tube of the unit. Into this tube a dipstick coated with salmonella antibodies is introduced and left in place for 20 minutes; at this time the antibodies will capture salmonella, if present. The dipstick, with salmonella attached, is then washed and placed into a tube containing growth medium. The dipstick is left in this tube for 4 hours. During this time, if salmonella is present, it replicates, and the newly produced salmonella are automatically trapped by the coated antibodies. Thus, after 4 hours of replication, the dipstick becomes saturated with trapped salmonella. The dipstick is then transferred to another tube containing a second antibody conjugated to enzyme, and the tube contents are allowed to react for 20 minutes. After this second antigen–antibody reaction, the dipstick is washed in the fifth tube and placed into the last tube for color development similar to other ELISA tests.

Development of a purple color on the dipstick indicates the presence of salmonella in the food. The entire process, from incubation of food sample to reading of the test results, requires about 22 hours, making it an attractive system for detection of salmonella. A similar system can now also detect listeria. An automated system is now being marketed.

The BioControl 1-2 test and the Unique Salmonella test are designed for laboratories with a low volume of tests. Thus, both the automatic systems and the hands-on unit systems have their place in different food testing laboratory situations.

A truly innovative development in applied microbiology is the immuno-magnetic separation system. Vicam (Somerville, MA) pioneered this concept by coating antibodies against listeria on metallic particles. Large numbers of these particles (in the millions) are added into a liquid suspected to contain listeria cells. The antibodies on the particles will capture the listeria cells while the mixture is rotated for about an hour. After the reaction has gone to completion, the tube is placed next to a powerful magnet which will immo-bilize all the metallic particles at the side of the glass test tube regardless of whether the particles have or have not captured the listeria cells. The rest of the liquid will be decanted. By removing the magnet from the tube, the metallic particles can again be suspended in a liquid. At this point, the only cells in the solution will be the captured listeria. By introducing a smaller volume of liquid (e.g., 10% of the original volume), the cells are now concentrated by a factor of 10. Cells from this liquid can be detected by direct plating on selective agar, ELISA tests, PCR reaction, or other microbiological procedures in almost pure culture state. Immunomagnetic capture can save at least one day in the total protocol of preenrichment and enrichment steps of pathogen detection in food.

Dynal (Oslo, Norway) developed this concept further by use of very homogeneous paramagnetic beads that can carry a variety of molecules such as antibodies, antigens, and DNA. Dynal has developed beads to capture *E. coli* O157, listeria, cryptosporidium, giardia, and others. Furthermore, the beads can be supplied without any coating materials, and scientists can tailor them to their own needs by coating with the necessary antibodies or other capturing molecules for detection of target organisms. Currently, many diagnostic systems (ELISA, PCR, etc.) are incorporating an immunomagnetic capture step to reduce incubation and increase sensitivity of the entire protocol.

Fluorescent antibody techniques have been used for decades for the detec-tion of salmonella and other pathogens. Similar to the DEFT test designed for viable cell count, fluorescent antibodies can be used to detect a great variety of target microorganisms such as *E. coli* O157:H7 in milk and juice.

One of the newest and fastest immunological methods to detect food-borne pathogens is the Pathatrix system (Matrix MicroScience, Golden, CO). Wu *et al.* [16] tested a same-day protocol for the detection of *Escherichia coli* O157:H7 by the Pathatrix system (which employs a novel immuno-capture method) and Colortrix (a rapid ELISA test). The Pathatrix system can circulate a 4.5 hour preenriched 250 ml sample (25 g of food in 225 ml of preenrichment

broth) over a sheet of paramagnetic beads coated with antibodies against
E. coli O157:H7 many times in 30 minutes to capture almost all target pathogens. This circulation system increased the concentration of *E. coli* O157:H7
from the population after 4.5 hours of enrichment to 1.2 to 2.6 log CFU/25 g
higher concentration in 30 minutes. After Pathatrix concentration the beads
with target pathogens are applied to the Colortrix system, a rapid ELISA
system that was able to detect *E. coli* O157:H7 in 15 minutes. The results
indicated an excellent correlation (100%) between positive Pathatrix/Colortrix
(5.25 hours) compared with a 30-hour conventional plating method. The
sensitivity of the system is from 0.7 to 2.1 log CFU/25 g as the initial concentration of *E. coli* O157:H7 in the sample before the 4.5 hours of enrichment.
This system is also able to detect *Listeria monocytogenes*, campylobacter,
and other pathogens.

Antigen–antibody reaction provides a powerful system for rapid detection of all kinds of pathogens and molecules. This section has described
some of the useful methods developed for applied food microbiology. Some
systems are highly automated, and others are exceedingly simple to operate.
It should be emphasized that many of the immunological tests described in
this section provide presumptive positive or presumptive negative screening test
results. For negative screening results, the food in question is allowed to be
shipped for commerce. For presumptive positive test results, the food will
not be allowed for shipping until confirmation of the positive is done by the
conventional microbiological methods.

25.6 INSTRUMENTATION AND BIOMASS MEASUREMENTS

As the field of rapid methods and automation has developed, the boundaries
between instrumentation and diagnostic tests have begun to merge. Instrumentation is now playing an important function in improving the efficiency
of diagnostic kit systems, and the trend will continue. The following discussions are mainly on instrumentation measuring signals related to microbial
growth.

Instruments can be used to monitor changes in a population such as
ATP levels, levels of specific enzymes, pH, electrical impedance, conductance
and capacitance, generation of heat, radioactivity, carbon dioxide, and others.
It is important to note that for the information to be useful, these parameters must be related to viable cell counts of the same sample series. In general,
the larger the number of viable cells in the sample, the shorter the detection time of these systems. A scattergram is then plotted and used for further
comparison of unknown samples. The assumption is that as the number of
microorganisms increases in the sample, these physical, biophysical, and
biochemical events will also increase accordingly. When a sample has 5 or 6 log
organisms/ml, detection time can be achieved in about 4 hours from the time
the sample is placed in the instrument.

All living things utilize ATP. In the presence of a firefly enzyme system (luciferase and luciferin system), oxygen, and magnesium ions, ATP will facilitate the reaction to generate light. The amount of light generated by this reaction is proportional to the amount of ATP in the sample. Thus, the light units can be used to estimate the biomass of cells in a sample. The light emitted by this process can be monitored by a sensitive and automated fluorimeter. Some instruments can detect as little as 100 to 1000 femtograms of ATP (1 femtogram, 1 fg, is $-15 \log$ g). The amount of ATP in one colony-forming unit has been reported as 0.47 fg with a range of 0.22 to 1.03 fg. Using this principle, many researchers have used ATP to estimate the number of microbial cells in solid and liquid foods.

Initially, scientists attempted to use ATP to estimate the total viable cell count in foods. The results are inconsistent due to the fact that (1) different microorganisms have different amounts of ATP per cell (e.g., a yeast cell can have 100 times more ATP than a bacterial cell); (2) even for the same organism, the amount of ATP per cell is different at different growth stages; and (3) background ATP from other biomass such as blood and biological fluids in the foods interferes with the target bacterial ATP. Only after much research and development will scientists be able to separate nonmicrobial ATP from microbial ATP and obtain reasonable accuracy in relating ATP to viable cell counts in foods. Since obtaining an ATP reading takes only a few minutes, the potential of exploring these methods further exists. To date, ATP has not been applied much to estimation of viable cell counts in food microbiology laboratories.

From another viewpoint, the presence of ATP in certain foods such as wine is undesirable regardless of the source. Thus monitoring ATP can be a useful tool for quality assurance in the winery.

There has been a paradigm shift in the field of ATP detection in recent years. Instead of detecting ATP of microorganisms, systems are now designed to detect ATP from any source for hygiene monitoring. The idea is that a dirty food processing environment will have a high ATP level, and a properly cleansed environment will have a low ATP level regardless of what contributed to the ATP in these environments. Once this concept is accepted by the food industry, there will be an explosion of ATP systems being used in the food industry for hygiene monitoring. In all of these systems, the key is to be able to obtain an ATP reading in the form of relative light units (RLUs) and to relate these units to the cleanliness of food processing surfaces. The scale of RLU readings obtained from different surfaces in food factories encompasses acceptable, marginal, and unacceptable levels. Since there is no standard as to what constitutes an absolutely acceptable ATP level in any given environment, these RLUs are quite arbitrary. In general, a dirty environment will have high RLUs, and after proper cleaning the RLUs will decrease. Besides the sensitivity of the instruments, an analyst should consider the following attributes in selecting a particular system: simplicity of operation, compactness of the unit, computer adaptability, cost of the unit, support from the company, and documentation of usefulness of the system.

Besides the above mentioned issues, Dreibelbis [17] in a study of five ATP instruments for hygiene monitoring of a food plant considered the following attributes to be important as selection criteria of the systems: the ability of the technicians in the microbiological laboratory to use the ATP bioluminescence hygiene monitoring system without supervision, the reputation of the ATP system in the industry, and the quality of services received from the manufacturer during the evaluation of the product.

Currently the following ATP instruments are available: Lumac (Landgraaf, the Netherlands), BioTrace (Plainsboro, NJ), Lightning (BioControl, Bellevue, WA), Hy-Lite (EM Science, Darmstadt, Germany), Charm 4000 (Charm Sciences, Malden, MA), Celsis system SURE (Cambridge, U.K.), Zylux (Maryville, TN), Profile 1 (New Horizon, Columbia, MD), and others.

As microorganisms grow and metabolize nutrients, large molecules are metabolized to smaller molecules in a liquid system and cause a change in electrical conductivity and resistance in the liquid as well as at the interface of electrodes. These changes can be expressed as impedance, conductance, and capacitance changes. When a population of cells reaches about 5 log CFU/ml, it will cause a change in these parameters. Thus, when a food has a large initial population, the time to make this change will be shorter than with a food that has a smaller initial population. The detection time of the test sample, the time when the curve accelerates upward from the baseline, is inversely proportional to the initial concentration of microorganisms in the food. In order to use these methods, a series of standard curves must be constructed by making viable cell counts in food with different initial concentrations of cells and then measuring the resultant detection time. A scattergram can then be plotted. Thereafter, in the same food system, the number of the initial population of the food can be estimated by the detection time on the scattergram.

The Bactometer (bioMerieux, Hazelwood, MO) has been in use for many years to measure impedance changes by microorganisms in foods, water, cosmetics, and similar products. Samples are placed in the wells of a 16-well module which is then plugged into the incubator to start the monitoring sequence. As the cells reach the critical number (5 to 6 log/ml), the change in impedance increases sharply, and the monitor screen shows a slope similar to the log phase of a growth curve. The detection time can then be obtained to determine the initial population of the sample. If one sets a cut-off point of 6 log CFU/g of food for acceptance or rejection of the product, and the detection time is 4 hours ± 15 minutes, then one can use the detection time as a criterion for quality assurance of the product. Food that exhibits no change of impedance curve after more than 4 hours and 15 minutes in the instrument is acceptable while food that exhibits a change of impedance curve before 3 hours and 45 minutes will not be acceptable. For convenience the instrument is designed such that the sample bar displayed on the screen for a food will flash red for an unacceptable sample, green if acceptable, and yellow for marginally acceptable. The rapid automated bacterial impedance technique (RABIT) is a similar system, marketed by Bioscience International (Bethesda,

MD) for monitoring microbial activities in food and beverages. Instead of the 16-well module used in the Bactometer, individual tubes containing electrodes are used to house the food samples.

The Malthus system (Crawley, U.K.) uses conductance changes of the fluid to indicate microbial growth; it generates conductance curves similar to impedance curves used in the Bactometer. The Malthus system uses individual tubes for food samples. Water heated to the desired temperature (e.g., 35°C) is used as the temperature control instead of heated air as with the previous two systems. All these systems have been evaluated by various scientists in the past 10 to 15 years with satisfactory results. All have their advantages and disadvantages depending on the type of food being analyzed. These systems can also be used to monitor targeted groups of organisms such as coliform or yeast and mold using specially designed culture media. In fact, the Malthus system has a salmonella detection protocol that was approved by AOAC International.

BacT/Alert Microbial Detection System (Organon Teknika, Durham, NC) utilizes colorimetric detection of carbon dioxide production by microorganisms in a liquid system using sophisticated computer algorithms and instrumentation. Food samples are diluted and placed in special bottles with appropriate nutrients for growth of microorganisms and production of carbon dioxide. At the bottom of the bottle there is a sensor that is responsive to the amount of carbon dioxide in the liquid. When a critical amount of the gas is produced, the sensor changes from dark green to yellow, and this change is detected by reflectance colorimetry automatically. The units can accommodate 120 or 240 culture bottles. Detection time of a typical culture of *E. coli* is about 6 to 8 hours.

BioSys (BioSys, Inc., Ann Arbor, MI) utilizes color changes of media (designed for specific target organisms) during the growth of cultures to detect and estimate organisms in foods and liquid systems. The uniqueness of the system is that the color compounds developed during microbial growth are diffused into an agar column situated at the bottom of the unit, and the changes are measured automatically without the interference of food particles in the chamber. Depending on the initial microbial load in the food, microbial information can be obtained during the same production shift that the sample was taken in a food processing operation. The system is easy to use and can accommodate 32 samples for one incubation temperature or 128 samples for 4 independent incubation temperatures in different models. The system is designed for bioburden testing and HACCP (hazard analysis critical control points) control and can test for indirect total viable cell, coliform, *E. coli*, yeast, mold, and lactic acid bacteria counts in swab samples and environmental samples.

Basically, any type of instrument that can continuously and automatically monitor turbidity and color changes of a liquid in the presence of microbial growth can be used for rapid detection of the presence of microorganisms. There will definitely be more systems of this nature on the market in years to come.

25.7 GENETIC TESTING

Rapid tests discussed earlier for detection and characterization of microorganisms were based on phenotypic expressions of genotypic characteristics of microorganisms. The phenotypic expressions are subject to growth conditions such as temperature, pH, nutrient availability, oxidation–reduction potentials, environmental and chemical stresses, toxins, and water activities. Phenotypic expression, even including immunological tests, depends on cells' ability to produce the target antigens to be detected by the available antibodies or vice versa. The conventional "gold standards" of diagnostic microbiology rely on phenotypic expression or traits that are inherently subject to variation.

Genotypic characteristics of a cell are far more stable than its phenotype. The natural mutation rate of a bacterial culture is about 1 in 100 million cells. Thus, there has been a push in recent years to make genetic test results the confirmative and definitive identification step in diagnostic microbiology. The debate is still continuing, and the final decision has not been reached by governmental and regulatory bodies for microbiological testing. Genetic-based diagnostic and identification systems are discussed in this section.

Hybridization of the deoxyribonucleic acid (DNA) sequence of an unknown bacterium by a known DNA probe is the first stage of genetic testing. The Genetrak system (Framingham, MA) provides a sensitive and convenient method to detect pathogens such as salmonella, listeria, campylobacter, and *E. coli* O157 in foods. Initially, the system utilized radioactive isotopes bound to DNA probes to detect complementary DNA of unknown cultures. The drawbacks of the first generation of this type of probes are (1) most food laboratories are not eager to work with radioactive materials in routine analysis and (2) there are limited copies of DNA in a cell. The second generation of probes uses enzymatic reactions to detect the presence of the pathogens and uses RNA as the target molecule. In a cell, there is only one complete copy of DNA; however, there may be 1,000 to 10,000 copies of ribosomal RNA. Thus, the new generation of probes is designed to detect target RNA using color reactions. After enrichment of cells (e.g., salmonella) in a food sample for about 18 hours, the cells (target cells as well as other microbes) are lysed by a detergent to release cellular materials (DNA, RNA, and other molecules) into the enrichment solution. Two RNA probes (designed to react with one piece of target salmonella RNA) are added into the solution. The capture probe with a long tail of a nucleotide (e.g., polyadenine tail or AAAAA) is designed to capture the RNA onto a dipstick with a long tail of thymine (TTTTT). The reporter probe, with an enzyme attached, will react with the other end of the RNA fragment. If salmonella RNA molecules are present, the capture probes will attach to one end of the RNA, and the reporter probes will attach to the other end. A dipstick coated with many copies of a chain of complementary nucleotide (e.g., thymine, TTTTT) will be placed into the solution. Since adenine (A) will hybridize with thymine (T), the chain (TTTTT) on the dipstick will react with the AAAAA and thus capture the target RNA complex onto the stick. After washing away debris and other molecules in the liquid, a chromagen is added. If the target RNA is captured, then the enzyme present in the second probe will react with the chromagen and will produce a color reaction indicating the presence of the pathogen in the food. In this case, the food is positive for salmonella. The system developed by Genetrak has been evaluated and tested for many years and has AOAC International approval of the procedure for many food types. More recently, Genetrak has adapted a microtiter format for more efficient and automated operation of the system.

PCR is now an accepted method to detect pathogens by amplification of the target DNA and detecting the target PCR products. Basically, a DNA molecule (double helix) of a target pathogen (e.g., salmonella) is first denatured at about 95°C to form single strands, then the temperature is lowered to about 55°C for two primers (small oligonucleotides specific for salmonella) to anneal to specific regions of the single stranded DNA. The temperature is increased to about 70°C for a special heat-stable polymerase, the TAQ enzyme from *Thermus aquaticus*, to add complementary bases (A, T, G, or C) to the single-stranded DNA and complete the extension to form a new double strand of DNA. This is called a thermal cycle. After this cycle, the tube will be heated to 95°C again for the next cycle. After one thermal cycle, one copy of DNA will become two copies. After about 21 cycles and 31 cycles, one million and one billion copies of the DNA will be formed, respectively. This entire process can be accomplished in less than an hour in an automatic thermal cycler. Theoretically, if a food contains one copy of salmonella DNA, the PCR method can detect the presence of this pathogen in a very short time. After PCR reactions, one still needs to detect the presence of the PCR products to indicate the presence of the pathogen. Four commercial kits for PCR reactions and detection of PCR products are briefly discussed in the following.

The BAX system (Qualicon, Inc., Wilmington, DE) for screening foodborne pathogens combines DNA amplification and automated homogeneous detection to determine the presence or absence of a specific target. All primers, polymerase, and deoxynucleotides necessary for PCR as well as a positive control and an intercalating dye are incorporated into a single tablet. The system works directly from an overnight enrichment of the target organisms. No DNA extraction is required. Assays are available for salmonella, *E. coli* O157:H7, listeria genus, and *Listeria monocytogenes*. The system uses an array of 96 blue LEDs as the excitation source and a photomultiplier tube to detect the emitted fluorescent signal. This integrated system improves the ease-of-use of the assay. In addition to simplifying the detection process, the new method converts the system to a homogeneous PCR test. The homogenous detection process monitors the decrease in fluorescence of a double-stranded DNA (dsDNA) intercalating dye in solution with dsDNA as a function of temperature. Following amplification, melting curves are generated by slowly ramping the temperature of the sample to a denaturing level (95°C). As the dsDNA denatures, the dye becomes unbound from the DNA duplex, and the fluorescent signal decreases. This change in fluorescence can be plotted against temperature to yield a melting curve waveform. This assay thus eliminates the need for gel-based detection and yields data amenable to storage and retrieval in an electronic database. In addition, this method reduces the hands-on time of the assay and reduces the subjectivity of the reported results. Further, melting curve analysis makes possible the ability to detect multiple PCR products in a single tube. The inclusivity and exclusivity of the BAX system assays reach almost 100% meaning that false positive and false negative rates are almost zero. The automated BAX system can now be used with assays for the detection of *Cryptosporidium parvum* and *Campylobacter jejuni/coli* and for the quantitative and qualitative detection of genetically modified organisms in soy and corn. The new BAX system is far more convenient than the old system in which a gel electrophoresis step was required to detect PCR products after thermal cycling.

The following two methods also have been developed to bypass the electrophoresis step to detect PCR products. These methods are called "real-time PCR" because they involve a solution in which a fluorescent signal increases if the target sequence is present in the solution. They rely on the use of fluorescent molecules and can directly measure the amplification products while amplification is in progress. The more target DNA in the solution, the sooner the number of PCR products will reach the detection threshold and can be detected since fewer thermal cycles are needed, compared to a solution with a smaller number of target DNA molecules. With the use of different fluorescent dyes in the same solution, several target DNA molecules can be studied simultaneously. This is called a multiplex PCR system.

The TaqMan system of Applied Biosystems (Foster City, CA) also amplifies DNA by a PCR protocol. However, during the amplification step a special molecule is annealed to the single-stranded DNA to report the linear amplification. The molecule has the appropriate sequence for the target DNA. It also has two attached particles. One is a fluorescent particle, and another one is a quencher particle. When the two particles are close to each other no fluorescence occurs. However, when the TAQ polymerase is adding bases to the linear single strand of DNA, it will break this molecule away from the strand (like the PacMan in computer games). As this occurs, the two particles will separate from each other, and fluorescence will occur. By measuring fluorescence in the tube, a successful PCR reaction can be determined. Note that the reaction and reporting of a successful PCR protocol occur in the same tube. The author's research team developed a TaqMan procedure to detect rapidly *Yersinia enterocolitica* in foods [18].

A new system called Molecular Beacon Technology (Stratagene, La Jolla, CA) was developed and can be used for food microbiology in the future [19]. In this technology, all reactions are again in the same tube. A Molecular Beacon is a tailor-made hairpin-shaped hybridization probe. The probe is used to attach to target PCR products. On one end of the probe there is attached a fluorophore, and on the other end a quencher. In the absence of the target PCR products the beacon is in a hairpin shape, and there is no fluorescence. However, during PCR reactions and the generation of target PCR products, the beacons will attach to the PCR products and cause the hairpin molecule to unfold. As the quencher moves away from the fluorophore, fluorescence will occur, and this can be measured. The measurement can be done as the PCR reaction is progressing, thus allowing "real-time" detection of target PCR products, and thus the presence of the target pathogen in the sample. This system has the same efficiency as the TaqMan system, but the difference is that the beacons detect the PCR products themselves, while in the TaqMan system they only report the occurrence of a linear PCR reaction and not the presence of the PCR product directly. By using molecular beacons containing different fluorophores, one can detect different PCR products in the same reaction tubes, and thus it is possible to perform "multiplex" tests of several target pathogens or molecules. The use of this technology is very new and not well known in food microbiology areas.

One of the major problems of PCR systems is contamination of PCR products from one test to another. Thus, if any PCR products from a positive sample (e.g., salmonella PCR products in a previous run) enter the reaction system of the next analysis, they may cause a false positive result. The Probelia system, developed by Institut Pasteur (Paris, France), attempts to eliminate PCR product contamination by substituting the base uracil for the base thymine in the entire PCR protocol. Thus, in the reaction tube there are adenine, uracil, guanine, and cytosine, and no thymine. During the PCR reaction, the resultant Probelia PCR products will be AUGC pairing and not the natural ATGC pairings. The PCR products are read by hybridization of known sequences in a microtiter plate. The report of the hybridization is by color reaction similar to an ELISA test in the microtiter system.

After one experiment is completed, a new sample is added into another tube for the next experiment. In the tube there is an enzyme, uracil-D-glycosylase (UDG), which will hydrolyze any DNA molecules that contain a uracil. Therefore, if there are contaminants from a previous run, they will be destroyed before the beginning of the new run. Before a new PCR reaction, the tube with all reagents is heated to 56°C for 15 minutes for UDG to hydrolyze any contaminants. During the DNA denaturization step, the UDG will be inactivated and will not act on the new PCR products containing uracil. Currently, Probelia can detect salmonella and *Listeria monocytogenes* from foods. Other kits under development include *E. coli* O157:H7, campylobacter, and *Clostridium botulinum*.

Theoretically, PCR systems can detect one copy of target pathogen DNA from a food sample (e.g., salmonella DNA). In practice, about 200 cells are needed to be detected by current PCR methods. Thus, even in a PCR protocol, bacteria in the food must be enriched for a period of time, e.g., overnight or at least 8 hours' incubation of food in a suitable enrichment liquid, so that there are enough cells for the PCR process to be reliable.

Besides the technical manipulations of the systems which can be complicated for many food product microbiology laboratories, two major problems need to be addressed: inhibitors of PCR reactions and the question of live and dead cells. In food, there are many enzymes, proteins, and other compounds that can interfere with the PCR reaction and result in false negatives.

These inhibitors must be removed or diluted. Since the PCR reaction amplifies target DNA molecules, even DNA from dead cells can be amplified, and thus food with dead salmonella can be declared as salmonella positive by PCR results. In this situation, food properly cooked but containing DNA of dead cells may be unnecessarily destroyed because of a positive PCR test. PCR can be a powerful tool for food microbiology once all the problems are solved, and analysts are convinced of its applicability in routine analysis of foods.

The aforementioned genetic methods are for detection of target pathogens in foods and other samples. They do not provide identification of the cultures to the species and subspecies level which is critical in epidemiological investigations of outbreaks or routine monitoring of occurrence of microorganisms in the environment. The following discussions will center around developments in the genetic characterization of bacterial cultures.

The RiboPrinter microbial characterization system (DuPont Qualicon, Wilmington, DE) characterizes and identifies organisms to genus, species, and subspecies levels automatically. To obtain a RiboPrint of an organism, the following steps are followed:

1. A pure colony of bacteria suspected to be the target organism (e.g., salmonella) is picked from an agar plate by a sterile plastic stick.
2. Cells from the stick are suspended in a buffer solution by mechanical agitation.
3. An aliquot of the cell suspension is loaded into the sample carrier to be placed into the instrument. Each sample carrier has space for eight individual colony picks.
4. The instrument will automatically prepare the DNA for analysis by restriction enzyme and lysis buffer to break the cell envelope, release and cut DNA molecules. The DNA fragments will go through an electrophoresis gel to separate DNA fragments into discrete bands. Lastly, the DNA probes, conjugate, and substrate will react with the separated DNA fragments, and light emission from the hybridized fragments is then photographed. The data are stored and compared with known patterns of the particular organism. The entire process takes eight hours for eight samples. However, at two-hour intervals, another eight samples can be loaded for analysis.

Different bacteria will exhibit different patterns (e.g., salmonella versus *E. coli*), and even the same species can exhibit different patterns (e.g., *Listeria monocytogenes* has 49 distinct patterns). Examples of numbers of RiboPrint patterns for some important food pathogens are: salmonella, 145; listeria, 89; *Escherichia coli*, 134; staphylococcus, 406; and vibrio, 63. Additionally, the database includes 300 lactobacillus, 43 lactococcus, 11 leuconostoc, and 34 pediococcus patterns. The current identification database provides 3267 RiboPrint patterns representing 98 genera and 695 species.

One of the values of this information is that in the case of a foodborne outbreak, scientists not only can identify the etiological agent (e.g., *Listeria monocytogenes*) but can pinpoint the source of the responsible subspecies. For example, in the investigation of an outbreak of *Listeria monocytogenes*, cultures were isolated from a slicer of the product and also from the drains of the plant. The question was: which source was responsible for the outbreak? By matching RiboPrint patterns of the two sources of *L. monocytogenes* against the foodborne outbreak culture, it was found that the isolate from the slicer matched the outbreak culture, thus determining the true source of the problem. The RiboPrinter system is a very powerful tool for electronic data-sharing worldwide.

These links can monitor the occurrence of foodborne pathogens and other important organisms as long as different laboratories utilize the same system for obtaining the RiboPrint patterns.

Another important system concerns the pulsed-field gel electrophoresis patterns of pathogens. In this system, pure cultures of pathogens are isolated and digested with restriction enzymes, and the DNA fragments are subjected to a system known as pulsed-field gel electrophoresis which effectively separates DNA fragments on the gel (DNA fingerprinting). For example, in a foodborne outbreak of *E. coli* O157:H7, biochemically identical *E. coli* O157:H7 cultures can exhibit different patterns. By comparing the gel patterns from different sources, one can trace the origin of the infection or search for the spread of the disease and thereby control the problem.

In order to compare data from various laboratories, the Pulse Net System was established under the National Molecular Subtyping Network for Foodborne Disease Surveillance at the Centers for Disease Control and Prevention (CDC). An extensive training program has been established so that all the collaborating laboratories use the same protocol and are electronically linked to share DNA fingerprinting patterns of major pathogens. As soon as a suspect culture is noted as a possible source of an outbreak, all the collaborating laboratories are alerted to search for the occurrence of the same pattern to determine the scope of the problem and share information in real time.

There are many other genetic-based methods, but they are not directly related to food microbiology and are beyond the scope of this review. It is safe to say that many genetic-based methods are slowly but surely finding their way into food microbiology laboratories, and they will provide valuable information for quality assurance, quality control, and food safety programs in the future.

25.8 BIOSENSORS

The use of biosensors is an exciting field in applied microbiology. The basic idea is simple, but the actual operation is quite complex and involves much instrumentation. Basically, a biosensor is a molecule or a group of molecules of biological origin attached to a signal recognition material. When an analyte comes in contact with the biosensor, the interaction will initiate a recognition signal which can be reported in an instrument.

Many types of biosensors have been developed, such as enzymes (a great variety of enzymes have been used), antibodies (polyclonal and monoclonal), nucleic acids, cellular materials, and others. Whole cells may also be used as biosensors. Analytes detected include toxins (staphylococcal enterotoxins, tetrodotoxins, saxitoxin, botulinum toxin, and others), specific pathogens (salmonella, staphylococcus, *Escherichia coli* O157:H7, etc.), carbohydrates (fructose, lactose, galactose, etc.), insecticides and herbicides, ATP, antibiotics (e.g., penicillins), and others. The recognition signals used include electro-chemical (potentiometry, voltage changes, conductance and impedance, light addressable, etc.), optical (such as UV, bioluminescence and chemilumin-escence, fluorescence, laser scattering, reflection and refraction of light, surface plasmon resonance, and polarized light), and miscellaneous transducers (such as piezoelectric crystals, thermistors, acoustic waves, and quartz crystals).

An example of a simple enzyme biosensor is the sensor for glucose. The reaction involves the oxidation of glucose (the analyte) by glucose oxidase (the biosensor) yielding the end products, gluconic acid and hydrogen peroxide. The reaction is reported by a Clark oxygen electrode which monitors the decrease in oxygen concentration amperometrically. The range of meas-urement is from 1 to 30 mM with a response time of 1 to 1.5 minutes and a recovery time of 30 seconds. The lifetime of the unit is several months. Some of the advantages of enzyme biosensors are their strong binding to the analyte, high selectivity and sensitivity, and rapid reaction time. Some of the disadvantages are expense, loss of activity when enzymes are immobilized on a transducer, and loss of activity due to deactivation. Other enzymes used include galactosidase, glucoamlyase, acetylcholinesterase, invertase, and lactate oxidase. Excellent review articles and books on biosensors are presented by Eggins [20], Cunningham [21], Goldschmidt [22], and others.

Recently much attention has been directed to the field of "biochips" and "microchips" development to detect a great variety of molecules including foodborne pathogens. Due to advancements in miniaturization technology, as many as 50,000 individual spots (e.g., DNA microarrays), with each spot containing millions of copies of a specific DNA probe, can be immobilized on a specialized microscope slide. Fluorescent labeled targets can be hybridized to these spots and be detected. An excellent article by Deyholos *et al.* [23] described the application of microarrays to discover genes associated with a particular biological process such as the response of a plant (arabidopsis) to NaCl stress and detailed analysis of a specific biological pathway such as one-carbon metabolism in maize.

Biochips can also be designed to detect all kinds of foodborne pathogens by imprinting a variety of antibodies or DNA molecules against specific pathogens on the chip for the simultaneous detection of pathogens such as salmonella, listeria, *Escherichia coli*, and *Staphylococcus aureus* on the same chip. According to Heron writing in 2000 [24], biochips are an exceedingly important technology in life sciences, and at that time the market value was estimated to be as high as $5 billion by the middle of the present decade. This technology is especially important in the rapidly developing field of proteomics which requires massive amount of data to generate valuable information.

Certainly, the development of these biochips and microarray chips is impressive for obtaining a large amount of information for biological sciences. As for foodborne pathogen detection, there are several important issues to consider. These biochips are designed to detect minute quantities of target molecule. The target molecules must be free from contaminants before being applied to the biochips. In food microbiology, the minimum requirement for pathogen detection is 1 viable target cell in 25 g of a food such as ground beef. A biochip will not be able to seek out such a cell from the food matrix without extensive cell amplification (either by growth or PCR) or sample preparation by filtration, separation, absorption, centrifugation, etc., as described in this chapter. Any food particle in the sample will easily clog the channels used in biochips. These preparations will not allow the biochips to provide "real-time" detection of pathogens in foods.

Another concern is viability of the pathogens to be detected by biochips. Monitoring the presences of some target molecule will only demonstrate the presence or absence of the target pathogen and will not show the viability of the pathogen in question. Some form of culture enrichment to ensure growth is still needed in order to obtain meaningful results. It is conceivable that the biomass of microbes can be monitored by biochips but instantaneous detection of specific pathogens such as salmonella, listeria, and campylobacter in a food matrix during food processing operations is still not possible. The potential of biochip and microarrays for food pathogen detection is great, but at present much more research is needed to make this technology a reality in applied food microbiology.

25.9 U.S., WORLD MARKET, AND TESTING TRENDS (1999–2008)

There is no question that many microbiological tests are being conducted nationally and internationally on food, pharmaceutical products, environmental samples, and water. The most popular tests are total viable cell count, coliform/*E. coli* count, and yeast and mold counts. A large number of tests are also performed on pathogens such as salmonella, listeria and *Listeria monocytogenes*, *E. coli* O157:H7, *Staphylococcus aureus*, campylobacter, and other organisms.

Applied microbiologists working in medical, food, environmental, and industrial settings in government, academia, and the private sector are interested in the numbers and kinds of microbiological tests being done annually on local, regional, national, and international scales.

Strategic Consulting, Inc. (phone: 802-457-9933; e-mail: weschler@ strategic-consult.com; Woodstock, VT) produced three major reports on the market for microbiological testing [25–27]. This group researched diagnostic testing companies through public records and interviews of hundreds of practitioners of applied microbiology by phone or other means to obtain estimated data to compile the reports. Readers are advised to contact Strategic Consulting, Inc. for details of these reports. Below is information that the author received permission to use for this article.

In 1998 the number of worldwide industrial microbiological tests was estimated to be 755 million with a total market value of US\$1.1 billion, assuming the average price per test to be US\$1.47. They also estimated that 56% of the tests were for food; 30% for pharmaceuticals; 10% for beverages; and 4% for environmental water tests [25]. Of these tests, 420 million were done in food laboratories with 360 million for "routine tests" (total viable cell counts, coliform counts, and yeast and mold counts) and 60 million for "specific pathogen tests" (salmonella, listeria, *Staphylococcus aureus*, *E. coli* O157:H7 tests). Approximately one third of all the tests were done in the U.S., another third in Europe, and the rest were performed in the rest of the world.

It was projected that from 1998 to 2003 there would be a 24.6% increase in the number of tests; 17% increase in the price per test, and 45.8% increase in the total revenue of the testing market by 2003. Of the 50 or so diagnostic companies reviewed, there seems to be no absolute dominance of the field by any one company, although there are clear leaders in the area [25]. The situation is quite fluid since some companies are constantly acquiring products from other companies. Many new companies are also emerging in this area as new technologies are developed.

The 2000 U.S. food industry market study [26] indicated that the total number of microbiological tests per year was 144.3 million, total number of tests for pathogens was 23.5 million, with a market value of US\$53.4 million, and the average selling price per test was US\$2.27. These data were obtained from a survey of 5,979 food processing plants with an average of 464 tests per plant per week, and 24,128 tests per plant per year. The percentage of microbiological testing performed on selected food categories was as follows: processed foods, 36.2%; dairy, 31.8%; meat, 22.3%; fruits/vegetables, 9.7%. The number of test to be done in the future for fruits and vegetables will certainly increase due to recent foodborne outbreaks related to these food commodities.

Another valuable set of data is the proportion of routine to pathogen tests which is 83.7% versus 16.3%. Further breakdown of these data revealed that the total viable count represented 37.2% of all tests; coliform/*E. coli*, 30.8%; yeast and mold, 15.7%; and pathogens, 16.3%. The percentage for pathogen testing is an increase from 15% reported in the 1998 review [25]. It is projected that this number will increase further in the years to come.

Estimation of the use of "rapid methods" versus "conventional methods" is hard to obtain. From the author's experiences, about 70% of microbial tests are currently done using manual or conventional methods and 30% using rapid methods. By 2008, for total testing, about 50% will be using conventional methods, and 50% will be using rapid tests. However, for pathogen testing 60 to 70% will be some form of rapid test, and 30 to 40% will the conventional tests. These projected changes are attributed to the current and future improvement of rapid methods.

The newest global test volume predictions for industrial microbiological testing (per year) for 2008 are: food, 715.6 million tests (47.5%, total); beverages, 137.0 million tests (9.1%); pharmaceuticals, 311.1 million tests (20.7%); personal care products, 249.1 million tests (16.5%); environmental, 55.9 million tests (3.7%); and material processing, 36.5 million tests (2.4%) [27].

It is safe to say that the field of rapid methods and automation in microbiology will continue to grow in number and kinds of tests to be done in the future due to the increased concern about food safety.

25.10 PREDICTIONS OF THE FUTURE

It is always difficult to predict the future development in any field of endeavor. In 1995 the author was honored to present a lecture at the annual meeting of the American Society of Microbiology as the Food Microbiology Divisional Lecturer concerning the current status and the future outlook of the field of rapid methods and automation in microbiology. The following is a synopsis of the ten predictions, with a look into the future made in 1995. A more detailed description of the predictions can be found in the paper by Fung published in 1999 [28].

1. Viable cell counts will still be used.
2. Real-time monitoring of hygiene will be in place.
3. PCR, ribotyping, and genetic tests will become a reality in food laboratories.
4. ELISA and immunological tests will be completely automated and widely used.
5. Dipstick technology will provide rapid answers (10 minutes).
6. Biosensors will be in place for HACCP programs in the future.
7. Instant detection of target pathogens will be possible by a computer-generated matrix in response to particular characteristics of pathogens (microarrays, biochips, microchips).
8. Effective separation and concentration of target cells will greatly assist in rapid identification.
9. Microbiological alert systems will be in food packages.
10. Consumers will have rapid alert kits for detection of pathogens at home.

Along with the prediction of the future of rapid testing methods, it is useful to describe the ten attributes and criteria for an ideal automated microbiology assay system as follows:

1. Accuracy for the intended purposes. Sensitivity, minimal detectable limits, specificity of test system, versatility, potential applications, comparison to reference methods.

2. Speed in productivity. Time in obtaining results, number of samples processed per run, per hour, per day.
3. Cost. Initial, per test, reagents, labor.
4. Acceptability by scientific community and regulatory agencies.
5. Simplicity of operation. Sample preparation, operation of test equipment, computer versatility.
6. Training. On-site, length of time, qualification of operator.
7. Reagents. Preparation, stability, availability and consistency.
8. Company reputation.
9. Technical services. Speed, availability, cost and scope.
10. Utility and space requirements.

The future looks very bright for the field of rapid methods and automation in microbiology. The potential is great and many exciting developments will certainly unfold in the near and far future.

ACKNOWLEDGMENT

This material is based upon work supported by the Cooperative State Research Education and Extension Service, United States Department of Agriculture, under Agreement No. 93-34211-8362. Contribution No. 05-85-B Kansas Agricultural Experimental Station, Manhattan, Kansas.

REFERENCES

1. Fung, D.Y.C., Rapid methods and automation in microbiology, *Compr. Rev. Food Sci. Food Saf.* (IFT), 1, 3, 2002.
2. Fung, D.Y.C. *et al.*, The Pulsifier: a new instrument for preparing food suspensions for microbiological analysis, *J. Rapid Methods Automat. Microbiol.*, 6, 43, 1998.
3. Kang, D.H., Dougherty, R.H., and Fung, D.Y.C., Comparison of Pulsifier and Stomacher to detach microorganisms from lean meat tissues, *J. Rapid Methods Automat. Microbiol.*, 9, 27, 2001.
4. Wu, V.C.H., Jitareerat, P., and Fung, D.Y.C., Comparison of the Pulsifier and the Stomacher for recovering microorganisms in vegetable, *J. Rapid Methods Automat. Microbiol.*, 11, 145, 2003.
5. Manninen, M.T., Fung, D.Y.C., and Hart, R.A., Spiral system and laser counter for enumeration of microorganisms, *J. Food Prot.*, 11, 177, 1991.
6. Chain, V.S. and Fung, D.Y.C., Comparison of Redigel, Petrifilm, Spiral Plate System, Isogrid and aerobic plate count for determining the numbers of aerobic bacteria in selected food, *J. Food Prot.*, 54, 208, 1991.
7. Fung, D.Y.C. and Lee, C.M., Double-tube anaerobic bacteria cultivation system, *Food Sci.*, 7, 209, 1981.
8. Ali, M.S. and Fung, D.Y.C., Occurrence of *Clostridium perfringens* in ground beef and turkey evaluated by three methods, *J. Food Prot.*, 11, 197, 1991.

9. Schmidt, K.A. *et al.*, Application of a double tube system for the enumeration of *Clostridium tyrobutyricum*, *J. Rapid Methods Automat. Microbiol.*, 8, 21, 2000.

10. Hartman, P.A., *Miniaturized Microbiological Methods*, Academic Press, New York, 1968.

11. Fung, D.Y.C. and Kraft, A.A., Microtiter method for the evacuation of viable cells in bacterial cultures, *Appl. Microbiol.*, 16, 1036, 1968.

12. Fung, D.Y.C. *et al.*, A collaborative study of the microtiter count method and standard plate count method on viable cell count of raw milk, *J. Milk Food Technol.*, 39, 24, 1976.

13. Fung, D.Y.C. and Kraft, A.A., Rapid evaluation of viable cell counts using the microtiter system and MPN technique, *J. Milk Food Technol.*, 32, 408, 1969.

14. Walser, P.E., Using conventional microtiter plate technology for the automation of microbiology testing of drinking water, *J. Rapid Methods Automat. Microbiol.*, 8, 193, 2000.

15. Irwin, P., Tu, S., Damert, W., and Phillips, J., A modified Gauss–Newton algorithm and ninety-six well micro-technique for calculating MPN using EXCEL spread sheets, *J. Rapid Methods Automat. Microbiol.*, 8, 171, 2000.

16. Wu, V.C.H. *et al.*, Rapid protocol (5.25 H) for the detection of *Escherichia coli* O157:H7 in raw ground beef by an immuno-capture system (Pathatrix) in combination with Colortrix and CT-SMAC, *J. Rapid Methods Automat. Microbiol.*, 12, 57, 2004.

17. Dreibelbis, S.B., Evaluation of Five ATP Bioluminescence Hygiene Monitoring Systems in a Commercial Food Processing Facility, Master's thesis, Kansas State University Library, Manhattan, KS, 1999.

18. Vishubhatla, A. *et al.*, A Rapid 5′ nuclease (TaqMan) assay for the detection of pathogenic strains of *Yersinia enterocolitica*, *Appl. Environ. Microbiol.*, 66, 4731, 2000.

19. Robinson, J.K., Mueller R., and Pilippone, L., New molecular beacon technology, *Am. Lab.*, 32, 30, 2000.

20. Eggins, B., *Biosensors: An Introduction*, John Wiley, New York, 1997.

21. Cunningham, A.J., *Bioanalytical Sensors*, John Wiley, New York, 1998.

22. Goldschmidt, M.C., Biosensors: scope in microbiological analysis, in *Encyclopedia of Food Microbiology*, Robinson, R., Batt, C., and Patel, P., Eds., Academic Press, New York, 1999, p. 268.

23. Deyholos, M., Wang, H., and Galbraith, D., Microarrays for gene discovery and metabolic pathway analysis in plants, *Life Sci.*, 2, 2, 2001.

24. Heron, E., Applied Biosystem: innovative technology for the life sciences, *Am. Lab*, 32, 35, 2000.

25. Strategic Consulting Inc., *Industrial Microbiology Market Review*, Strategic Consulting Inc., Woodstock, VT, 1998.

26. Strategic Consulting Inc., *Pathogen Testing in the U.S. Food Industry*, Strategic Consulting Inc., Woodstock, VT, 2000.

27. Strategic Consulting Inc., *Industrial Microbiology Market Review*, 2nd ed., Strategic Consulting Inc., Woodstock, VT, 2004.

28. Fung, D.Y.C., Prediction in the future of rapid methods and microbiology, *Food Test. Anal.*, 5, 18, 1999.

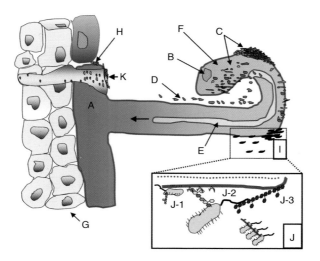

FIGURE 2.1 Anatomy of a root hair. Schematic representation of structures that are part of the anatomy of a plant root hair with attached bacteria. (A) Root hair epidermal cell; (B) nucleus; (C) bacteria bound to epidermal cell surface in aggregates and as biofilm; (D) rhizobial bacteria; (E) root hair infection thread initiated by rhizobial bacteria; (F) curling root hair tip; (G) cortex cells; (H) junction between root epidermal cells with attached bacteria; (I) bacteria binding as single cells, then aggregating; (J) magnification of I (not drawn to scale): J-1, single bacterial cell binding by pili/fimbriae; J-2, plant lectins interacting with bacterial carbohydrate (e.g., EPS, LPS, CPS, cellulose fibrils); J-3, bacterial flagellin interacting with plant receptor (e.g., polysaccharide); (K) lesion produced by plant pathogen.

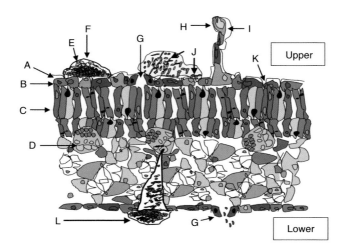

FIGURE 2.2 Anatomy of the cross-section of a leaf. Schematic representation of structures that are part of the anatomy of most plant leaves and attached microorganisms. (A) Cuticle layer; (B) upper epidermis; (C) palisade parenchyma; (D) vascular bundle composed of phloem and xylem; (E) biofilm composed of bacteria and other microorganisms; (F) EPS; (G) stomates within upper and lower epidermis; (H) trichome; (I) cuticle; (J) free bacteria and other microorganisms within water droplet; (K) recessed area between epidermal cells; (L) biofilm on underside of leaf forming a lesion into the vascular system.

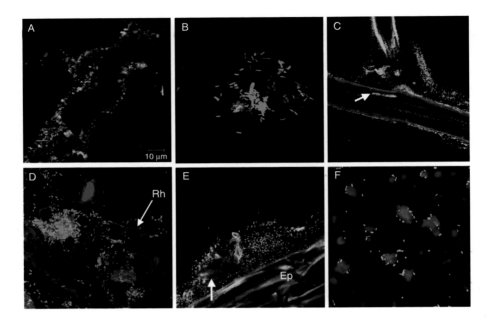

FIGURE 2.3 Confocal micrographs of bacteria on plant leaf, stem, and root tissues, and bacteria bound to material extracted from leaves. (A) Natural microorganisms, mostly bacteria, bound to junction of epidermal cells on a lettuce leaf. The bacteria were stained with LIVE *Bac*Light Gram stain (Molecular Probes, OR). (B) GFP-labeled *S. enterica* and dsRed-labeled *P. agglomerans* cells bound singly and in aggregates after their inoculation and incubation on the leaves of cilantro plants. Natural epiphytic bacteria were stained with SYTO® 62 (Molecular Probes) and were detected in the close vicinity of the inoculated strains. The SYTO® 62 signal was assigned the pseudocolor blue. (C) GFP-labeled *Ec*O157:H7 bound in the region of a lateral root emerging from an *Arabidopsis thaliana* plant. The arrow points to a region where the *Ec*O157 cells have become internalized. (D) GFP-labeled *S. enterica* bound to the root hairs (Rh) of an alfalfa sprout. (E) A thick biofilm of natural microorganisms colonizing the root of an alfalfa sprout and stained with LIVE *Bac*Light Gram stain (arrow). (F) GFP-labeled *S. enterica* cells attached to a dried compound extracted from cilantro leaves and identified as stigmasterol. (Brandl and Mandrell, unpublished data.)

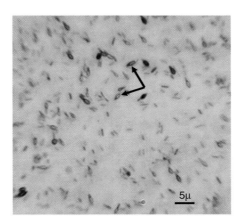

FIGURE 7.1 Gram stain of *Alicyclobacillus acidoterrestris*. Note swollen sporangia at arrows. (Magnification ×1000.)

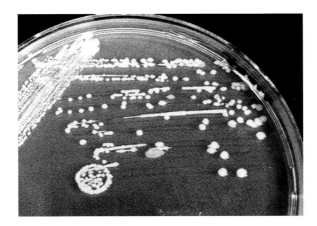

FIGURE 7.2 Colonies of *Alicyclobacillus acidoterrestris* ATCC 49025 on Ali agar after 24 hours at 45°C. Key characteristics: white/cream color, smooth to irregular edges. Older, larger colonies may develop translucent quality with slightly raised margins.

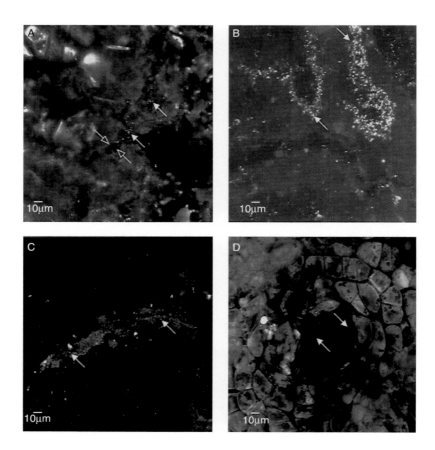

FIGURE 24.1 Confocal laser scanning microscopy (CSLM) images showing attachment of *E. coli* O157:H7 to various sites on the surface of apples. (A) Bruised tissue of unwashed, unrubbed, bruised apple at a junction (4.8 μm depth) between wax platelets: edge of wax platelets (open arrow); most cells attached to the edge of the wax platelets (filled arrow). (B) Bruised tissue of unwashed, unrubbed, bruised apple at a junction (6.6 μm depth) between wax platelets: heavy colonization of junctions between wax platelets (arrow). (C) Bruised tissue of unwashed, rubbed, bruised apple showing a cuticular crack on surface of apple (16.6 μm depth): cells are trapped within the cuticular crack (arrow). (D) Bruised tissue of unwashed, unrubbed, bruised apple showing a lenticel (9.2 μm depth): cells are within lenticel (arrow) at a depth of 20.6 μm below the surface of the apple. (From Kenney, S.J., Burnett, S.L., and Beuchat, L.R., *J. Food Prot.*, 64, 132, 2001. With permission. Copyright International Association for Food Protection, Des Moines, IA.)

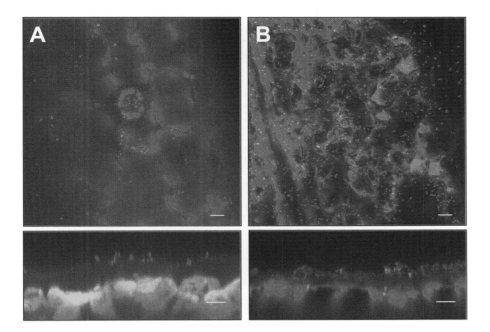

FIGURE 26.2 CLSM micrograph of GFP-labeled *S. enterica* serovar Thompson cells after their inoculation and incubation on the leaves of cilantro plants. *S.* Thompson cells were observed on healthy (A) and diseased (B) leaf tissue. The lower panels show cross-sections of the tissue in the top panel that were acquired by optical sectioning in the *xz* plane. They reveal that the bacterial cells are located on top of the cuticle on the healthy leaf (A), but have invaded the damaged tissue of the diseased leaf (B). Bars, 10 μm. (From Brandl, M.T. and Mandrell, R.E., *Appl. Environ. Microbiol.,* 68, 3614, 2002.)

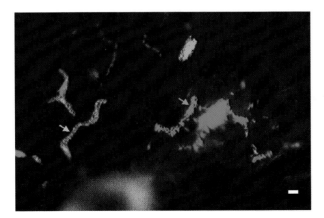

FIGURE 26.4 Epifluorescence micrograph of *P. agglomerans* and *P. fluorescens* cells labeled with CFP (blue arrow) and GFP (green arrow), respectively, and inoculated onto the leaves of bean plants incubated under humid conditions. The dual labeling with fluorescent proteins provided a good means to quantify the degree of spatial segregation between aggregates of these two bacterial species by image analysis. Bar, 20 μm.

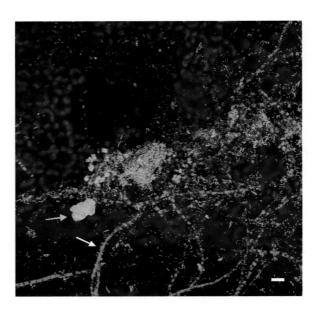

FIGURE 26.5 CLSM micrograph of the microbial community on a leaf section of field-grown bean plants. The microbes were stained with LIVE *Bac*Light Gram stain. Single cells and large mixed aggregates composed of fungi (bright green filaments, white arrow), and putative Gram-negative (green cells) and Gram-positive (red/orange cells) bacterial cells are present on the red autofluorescent leaf and in the vicinity of a glandular trichome (yellow arrow). Bar, 20 μm.

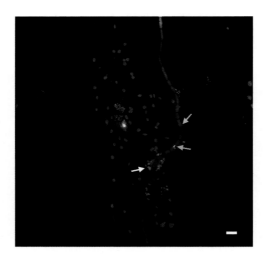

FIGURE 26.6 CLSM micrograph of GFP-labeled *S. enterica* cells (yellow arrow) and DsRed-labeled *P. agglomerans* cells (white arrow) attached to a SYTO 62-stained fungal hypha (blue arrow) in the phylloshere of cilantro plants. The bright blue round objects are the chloroplasts of the leaf vein epidermal cells. The SYTO 62 stain which emits at 680max nm was assigned the pseudocolor blue. The image was acquired by excitation with argon, krypton, and He/Ne lasers (Leica Microsystems, Wetzlar, Germany). Bar, 20 μm.

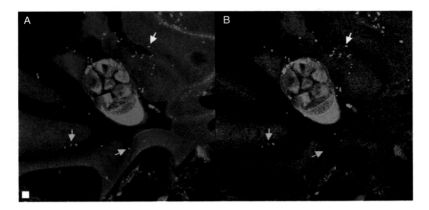

FIGURE 26.8 CLSM micrograph of the spatial distribution of *E. herbicola* cells harboring an *ipdC-gfp* fusion in the vicinity of a trichome (orange arrow) on the surface of bean leaves. The same field of view was imaged sequentially with a rhodamine (A) and a GFP (B) emission filter. *E. herbicola* cells were detected by FISH with a rhodamine-labeled 16S rRNA probe and performed on leaf discs mounted in agar (A). Large variations in the expression of *ipdC-gfp* were detected among the population of rhodamine-labeled cells (B). White and yellow arrows show *E. herbicola* cells with high and low levels of *ipdC-gfp* expression, respectively. Width of white square, 5 μm. (From Brandl, M.T., Quinones, B., and Lindow, S.E., *Proc. Natl. Acad. Sci. USA*, 98, 3454, 2001. Copyright 2001 National Academy of Sciences, USA.)

TABLE 26.1
List of Non-Conjugated Fluorochromes Commonly Used in Bacteriology.

Fluorescent Proteins	Ex. (nm)	Em. (nm)	Ref.		Biological Parameters	Ex. (nm)	Em. (nm)	Ref.
BFP (Blue GFP); eBFP	384	448	71	Kits	LIVE-DEAD BacLight			
CFP (Cyan GFP); eCFP	434	476	71		SYTO® 9 (total cells)	485	498	73
WtGFP	393/473	504	71		Propidium Iodide (dead cells)	536	617	73
GFPuv	399	511	71		LIVE BacLight Gram Stains			
GFP (S65T), eGFP	488	508	71		SYTO® 9 (total cells)	485	498	73
YFP (Yellow GFP); eYFP	514	527	71		Hexidium iodide (gram+ cells)	518	600	73
DsRed	558	583	71		ViaGram Red$^+$			
HcRed	592	645	71		DAPI (viable cells)	359	461	73
					SYTOX® Green (dead cells)	504	523	73
					Texas Red®-X-WGA (gram+ cells)	595	615	73

Fluorescent Dyes	Ex. (nm)	Em. (nm)	Ref.			Ex. (nm)	Em. (nm)	Ref.
Proteins Nile Red	540	600	73	Viability	CTC	450^a	630^a	
Sypro® Ruby	280, 450	610	73		Ethidium Bromide	545	610	9
Sypro® Orange	300, 470	570	73		Propidium Iodide	530	615	9
Sypro® Rose	350	610	73		SYTOX® Blue	431	480	73
Sypro® Red	300, 550	630	73		SYTOX® Green	504	523	73
Sypro® Tangerine	300, 490	640	73		SYTOX® Orange	547	570	73
Nucleic acids Acridine Orange (+DNA)	500	526	73		TO-PRO™-1	515	531	73
Acridine Orange (+RNA)	460	650	73		TO-PRO™-3	642	661	73
DAPI	358	461	73		TO-PRO™-5	747	777	73
Blue SYTO® dyes (40–45)	420–455	441–484	73	pH	BCECF	500	530/620	9
Green SYTO® dyes (11–25)	488–521	509–556	73		SNAFL®-1	508/540	540/623	73
Orange SYTO® dyes (80–85)	530–567	544–583	73		SNARF® (high pH)	574	630	9
Red SYTO® dyes (17, 59–64)	559–657	619–678	73		SNARF® (low pH)	548	587	9
SYTO® RNASelect	490	530	73					
Polysaccharides DTAF	492	516	73					

^{a} Data provided by Polysciences, Inc, Warrington, PA.

26 Methods in Microscopy for the Visualization of Bacteria and Their Behavior on Plants

Maria T. Brandl and J.-M. Monier

CONTENTS

26.1 INTRODUCTION

Since the discovery of microbes by Robert Hooke and Antonie van Leeuwenhoek in the 17th century, microscopy has made great strides to enhance our understanding of the microbial world, including the microflora of plants. However, despite our increasing ability to probe the minuscule at high resolution, with instruments such as the electron microscope, bacteria have remained relatively anonymous because of their lack of morphological diversity at the cellular scale. In addition, most types of electron microscopy involve extensive sample preparation that may dislodge or alter bacterial cells, leaving the microscopist in doubt about the interpretation of the observations made. It was the discovery of confocal microscopy, and the green fluorescent protein (GFP) as an intrinsic bacterial label, that spurred a new revolution, starting in the 1990s, in the use of fluorescence microscopy to study bacteria in their natural habitat. Because most bacterial species or strains cannot be distinguished from each other microscopically, intrinsic labeling of bacteria with GFP or other fluorescent proteins has been used widely to track specific bacteria in complex environments, including plants.

This chapter focuses on novel experimental approaches in fluorescence microscopy to detect bacteria and investigate their behavior on plants. Recent advances in microscope technologies that may be applied to plant microbiology research are also discussed.

26.2 VISUALIZATION OF BACTERIA ON PLANTS: AVAILABLE TOOLS

26.2.1 LABELING OF BACTERIA WITH FLUORESCENT PROTEINS

The recent renaissance in the application of microscopy to the study of bacterial behavior on plants is largely attributable to the discovery of GFP. The popularity of GFP as a fluorophore lies in its bright and relatively stable fluorescence, the expression of *gfp* in most bacterial species, and the ability to use it as an intrinsic label without the need for a substrate. The latter property means that samples can be mounted without prior processing for visualization under the microscope, and thus disruption of bacterial cells in the environment under study is minimal. This is in sharp contrast with the potential artifacts generated during sample preparation for immunofluorescence or electron microscopy. Additionally, mutants of GFP that have enhanced fluorescence intensity and emit at various wavelengths of the visible spectrum have provided investigators with versatile tools to circumvent problems with detecting the GFP signal against autofluorescent backgrounds. This problem is especially prevalent in plant tissues, which emit with various intensities in different regions of the visible spectrum [1]. It is exacerbated also by contamination of the image with stray fluorescence, but may be minimized by the use of confocal microscopy. Because the confocal laser scanning microscope (CLSM) collects the fluorescent signal solely from the focal plane by rejecting scattered light with its pinhole, the CLSM can improve greatly the detection of fluorescently tagged bacterial cells on plants.

Despite the availability of GFP mutants with enhanced fluorescence intensity, such as the widely used S65T mutant [2], the detection of GFP fluorescence at low levels of expression in bacterial cells remains difficult. Therefore, intrinsic labeling with GFP is commonly achieved by transforming the bacterial strain of interest with a plasmid that is maintained stably at a moderate copy number per cell and that harbors gfp expressed from strong promoters. Adverse effects of GFP on the fitness of bacteria that were transformed with high copy number plasmids encoding GFP have been reported [3], presumably because the resultant high concentrations of GFP overload bacterial metabolism or disrupt cellular functions, and thus use of such plasmids should be avoided.

GFP production from single insertion of gfp into the bacterial chromosome may circumvent problems associated with excessive GFP concentrations or plasmid instability, but may still decrease the competitiveness of the tagged strain compared to its parental strain under stressful conditions such as carbon-substrate limitation [4]. In addition, gfp expression from a single location on the bacterial chromosome may yield a signal-to-noise ratio that approaches the lower limit of detection for GFP in a strong fluorescent background. In such cases, GFP fluorescence may be imaged only by high signal gain, which results in poor definition of the bacterial cell profile and in grainy images [5]. This is particularly true for imaging of dim GFP-tagged bacteria on plant tissue that emits in the green range, e.g., the roots of certain plant species.

Expression of gfp from plasmids requires that plasmid stability in the transformed bacterial strain be assessed in the plant environment under study. By comparison of population sizes of $E.$ $coli$ O157:H7 cells that showed green fluorescence to those that were detected by immunolabeling, Takeuchi and Frank demonstrated that $E.$ $coli$ O157:H7 pEGFP retained pEGFP, or GFP per se, at higher frequency on lettuce leaves and cauliflower florets than on tomato [6]. Since few broad-host-range plasmids are completely stable in any bacterial species, it is expected that the frequency at which a GFP-encoding plasmid is lost in a cell population will increase as the bacterial growth rate increases, or as physiological stress impacts that population. Therefore, the retention of a GFP plasmid among a bacterial population may vary greatly depending on the experimental conditions and on the plant host tissue or species studied.

In addition to GFP and its color variants, intrinsic fluorophores emitting in the near red (DsRed) and far red (HcRed) have been cloned from $Discoma$ and $Heteractis$ $crispa$, respectively [7,8]. The availability of fluorophores that span the visible range allow for multispectral imaging of bacteria within a same sample. The excitation and emission spectra of the red fluorescent proteins and of the most widely used variants of GFP are shown in Figure 26.1. The peaks of their excitation and emission spectra are listed in Table 26.1.

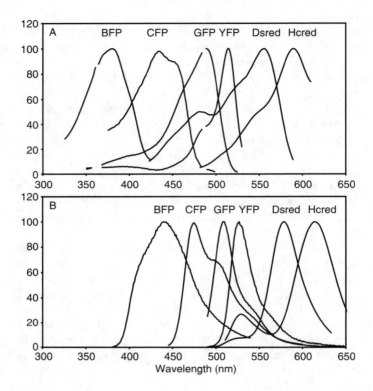

FIGURE 26.1 Excitation (A) and emission (B) spectra of blue (BFP), cyan (CFP), green (GFP), and red (DsRed and HcRed) fluorescent proteins suitable for intrinsic labeling of bacteria. The Y-axis represents the normalized fluorescence intensity. (Fluorescence data courtesy of BD Biosciences Clontech, Palo Alto, CA.)

26.2.2 LABELING OF BACTERIA WITH DYES AND FLUORESCENT CONJUGATES

In addition to fluorescent proteins, a wide range of fluorescent dyes and bioconjugates are available for the visualization of bacteria on plants. Exhaustive lists of fluorescent probes and their applications in biological microscopy are beyond the scope of this chapter, but can be found in an excellent review by Kasten [9]. The DNA-intercalating stain acridine orange (AO) and the newer SYTO® dyes (Molecular Probes, Eugene, OR), which are available in a wide range of the visible spectrum, are particularly useful to visualize bacteria against the autofluorescent plant background. 4′,6-Diamidino-2-phenylindole (DAPI), also a nucleic acid stain, requires ultraviolet (UV) illumination, and thus may cause extensive damage to plant cells. Other dyes, such as Sypro® Orange, a general protein stain, and Nile Red (Molecular Probes), which stains lipid-hydrophobic sites , have been applied to the study of bacterial biofilms [10].

TABLE 26.1
List of Non-Conjugated Fluorochromes Commonly Used in Bacteriology (Color Insert Follows Page 594)

Fluorescent Proteins		Ex. (nm)	Em. (nm)	Ref.
	BFP (Blue GFP); eBFP	384	448	71
	CFP (Cyan GFP); eCFP	434	476	71
	WtGFP	393/473	504	71
	GFPuv	399	511	71
	GFP (S65T), eGFP	488	508	71
	YFP (Yellow GFP); eYFP	514	527	71
	DsRed	558	583	71
	HcRed	592	645	71
Fluorescent Dyes		**Ex. (nm)**	**Em. (nm)**	**Ref.**
Proteins	Nile Red	540	600	73
	Sypro® Ruby	280, 450	610	73
	Sypro® Orange	300, 470	570	73
	Sypro® Rose	350	610	73
	Sypro® Red	300, 550	630	73
	Sypro® Tangerine	300, 490	640	73
Nucleic acids	Acridine Orange (+DNA)	500	526	73
	Acridine Orange (+RNA)	460	650	73
	DAPI	358	461	73
	Blue SYTO® dyes (40–45)	420–455	441–484	73
	Green SYTO® dyes (11–25)	488–521	509–556	73
	Orange SYTO® dyes (80–85)	530–567	544–583	73
	Red SYTO® dyes (17, 59–64)	559–657	619–678	73
	SYTO® RNASelect	490	530	73
Polysaccharides	DTAF	492	516	73

	Biological Parameters	Ex. (nm)	Em. (nm)	Ref.
Kits	LIVE-DEAD *Bac*Light			
	SYTO® 9 (total cells)	485	498	73
	Propidium Iodide (dead cells)	536	617	73
	LIVE *Bac*Light Gram Stains			
	SYTO® 9 (total cells)	485	498	73
	Hexidium iodide (Gram+ cells)	518	600	73
	ViaGram Red$^+$			
	DAPI (viable cells)	359	461	73
	SYTOX® Green (dead cells)	504	523	73
	Texas Red®-X-WGA (Gram+ cells)	595	615	73
Viability	CTC	450^a	630^a	9
	Ethidium Bromide	545	610	9
	Propidium Iodide	530	615	73
	SYTOX® Blue	431	480	73
	SYTOX® Green	504	523	73
	SYTOX® Orange	547	570	73
	TO-PRO™-1	515	531	73
	TO-PRO™-3	642	661	73
	TO-PRO™-5	747	777	73
pH	BCECF	500	530/620	9
	SNAFL®-1	508/540	540/623	73
	SNARF® (high pH)	574	630	9
	SNARF® (low pH)	548	587	9

^{a} Data provided by Polysciences, Inc, Warrington, PA.

In contrast to general DNA dyes, immunostaining with fluorescent-labeled antibodies [11] or fluorescence *in situ* hybridization (FISH) with 16S rRNA probes [12–16] may provide high specificity to detect a given bacterial species or strain among the plant microflora. Also, fluorescently labeled lectins have proven useful to probe bacteria associated with the exopolysaccharide matrix of aggregates on plants and in biofilms in general [17,18]. While fluorescein and rhodamine bioconjugates have been popular fluorochromes for the visualization of bacteria on plant tissue, newer products for labeling of bioconjugates, such as Alexa Fluor® (Molecular Probes) and Cy™ dyes (Amersham Biosciences Corp., Piscataway, NJ), exhibit brighter fluorescence and enhanced photostability.

General stains and bioconjugates can be used alone or in combination with fluorescent proteins to probe complex microenvironments, processes, or biological parameters governing bacterial behavior on plants. In the sections below, we discuss various applications of fluorescence microscopy to the study of the ecology of plant-associated bacteria and human enteric pathogens in the plant environment. Table 26.1 lists the excitation and emission peaks, as well as specific applications, of a variety of fluorescent proteins and stains, some of which are mentioned in this chapter.

26.3 APPLICATIONS

26.3.1 SPATIAL DISTRIBUTION

The microscopic investigation of the localization of human or plant pathogenic bacteria on plant surfaces has provided new insights into their ecology in that environment. The observations by confocal microscopy that GFP-labeled *Salmonella enterica* formed microcolonies on the leaves of cilantro plants provided evidence that this enteric pathogen has the ability to colonize plants in a preharvest environment and may cause outbreaks due to contamination that occurred in the field [19]. Furthermore, *S. enterica* developed high-density populations and large heterogeneous aggregates in the vein area of the leaf, thus following spatial colonization patterns similar to those of the natural plant microflora on uninoculated plants [20]. In this same study, Brandl and Mandrell used confocal scans in the *xz* plane to obtain optical cross-sections of healthy and diseased cilantro leaves, and demonstrate that *S. enterica* cells had gained access to internal tissue while growing in the plant lesion whereas they remained on the cuticle layer while colonizing a healthy leaf [19] (Figure 26.2). Optical cross-sectioning by CLSM is an effective way to demonstrate unequivocally the internalization of bacteria in plant openings or tissue without the potential artifacts created by mechanical sectioning, which may contaminate internal tissue with external bacteria.

Confocal microscopy of GFP-labeled cells was used in several studies to demonstrate that enteric pathogens attach to and multiply at high densities in the damaged tissue of a variety of fruits and vegetables [6,19,21]. Burnett *et al.* combined CLSM and digital image analysis to count GFP-tagged *E. coli*

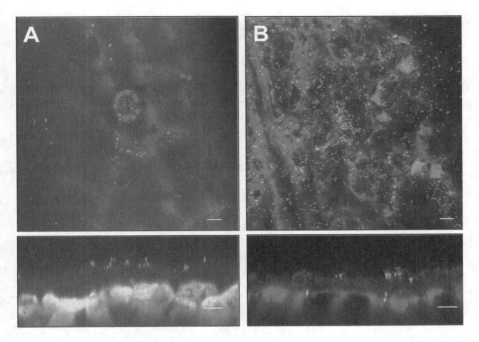

FIGURE 26.2 (Color insert follows page 594) CLSM micrograph of GFP-labeled *S. enterica* serovar Thompson cells after their inoculation and incubation on the leaves of cilantro plants. *S.* Thompson cells were observed on healthy (A) and diseased (B) leaf tissue. The lower panels show cross-sections of the tissue in the top panel that were acquired by optical sectioning in the *xz* plane. They reveal that the bacterial cells are located on top of the cuticle on the healthy leaf (A), but have invaded the damaged tissue of the diseased leaf (B). Bars, 10 μm. (From Brandl, M.T. and Mandrell, R.E., *Appl. Environ. Microbiol.*, 68, 3614, 2002.)

O157:H7 cells that attached at various depths in healthy or punctured apple tissue during inoculation [21]. From this approach, they concluded quantitatively that the pathogen infiltrated intact tissue and natural openings to a greater extent under negative than positive temperature differential.

The ability of plant pathogens and human enteric pathogens to become internalized in plant tissue, where they are shielded from the adverse effects of environmental conditions or chemical control agents, has been the focus of much interest in plant microbiology. Confocal microscopy has been a valuable means of probing the internal tissue of plants inoculated with GFP-tagged bacteria. Hallmann *et al.* were among the first to use the CLSM to observe the endophytic localization of a bacterial species, *Rhizobium etli*, within *Arabidopsis thaliana* roots [22]. Using a similar approach, endophytic colonization of *A. thaliana* and alfalfa roots by *E. coli* O157:H7 and *S. enterica*, respectively, has been demonstrated in plant systems *in vitro* [23,24]. *A. thaliana* and various other plant species have transparent roots, which lend themselves well to optical sectioning. In contrast, opaque plant tissues, e.g., potato tubers, make visualization of internalized GFP-labeled bacteria more challenging. Although opaque plant tissue can be cleared with various techniques, these generally inhibit or destroy GFP fluorescence [22]. In an experimental soil system that attempted to simulate field conditions, albeit with high inoculum levels, Solomon *et al.* found evidence for the transmission of GFP-tagged *E.coli* O157:H7 from contaminated manure to the internal tissue of lettuce leaves [25]. However, the low resolution of their published confocal micrograph reveals the difficulty of imaging fluorescent bacterial cells located under several layers of opaque leaf tissue. Also challenging is the localization of the internalized bacterial cells within the plant tissue in the epifluorescence mode during browsing before image acquisition in the confocal mode; browsing large numbers of fields of view or samples for the visualization of rare events is impractical by confocal microscopy. In our experience, this problem occurs even with brightly fluorescent bacterial cells.

Fluorescence microscopy has been instrumental to the discovery that bacteria can form biofilms on plants. Morris *et al.* used epifluorescence microscopy and AO staining of microbial cells to demonstrate the presence of natural biofilms on leaves of a variety of vegetables [26]. Since this first report, biofilm formation on plants has been shown in several studies using various microscopy techniques. Monier and Lindow developed a method to study the frequency, size, and localization of bacterial aggregates *in situ* on leaf surfaces [27]. Quantitative data were obtained by digital image analysis of epifluorescence micrographs of AO-stained bacteria present on leaf samples. The analysis was performed based on the outlining of the profiles of single bacteria or bacterial aggregates, by thresholding on their bright fluorescence intensity against the less fluorescent plant background. This method was applied to the quantification of the size of aggregates formed by *S. enterica* in the cilantro phyllosphere using GFP as an intrinsic fluorescent bacterial label instead of AO [28], and is illustrated in Figure 26.3.

26.3.2 CELL–CELL INTERACTIONS

The characterization of individual microbial aggregates isolated from plants revealed that they harbor a wide range of microorganisms including numerous species of Gram-negative and Gram-positive bacteria, as well as yeasts and filamentous fungi [26,29]. The heterogeneous composition of aggregates suggests that a complex pattern of microbial interactions is possible even at such small scales in the plant environment. While the spatial organization of epiphytic bacterial populations had remained obscure until recently, the use of marker genes conferring the production of fluorescent proteins combined with fluorescent stains has proven to be a valuable tool and has provided new insight into our understanding of bacterial interactions on plant surfaces.

Despite the heterogeneity of the plant surface habitat, which makes it a difficult task to study bacterial interactions *in situ*, a few studies attempting to decipher the factors shaping the structure of epiphytic communities have

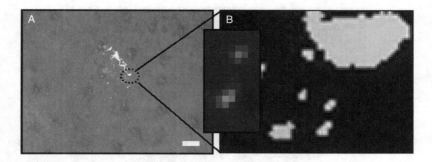

FIGURE 26.3 Schematic diagram illustrating the basis for digital analysis of fluorescence images. (A) Epifluorescence micrograph of cellular aggregates of GFP-labeled *S. enterica* on a cilantro leaf. Each GFP-labeled *S. enterica* single cell or aggregate in the image can be identified by thresholding on the bright pixels originating from the GFP fluorescence, which is of higher intensity than the background pixels originating from the leaf surface (inset). This thresholding yields objects (B) for which a variety of parameters, such as total number of pixels or mean pixel intensity, can be automatically measured with the image analysis software. Because of the highly heterogeneous spatial distribution of bacteria on plants, as well as variations between plants, this type of analysis requires the acquisition of a large number of images from random fields of view of multiple plant samples in order to yield unbiased data.

been reported. Normander *et al.* have reported the significance of bacterial distribution on genetic exchange in the phyllosphere using GFP as an indicator of plasmid transfer [30]. Conjugation was observed under CLSM to occur primarily in the interstitial spaces of epidermal cells and vein cells, and in stomata; bacterial aggregation had a great stimulatory effect on plasmid transfer. Such data are pertinent to assessing the risk associated with the dissemination of antibiotic resistance genes among bacterial cells on plants.

Monier and Lindow tested the spatial partitioning of cells within aggregates on leaf surfaces by establishing different pairwise mixtures of three different epiphytic bacterial species that were tagged either with GFP or CFP [31]. The spatial structure of the resulting aggregates was studied *in situ* on leaves by epifluorescence microscopy. Digital image analysis was employed to quantify the degree of segregation of the GFP- and the CFP-marked strains and revealed that the fraction of cells in direct contact ranged from 0.2 to 8.0%. The highest segregation occurred between two bacterial species exhibiting negative interactions (Figure 26.4).

Fluorescence microscopy has proven useful also for the assessment of various bacterial genes hypothesized to have a role in cell–cell interactions on plants. For example, by comparing the behavior of GFP-labeled parental and mutant strains *in situ* on plants under the microscope, the function of the adhesin encoding *hecA* gene in the attachment and aggregation of *Erwinia chrysanthemi* on plants was confirmed [32]. In contrast, comparison of GFP-labeled *S. enterica* parental and LuxS⁻ mutant strains by digital image

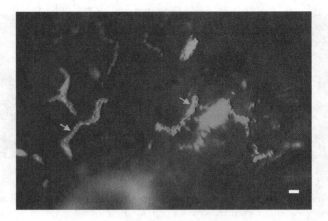

FIGURE 26.4 (Color insert follows page 594) Epifluorescence micrograph of *P. agglomerans* and *P. fluorescens* cells labeled with CFP (blue arrow) and GFP (green arrow), respectively, and inoculated onto the leaves of bean plants incubated under humid conditions. The dual labeling with fluorescent proteins provided a good means to quantify the degree of spatial segregation between aggregates of these two bacterial species by image analysis. Bar, 20 μm.

analysis of their aggregate sizes on leaves revealed that production of the autoinducer-2 molecule for cell–cell signaling had no detectable role in aggregate formation by this human pathogen in the cilantro phyllosphere [28].

Besides the fluorescent proteins, or in combination with them, fluorescent dyes can provide a means to view the microbial composition and possible interactions on plant surfaces. The LIVE *Bac*Light™ bacterial Gram stains (Molecular Probes) impart green and red fluorescence to Gram-negative and Gram-positive bacterial cells, respectively. The DNA dye SYTO® 9, which is included in the assay, also stains fungi, and therefore the assay allows for easy visualization of the fungal and bacterial composition of the plant microflora (Figure 26.5) [33]. In a study of the interaction of *S. enterica* with the common bacterial epiphyte *P. agglomerans* in the cilantro phyllosphere, the SYTO® 62 dye, which emits in the red region of the spectrum, enabled the observation under CLSM that these two species were part of larger bacterial aggregates (Chapter 2, Figure 2.3B), and that they attached to fungal hyphae (Figure 26.6).

26.3.3 MEASUREMENT OF BIOLOGICAL PARAMETERS

The assessment of bacterial cell viability, although complex in its interpretation, is central to our understanding of how bacteria survive in a given habitat. Also, there is an increasing need to understand the physiology of bacteria in the plant environment in order to design efficient strategies to control plant colonization by human or plant pathogens, or to sanitize fruits and vegetables. Fluorescent reporters may provide information about the

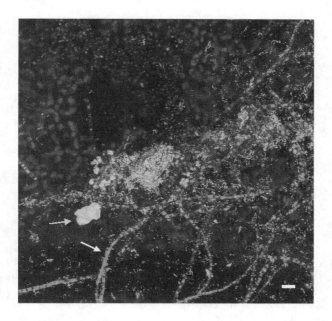

FIGURE 26.5 (Color insert follows page 594) CLSM micrograph of the microbial community on a leaf section of field-grown bean plants. The microbes were stained with LIVE *Bac*Light Gram stain. Single cells and large mixed aggregates composed of fungi (bright green filaments, white arrow), and putative Gram-negative (green cells) and Gram-positive (red/orange cells) bacterial cells are present on the red autofluorescent leaf and in the vicinity of a glandular trichome (yellow arrow). Bar, 20 μm.

physiological status of bacterial cells in complex ecosystems. They can be used to determine cell viability via the assessment of basic cell functions such as reproductive ability, membrane integrity, and respiration, or to measure cellular parameters, such as pH and levels of various ions.

26.3.3.1 Kogure Assay for Cell Viability

The ability of bacterial cells to grow and multiply has been the gold standard to demonstrate cell viability. In an approach based on the Kogure assay [34], Wilson and Lindow used a direct viable count method to examine the viability of epiphytic populations of *Pseudomonas syringae* on bean plants under desiccation stress [35]. The method consisted of incubating cells recovered from bean leaves in low-percentage yeast extract and in nalidixic acid to provide substrates for growth and to prevent cell division, respectively. The cells were then stained with DAPI, and cells that were fluorescent and elongated (growing cells in which division is inhibited by nalidixic acid) were counted as viable cells under the epifluorescence microscope. The increasing frequency of viable but nonculturable cells of the pathogen *R. solanacearum* during infection of tomato

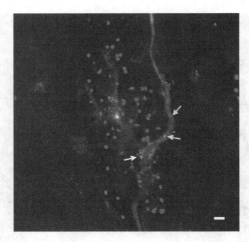

FIGURE 26.6 (Color insert follows page 594) CLSM micrograph of GFP-labeled *S. enterica* cells (yellow arrow) and DsRed-labeled *P. agglomerans* cells (white arrow) attached to a SYTO® 62-stained fungal hypha (blue arrow) in the phyllosphere of cilantro plants. The bright blue round objects are the chloroplasts of the leaf vein epidermal cells. The SYTO® 62 stain which emits at 680_{max} nm was assigned the pseudocolor blue. The image was acquired by excitation with argon, krypton, and He/Ne lasers (Leica Microsystems, Wetzlar, Germany). Bar, 20 μm.

plants was reported in a similar study [36]. This method has the advantage of examining viability directly via the ability of the cells to grow, but provides little information about the spatial distribution of the viable and nonviable cells at the microscale *in situ* on plants.

26.3.3.2 Indicators of Membrane Integrity

Biological stains that report on bacterial membrane activity can be useful to probe bacterial cell viability directly on plants. These stains penetrate only cells that have a compromised cytoplasmic membrane, and therefore are presumably nonviable. They include the DNA dyes ethidium bromide, TO-PROTM-3 and SYTOX® Green (Molecular Probes), and propidium iodide (PI), which is probably the most commonly used. The LIVE/DEAD *Bac*Light bacterial viability assay (Molecular Probes) is a popular and simple method for determination of bacterial cell viability in which live cells fluoresce green due to staining with SYTO® 9, and dead cells fluoresce red due to staining with PI. The spectra of both stains are sufficiently close to detect both the green and red cells with a fluorescein filter. Using this assay, Warriner *et al.* reported the differential survival of *E. coli* and *S. enterica* cells inside and on the surface of bean sprouts after their treatment with sodium hypochlorite [37]. Also, PI staining was combined with immunostaining to detect specifically viable and nonviable cells of *E. coli* O157:H7 introduced onto cut lettuce and exposed to sodium hypochlorite [11].

Monier and Lindow developed a protocol based on PI staining, epifluorescence microscopy, and digital image analysis to determine the viability of individual bacterial cells directly on plants [38]. In their studies, PI was used to map the distribution of viable and nonviable GFP- or CFP-labeled cells on leaves (1) before and after they were exposed to desiccation stress, (2) in homogenous versus heterogeneous aggregates, and (3) after they landed in an aggregate of the same species versus in that of another species. These studies demonstrated quantitatively the importance of aggregation in the survival of epiphytic bacteria [38], the existence of antagonistic interactions at the bacterial scale within mixed aggregates [31], and the differential fate of immigrant bacteria to leaf surfaces depending on resident bacteria and on leaf anatomical features at the landing site [39].

Because the correlation between membrane integrity and physiological status of the cell is still controversial, cell viability data obtained with stains such as PI should be interpreted carefully. It is important to emphasize that the choice of a fluorescent viability probe is critical, and whether a certain fluorescent probe is suitable for viability assessment under the conditions tested has to be assessed.

26.3.3.3 GFP Fluorescence and Cell Viability

With the general excitement about GFP as an intrinsic label for bacteria, a thorny issue that has received little attention is whether GFP or its variants can serve as an indicator of cell viability. That is, can all fluorescent GFP-tagged cells observed under the microscope be considered as viable cells? Lowder *et al.* reported a strong correlation between *Pseudomonas fluorescens* cell death and leakage of GFP from cells; most dead cells (as assessed by viability staining) were not GFP fluorescent, but a small percentage were dead and retained green fluorescence [40]. We have made similar observations with cultured GFP-labeled *S. enterica* cells that were exposed to the lethal stress of high temperature, desiccation, or calcium hypochlorite treatment [41]. Thus, it appears that GFP, which is considered as a stable fluorochrome in living cells, is lost rapidly upon cell death in bacteria. However, because of the small percentage of fluorescent GFP cells for which the cell status is unclear, it is preferable to confirm cell viability with an additional method when inferring specifically on the viability of GFP-tagged cells.

26.3.3.4 Other Fluorescent Indicators of Bacterial Physiology

Although many studies have investigated the detection of metabolically active bacteria in aquatic environments with fluorogenic substrates as indicators of bacterial respiratory or enzymatic activity [42,43], few of these dyes have been used so far to investigate bacterial activity on plants. Similarly, the assessment of intracellular pH has been performed at the single bacterial cell level by fluorescence ratio imaging with fluorescent pH indicators (e.g., 5- (and 6-) carboxyfluorescein, BCECF, SNAFL, SNARF) [42,43], and with GFP [44], but their utility for *in situ* probing of bacterial pH on plants has not been explored. With the increasing interest in bacterial survival to acid stress in fresh-cut fruits and vegetables, such an approach may prove to be almost essential. Additionally, the use of fluorescent probes in combination with flow cytometry in antimicrobial research has been widely reported. Fluorescent probe technology and microscopy may be applied successfully to quantify the effect of decontamination agents on human or plant pathogens on agricultural plants.

26.3.4 BACTERIAL GENE EXPRESSION *IN SITU* ON PLANTS

26.3.4.1 GFP as a Reporter of Gene Expression

The combination of fluorescent markers with reporter gene technology has proven to be a powerful tool to study the behavior of bacteria on plants. The fusion of fluorescent reporter genes to bacterial genes of interest allows for the measurement of the transcriptional activity of that gene at the single bacterial cell level under the microscope, rather than at the population level. Thus, the distribution of transcriptional activity of a gene, and, potentially, the role of its phenotype, can be assessed in particular environments.

Unless the experiments are performed in a gnotobiotic system, an additional marker is required to distinguish the bacterial cells under study from those belonging to the indigenous microflora. This ensures that the bacterial cells in which transcriptional activity of the gene of interest is low or off, and therefore in which the reporter signal is low, will be detected. In the first study of this type on plants, Brandl *et al.* used a transcriptional *gfp* fusion to an auxin (IAA) biosynthetic gene of *E. herbicola* in combination with FISH to assess the distribution of IAA synthesis in this bacterial species on bean leaves [12] (Figure 26.7). Plants were inoculated with a strain of *E. herbicola* transformed with the *gfp* fusion and incubated to allow for colonization to occur. Then, the bacterial cells were washed off the leaves and subjected to FISH on microscope slides with tetramethylrhodamine-labeled 16S rRNA probe specific to *E. herbicola*. The green fluorescence intensity of bacterial cells that were stained red by FISH was measured by analysis of digital images acquired under the epifluorescence microscope. The frequency distribution of GFP fluorescence intensity per cell revealed that a small proportion of the *E. herbicola* cells on the leaves expressed the IAA gene at very high levels, suggesting that there were microsites on the leaf that were conducive to high production of IAA. With the same approach, subsequent studies demonstrated the heterogeneous distribution of the availability of sucrose [15] and fructose [14] to *E. herbicola*, and of iron to *Pseudomonas syringae* [13], on plant surfaces. In all of the above studies, FISH enabled the specific labeling of bacterial cells at the strain level

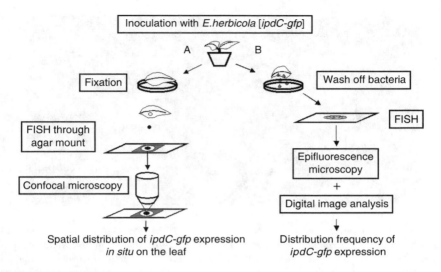

Spatial distribution of *ipdC-gfp* expression
in situ on the leaf

Distribution frequency of
ipdC-gfp expression

FIGURE 26.7 Schematic diagram of fluorescence microscopy strategies to investigate the distribution of gene expression at the bacterial cell level on plants. The protocols make use of dual labeling with GFP as a reporter of transcriptional activity, and with rhodamine as a marker for FISH to identify specific bacterial cells among the natural plant microflora. Spatial distribution of *ipdC-gfp* expression is assessed by visualization under CLSM of the GFP fluorescence in bacterial cells that were identified by FISH, performed directly on plant samples (A). Frequency distribution data of the activity of transcriptional fusions to *gfp* are acquired by measuring the fluorescence of individual bacterial cells that were washed off the plant surface and identified by FISH (B). (From Brandl, M.T., Quinones, B., and Lindow, S.E., *Proc. Natl. Acad. Sci. USA,* 98, 3454, 2001.)

within a given species. Additionally, the rhodamine label of the 16S rRNA probe provided a fluorescent signal with an emission spectrum sufficiently distinct from that of GFP to prevent misinterpretation of the fluorescent signals in each microscope filter set. Optical crosstalk is a major issue in multilabeling experiments.

Frequency distribution analysis of bacterial gene expression has been performed mostly by epifluorescence microscopy to allow for fluorescence measurements of a large number of bacterial cells that were recovered from plant tissue. Brandl *et al.* developed a method for the assessment of bacterial GFP fluorescence *in situ* on leaves under the CLSM (Figure 26.7) [12]. The method consisted in performing FISH on leaf disks that were fixed in paraformaldehyde and then covered with a thin film of low-percentage agar to prevent disruption of the spatial distribution of the bacterial cells during hybridization procedures. The transcriptional activity of GFP reporter fusions was assessed subsequently in 16S rRNA-labeled *E. herbicola* cells, through the agar, by confocal microscopy (Figure 26.8). In this manner, spatial patterns of gene expression in a specific bacterial population could be established.

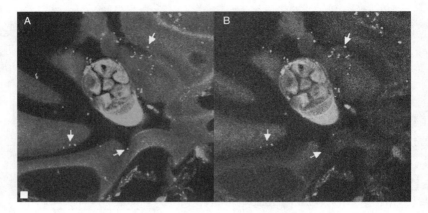

FIGURE 26.8 (Color insert follows page 594) CLSM micrograph of the spatial distribution of *E. herbicola* cells harboring an *ipdC-gfp* fusion in the vicinity of a glandular trichome on the surface of bean leaves. The same field of view was imaged sequentially with a rhodamine (A) and a GFP (B) emission filter. *E. herbicola* cells were detected by FISH with a rhodamine-labeled 16S rRNA probe and performed on leaf disks mounted in agar (A). Large variations in the expression of *ipdC-gfp* were detected among the population of rhodamine-labeled cells (B). White and yellow arrows show *E. herbicola* cells with high and low levels of *ipdC-gfp* expression, respectively. Width of white square, 5 μm. (From Brandl, M.T., Quinones, B., and Lindow, S.E., *Proc. Natl. Acad. Sci. USA*, 98, 3454, 2001. Copyright 2001 National Academy of Sciences, USA.)

The discovery of the fluorescent protein DsRed caused much excitement because of its potential to be used as an intrinsic bacterial label along with a transcriptional fusion to GFP in gene expression studies. Despite reports that GFP and DsRed can be detected in a single cell upon simultaneous excitation [45], it appears that DsRed is also a good acceptor molecule, with GFP or CFP as a donor, in fluorescence resonance energy transfer (FRET) imaging [46]. The interaction between these fluorescent proteins may confound the quantitative interpretation of the fluorescent signals, and therefore the usefulness of DsRed for dual-labeling with GFP or its variants still needs to be ascertained.

26.3.4.2   Practical Note on the Use of GFP for Gene Expression Measurements

Because of the great stability of GFP, unstable GFP variants with a short half-life have been constructed to measure transient bacterial gene expression in time-course experiments on plants [47–49]. These destabilized variants prevent the accumulation of GFP under basal or noninduced gene expression conditions and are more accurate reporters of transcriptional activity at a given time point. Also, GFP fluoresces poorly under low oxygen conditions, and some variants like EGFP have decreased fluorescence at a pH between 7.0 and

4.5, with only 50% of its fluorescence at pH 6.0; other factors affecting GFP chromophore formation include temperature, chemical denaturants, and certain solvents [50]. In addition, an effect of bacterial growth rate on GFP fluorescence intensity of individual cells has been reported [49]. Therefore, the effect of experimental conditions on the accumulation and the function of GFP as a chromophore *per se* should be tested with a constitutively expressed *gfp* to avoid misinterpretation of GFP fluorescence data.

26.3.4.3 FISH for the Detection of Bacterial mRNA

Although not reported in plant studies at the present, bacterial mRNA detection and quantification by FISH has been successfully performed in single bacterial cells in environmental samples. With as many as five fluorescently labeled oligodeoxynucleotide probes (depending on abundance of the transcript), the mRNA of two enterobacterial genes that are induced at different stages of growth was targeted to profile the physiological activity of Enterobacteriaceae in a waste water microbial community [51]. The mRNA profile was obtained in conjunction with rRNA FISH for the identification of the bacterial cells at the taxon level within the community. In other cases where abundance of target mRNA is low, signal amplification is necessary to detect fluorescence in single cells. Pernthaler and Amann have developed a FISH protocol for the sensitive detection of low-abundance mRNAs at the single bacterial cell level by enzymatic amplification of the fluorescence signal emitted from long oligonucleotide probes that were labeled at high density [52]. This enabled the detection of the expression of a single gene in methanotrophic bacteria present in sediment samples. The application of such powerful reporter systems to the detection of mRNA in single bacterial cells may offer a means of probing specific bacterial functions in natural microbial consortia in the plant environment. It remains to be determined, however, whether detection of the hybridization signal can be achieved against the often autofluorescent background on plants.

26.3.4.4 Immunolabeling of Gene Products

Immunofluorescence labeling represents an alternative method to FISH to investigate the localization of specific bacteria on plants, but it is suitable also for quantitative analysis of proteins or their enzymatic products. For example, patterns of regulation of a *Ralstonia solanacearum* virulence gene (*eps*) were determined by quantifying the amount of β-galactosidase protein present in single cells of a transformant of this plant pathogen that carried an *eps-lacZ* reporter fusion [53]. Quantitative measurements were performed by digital analysis of the fluorescence of single *R. solanacearum* cells recovered during infection of tomato plants, and then immunolabeled against β-galactosidase. Immunofluorescence microscopy with antibodies against the *R. solanacearum* exopolysaccharide EPS I54 and a specific *Xanthomonas axonopodis* lipopolysaccharide [55] proved useful also to determine the spatiotemporal production of these virulence factors during progression of disease in their respective plant host. Such an approach may have great potential in the investigation of the surface components of plant-associated or human bacterial cells while they grow or survive in the plant environment. The increasing availability of very bright fluorescent antibody conjugates that allow for the detection of even small amounts of molecules at the single cell level makes this method worthy of consideration.

Recent breakthroughs in the development of fluorescent bioconjugates that are bright and span the visible range have contributed to the arsenal of strategies that microbiologists can employ to investigate the molecular biology of bacteria on plants. However, both FISH and immunofluorescence labeling *in situ* on plant surfaces require great attention to avoid the perturbation of the bacterial cells during the many washes involved in these procedures.

26.4 OTHER TYPES OF MICROSCOPY

26.4.1 MULTIPHOTON EXCITATION FLUORESCENCE MICROSCOPY

In multiphoton excitation microscopy, a fluorophore is excited simultaneously by two or more photons of longer wavelength than that of the emitted light. Because excitation occurs only at the focal point of the microscope, there is little out-of-focus absorption, and therefore more excitation light reaches the focal plane. In principle, there are three main advantages to this type of microscopy: greater penetration through thick samples, minimized photodamage of living cells by excitation with infrared light, and decreased photobleaching outside the focal plane [56].

Multiphoton microscopy may have potential for improved imaging of microbial cells embedded deep in plant tissue. In addition, since UV illumination in other types of microscopy causes considerable photodamage of plant cells, multiphoton microscopy may be useful for visualization of bacterial cells labeled with UV-absorbing fluorophores on plant surfaces. This may broaden the range of fluorophores available to microbial ecologists to investigate the behavior of bacteria in the plant environment. However, there is evidence that photobleaching is actually more acute within the focal volume than with one-photon excitation, particularly with thin samples [57]. This was confirmed even with GFP, a fluorescent molecule considered relatively stable [58]. However, perhaps more limiting to its application to plant microbiology is the fact that multiphoton microscopy provides inherently less resolution than single-photon microscopy such as CLSM [59]. Thus, although this technology has improved greatly three-dimensional imaging of plant tissue, it actually may be suited only for the effective visualization of its larger microbial inhabitants and not for bacteria. To our knowledge, multiphoton microscopy has been applied to microbiology so far mostly for the study of biofilms [60,61]. As less expensive multiphoton systems become available, time will tell if this technology is advantageous for the imaging of bacterial cells in or on plant tissue.

26.4.2 FLUORESCENCE STEREOMICROSCOPY

The fluorescence stereomicroscope provides little resolution of single bacterial cells, and thus is not widely used in microbiology. However, it may be useful for preliminary observations of microbial assemblages on plant surfaces, such as biofilms and fungi, or to select tissue samples for further high-resolution microscopy. For example, GFP-labeled $S.$ $enterica$ and $Listeria$ $monocytogenes$ cells were observed by stereomicroscopy as large aggregates on the seed coat edge and the root hairs of alfalfa sprouts [62,63]. In contrast, GFP-labeled $E.$ $coli$ O157:H7 cells were located in these areas at significantly lower densities [62]. These observations under the fluorescence stereomicroscope were possible because of the intense green fluorescence emitted by large aggregates of brightly fluorescent bacterial cells and on portions of the sprouts that have relatively little autofluorescence in the green range. When inoculated onto the roots of growing lettuce plants at low cell concentrations in irrigation water, GFP-labeled $E.$ $coli$ O157:H7 was present as single cells or small colonies scattered at distant locations on the root, and detectable by confocal microscopy only [64].

26.4.3 IMMUNOELECTRON MICROSCOPY

Despite the important role of fluorescence microscopy in cellular imaging, its resolution is still well below that attained with electron microscopy (EM). However, with the exception of biofilms on plant surfaces, which were revealed by scanning EM (SEM) as complex assemblages of diverse microbes embedded in organic material [26,65,66], most bacterial cells that are imaged on plants under the EM remain disappointingly anonymous to the investigator.

Fortunately, the discovery of colloidal gold as a label in immunoelectron microscopy has provided new opportunities to detect specific bacteria and to unravel the more complex biology of bacterial cells in their natural environment. This approach has the advantage of combining the highly specific localization of molecules in $situ$ with the high resolution of EM, and has been used in many studies in plant pathology. For example, some elegant experiments were performed under EM to propose a model for the role of the Hrp pilus and effector proteins for type III secretion during the interaction of $Erwinia$ $amylovora$ and $Pseudomonas$ $syringae$ with plant cells. These studies involved single and double labeling with gold particles of different sizes to detect two types of protein, and observations under transmission EM in plants [67,68]. Using SEM in combination with somatic and flagellar gold-labeled monoclonal antibodies specific to $Salmonella$ $enterica$ serovar Thompson, this pathogen was visualized at high resolution on leaf surfaces after its inoculation onto cilantro plants; more importantly, the immunodetection of flagellar components allowed for the observation that $S.$ Thompson cells produced flagella that appeared anchored to the leaf surface, suggesting that they may serve as attachment factor to plants (Figure 26.9) [69]. Besides immunocytochemistry, other cytochemical approaches have been developed for gold

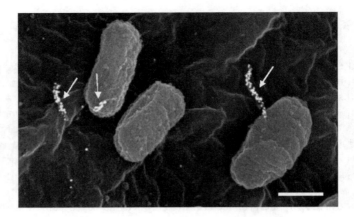

FIGURE 26.9 Backscattered electron image of *S. enterica* serovar Thompson cells on a cilantro leaf after their inoculation onto cilantro plants. Gold-labeled flagellar antibodies that are specific to this serovar are shown binding to the flagellum, which appears to be anchored to the plant surface (arrows). Bright dots are 10 nm gold particles. Bar, 500 nm. (Micrograph courtesy of Delilah F. Wood.)

labeling in EM based on the binding affinity of lectins, enzymes, or proteins to specific molecules. For a discussion and description of methodology regarding the use of gold labeling and EM to investigate plant–microbe interactions, the reader is referred to an excellent review by Benhamou and Belanger [70].

26.5 CONCLUDING REMARKS

Plant microbial ecologists face the challenge of investigating the behavior of their subjects at the relevant spatial scale, that of the bacterial cell. Although the discovery of fluorescent proteins and confocal microscopy has propelled the use of cell imaging to study bacteria in the plant environment, it is evident that this approach remains in its infancy compared to the recent advances in fluorescence imaging of protein dynamics in living eukaryotic cells, such as FRET, FRAP (fluorescence recovery after photobleaching), FLIP (fluorescence loss in photobleaching), and FCS (fluorescence correlation spectroscopy) [71]. The intense fluorescent signal required in this type of imaging due to the small size of bacterial cells, combined with the difficulty of performing time-lapse studies on plants under the fluorescence microscope without altering the physicochemical microenvironment of the bacterial cells, may have limited the application of these technologies to plant microbiology research.

Other types of microscopy are emerging that can be applied to fully hydrated cells, and thus have great potential to impact our ability to probe the behavior of bacteria in their natural habitat. Atomic force microscopy (AFM), which measures the force between a sharp tip and the surface of a sample, has extended our capability to view the minuscule. AFM has been used already to map the surface of individual bacterial cells and various bacterial attachment factors at unparalleled resolution, to map and quantify the adhesion force of microbes to a substratum, and to measure bacterial cell wall elasticity [72]. Scanning transmission X-ray microscopy (STXM), which uses soft X-ray absorption spectra to provide detailed quantitative chemical information about a sample at high resolution, is another new technology that has been used recently to investigate the distribution of proteins, lipids, saccharides, and nucleic acids in a biofilm [10]. STXM may be useful to map the biochemistry of bacteria and their chemical environment on plant surfaces, and thereby gain a better understanding of their physiology in this habitat. Thus, developments in microscopy keep providing powerful tools to explore fundamental questions regarding the biology of bacteria on plants, and their interactions with other plant microflora and with their plant host.

REFERENCES

1. Rost, F.W.D., Autofluorescence in plants, fungi and bacteria, in *Fluorescence Microscopy*, Rost, F.W.D., Ed., Cambridge University Press, New York, 1995.

2. Heim, R., Cubitt, A.B., and Tsien, R.Y., Improved green fluorescence, *Nature*, 373, 663, 1995.

3. Oscar, T.P., Comparison of predictive models for growth of parent and green fluorescent protein-producing strains of *Salmonella*, *J. Food Prot.*, 66, 200, 2003.

4. Fuchslin, H.P. *et al.*, Effect of integration of a GFP reporter gene on fitness of *Ralstonia eutropha* during growth with 2,4-dichlorophenoxyacetic acid, *Environ. Microbiol.*, 5, 878, 2003.

5. Gandhi, M. *et al.*, Use of green fluorescent protein expressing *Salmonella* Stanley to investigate survival, spatial location, and control on alfalfa sprouts, *J. Food Prot.*, 64, 1891, 2001.

6. Takeuchi, K. and Frank, J.F., Expression of red-shifted green fluorescent protein by *Escherichia coli* O157:H7 as a marker for the detection of cells on fresh produce, *J. Food Prot.*, 64, 298, 2001.

7. Matz, M.V. *et al.*, Fluorescent proteins from nonbioluminescent *Anthozoa* species, *Nat. Biotechnol.*, 17, 969, 1999.

8. Gurskaya, N.G. *et al.*, GFP-like chromoproteins as a source of far-red fluorescent proteins, *FEBS Lett.*, 507, 16, 2001.

9. Kasten, F.H., Introduction to fluorescent probes: properties, history and applications, in *Biological Techniques Series; Fluorescent and Luminescent Probes for Biological Activity: A Practical Guide to Technology for Quantitative Real-Time Analysis*, 2nd ed., Mason, W.T., Ed., Academic Press, New York, 1999.

10. Lawrence, J.R. *et al.*, Scanning transmission X-ray, laser scanning, and transmission electron microscopy mapping of the exopolymeric matrix of microbial biofilms, *Appl. Environ. Microbiol.*, 69, 5543, 2003.

11. Takeuchi, K. and Frank, J.F., Penetration of *Escherichia coli* O157:H7 into lettuce tissues as affected by inoculum size and temperature and the effect of chlorine treatment on cell viability, *J. Food Prot.*, 63, 434, 2000.

12. Brandl, M.T., Quinones, B., and Lindow, S.E., Heterogeneous transcription of an indoleacetic acid biosynthetic gene in *Erwinia herbicola* on plant surfaces, *Proc. Natl. Acad. Sci. USA*, 98, 3454, 2001.

13. Joyner, D.C. and Lindow, S.E., Heterogeneity of iron bioavailability on plants assessed with a whole-cell GFP-based bacterial biosensor, *Microbiology*, 146, 2435, 2000.

14. Leveau, J.H.J. and Lindow, S.E., Appetite of an epiphyte: quantitative monitoring of bacterial sugar consumption in the phyllosphere, *Proc. Natl. Acad. Sci. USA*, 98, 3446, 2001.

15. Miller, W.G. *et al.*, Biological sensor for sucrose availability: relative sensitivities of various reporter genes, *Appl. Environ. Microbiol.*, 67, 1308, 2001.

16. van Overbeek, L.S. *et al.*, A polyphasic approach for studying the interaction between *Ralstonia solanacearum* and potential control agents in the tomato phytosphere, *J. Microbiol. Methods*, 48, 69, 2002.

17. Carmichael, I. *et al.*, Bacterial colonization and biofilm development on minimally processed vegetables, *J. Appl. Microbiol. Symp. Suppl.*, 85, 45S, 1999.

18. Neu, T., Swerhone, G.D., and Lawrence, J.R., Assessment of lectin-binding analysis for *in situ* detection of glycoconjugates in biofilm systems, *Microbiology*, 147, 299, 2001.

19. Brandl, M.T. and Mandrell, R.E., Fitness of *Salmonella enterica* serovar Thompson in the cilantro phyllosphere, *Appl. Environ. Microbiol.*, 68, 3614, 2002.

20. Leben, C., Relative humidity and the survival of epiphytic bacteria with buds and leaves of cucumber plants, *Phytopathology*, 78, 179, 1988.

21. Burnett, S.L., Chen, J., and Beuchat, L.R., Attachment of *Escherichia coli* O157:H7 to the surfaces and internal structures of apples as detected by confocal scanning laser microscopy, *Appl. Environ. Microbiol.*, 66, 4679, 2000.

22. Hallmann, J. *et al.*, Endophytic colonization of plants by the biocontrol agent *Rhizobium etli* G12 in relation to *Meloidogyne incognita* infection, *Phytopathology*, 91, 415, 2001.

23. Cooley, M.B., Miller, W.G., and Mandrell, R.E., Colonization of *Arabidopsis thaliana* with *Salmonella enterica* and enterohemorrhagic *Escherichia coli* O157:H7 and competition by *Enterobacter asburiae*, *Appl. Environ. Microbiol.*, 69, 4915, 2003.

24. Dong, Y. *et al.*, Kinetics and strain specificity of rhizosphere and endophytic colonization by enteric bacteria on seedlings of *Medicago sativa* and *Medicago truncatula*, *Appl. Environ. Microbiol.*, 69, 1783, 2003.

25. Solomon, E.B., Yaron, S., and Matthews, K.R., Transmission of *Escherichia coli* O157:H7 from contaminated manure and irrigation water to lettuce plant tissue and its subsequent internalization, *Appl. Environ. Microbiol.*, 68, 397, 2002.

26. Morris, C., Monier, J., and Jacques, M., Methods for observing microbial biofilms directly on leaf surfaces and recovering them for isolation of culturable microorganisms, *Appl. Environ. Microbiol.*, 63, 1570, 1997.

27. Monier, J.M. and Lindow, S.E., Frequency, size, and localization of bacterial aggregates on bean leaf surfaces, *Appl. Environ. Microbiol.*, 70, 346, 2004.

28. Brandl, M.T. *et al.*, Production of autoinducer-2 in *Salmonella enterica* serovar Thompson contributes to its fitness in chickens but not on cilantro leaf surfaces, *Appl. Environ. Microbiol.*, 71, 2653, 2005.

29. Morris, C.E., Monier, J.M., and Jacques, M.A., A technique to quantify the population size and composition of the biofilm component in communities of bacteria in the phyllosphere, *Appl. Environ. Microbiol.*, 64, 4789, 1998.

30. Normander, B. *et al.*, Effect of bacterial distribution and activity on conjugal gene transfer on the phylloplane of the bush bean (*Phaseolus vulgaris*), *Appl. Environ. Microbiol.*, 64, 1902, 1998.

31. Monier, J.M. and Lindow, S.E., Spatial organization of dual-species bacterial aggregates on leaf surfaces, *Appl. Environ. Microbiol.*, in press.

32. Rojas, C.M. *et al.*, HecA, a member of a class of adhesins produced by diverse pathogenic bacteria, contributes to the attachment, aggregation, epidermal cell killing, and virulence phenotypes of *Erwinia chrysanthemi* EC16 on *Nicotiana clevelandii* seedlings, *Proc. Natl. Acad. Sci. USA*, 99, 13142, 2002.

33. Lindow, S.E. and Brandl, M.T., Microbiology of the phyllosphere, *Appl. Environ. Microbiol.*, 69, 1875, 2003.

34. Kogure, K., Simidu, U., and Taga, N., A tentative direct microscopic method for counting living bacteria, *Can. J. Microbiol.*, 25, 415, 1979.

35. Wilson, M. and Lindow, S.E., Relationship of total viable and culturable cells in epiphytic populations of *Pseudomonas syringae*, *Appl. Environ. Microbiol.*, 58, 3908, 1992.

36. Grey, B.E. and Steck, T.R., The viable but nonculturable state of *Ralstonia solanacearum* may be involved in long-term survival and plant infection, *Appl. Environ. Microbiol.*, 67, 3866, 2001.

37. Warriner, K. *et al.*, Internalization of bioluminescent *Escherichia coli* and *Salmonella* Montevideo in growing bean sprouts, *J. Appl. Microbiol.*, 95, 719, 2003.

38. Monier, J.M. and Lindow, S.E., Differential survival of solitary and aggregated bacterial cells promotes aggregate formation on leaf surfaces, *Proc. Natl. Acad. Sci. USA*, 100, 15977, 2003.

39. Monier, J.M., Aggregates of resident bacteria facilitate the survival of immigrant bacteria on leaf surfaces, *Microbial Ecol.*, 49(3), 2005.

40. Lowder, M. *et al.*, Effect of starvation and the viable-but-nonculturable state on green fluorescent protein (GFP) fluorescence in GFP-tagged *Pseudomonas fluorescens* A506, *Appl. Environ. Microbiol.*, 66, 3160, 2000.

41. Brandl, M.T., unpublished data.

42. Breeuwer, P. and Abee, T., Assessment of viability of microorganisms employing fluorescence techniques, *Int. J. Food Microbiol.*, 55, 193, 2000.

43. Shapiro, H.M., Microbial analysis at the single-cell level: tasks and techniques, *J. Microbiol. Methods*, 42, 3, 2000.

44. Olsen, K.N. *et al.*, Noninvasive measurement of bacterial intracellular pH on a single-cell level with green fluorescent protein and fluorescence ratio imaging microscopy, *Appl. Environ. Microbiol.*, 68, 4145, 2002.

45. Maksimow, M. *et al.*, Simultaneous detection of bacteria expressing GFP and DsRed genes with a flow cytometer, *Cytometry*, 47, 243, 2002.

46. Erickson, M.G., Moon, D.L., and Yue, D.T., DsRed as a potential FRET partner with CFP and GFP, *Biophys. J.*, 85, 599, 2003.

47. Ramos, C., Molbak, L., and Molin, S., Bacterial activity in the rhizosphere analyzed at the single-cell level by monitoring ribosome contents and synthesis rates, *Appl. Environ. Microbiol.*, 66, 801, 2000.

48. Miller, W.G., Leveau, J.H., and Lindow, S.E., Improved *gfp* and *inaZ* broad-host-range promoter-probe vectors, *Mol. Plant Microbe Interact.*, 13, 1243, 2000.

49. Leveau, J.H. and Lindow, S.E., Predictive and interpretive simulation of green fluorescent protein expression in reporter bacteria, *J. Bacteriol.*, 183, 6752, 2001.

50. Anon., *Living Colors User Manual,* BD Biosciences Clontech, 2001.

51. Chen, H. *et al.*, Culture-independent analysis of fecal enterobacteria in environmental samples by single-cell mRNA profiling, *Appl. Environ. Microbiol.*, 70, 4432, 2004.

52. Pernthaler, A. and Amann, R., Simultaneous fluorescence *in situ* hybridization of mRNA and rRNA in environmental bacteria, *Appl. Environ. Microbiol.*, 70, 5426, 2004.

53. Kang, Y. *et al.*, Quantitative immunofluorescence of regulated *eps* gene expression in single cells of *Ralstonia solanacearum*, *Appl. Environ. Microbiol.*, 65, 2356, 1999.

54. Mc Garvey, J.A., Denny, T.P., and Schell, M.A., Spatial-temporal and quantitative analysis of growth and EPS I production by *Ralstonia solanacearum* in resistant and susceptible tomato cultivars, *Phytopathology,* 89, 1233, 1999.

55. Boher, B. *et al.*, Extracellular polysaccharides from *Xanthomonas axonopodis* pv. *manihotis* interact with cassava cell walls during pathogenesis, *Mol. Plant Microbe Interact.*, 10, 803, 1997.

56. Piston, D.W., Imaging living cells and tissues by two-photon excitation microscopy, *Trends Cell Biol.*, 9, 66, 1999.

57. Patterson, G.H. and Piston, D.W., Photobleaching in two-photon excitation microscopy, *Biophys. J.*, 78, 2159, 2000.

58. Drummond, D.R., Carter, N., and Cross, R.A., Multiphoton versus confocal high resolution z-sectioning of enhanced green fluorescent microtubules: increased multiphoton photobleaching within the focal plane can be compensated using a Pockels cell and dual widefield detectors, *J. Microsc.*, 206, 161, 2002.

59. Cox, G. and Sheppard, C., Multiphoton fluorescence microscopy, in *Biological Techniques Series; Fluorescent and Luminescent Probes for Biological Activity: A Practical Guide to Technology for Quantitative Real-Time Analysis*, 2nd ed., Mason, W.T., Ed., Academic Press, New York, 1999.

60. Neu, T.R., Kuhlicke, U., and Lawrence, J.R., Assessment of fluorochromes for two-photon laser scanning microscopy of biofilms, *Appl. Environ. Microbiol.*, 68, 901, 2002.

61. Neu, T.R., Woelfl, S., and Lawrence, J.R., Three-dimensional differentiation of photo-autotrophic biofilm constituents by multi-channel laser scanning microscopy (single-photon and two-photon excitation), *J. Microbiol. Methods,* 56, 161, 2004.

62. Charkowski, A.O. *et al.*, Differences in growth of *Salmonella enterica* and *Escherichia coli* O157:H7 on alfalfa sprouts, *Appl. Environ. Microbiol.*, 68, 3114, 2002.

63. Gorski, L., Palumbo, J.D., and Nguyen, K.D., Strain-specific differences in the attachment of *Listeria monocytogenes* to alfalfa sprouts, *J. Food Prot.*, 67, 2488, 2004.

64. Wachtel, M.R., Whitehand, L.C., and Mandrell, R.E., Association of *Escherichia coli* O157:H7 with preharvest leaf lettuce upon exposure to contaminated irrigation water, *J. Food Prot.*, 65, 18, 2002.
65. Fett, W.F. and Cooke, P.H., Scanning electron microscopy of native biofilms on mung bean sprouts, *Can. J. Microbiol.*, 49, 45, 2003.
66. Rayner, J., Veeh, R., and Flood, J., Prevalence of microbial biofilms on selected fresh produce and household surfaces, *Int. J. Food Microbiol.*, 95, 29, 2004.
67. Brown, I.R. *et al.*, Immunocytochemical localization of HrpA and HrpZ supports a role for the Hrp pilus in the transfer of effector proteins from *Pseudomonas syringae* pv. *tomato* across the host plant cell wall, *Mol. Plant Microbe Interact.*, 14, 394, 2001.
68. Jin, Q. *et al.*, Visualization of secreted Hrp and Avr proteins along the Hrp pilus during type III secretion in *Erwinia amylovora* and *Pseudomonas syringae*, *Mol. Microbiol.*, 40, 1129, 2001.
69. Brandl, M.T. and Wood, D.F., unpublished data.
70. Benhamou, N. and Belanger, R., Immunoelectron microscopy, in *Molecular Methods in Plant Pathology*, Singh, R.P. and Singh, U.S., Eds., CRC Press, Boca Raton, FL, 1995.
71. Lippincott-Schwartz, J. and Patterson, G.H., Development and use of fluorescent protein markers in living cells, *Science*, 300, 87, 2003.
72. Dufrene, Y.F., Recent progress in the application of atomic force microscopy imaging and force spectroscopy to microbiology, *Curr. Opin. Microbiol.*, 6, 317, 2003.
73. Haugland, P.H., *Handbook of Fluorescent Probes and Research Products*, 9th ed., Molecular Probes, Inc., Eugene, 2002.

Index